LASER MICROFABRICATION
Thin Film Processes and Lithography

LASER MICROFABRICATION
Thin Film Processes and Lithography

Edited by

Daniel J. Ehrlich

Lincoln Laboratory
Massachusetts Institute of Technology
Lexington, Massachusetts

Jeffrey Y. Tsao

Sandia National Laboratories
Albuquerque, New Mexico

ACADEMIC PRESS, INC.

Harcourt Brace Jovanovich, Publishers

Boston San Diego New York
Berkeley London Sydney
Tokyo Toronto

ACADEMIC PRESS, INC.
1250 Sixth Avenue, San Diego, CA 92101

United Kingdom Edition published by
ACADEMIC PRESS INC. (LONDON) LTD.
24–28 Oval Road, London NW1 7DX

Designed by Joni Hopkins

Library of Congress Cataloging-in-Publication Data

Laser microfabrication : thin film processes and lithography / edited by Daniel J. Ehrlich, Jeffrey Y. Tsao.
p. cm.
Includes bibliographies and index.
ISBN 0-12-233430-2
1. Lasers—Industrial applications. 2. Thin film devices—Design and construction. 3. Microlithography. I. Ehrlich, Daniel J. II. Tsao, Jeffrey Y.
TA1677.L366 1989
621.36′6—dc 19 88-24215
CIP

Printed in the United States of America
89 90 91 92 9 8 7 6 5 4 3 2 1

CONTENTS

CONTRIBUTORS

Numbers in parentheses indicate the pages on which the authors' contributions begin.

C.I.H. Ashby (231), *Organization 1126, Sandia National Laboratories, PO Box 5800, Albuquerque, New Mexico 87185*

M.R. Aylett (453), *British Telecom Research Laboratories, Martlesham Heath, Ipswich, United Kingdom*

T.H. Baum (385), *IBM-Almaden Research Center K91/802, 650 Harry Road, San Jose, California 95120-6099*

Ian W. Boyd (539), *Department of Electrical Engineering, University College London, Torrington Place, London WC1E 7JE, United Kingdom*

T.J. Chuang (87), *IBM-Almaden Research Center K33/801, 650 Harry Road, San Jose, California 95120-6099*

P.B. Comita (385), *IBM-Almaden Research Center, 650 Harry Road, San Jose, California 95120-6099*

D.J. Ehrlich (285, 385), *Lincoln Laboratory, Massachusetts Institute of Technology, 244 Wood Street, PO Box 73, Lexington, Massachusetts 02173*

J. Haigh (453), *British Telecom Research Laboratories, Martlesham Heath, Ipswich, United Kingdom*

S.J.C. Irvine (503), *Royal Signals and Radar Establishment, St. Andrews Road, Great Malvern, Worcester WR14 3PS, United Kingdom*

R.L. Jackson (385), *IBM-Almaden Research Center, 650 Harry Road, San Jose, California 95120-6099*

T.T. Kodas (385), *IBM-Almaden Research Center, 650 Harry Road, San Jose, California 95120-6099*

Y.S. Liu (1), *General Electric Research and Development Center, Building KW-B1307, PO Box 8, Schenectady, New York 12301*

John J. Ritsko (333), *IBM-T.J. Watson Research Center, Mail Stop 39-144, PO Box 218, Yorktown Heights, New York 10598*

M. Rothschild (163), *Lincoln Laboratory, Massachusetts Institute of Technology, 244 Wood Street, PO Box 73, Lexington, Massachusetts 02173*

J.Y. Tsao (231, 285), *Organization 1141, Sandia National Laboratories, PO Box 5800, Albuquerque, New Mexico 87185*

H.J. Zeiger (285), *Lincoln Laboratory, Massachusetts Institute of Technology, 244 Wood Street, PO Box 73, Lexington, Massachusetts 02173*

PREFACE

Since the initial studies at the beginning of this decade, interest in laser microfabrication processes, especially those based on laser-stimulated thin-film chemistry, has undergone an explosive and sustained growth. An intriguing aspect of this growth is that it was (and continues to be) generated by scientists and engineers from a wide range of fields and disciplines, among them quantum electronics, surface science, materials science, chemical physics, and microelectronics engineering. Indeed, it is perhaps *because* of the interdisciplinary nature of laser microfabrication that it has grown so rapidly.

From an applications point of view, selective and (in many cases) non-thermal chemistry, accompanied by micrometer spatial confinement, was seen to have wide potential for microelectronics. An entirely new set of technologies was spawned, based on laser direct-write "microchemistry." From a basic science point of view, there has been interest in developing a fundamental understanding of, *inter alia,* energy transfer between adsorbates and surfaces, surface electromagnetism, surface photochemistry, catalysis, and chemical kinetics in small reaction zones.

Our intent is that this volume will serve three distinct functions. First, we review in a convenient format the fundamental scientific knowledge on laser-stimulated surface chemistry, which has accumulated over the past few years. It can thus serve as a starting point for students and researchers to learn about the field. Second, we provide a capsule summary of the current state-of-the-art in the technology of these processes. It should thus be useful to applications engineers as well. Third, we provide up-to-date reference data (photodissociation cross-sections, thermochemical constants, etc.) compiled from many scattered sources. It should therefore be useful to the active researcher in the field.

We are indebted to numerous colleagues throughout the years for discussions and collaborations. We especially thank Carol Ashby, Jerry Black, Steve Brueck, Tom Deutsch, Yung Liu, Alan McWhorter, Bob Mountain, Rick Osgood, Richard Reynolds, Mordy Rothschild, Jan Sedlacek, Don Silversmith and Howard Schlossberg. Many thanks also to the contributors to this volume!

Finally, we dedicate this volume to our (extended) families, with love and affection.

J.Y. Tsao
D.J. Ehrlich

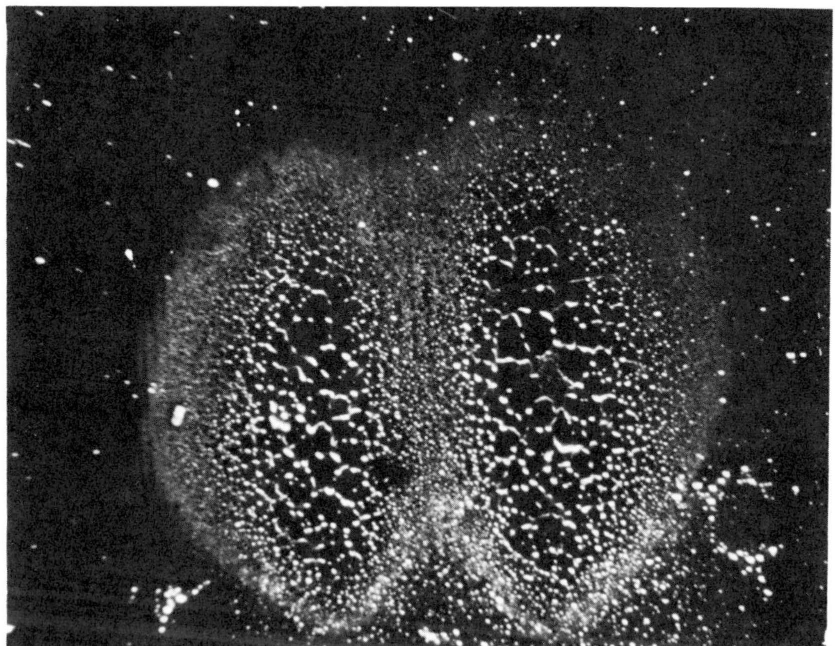

Result from the first day's experiments. Iodine, photodeposited from CF_3I using a 257-nm laser beam that had passed around a vertical wire (note the fringed far-field diffraction pattern). Laser was on loan from H. Schlossberg (then at RADC) and CF_3I was on loan from T.F. Deutsch, both deflected for the afternoon from fusion laser studies (D.J. Ehrlich, unpublished).

PART I

Technology

CHAPTER 1

Sources, Optics, and Laser Microfabrication Systems for Direct Writing and Projection Lithography

Y.S. LIU

GE Research and Development Center
Schenectady, New York

0-12-233430-2

1. Introduction

Since the first coherent laser radiation from ruby was demonstrated in 1960, coherent radiation has been observed from the UV to the far-IR from literally thousands of materials. During this period, lasers have been used in various material processing and device fabrication applications (Von Allmen, 1987, Bäuerle, 1986). The recent increasing interest in laser processing has been driven by future needs for application specific integrated circuits designs, circuitry restructuring, and yield enhancements, all of which can benefit from an adaptive processing technology such as laser processing for a higher degree of precision, resolution, and process automation, and for quick turnaround (Smart, et al., 1987).

Similar to other particle-beam technologies, e.g., ion beam and electron beam, a laser is either used as a focused beam to deliver photon energy onto a small area or as a broad beam for larger area processing. Unlike other beam processing technologies, however, lasers are not required to operate in a vacuum, thus providing greater flexibility in beam delivery and transporting. In-situ process monitoring, diagnostics, and controlling are additional capabilities in laser processing (Boyd, 1987).

The purpose of this chapter is to give readers the essential knowledge on optical considerations useful for laser microfabrication as well as to review several state-of-the art laser microfabrication systems and some instrumental design parameters. The organization of this chapter follows.

In the next section, we review a selected number of coherent and incoherent sources useful for laser microprocessing. In addition, some useful nonlinear optical techniques such as harmonic generation and stimulated Raman scattering for visible and UV generation are discussed.

In Section 3, we discuss optical considerations relevant to focused laser-beam technology, in which a scanning visible or UV laser, such as argon-ion and krypton-ion lasers, or high-repetition-rate, pulsed solid-state YAG lasers are frequently employed. Parameters such as spatial mode structures, Gaussian beam propagation, focusing, depth of focus, collimation and beam shaping, modulation, and monitoring are discussed. Several state-of-the-art laser direct-write systems for interconnection are reviewed.

In Section 4, we discuss broad-beam laser projection technology in which an excimer laser source is typically used. In addition to the source (excimer laser) properties, we review important considerations such as image formation, coherent effects, resolution, modulation transfer function, depth of field, as well as other practical constraints such as source life, cost, and reliability. Several recently developed excimer-laser-based microfabrication systems are discussed.

2. Coherent and Incoherent Sources

2.1. Coherent Sources

A laser device generally consists of two components: an active medium that produces optical gain under proper pumping conditions and an optical resonator that consists of a pair of mirrors with specific radii of curvature and reflectivities to provide optical feedback and output coupling (Yariv, 1975; Svelto, 1982; Siegman, 1971). The laser medium can be pumped by various excitation mechanisms such as gas discharge, optical pumping (coherently or incoherently), or electrical current to produce a population inversion. Several selected laser sources useful for microfabrication are given in Table 1.1.

The laser transitions in a gas laser involve electronic, vibrational, and rotational transitions in atoms, molecules, or ions (e.g., argon laser) and are commonly pumped with electrical discharges (DC, RF, or pulsed).

Table 1.1 General Characteristics of a Selected Number of Laser Sources Useful for Laser Processing.

Laser	Emission Wavelength (μm)	Power (W)	Pulse Energy (J)	Pulse Width (ns)	Beam Diameter (mm)	Beam Divergence (mrad)
CW LASERS						
Argon (UV)	0.35–0.65	4	na	na	1.5	0.4
(Visible)	0.45–0.53	20	na	na	1.5	0.6
Krypton	0.33–0.8	4	na	na	2	0.5
He-Ne	0.63	0.01	na	na	1.5	0.5
GaAs	0.82	0.2	na	na	0.1 × 0.001	100 × 350
Nd:YAG	1.06	400	na	na	8	15
CO_2	9–11	1000	na	na	10	2
PULSED LASERS						
Excimer Lasers						
ArF	0.193	50	0.4	20	20 × 10	2 × 3
KrF	0.248	100	1	20	20 × 10	2 × 3
XeCl	0.308	50	1	20	20 × 10	2 × 3
XeF	0.351	50	0.5	20	20 × 10	2 × 3
Argon	0.45–0.51	5	10^{-8}	20	1.2	0.7
Copper	0.51–0.58	20	0.005	20	20	5
Nd:YAG	1.06	100	0.2	30	10	2
Nd:Glass	1.06	100	10	50	10	4
CO_2	9–11	1000	1	200	10	2

The numbers used here are representative and do not indicate the maximum values available.

Since the gain volume in a gas laser can be scaled up, some of the most powerful lasers are based on gaseous media, including molecular CO_2 (10.6 μm) and CO lasers (5 μm), HF and HCl chemical lasers (2–3 μm), iodine lasers (1.3 μm), argon-ion lasers (0.5 μm), and excimer lasers (0.2–0.35 μm). Among these gaseous lasers, all but excimer lasers can be operated both continuously (cw) or pulsed. The efficiency of a gaseous laser is typically a few percent, higher than that typically obtained in a solid-state laser. Beside excimer lasers, which have become a popular UV source for high-resolution lithography, CO_2 lasers are the most useful laser source for a variety of material processing applications requiring high power (~kW) such as welding, cutting, and drilling (Ready, 1971; Banas and Webb, 1982). However, the resolution achievable with a CO_2 laser is limited to tens of microns because of the long wavelength. Liquid lasers including dye lasers are tunable over a wide spectral bandwidth in visible and near IR. They are the most popular coherent source for spectroscopic applications. Their tunability, however, has not yet been explored for microfabrication and electronic material processing.

The solid-state lasers, noticeably Nd:YAG lasers, are perhaps the most useful laser sources in material processing applications requiring less power, such as trimming, marking, and circuit and mask repairs (Cohen et al., 1982). The YAG laser is pumped with a flashlamp or with other lasers whose emission wavelengths match the absorption bands of YAG at around 0.8 μm (Weber, 1979). The solid-state laser is generally more compact than a gas laser. Optical components are readily available for use at 1.06 μm emitted from solid-state YAG lasers. Furthermore, YAG lasers can be operated cw, pulsed, Q-switched, and mode-locked with a pulse duration from picoseconds, nanoseconds, and microseconds to milliseconds. The flexibility found in YAG lasers contributes to its popularity. In addition, the availability of many efficient nonlinear optical crystals for YAG lasers provides a frequency agility that far exceeds almost all other kinds of lasers (Hon, 1979).

Another important solid-state laser is the GaAs semiconductor laser, which emits radiation in the near IR. The impact of semiconductor lasers is apparent in optical communication, data storage, and computer I/O devices such as laser printers. The potential use of diode lasers for microfabrication has already been demonstrated in read-and-write optical memory devices. As the output power from semiconductor lasers continuously improves, diode lasers will be potentially very useful for material processing and device fabrication (Arjavalingam et al., 1988). Table 1.1 summarizes general characteristics of selected lasers sources useful for laser microprocessing.

The parameters that are used to characterize a laser source are emission wavelength, spectral bandwidth, average power, beam size, and beam divergence. For a pulsed laser, pulse energy and pulse width are also important parameters. Other parameters such as efficiency, size, cost, and reliability are important when a laser is used in an industrial environment. Table 1.2 summarizes the main laser parameters and the related processing parameters that are effected by the corresponding source parameters either directly or indirectly. In the following section, our discussions will focus on a selected number of laser sources and devices that are most useful in microfabrication, including argon and krypton, YAG, and excimer lasers.

Figure 1.1 shows the temperature range achieved using different types of laser sources plotted as a function of the interaction time in various laser-based chemical processes. As is shown in the figure, the enormous dynamic range (on the order of 10^{16}) in processing time explains why lasers are such a unique processing tool.

Table 1.2 Laser Parameters and Related Processing Parameters.

Laser Parameters	Processing Parameters
Power (average) (W)	Temperature (steady state)
	Process throughput
Wavelength (μm)	Optical absorption, reflection, and transmission
	Resolution
	Photochemical effects
Spectral linewidth (nm)	Temporal coherence
	Chromatic aberration
Beam size (mm)	Focal spot size
	Depth of focus
	Beam transformation characteristics
	Intensity
Mode structures	Intensity distribution
	Spatial uniformity
	Speckles
	Spatial coherence
	Modulation transfer function
Peak power (W)	Peak temperature
	Damage
	Nonlinear effects
Pulsewidth (sec)	Interaction time
	Transient processes
Stability (%)	Process latitude
Efficiency (%)	Cost
Reliability	Cost

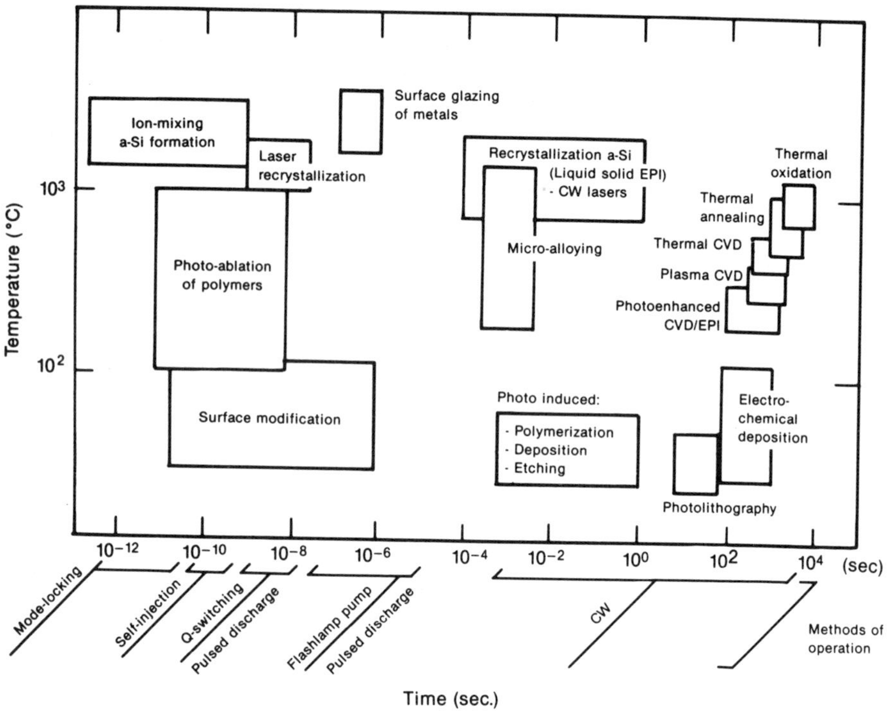

Fig. 1.1 Temperature ranges and processing times observed in various laser-assisted processes employing different types of lasers and modes of operation. Also shown in the figure are those of several conventional IC processing techniques such as oxidation, annealing, and chemical vapor deposition (CVD). [Refs.: self-injection (Liu, 1979), Q-switching (Koechner, 1976), mode locking (Smith et al., 1974).]

2.2. Ion Lasers

The high-power coherent radiation in the visible and UV from ion lasers makes these lasers (especially argon) the most popular laser sources used in direct-writing applications. The gain in the argon laser is provided by argon ions emitting visible radiation between 454.5 and 514.5 nm and UV radiation between 351.1 and 363.8 nm (Dunn and Ross, 1977). The krypton laser has visible output between 413.1 and 799.3 nm and a UV output between 337.4 and 356.4 nm and between 406.7 and 415.4 nm. The ion lasers can be operated on multilines consisting of 457.9- to 514.5-nm lines for argon laser and 647.1- to 676.4-nm lines for krypton laser. Both lasers can also be operated single line using a dispersive element inside the cavity. The strongest laser lines from an argon laser are at 514 nm and 488 nm, and from a krypton laser there is one at 647.1 nm. The argon-ion laser is available up to 20-W multiple lines in the visible, about 10-W single line at 514.5 nm and 7 W at 488 nm. UV argon lasers with multiline UV outputs of 4 W (351–364 nm) are now available. All powers given here are TEM_{00} mode with a typical beam diameter about 1 mm and a beam divergence less than 1 mrad.

In most cases, argon-ion lasers are operated continuously. The laser can also be operated in the pulsed mode by pulsed discharging. Argon lasers can be mode locked to produce pulses of about 100 psec duration at a repetition rate equal to $c/2L$, where L is the laser cavity length and c is the speed of light. A mode-locked argon laser is usually used in pumping a dye laser to generate tunable visible or UV pulses with durations in the subpicosecond region for spectroscopic applications.

A visible argon laser can be frequency doubled to generate cw UV output at 257 nm using the nonlinear optical crystals KDP, ADP, or BBO. Efficient doubling requires phase matching, which is accomplished either by properly orienting the crystal or by adjusting the crystal temperature (Akhmanov et al., 1975). Because frequency doubling is a nonlinear process and the conversion efficiency depends upon the incident peak power, the output from a frequency-doubled cw argon laser is usually low, typically on the order of 10 mW. Very recently, a 100-mW UV frequency-doubled argon laser at 257 nm using an intracavity frequency doubling technique has become commercially available (Spectra Physics Inc., Mountain View, CA.). The UV output has an amplitude stability of about ±3% using feedback control.

2.3. Solid-State Lasers

The solid-state Nd:YAG, Nd:glass, ruby, and tunable Alexandrite lasers (Walling, 1987) are useful laser sources for electronic materials processing (Weber, 1979; Findlay and Goodwin, 1970). Among these solid-state lasers, the Nd:YAG laser is by far the most popular for cutting, welding, drilling, marking, resistor trimming, repairing masks, scanning, and micro-soldering (Cohen et al., 1982). The YAG laser is a very versatile laser source, operable continuously, pulsed, Q-switched, or mode locked, with pulse widths from picoseconds to milliseconds, and with pulse energies varying from a few millijoules to several hundred joules (Danielmeyer, 1976). The average power generated from a slab-geometry YAG laser has been up to 500 W in cw operation. A YAG laser can be miniaturized and pumped by a diode laser to produce output with a linewidth narrowed to several hundred hertz with excellent amplitude stability (Zhou et al., 1985). Furthermore, nonlinear optical crystals for the YAG laser at 1.06 μm are well developed to provide a frequency agility that far exceeds any other lasers.

The strongest YAG laser emission is at 1.064 μm at room temperature. The YAG laser is a four-level system with its lower lasing level at 2,111 cm^{-1} above the ground state, and, as a result, the YAG laser has a low oscillation threshold and is operable continuously. A cw YAG laser can be easily converted into a pulsed laser by inserting an acousto-optical modulator inside the cavity. The modulator is used to generate Q-switched pulses at a repetition rate up to several kilohertz and an average power of 10 to 50 W. This type of YAG laser is used for scribing, marking, and circuit and mask repairing. The laser can be operated very reliably, with the operating life limited by the pump lamp to a few hundred hours. The pump lamp can be easily replaced without perturbing the laser cavity.

A cw-pumped, acousto-optically Q-switched YAG can be frequently doubled to produce visible output at 0.53 μm. Commonly used doubling crystals include KTP, KDP, CDA, and $LiNbO_3$. An acousto-optically Q-switched YAG laser can be efficiently doubled using the intracavity doubling technique. An intracavity doubled YAG laser with a significantly improved average power has been recently developed by Liu et al. (1984) using the nonlinear crystal potassium titanyl phosphate (KTP). Using the cavity configuration shown schematically in Fig. 1.2, an average power up to 20 W has been generated at 0.53 μm at a repetition rate of 5 kHz from the YAG oscillator (Fahlen and Perkins, 1984).

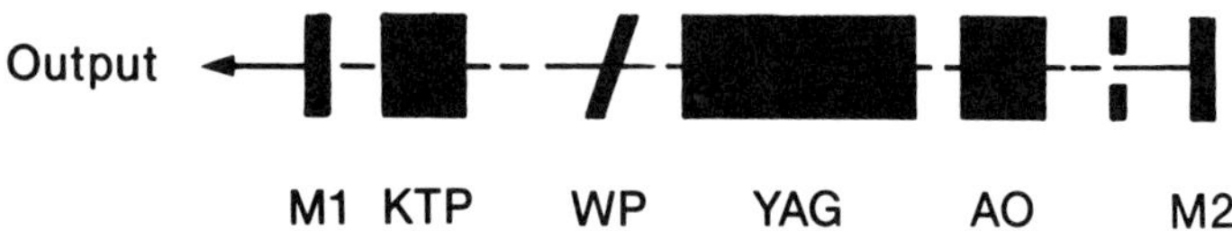

(a)

Fig. 1.2 (a) Experimental arrangement for intracavity second-harmonic generation in an acousto-optical Q-switched Nd:YAG oscillator using KTP. M1, dichroic-coated output mirror with maximum transmission for 532 nm and maximum reflection for 1.06 μm; WP, half-wave plate for 1.06 μm; AO, AO modulator; M2, rear mirror with maximum reflection at 1.06 μm; YAG, Nd:YAG (3 mm diameter and 79 mm long); KTP, the doubling crystal (5.8 mm × 4.1 mm in cross section and 3.5 mm in length) angle-tuned for type II interaction. (b) Intracavity second-harmonic generation in KTP versus $LiIO_3$ in the acousto-optically Q-switched Nd:YAG oscillator at 5 kHz. Second-harmonic outputs for KTP and for $LiIO_3$ plotted as a function of the lamp current (Liu et al., 1984).

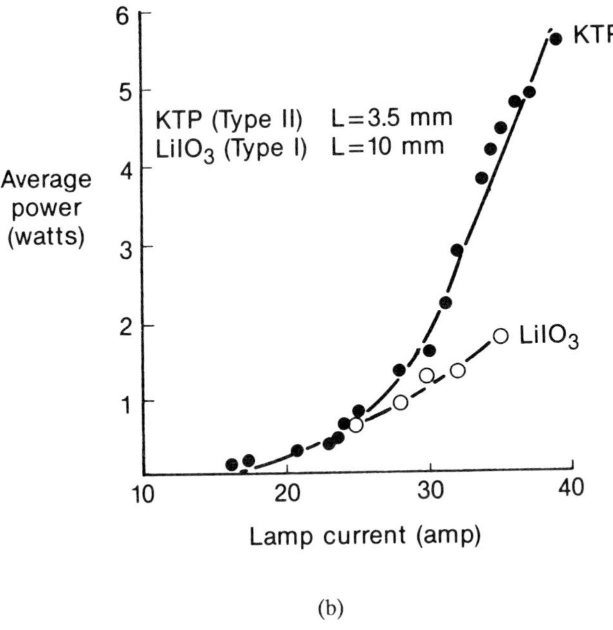

(b)

Fig. 1.2 (*continued*)

2.4. *Slab-Geometry Solid-State Lasers*

The average power from a solid-state laser, fabricated in the conventional rod shape, is limited by thermally induced focusing (defocusing) and birefringence resulting from the nonuniform radial temperature distribution when the gain medium is exposed to intense flashlamp pumping. These distortions reduce the output power and increase the beam divergence (Koechner, 1970). A way to optically compensate for these distortions is to fabricate the solid gain medium in the shape of a slab geometry, as is shown schematically in Fig. 1.3 (Martin and Chernoch, 1972). In this configuration, an optical beam propagating through a zigzag path via total internal reflection between the two parallel slab faces sees no net thermal lensing effect with minimum depolarization (Jones et al., 1972; Kane et al., 1983). As a result, thermally induced lensing and birefringence, which limit the average power in a rod-type solid-state laser, are significantly reduced. Applying the slab-geometry design, several prototype high-average power solid-state YAG and glass slab-lasers have been developed at GE (Jones et al., 1978; Liu et al., 1981b),

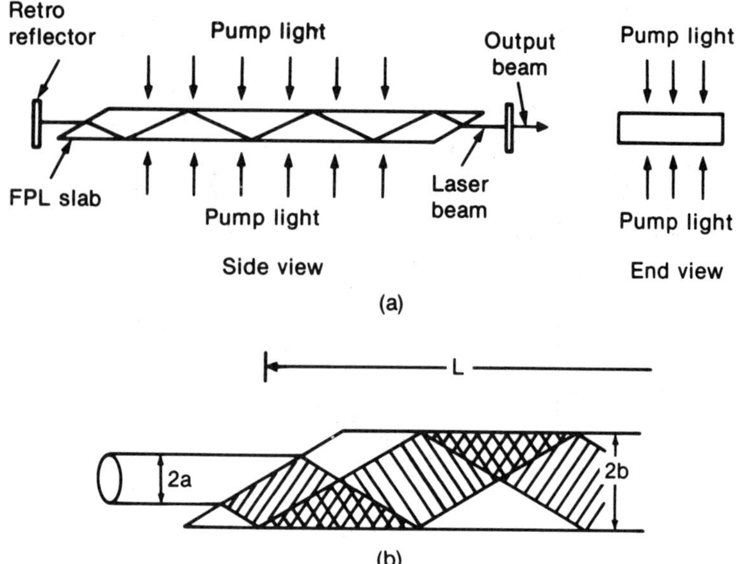

Fig. 1.3 (a) A schematic of a total-internal-reflective slab solid state laser. (b) Brewster-angle entrance face of a slab laser (Liu et al., 1981).

including a cw YAG slab laser at 400-W average power for industrial material processing applications; a Q-switched YAG slab laser with an average power output over 80 W at 400 Hz; a frequency-doubled slab YAG laser with 30-W output (400 Hz) at 0.53 μm (Hulme and Jones, 1975); and a Q-switched glass slab laser with an average output of over 30 W at 10 Hz and a frequency-doubled output at 10 W using type II KDP doubling (Liu et al., 1981b; Jones and Hulme, 1978). Recent developments at Stanford University (Eggleston et al., 1982, 1984; Basu et al., 1986) and at other laboratories have further advanced the slab-laser technology. The ability to generate near-diffraction-limited, high peak power (Brown, 1981), and high-average power output from a slab-laser makes the solid-state laser ideal for nonlinear frequency conversion. Efficient frequency doubling into UV at 266 nm was reported by Liu et al. (1976). The spectral bandwidth of a Nd:glass laser (about 200 cm^{-1}) is comparable to that of excimer lasers, and speckle effects for a quadrupled Nd:glass are expected to be minimal. High-power glass slab lasers are therefore potentially useful for high-resolution lithography applications. Recently, laser-generated plasma x-ray sources pumped by a solid-state slab laser have become commercially available and are an attractive candidate for the next-generation plasma x-ray source for x-ray lithography (Hoffman et al., 1985; Pepin et al., 1987; Heuberger, 1986; Peters et al., 1988).

2.5. *Excimer Lasers*

Excimer lasers have emerged as the most important UV source for microfabrication applications including lithography, surface modification, and micromachining of both organic and inorganic materials (Reintjes, 1985). The rare-gas halide excimer lasers as a coherent UV source have many attractive features: high power, high efficiency, and high brightness. The emission from an excimer laser originates from transitions from electronically excited bound states of a rare-gas dimer to a repulsive ground state. The term, *excited dimer,* refers to a molecule that is bound only in an excited electronic state, but unbound in the ground state. As a result, emission from an excimer laser has a relatively broad emission bandwidth. This reduces the temporal coherence of radiation from an excimer laser (see Section 4.5). In addition, the presence of a large number of transverse modes in an excimer laser cavity scrambles the spatial coherence and thus eliminates the speckle noise that frequently is observed in images formed from coherent illumination (see Section 4.4). Figure 1.4 shows available UV powers obtained from excimer lasers and several other UV lasers in comparison with that from a 1-kW medium-pressure incoherent mercury arc lamp measured at 1 m distance away from the lamp.

Most excimer laser development activities took place in the late seventies, and their potential applications to microlithography were recognized in the early eighties (Jain et al., 1982). Several good review papers on excimer lasers are available (Brau, 1984; Hutchinson, 1987; Pummer et al., 1987). Here we review some properties of excimer lasers of relevance to thin-film processing and lithography applications.

2.5.1 Temporal Characteristics

Excimer lasers are pumped by electrical discharge or by e-beams (Hutchinson, 1987) with UV or x-ray preionization schemes. Typical pulsewidths are 10 to 20 nsec at a repetition rate to several hundred hertz. The pulse-to-pulse amplitude stability is typically ±3% to ±5%, and a pulse-to-pulse temporal jitter from a reference signal is about ±2 nsec. The requirement for a uniform discharge in the plasma tube dictates that the pulse duration be on the order of tens of nanoseconds. Excimer lasers with pulsewidths to several hundred nanoseconds have been developed using a distributed resistive ballasting of discharge

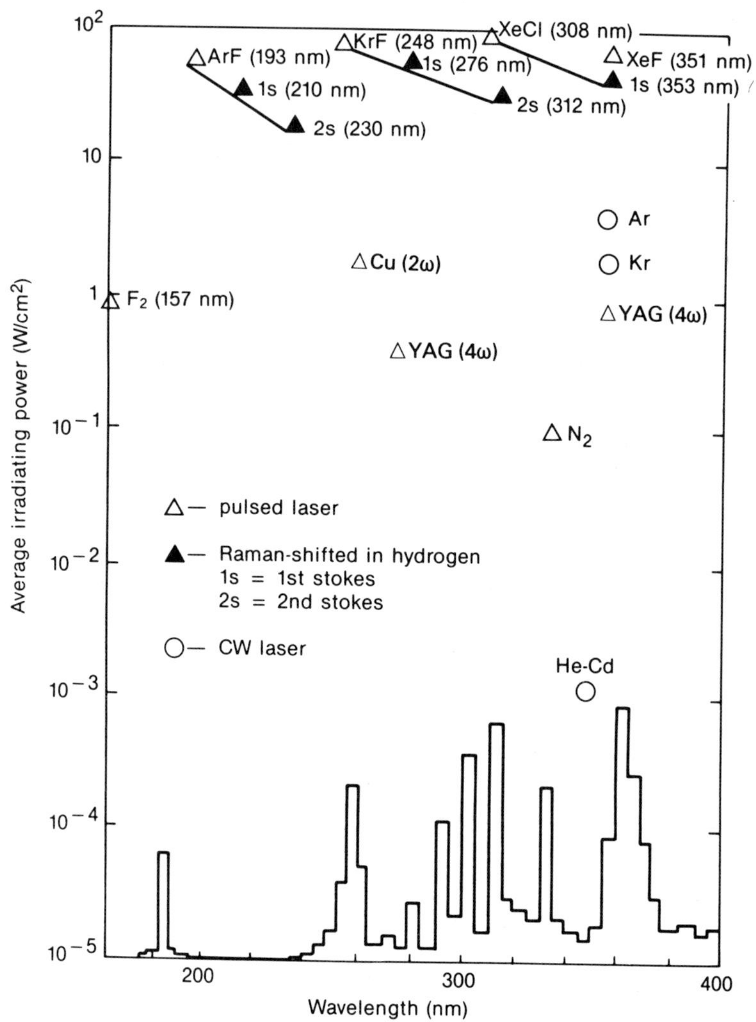

Fig. 1.4 Available irradiating power from various coherent and incoherent UV sources. The first and second Stokes lines using stimulated Raman scattering in hydrogen pumped with ArF, KrF, and XeCl lasers are marked as 1*s* and 2*s*. The irradiating power from a medium-pressure 1-kW He-Hg lamp, shown at the bottom, is measured at a distance 1 m away from the lamp.

electrodes or inductive ballasting (Sze, 1983). Recently long-pulse excimer lasers with pulsewidths to 300 nsec have become commercially available (Lambda Physik, Acton, MA). Microsecond excimer lasers have also been reported using RF pumping (Wisoff et al., 1982). Long pulse duration reduces the maximum peak power. This is an important consideration when an optical fiber is used as the optical delivery system in that high-peak UV power causes unwanted damage or nonlinear processes. At the other extreme, high-power picosecond KrF (Bucksbaum et al., 1982) and ArF (Egger et al., 1982) lasers have been achieved using a picosecond seed pulse that was generated from a dye laser.

2.5.2 Spectral Characteristics

The spectral bandwidth is an important design parameter for high-resolution lithography because of chromatic aberration. The spectral linewidths of free-running excimer lasers are typically several tenths of nanometers. Typical fluorescence and stimulated emission spectra from free-running ArF, KrF, XeCl, and XeF lasers are shown in Fig. 1.5.

The spectral linewidth of an excimer laser can be narrowed using a variety of methods including intracavity etalons, prisms, gratings, and apertures. A quartz prism or a set of quartz prisms have been used to reduce linewidths of KrF and ArF lasers to about 0.1 nm, with a tuning range of about 2 nm and an output energy of 30 mJ (Loree et al., 1978). Since the resolution of a prism depends upon the beam divergence, use of a set of prisms for beam expansion in conjunction with dispersive prisms improves the resolving power and reduces the linewidth. Insertion of a small aperture inside the laser cavity to limit the transverse modes can further reduce the linewidth. The operation of a line-narrowed excimer laser oscillator using multiple prisms as the tuning element requires a very critical alignment and the procedure could be tedious. Intracavity etalons provide an alternate and simpler tuning method, and are particularly useful in obtaining oscillation linewidths to a few 10^{-3} nm. Linewidths of 0.001 nm in KrF and <0.002 nm in XeCl using three etalons of different free spectral ranges were reported by Goldhar et al. (1980). Diffraction gratings provide another simple tuning method for obtaining very narrow linewidths. Linewidths less than 0.002 nm in KrF (Caro et al., 1982) and less than 0.002 nm in ArF (Gower, 1983) were obtained using the grazing angle grating technique.

2.5.3 Spatial Characteristics

An excimer laser is typically operated at a high Fresnel number (> several thousands); a stable resonator would therefore produce a

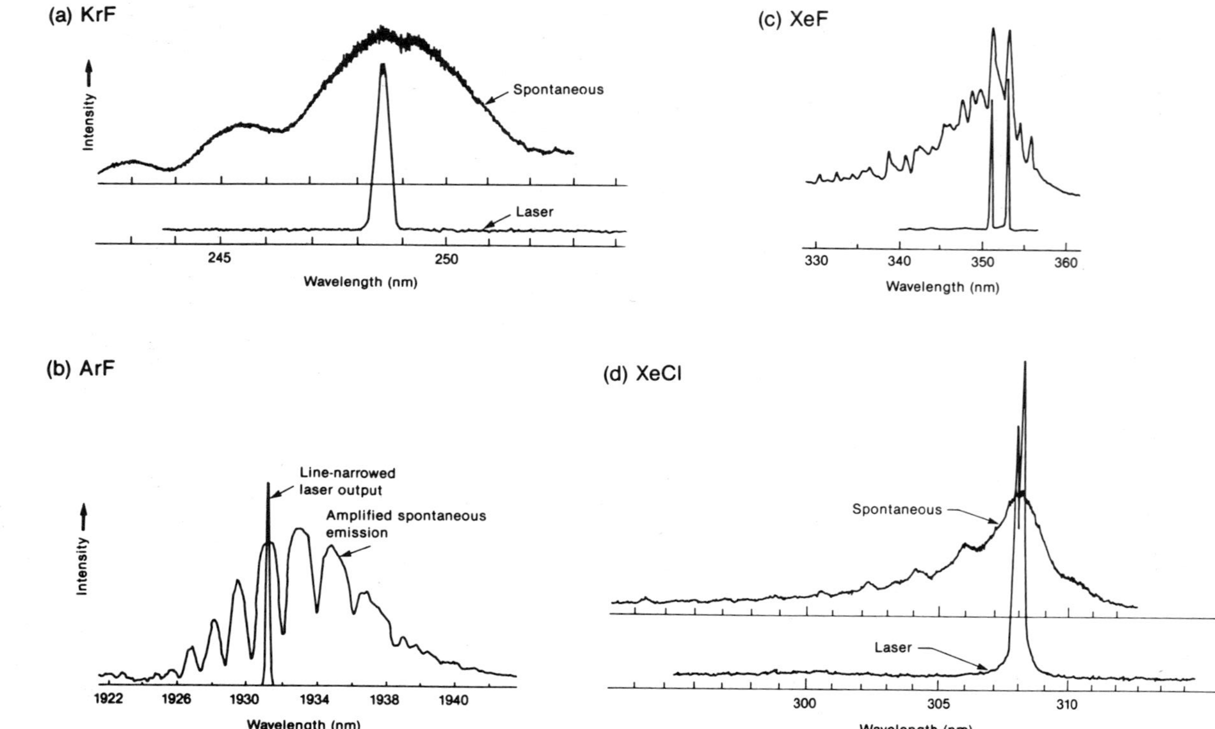

Fig. 1.5 Typical fluorescence and free-running laser emission spectra of (a) KrF (Ewing and Brau, 1975), (b) ArF (Grower, 1983), (c) XeF (Golden et al., 1978), and (d) XeCl (Ewing and Brau, 1975). Structures in the ArF spectrum are due to atmospheric oxygen absorption.

multimode beam with a large beam divergence. A low divergence TEM_{00} mode in a stable resonator would require a beam diameter less than 1 mm in a 1-m long cavity. This can be achieved in an excimer laser cavity at a greatly reduced power. To obtain a high output energy and a low divergence, unstable resonator configurations are commonly used in excimer lasers (Siegman, 1974; Goldhar and Murray, 1977). An unstable resonator using a diffraction-coupled output mirror results in a donut-shape mode in the near field, and therefore beam homogenization is required for obtaining a spatially uniform beam (Section 3.4). An alternative is to use a transmission-coupled mirror in an unstable resonator in which the output mirror is dielectric coated. Beam uniformity can be substantially improved with some reduction in output power. Several excimer lasers using this technique to produce a diffraction-limited, high pulse energy with a uniform beam intensity distribution are now available. Very recently, x-ray preionization has been used to replace the commonly used UV preionization scheme to produce more uniform preionization and therefore excellent spatial beam uniformity. A schematic of the x-ray preionized excimer laser and the output beam profile are shown in Fig. 1.6 (Lambda Physik, Göttingen, West Germany).

2.6. *Nonlinear Optical Techniques*

Nonlinear optical techniques such as harmonic, sum, or difference frequency generation, parametric oscillators, and stimulated Raman scattering (SRS) are useful for extending the spectral range from a laser source at a fixed wavelength. By properly designing a nonlinear frequency converter, one can achieve a high conversion efficiency of 30% to 50%. For example, a Nd:YAG laser at 1.06 μm can be efficiently doubled, tripled, or quadrupled to generate 532-, 355-, and 266-nm radiation. The most useful nonlinear conversion processes discussed in this section are second harmonic generation and stimulated Raman scattering. Readers who are interested in further discussions on the subject will find the following references useful: Shen (1984), Byer (1975), Reintjes (1985), and Akhmanov et al. (1975).

The response of an optical material to an incident laser beam at frequency ω is described by $P = \chi E$, where E is the electrical field of the optical wave at the frequency ω and χ is the polarizability of the medium. When the optical field is intense enough, the higher order polarizability becomes significant in a nonlinear medium; then $P = \chi E$ becomes

$$P = \chi^{(1)}E + \chi^{(2)}EE + \chi^{(3)}EEE + \cdots$$

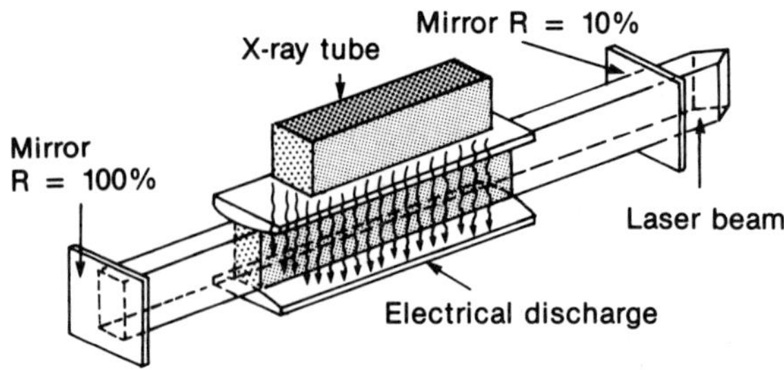

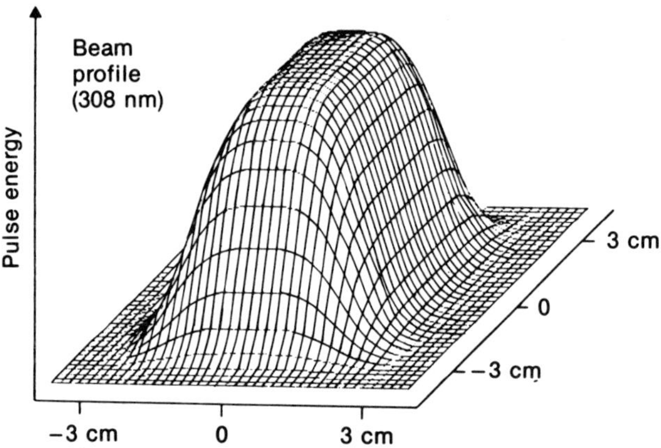

Fig. 1.6 The spatial beam intensity distribution of an x-ray preionized XeCl laser and the cavity schematic. From Lambda Physik, Göttingen, West Germany.

where $\chi^{(2)}$, $\chi^{(3)}$ are the second- and third-order nonlinear susceptibilities and are characteristics of the optical material and types of the nonlinear interactions employed.

2.6.1 Harmonic Generation

Second-harmonic generation is produced via the second-order term, $\chi^{(2)}$ (Bloembergen, 1965). Figure 1.7 illustrates techniques of producing various harmonics of Nd lasers using second-harmonic and sum frequency generation processes. In second-harmonic generation, two electromagnetic waves at frequencies ω are coupled to induce a polarization wave at

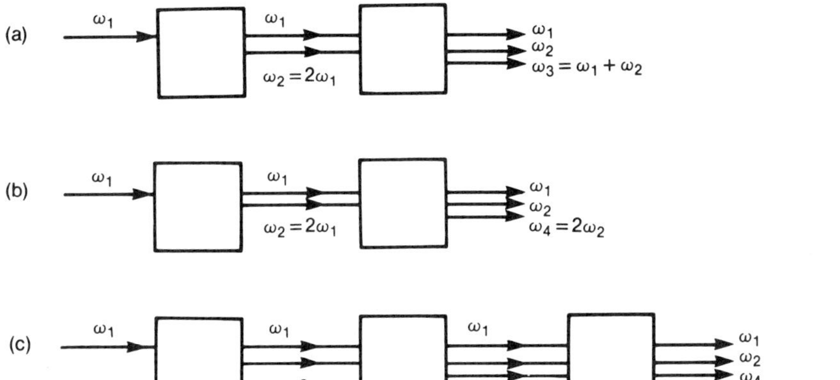

Fig. 1.7 Schematic illustration of techniques for generation of various harmonics of a laser at ω_1 using second-order processes: (a) Third-harmonic generation via sum-frequency mixing of ω_1 and ω_2. (b) Fourth-harmonic generation via twice frequency doubling in tandem. (c) Fifth-harmonic generation via sum-frequency mixing of ω_1 and ω_4. From Reintjes, 1985.

2ω inside the nonlinear medium and to generate the harmonic wave at the frequency 2ω. In order to efficiently convert energy from the fundamental waves to the harmonic, the phase velocities of these coupled waves have to be equal, a condition called *phase matching*. Generally speaking, there are two types of phase matching: Type I, in which the two incident waves have the same polarization, and type II, in which the two incident waves have orthogonal polarizations. Depending upon the structure of a nonlinear crystal, phase matching is achieved by adjusting temperature or orientation of the nonlinear crystal. The second harmonic conversion efficiency can be described as

$$\eta = \frac{P(2\omega)}{P(\omega)} = \eta_0 \operatorname{sinc}^2\left(\frac{\Delta kL}{2}\right),$$

where η_0 is the power conversion efficiency and is a function of the incident power density and the effective nonlinear coefficient of the material, L is the crystal length, and $\Delta k = k_2 - 2k_1$ determines the degree of phase matching via the function

$$\sin c\left(\frac{\Delta kL}{2}\right) = \left[\sin\left(\frac{\Delta kL}{2}\right) \Big/ \left(\frac{\Delta kL}{2}\right)\right]$$

and therefore the nonlinear conversion efficiency. Double refraction, temperature nonuniformity, beam divergence, and crystal dispersion contribute to phase mismatching and thus reduce the conversion efficiency. The design of an efficient harmonic converter for a given laser has to consider source characteristics such as the beam divergence and spectral bandwidth, focused spot size, and crystal selection. As an example, the spectral phase-matching bandwidths for harmonic generation at 1.06 μm in several commonly used nonlinear crystals are shown in Fig. 1.8 (Liu, 1977). The phase-matching spectral bandwidth determines the spectral component that can be efficiently converted into the second harmonic. Similar relationships exist for angle and temperature phase matching. In Fig. 1.9, the optical transmission ranges of a number of the commonly used nonlinear optical crystals are shown. Among these nonlinear crystals, potassium titanyl phosphate (KTP) and beta barium borate (BBO), discussed below, are of particular interest for laser microfabrication.

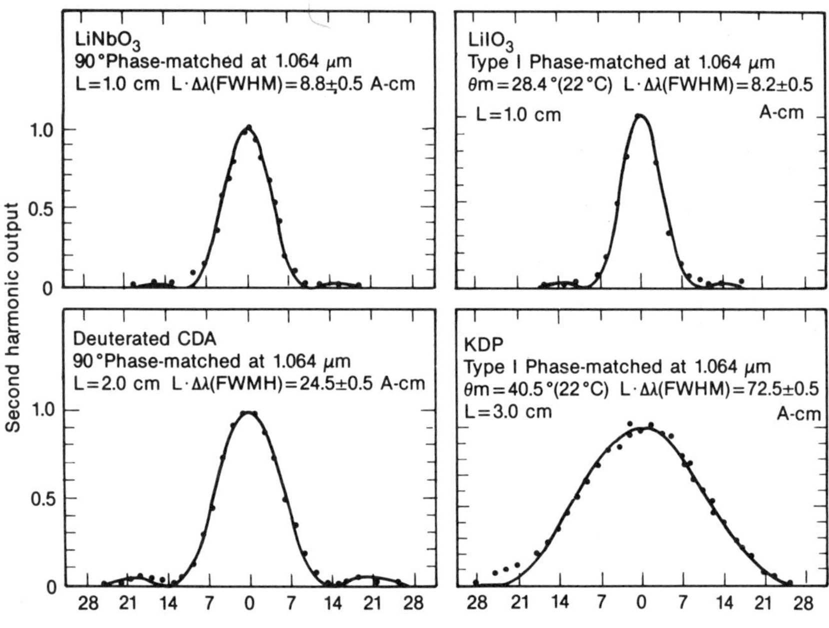

Fig. 1.8 Spectral phase-matching bandwidths for $LiNbO_3$, $LiIO_3$, KDP, and deuterated CDA measured at 1.06 μm using various types of phase matching. The bandwidth and length product is a characteristic of the crystal, and the type of nonlinear interactions. From Liu 1977.

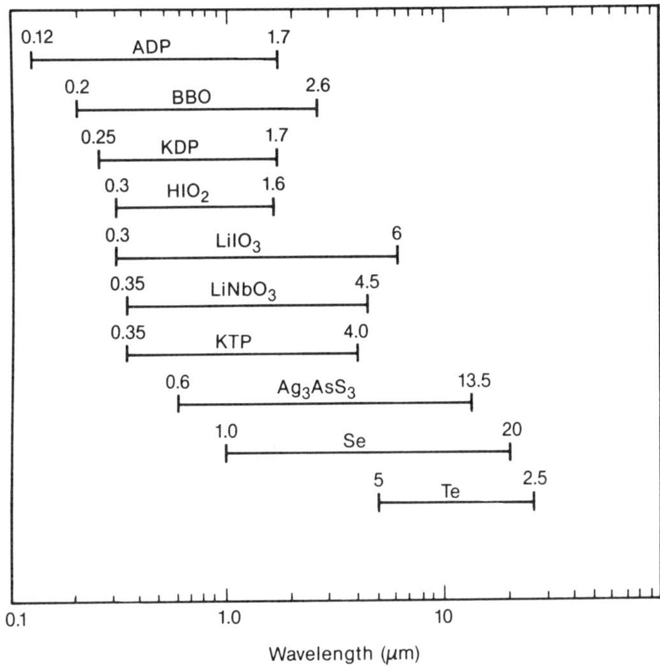

Fig. 1.9 Optical transmission range of several commonly used nonlinear crystals.

KTP has been commercially available since 1983 (Belt et al., 1985) and is the most efficient doubling crystal for 1.06 μm (Liu et al., 1982). It is transmissive from 0.35 μm to about 4 μm and can be phased matched for second-harmonic and sum frequency generation from 458 nm to 550 nm (Kato, 1988). The crystal has been used for intracavity doubling an acousto-optically Q-switched Nd:YAG laser to generate 6- to 20-W output at 532 nm at a pulse rate of 10 kHz and pulse duration of 100 to 200 nsec (Liu et al., 1984; Fahlen and Perkins, 1984). The crystal has also been used for the generation of sum frequency at 459 nm by mixing a 1.06 μm YAG output with a semiconductor laser output at 809 nm (Baumert et al., 1987; Risk et al., 1987).

BBO crystal was first reported in early 1984 by Chen et al. (1984) and is optically transparent to 189 nm. The crystal is phase matched at the fifth harmonic of Nd:YAG by mixing the fundamental (1.06 μm) and its fourth harmonic (266 nm) or by mixing the second harmonic (532 nm) with the third harmonic (355 nm) to generate UV radiation at 212 nm

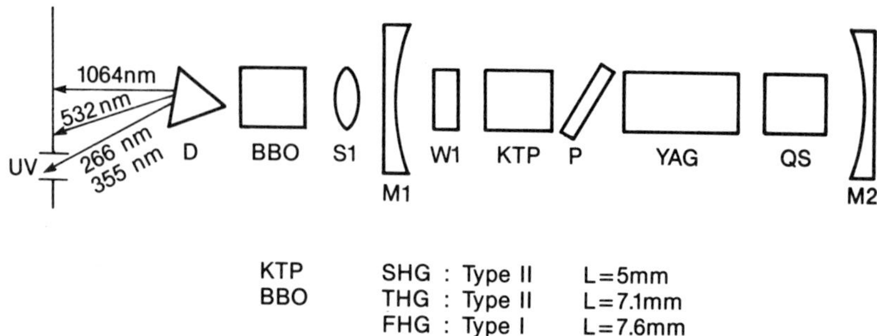

Fig. 1.10 Experimental arrangements for generating second (SHG), third (THG) and fourth (FHG) harmonic generation from a Q-switched YAG laser using KTP and BBO crystals. D = prims; S1 = lens, M1 and M2 = mirrors, W1 = wave plate, P = polarizer, QS-Q = switch cell, and YAG = gain medium. From Liu and Rhodes, 1987.

(Liu and Rhodes, 1987). Kato (1987) has further extended the wavelength to as short as 204.8 nm by second-harmonic generation of 409.6-nm pulses with an efficiency to 17%. Both KTP and BBO crystals have relatively high damage thresholds and can be used in tandem to generate high-power UV radiation. These two new nonlinear optical crystals hold great promise for future microfabrication applications such as direct writing, photochemically induced CVD for mask repair, and recticle fabrication. An experimental schematic used by Liu and Rhodes (1986) for second-, third-, and fourth-harmonic generation from a Q-switched YAG oscillator is shown in Fig. 1.10. Using this setup, they obtained 3.8 W at 532 nm, 0.7 W at 355 nm, and 1 W at 266 nm from a Q-switched YAG oscillator, and 3 W at 532 nm, 0.6 W at 355 n, and 1 W at 266 nm from a mode-locked oscillator.

2.6.2 Raman Frequency Shifting

Stimulated Raman scattering (SRS) is an efficient nonlinear conversion process involving the third-order nonlinear susceptibility $\chi^{(3)}$ (Wilke and Schmidt, 1979). When an incident beam I_0 at frequency ν_0 interacts with a Raman active medium, a scattered beam, called the *first Stokes signal,* is generated at the frequency $\omega_1 = \omega_0 - \omega_s$, where ν_s is the Raman frequency, a characteristic frequency of the medium. When the pump beam intensity is sufficiently high, the medium will exhibit a net gain and the intensity of the first Stokes beam I_1 increases exponentially as it propagates along the gain medium and the process is called *stimulated Raman scattering* (SRS). When the intensity of the stimulated Stokes beam is sufficiently intense, higher-order Stokes beams are generated at $\omega_2 = \omega_0 - 2\omega_s$, $\omega_3 = \omega_0 - 3\omega_s, \ldots,$ called the second Stokes, the third

Stokes, and so forth. SRS has been observed in various solid, liquid, and gaseous media. Hydrogen and methane gases operated at several tens of atmospheric pressure range are efficient Raman media. Conversion efficiencies over 30–40% are readily achievable in these materials using an excimer laser as the pump beam. Figure 1.11a shows the schematic of an experimental setup using a high-pressure hydrogen cell as the Raman converter for shifting an excimer laser to longer wavelengths, and Fig. 1.11b shows the relative intensities of different-order Stokes signals

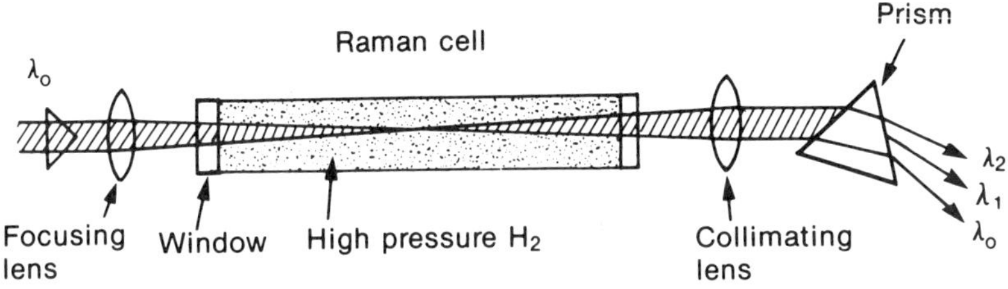

(a) Schematic of a Raman converter

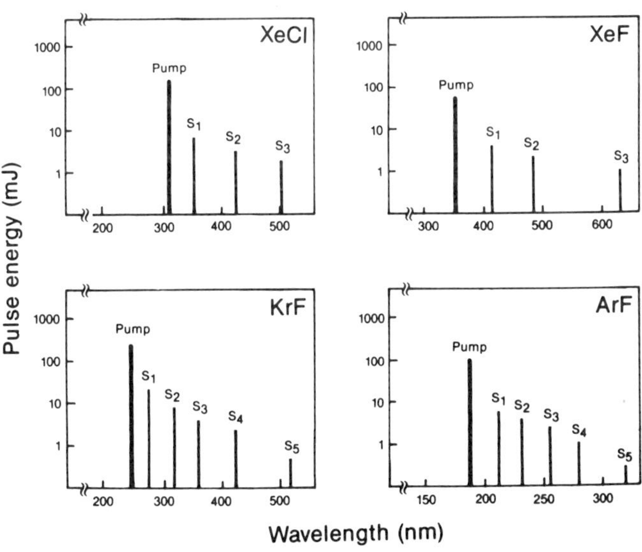

(b) Spectral output of Raman-shifted excimer lasers

Fig. 1.11 (a) the schematic of a Raman converter using a high-pressure hydrogen gas and (b) spectra outputs of Raman-shifted excimer lasers and the relative output powers at various Stokes lines.

generated from hydrogen pumped with ArF, KrF, XeCl, and XeF excimer lasers (Brink et al., 1982; Loree et al., 1979). A good review paper on stimulated Raman scattering is given by White (1987).

2.7. Incoherent UV Sources

Commonly used incoherent UV sources are mercury arc lamps, deuterium or hydrogen lamps, and high-pressure xenon flashlamps. For the low-pressure mercury lamp, the strongest emission lines are atomic-mercury $^3P_1$–$^1S_0$ and $^1P_1$–$^1S_0$ transitions at 253.7 nm and 184.9 nm, respectively. Because of strong self-absorption, these resonant lines occur only in the low-pressure mercury lamp. The low-pressure mercury lamp, operated by electrode or electrodeless discharge, is the most useful deep-UV source for inducing photosensitized reactions (Calvert and Pitts, 1967). Inert gases such as krypton or xenon are often used to mix with mercury to improve the irradiating intensity.

When a low-pressure Hg-inert-gas discharge operates at high-current densities over 1 A/cm^2, the Hg-ion resonance transition ($6p^2P^0_{1/2}$ to $6s^2S_{1/2}$) at 194.2 nm becomes the dominant deep-UV emission, and the UV output increases superlinearly up to several tens of A/cm^2, whereas the intensities of the Hg-atom lines at 253.7 nm and 184.9 nm are saturated at a lamp current density around 1 A/cm^2 (Johnson, 1971). This

Table 1.3 Dominant Spectral Line Energy Distribution in Low- and Medium-Pressure Mercury Lamps (max = 100).

Wavelength (nm)	Low-Pressure Mercury Arc	Medium-Pressure Mercury Arc
222.4	—	14
238	—	9
240	—	7
248.2	—	9
253.7	100	17
265.2–265.5	—	15
280.4	—	9
296.7	0.2	17
302.2–302.8	—	24
312.6–313.2	0.6	50
334.1	—	9
365–366	0.5	100
404.5–407.8	0.4	42

Calvert and Pitts, 1967

type of high-current density Hg-inert-gas discharged lamp is useful as a deep-UV source for inducing photochemical reactions.

When a mercury lamp is operated at medium (subatmosphere) to high pressure (several to several hundred atmospheres), the emission below 200 nm disappears because of self-absorption. Meanwhile, stronger emission lines in the spectral range between 222 nm and 580 nm occur. These lines radiate from mercury excited-states other than $^3P_1$. The medium-pressure mercury lamps are commonly used for photolysis studies.

Typical deep-UV Hg-Xe lamps with rated powers of 1 kW have a spectral intensity of about 100 μW/cm^2 in 180–235 nm, about 1 mW/cm^2 in 180–300 nm, about 380 μW/cm^2 in 360–370 nm, and about 130 μW/cm^2 in 400 to 410 nm, measured on a surface irradiated normally at a distance 1 m away from the source. Table 1.3 summarizes the energy distribution

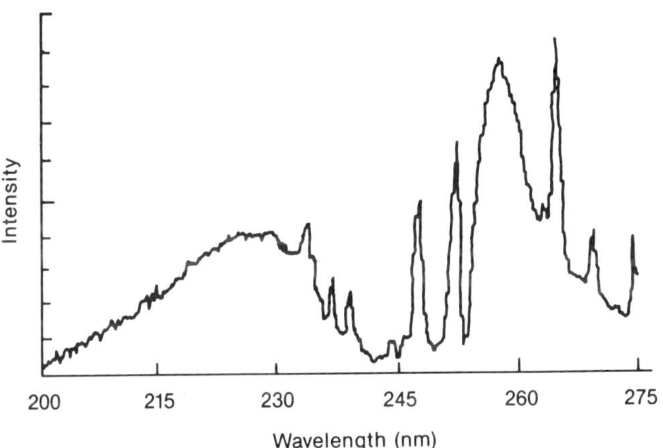

Interval (nm)	Power (Watts)
200-215	11.15
215-230	33.27
230-245	19.45
245-260	40.79
260-275	35.71
275-290	17.50
290-305	12.04

Fig. 1.12 The UV spectrum and output power of a microwave-discharged high-pressure mercury lamp operated at 1400 W input power with 500 hours operating life. From Matthews et al., 1983.

of the dominant emission lines of interest to chemical processing from low- and medium-pressure mercury lamps.

A more recently developed microwave discharged mercury lamp generates 114 W of deep UV in 200 to 260 nm, with an efficiency greater than 9% and 500 hours lifetime (Matthews et al., 1983). The microwave source is a magnetron operated at 1400 W and beamed into a spherical cavity. The microwave heats the argon gas, which in turn vaporizes the mercury, producing 1 to 2 atm pressure. Over 85% of the microwave power is coupled to the bulb and about 275 W are emitted in the UV range (200 to 400 nm) and 225 W in the visible range (400 to 700 nm); the rest is dissipated as radiant and convected heat. Figure 1.12 shows the emission and power output from a microwave-discharged mercury lamp with 500-hours operating life.

Other incoherent UV sources include deuterium- or hydrogen-discharged lamps. These lamps are of low UV power and are useful mainly for spectroscopic applications in the vacuum UV below 160 nm.

3. Optical Considerations for Direct Writing

The increasing complexity in VLSI design and fabrication has increased the demand for discretionary processing techniques for application specific designs, circuitry restructuring, yield enhancement, and localized masking and coating. Among various laser processing techniques, laser direct writing for circuit interconnection in VLSI and packaging applications is perhaps one of the most likely technologies to have a major impact on next-generation circuit and device fabrication. In this section we will discuss the key optical considerations in applying laser microchemical processing to direct writing: resolution, writing speed, Gaussian beam propagation and transformation, beam shaping, measuring, and scanning. Several state-of-the-art laser direct-write systems that have been developed for discretionary interconnect in gate array and packaging applications are reviewed.

The key considerations in applying laser microchemical processing to direct writing are resolution, writing speed, physical and chemical properties (e.g., morphology, dielectric or electrical properties), and mechanical properties (e.g., adhesion). Table 1.4 summarizes these key process considerations and corresponding physical, chemical, and optical mechanisms. In this section, we shall consider the first two process parameters that are optics related: resolution and writing speed. The rest of the more material- and process-related considerations are discussed in various other chapters in this book.

Table 1.4 Process Considerations in Laser Direct Writing.

Process Considerations	Related Mechanisms
Resolution	Wavelength Beam spot size Thermal diffusivity of substrates Nonreciprocity
Writing speed	Beam size Film growth rates
Resistivity	Composition Impurity Physical structures
Morphology	Coherence effects Nonuniform heating Instabilities
Adhesion	Interfaces Thermal mismatch

3.1. Resolution

The spatial resolution in a laser direct writing process depends upon the laser spot size, the thermal property of the substrate, and the type of chemical reactions involved. When a laser beam with a Gaussian intensity distribution is incident upon an absorbing substrate, the local temperature increases as a result of absorption of the incident laser energy. A unique characteristic in laser microchemical processing is the nonreciprocal property caused by the strong nonlinear temperature dependence of the local reaction rate (Ehrlich and Tsao, 1984). Depending upon the activation energy of the reactions employed, localization of the chemical reaction can result in an improvement of the writing resolution such that the structure produced is smaller than the diffraction-limited focused spot size. For example, Fig. 1.13 shows the temperature profiles induced by an argon laser with a Gaussian intensity, $I(r) = I_o \exp(-2r^2/\omega_o^2)$ for 1-μm spot size. The calculated relative reaction rate and the laser-beam profile, normalized to the peak surface temperature, are also shown. The one-dimensional thermal calculation assumed an Arrhenius-type rate with an activation energy of 2 eV. The reaction as shown is confined to a width less than the laser-beam size. The degree of confinement depends upon the values of the activation energy, thermal diffusivity of the substrate, and the dwell time. Higher activation energies result in a higher degree of confinement and reduced linewidth. Fabrication of submicron structures have been reported in a variety of systems including

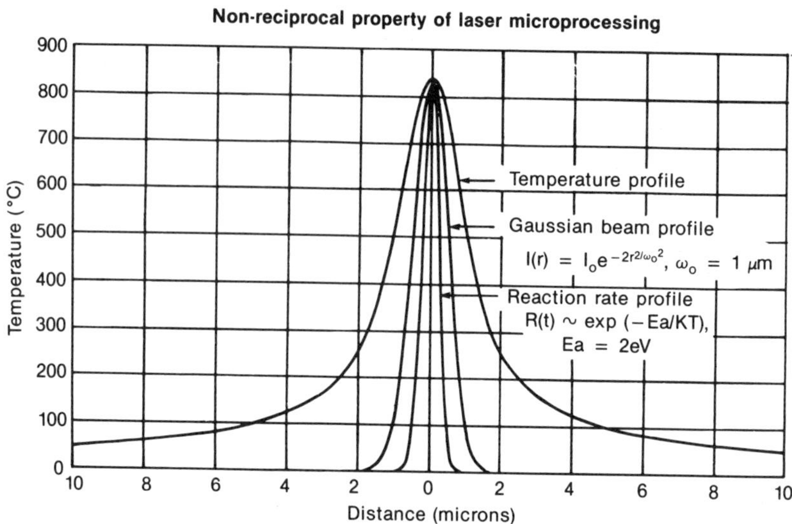

Fig. 1.13 Temperature profile, the relative local reaction rate and beam profile (normalized to the peak temperature at the beam center) for an argon laser beam with a Gaussian intensity distribution $I(r) = I_o \exp(2r^2/\omega_o^2)$ with $\omega_o = 1\ \mu$m, incident upon an amorphous silicon layer deposited on a SiO_2 substrate and scanned at 1 cm/s. In the calculation, one-dimensional thermal conduction is assumed and the reaction is Arrhenius type, with an activation energy of 2 eV. The temperature profile is plotted in the direction transverse to the scanning direction. From Liu, 1985.

photoinduced deposition (Ehrlich et al., 1982), photoelectro-chemical etching of GaAs (Podlesnik et al., 1983), pyrolytically deposited polysilicon (Ehrlich et al., 1981), and laser-induced tungsten deposition via silicon surface reduction reaction (Liu et al., 1985). Ehrlich and Tsao (1983) have derived the following expressions for the minimum resolutions, L_o, obtained in

a. a photoresist process

$$\frac{L_o}{2\omega_o} = \frac{\log(2) + 1/\gamma^*}{[2\log(e)]^{1/2}} \qquad \text{and}$$

b. an Arrhenius-type process

$$\frac{L_o}{2\omega_o} = \frac{1}{[2\log(e)\gamma^*]^{1/2}},$$

where $2\omega_o$ is the beam diameter and γ^*, analaglous to the γ factor used commonly to describe the contrast factor in the photoresist process

(King, 1981), is defined as

$$\gamma^*(I_p) = \ln(10)\left[\frac{d(\ln(f))}{d(\ln I)}\right] \quad \text{for} \quad I = I_p,$$

where f is the reaction rate $[f = f(I)]$ evaluated at the operating intensity. It can be shown that for an Arrhenius-type reaction at temperature T, the γ^* is

$$\gamma^* = \frac{\ln(10)E_a T}{k(T + T_s)^2},$$

where E_a is the activation energy of the process and T_s is the substrate temperature. This result suggests that in the conventional photoresist process the resolution limits are about $0.6 \times (2\omega_0)$ for γ^* exceeding 10, and saturate more or less for γ^* much larger than 10. In the microchemical process, however, the resolution increases steadily to about $0.15 \times (2\omega_o)$ with increasing γ^*. For γ^* above 20, a linewidth smaller than $0.25\omega_o$ is achievable. The minimum resolution achievable in a project-patterning process and in a laser microchemical direct-write process are shown in Fig. 1.14 (Rothschild and Ehrlich, 1988). Experimentally, Podlesnik et al. (1983) achieved submicron grating structures using laser etching of GaAs, and Ehrlich and Tsao (1984) demonstrated 0.25-μm linewidth in laser-induced polysilicon deposition. Table 1.5 shows the effective γ^* for various laser-writing schemes.

3.2. Writing Speed

Another unique property observed in a laser microchemical process using gas-phase deposition is the enhancement of the available reaction flux. The reaction rate in a heterogeneous reaction at a gas-solid interface is generally limited either by diffusion of reactants and/or products or by reaction rates on the solid surface. The reaction flux channeled into the reaction zone increases at high pressure as the dimension of a reaction zone decreases to a value small compared with the gas diffusion distance. On the other hand, for an extended heated surface, the reaction flux at high pressure is usually limited by gas-phase diffusion. From a geometrical scaling consideration, net reaction rates in a microchemical process induced by a focused laser beam can be several orders of magnitude faster than those observed in a diffusion-limited case such as a furnace-heated low-pressure CVD process. The steady-state molecular reaction flux as a function of pressure for various values of beam-spot radii ranging from 1 μm to 1000 μm is calculated and shown in Fig. 1.15 (Ehrlich and Tsao, 1983). It shows that a reaction flux of 1.0×10^{21} cm^{-2} s^{-1} is available at a pressure of 100 torr for a 10-μm focused

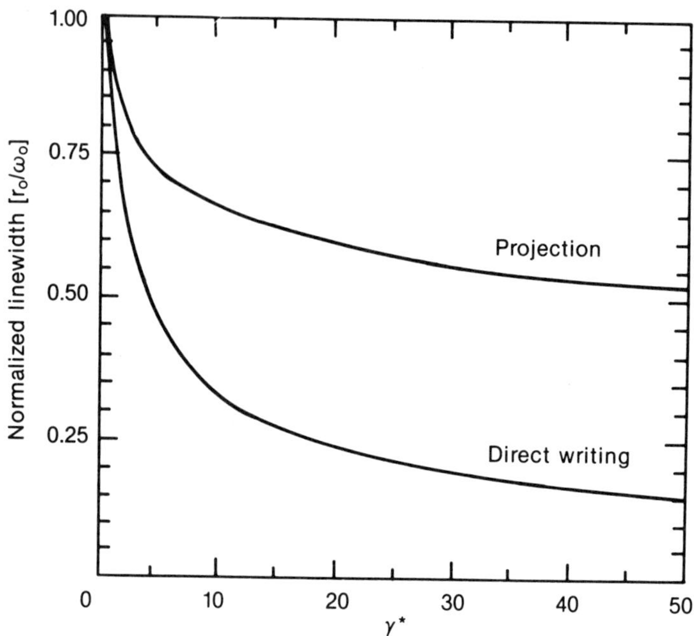

Fig. 1.14 The calculated resolution normalized to the laser beam diameter, ω_o, is plotted as a function of γ^* for direct writing and projection patterning. The resolution is defined as the smallest separation between two parallel exposed lines. From Rothschild and Ehrlich, 1988.

Table 1.5 Effective γ^* for Various Thermal Processes in Laser Direct Writing (Ehrlich and Tsao, 1983).

	E_a(eV)	T_m(K)	γ^*
(I) Etching			
Si/Cl_2	1.7	1680	22
Ge/Cl_2	1.1	1200	18
Si/HCl	0.43	1680	5.8
(II) Thermal decomposition			
SiH_4	0.74	1400	11
$SiCl_4/H_2$	1.9	1400	30
(III) Doping			
Si	3.69	1680	48

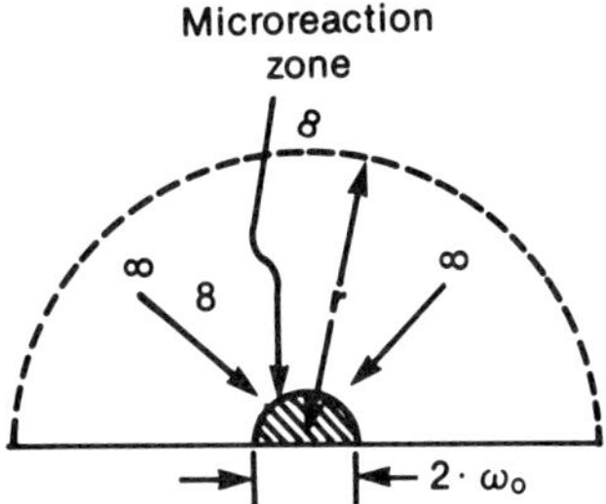

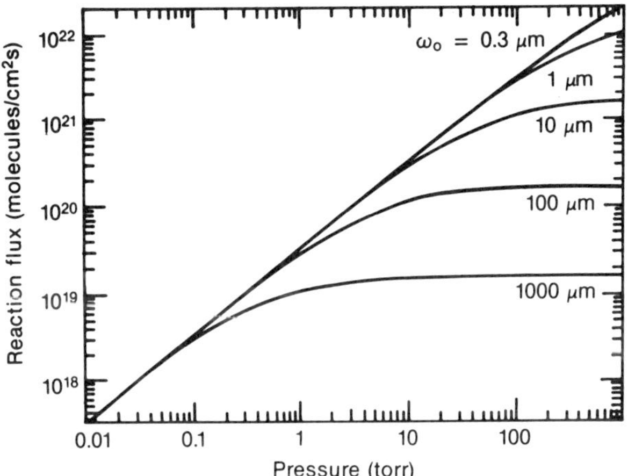

Fig. 1.15 Available reaction flux as a function of pressure for various laser beam spot sizes. From Ehrlich and Tsao, 1983.

Gaussian beam, and the value reduces to $1.0 \times 10^{19}\,\mathrm{cm}^{-2}\,\mathrm{s}^{-1}$ for a Gaussian beam of 1-mm radius.

The writing speed in a laser direct-writing process is further dictated by the beam size, growth speed, and the final film thickness required for a specific application. The ultimate writing speed can be calculated using a simple model shown in Fig. 1.16 in which D is the beam diameter, d the film thickness, V_g the film growth velocity, and V_S is the lateral writing speed. The values D and d are dictated by the resolution and conductance required. The growth velocity V_g further depends upon the type of reaction, pressure, temperature, and the mass transport discussed earlier.

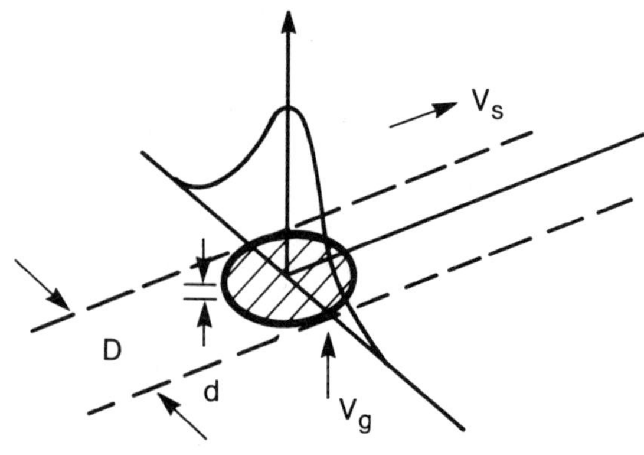

where:

$$V_s = \frac{D}{d} V_g$$

V_s = Scanning speed

V_g = Growth velocity

D = Beam diameter

d = Film thickness

Fig. 1.16 A simple model for relating the writing speed, film growth speed, thickness, and width. From Liu, 1985.

For example, a writing speed of 1 mm/s is required for writing a 1-μm thick and 10-μm wide line at a film growth rate of 100 μm/s (Liu, 1985). At this writing speed, a 1-m long interconnect requires 16.6 min to complete.

3.3. Gaussian Beam Propagation

In direct-writing applications, a focused beam with a Gaussian radial intensity distribution is most commonly employed. A laser resonator supports many transverse modes, and the Gaussian beam is usually the lowest-order mode (TEM_{00}) and is characterized by its unique propagation and transformation properties. These properties are discussed in this section. The radial intensity distribution of a Gaussian beam is described by

$$I(r) = I_o \exp\left(\frac{-2r^2}{\omega^2}\right), \tag{1.1}$$

where ω is the beam radius at which the intensity falls to e^{-2} of its value I_o on the axis. The spot size of a Gaussian beam is defined as 2ω. The total power of a Gaussian beam is obtained by integrating Eq. 1.1 and is equal to $\pi\omega^2 I_o/2$.

A Gaussian beam propagating in a diffraction-limited optical system remains always a Gaussian beam when the aperture effect is not considered. A Gaussian beam can be described by the radius function $\omega(z)$ and the curvature function $R(z)$ along the propagation direction z. The function $\omega(z)$ and $R(z)$ are given by (Kogelnik and Li, 1966; Dickson, 1970).

$$\omega(z) = \omega_o\left[1 + \left(\frac{2z}{b}\right)^2\right]^{1/2} \tag{1.2}$$

$$R(z) = z\left[1 + \left(\frac{b}{2z}\right)^2\right], \tag{1.3}$$

where λ is the wavelength, z is the propagation distance away from the beam waist ω_0 at $z = 0$, and

$$b = \frac{2\pi\omega_o^2}{\lambda} \tag{1.4}$$

is called the *confocal parameter* and is the distance within which the diameter of a focused beam remains almost constant (i.e., depth of field). The region, $-b/2 < z < +b/2$, is sometimes called the *Rayleigh range*. Equations 1.2 and 1.3 show that the propagation properties of a Gaussian beam are determined once the beam waist is given. The beam radius varies from ω_o at $z = 0$ to $\sqrt{2}\omega_o$ at $z = b/2$, while the curvature of the beam changes from infinite (plane wave) at $z = 0$ to b, the minimum value, at $z = b/2$. For $z \gg b$, the beam radius increases linearly as a function of z, with a divergence angle equal to

$$\theta = \frac{\lambda}{\pi\omega_o} = \frac{2\omega_o}{b}. \tag{1.5}$$

This relationship shows that the beam waist ω_o, beam divergence θ, and confocal parameter b are related to each other. When a laser beam emerges from a cavity, the minimum beam radius (beam waist) is usually (but not always) located at the output mirror. For example, a He-Ne laser (638 nm) with a 0.5-mm beam waist has a confocal parameter b of 2.46 m and a 0.4-mrad beam divergence. The variation of a Gaussian beam near its beam waist is shown in Fig. 1.17, including the definitions of beam waist, confocal parameter, and beam divergence.

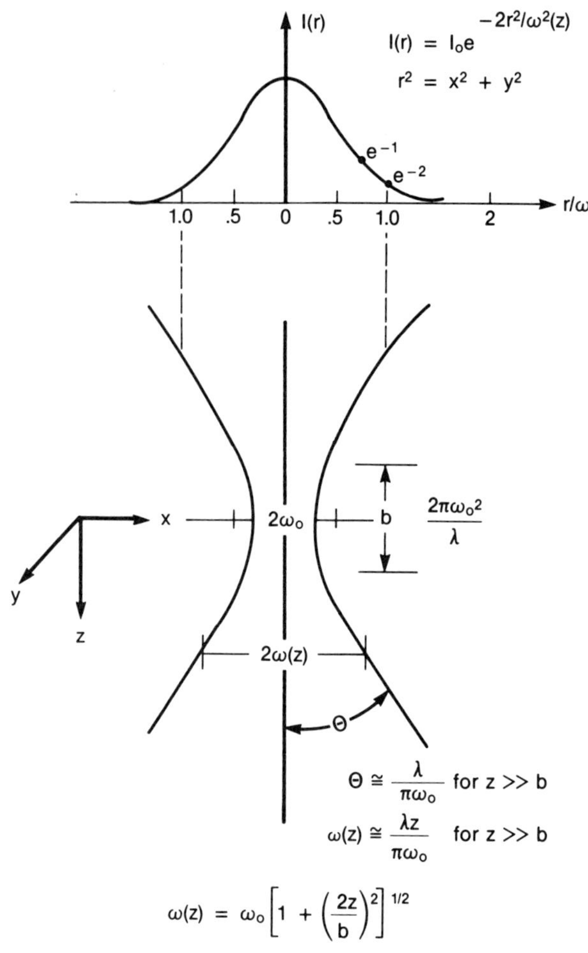

Fig. 1.17 Gaussian beam profile at near the beam waist ω_o and definitions of confocal parameter b and beam divergence angle θ.

3.4. *Transformation of a Gaussian Beam*

A Gaussian beam can be transformed from one set of beam parameters (ω_1, b_1, or θ_1) to a new set of beam parameters (ω_2, b_2, or θ_2) using a lens system. The transformations such as focusing, collimating, or expansion depend upon the specific application desired. Figure 1.18 shows the transformation of a Gaussian beam by placing a focusing lens of focal length f at a distance d_1 from the initial beam waist ω_1. A new

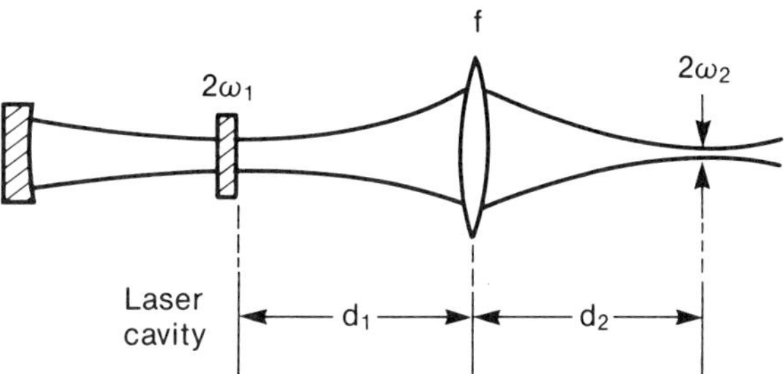

Fig. 1.18 Transformation of a Gaussian beam with a lens. (f is the focal length, d_1 and d_2 are the distances between the lens and the location of the original beam waist at ω_1, and the new beam waist at ω_2, respectively.)

beam waist ω_2 is formed at a distance d_2. the relationships between d_1, d_2, and f are given by the following equations, (Kogelnik and Li, 1966)

$$d_1 = f \pm \frac{b_1}{2}\left[\left(\frac{f}{f_0}\right)^2 - 1\right]^{1/2} \tag{1.6}$$

$$d_2 = f \pm \frac{b_2}{2}\left[\left(\frac{f}{f_0}\right)^2 - 1\right]^{1/2}, \tag{1.7}$$

where

$$b_1 = \frac{2\pi\omega_1^2}{\lambda},$$

$$b_2 = \frac{2\pi\omega_2^2}{\lambda},$$

and

$$f_0 = \sqrt{\frac{b_1 b_2}{4}}.$$

Here, f_0 is the minimum focal length such that any lens with a focal length $f > f_0$ can be used to give the desired transformation. In practice, usually the values of b_1 (beam waist of the laser), f (focal length of a lens), and d_1 (the location of the lens from the beam waist) are given, and one wants to know the values of b_2 and d_2; then the following formulas are useful:

$$b_2 = \frac{b_1 f^2}{(d_1 - f)^2 + (b_1/2)^2} \tag{1.8}$$

$$d_2 = f + \frac{(d_1 - f)f^2}{(d_1 - f)^2 + (b_1/2)^2}. \tag{1.9}$$

The two most commonly used focusing configurations are:

a. For $d_1 \ll b_1$, i.e., the lens is placed in the near field, we have

$$\omega_2 = f \cdot \theta = \frac{f\lambda}{\pi\omega_1} \tag{1.10}$$

b. For $d_1 \gg b_1$, i.e., the lens is placed far out of the Rayleigh range,

$$\omega_2 = \frac{\omega_1 f}{d_1}. \tag{1.11}$$

Equation 1.10 shows that in order to achieve a smaller focused spot, one has to either increase the initial beam waist ω_1 or reduce f or λ. Figure 1.19 shows that the diffraction-limited focused spot diameters and the corresponding depths of focus that can be achieved as a function of focal length for lasers at 0.19, 0.25, 0.50, 0.75, and 1.0 μm, using an incident beam diameter equal to 10 mm.

3.4.1 Collimation

From Eq. 1.10, it follows that the minimum spot achievable with a focusing lens of focal length f is limited by the beam divergence (θ). To obtain a smaller spot size, the beam divergence θ has to be reduced through collimation. In Figs. 1.20a and 1.20b, the normalized values of d_2/f, and b_2/f are plotted as a function of d_1/f for various values of b_1/f (Eq. 1.8 and Eq. 1.9). d_2 is maximum when $d_1 = f + (b_1/2)$, namely, when the collimating lens is placed at a distance equal to its focal length plus half a Rayleigh distance. Under this optimum collimation condition, the new beam waist locates at a distance equal to

$$d_2 = f + f^2/b_1.$$

These results show that the best collimation, i.e., the maximum value for d_2, is achieved by using a small b_1. This suggests that one can use a pair of short and long focal length lenses to achieve an excellent collimation of a laser beam to minimize the beam divergence. Figure 1.21 shows two commonly used collimating optical systems: a Keplerian telescope and a Galilean telescope (O'Shea, 1985, Kingslake, 1983).

3.5 Beam Shaping

In many applications, it is desirable to have a beam with an intensity distribution that is not Gaussian. For example, in laser projection printing, one would like to have a uniform intensity at the focal plane of the lens. For laser recrystallization, it is crucial to control the melt front

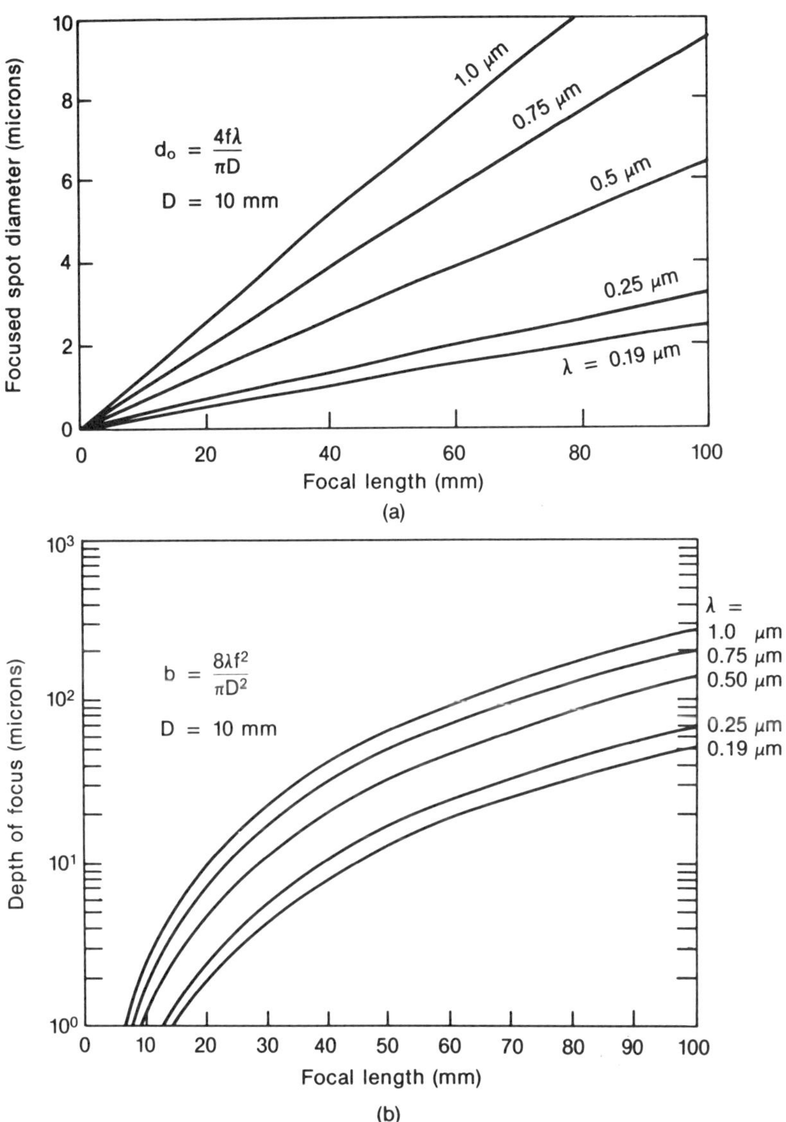

Fig. 1.19 (a) Shows the diffraction-limited focused spot diameters achievable using various focal length lens for lasers at $\lambda = 0.19$, 0.25, 0.50, 0.75, and 1.0 μm, (beam diameter $D = 10$ mm) and (b) shows depths of focus plotted as a function of the focal length for each of the corresponding wavelengths.

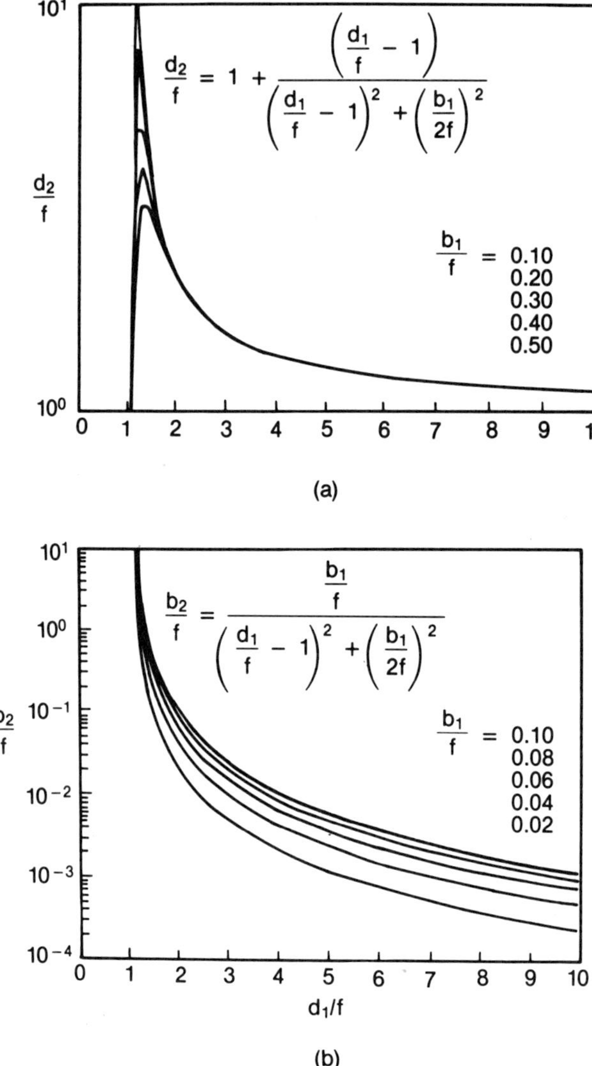

Fig. 1.20 (a) The normalized values of d_2/f, and (b) the normalized values of b_2/f are plotted as a function of d_1/f for various values of b_1/f shown. The curves are calculated from Eqs. 1.8 and 1.9.

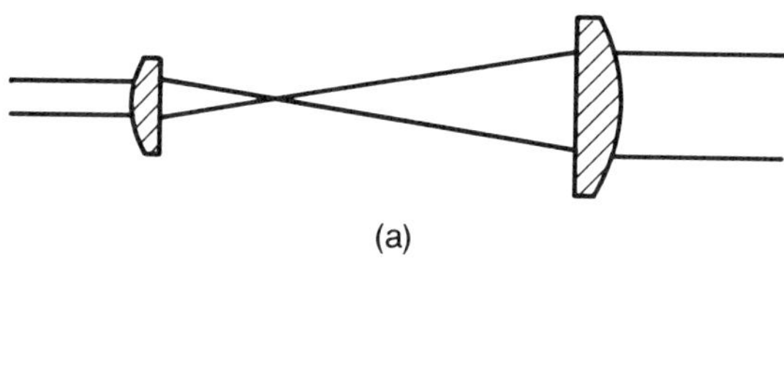

(a)

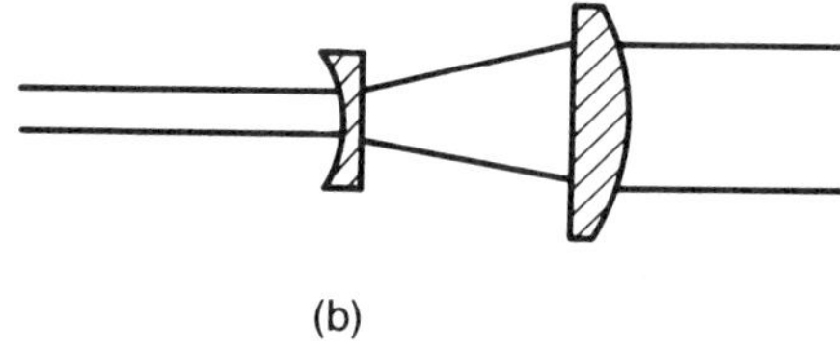

(b)

Fig. 1.21 Collimation of a Gaussian beam: (a) Keplerian telescope and (b) Galilean telescope.

using beam shaping to prevent the growth of unwanted grain boundaries (Biegelsen et al., 1981). In laser annealing, a spatially uniform beam is required (Hill, 1982). In general, techniques for shaping a Gaussian beam into a uniform intensity are a) redistribution of the intensity spatially via absorptive (apodization) (Ih, 1972), reflective (homogenization) (Grojean et al., 1980; Cullis et al., 1979), or refractive (aspherical optics) (Rhodes and Shealy, 1980) optical methods and b) diffraction methods such as phase plates and holographic techniques (Veldkamp, 1982). In the following, we discuss two of the more useful beam-shaping techniques using phase plates and beam homogenization.

The principle of using a phase plate to shape a beam intensity distribution derives from the fundamental relationship that the amplitude of a beam profile in the image plane is the Fourier transform of that in the object plane (Champeney, 1973). Therefore, by spatial modulating the phase of a laser beam, the beam intensity at its focal place can be redistributed to the desired form. For example, a transmissive or reflective binary phase grating can be used to shape a beam intensity in one or two dimensions from a centrosymmetric Gaussian distribution into a far-field, flat-top profile (Veldkamp and Kastner, 1982). A binary diffraction grating is used to disperse beam energy to the shoulders of the Gaussian bean profile, where it constructively interferes with natural beam divergent components of the zeroeth diffraction order, whereas in

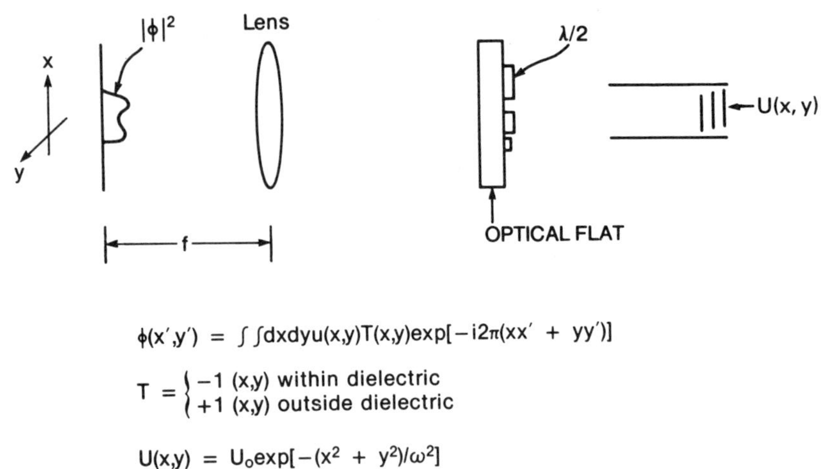

Fig. 1.22 Transformation of a Gaussian beam into a dumbell-shape intensity distribution at the focal plane using a binary phase grating. From Possin et al., 1983.

the tails, destructive interference takes place to produce a flat-top profile with a steep intensity dropoff. Figure 1.22 is a binary transmissive phase grating used to transform a Gaussian beam into a dumbbell-shaped intensity distribution at the focal plane (Possin et al., 1983).

A Gaussian intensity profile can be transformed into a uniform profile using a computer-generated holographic technique. Han et al. (1983) used a computer to generate a pair of holograms. The first hologram was designed to deflect the incident rays to convert the incoming Gaussian distribution into a uniform one when the laser beam reaches the second holographic plane, while the second hologram deflects the rays into a collimated beam with a uniform distribution. Since the diffraction efficiency of a volume phase hologram is high, this technique of beam shaping offers a high transmitted efficiency. The holographic method for beam shaping has been applied for transformation of a laser beam into a line source of required curvature and numerical aperture (Jain et al., 1984a, 1984b). The technique was developed for shaping the output from an excimer laser into a curved geometry to match the projection optics in a lithography system.

In laser recrystallization of polysilicon island structures, it is critical to shape the beam profile to match the melt front in order to control nucleation. Since crystal growth is driven by the local temperature gradient, irradiation with a Gaussian beam leads to competitive grain growth and spontaneous solidification. A way to stabilize the crystal growth is to use a shaped beam, which provides a concave trailing edge. Possin et al. (1983) used a phase plate for beam shaping to produce a concave melt front. The phase plate was fabricated by patterning a dielectric film that was about one-half wave thick onto a fused quartz flat. The phase plates were patterned using sputtered indium tin oxide (ITO). The phase plate was inserted in the beam path before the objective lens. The incident field amplitude $U(x, y)$ is Gaussian. The field amplitude at the focal plane ϕ is the Fourier transform of $U(x, y)$ (Fig. 1.22). Figure 1.23 shows the shaped beam intensity at the focal plane measured using an IR imaging technique and compared with a calculated intensity distribution. This beam-shaping technique has flexibility in modifying the beam distribution in the focal plane by changing the position of the plate

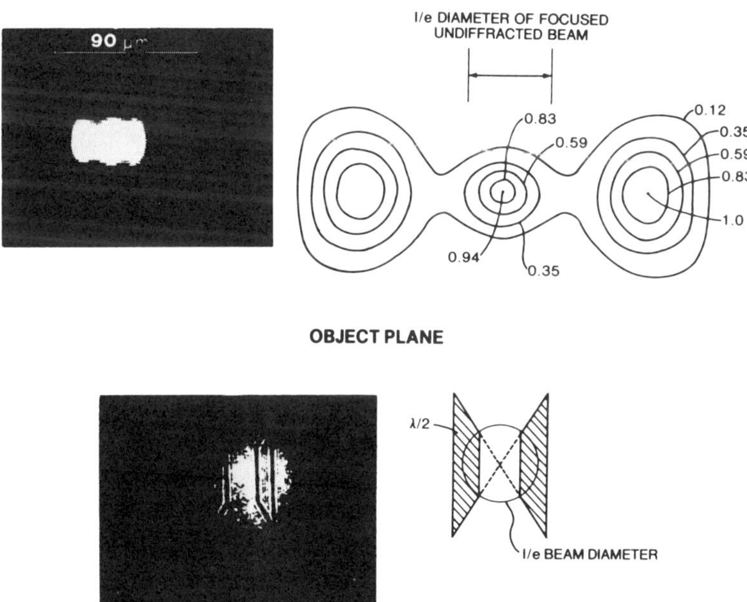

Fig. 1.23 The photographs of the IR images of the original Gaussian beam intensity incident on a phase plate (object plane, bottom) and of the shaped-beam intensity at the focal plane. (image plane, top). A calculated intensity distribution at the image plane using Fourier transformation is shown at the top right. From Possin et al., 1983.

in the beam path, by varying the divergence of the beam using a telescope, or by varying the focus of the objective. This beam-shaping technique is simple and power efficient.

3.6. Beam-Homogenization

In many incidences, the laser beam from a laser cavity is not Gaussian due to the presence of higher order modes, distortion, or diffraction. Under these circumstances, beam homogenization is required to produce a spatially uniform beam for device fabrication or material processing. Multiple scanning can be used to spatially integrate over a scanned area larger than the beam spot size. Another simple method is to place a diffuser such as sand-blasted fused-quartz plates or ground-glass sheets in front of the sample to spatially homogenize a nonuniform beam. This method, however, suffers large transmission loss, and, depending upon the degree of coherence of the source and the structure of the diffuser used, considerable speckle noises can be introduced (Lahart and Marathay, 1975). Cullis et al. (1978, 1979) used a diffuser and ligh-guide structure to homogenize a nonuniform beam, while eliminating the lossy scattering (Fig. 1.24). The laser output is directed onto the input face of an optical light guide with a flat entrance surface finished by grinding with 14-μm diamond powder on a lapping plate. The beam that enters the optical light guide is therefore strongly diffused. Most of the scattered light beams are in the forward direction within the light cone confined by the acceptance angle of the light guide. Only a relatively small fraction of the incident light is scattered through large angles at the input entrance face and is lost. When the collected light propagates down the light guide, repeated internal reflection further diffuses the light and homogenizes the beam uniformly. A 90° bending is introduced into the light guide to homogenize the on-axis component, which otherwise does not undergo internal reflection in a straight light guide. Such a bending, however, introduces caustic patterns and requires a further length of the light guide to rehomogenize the beam. In addition, a taper section was introduced before the light exit of the light guide to increase the beam power density. The exit face of the light guide is highly polished to reduce the loss. The intensity distribution near the exit face of the light guide can be highly uniform. The divergence of the output beam is relatively large so that specimens to be irradiated have to be placed as close to the exit face as possible (within 1 mm). To project onto a remote specimen, a collimating lens system with a large numerical aperture is required. The light guide is made of optical transparent materials; among them silica is the most commonly used. With a transmission spectral

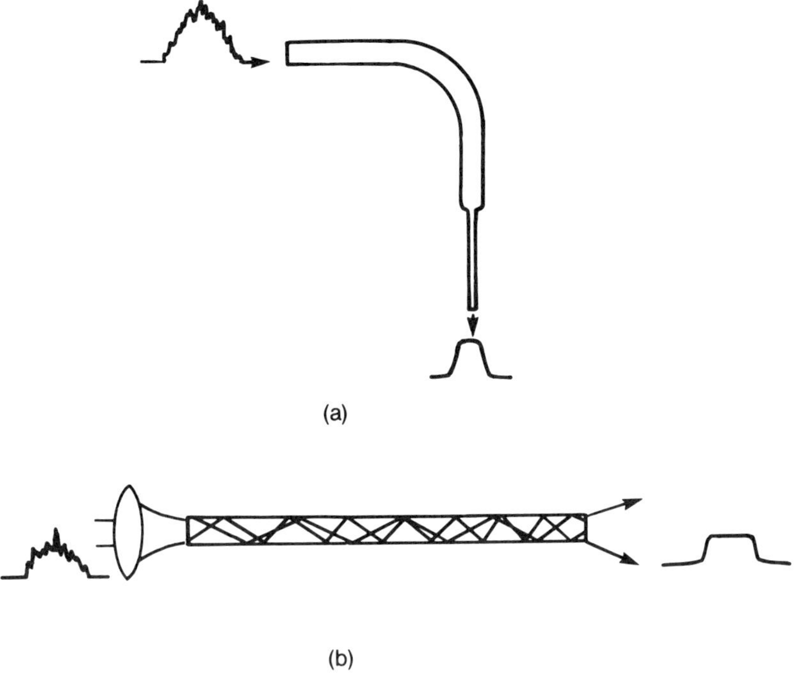

Fig. 1.24 Beam homogenizer (a) consisting of a tapered quartz rod bent 90° with both input and output faces polished and (b) a straight light pipe configuration. Figure 11.24a is from Cullis et al., 1979.

range between 190 nm to about 3.5 μm and a refractive index of 1.46, silica is useful for UV as well as for near-infrared lasers. The diameter of the light guide should be chosen to match the laser-beam diameter for maximum efficiency. Fused silica can handle cw power up to several hundred watts without damage. A slightly modified version of this design has been reported using a double bend in the output section to make the output less sensitive to the exact length of the output section compared to the original design (Hill, 1982). In both cases, a uniformity of 5% in intensity distribution over the output beam diameter can be obtained with a total loss of about 40%.

3.7. *Beam Profile Measurements*

Since the measurement of a beam profile is critical for determining the energy or power density irradiated upon a sample surface, it is desirable to have a simple and reliable technique to measure the spot size accurately. Means such as vidicon and silicon detector arrays provide real-time monitoring of a laser-beam profile. Measuring the submicron beam profile at the focused spot, however, requires more sophisticated techniques.

A commonly used method for measuring the beam profile of micron dimension is the knife-edge technique, in which a sharp edge is used to scan across the focused spot, and the transmitted or reflected beam intensity is monitored. For a Gaussian spot, the amount of light transmitted or reflected can be calculated and compared with the measured values to yield the beam diameter (Suzaki and Tachibana, 1975; Khosrofian and Garetz, 1983). For a Gaussian beam,

$$I(x) = \left(\frac{2P_o}{\pi\omega^2}\right) \exp\left[\frac{-2x^2}{\omega^2}\right], \tag{1.12}$$

where P_o is the total power and ω is the e^{-2}-beam radius. When the laser beam scans across a knife edge, the transmitted or reflected light intensity is given by the error function,

$$P(x) = \sqrt{\frac{2}{\pi}}\frac{P_0}{\omega}\int_{-\infty}^{x} e^{-2x^2/\omega^2}\, dx. \tag{1.13}$$

Using the knife-edge technique, one should be aware that when the spot size is comparable to the diffraction-limited beam diameter, the knife-edge method may yield larger beam radii than the actual values because of diffraction effects caused by the knife edge itself. Since the interference is proportional to λ/ω, where ω is the beam radius, the effect becomes important for beam sizes on the order of wavelengths.

For measuring spots of submicron dimensions, more elaborate methods are required. Schneider and Webb (1981) discussed several useful methods for measuring submicron laser spots, including the point scan method in which a fluorescent particle, which is much smaller than the beam waist, was used to scan through the focused spot and the fluorescence signal was monitored to yield the beam profile. They used 0.038-μm latex beads labeled with fluorescent isothiocyanate (FITC) and monitored the fluorescence when a bead is moving through the beam waist of a focused laser. Since the fluorescence signal is proportional to the beam intensity, the technique allows a direct measurement of the focused Gaussian spot with a submicron resolution. A drawback of this technique is that some degree of proficiency at fluorescence labeling is required before one can reliably perform the measurement.

One other method that is relatively easy to carry out is the expanded image method. The image of the focused laser spot (to be measured) formed in the image plane of the objective lens by reflection from a smooth surface in the focal plane is recorded on a film or by a vidicon imager. Knowing the magnification of the optics, one can determine the spot diameter by measuring the dimensions of the expanded image of the laser beam. Since typically the magnification is on the order of 10^2 to 10^3, the expanded image is on the order of 0.1 to 1 mm and is easy to resolve. This expanded-image technique for measuring focused-spot profiles is compatible with most laser direct-writing experiments in which a vidicon camera is usually available for viewing and monitoring.

3.8. Beam Scanning

Pattern generation is an essential element in laser direct writing. This can be accomplished by moving the sample using a positioning table, while keeping the laser beam stationary. The recent development of high precision stages provides a positioning resolution better than 0.5 μm with a high degree of position accuracy and a large scanning distance. In addition, a stationary optical beam is simpler to transfer and easier to focus into a diffraction-limited submicron spot. Two disadvantages of this scanning configuration are a limitation on its scanning speed and an inability to manipulate the beam in a nonraster manner.

An alternate approach is to keep the sample stationary while scanning the beam using a scanner such as a rotating polygon (Sherman, 1985), galvanometer (Montagu, 1985), or a holographic (Sincerbox, 1985) or acousto-optical scanner (Young and Yao, 1969, Chang, 1976). This scan configuration requires more complicated optical designs including tradeoffs between speed, resolution, field size, and accuracy (Zook, 1973). In this section, we shall discuss several scanning techniques and some important considerations for designing a scanning system for direct writing applications.

3.8.1 Resolution

The resolution of a scanning device is defined as the number of resolvable spots per scan line. In a diffraction-limited system, the resolution is determined by the ratio of the total scanning angle of the scanning device to the minimum diffraction-limited angle of the scanned beam, i.e.,

$$\text{Resolution} = \frac{\text{total scan angle } (\Theta)}{\text{diffraction-limited angle}}$$

$$= \frac{\Theta \cdot D}{a\lambda}, \tag{1.14}$$

where D is the aperture diameter of the scanner, λ is the wavelength of laser, and a is a constant dependent upon the aperture geometry ($a = 1.22$ for a circular beam, and 1 for a rectangular beam). From Eq. 1.14, it follows that the resolution of a scanning system is determined both by the scanning device and by the intrinsic divergent angle of the optical beam itself. High-resolution scanners used in reconnaissance and graphic arts with polygon or holographic scanners usually have over 10,000 resolvable spots per scan. Medium-resolution scanners found in business graphics typically have resolutions between 2000 and 10,000 spots per scan. A low-resolution scanner with a resolution less than 2000 is used in bar code and video scanning. In addition to resolution, other scanner parameters are linearity, speed, random accessibility, and cost. Commonly used scanners are discussed in the next few sections.

3.8.2 Galvanometer-Type Scanner

This type of scanner is the simplest and the least expensive scanning device with high resolution. It employs a single-element scanner over a limited range of angles; the scan mirror used in a galvanometer typically has a low inertia with a resolution of >1000 spots/scan and is commonly used in scanning laser acoustic microscopes, color separators, facsimile transmitters and receivers, laser trimmers, laser markers, laser pattern generators, and optical disk readers. A galvanometer scanner can be operated in a resonant mode when the driving frequency is near the resonant frequency of the device, typically 1000 to 3000 Hz. At the resonant mode, the device produces the maximum scanning amplitude at a high scanning rate with minimum driving power, but it cannot be randomly accessed. If a galvanometer scanner is operated below its resonant frequency, it is considered to be operating in a broadband mode. In the broadband mode, the galvanometer can be driven with sinusoidal or nonsinusoidal signals such as linear ramp and sawtooth signals, and it is randomly accessible. Random accessibility is an important consideration when the writing time, and hence the throughput, has to be minimized. The position accuracy of this type of scanner is about 50 mrad, limited by wobble, and the repeatability is about 10 mrad. A major drawback of this type of scanner is the difficulty in maintaining system rigidity while having a low-inertia armature and mirror system. Wobble-free operation is difficult to achieve. Raamont and Zaleckas (1973) developed a YAG-laser-based laser pattern generator using galvanometers as deflectors for fabrication of thin film masks. The system had 4000 by 4000 spot resolution, with a 25-μm spot size and a 142-mm diameter field. The system can be operated in a random access mode, using optical beam position detection with digital computer control, or it can be operated as a flying spot scanner in a raster-type format. Many laser pattern generators available in the market today are based on this type of scanning technology.

3.8.3 Polygon Scanner

The polygon scanner is a mechanical device providing high-speed, wide scan angle, and high-resolution scanning capability. The polygonal scanner is a unidirectional scanner with a fixed angular velocity. The scanning is achieved by rotating a polygonal mirror at a fixed angular velocity. Because of its relatively high inertia, the polygon scanner has good mechanical stability.

The basic parameters of a polygon scanner are the number of facets M and the radius of rotation R. If the facet length is L and the beam diameter is D, then the duty cycle during which the beam is fully on a single facet is $(L-2D)/L$ (Fig. 1.25). To choose a polygon scanner, one has to consider the required resolution and the beam size, then apply the scan equation (Eq. 1.14) to decide the number of facets needed. More

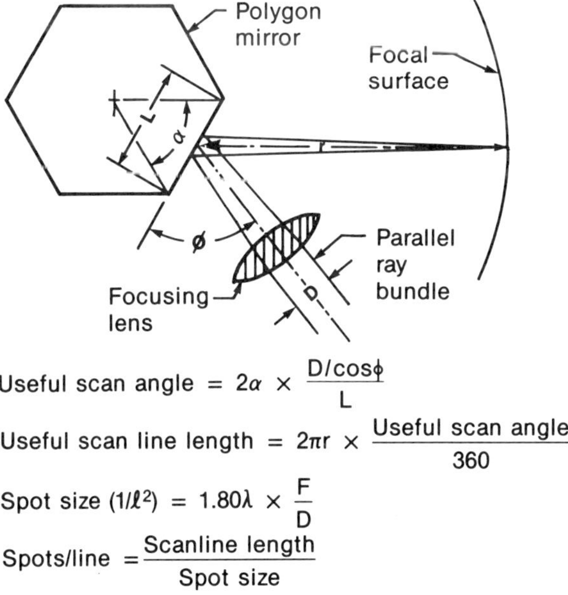

Fig. 1.25 Rotating polygon scanner. From Sherman, 1985. (F = focal length of the lens.)

facets reduce the windage noise and the scan angle, and thus the f-number of the system. The precision in fabrication of oriented facets determines the performance of a polygon scanner. Since it is achromatic, as is that for a galvanometer scanner, the polygon scanner can be used over a wide spectral range from UV to IR. Commercially available polygon scanners have a facet number as high as 60 and can accommodate beam diameters up to 70 mm. The unidirectional scanning rate varies from 30 to 20,000 Hz with a scan resolution over 20,000 spots per scan line. Current commercial polygon scanners have facet-to-facet angular error to within a few arc-sec. This type of scanner is commonly found in nonimpact printers and graphic facsimile systems (Marshall, 1985).

There are two types of scanning geometries used: prelens scanning and postlens scanning, illustrated in Fig. 1.26. In the postlens scanning geometry (also called *exit scanning*), the scanner is used to deflect a focused light beam over a scanned field that is slightly curved, with its curvature approximately equal to the distance between the scanned field and the mirror facet. When the depth of focus of the scanning beam is large, postlens scanning provides a simple scanning configuration with minimum demand on the optical components. In prelens scanning, also called *entrance scanning,* the beam is deflected via a rotating polygon, then focused through the image lens, typically a f-θ lens. The f-θ lens provides a flat field and uniform spot size over the scanned field. The type of scanning therefore requires more complicated and costly optical components.

3.8.4 Acousto-Optical Scanner

AO scanners are commonly used in IC processing for interconnects and mask repair and in the laser microscope. In these applications, the scan field is small, while a high scan rate and the random accessibility are important. An acousto-optical scanner deflects a light beam by the phase grating generated by an acoustic wave inside a solid medium. The compression and rarefractions of medium generated by the acoustic wave produce a periodic index variation and form a phase grating. A laser beam is diffracted according to the grating formula (Fig. 1.27))

$$m \cdot \lambda = 2d \sin \theta \tag{1.15}$$

where m is an integer, λ the wavelength of the laser, and d is the grating spacing. If the wavelength of the acoustic wave is equal to d, the diffraction condition for $m = 1$ becomes

$$\sin \theta = \frac{\lambda}{2\lambda_s}.$$

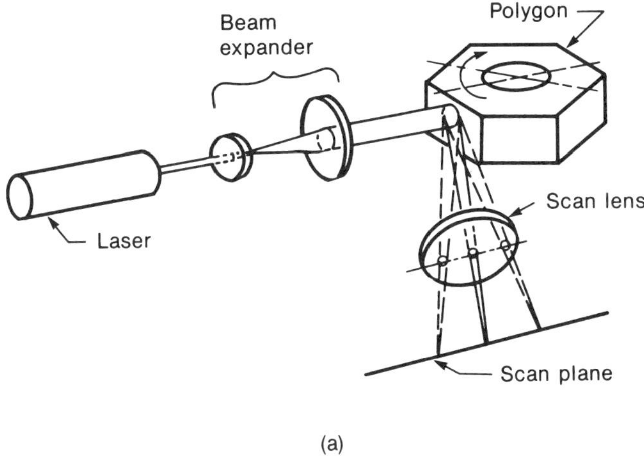

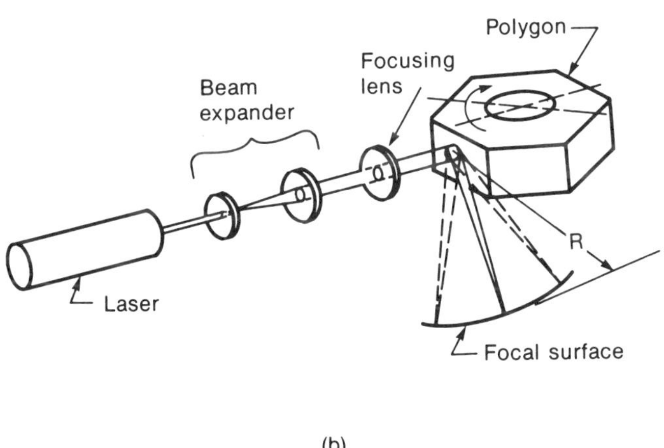

Fig. 1.26 Prelens scanning and postlens scanning. From Sherman, 1985.

For an acoustic wavelength of about 10^3 μm, one can immediately see from Eq. 1.15 that the diffraction angle of an acoustic wave is typically on the order of 10^{-3} rad. The diffraction efficiency into the first order is usually about 50% to 60%. When the frequency of the acoustic signal is swept, then the deflection angle of the laser beam is scanned accordingly,

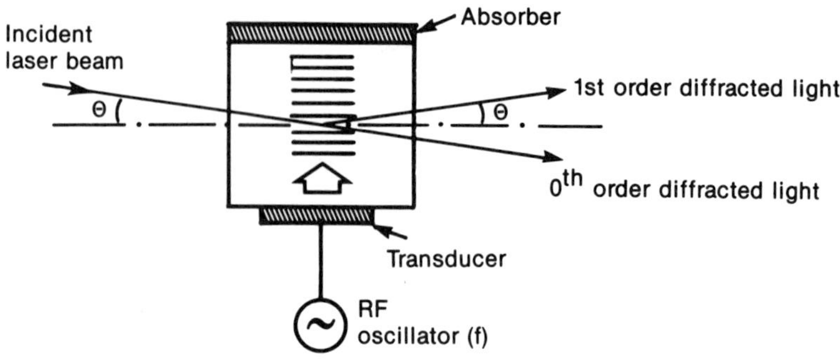

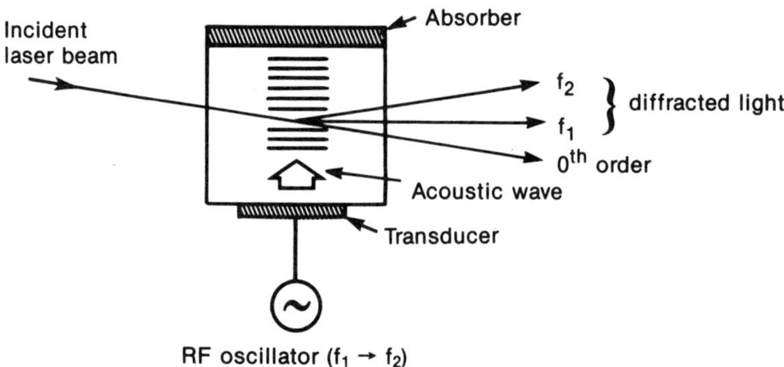

Fig. 1.27 Acousto-optical modulator and deflector.

with an angular displacement determined by the bandwidth of the frequency sweep. The resolution of an AO deflector is proportional to the product of the acoustic frequency bandwidth and the transit time of the acoustic wave across the laser beam (assuming the laser beam is uniformly illuminating the aperture). Beam shaping can be used to increase the beamwidth in the direction in which the acoustic wave propagates to increase the transit time and thus the resolution. A higher resolution is achieved, however, at the expense of frequency response. For further discussion on laser scanning devices, readers are referred to the following articles: Beiser (1986), Marshall (1985), O'Shea (1985), Zook (1974), and Dickson et al. (1982).

3.9. Laser Direct-Write Systems

3.9.1 Direct Write for Gate Array Interconnects

Early work on laser direct writing of polysilicon lines from laser-induced gas-phase decomposition of silane was reported by Bauerle et al. (1982) and Ehrlich and Tsao (1984) at scan speeds up to 100 μm/sec and linewidths as narrow as 0.2 μm. However, at these writing speeds applications for interconnect are limited to short runs such as minor circuit alterations, local restructuring, and mask repairs.

Further advances were made by Ehrlich and coworkers (1984) in directly depositing heavily doped polysilicon interconnection by local pyrolysis of SiH_4 and a dopant source gas, and applying this to prototyping CMOS gate arrays. This work has demonstrated the feasibility of using laser direct write for device customization. Direction writing of tungsten metal at speeds up to several cm/s was demonstrated by Liu (1985) using surface reduction reaction of WF_6 by silicon. Smooth tungsten lines of several tens of centimeters long, with linewidths of 2 to 15 μm, were deposited at writing speeds up to several cm/s. These results were achieved with a 514-nm argon laser at a power of about 50 mW, with a focused spot size of about 20 μm. The fabricated tungsten lines have excellent surface morphology, with thickness self-limited to about 0.2 μm. Black et al. (1986) applied selective tungsten deposition technique on laser-direct written polysilicon to significantly reduce the resistivity of laser-written interconnects to less than 10 μΩ-cm and made the laser process useful to interconnect complex gate arrays using less than 2-μm CMOS technology.

These earlier results have led to the development of a laser-based direct-write system by LASA (LASA Industries, Inc., San Jose, CA) using chemical vapor-phase (CVD) deposited tungsten as the interconnect metal for fabircation of application-specific IC (ASIC). The system is designed to interconnect up to 50,000 gate arrays with two levels of metal, for a total interconnect run about 1.5 m long. The processing time for 10,000 gates is about 30 min. The CVD reactor contains a gas mixture made up of tungsten hexafluoride, chlorine, nitrogen, and silane. An argon laser is focused to a 1-μm spot at about 1 cm/s writing speed. The process includes laser direct-writing a first metal layer of tungsten, deposition of silicon-dioxide for dielectric isolation interlayer, direct-etching vias for interconnect, direct-writing a second metal level of

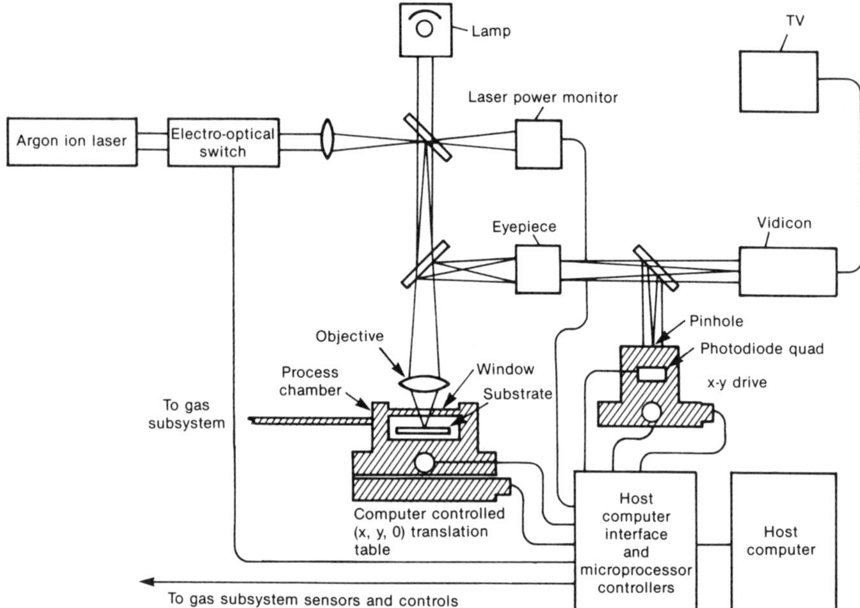

Fig. 1.28 The LASA QT-GA system developed for direct writing of tungsten lines for gate arrays interconnect. From Burggraff, 1988.

tungsten, and final sealing of the package. A schematic of the QT-GA (quick tune gate array) system developed by LASA is shown in Fig. 1.28.

3.9.2 Laser Pantography System

The laser pantography system for interconnection of customized circuits was developed at Livermore National Laboratory (McWilliams et al., 1984; Whitehead et al., 1986; Berhhardt et al., 1987) and has been applied to commercially available CMOS gates arrayed with 5-μm design rules and single-level metallization. The reaction gas used in this process is a mixture of silane, disilane, and dopant gas—phosphine for n-type doping and diborane for p-type doping—with a total reactant gas pressure of about 1 atm. The apparatus used in this work is shown in Fig. 1.29, in which an argon laser at 514 nm was focused through the reaction chamber window by a 20X microscope object to a ca. 1-μm spot on the substrate. The beam is scanned by an orthogonal pair of acousto-optic crystals to provide a 512-μm by 512-μm field. The direct writing is

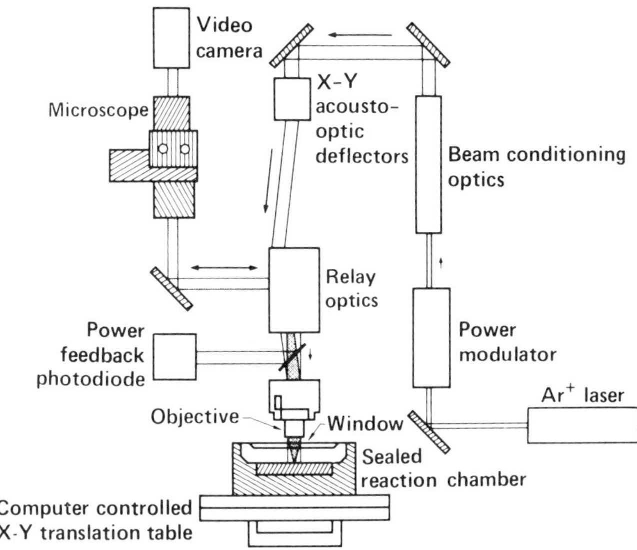

Fig. 1.29 Livermore laser pantography system layout for polysilicon direct write by solid-state acousto-optic scan over a 512-μm by 512-μm field. From Bernhardt et al., 1987 and McWilliam et al., 1984.

accomplished using programmable hardware to control beam position, writing speed, and beam power. The beam power is set as required by closed-loop feedback control of an acousto-optical modulator. The Livermore group has developed two software packages for automatically generating the command sequences required for discretionary laser writing of complex patterns: the Assembler translates an industry-standard CIF file, which contains a geometric description of a circuit design into a set of instructions that control the beam scan code for controlling position, velocity, direction, and laser power. The scan code also includes the information for positioning and aligning the substrate after each 512-μm square scan field has been written. The second software package is the Monitor for providing a user interface to the apparatus. The program accepts the scan code produced by the Assembler and uses it to drive the scanner, beam power controller, and X-Y translation stages. Once a circuit is designed and physical placement and routing are complete, the Livermore laser direct-write system allows a 16-bit minicomputer running the software described to generate the scan code in less than 1 hour. With this direct-write system, poly-Si lines of 3 to 4 μm wide and 1 μm high were deposited at a lateral writing speed of

1 mm/s and a laser power of 100 mW. The deposited material has a resistivity of 3×10^{-3} Ω-cm.

3.9.3 Other Laser-Based Interconnect Systems

In addition to the Livermore and LASA systems using the direct-writing technique discussed above, several other laser-based interconnect systems using different approaches have been developed. Among them, Lasarray S.A. has developed a laser pattern generator for photoresist exposure used in device customization such as ASIC prototyping (Fitzgibbons et al., 1987). The laser pattern generator is optimized for a 2400-gate CMOS base array and is fabricated using 2-μm double-layer metal technology. The system block diagram and the upper layer interconnect metallization schematic are shown in Fig. 1.30, in which a He-Cd laser (442 nm) is used to write on positive photoresist. The wafer is mounted on top of a translation stage traversed at a speed of 30 cm/s in the x direction and stepped approximately 8 micron in the y direction. The UV laser is operated at about 10 mW at the entrance pupil of the system and is attenuated to about 0.5 mW at the wafer surface. An autofocus optic is used to keep the laser-focused spot to about 1.8 μm in diameter. Another He-Ne laser is employed to scan the surface to supply data for the pattern recognition without exposing the photoresist. The accuracy and repeatability for the 8-μm metal pitch require the total positional error to be within ±0.75 μm over a 100-mm wafer and to have a total scan distance in excess of 1 Km. Sophisticated pattern recognition was required in order to maintain the alignment accuracy.

A different approach for gate array interconnection is to use lasers to cut links that have been prefabricated on the wafer (Raffel, 1987). This approach employs the technology that is relatively mature and is similar to cutting the silicide or aluminum link structures in repairing memory devices with redundant circuitry. The link cutting is performed using a short-pulse laser, such as a YAG laser, with a pulse energy of 1 mJ. Typically 4- to 6-μm laser spots are used to cut 2-μm links, with a travel distance between links of about 20 μm and at a throughput of about 60 link cuts per second. For a device of 100,000 links, the total link cut time is about 32 min. Compared to other existing programming techniques used for ASIC device fabrications, including application-specific final masks, electrical fuses, EPROMs, and EEPROMs, laser links are far simpler and smaller. Laser interconnect systems that utilize the cut-link approach are found in systems developed by Teradyne, Laserpath, and Elron Electronics Industries.

Laser pattern generators for photoresist exposures have also been

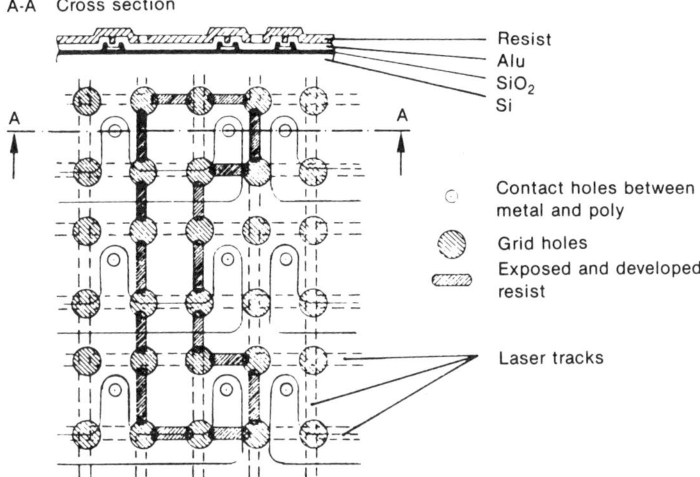

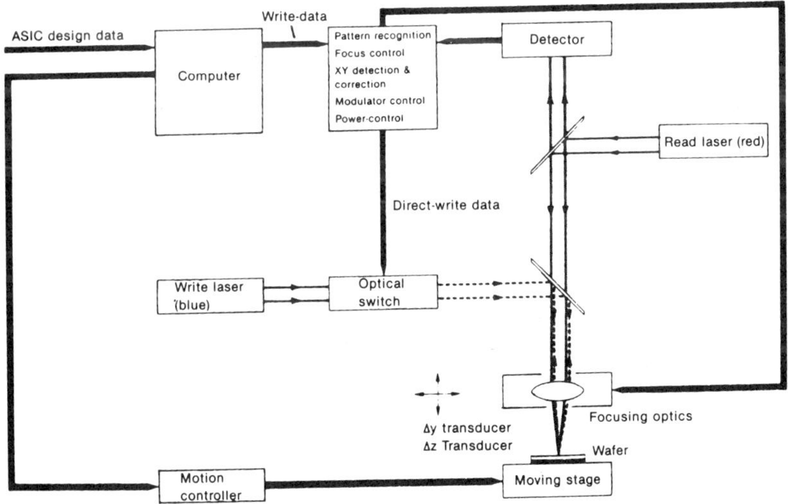

Fig. 1.30 Schematic of Lasarray direct-write laser pattern generator and the upper layer interconnect metallization methodology. From Fitzgibbons et al., 1987.

developed for retical and printed circuit board fabrications. Typical systems accept input from the magnetic tape generated from CAD/CAM stations and convert them into a rasterized image to drive a UV laser. The laser imaging system then generates printed circuit board patterns onto photoresist-coated copper-clad substrates or onto UV-sensitized media such as silver-halide-coated film or glass. A UV argon laser is commonly used as the exposure source scanned over an image field as large as 45 cm × 60 cm, at an accuracy better than 25 μm and a resolution of about 75 μm. To achieve the large scan field and speed, mechanical scanners are used in this type of laser pattern generator. High-resolution systems have been developed by ATEQ using a single Ar-ion laser (363.8 nm) split into eight beams simultaneously scanned by a high-speed rotating polygon and independently modulated at 50 MHz by an acousto-optical modulator. The system is designed for the high-speed writing of 5X and 10X reticles (Warkentin and Schoeffel, 1986).

3.9.4 Laser Direct-Write Lithography System for High-Density Packaging

Polymeric materials have many desirable features for packaging applications in VLSI and multichip circuitry, and as interconnect substrates. Their low dielectric constants reduce capacitance coupling between metallization stripes and thus improve the ultimate speed of the packaged device for high-speed interconnection. Laser processing of polymeric materials is a key technology and requires a very different set of processing parameters from those commonly used in laser processing of semiconductors. Several studies have been recently reported on laser selective metallization of copper on polyimide (Cole et al., 1986), photoetching of polymers (Liu et al., 1987), and measurements of UV and VUV absorptive and dispersive optical properties of various polymers (Philipp et al., 1986).

A laser adaptive lithography system was developed at GE for high-density interconnects in electronic packaging (Eichelberger et al., 1988). The approach uses polymer overlays laminated over bare chips mounted on a substrate. The photoresist is spun on the overlays and laser patterned, then copper is deposited to connect the chips and I/O. In this approach, both vias and interconnect lines are laser fabricated without any patterning mask. Chip misalignment is accommodated by laser adaptive writing. An interconnect density of 2-mil pitch has been achieved. In addition, the polymer overlay can be removed and replaced without damage to the bare chips.

Figures 1.31a and 1.31b show the basic concept of the high-density

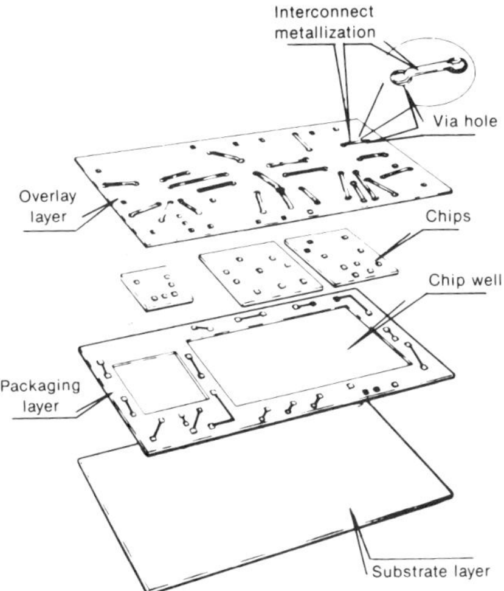

(a) High-density interconnect (exploded view)

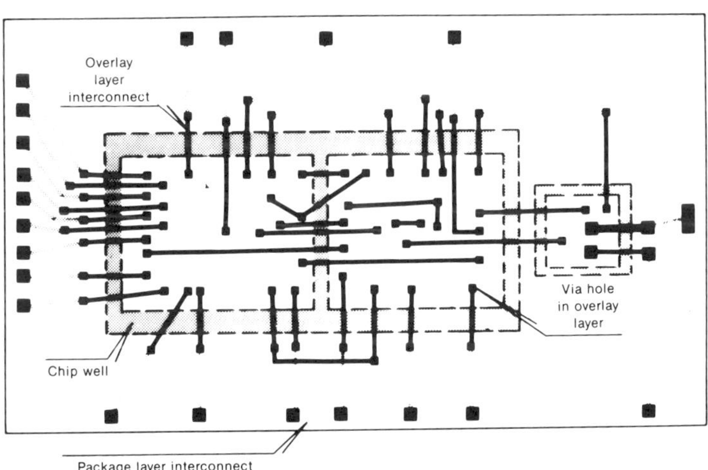

(b) Top view, high-density interconnect. Chip well usually accommodates many very closely spaced chips.

Fig. 1.31 High-density multilayer interconnect structures: (a) an exploded view and (b) the top view of overlay polymer layer on which vias and copper inter-connect lines are fabricated using an adaptive laser lithography system. From Eichelberger et al., 1988.

interconnect structure, which consists of three layers: substrate, packaging layer, and overlayer. The integrated circuits are bonded to the base substrate, which also provides the support structure for the interconnect to the outside world. The next section in the structure is the packaging layers for power distribution and ground, and for providing interconnects between chips separated by a large distance and interconnects to the pin system of the substrate. The final portion of the structure is the overlay polymer layer with different wiring configurations. The use of this overlay offers performance advantages as well as partitioning, testing, and repair options. The adaptive laser lithography system eliminates the need for exact placement. Via holes are etched using a laser ablation technique, and metallization paths are patterned using an adaptive laser lithography technique of exposing photoresist to form the desired interconnect patterns. The laser lithography system for high-density interconnects provides an ideal process for prototyping and customization in hybrid packaging.

A different approach using laser-patterned interconnect for thin-film hybrid wafer-scale packaging was developed at Livermore (Tuckerman, 1987). In this approach, the chips are gold bonded to a silicon substrate and thin-film wires are patterned down the edges of a silicon chip using a laser-etching technique. The novel etching process employs an argon-laser etch via in an amorphous silicon layer in a Cl_2 gas environment. The laser, at 300-mW power in a 5-μm spot, is acousto-optically scanned at 3 mm/s. The via pattern is then transferred to the underlying SiO_2 by reactive ion etching to form vertical interconnect paths. This approach provides a potentially very high interconnect density of 1600/chip.

4. Laser Projection Optics

One of the most significant technological advancements in laser microfabrication is in the area of pattern replication. Specifically, the use of excimer lasers as deep-UV sources for contact, proximity, and projection lithography has extended the capability of optical lithography to the sub-half-micron region (Jain, 1987; Rothschild and Ehrlich, 1988). Only a few years ago, this kind of resolution was considered not achievable using optical lithography. The high-intensity and high-energy deep-UV photon sources made available by excimer lasers have opened up many new opportunities in patterning, chemical processing, and modifications of surfaces. These new applications include, to name a few, shallow junction formation (Carey et al., 1985), photoablation of polymers (Srinivasan and Mayne-Banton, 1982), photo-catalyzed polymerization (Ehrlich and Tsao, 1985), and laser-induced oxidation (Liu et al., 1981a; Orlowski and

Richter, 1984). [For other examples, readers are referred to a recent review paper by Rothschild and Ehrlich (1988).] In this section, we discuss some key parameters in this rapidly developing technology in using excimer lasers for patterning and surface processing. The main optical considerations, such as source coherence and its relationship with image formation, resolution, contrast, and depth of focus, are discussed. Concepts such as modulation transfer function and some recent developments of excimer-laser-based high resolution lithography systems are reviewed.

4.1. *Resolution and Depth of Focus*

The resolution of a diffraction-limited imaging system is given by (Hopkins, 1957; Thompson and Bowden,1983)

$$R = K_1 \frac{\lambda}{\mathrm{NA_o}}, \tag{1.16}$$

where K_1 is a constant whose value depends upon degree of coherence and varies usually between 0.6 and about 0.8, and $\mathrm{NA_o}$ is the numerical aperture of the imaging space.

Equation 1.16 shows that the resolution of an imaging systems can be improved by using a shorter wavelength light source or by increasing the value of NA. However, since the depth of focus of an optical system decreases as the square of NA as given by (Born and Wolf, 1970)

$$z = \pm \frac{\lambda}{2\mathrm{NA}^2}, \tag{1.17}$$

any improvement of resolution by increasing NA would have to reduce the depth of focus of the optical system. For example, for a G-line stepper at 436 nm with a half-micron resolution, the depth of field is about 0.7 μm at a relatively high NA of 0.55. The depth of focus of an imaging system is a critical design consideration. Further improvement in resolution by increasing NA is difficult to accomplish, as the focus budget is limited by wafer flatness variation over the field size, circuit topology, and focusing error (Markle, 1987; Lin, 1987).

Recent efforts in developing projection steppers using shorter wavelength sources by going from the G-line (436 nm), to H-line (405 nm), then I-line (365 nm), and finally KrF excimer laser at 248 nm suggests a logical development sequence in improving the resolution by using shorter-wavelength sources. This has been the driving force behind the development of an excimer-laser-based high-resolution lithography stepper. Figure 1.32 shows resolution (Eq. 1.16) and depth of focus (Eq.

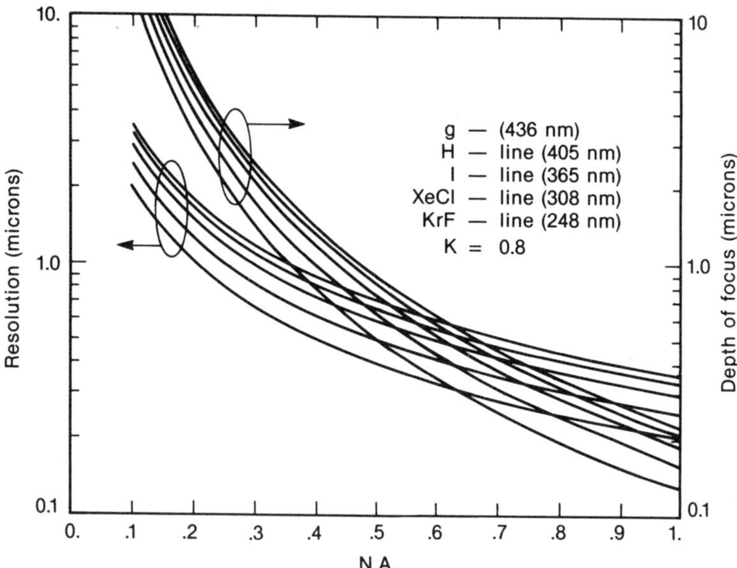

Fig. 1.32 The theoretical resolution and the depth of focus of a diffraction-limited optical imaging system plotted as function of NA for several light sources at 436 nm (G-line), 405 nm (H-line), 365 nm I-line). 308 nm (XeCl line), and 248 nm (KrF line) ($K_1 = 0.8$).

1.17) as functions of NA for G-line (436 nm), H-line (405 nm), I-line (365 nm), XeCl-line (308 nm) and KrF-line (248 nm) respectively. From the figure, it can be seen that a resolution of 0.55 μm and a depth of focus of 1 micron can be achieved using a KrF laser stepper of 0.35 NA (Nakase, 1985; Wilczynski, 1987). In the following sections, we will discuss the effect of source coherence on the depth of focus in a lithography system. Before this discussion, we first review the concepts of mutual coherence function and modulation transfer function (MTF).

4.2. *Linear Systems and Coherence*

In a photolithography system, source and system coherence are a major determinant of image quality. In such a system, the image is formed through diffraction of light. The image qualities, including resolution and contrast, depend therefore on the degree of coherence of illumination, which can be described by the mutual coherence function $G_{12}(x)$, defined as follows. For a monochromatic optical field $U(x, t)$ at frequency v,

$$U(x, t) = \phi(x) \exp(-i2\pi vt), \tag{1.18}$$

where $\phi(x)$ is complex amplitude of the field. The measured intensity of the optical field is

$$I(x) = \langle U(x, t_1)U^*(x, t_2)\rangle_t \tag{1.19}$$

where $\langle\ \rangle_t$ represents the time average and U^* is the complex conjugate of U. Where two optical fields are added together, the resultant intensity is

$$\begin{aligned} I_{\text{total}}(x) &= \langle U_1(x, t_1)U_2^*(x, t_2)\rangle \\ &= \langle U_1 U_1^*\rangle + \langle U_2 U_2^*\rangle + \langle U_1 U_2^*\rangle + \langle U_1^* U_2\rangle \\ &= I_1 + I_2 + 2\,\text{Re}\langle U_1(x, t_1)U_2^*(x, t_2)\rangle. \end{aligned} \tag{1.20}$$

The third term is a cross-correlation function and is a measure of the degree of coherence between the two optical fields U_1 and U_2. The *mutual coherence function* $G_{12}(x)$ between U_1 and U_2 is (Born and Wolf, 1970)

$$G_{12}(x, t) = \langle U_1(x, t_1)U_2(x, t_1 + t)^*\rangle, \tag{1.21}$$

and the normalized coherence function between the two optical fields is

$$g_{12}(x, t) = \frac{\langle U_1(x, t_1)U_2^*(x, t_1 + t)\rangle}{[I_1(x)I_2(x)]^{1/2}}. \tag{1.22}$$

For an incoherent illumination system, the mutual coherence function $g_{12} = 0$, $I_{\text{total}} = I_1 + I_2$, and the system is considered linear (in intensity). In such a linear system, the image intensity is a convolution of the object intensity and the intensity point-spread function of the optical imaging system. For a partially coherent system such that $0 < g_{12} \leq 1$, the system is no longer a linear function in intensity and becomes sensitive to object contrast and structure (Swing and Clay, 1967).

Thompson and Wolf (1957) and Thompson (1958) measured the visibility of two beam-interference fringe patterns to determine the degree of the source coherence. Their studies clearly demonstrated that under quasimonochromatic illumination the degree of spatial coherence depends upon the source size, while the degree of temporal coherence depends upon the spectral bandwidth of the source. The spatial coherence of a laser source is related to the transverse mode structure of the laser, and the temporal coherence depends on the longitudinal mode property. In principle, the degree of coherence of an optical field within a volume element is dependent upon both spatial and temporal coherent properties of the source (Yu, 1985). In practice, the two are often separated. Aside from speckle effects, image resolution in a system with nonzero chromatic aberration is degraded by increased spectral bandwidth and hence by a decreased temporal coherence. This kind of

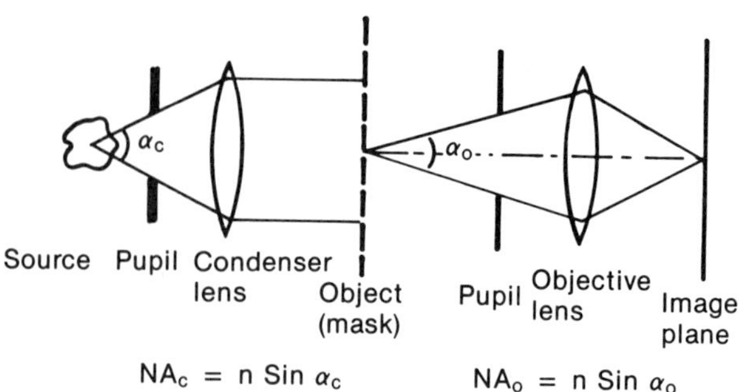

Fig. 1.33 A schematic illustrating Na_c and Na_o in a refractive imaging system illuminated with an incoherent source.

degradation is discussed in Section 4.5. Even in a system free of chromatic aberration, however, image resolution is strongly affected by spatial coherence.

In order to treat only spatial coherence, it is usual to use a measure simpler than that defined by the coherence function in Eq. 1.22. In particular, the degree of coherence in an illumination system is commonly described by the parameter s, defined as

$$s = \frac{NA_c}{NA_o}, \tag{1.23}$$

where the numerical aperture, NA_c, of the condenser lens is defined as $NA_c = n \sin(\alpha_c)$, and that for the objective lens is $NA_o = n \sin(\alpha_o)$, where n is the refractive index and α_c and α_o are the maximum collection angles of the condenser and objective lenses (defined in Fig. 1.33). The parameter s therefore defines the degree of filling of the entrance pupil of a projection lens by the condenser. For a source of infinite extent, $s = \infty$, the system is incoherent, whereas for a point source, $s = 0$, illumination is considered coherent.

4.3. Modulation Transfer Function

The quality of an imaging system is frequently described in terms of the modulation transfer function (MTF). In an incoherent illumination system, the image intensity $I_i(x, y)$ is obtained by the convolution of the intensity distribution $I_o(x, y)$ in the object plane and the point spread function $K(x, y; x', y')$ of the lens system (Born and Wolf, 1970).

$$I_i(x, y) = \iint I_o(x', y')K(x, y; x', y')\, dx'\, dy'. \tag{1.24}$$

This equation indicates that the intensities from different elements of the object plane are additive, namely, the optical system is linear in intensity. In using Eq. 1.24 to describe an optical system, we usually imply that the point spread function $K(x, y; x', y')$ does not vary around the image point under consideration, namely,

$$K(x, y; x', y') = K(x - x', y - y').$$

An imaging area with this property is called an *isoplanatic region* (Born and Wolf, 1970). The Fourier transform of an intensity distribution function $I(x, y)$ gives the respective Fourier amplitude associated with each individual sinusoidal spatial frequency component,

$$I_i(v, \mu) = \iint I_i(x, y) \exp[i2\pi(vx + \mu y)]\, dx\, dy$$

$$I_o(v, \mu) = \iint I_o(x, y) \exp[i2\pi(vx + \mu y)]\, dx\, dy \tag{1.25}$$

$$\text{OTR}(v, \mu) = \iint K(x,y) \exp(i2\pi(vx + \mu y)]\, dx\, dy.$$

From the convolution theorem, we have

$$\text{OTR}(v, \mu) = \frac{I_i(v, \mu)}{I_o(v, \mu)}, \tag{1.26}$$

where the optical transfer function, $\text{OTR}(v, \mu)$, contains modulus and phase terms, with the modulus component commonly referred to as the modulation transfer function (MTF). Therefore, the value of MTF at each frequency component is the ratio of the amplitude of a sinusoidal function in the image plane to that in the object plane.

For a one-dimensional sinusoidal object with $I_o(x)$ given as

$$I_o = \frac{[1 + \sin(2\pi vx)]}{2}, \tag{1.27}$$

$$\text{MTF}(v) = \frac{I_{max} - I_{min}}{I_{max} + I_{min}}, \tag{1.28}$$

where I_{max} and I_{min} are the maximum and minimum intensities of the image in the image plane. As the spatial frequency increases, the image contrast reduces and the MTF becomes smaller. The value of the MTF at a given spatial frequency of an optical system can be determined by using a periodic space bar as the object and measuring the contrast using Eq. 1.28. Thus the MTF of an optical system can be determined by plotting the contrast as a function of spatial frequency. Since MTF is linear, the overall MTF of a lithography system is simply the product of the MTFs of each individual optical component. In practice, the MTF provides an optical designer with a method for measuring the performance of an optical lens (system) in a manner similar to knowing the frequency response of an audio component in an audio hi-fi system.

4.4. *Spatial Coherence*

Strictly speaking, MTF is defined only for an incoherent imaging system (Swing and Cley, 1967). Nevertheless, MTF serves as a useful parameter in describing the image quality obtainable in an optical system, even in the case of partially coherent illumination. The significance of the spatial coherence to the depth of focus can be illustrated by a calculation of the MTFs for a one-dimensional object with a sinusoidal amplitude distribution function (Rothschild and Ehrlich, 1988). The result, shown in Fig. 1.34, illustrates the effects of spatial coherence on resolution, as determined by plotting MTFs for three different values of coherence: $s = 0.2$, 0.5, and 2.0, corresponding to highly coherent to highly incoherent illumination cases. The results are plotted for three different defocusing conditions to show the effects of degree of coherence on defocusing error in a diffraction-limited system. In the perfect focusing condition, coherent illumination yields a higher-value MTF for $\nu < \mathrm{NA}/\lambda$ and a lower-value MTF for $\nu > \mathrm{NA}/\lambda$ and has a larger depth of focus. For example, at $\nu = \mathrm{NA}/\lambda$, a defocusing error of $\lambda/2(\mathrm{NA})^2$ reduces MTF from 0.65 to 0.32 (50% reduction) for incoherent illumination ($s = 2$), while it only reduces MTF by 5% (0.65 to 0.62) under the coherent illumination condition.

Another phenomenon in imaging due to spatial coherence is the "ringing effect" usually seen at the edge of a sharply defined object. In addition, the ringing causes an apparent shift in the position of the edge. For an open slit, the shift is toward the illuminated region. Similar effects are observed for an object consisting of a bar or a single edge. It is interesting to note that the slope of the edge response for a coherent image is steeper than that for the corresponding incoherent image. This property can be used to improve the image constrast (Thompson, 1977). A calculation of the intensity distribution in the image plane of a sharp edge for several values of the coherence factor is shown in Fig. 1.35 (Rothschild and Ehrlich, 1988).

Finally, when coherent light illuminates a diffuse object, speckle patterns are observed as bright and dark spots caused by constructive and destructive interferences originating from various scattering points

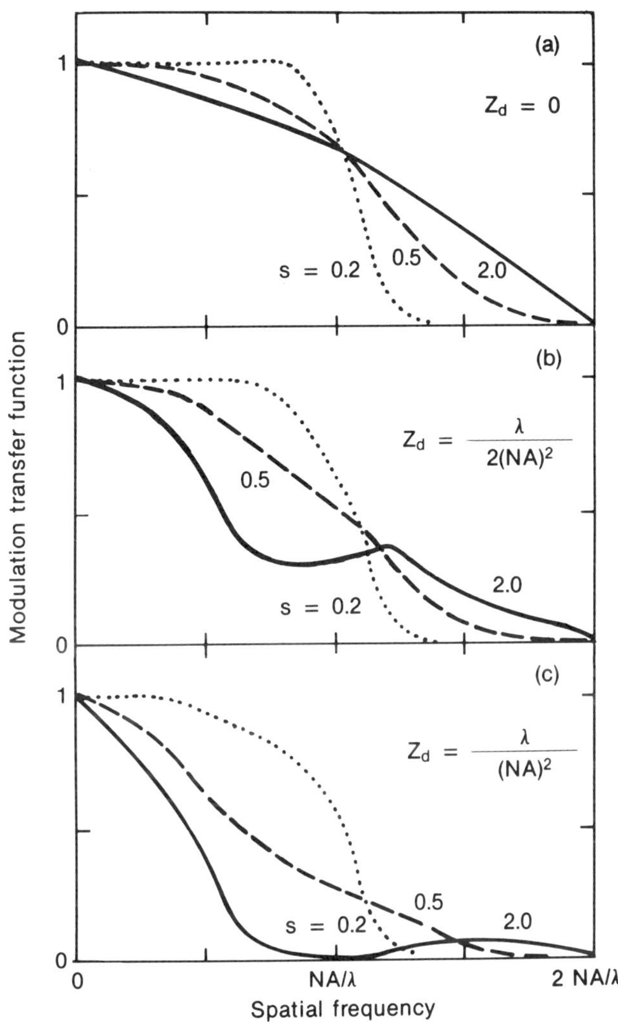

Fig. 1.34 The effects of spatial coherence and defocusing on resolution, as determined by the modulation transfer function for a one-dimensional object with a sinusoidal amplitude distribution function. The MTF functions were calculated for three values of spatial coherence, $s = 0.2$ (highly coherent), 0.5 (intermediate), and 2.0 (highly incoherent). In (a) perfect imaging is assumed and defocusing tolerance is zero, and in (b) and (c) defocusing of one and two Rayleigh ranges is introduced in the calculation. From Rothschild and Ehrlich, 1987.

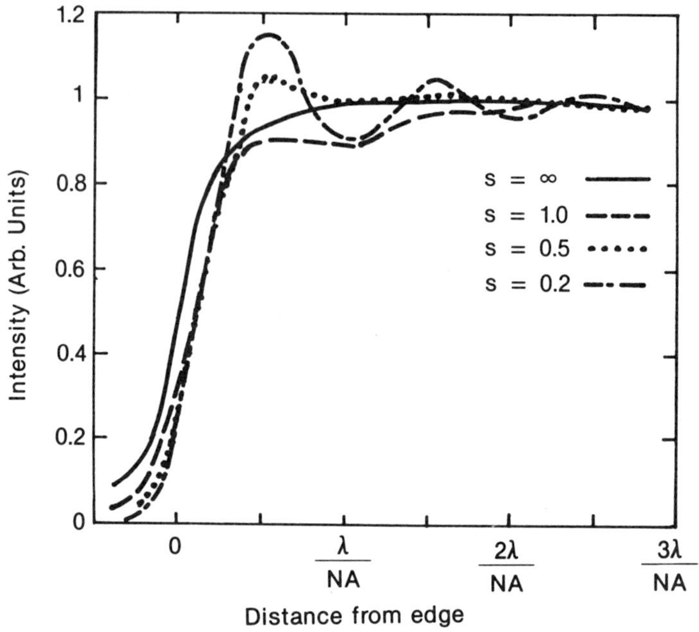

Fig. 1.35 The intensity distribution in the image plane of a sharp edge for several values of the coherence parameter s. For coherent illumination, note the $\pm 15\%$ ringing in the illuminated area as well as the lateral shift of $\sim\lambda/4\,\text{NA}$ in the position of the edge (defined as the point of intensity 0.5). On the other hand, incoherent illumination has a significant "toe," i.e., intensity extending $\sim\lambda/4\,\text{NA}$ into the geometrical shadow. From Rothschild and Ehrlich, 1988.

(Goodman, 1977). The speckle has adverse effects on image resolution. Using a plane-polarized single-line argon laser at 514 nm as the coherent illumination source, and comparing with incoherent illumination achieved by placing a diffuser moving continuously between the source and the stationary diffused object, Kozma and Christensen (1976) observed that the aperture of a coherent system must be several times larger than that of an incoherent system in order to produce a comparable resolution. When imaging a continuous-tone object, this factor increases to five. In order for a coherent system to accomplish the resolution that is comparable to an incoherent system of equal aperture, the signal-to-noise ratio of a coherent image has to increase by a factor about 10. The signal-to-noise ratio of the image is improved to $\sqrt{N}$ if N independent exposures are added together.

An excimer laser is operated with a high Fresnel number, greater than several hundreds, and therefore has relatively poor spatial coherence. However, when an aperture is inserted inside the excimer laser cavity for spectral narrowing, the transverse mode number is also reduced and the control of spatial coherence becomes an important consideration. Several mode scrambling methods used for excimer-laser illumination have been reported including use of quartz diffusers (Kerth et al., 1986), "fly's-eye" lenslets (Horiike et al., 1986), and multiple scanning of the source to control the filling factor s in the entrance pupil (Pol et al., 1986).

4.5. Spectral Bandwidth Considerations

For sub-half-micron lithography using a single-material lens system, the spectral bandwidth of a free-running excimer laser is generally too large to be tolerable, due to chromatic aberration. For example, in fused silica, the dispersion $dn/d\lambda$ in deep UV between 180 nm and 350 nm has values of $8.9 \times 10^{-4}\,\text{nm}^{-1}$ at 193 nm (ArF), $6 \times 10^{-4}\,\text{nm}^{-1}$ at 248 nm (KrF), $3.2 \times 10^{-4}\,\text{nm}^{-1}$ at 308 nm (XeCl), and $1.8 \times 10^{-4}\,\text{nm}^{-1}$ at 351 nm (XeF). For a point object imaged through a lens of focal length f, chromatic aberration will introduce a defocusing error, df, as a function of source bandwidth,

$$df = \frac{f}{1-n}\left(\frac{dn}{d\lambda}\right)\Delta\lambda. \tag{1.29}$$

To achieve a defocusing error less than 1 micron in the image plane using a $f = 1$ cm lens, from Eq. 1.29, the spectral linewidth of the source has to be no greater than 6×10^{-2} nm for an ArF laser and 8×10^{-2} nm for a KrF laser, respectively, in a single-material lens. It is difficult to design and manufacture an achromatic projection lens for the excimer laser in deep UV due to the small difference in chromatic dispersion between fused quartz and fluorite, in addition to other material-related problems, such as UV-induced defect formation. Therefore, most of the current excimer-laser-based lithography systems are taking the line-narrowing approach to minimize the stringent optical design required for chromatic correction (Pol et al., 1986; Walsh et al., 1987; Endo et al., 1987). To take this approach, however, one has to impose spectral control and wavelength stabilization on the laser source. Jewell et al. (1987) showed that the image contrast depends on the spectral bandwidth stability. The wavelength drift for a KrF laser, either single-stage line narrowed or injection locked, is typically on the order of 0.01 nm over a period of several hours.

An alternate approach to eliminating chromatic aberration problems is to use an all-reflective system. Ehrlich et al. (1985) applied a reflective two-mirror Schwartzchild microscope objective lens (Fig. 1.36) in the

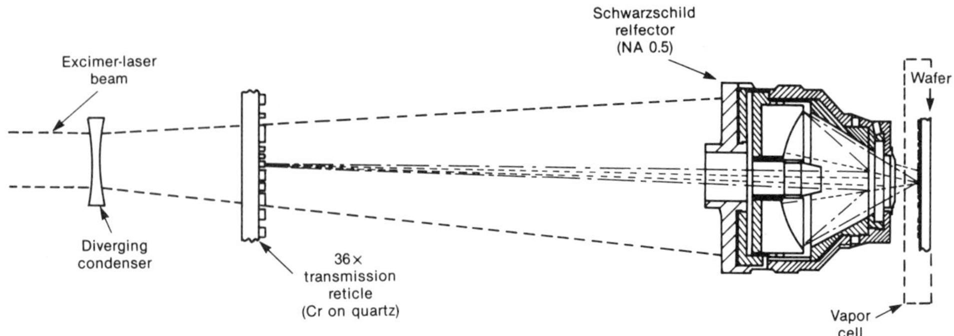

Fig. 1.36 The Schwarzschild object-imaging lens system using all-reflective optics for submicron excimer projection applications. From Ehrlich et al., 1985.

deep UV down to the wavelength of the F_2 laser at 157 nm. The lens was used to image a transmission mask with a demagnification of 36X. Using this reflective projection lens system, 0.13-μm lines and spaces patterns were produced using 193-nm pulses on a hard carbon film on GaAs. The NA of the lens was 0.5. Apart from its achromatism, the Schwartzchild lens provides a large working distance (Born and Wolf, 1970). The drawback of this type of lens is a very limited field of view (<2 mm).

An excimer scanning projection lithography system using reflective optics was reported by Jain and Kerth (1984a) by retrofitting a XeCl laser at 308 nm to a Perkin-Elmer Model 111 full-field 1:1 scanner projected by a set of mirrors. The large bandwidth tolerance (typically 1 nm for a reflective optics) allows the operation of a broadband excimer laser. The numerical aperture in the condenser was 0.14 and with $f/3$ projection optics gives a value $s = 0.86$ for the partial coherence factor and achieves a 1-μm resolution. With an optical transmission efficiency from laser to wafer of about 6%, a laser output of 5–10 W is sufficient to give a wafer exposure throughput of about one hundred 125-nm-diameter wafers/hour (Kerth et al., 1986).

Another design approach using a catadioptric system (a mixture of refractive and reflective components) to accommodate a broadband excimer laser source is the 1:1 Wynne Dyson lens developed at Ultratech (Markle 1987). This system is shown schematically in Fig. 1.37 (Goodall et al., 1986). In this configuration, the image and object are separated by two 45° LiF prisms, which are in contact with fused-silica windows and planoconvex LiF lenses. The LiF lens is contacted around the periphery of the fused-silica meniscus lens. Using the Wynne–Dyson lens, the

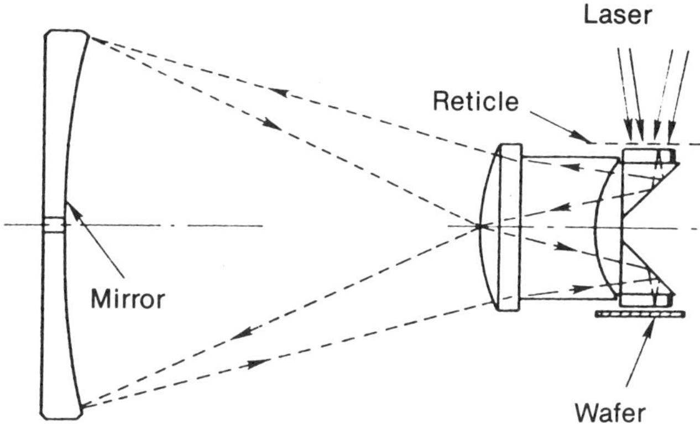

Fig. 1.37 A schematic of the 1:1 Wynne Dyson lens design. From Goodall et al., 1986.

projection system can be designed to operate at the 0.3-nm bandwidth with the KrF laser. The system has a 0.3 NA, a partial coherence factor of 0.6, and a field size of 60 mm in diameter. The optical projection system can achieve a depth of focus of 1 μm with minimum resolution of 0.5 μm.

4.6. Excimer Laser Projection Systems

Since the first demonstration of excimer laser projection for high-resolution patterning in 1982 (Dubreucq and Zahorsky), progress in this field has been impressive. Within four to five years, several excimer-laser-based projection systems have been developed for pattern generation and sub-micron lithography. Hafner (1988) reported the development of a XeCl-laser-based pattern generator for making 5X reticles used in VLSI device fabrication. The system was based on a commercial GCA M3600F pattern generator modified to fit a XeCl laser (Fig. 1.38). The laser output was transferred through an optical assembly including: a) a laser-to-fiber coupling unit for imaging an 8 mm × 22 mm excimer output onto a fiber optics entrance of 5 mm diameter, b) a 1.8-m long flexible UV fiber optics for beam homogenization, c) a condenser lens, and d) a 20X reduction lens. The system, operated in a flash-on-the-fly mode, has been operated successfully since 1985 for fabrication of 134 reticles, with accumulation of a total of 50 million pulses at an average 373,000 pulses per reticle. This development has shown that excimer laser pattern generators are promising tools for the production of 5X and 10X reticles

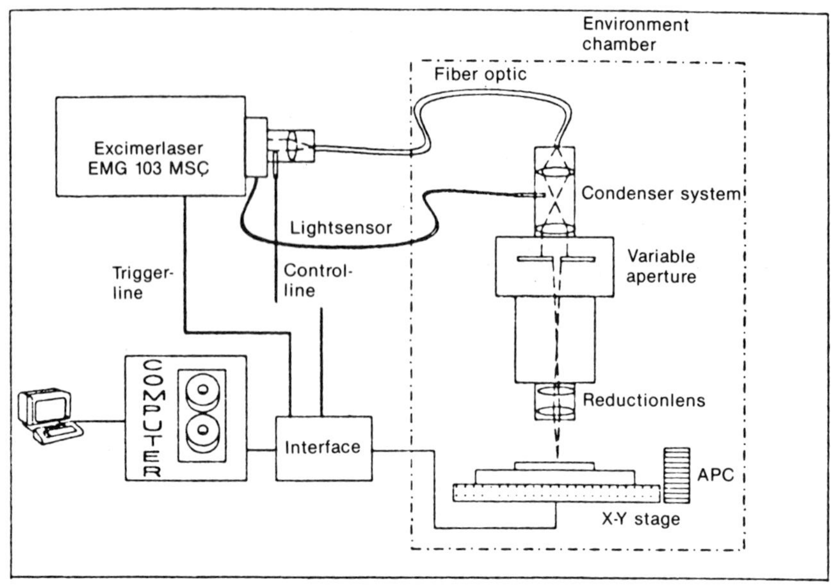

(a)

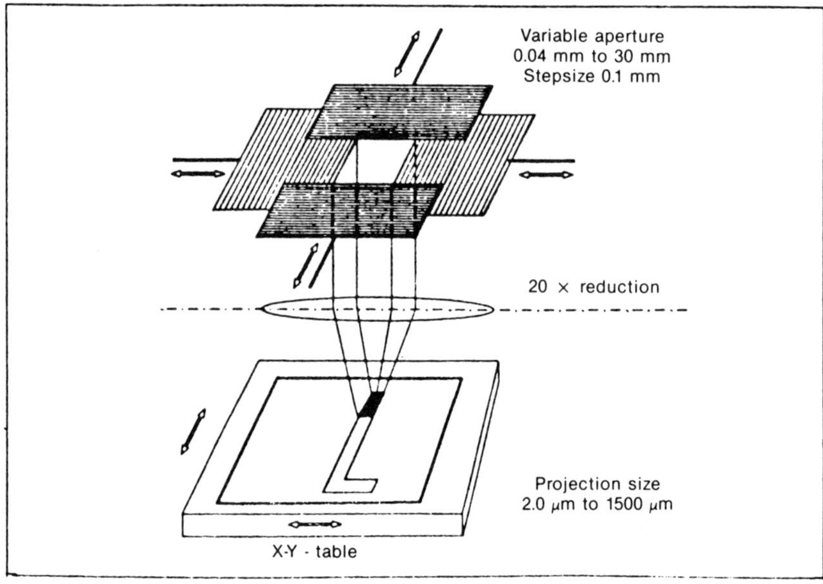

(b)

Fig. 1.38 (a) The schematic of a XeCl-excimer-laser-based pattern generator system and (b) the variable aperture used in the exposure system. From Hafner, 1988.

Table 1.6 Reported Performances of Excimer Laser-Based Lithography Systems. From Higashikawa et al. 1988.

	AT & T Bell Lab.	GCA	Matsushita	Toshiba	Nikon	Nikon	Canon	Mitsubishi	Sony
NA	0.38	0.35	0.36	0.37	0.4	0.37	0.35	0.37	0.35
Wavelength (nm)	248	248	248	248	248	248	248	248	248
$\Delta\lambda$ (FWHM) (nm)	0.005	—	0.007	~0.4	—	~0.4	—	~0.4	0.002
Type	Chromatic	Chromatic	Chromatic	Achromatic	Chromatic	Achromatic	Chromatic	Achromatic	Chromatic
Reduction ratio	1/5	1/5	1/5	1/10	1/5	1/10	1/5	1/10	1/5
Field size (mm)	20ϕ	20ϕ	15×15	5×5	15×15	5×5	5ϕ	5×5	4×4
Resolution ($k = 0.8$)(μm)	0.52	0.57	0.55	0.54	0.50	0.54	0.57	0.54	0.57
Alignment Precision (μm)	±0.25 (2σ)	±0.3 (3σ)	±0.3 (3σ)	±0.2 (3σ)	±0.15 (3σ)	—	—	—	—

containing up to 5 million flashes for VLSI applications. Several commercial optical pattern generators using conventional Hg lamp sources (manufactured by GCA and ASET) have been retrofitted to an excimer laser source (Burggraaf, 1988).

In the submicron laser-based stepper development, the progress has been very rapid. Table 1.6 summarizes the excimer-laser-based lithography systems that have been reported to date (Kameyama and Ushida, 1987; Higashikawa et al., 1988), in addition to the several experimental systems discussed in the earlier section. Recently, Nakagama et al. (1988) reported the development of a compact KrF-laser-based step-and-repeat system for 0.4-μm VLSI (16 MDRAM) device fabrication. The lithography system has a 15-mm × 15-mm field and 0.37 NA. The laser has a spectral bandwidth of 0.005 nm.

Finally, we briefly summarize some of the instrumental features in the excimer-laser-based stepper developed by Pol and his coworkers at Bell Labs (1987). The system, shown schematically in Fig. 1.39, includes a KrF laser line-narrowed to 0.007 nm using two etalons, with 1 mJ/pulse output at 300 Hz, and with about 50% transmitted onto the wafer. Another approach employed by the Bell Labs group used an injection-

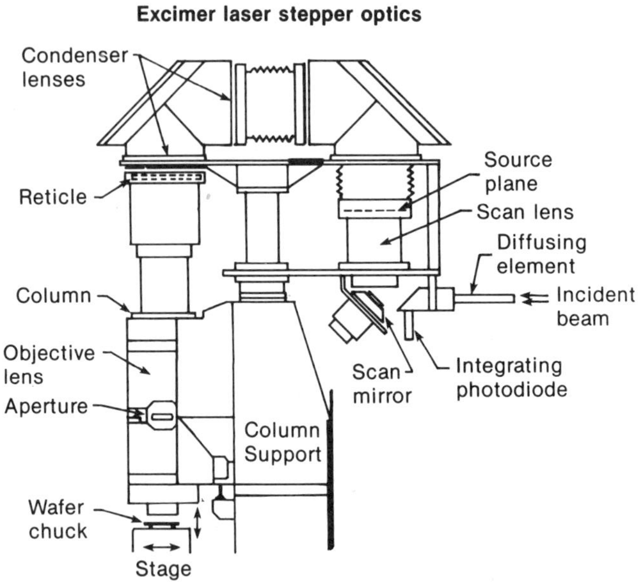

Fig. 1.39 The excimer-laser stepper illumination and projection optics developed at Bell Labs. From Pol et al., 1986.

locked excimer laser to produce a 0.004-nm bandwidth with 7-W output (Bennewitz et al., 1986). A novel illumination assembly was developed utilizing a scanning device to deflect the incoming pulses and to refocus them at the different points in the source plane, thus effectively controlling the filling factor s. The excimer laser source was retrofitted into an GCA Model 4800 DSW wafer stepper with a 0.38 NA and a field-size diameter of 14.5 mm to 20 mm. With an average energy density exposing the wafer of about 100 mJ/cm^2, a total number of 200 pulses per exposure were required. The system achieved a resolution of 0.5 μm across the field, with a depth of focus of ± 0.5 μm. Effects of laser source characteristics such as spectral linewidth, wavelength stability, and background emission on stepper performances have been studied by Jewell et al. (1987).

4.7. *Practical Constraints*

A unique characteristic in excimer-laser lithography is the pulsed nature of the exposure source used. The advantages of a pulsed source include high brightness, high intensity, and short duration. The high intensity allows new processing opportunities for exploring nonreciprocal processes, and the nanosecond pulse duration makes the flash-on-the-fly lithography system possible. On the other hand, a pulsed system creates new constraints for system design. Such constraints include the source lifetime, which inevitably becomes shortened in a pulsed operation, and shot-to-shot fluctuations and considerations of the optimum pulse repetition rate. Tsao et al. (1987) considered the trade-offs in terms of four operating parameters: a) shot-to-shot stability, b) source life in terms of shots, c) single pulse fluence, and d) repetition rate, and the results of their analysis for a broadband and narrowed-band excimer laser are shown in Fig. 1.40. In this analysis, the trade-off between shot-to-shot stability and life-time, and between single-pulse fluence and repetition rate, are considered within the constraints of resist damage, coherence effects, mask damage, exposure latitude, statistical dose control, and stepper failure. The first trade-off defines a range of source lifetimes and shot-to-shot fluctuations compatible with practical lithography. The second trade-off defines the range of single pulse fluences and repetition rates compatible with practical lithography. Significant progress has been made in developing reliable excimer lasers for industrial applications, including lithography, pattern generation, printed circuit board fabrication, and surface treatments. These industrial applications demand significant improvement in laser reliability and ease of operation (Znotins et al., 1988; Austin, et al., 1987; Sengupta, 1988).

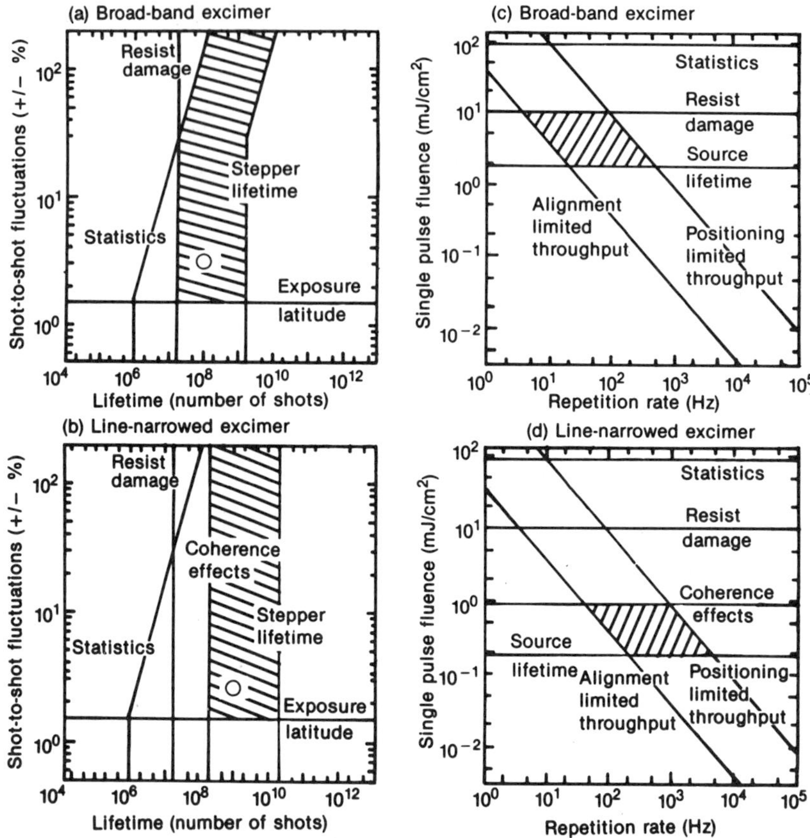

Fig. 1.40 Shot-to-shot stability versus lifetime trade-offs (a) for a broadband excimer laser and (b) for a line-narrowed excimer laser. Pulse fluence versus repetition rate trade-offs (c) for a broadband excimer laser and (d) for a line-narrowed excimer laser. Assume: pulse duration = 20 nsec, wavelength = 248 nm, exposure fluence = 10 mJ/cm^2, resist sensitivity = 180 mJ/cm^2, field size = 1.5 cm × 1.5 cm, wafer size = 6 in., exposure latitude = ±3%, exposure time = 5 sec/field, and positioning time = 5 sec/field. The shaded areas are the desired operating regions. The open circles represent sources with (2σ) shot-to-shot fluctuation and lifetimes of (a) ±3% and 10^8, and (b) ±3% and 10^9. These values are assumed the trade-offs shown in (c) and (d), respectively. From Tsao et al., 1987.

5. Conclusion

In this paper we have reviewed the status of laser-based microfabrication technology and discussed optical considerations critical to laser direct writing and excimer-laser projection lithography. Laser processing, being maskless and adaptive, is a particularly attractive technique for fabrication of application specific ICs and for circuit restructuring. In the area of laser direct-write technology, the first commercial laser direct-write two-level metal interconnect system has recently become available for interconnects of up to 50,000 gate arrays using 1.5-μm design rules. In the area of laser projection lithography, significant efforts are underway in many laboratories to develop the next-generation submicron lithography tool. Efforts in exploring the feasibility of using optical lithography for fabrication of sub-half-micron devices have already begun to draw much research attention. Excimer lasers provide not only a new UV source for submicron lithography, the high peak power and high photon energy further offer tremendous opportunities in developing totally new methods for patterning, imaging, and area-selective processing. In conjunction with other processing techniques, such as laser-induced etching, doping, and surface modifications for area-selective thin film deposition, laser processing offers the potential for the development of a fully adaptive processing technology. In the area of electronic packaging, laser processing is particularly useful for interconnecting multilevel structures using polymeric materials for next-generation high-performance packaging module.

Several key technical issues, which are related to the topics discussed in this chapter but could not be included, are summarized as follows and listed with further references: a) The optimum optical design requires further considerations of the trade-off between shorter wavelength versus larger numerical aperture, bandwidth control versus defocus tolerance, and related alignment and overlay technologies for the sub-half-micron excimer-laser lithography system (Markle, 1987; Kameyama and Ushida, 1987). b) More studies are needed on the long-term reliability of optical materials used as optical components in various optical transport systems that are exposed to high-energy and high-peak-power deep-UV radiation (Rothschild and Ehrlich, 1988). c) Continued improvements are required on excimer-laser sources in both the operational characteristics, such as pulse-to-pulse stability, temporal, spectral, and spatial mode control, as well as the long-term reliability and life (Znotins, 1986). d) Further understanding on material responses upon exposure to high-power deep-UV laser irradiation is required in areas such as reciprocity properties (Rice and Jain, 1984a; Davis and Gower, 1987; Polasko et al., 1984), defect generation (Rice and Jain, 1984b; Danielzik et al., 1986; Cole et al., 1986; Koren, 1987), etching properties (Bäuerle et al., 1988; Ehrlich et al., 1985; Andrew et al., 1983; Koren et al., 1986; Rothschild et al., 1986), surface modifications (Tsao and Ehrlich, 1984; Stuke et al., 1988), and new patterning techniques via self-developing resists (Deutsch and Geis, 1983; Rothschild and Ehrlich, 1987). The area of material interactions with high-power UV laser irradiation offers enormous potential for developing totally new processes for patterning and for fabrication of high-resolution structures.

Acknowledgements

The author wishes to acknowledge many useful discussions with Dr. J.Y. Tsao and Dr. D.J. Ehrlich during the the course of preparing this manuscript and to thank the following persons for providing original figures to the author: Dr. M. Rothschild (Figs. 1.34 and 1.35), Dr. A.F. Bernhardt (Fig. 1.29), Dr. K.C. Liu (Fig. 1.10), Dr. Y. Horiike and I. Higashikawa (Table 1.6), and Dr. H.L. Witting for data on mercury lamps. Many stimulating interactions over the years with my colleagues are also acknowledged, including H.S. Cole, R. Guida, L.M. Levinson, H.R. Philipp, G.E. Possin, and J.W. Rose. Thanks are also due to Dave Raycroft for preparing this manuscript.

References

Akhmanov, C.A., Kovrygin, A.I., and Sukhorukov, A.P. (1975). Optical harmonic generation and optical frequency multipliers. In *Quantum Electronics: A Treatise* (Rabin, H., and Tang, C.L., eds.). Academic Press, vol. I, Part A, 476–586.

Andrew, J.E., Dyer, P.E., Greenough, R.D., and Key, P.H. (1983). Metal films removal and patterning using XeCl laser. *Apply Phys. Lett.* **43,** 1076–1078.

Arjavalingam, G., Oprysko, M.M., and Hurst, J.E. Jr. (1988). Compact laser source for metal deposition. In *Laser and Particle-Beam Chemical Processing for Microelectronics* (Ehrlich, D.J., Higoshi, G.S., and Oprysko, M.M., eds.). Materials Res. Soc., Pittsburgh, Pennsylvania, 81–88.

Austin, L., Basting, D., Kablert, H.D., Mucherheim, W., and Rebhan, U. (1987). Industrial excimer lasers, *Lambda Physik Inc.,* Tech. Note.

Banas, C.M., and Webb, R. (1982). Macro-materials processing. *Proc. IEEE* **70,** 556–565.

Basu, S., Kane, T.J., and Byer, R.L. (1986). A proposed 1 KW average power moving slab Nd:glass laser. *IEEE J. Quant. Elect.* **QE-22,** 2052–2057.

Bäuerle, D. (1986). *Chemical Processing with Lasers.* Springer Series of Materials Science, Vol. I, Springer-Verlag, New York.

Bäuerle, D., Eyett, M., Kolzer, U., Kullmer, R., Mogyorosi, P., and Piglmayer, K. (1988). Laser-induced surfaces modification and etching of materials. In *Proc. of Laser and Particle Beam Chemical Processing for Microelectronics.,* Mat. Res. Soc., 411–420.

Baumert, J.C., Schellenberg, F.M., Lenth, W., Risk, W.P., and Bjorklund, G.C. (1987). Generation of blue CW coherent radiation by sum frequency mixing in KTP. *Appl. Phys. Lett.* **51,** 2192–2194.

Beiser, L. (1986). Imaging with laser scanners. *Optics News.* Opt. Soc. of Am., 10–16.

Belt, R., Gashurov, G., and Liu, Y.S. (1985). KTP as a harmonic generator for Nd:YAG lasers. *Laser Focus,* Oct., 110–122.

Bennewitz, J.H., Escher, G.C., Feldman, M., Firtion, V.A., Jewell, T.E., Pol, V., Wilcomb, B.E., and Clemens, J.T. (1986). Excimer-laser-based lithography for 0.5 μm device technology. Proc. of Intern. Elect. Dev. Meeting, IEEE, Piscataway, New Jersey, 312–315.

Bernhardt, A.F., McWilliams, B.M., Mitlitsky, B.M., and Whitehead, J.C. (1987). Laser microfabrication technology and its applications to high speed interconnect of gate arrays. In *Photon, Beam and Plasma Stimulated Chemical Processes at Surfaces* (Donnelly, V.M., Herman, I.P., and Hirose, M., eds.), Vol. 75, Materials Res. Soc. Proc., Pittsburgh, Pennsylvania.

Biegelsen, D.K., Johnson, N.M., Bartelink, D.J., and Moyer, M.D. (1981). Laser-induced crystallization of silicon islands on amorphous substrates. *Appl. Phys. Lett.* **38,** 150–152.

Black, J.G., Ehrlich, D.J., Sedlacek, J.H.C., Feinerman, A.D., and Busta, H.H. (1986). Rapid low-resistance interconnect by selective tungsten deposition on laser direct written polysilicon. *IEEE Elect. Dev. Letts.* **EDL-7,** 422–424.

Bloembergen, N. (1965). *Nonlinear Optics*. Benjamin, New York.

Born, M., and Wolf, E. (1970). *Principles of Optics,* 2nd ed., Pergamon Press, Oxford.

Boyd, I.W. (1987). *Laser Processing of Thin Films and Microstructures*. Springer Series in Materials Science, Vol. 3, Springer-Verlag, New York.

Brau, C.A. (1984). In *Excimer Lasers,* 2nd ed. (Rhodes, C.K., ed.). *Topics Apply. Phys.,* Vol. 30. Springer-Verlag, New York, 87–138.

Brink, D.J., Proch, D., Basting, D., Hohla, K., and Lokai, P. (1982). Raman-frequency shifting: A simple efficiency way to VUV and IR generation. *Laser and Optoelektronik* **3**.

Brown, D.C. (1981). *High Peak Power ND:Glass Laser Systems*. Springer Verlag, New York.

Bucksbaum, P.H., Bokor, J., Storz, R.H., and White, J.C. (1982). Amplification of ultrashort pulses in KrF laser at 248 nm. *Opt. Lett.* **7,** 399–401.

Burggraaf, P. (1988). Laser-based pattern generation. *Semiconductor International,* May, 116–121.

Burnham, R., Powell, F.X., and Djeu, N. (1976). Efficient electrical discharge lasers in XeF and KrF. *Appl. Phys. Lett.* **29,** 30–32.

Byer, R.L. (1975). Optical parametric oscillators. In *Quantum Electronics: Treatise* (Rabin, H., and Tang, C.L., eds). Academic Press, 588–702.

Calvert, J.G., and Pitts, Jr., J.N. (1967). *Photochemistry*. John Wiley & Sons, New York, 686–722.

Carey, P.G., Sigmon, T.W., Press, R.L., and Fahlen, T.S. (1985). Ultra-shallow high concentration boron profiles for CMOS processing. *IEEE Elect. Dev. Lett.* **EDL-6,** 291–293.

Caro, R.G., Gower, M.C., and Webb, C.E. (1982). *J. Phys.* **D15,** 767.

Champeney, D.C. (1973). *Fourier Transforms and Their Physical Applications*. Academic Press, London and New York.

Chang, I.C. (1976). Acousto-optic devices and applications. *Proc. IEEE,* **SU-23,** 2–22.

Chen, C., Wu, B., Jiang, A., and You, G. (1984). *Sci. Sinica Ser.* **B28,** 235.

Cole, H.S., Liu, Y.S., and Levinson, L.M. (1986). Laser patterning of polymer for electronic packaging. In *Proc. of the Third U.S. and Japan Seminar on Dielectric and Piezoelectric Ceramic* (Yamaguchi, T., ed.). Keio University, Japan, 118–119.

Cole, H.S., Liu, Y.S., Phillipp, H.R. and Guida, R. (1986). Laser etching of polymers. In *Electronic Packaging Materials Science II* (Jackson, K.A., Pohanka, R.C., Uhlmann, D. R., and Ulrich, D.R. eds.). Mat. Res. Soc., Pittsburgh, Pennsylvania.

Cohen, M.C., Kaplan, R.A., and Arthurs, E.G. (1982). Micro-materials processing. *Proc. of IEEE* **70,** 545–555.

Cullis, A.G., Weber, H.C., and Bailey, P. (1978). British Patent No. 46015.

Cullis, A.G., Weber, H.C., and Bailey, P. (1979). A device for laser beam diffusion and homogenization. *J. Phys. E. Sci. Instrum.* **12,** 688–689.

Danielmeyer, H.G. (1976). In *Lasers* (Levine, A.K., and DeMaria, A.J., eds.). Marcel Dekker, New York, Vol. 4, Chapter 1.

Danielzik, B., Fabricius, N., Rowekamp, M., and Von der Linde, D. (1986). Velocity distribution of molecular fragments from polymethacrylate irradiated with UV laser pulses. *Appl. Phys. Lett.*, **48,** 212–214.

Davis, G.M., and Gower, M.C. (1986). Excimer laser lithography: Intensity dependent resist damage. *IEEE Elect. Dev. Lett.*, **EDL-7,** 543–545.

Deutsch, T.F., and Geis, M.W. (1983). *J. Appl. Phys.* **54,** 7201.

Dickson, L.D. (1970). Characteristics of a propagating gaussian beam. *Appl. Opt.* **9,** 1854–1861.

Dickson, L.D., Sincerbox, G.T., and Wolfheimer, A.D. (1982). Holography in IBM 3687 supermarket scanner. *IBM J. Res. Dev.* **26,** 228–234.

Dubreucq, G. M., and Zahorsky, D. (1982). KrF excimer laser as a future deep UV source for projection printing. In *Proceedings of the International Conference on Microcircuits Engineering,* 73–78.

Dunn, M.H., and Ross, J.N. (1977). In *Progress in Quantum Electronics* (Sanders, J.H., and Stenholm, S., eds.). Pergamon Press, London, 233–270.

Egger, H., Pummer, H., and Rhodes, C.K. (1982). *Laser Focus* **18,** 59.

Eggleston, J.M., Kane, T.J., Kuhn, K., Unternahrer, J., and Byer, R.L. (1984). The slab geometry laser—part I: Theory. *IEEE J. of Quant. Elect.* **QE-20,** 289–301.

Eggleston, J.M., Kane, T.J., Unternahrer, J., and Byer, R.L. (1982). Slab geometry Nd:glass laser performance studies. *Opt. Lett.* **9,** 405–407.

Ehrlich, D.J., Osgood, R.M., and Deutsch, T.F. (1981). Laser microreaction for deposition of doped silicon. *Appl. Phys. Lett.*, **39,** 957–960.

Ehrlich, D.J., Osgood, R.M. Jr., and Deutsch, T.F. (1982). Photodeposition of metal films with UV laser light. *J. Vac. Sci. and Tech.* **21,** 23–27.

Ehrlich, D.J., and Tsao, J.Y. (1983). A review of laser microchemical processing. *J. Vac. Sci. Tech.* **B1**(4), 969–984.

Ehrlich, D.J., and Tsao, J.Y. (1984). Nonreciprocal laser-microchemical processing: Spatial resolution limits and demonstration of 0.2 μm linewidths. *Appl. Phys. Lett.* **44,** 270–272.

Ehrlich, D.J., and Tsao, J.Y. (1985). UV laser photodeposition of patterned catalyst films from adsorbate mixtures. *Appl. Phys. Lett.* **46,** 198–200.

Ehrlich, D.J., Tsao, J.Y., and Bozler, C.O. (1985). Submicrometer patterning by projected excimer laser-based-beam induced chemistry. *J. Vac. Sci. Technol.* **B3,** 1–5.

Ehrlich, D.J., Tsao, J.Y., Silversmith, D.J., Sedlacek, J.H.C., Mountain, R.W., and Graber, W.S. (1984). Direct write metallization of silicon MOSFET's using laser photodeposition. *IEEE Elect. Dev. Lett.* **EDL-5,** 32–34.

Eichelberger, C.W., Wojnarowski, R.J., Carlson, R.O., and Levinson, L.M. (1988). High density interconnects for electronic packaging. SPIE Symp. Innovative Sci. and Technol. paper #877–15.

Endo, M., Sasago, M., Hirai, Y., Ogawa, K., and Ishihara, T. (1987). Half-micron KrF excimer laser stepper lithography with new resist and water soluble contrast enhanced materials. *SPIE* **774,** 138–146.

Ewing, J.J., and Brau, C.A. (1975). Laser actions in KrF and ZeCl. *Appl. Phys. Lett.* **27,** 350–352.

Fahlen, T.S., and Perkins, P. (1984). Materials and medical applications using a 20 W frequency-doubled Nd:YAG laser. *Digest of Conf. on Lasers and Electro-optics,* Opt. Soc. Am., 138.

Findlay, D., and Goodwin, D.W. (1970). In *Advances in Quantum Electronics* (Goodwin, D. W., ed.). Academic Press, New York, 77–128.

Fitzgibbons, E.T., Kempter, M., and Walther, R. (1987). A direct write laser pattern generator for rapid semiconductor device customization. In *Proc. of Lasers in Microlithography,* Vol. 774 (Ehrlich, D.J., Batchelder, J.S., and Tsao, J.Y., eds.), SPIE, 82–87.

Golden, J., Eden, J.G., Mahaffey, R.A., Pasour, J.A., and Waynant, R.W. (1978). *Proc. Intern. Conf. on Lasers '78,* 18.

Goldhar, J., and Murray, J.R. (1977). Injection-locked narrow-band KrF discharged laser using an unstable resonator cavity. *Opt. Lett.* **1,** 199.

Goldhar, J., Rapaport, W.R., and Murray, J.R. (1980). An injection-locked unstable resonator rare-gas halide discharge laser of narrow linewidth and high spatial quality. *IEEE J. Quant. Elect.* **QE-16,** 235.

Goodall, F., Lawes, R.A., and Sharp, P.H. (1986). Excimer lasers as deep UV sources for photolithographic system. *Microelectron. Eng.* **5,** 445–452.

Goodman, J.W. (1977). Some fundamental properties of speckle. In *Coherent Optical Engineering* (Arecchi, F.T., and Degiorgio, V., eds.), North-Holland Company, New York, 29–48.

Grojean, R.E., Feldman, D., and Roach, J.F. (1980). Production of flat-top beam profiles for high energy lasers. *Rev. Sci. Instrum.* **51,** 375–376.

Grower, M.C. (1983). Phase conjugation at 193 nm. *Opt. Lett.* **8,** 70–72.

Han, C.Y., Ishii, Y., and Murata, K. (1983). Reshaping collimated laser beams with gaussian profile to uniform profiles. *App. Opt.* **22,** 3644–3647.

Hafner, B.F. (1988). Optical pattern generation using excimer laser. In *Optical/Laser Microlithography.* SPIE, **922,** 417–423.

Heuberger, A. (1986). X-ray lithography. *Solid State Technology,* Feb., 93–107.

Higashikawa, I., Nonaka, M., Sato, T., Nakase, M., Ito, S., Horioka, K., and Horiike, Y. (1987). Recent progress in excimer laser lithography. In *Proc. of Laser and Particle-Beam Chemical Processing for Microelectronics* (Ehrlich, D.J., Higashi, G.S., and Oprysko, M.M., eds.). Materials Res. Soc., Pittsburgh, Pennsylvania, 3–12.

Hill, C. (1982). Factors influencing applications. In *Laser Annealing of Semiconductors* (Poate, J.M., and Mayer, J.W., eds.). Academic Press, New York, 479–552.

Hoffman, A.L., Albrecht, G.F., Crawford, E.A., and Rose, P.H. (1985). High brightness laser/plasma source for high throughput sub-micron X-ray lithography. In *Electron-beam, X-ray and Ion-beam Techniques for Sub-micron Lithography,* SPIE, Vol. 537, 198–205.

Hon, D. (1979). High average power efficient second harmonic generation. In *Laser Handbook* (Stitch, M.L., ed.). North Holland, New York, 423–484.

Hopkins, H.H. (1957). Applications of coherence theory in microscopy and interometry. *J. Opt. Soc. Am.* **47,** 508–526.

Horiike, Y., Yoshikawa, R., Okano, H., Nakase, M., Komano, H., and Takigawa, T. (1986). Microfabrication technologies for advanced VLSI devices. In *Proc. of Mat. Res. Soc. Fall Meeting,* paper #B1.

Hulme, G.J., and Jones, W.B. (1975). Total internal reflection face-pumped lasers. *SPIE* **69,** 38–45.

Hutchinson, M.H. (1980). Excimers and excimer lasers. *Appl. Phys.* **21,** 95–114.

Hutchinson, M.H. (1987). Excimer lasers, in tunable lasers. In *Topics Appl. Phys.,* Vol. 59. (Mollenauer, L.F., and White, J.C., eds.). Springer-Verlag, New York, 19–56.

Ih, C.S. (1972). Absorption lens for producing uniform laser beams. *Appl. Opt.* **11,** 694–695.

Jain, K. (1987). Advances in excimer laser lithography. In *Proc. of SPIE* **774,** 115–124.

Jain, K., and Kerth, R.T. (1984a). Excimer laser projection lithography. *Appl. Opt.* **23,** 648–649.

Jain, K., Latta, M.R., and Sincerbox, G.T. (1984b). Holographic method and apparatus for transforming of a light beam into a line source of required curvature and finite numerical aperture. U.S. Patent 4,444,456 and Patent 4,516,832.

Jain, K., Wilson, C.G., and Lin, B.J. (1982). Ultra-high resolution contact lithography with excimer lasers. *IBM Res. Dev.* **26,** 151.

Jewell, T.E., Bennewitz, J.H., Escher, G.C., and Pol, V. (1987). Effect of laser characteristics on the performance of a deep UV projection system in lasers. In *Microlithography,* SPIE. **774,** 124–132.

Johnson, P.D. (1971). Excitation of Hg^+ 194.2 nm in the high current low pressure discharge. *Appl. Phys. Lett.* **18,** 381–382.

Johnson, P.D. (1971). Mercury resonance radiation in the high current low pressure discharge. *J. Opt. Soc. Am.* **61,** 1451–1453.

Jones, W.B., Goldman, L.M., Chernoch, J.P., and Martin, W.S. (1972). The mini-FPL—a face pumped laser: Concept and implementation. *IEEE J. Quant. Elect.* **QE-8,** 534.

Jones, W.B., and Hulme, G.J. (1978). ND:slab face-pumped lasers. In *Proc. Electro-optics/Lasers '78. Conf. Ind. Sci. Manage.,* Chicago, Illinois, 515.

Kameyama, M., and Ushida, K. (1987). Excimer laser stepper for sub-micron lithography. *Proc. of SPIE* **774,** 147–154.

Kane, T.J., Eckardt, R.C., and Byer, R.L. (1983). Reduced thermal focusing and birefringence in zig-zag slab geometry crystalline lasers. *IEEE J. of Quant. Elect.* **QE-19,** 1351–1354.

Kato, K. (1987). Second harmonic generation to 204 nm in BBO. *IEEE J. Quant. Elect.* **QE-22,** 1013–1014.

Kato, K. (1988). Second harmonic and sum frequency generation to 495 nm and 458 nm in KTP. *IEEE J. Quant. Elect.* **QE-24,** 3–4.

Kerth, R.T., Jain, K., and Latta, M.R. (1986). Excimer laser projection lithography on a full field scanning projection system. *IEEE Electron. Dev. Lett.* **EDL-7,** 299–231.

Khosrofian, J.M., and Garetz, B.A. (1983). Measurement of a gaussian laser beam diameter through the direct inversion of knife-edge data. *Appl. Opt.* **22,** 3406–3410.

King, M.C. (1981). In *VLSI Electronics: Microstructures Science* (Einspruch, N.G. ed.). Academic Press, New York, Vol. 1, 41.

Kingslake, R. (1983). *Optical System Design.* Academic Press, New York.

Koechner, W. (1970). Thermal lensing in a Nd:YAG laser rod. *Appl. Opt.* **9,** 2548–2553.

Koechner, W. (1976). *Solid State Laser Engineering.* Springer Series in Optical Sciences, Springer Verlag, New York.

Kogelnik, H., and Li, T. (1966). Laser beams and resonator. *Proc. IEEE* **54,** 1312–1329.

Koren, G. (1987). Temporal measurements of photofragment attenuation at 248 nm in the laser ablation of polyimide in air. *Appl. Phys. Letts.* **50,** 1030–1032.

Koren, G., Ho, F., and Risko, J.J. (1986). XeCl laser controlled chemical etching of Al in chlorine gas. *Appl. Phys.* **A40,** 13–23.

Kozma, A., and Christensen, C.R. (1976). Effects of speckle on resolution. *J. Opt. Soc. Am.* **66,** 1257–1260.

Lahart, M.J., and Marathay, A.S. (1975). Image speckle patterns of weak diffusers. *J. Opt. Soc. Am.* **65,** 769–778.

Lin, B. (1987). The future of sub-half-micrometer optical lithography. *Microelect. Eng.* **6,** 31–51.

Liu, K.C., and Rhoades, M. (1987). High average power UV generation at 266 and 355 nm in BBO. *Digest of Conf. of Lasers and Electro-Optics,* paper #ThA2, 208.

Liu, Y.S. (1977). Spectral phase-matching properties for second harmonic generation in nonlinear crystals. *Appl. Phys. Lett.* **31,** 187–189.

Liu, Y.S. (1979). Generation of high power nanosecond pulses from a Q-switched Nd:YAG oscillator using intracavity-injection technique. *Opt. Lett.* **4,** 372–374.

Liu, Y.S. (1985). Laser direct writing of tungsten lines for VLSI applications. In *Tungsten and Other Refractory Metals Deposition for VLSI Applications* (Blewer, R.L., ed.). Materials Research Society, 43–52.

Liu, Y.S., Chiang, S.W., and Bacon, F. (1981a). Rapid oxidation via adsorption of oxygen in laser-induced amorphous silicon. *Appl. Phys. Lett.* **38,** 1005–1007.

Liu, Y.S., Cole, H.S., Philipp, H.R., and Guida, R. (1987). Photoetching of polymers with excimer lasers. In *Lasers in Microlithography,* SPIE **774,** 133–137.

Liu, Y.S., Dentz, D., and Belt, R. (1984). High average power intracavity second harmonic generation using KTP in an acousto-optically Q-switched Nd:YAG laser oscillator at 5 kHz. *Opt. Lett.,* 76–78.

Liu, Y.S., Drafall, L., Dentz, D., and Belt, R. (1982). *Nonlinear Optical Phase Matching Properties of KTP.* General Electric TIS Report 82CRD016, Schenectady, New York.

Liu, Y.S., Jones, W.B., and Chernoch, J.P. (1976). High-efficiency high-power coherent UV generation at 266 nm in 90° phase matched deuterated KDP. *Appl. Phys. Lett.* **29,** 32–34.

Liu, Y.S., Jones, W.B., and Chrenoch, J.P. (1981b). *Recent Development of High Power Visible Laser Sources Employing Solid State Slab Lasers and Nonlinear Harmonic Conversion Techniques.* General Electric TIS Report 81CRD104, Schenectady, New York.

Liu, Y.S., Yakymyshyn, C.P., Philipp, H.R., Cole, H.S., and Levinson, L.M. (1985). Laser-induced selective deposition of tungsten on silicon. *Vac. Sci. and Tech.* **B3**(5), 1441–1444.

Loree, T.R., Sze, R.C., Barker, D.L., and Scott, P.B. (1979). New lines in the UV: Stimulated Raman scattering of excimer laser wavelengths. *IEEE J. Quant. Elect.* **QE-15,** 551–553.

Markle, D.A. (1984). The future and potential of optical scanning systems. *Solid State Technology,* 159–166.

Markle, D.A. (1987). Deep UV lithography: Problems and potential. In *Proc. SPIE* **774,** 108–114.

Marshall, G.F., ed. (1985). *Laser Beam Scanning.* Marcel Dekker, New York.

Martin, W.S., and Chernoch, J.P. (1972). Multiple internal reflection face pumped laser. U.S. Patent 3,633,126.

Matthews, J.C., Ury, M.G., Birch, A.D., and Lashman, M.A. (1983). Microlithography techniques using a microwave powered deep UV source. *SPIE Conf. on Microlithography,* Mar. 13–17.

McWilliam, B.M., Chin, H.W., Herman, I.P., Hyde, R.A., Mitlitsky, F., Whitehead, J.C., and Wood, L.L. (1984). Wafer scale laser pantography: Direct write interconnection of VLSI gate arrays. *SPIE Proc.* Vol. 459, *Laser Assisted Deposition, Etching and Doping,* 22–27.

McWilliams, B.M., Herman, I.P., Mitlitsky, F., Hyde, R.A., and Wood, L.L. (1983). Wafer scale laser pantography: Fabrication of n-MOS transistors and small scale IC's by direct write laser induced pyrolytic reactions. *Appl. Phys. Lett.* **43,** 946–948.

Montagu, J. (1985). Galvanometric and resonant low inertia scanners. In *Laser Beam Scanning* (Marshall, G. F., ed.). Marcel Dekker, New York, 193–288.

Nakagawa, H., Sasago, M., Endo, M., Hirai, Y., Ogawa, K., and Ishihara, T. (1988). An advanced KrF excimer laser stepper for production of 16MDRAMs. In *Proc. of Optical and Laser Microlithography,* SPIE, **922,** 400–408.

Nakase, M. (1985). The potential of optical lithography, *SPIE* **537,** 160–167.

Orlowski, T.E., and Richter, H. (1984). Ultrafast laser-induced oxidation of silicon: A new approach towards high quality, low temperature, patterned oxide. *Appl. Phys. Lett.* **45,** 241–243.

O'Shea, D.C. (1985). *Elements of Modern Optical Design.* John Wiley & Sons, New York.

Polasko, K.J., Ehrlich, D.J., Tsao, J.Y., Pease, R.F., and Martinero, E.E. (1984). Deep UV exposure of $Ag_2Se/GeSe_2$ utilizing an excimer laser. *IEEE Elect. Dev. Lett.,* **EDL-5,** 24–26.

Pepin, H., Alaterre, P., Chaker, M., Fabro, R., Faral, B., Toubhans, I., Nagel, D.J., and Peckerar, M. (1987). X-ray sources for microlithography created by laser radiation at 0.266 μm. *J. Vac. Sci. Technol.* **B5,** 27–32.

Peters, D.W., Drumheller, J.P., Frankel, R.D., Kaplan, A.S., Preston, S.M., and Tomes, D.N. (1988). Application and analysis of production suitability of a laser-based plasma x-ray stepper. *SPIE* **923,** 5.

Philipp, H.R., Cole, H.S., Liu, Y.S., and Stitnik, T.A. (1986). Optical properties of polymers in the vacuum UV 150–250 nm region. *Appl. Phys. Lett.* **48,** 192–194.

Podlesnik, D.V., Gilgen, H.H., Osgood, R.M. Jr., and Sanchez, A. (1983). Maskless, chemical etching of sub-micrometer gratings in single crystal GaAs. *Appl. Phys. Lett.* **43,** 1083–1085.

Pol, V., Bennewitz, H., Escher, G.C., Feldman, M., Firtion, V.A., Jewell, T.E., Wilcomb, B.E., and Clemens, J.T. (1986). Excimer laser-based lithography: A deep UV wafer stepper. In *Optical Microlithography V,* SPIE Proc. No. 633, 6–16.

Possin, G.E., Parks, H.G., Chiang, S.W., and Liu, Y.S. (1983). The effects of selectively absorbing dielectric layers and beam shaping on recrystallization and FET characteristics in laser recrystallized silicon on amorphous substrates. In *Materials Res. Soc. Sym. Proc.* **13,** 549–555.

Pummer, H., Egger, H., and Rhodes, C.K. (1987). High spectral brightness excimer systems. In *Excimer Lasers* (Rhodes, C.K., ed.). *Topics Appl. Phys.,* Vol. 30. Springer-Verlag, New York, 217–228.

Raamot, J., and Zaleckas, V.J. (1973). Laser pattern generation using x-y beam deflection. *Appl. Opt.* **13,** 1179–1183.

Raffel, J.I. (1987) *Laser Linking for Defect Avoidance and Customization. in Lasers in Microlithography* (Ehrlich, D.J., Tsao, J.Y., and Batchelder, J.S., eds.). Proceedings of SPIE **774,** 93–100.

Ready, F.J. (1971). *Effects of High Power Laser Radiation,* Academic Press, New York.

Reintjes, J.F. (1985). Coherent UV and vacuum UV sources. In *Laser Handbook* (Bass, M., and Stitch, M.L., eds.). North-Holland, New York, 1–202.

Rice, S., and Jain, K. (1984a). Reciprocity behavior of photoresist in excimer laser lithography. *IEEE Trans. Elect. Dev.,* **ED-31,** 1–3.

Rice, S., and Jain, K. (1984b). Direct high-resolution excimer laser photoetching. *Appl. Phys.* **A33,** 195–198.

Rhodes, P., and Shealy, R.L. (1980). Refractive optical systems for irradiance redistribution of collimated radiation: Their design and analysis. *Appl. Opt.* **19,** 3545–3549.

Risk, W.P., Baumert, J.C., Bjorklund, G.C., Schellenberg, F.M., and Lenth, W. (1987). Generation of blue light by intracavity frequency mixing of the laser and pump radiation of a miniature Nd:YAG laser. *Appl. Phys. Letts.,* **51,** 2192–4.

Rothschild, M., Arnone, C., and Ehrlich, D.J. (1986). Excimer laser etching of diamond and hard carbon films by direct writing and optical projection. *J. Vac. Sci. Technol.* **B4,** 310–314.

Rothschild, M., and Ehrlich, D.J. (1987a). Attainment of 0.13 μm lines and spacings by excimer laser projection lithography in diamond-like carbon resist. *J. Vac. Sci. Technol.* **B5,** 389.

Rothschild, M., and Ehrlich, D.J. (1987b). Optical considerations for excimer laser lithography. In *Proc. of Laser and Particle-beam Chemical Processing for Microelectron-*

ics (Ehrlich, D.J., Higashi, G.S., and Oprysko, M.M., eds.). Materials Res. Soc., Pittsburgh, Pennsylvania, 13–20.

Rothschild, M., and Ehrlich, D.J. (1988). A review of excimer laser projection lithography. *J. Vac. Sci. Technol.* **B6,** 1–17.

Sengupta, U. (1988). Private communication.

Schneider, M.B., and Webb, W.W. (1981). Measurement of submicron laser beam radii. *Appl. Opt.* **20,** 1382–1388.

Shen, Y.R. (1984). *The Principles of Nonlinear Optics.* John Wiley & Sons, New York.

Sherman, R.J. (1985). Polygon scanners. In *Laser Beam Scanning* (Marshall, G.F., ed.). Marcel Dekker, New York, 63–123.

Siegman, A.E. (1971). *An Introduction to Lasers and Masers.* McGraw Hill, New York.

Siegman, A.E. (1974). Unstable optical resonator. *Appl. Opt.* **13,** 353–378.

Sincerbox, G.T. (1985). Holographic scanners: Applications, performance and design. In *Laser Beam Scanning* (Marshall, G.F., ed.). Marcel Dekker, New York, 1–62.

Smart, D.V., and Stewart, D.M. (1987). Laser processing for application specific integrated circuits. In *Lasers in Microlithography* (Ehrlich, D.J., Batchelder, J.S., and Tsao, J.Y., eds.). SPIE, Vol. 774, 88–92.

Smith, P.W., Duguay, M.A., and Ippen, E.P. (1974). *Progress in Quantum Electronics* (Sanders, J.H., and Stevens, K.W., eds.). Pergamon Press, Oxford, 107–229.

Srinivason, R., and Mayne-Banton (1982). *Appl. Phys. Lett.* **41,** 576.

Stuke, M., Zhang, Y., and Kuper, S. (1988). Fundamentals of laser photochemistry for surface modification. In *Proc. of Laser and Particle Beam Chemical Processing for Microelectronics.* Mat. Res. Soc., Pittsburgh, Pennsylvania, 139–150.

Suzaki, Y., and Tachibana, A. (1975). Measurement of the micron sized radius of gaussian laser beam using the scanning knife edge. *Appl. Opt.* **14,** 2809–2810.

Svelto, O. (1982). *Principles of Lasers.* (Hanna, D.C., translator). Plenum Press, New York.

Swing, R.E., and Cley, J.R. (1967). Ambiguity of the transfer function with partially coherent illumination. *J. Opt. Soc. Am.* **57,** 1180–1189.

Sze, R.C. (1983). AIP Conf. Proc. No. 1000. Excimer Lasers — 1983, 73.

Thompson, B.J. (1958). Illustration of the phase change in two-beam interference with partially coherence light. *J. Opt. Soc. Am.* **48,** 95.

Thompson, B.J. (1977). Image formation with coherent light—a tutorial review. In *Coherent Optical Engineering* (Arecchi, F.T. and Degiorgio, V. eds.). North-Holland, New York, 49–62.

Thompson, B.J., and Wolf, E. (1957). Two-beam interference with partially coherent light. *J. Opt. Soc. Am.* **47,** 895.

Thompson, L.F., and Bowden, M.J. (1983). The lithography process: Physics. In *Introduction to Microlithography, Theory, Materials and Processing* (Thompson, L.F., Wilson, C.G., and Bowden, M.J., eds.). ACS Symposium Series, Vol. 219, 15–86.

Tsao, J.Y., and Ehrlich, D.J. (1984). Patterned photonucleation of chemical vapor deposition of Al by UV-laser photodeposition. *Appl. Phys. Lett.* **45,** 617–619.

Tsao, J.Y., Picraux, S.T., Light, R.W., and Hsing, W.W. (1987). Practical constraints on sources for pulsed beam lithographies. *Sandia National Laboratory Report,* SAND87-1606.

Tuckerman, D.B. (1987). Laser-patterned interconnect for thin film hybrid wafer scale circuits. *IEEE Elect. Dev. Letts.* **EDL-8**(11), 540–544.

Veldkamp, W.B. (1982). Laser beam profile shaping with interlaced binary diffraction gratings. *App. Opt.* **21,** 3209–3212.

Veldkamp, W.B., and Kastner, C.J. (1982). Beam profile shaping for laser radars that use detector arrays. *App. Opt.* **21,** 345–356.

Von Allman, M. (1987). Laser beam interactions with materials. In *Springer Series in Materials Science,* Vol. 2, Springer-Verlag, New York.

Walling, J.C. (1987). Tunable paramagnetic-ion solid state lasers. In *Tunable Lasers* (Mollenauer, L.F., and White, J.C., eds.). Springer-Verlag, New York, 331–398.

Walsh, K.F., Dunn, M.M., Holbrook, D.S., and Brunning, J.H. (1987). Performance evaluation of a practical 248 nm wafer stepper. In *SPIE.* **774,** 155–160.

Warkentin, P.A., and Schoeffel, J.A. (1986). Scanning laser technology applied to high speed reticle writing. In *Optical Microlithography,* SPIE, Vol. 633, 286–291.

Weber, M.J. (1979). Solid state lasers. In *Methods of Experimental Physics,* Vol. 15A. Academic Press, New York, 167.

Weber, M.J. (1985). In *Handbook of Laser Science and Technology, Vol. I, Lasers and Masers.* CRC Press, Boca Raton, Florida.

White J.C. (1987). Stimulated Raman Scattering. In *Tunable Lasers* (Mollenauer, L.F., and White, J.C., eds.). Springer-Verlag, New York, 115–208.

Whitehead, J.C., Mitlitsky, F., Ashkenas, D.J., Bernhardt, A.F., Farmwald, Kaschmitter, J.L., and McWilliams, B.M. (1986). Laser fabrication of interconnect structures on CMOS gate arrays. In *Manufacturing Applications of Lasers,* SPIE, Vol. 621, 62–70.

Wilczynski, J.S. (1987). Optical lithographic tools: Current status and future potential. *J. Vac. Sci. Technolog.* **B5,** 288–292.

Wilke, V., and Schmidt, W. (1979). Tunable coherent radiation source covering a spectral range from 185 nm to 880 nm. *Appl. Phys. Lett.* **18,** 177–181.

Wisoff, P.J.K., Mendelsohn, A.J., Harris, S.G., and Young, J.F. (1982). Improved performance of the microwave-pumped XeCl laser. *IEEE J. Quant. Elect.* **QE-18,** 1839.

Yariv, A. (1975). *Quantum Electronics,* 2nd ed. John Wiley & Sons, New York.

Young, E.H., and Yao, S.K. (1976). Design considerations for acousto-optic devices. *Proc. IEEE* **69,** 54–64.

Yu, F.T.S. (1985). *White Light Optical Signal Processing.* John-Wiley & Sons, New York.

Zhou, B., Kane, T.J., Dixon, G.J., and Byer, R.L. (1985). Efficient frequency-stable laser-diode-pump Nd:YAG laser. *Opt. Lett.* **10,** 62–65.

Znotins, T. (1986). Excimer lasers: An emerging technology in semiconductor processing. *Solid State Technology,* Sept., 99–104.

Znotins, T., McKee, T., Gutz, S., and Tan, K. (1988). The design of excimer lasers for use in microlithography. In *Symposium in Microlithography,* SPIE, **922,** 454–460.

Zook, J.D. (1973). Light beam deflector performance: A comparative analysis. *Appl. Opt.* **13,** 875–887.

PART II

Fundamentals

CHAPTER 2

Laser-Stimulated Molecular Processes on Surfaces

T.J. CHUANG
IBM Almaden Research Center
San Jose, California

1. Introduction

This chapter consists of excerpts from two review articles by the author: "Laser-induced gas-surface interactions," published in *Surface Science Reports*, Vol. 3, pp. 1–105 (1983) and "Laser-induced molecular processes on surfaces," *Surface Science*, Vol. 178, pp. 736–786 (1986).

Photochemistry is usually considered as a means for carrying out

"Laser-Induced Gas-Surface Interactions" from *Surface Science Reports* 3, 1983, pp. 1–105, by T.J. Chuang and "Laser-Induced Molecular Processes on Surfaces" from *Surface Science* 178, 1986, pp. 763–786 by T.J. Chuang, adapted and reprinted by permission of North-Holland Physics Publishing, a division of Elsevier Science Publishers B.V.

molecular or even mode-selective chemical processes for chemical synthesis and purifications, isotope enrichments, and chemical fabrications of microelectronic and micromechanical devices. For a given chemical system, whether or not the photon energy can be effectively utilized for chemical processes depends upon the rates of energy acquisition, storage, and the chemical reactivity of the system. These subjects have been extensively investigated in both gaseous and condensed phases for more than a decade. These studies have led to tremendous progress in our understanding of molecular structure and dynamics. This progress is made possible because of the advances in both theoretical and experimental techniques. The latter is due primarily to the development of lasers, which have been utilized both as optical pumping and as probing tools. Lasers, because of the inherent monochromaticity and coherent characteristics, are particularly suitable for studying the properties of chemical systems. The fine resolution of the photon source can selectively excite a set of atoms or molecules with a bandwidth as small as 1 kHz. The light source can also be manipulated to excite matter and probe its dynamic behavior on time scales shorter than a picosecond. Clearly, the same powerful tool can be employed to investigate surface structures and heterogeneous chemical interactions. Although the applications of lasers to studying surface photochemistry are still rather limited, definite progress has been made in recent years. In this chapter, the basic molecular processes on surfaces stimulated by laser radiation are outlined and discussed.

In considering gas-surface chemistry, even the simplest gas-surface interactions involve several steps that begin with the collision of the incident particle with the solid surface. Depending on the particle-surface potential, which depends on both the electronic and vibrational states of the gaseous species and the electronic structure of the solid, a certain fraction of the incident particles is trapped in the attractive surface potential. Once trapped, they can move along the surface by diffusion. The adsorbed species may desorb from the surface if sufficient energy is imparted to it by the rather efficient adsorbate-substrate energy transfer processes. It is also possible for the adspecies to decompose into fragments if the attractive force to the surface is sufficiently strong. Dissociative chemisorption may then lead to further adsorbate-adsorbate and adsorbate-adsorbent interactions. New species may be formed and desorbed into the gas phase. When a photon beam is present at the gas-solid interface, it can affect many of these surface interaction steps. For instance, the photon can resonantly excite the incident gaseous species and influence its surface sticking probability. For the adsorbed species, resonant photoexcitation can activate surface reaction and desorption. If the photon beam is not absorbed by the gaseous or the adsorbed species, it can be absorbed by the solid substrate to promote electronic excitation in the valence band and the conduction band of the material. To begin with our discussion, it is useful to divide the photon-simulated surface interactions into three basic processes, namely, the interaction of laser photons with a) the gaseous species, b) the adsorbed species, and c) the solid substrate. In the first two processes, both electronic and lattice phonon (thermal) excitation of the solid substrate can take place. Depending on the chemical system, the optical arrangement, and the laser wavelength, one or more of these three processes may be involved. Detailed classification and discussion of these basic types of laser excitation are given in the following section, emphasizing primarily on phenomenological observations. In Section 3, the fundamental surface processes and basic interaction mechanisms involved in adsorption, surface reaction, and desorption affected by the laser radiation will be more thoroughly examined. Both important experimental observations and theoretical considerations will be discused, with the goal of clarifying the relative importance of the interaction steps and extracting useful concepts from these studies. Energy acquired from photoexcitation would be wasted if it could not be efficiently used for chemical purposes. Therefore, the important electronic and vibrational energy transfer and relaxation processes will also be closely examined. Photon-stimulated desorption is a distinctive surface phenomenon and the subject has been quite extensively investigated in recent years. Both electronic excitation by ultraviolet-visible light and vibrational activation with infrared photons have been investigated. Because of the relatively large amount of photodesorption studies to be covered, and in order to preserve the continuity of discussion on important surface processes in Section 3, we have divided the discussion on photodesorption into two parts. The basic concepts, theoretical treatments, and computational illustrations will be outlined in Section 3, and the experimental studies will be presented in Section 4.

For our discussion, we will emphasize studies that have been carried out on well-characterized solid surfaces under relatively well-controlled experimental conditions. Subjects that are of peripheral interest to the present topic, such as the physical effects induced by high-power laser radiation, semiconductor photoconductive and photovoltaic properties, early studies (before 1975) on photocatalysis on semiconductor surfaces and surface-enhanced Raman scattering by lasers, etc., are not included in this chapter. Relevant studies and concepts that have been developed from these studies naturally are not excluded and suitable references will be given wherever appropriate.

2. Basic Modes of Laser Excitation

2.1. Gas-Phase Excitation

2.1.1 Electronic Excitation

Laser-induced chemical reactions due to electronic excitation of one of the reactants have been extensively investigated in the gas phase as well as in condensed phases (see, e.g., Knebe and Wolfrum, 1980; Steinfield, 1981). These photoexcited reactions occur either because of the enhanced reactivity for excited electronic states or because of the reactive radicals generated by photodissociation. Both reaction mechanisms have also been demonstrated in heterogeneous chemical systems. The possible optical excitation schemes for gas-surface systems are shown schematically in Fig. 2.1. In a recent experiment, Umstead et al. (1980) used a cw Ar^+ ion laser to study the radiation effect on C_2H_4 oxidation by NO_2 catalyzed by a polycrystalline Pt surface maintained at 250°C. They found that when NO_2 molecules were excited into the first excited electronic state by a 1-W laser beam at 488 nm, the CO_2 production compared to the yield in the absence of radiation increased by a factor of four. The laser beam was incident collinearly with the Pt coil and did not have direct contact with the catalyst surface, similar to the arrangement shown

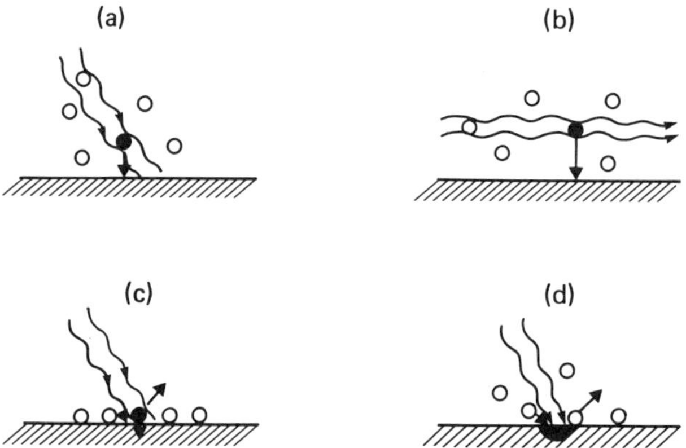

Fig. 2.1 Laser excitation of gas phase species with (a) or without (b) the radiation on the surfaces may enhance chemisorption. Resonant excitation of the adsorbate may promote adsorbate-adsorbate and adsorbate-adsorbent interaction as well as molecular desorption (c). Direct laser excitation and heating of the substrate may also induce surface reactions and desorption (d). Resonant excitation of chemical species in the gas phase or in the adsorbed state is indicated by solid dots.

in Fig. 2.1b. The authors believed that the laser enhancement of CO_2 production in this catalytic bimolecular reaction was due to the enhanced reaction of vibrationally excited NO_2^* with chemisorbed C_2H_4 or its related surface species. Their interpretation was based on the rapid collisional energy conversion from the electronically excited state into vibrationally excited states in the pressure range, i.e., 0.1–1.0 torr, that they had used. For better understanding of the initial electronic excitation effects, it would be desirable to study the reaction yield dependence on the laser wavelength and over a larger range of gas pressures. Simultaneous surface analyses would also be very helpful for further elucidation of the reaction mechanism.

Because radicals frequently can react with surfaces that appear inert to the parent molecules, surface reactions can often be greatly enhanced due to radicals generated by photofragmentation. The prominent examples are Ge-Br_2 and SiO_2-Cl_2 reactions enhanced by an Ar^+ ion laser. In the Ge-Br_2 experiment reported by Beterov et al. (1978) and Baklanov et al. (1974), the collimated laser beam was incident parallel to Ge surfaces, as shown in Fig. 2.1b. The 488-nm photons could photolyze Br_2 to produce Br atoms, which reacted with Ge to form germanium bromide products. In the SiO_2-Cl_2 reaction investigated by Chuang (1982c), the laser beam was incident perpendicular to the quartz substrate, as shown in Fig. 2.1a. Since the SiO_2 substrate did not absorb the laser light at 458 nm, the dominant effect of the photon radiation was to excite Cl_2 molecules into the first dissociative state. The photon-generated Cl atoms then reacted with SiO_2 surface to form volatile products. This SiO_2-Cl_2 photoreaction was further studied with a laser beam at 515 nm. It was found that the 515-nm laser beam induced a much lower surface reaction rate than the 458-nm light, evidently because Cl_2 photodissociation yield was much reduced at this longer wavelength (Herzberg, 1950). Without photofragmentation, Cl_2 molecules do not attack a SiO_2 surface spontaneously. It is interesting to note that depending on the specific chemical interaction of a gas-solid system, the photofragments can either react with or condense on the solid. Thus in contrast to photon-induced chemical etching reactions, lasers have also been used to induce deposition of materials on solid surfaces. One such example was given by Deutsch et al. (1979), who showed that Al or Cd metal could be deposited on a SiO_2 substrate by UV-laser photolysis of an alkyl $Al(CH_3)_3$ or $Cd(CH_3)_2$ compound at 257 or 195 nm. In subsequent studies by Ehrlich et al. (1981), it was further demonstrated that direct metal film growth could be accomplished by first "prenucleating" the desired region of growth *via* photodissociation of a thin surface layer of adsorbed alkyl molecules. Further growth of the predeposited film was achieved by a photon-generated spatially uniform atom source. With this approach metal patterns could be produced efficiently and with high spatial resolutions. Both laser-induced chemical etching and chemical vapor deposition can have important implications for microelectronic fabrication.

2.1.2 Vibrational Excitation

The similarity of the vibrational energy of a molecule to the activation barrier for chemical reaction has inspired numerous studies to search for a correlation between surface reactivity and vibrational activation. With these efforts, a variety of phenomena have been observed, revealing many facets of the effects of vibrational excitation on surface chemical processes. It is interesting to observe that in some instances surface sticking coefficients are found to decrease when certain molecules are vibrationally excited. In other systems, the probabilities for dissociative chemisorption are shown to be enhanced because of vibrational activation. In an experiment carried out by Gochelashvili et al. (1976), it was reported that when the ν_3 mode of a $^{11}BCl_3$ molecule was vibrationally excited by a CO_2 laser at 10.6 μm, the sticking coefficient of the molecule on stainless steel walls at 160°K was decreased. It was further reported that by this infrared radiation, an enrichment of ^{11}B species in the $^{11}BCl_3$ and $^{10}BCl_3$ gas mixture could be obtained. In other non-photon-related experiments, CO_2 (Basov et al., 1975; Brzhazovskii et al., 1976) molecules were found to be more difficult to condense on cool surfaces when the molecules were internally (mainly vibrationally) excited. These experiments involved interaction of excited gaseous species with multilayers of condensate on cool surfaces. The solid surfaces were not characterized. By now, the phenomenon of infrared-laser-excited desorption is rather well established and this phenomenon will be discussed in detail in Sections 3 and 4.

There have been a number of experiments illustrating heterogeneous decomposition of polyatomic molecules enhanced by infrared laser excitation of the gaseous species. For instance, Bass and Franchi (1976) studied N_2O decomposition on Cu surfaces. In this experiment, N_2O molecules were excited from (001) to (100) vibrational level by a N_2O laser at 10.8 μm. The rate of N_2O-Cu reaction that presumably formed N_2 and CuO was determined by measuring the pressure increase of noncondensible gaseous product (i.e., N_2) at 77°K. The surface reaction rate was found to be enhanced by a factor of 5000 when the gas-solid system was irradiated by the IR laser. By examining the laser heating effects on the cell windows, copper insert, and the thermal gradient of the gas, the authors concluded that the enhanced reaction was due to laser vibrational excitation of N_2O molecules. The study was followed by other catalytic decomposition experiments by other researchers, e.g., NH_3 on Pt (Khmelev et al., 1977), HCOOH on Pt (Umstead and Lin, 1978), $BCl_3 + H_2$ on Ti and Pb (Lin et al., 1977; Lin and Atvars, 1978) and 2-propanol on CuO surfaces (Farneth et al., 1983).

For many chemical systems, the activation barriers for reaction can be much higher than a single vibrational quantum level, and multiple photon excitation is necessary to promote surface reactions. Depending on the laser excitation conditions, absorption of multiple infrared photons can pump the gaseous molecules into highly vibrational excited states or drive the molecules beyond the threshold for dissociation in the ground electronic state. The highly vibrationally excited molecules or the multiple-photon-dissociated (MPD) radicals can readily interact with solid surfaces. One such example is the SF_6 interaction with silicon. In a study by Chuang (1980b, 1981a), it was found that although SF_6 molecules were inert to Si at 25°C, the species could be induced by multiple photon excitation with a pulsed CO_2 laser to react with the solid, causing Si atoms to be chemically removed from the surface. A systematic investigation, including the determination of surface reaction yields as a function of the laser wavelength, the laser intensity, the gas pressure, and the addition of buffer gases, had led to the conclusion that vibrationally excited SF_6 could directly react with Si to form a volatile SiF_4 product. The level of molecular excitation was estimated to involve absorption of 3–4 CO_2 laser photons. It was further shown that by focusing the CO_2 laser beam to obtain a very high laser intensity, SF_6 molecules could be photodecomposed because of coherent and consecutive absorption of more than 30 photons to produce SF_4 and F atoms that could subsequently react with Si to form SiF_4. Multiple photon dissociation with a pulsed CO_2 laser was also used by Hanabusa et al. (1979) to decompose SiH_4 molecules for depositing silicon on quartz or glass substrates.

2.2. *Adsorbate Excitation*

When a photon beam is incident on an optically transparent or a weakly light absorbing substrate with an adsorbate on its surface, the photons, in a suitable wavelength region, can be absorbed by the adsorbed species and induce the species to break the surface bond and desorb into the gas phase. This photon-stimulated desorption process will be discussed separately later. Another possibility is that the resonant molecular excitation may enhance chemical interactions between two different adsorbates or between the adsorbate and substrate atoms, as shown schematically in Fig. 2.1c. An example is the Si-SF_6 interactions, which have been studied in both gas-phase and condensed states. Whereas SF_6 molecules do not chemisorb on Si at room temperature, the gas can be physisorbed on the solid at 90°K. At this temperature, solid SF_6 can still vaporize so that the gas does not condense permanently on Si. Nevertheless, a few monolayers of SF_6 molecules can be adsorbed for a period of time sufficient for pulsed-laser mass spectrometric measurements. When the Si surface at 90°K covered with 1–2 monolayers of SF_6 is irradiated by CO_2 laser pulses at $\nu = 942\ \text{cm}^{-1}$, with the laser intensity $I = 0.8\ \text{J/cm}^2$, a substantial amount of SiF_3^+ ions is detected with a mass spectrometer in an ultrahigh vacuum chamber, indicating that SiF_4 is produced from the laser-induced Si-SF_6 reaction (Chuang, 1982a, 1983a). In addition to SiF_4, SF_6 desorption is also observed (detected as SF_5^+ ions). When the laser is tuned to $978\ \text{cm}^{-1}$, where the laser photons are not absorbed by the physisorbed SF_6, no SiF_4 product is observed. The typical experimental results are shown in Fig. 2.2. The reaction yields in the $930–950\ \text{cm}^{-1}$ region coincide with the infrared absorption spectra of solid SF_6 (Brueck et al., 1979) and of SF_6 molecules adsorbed on silver surfaces that were determined with surface photoacoustic spectroscopic technique (Träger et al., 1982; Chuang et al., 1983c). The results suggest that like the gas-phase excitation, vibrationally excited SF_6 due to multiple IR photon absorption is also quite reactive to Si at a relatively low substrate

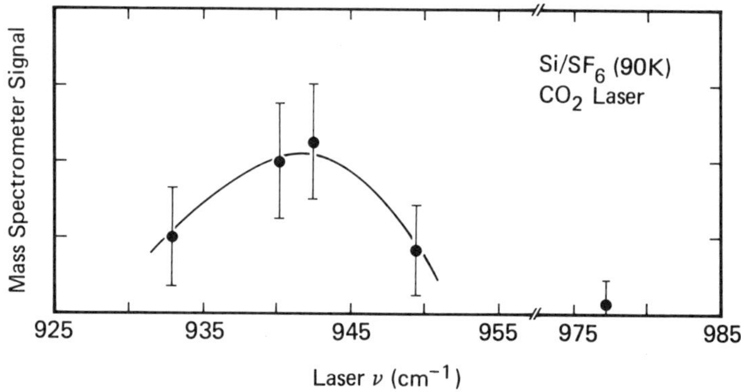

Fig. 2.2 Relative SiF_4 production yields per laser pulse determined as SiF_3^+ ions with the mass spectrometer as a function of the laser frequency. The laser intensity is fixed at $0.8\ \text{J/cm}^2$, p-polarized, and incident at 75° from surface normal. SF_6 surface coverage is about two monolayers. Each data point is an average of mass peak heights due to 10 laser pulses. Data according to Chuang (1983a).

temperature. It should be noted that silicon is quite transparent to CO_2 laser light. The laser excitation and heating of the Si substrate alone, i.e., without resonant SF_6, excitation, does not cause the surface reaction to occur. The laser-activated reaction has been further studied with x-ray photoemission spectroscopy (XPS) in the same UHV system. When the laser-irradiated Si-SF_6 sample at 90°K (200 pulses at $942\,cm^{-1}$, $I = 0.8\,J/cm^2$ with 100-nsec pulse duration) is examined with XPS, a small shoulder about 2.5 eV higher in binding energy appears in the Si(2p) and Si(2s) spectra, as shown in Fig. 2.3. This new band is very similar to that attributed to the "SiF_2"-like species observed in the Si-XeF_2 reaction also investigated with XPS (Chuang, 1980c). It thus appears that the laser-induced Si-SF_6 interaction can also transform the clean Si into a fluorinated surface. In a separate experiment, about 1–2 monolayers of SiF_4 were condensed on Si at 90°K and irradiated with the same CO_2 laser at $942\,cm^{-1}$. No significant SiF_4 desorption was observed. Clearly, the small substrate heating effect alone is incapable of thermally desorbing ground state SiF_4. The fact that SiF_4 molecules produced in the laser-induced Si-SF_6 reaction desorb from Si at 90°K suggests that the product molecules may be formed in excited states, which can have a higher desorption rate. There is evidence to show that vibrational excited

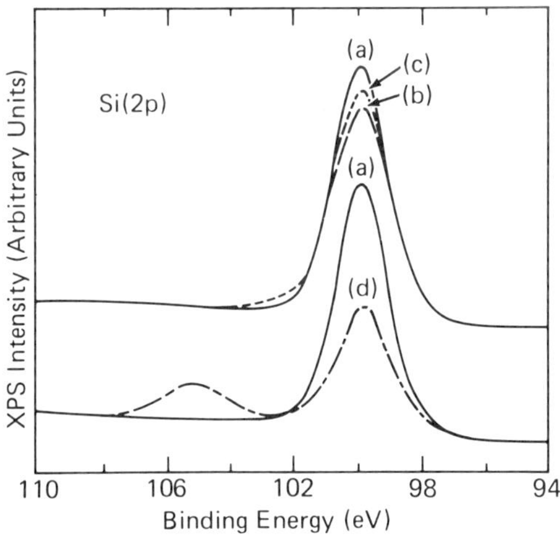

Fig. 2.3 Si(2p) XPS spectra for a silicon crystal exposed to SF_6 at 90°K substrate temperature: (a) Si clean; (b) about two monolayers of SF_6 adsorbed on Si before laser irradiation; (c) after irradiation by 200 CO_2 laser pulses at $\nu = 942.4\,cm^{-1}$, $I = 0.8\,J/cm^2$; (d) a SiF_4 layer condensed on Si for comparison. Data according to Chuang (1983b).

species are formed in the Si-XeF_2 reaction, as observed by infrared chemiluminescence (Chuang, 1979, 1980a).

When a solid is exposed to a gaseous chemical at certain gas pressure, depending on the adsorption isotherm of the system, a substantial amount of the gas can be physisorbed on the solid surface. Again using Si-SF_6 as an example, about half of a monolayer of molecules can be adsorbed on Si at room temperature when the gas pressure is maintained at 1 torr. As the gas pressure is increased to 20 torr, the surface coverage increases slightly but does not exceed one monolayer, as determined with a silicon quartz crystal microbalance (Chuang, 1981a). If a CO_2 laser pulse is incident on the gas-solid interface, SF_6 vibrational excitation can take place both in the gas phase and in the adsorbed state at 25°C, and the Si-SF_6 reaction can occur. Similarly, photochemical vapor deposition has been demonstrated for metal alkyl molecules physisorbed on solid surfaces. Ehrlich and Osgood (1981) showed that $Al_2(CH_3)_6$ and $Cd(CH_3)_2$ could be condensed at a gas pressure greater than 1 torr to form an overlayer on SiO_2. Photolysis of this molecular thin layer by a UV laser could produce a metal deposit at high spatial resolution. Other adsorbate excitation-promoted surface reactions include the dehydroxylation of OH groups adsorbed on SiO_2 surfaces reported by Djidjoev et al. (1976) using a CO_2 laser in the 950–970 cm^{-1} region.

2.3. *Solid Excitation*

In a gas-solid system, if the incident photon beam is also present on the solid surface (Fig. 2.1d), the radiation effects on the substrate have to be considered. Unlike gaseous atomic and molecular species, which have a relatively small number of allowed optical transitions in the UV, visible, and infrared regions, most solids have rather wide ranges of optical absorption bands with sufficiently large optical extinction coefficients. For instance, the optical absorption coefficient of Si is small in the 1 to 12 μm region. Below 1 μm (band gap), the absorption band is continuous and reaches a maximum around 265 nm. Polished copper is highly reflective in the infrared but absorbs UV photons quite well. For illustrative purposes, the optical absorption characteristics of Si and Cu are shown in Fig. 2.4. The original optical data and references can be found in the *American Institute of Physics Handbook* (Gray, 1972). Some optical absorption depths are also indicated in the figure. It is worth noting that practically all metals are opaque from UV to IR and can be readily excited by commonly available lasers.

Mechanistically, the initial interaction of the laser light with a solid is almost always with its electrons, except in the medium- and far-IR

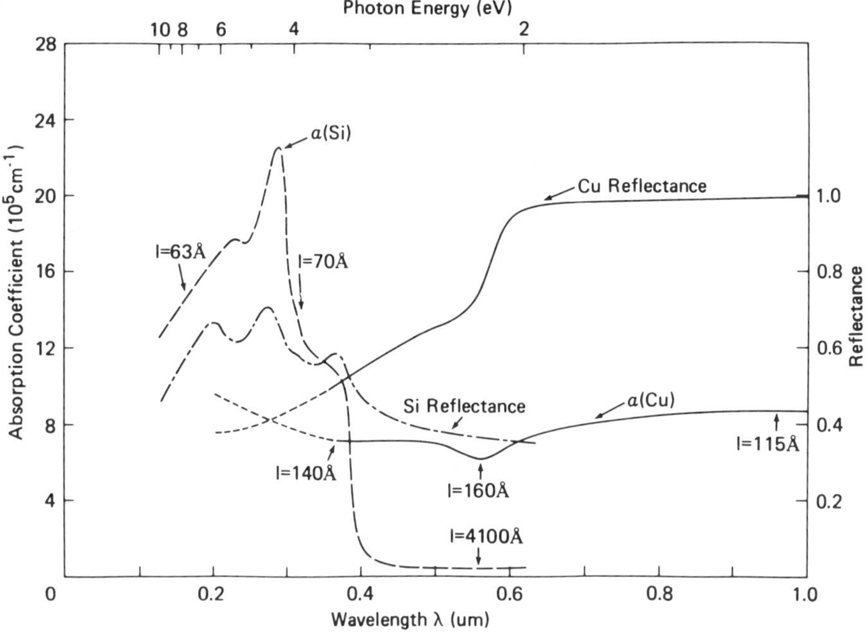

Fig. 2.4 Absorption coefficients and reflectances for silicon and copper in the ultraviolet, visible, and near-infrared region. Some optical absorption depths (l) are also indicated.

regions where direct lattice phonon excitation can occur. The optical absorption can result in band-gap excitation, interband transitions, and excitation of free electrons. The energy absorbed by the electrons is ultimately shared with the atoms of the solid as electronic excitation is transformed into lattice excitation (heat). The exact sequence of photon-electron-phonon interaction depends strongly on the detailed electronic structure of the solid and the intensity of the laser radiation field. Generally, because of the high photon flux, the initial absorption of a laser beam can often alter the electronic structure of the material, which in turn can lead to a self-induced increase or reduction of the degree of absorption. This complex coupling phenomenon is not generally understood for most materials. The lack of such physical understanding has also impeded the development of a more detailed understanding of the chemical interactions involved in the laser-stimulated surface processes. While a complete formulation of a given gas-solid reaction is not possible at this time, we can, nevertheless, estimate certain limits in terms of the rates of interactions and gain some insight into the chemical processes.

Using Si as an example, Brown (1980) illustrated that if 2-eV photons with 0.2 J/cm^2 intensity were suddenly absorbed in the top 1000 Å layer of the solid (approximately the photon energy and the absorption depth for 500- to 600-nm light in Si), it would produce a carrier concentration $6 \times 10^{22}/\text{cm}^3$ in the layer. The Auger process associated with the electron-hole recombination would decrease the carrier concentration to $6 \times 10^{20}/\text{cm}^3$ in less than 10^{-11} sec. Simultaneously, the electron-phonon collisions would transfer the excess energy of the electron-hole pair to lattice thermal excitation at the rate of 10^{11}–10^{12}/sec. Thus, in about 10^{-11} sec, 99% of the light energy absorbed by the solid would have been transformed into heat. The chemical interaction rates of a gas-phase species with the photoexcited Si surface can be estimated from the gas-surface collision rates. For a gas at $p = 500$ torr, the gas-surface collision rate is about $1 \times 10^{23}/\text{cm}^2$-sec at 25°C. Thus for each Si surface atom, the rate of collisions by gaseous particles is about 1×10^8/sec. If there were no molecules adsorbed on the Si surface before the arrival of the laser pulse and if the surface reaction were to depend entirely on the gas-surface collision rate, then the time involved in the surface chemical process would be about 1×10^{-8} sec or longer. On this time scale, as suggested by Brown, the speed by Auger recombination for carrier concentration $n \geq 3 \times 10^{19}$ would not allow the carrier concentration to be far out of equilibrium with the temperature of the lattice. Under this condition, the chemical process would be dominated by the thermal effect. Figure 2.5 illustrates the relevant time scale discussed here. Also included in the figure are the electron-electron ($\tau_{\text{e-e}}$) and electron-plasmon ($\tau_{\text{p-plas}}$) collision times, which are in the 10^{-13}–10^{-14} sec region and can be important if the laser intensity is $\geq 10^8$ W/cm^2 (von Allmen, 1980). If, on the other hand, the chemical species is initially adsorbed on the solid surface before the laser impulse, the time involved ($\tau_{\text{ads-s}}$ in Fig. 2.5) in the chemical interaction between the adsorbate and the adsorbent induced by the laser pulse can be shorter than 10^{-9} sec. If the activation barrier is low, this $\tau_{\text{ads-s}}$ could even be $\leq 10^{-12}$ sec. In this situation, the chemical reaction rate clearly does not have to depend on the rate of chemisorption from the gas phase because the reactant is already present on the surface. Thus the surface reaction can kinetically be nonthermal in nature. The time scales for chemical interactions shown in Fig. 2.5 are particularly useful for differentiating the photon-induced chemistry originated in the gas-phase excitation from that initiated by adsorbate-substrate excitation, when a pulsed laser is used. At a given gas pressure, if the gas-surface collision time is substantially longer than the laser pulse duration, the laser radiation effect is most likely associated with the adsorbate and/or adsorbate-substrate excitation.

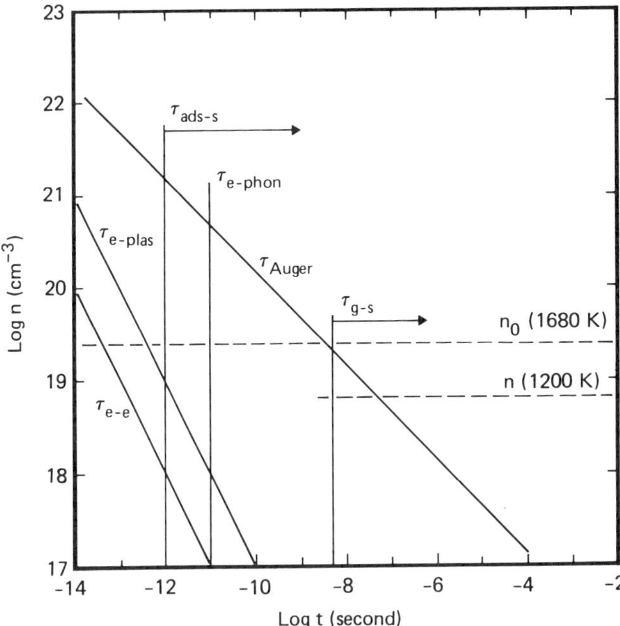

Fig. 2.5 Decay of carrier concentration (n) by Auger recombination (τ_{Auger}) as a function of time following a UV or visible laser impulse excitation on Si. Also shown are the time scales for electron-plasmon ($\tau_{e\text{-}plas}$), electron-electron ($\tau_{e\text{-}e}$), and electron-phonon ($\tau_{e\text{-}phon}$) collisions; the gas-surface ($\tau_{g\text{-}s}$) collisions time for $p \leq 760$ torr; and the time for adsorbate-surface chemical interactions ($\tau_{ads\text{-}s}$). The equilibrium carrier concentrations at 1680°K (Si melting point) and 1200°K are also indicated.

For gas-surface reactions initiated by photo-excitation of the solid, direct substrate heating is always an important factor to consider. Indeed, laser excitation and heating has been utilized for annealing of semiconductors (see, e.g., Van Vechten, 1980), desorption of adsorbates for surface cleaning and characterization (Bedair and Smith, 1969; Allen, 1982), and studies of thermal desorption kinetics (Levine et al., 1967, Christmann et al., 1974; Cowin et al., 1978). Nonthermal effects due to solid excitation have been investigated in some chemical systems. In a recent photooxidation study on silicon, Schafer and Lyon (1981, 1982) found that visible laser radiation ($\lambda = 413$–647 nm) could increase the Si thermal oxidation rate by about 40% and UV radiation at 350 nm by about 60%, in the temperature range of 770–900°C. At a given photon flux, the photoeffect induced by the visible light is insensitive to the wavelength. For the greater enhancement of the oxidation rate induced by UV photons, the authors suggested that the photoeffect was due partly to the thermal heating of the substrate and partly to the electronic excitation of the Si-SiO_2 interface. The excitation could involve the transfer of an electron from the conduction band of Si into the conduction bands of SiO_2 where charged oxygen species could be formed. Such an excitation, according to Goodman (1966), would require an activation energy of 3.3 eV, i.e., an energy gap between the two conduction band edges. The energy of UV photons used by Schafer et al. was about 3.5 eV. If this is the principal mechanism, it is not clear why a 2-eV photon in the visible region can also cause enhancement in the Si oxidation rate. Although the precise nature of the laser radiation effect is not yet certain, the nonthermal photon-induced behavior has been confirmed by Young and Tiller (1983). These authors further showed that as the laser intensity was increased, the thermal effect became the major factor in determining the Si oxidation rate. The nonthermal oxidation effects induced by low-power laser radiation were also observed by other researchers and found to be much more pronounced on GaAs surfaces (Schafer and Lyon, 1981; Petro et al., 1982).

The nonthermal effect in the laser-enhanced surface reaction has also been observed in silicon-fluorine interactions. In the Si-XeF_2 reaction, it was found that the spontaneous Si etching reaction at 25°C (Winters and Coburn, 1979) could be greatly enhanced not only by an ion beam (Tu et al., 1981), but also by a pulsed CO_2 laser (Chuang, 1981b) incident on the solid. From an XPS investigation of the reaction and the mass spectrometric determination of SiF_4 chemisorption and desorption behavior at both 25°C and −150°C, it was proposed (Chuang, 1981b) that the primary effect of the external radiation was to promote fluorine atoms chemisorbed in the surface region as "SiF_2"-like species to react *via* a disproportionation reaction to form the SiF_4 product. By comparing the interactions at −150°C with those at 25°C, Chuang concluded that the laser-enhanced Si-XeF_2 reaction was not due entirely to the thermal effect. The nonthermal interaction was more clearly shown later by Houle (1983), who used a cw Ar^+ ion laser in the visible region that could promote Si band-gap excitation. In this study, it was found that the SiF_4, SiF_3, SiF_2, and SiF product distribution, with SiF_4 being the dominant species (Tu et al., 1981; Winters and Houle, 1983), in the laser-enhanced reaction was very different from that produced in the absence of light. Furthermore, the reaction rate enhanced by the low-intensity Ar^+ laser beam could not be explained by the increase in substrate temperature. Like the laser-enhanced Si oxidation, the detailed mechanism for the nonthermal effect in the Si-XeF_2 etching reaction enhanced by laser radiation is yet to be elucidated. In view of the chemical similarity between the halogens and oxygen, e.g., both being highly electronegative species, as pointed out by Winters et al. (1983), it is perhaps not very surprising to see that the chemical interactions of these species with silicon are also quite similar, including the nonthermal photon-induced effects.

It should be clearly noted that in our discussion of the laser excitation of solid substrates, our primary interest has been focused on the surface *chemical* processes and not the *physical* effects such as those induced by the high-power laser radiation. The physical effects produced by lasers at very high light intensity, including surface melting, solid vaporization, surface damage, plasma formation, and ion and electron emission, have been quite thoroughly dealt with by Ready (1971). Photon-induced particle emission or "sputtering" involving both thermal and nonthermal excitation mechanisms have also been investigated by many researchers (Townsend, 1979; Hanabusa et al., 1981; Itoh and Nakayama, 1982). Unless otherwise noted, the vast majority of studies on laser-stimulated gas-surface chemistry that we describe in this chapter have been performed with a laser intensity of less than 10^6 W/cm^2 so that such physical effects as surface melting or solid vaporization are minimized or unimportant.

3. Photon-Stimulated Surface Processes

In the preceding section, we have given a phenomenological description of some laser-induced gas-surface interactions. In this section, we will discuss some fundamental aspects of the physicochemical processes involved in the radiation-enhanced surface chemistry. When a photon beam is present at the gas-solid interface, many effects can take place, such as the increase in local electromagnetic fields, excitation of surface plasmons, the electronic excitation of the adsorbate-substrate complex, the vibrational excitation of adspecies, phonon excitation, electron-hole pair excitation, photocarrier generation and diffusion, charged particle emission, etc. In general, the interactions among photons, gaseous particles, adspecies, and solid atoms are rather complex. Without adequate experimental data and detailed analyses, it is often difficult to sort out the important mechanisms directly influencing the surface chemical process of interest. In the following, a few subjects are chosen for further discussion, not only because physical understanding of these processes are important, but also because a substantial amount of experimental and theoretical studies have been performed in these areas. Except for photon-stimulated desorption, where the experimental studies will be discussed in Section 4, both experimental observations and theoretical considerations, whenever available, are discussed together.

3.1. Photon-Enhanced Adsorption

Physical and chemical adsorption in a gas-surface system can be enhanced by photon excitation in several different ways, e.g., by photodissociation, vibrational excitation, electronic excitation of the adsorbate-substrate complex, substrate electron-hole pair excitation, and, naturally, pure thermal activation. Among the nonthermal effects, enhanced adsorption due to photofragmentation is the easiest to understand. In elementary surface studies, it is well established that radicals frequently can chemisorb on surfaces that are inert to parent molecules. One prominent example is nitrogen interaction with metal surfaces. It has been shown the N atoms are readily chemisorbed on copper (Tibbetts, 1979; Lee and Farnsworth, 1965), iridium, and rhodium (Nimeault and Hansen, 1966), whereas N_2 molecules are not chemisorbed on any of these surfaces. As elucidated by Winters et al. (1983), there are two mechanisms that can allow a radical to be more reactive than its parent molecule. First, the reaction between a parent molecule and surface may be exothermic but require a large activation energy to proceed. When the need for activation energy is avoided by dissociating the molecule while it is in the gas phase, the resulting fragments can spontaneously react with the surface. Both single and multiple photon absorption can be effective for generating gaseous radicals. The multiple photon dissociated SF_6 interaction with silicon very likely occurs by this mechanism. The formation of a Si-F bond in SiF_4 should release more chemical energy than that is required to break a S-F bond in SF_6 (Cottrell, 1958), and yet SF_6 is inert to Si at 25°C. The single-photon photolyzed Br_2 interaction with germanium may also belong to this category. The second mechanism is related to adsorbed states that are endothermic; namely, the sum of binding energies of radicals to the surface is smaller than the dissociation energy of the parent molecule. In this case, the parent molecule is inert to the surface in the absence of external radiation, whereas the radicals can be quite reactive. Clearly, fragmentation by photons can also promote endothermic surface reactions.

The determination of vibrational excitation for enhancing surface chemisorption is less straightforward. Although a number of experiments as described in Sections 2.1 and 2.2 have demonstrated that highly vibrationally excited molecules can be quite reactive for surface reactions, to our knowledge, there are still very few clear-cut experiments to illustrate that surface reactivity can be enhanced due to excitation of a single vibrational quantum state. In prior experiments, there was exclusive usage of lasers in the 10-μm region corresponding to about 3 kcal/mole of photon energy. Three or more such infrared photons were usually needed to promote surface chemisorption. A possibly single IR photon excited dissociative chemisorption was demonstrated by Umstead and Lin (1978). They studied the effect of CO_2 laser radiation on the catalytic decomposition of formic acid on a platinum surface. The molecule was known to be catalytically decomposed into $CO_2 + H_2$ and $CO + H_2O$, with CO_2 production being the dominant process. With a 10-W cw CO_2 laser, the authors observed a change in the CO_2/CO production ratio by as much as 50% when HCOOH gas was irradiated by the laser at 9.6 μm in the presence of a Pt filament maintained at 250°C. If the laser was tuned off resonance from the formic acid absorption bands, the CO_2/CO ratio was the same as in the absence of the laser radiation. It was also found that the laser enhancement effect disappeared after a certain period of time, apparently because of catalyst surface poisoning. For other single IR photon excitation, there were three independent efforts in the 1978–1979 period trying to determine the role played by vibrational activation of methane molecules in their chemisorptive interactions with rhodium (Brass et al., 1979, Yates et al., 1979) and tungsten surfaces (Chuang and Winters, 1979). He-Ne lasers (cw) at 3.39 μm were used in these three cases. These studies were stimulated by the earlier investigations of CH_4 chemisorption behavior on Rh by Steward and Ehrlich (1975) and on W by Winters (1975, 1976). In the Rh-CH_4 experiments, the surface temperature was maintained at 245°K, while the gas temperature was varied in the 600–710°K range. The rate of CH_4 dissociative chemisorption was found to be independent of the gas translational energy, but rather sensitive to internal molecular temperature. Winters, on the other hand, varied the tungsten temperature from 600–2500°K while maintaining the initial gas temperature at 300°K. Both studies showed strong isotope effects, namely, the sticking probabilities decreased when H atoms in methane were replaced by deuterium atoms. Both studies suggested that vibrational excitations were the major mechanism responsible for the observed surface dissociative chemisorption. In the effort to directly verify such reaction mechanisms, Brass et al. (1979) measured the reaction rate by following the pressure decrease in a closed system containing CH_4 in contact with Rh films. Photoexcitation of the CH_4-ν_3 mode was accomplished with a He-Ne laser, while an arc light source was used for exciting the $2\nu_3$ mode. $2\nu_4$ mode excitation from the ν_3 mode *via* collisional energy transfer was also tested. None of these excitations led to any detectable changes in the Rh-CH_4 surface reaction rate. The separate effort by Yates et al. (1979) involved hydrogen thermal desorption from the Rh(111) surface to determine the laser radiation effect. Chuang and Winters (1979) used an Auger spectrometer to monitor the amount of hydrocarbon on a tungsten ribbon exposed to

CH_4 in the presence and absence of the 3.39-μm light (laser power about 15 mW). No discernible photoeffects were observed in either study. These experiments, of course, did not rule out the possibility that higher vibrational states, other than those explored by these studies, might be important in methane chemisorption on metal surfaces.

Little is known experimentally or theoretically about the direct effects of electronic excitation by laser radiation on gaseous adsorption behavior on metal surfaces. The relatively strong interaction between the adspecies and metal atoms can result in complex surface electronic structure different from the bulk states of the solid and the electronic structure of the isolated gaseous species. The interaction of a photon radiation field with such an adsorbate-substrate complex necessarily involves time-dependent wavefunctions of the system, making the detailed theoretical calculation on the chemisorption behavior more difficult to perform than the chemisorption in the absence of radiation. As qualitatively illustrated by Wautelet (1980), the optical effects on adatoms would not only modify the bonding between the adspecies and substrate atoms and affect their surface migration and desorption and therefore the surface sticking probability, but also would excite this bonding in many different ways. As an example, a situation where the adatom possessed two states, E_a (bonding) and E_b (antibonding), respectively, below and above the Fermi level, was considered with the photon energy less than the surface work function. The electron might come from a bulk state and be excited to E_b state, resulting in the formation of a negatively charged adatom. On the other hand, the adatom electron might be promoted from E_a to bulk states and a positively charged adatom could be produced. Another possibility was that an electron might be excited from the occupied E_a state into the unoccupied E_b state. Adatom migration, surface diffusion (Slutsky and George, 1979; George et al., 1980), and desorption might be enhanced by these processes. As further pointed out by Wautelet, gas-surface interactions could also be modified by the indirect radiation effects such as the generation of a static electric field in the surface region. As calculated by Gauthier and Guittard (1976), electric fields as high as 10^{-4} to 10^{-1} V/Å could be produced in the laser-irradiated area. The charge generation, the subsequent separation of electron and hole pairs due to the different carrier mobilities, and the formation of a surface dipole layer could lead to these high fields. Such electric fields could change surface electronic structure and shift chemisorption equilibrium, and even cause band bending on semiconductor surfaces (Tsong, 1979). These indirect field effects might be particularly important on metal surfaces with a chemisorbed layer and on oxidized metal and semiconductor surfaces. This mechanism may have some relevance to the nonthermal effects observed in the photon-enhanced Si-XeF_2 and Si-O_2 reactions mentioned earlier.

The effects of electromagnetic fields on rough surfaces have been considered by Nitzan and Brus (1981). The proposed classical model included a polarizable point dipole representing a protrusion on a rough surface. A similar approach was used earlier by Gersten and Nitzan (1980) in the attempt to explain the phenomenon of surface-enhanced Raman scattering. The calculation by these authors indicated that enhancement of resonant excitation was possible depending on the overlap between the molecular resonance and the excitable surface resonances, such as surface plasmons, which were determined by the dielectric function and the shape of the solid. The authors also numerically demonstrated that for very fast reactions such as I_2 photodissociation on a silver sphere (diameter ~1000 Å), the photon absorption cross section, and therefore the photodissociation yield, could be greatly enhanced by visible photons at $\lambda = 450$ nm incident on the metal surfaces. For processes that required finite time for energy accumulation to facilitate the chemical reaction, energy damping because of the rapid energy transfer to the substrate would become a severe competitive process. In this case, enhanced optical absorption would not necessarily imply enhanced photochemical yield. In order to determine whether a polyatomic molecule could absorb enough photon energy and store it for a sufficient period for a chemical process to take place, they calculated the internal energy of SF_6 molecules as a function of time when a CO_2 laser beam with the intensity of 10^8 W/cm^2 was incident on the SF_6-InSb interface. The n-type semiconductor was chosen because it had a free carrier plasmon resonance near the CO_2 laser line at $\nu = 967$ cm^{-1}, which was also close to the ν_3 vibrational mode of SF_6. Assuming an energy relaxation rate of 3×10^{12} sec (bandwidth 16 cm^{-1}), the calculated results showed that rapid excitation and efficient energy storage could be achieved if the molecule was a few angstroms (5–10 Å) away from the surface. If the separation between the molecule (considered as a point dipole) and the surface was 2 Å, i.e., SF_6 in direct contact with the solid, energy damping could be so fast that energy storage would become very difficult. The model calculation seems to indicate that the first molecular layer in direct contact with the solid surface could not store energy efficiently, but the second or third monolayer might be able to. The calculation is quite illuminating considering that we have observed the enhanced dissociative chemisorption of SF_6 on silicon both at 300°K and 90°K when the systems are excited by CO_2 laser pulses near resonance with the ν_3 vibration. How important this mechanism is for Si-SF_6, however, cannot be clearly determined. For a basic understanding of the photon interaction with a gas-metal system, it should be noted that the quantum mechanical approach used by Lundquist et al. (1979) and Norskov et al. (1979), by considering the electronic factors affecting chemisorption, sticking probability, and radiative surface process, appears to be very useful. The proposed electron-jump mechanism in the nonadiabatic chemisorptive step would seem to be of particular relevance, not only to photon emission (chemiluminescence), but also to photon-enhanced adsorption processes.

In contrast to the gas-metal system, photon-enhanced adsorption on insulator and semiconductor surfaces has been investigated quite extensively. This is particularly so for photon-enhanced adsorption and desorption of oxygen on oxide surfaces such as ZnO, TiO_2, etc. The subject has been thoroughly discussed by Volkenstein and Nagaev (1973, 1975), and Morrison (1977). The basic mechanism appears to involve the photoexcitation of electrons from a filled band to a conduction band of the solid and the electron transition to or from an adspecies. Following the photon absorption, two classes of processes can occur as the electrons and holes diffuse to the surface. The first is simple electron-hole recombination to generate heat, and the second is recombination with a chemical reaction by the electrons or holes being captured by surface groups. Depending on the detailed electronic characteristics of the system, it is suggested that an oxygen molecule can be photoadsorbed by capturing an electron *via*

$$e^- + O_2(g) \rightarrow O_2^-(ads) \tag{2.1}$$

or photodesorbed by capturing a hole *via*

$$h^+ + O_2^-(ads) \rightarrow O_2(g). \tag{2.2}$$

In early studies by Volkenstein and those cited by Morrison, experiments were not performed under very high vacuum conditions and the oxide surfaces were not characterized. While the proposed primary mechanisms involving photon-generated electrons and holes interacting with the adsorbate are physically reasonable, detailed interpretation of experimental results may not be very straightforward because of the experimental uncertainties such as the presence of surface impurities. This is particularly the case for photodesorption from semiconductor surfaces, which has been critically reviewed by Lichtman and Shapira (1978).

3.2. Photon-Induced Surface Reactions

In addition to promoting chemisorption, photon radiation may in a number of cases induce further surface reactions. By multiple CO_2-laser photon excitation, SF_6 molecules could be excited to high vibrational levels for dissociative chemisorption on both Si and Ta surfaces to form fluorinated surface layers. The photoactivated species could further react with the fluorinated surface layers to form silicon and tantalum fluoride products (Chuang, 1981a, 1982c). CO_2 lasers have also been used for catalytic decomposition reactions. For instance, Khemelev et al. (1977) observed that when NH_3 molecules were vibrationally excited with a pulsed CO_2 laser at 948 cm^{-1}, the molecules decomposed into N_2 and H_2 in the presence of the Pt catalyst at 25°C. When the experiments were repeated with copper or in the absence of the Pt film, no NH_3 decomposition was detected by a mass spectrometer. The authors suggested that the major effect of the vibrational excitation was to enhance initial molecular adsorption on Pt surfaces, which was followed by molecular decomposition and recombination reactions in the adsorbed state. No direct experimental confirmation of the surface reaction was available. Another example is the catalytic decomposition of HCOOH influenced by CO_2 laser radiation (Umstead and Lin, 1978) that has just been described above. Apart from the vibrational excitation of the gaseous species, the details of the surface chemical process were not determined.

An ultraviolet photon excitation effect on the gas-metal surface reaction was investigated by Baddour and Modell (1970). In the oxidation study, they reported that the rate of oxidation on polycrystalline Pd surfaces could be enhanced by a factor of 10 at 135°C and a factor of three at 165°C when the metal substrate was irradiated by a mercury arc lamp producing 300-nm photons. The photoreaction was also investigated by Chen et al. (1977), with particular emphasis on separating the photon-induced effects from the thermal effect. In this study, it was found that at low gas pressure near 10^{-5} torr, no photoenhancement was measurable for a wide range of experimental conditions. At a higher pressure near 20 torr (CO and O_2, respectively, at 5 and 15 torr), CO oxidation rate was greatly enhanced by UV photons at 254 nm. Furthermore, the temperature dependences of the thermal and photon-induced rates showed sharp changes between 413° and 443°K. Between 333° and 413°K, the photooxidation rate had an apparent activation energy of 45 kJ/mole, but above 413°K, the oxidation rate decreased. In contrast, the thermal oxidation rate increased drastically above this temperature. From the pressure dependence, the authors suggested that the photoeffect was related to the presence of weakly bound CO on Pd surfaces. Surface infrared studies showed that although CO interaction with the Pd surface was rather complex, there were two general types of CO adsorbed on Pd depending on the surface coverage. The more weakly bound species with the vibrational frequency (ν) near 2100 cm^{-1}, which generally appeared at a high surface coverage, could be removed at least partially by evacuation at 300°K or by heating at a high gas pressure. At a certain temperature between 413° and 493°K, this weakly bound CO could not remain on the metal surface. The tightly bound species at $\nu = 1900–1970\,\text{cm}^{-1}$ had a desorption temperature higher than 473°K. Thus the authors suggested that the decrease in the photooxidation rate for the Pd temperature above 413°K might be due to the disappearance of the weakly bound CO species. The question as to how the UV photons affect the CO adsorption and desorption, and the surface oxidation processes remains unanswered. Overall, it is fair to say that our understanding of chemical reactions on metal surfaces affected by photon radiation is primitive and incomplete.

On oxide and semiconductor surfaces, there have been a number of studies related to photocatalytic reactions (see, e.g., Volkenshtein and Nagaev, 1973; Morrison, 1977), again involving photogeneration of electrons and holes in the solids. For illustration, the work by Hemminger et al. (1980) on the photoeffects on the $CO_2 + H_2O$ reaction on Pt-$SrTiO_3$ surfaces is briefly described below. In order to investigate the mechanisms and the elementary steps involved in CH_4 production from the bimolecular surface reaction, the authors carried out a series of surface studies with a combination of analytical techniques including electron energy loss spectroscopy (ELS), ultraviolet photoelectron spectroscopy (UPS), and Auger and low-energy electron diffraction (LEED) in specially designed UHV chambers. The chemisorption behavior of H_2O, CO_2, CO, and O_2 on both Pt and $SrTiO_3$ single crystal faces, and the thermal and photo production yields of CH_4 were examined quite thoroughly. Photoexcitation across the band gap of the metal oxide (>3 eV) was accomplished with UV light from a mercury arc lamp. The important results included the observation that both $SrTiO_3$ oxide and the metal surfaces were needed for CH_4 production; H_2O dissociatively chemisorbed on the oxide surface while it remained molecular on the Pt surface; CO was oxidized to CO_2 on the oxide surface; O_2 chemisorbed on both the oxide and metal surfaces; part of the chemisorbed O_2 that was used for the oxidation of Ti^{+3} to Ti^{+4} was photodesorbed and the Ti^{+3} sites were regenerated; band-gap excitation was necessary to carry out the photochemical reactions. The proposed sequence of surface reactions was started by H_2O dissociative adsorption on $SrTiO_3$ surfaces *via*

$$Ti^{+3} + H_2O \rightleftharpoons Ti^{+4} + OH^- + H. \tag{2.3}$$

The photoelectrons and holes generated by UV photons would participate in the production of O_2 and regeneration of Ti^{+3} sites *via*

$$Ti^{+4} + 2OH^- + h^+ \rightleftharpoons Ti^{+3} + \tfrac{1}{2}O_2 + H_2O. \quad (2.4)$$

In other experiments, OH^- groups were similarly suggested as hole traps (Bickley and Jayanty, 1974). The reaction was followed by partial O_2 photodesorption, presumably *via* Eq. 2.2. H atoms produced on the oxide surface would migrate to the Pt surface where the hydrogenation of CO_2 could produce HCOOH and H_2CO types of reaction intermediates that were not experimentally identified. Stepwise reactions appeared to be necessary because the endothermic production of CH_4 from CO_2 and H_2O required an energy much greater than the band-gap energy of $SrTiO_3$. The authors also pointed out that in addition to CH_4 other molecules might be produced by using different oxide-metal contacts, photons of different wavelengths, and different gas mixtures. It is possible that photon-assisted reactions over oxide-metal contacts may provide a new route for the production of many different small molecules.

3.3. *Photon-Stimulated Desorption*

Like electron-stimulated desorption (ESD) (Madey, 1981), the subject of photon-simulated desorption (PSD) has been studied quite extensively (see, e.g., Menzel, 1982) for more than 15 years. This may be due to the fact that among the major surface processes, i.e., adsorption, surface reaction, and desorption, desorption seems to be a relatively simple process to investigate both theoretically and experimentally. Moreover, such studies have yielded important information for understanding not only the detailed interactions between the adsorbate and the adsorbent but also the surface excitation and energy transfer mechanisms. In PSD, most studies have been performed with high-energy photons (>10 eV) where both valence excitation and core ionization can participate in promoting desorption. In contrast, studies on photodesorption for photon energies less than 6 eV are still rather limited. This lower energy regime should also be of great interest because for many gas-surface systems, the adsorbate-surface bonds have bond energies in the 0.1–5 eV range. If the photon energy from UV-visible to infrared can be effectively utilized, breaking of the surface bond resulting in desorption should be an energetically feasible and perhaps even an efficient process. From mechanistic and fundamental points of view, it should be very interesting to find out how the photons would interact with an adsorbate-surface bond, which has a rather low vibrational frequency, e.g., about $1000\ cm^{-1}$ or less for adsorbed H atoms and $600\ cm^{-1}$ or less for other adspecies, and induce the bond to rupture. In the past decade, practically all the major studies on low-energy photodesorption were centered around the subject of separating quantum effects and the thermal effect. Basically, photon-induced desorption *via* a quantum process may be defined as the absorption of a photon by the adsorbate-substrate complex followed by the adsorbate desorption without equilibration of the absorbed energy in the surrounding solid lattice. In the UV-visible spectral region, the photon can be absorbed by the electron density of the gas-solid system. In the infrared region, the photon can be absorbed by the discrete intramolecular vibrational bands or the vibrational modes involving the adsorbate-surface bonds. Desorption may be able to take place before the energy is relaxed to lattice phonons. At the other extreme, the photon energy is simply absorbed by the solid atoms and heats the lattice locally and by conduction throughout. Thermal desorption may occur *via* a local heating effect. Conceptually, the quantum effects and the thermal effect may happen on different time scales. Since optical transitions occur in less than 10^{-14} sec, photodesorption due to the quantum effects could presumably take place in a very short time period, whereas the thermal process involving thermal conduction should occur at a substantially slower rate. For clarity, we will divide our discussion on photodesorption into two parts. In this section, the basic concepts and theoretical considerations will be outlined. Experimental studies on photon-induced thermal desorption and desorption due to UV-visible and infrared photon absorption resulting in electronic and vibrational excitations will be described in Section 4. Comparison of theories with experiments whenever available will be emphasized.

3.3.1 Electronically Excited Desorption

For UV-visible-photon-induced desorption studies up to 1979, the experiments and results have been critically reviewed by Lichtman and Shapira (1978), and Koel et al. (1980). The general conclusion was that for gas-metal systems such as CO on nickel and tungsten, the observed photodesorption was mainly thermal in nature. The quantum effect, if present at all, was very small, with the quantum efficiency less than 10^{-8} molecules desorbed per photon, corresponding to a cross section of less than $10^{-22}\,\mathrm{cm}^2$. A noted exception was that reported by Kronauer and Menzel (1972), who studied CO desorption from tungsten surfaces and observed relatively high photon desorption yield for the photon wavelength below 300 nm. At $\lambda = 250$ nm (4.96 eV), they reported a quantum yield of 4×10^{-7}, which was attributed to a quantum effect based on the careful compensation of photo-induced temperature increases. To account for this observed effect, the authors proposed two possible mechanisms. The first mechanism involved the photon absorption by the metal, resulting in the excitation of an electron from the conduction band to an unoccupied state above the Fermi energy. The excitation energy was then transferred to the adsorbate, possibly *via* elastic electron tunneling through the surface barrier into an empty state of the adsorbate above the Fermi level. The threshold for such a process would be the minimum energy required to lift an electron from the top of the conduction band to the empty adsorbate level. The second mechanism involved the electronic excitation of the adsorbate-metal complex, leading either directly to a repulsive or to a metastable state that could cross into a repulsive state. The state involved in the absorption of photons with 5 eV or less could not be derived from the slightly perturbed molecular CO upon chemisorption, because the lowest excited state of CO lies 6 eV above the ground state. To account for the low threshold energy, the authors suggested that the CO-W surface bond formation might lead to new, low-lying unoccupied molecular orbitals. So far for CO on tungsten, the proposed low-lying antibonding states from which desorption can take place have not been theoretically calculated, and the question as to why a photon with an energy of 5 eV or less can induce CO desorption remains unsolved.

The second possible mechanism mentioned above is very similar to a mechanism proposed by Menzel and Gomer (1964) and by Redhead (1964) for interpreting electron-stimulated desorption. This is an important concept, whether or not it is directly applicable to CO desorption from tungsten. More specifically, the MGR model suggested a two-step mechanism that was started with an optical or electron-induced excitation, causing a Frank-Condon transition from the ground state of the adsorbate-substrate complex to a neutral or ionic antibonding state of the system. As a consequence of this excitation, the neutral or ionic particle would begin to move away from the metal surface. If de-excitation did not occur, either ions or neutrals could be desorbed. If, on the other hand, the excited state was recaptured by an electron tunneling to fill the orbital emptied by the initial excitation, the system would relax and no desorption would take place. Even if not applicable for CO on tungsten, this model may still be relevant to photodesorption in other gas-surface systems that have low-lying antibonding states that can be reached by UV-visible photons. A different mechanism has been proposed by Knotek and Feibelman (1978) for ESD involving core-hole excitation. This mechanism is not expected to be applicable to photodesorption in the low photon energy region (<6 eV). Another desorption mechanism for PSD and ESD was suggested by Antoniewiez (1980). In the proposed model, a neutral particle (A) interacted with the substrate (M) *via* a

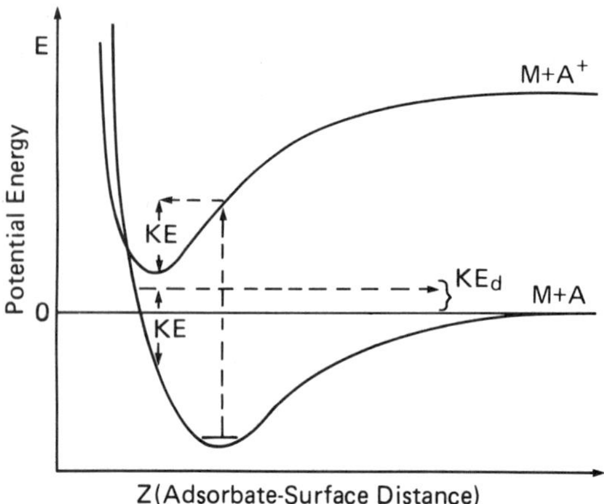

Fig. 2.6 A sequence leading to the desorption of a neutral adspecies according to Antoniewiez (1980). The potential energy curves represent the total energy of the system consisting of the substrate (M) and the adsorbate in either the neutral (A) or ionic state (A^+). After excitation from the ground M-A state into the upper M-A^+ state, the adsorbate is accelerated toward the substrate, gaining kinetic energy. The adsorbate is subsequently de-excited to the lower curve by interaction with the substrate. It still retains the kinetic energy gained in the above, and now its total energy is greater than the M-A binding energy. Consequently, the adspecies is desorbed with a kinetic energy KE_d.

potential energy curve that had an attractive part and a steeply repulsive part as the distance between the adspecies and the substrate surface became small. This energy diagram is shown schematically in Fig. 2.6. The position of the minimum energy in the M-A potential curve was approximately given by the sum of the atomic radii of the adspecies and substrate atoms. In the desorption sequence an adsorbed atom or molecule was ionized by an electron or photon. Since the ion was substantially smaller than the neutral particle, the ion-substrate (M-A^+) potential energy curve had its minimum nearer the substrate surface. Therefore, the ion would be forced to move toward the substrate and gain kinetic energy. As the ion came nearer the solid surface, the probability for an electron to tunnel from the substrate to neutralize the ion was increased and the ion could be neutralized. The neutralized particle had gained the kinetic energy and was nearer the surface than the initial equilibrium position, and therefore higher up on the repulsive part of the M-A curve. If the sum of the new potential energy and the acquired kinetic energy could be more than the desorption energy, the particle would desorb according to Fig. 2.6. In a similar manner, the author also proposed a scheme for ion desorption. In this case, three potential energy curves were needed: one for the neutral species, one for the singly-ionized ion, and one for the doubly-ionized ion. The essential feature of this proposal is the creation of an adsorbate ion in the initial electron or photon excitation process so that a neutral can be desorbed. The minimum energy for this step is the energy required to transfer an electron from the adsorbate to the Fermi level of the substrate. For some gas-surface system, this energy might be provided by low-energy photons. The applicability of the theory to photodesorption in the UV-visible region remains to be tested.

For photodesorption from oxide and sulfide semiconductor surfaces, the mechanism proposed by Baidyaroy et al. (1971) and Shapira et al. (1977) can account for most of the observed phenomena. In early experiments, although the solid surfaces were exposed to oxygen, there was confusion and uncertainty about the desorbed species when the systems were subjected to UV-visible photon radiation (Lichtman and Shapira, 1978). Later experiments showed that neutral CO_2 was the dominant or the only species photodesorbed from ZnO (Shapira et al., 1975, 1977), TiO_2 (Van Hieu and Lichtman, 1981a), V_2O_5 (Van Hieu and Lichtman, 1981b), and other semiconductor surfaces (Lichtman, 1979). The essential feature of this class of photodesorption is the requirement of a threshold photoenergy equal to the band gap of the solid substrate. For ZnO, TiO_2, V_2O_5, the threshold photon energies are 2.7 eV, 3.1 eV, and 2.35 eV, respectively. According to Shapira et al. (1977), using ZnO as an example, the sequence of desorption involved the oxygen adsorption on ZnO surfaces at or near the surface impurity carbon atoms. The carbon impurities were oxidized in a process including the capture of electrons from the ZnO conduction band to form CO_2^- species. The removal of electrons from the conduction band resulted in a decrease in surface conductivity. When the solid was excited by photon absorption, the holes generated in the surface region migrated to the surface, where they could recombine with and neutralize CO_2^- ions, very similar to Eq. 2.2. The neutralized species subsequently desorbed and the surface conductivity was increased. Photodesorption from such semiconductor surfaces was found to be a very efficient process, with a cross section of about $10^{-17}\,cm^2$. From the experimental observation that the photodesorption yield decreases as the surface carbon is depleted, it is apparent that in addition to band-gap excitation, the presence of carbon impurity on oxide surfaces is essential based on this photodesorption model.

Another photodesorption scheme for semiconductor surfaces was proposed by Gauthier and Guittard (1976), which was briefly mentioned earlier. The model was based on Volkenshtein's theory (Volkenshtein and

Nagaev, 1973, 1975) in which the adsorbed particle was considered as an impurity or a structural defect and would form a quantum system with the crystal lattice. The ability of the adspecies for transferring an electron or a hole to the surface would determine the existence of various adsorbate-surface bond strengths. An adspecies with an energy level near the conduction or valence bands of the semiconductor could either behave as an electron acceptor or donor, or remain neutral, depending on the chemical nature of the system. The neutral species was considered to be weakly bound and could be easily desorbed. The chemical adsorption could create charged superficial zones and promote dipolar layer formation and consequently could induce semiconductor band bending. According to this proposal, the effect of the laser radiation was to promote a surface imbalance between the irradiated and dark regions, which would cause local concentration variations of the electrons and holes. Because of the difference in electron and hole mobilities, a large electric field could be generated from the photon-generated dipolar layers. The authors demonstrated that with a ruby laser ($\lambda = 694$ nm) of 10^{-3} to 4-J pulse energy and 0.5-msec pulse duration irradiated on a Ge surface, an electric field as high as 10^4–10^7 V/cm in the surface layer could be produced. Such a high electric field could change the surface chemical equilibrium. Furthermore, the surface electrons under high local fields could acquire sufficient kinetic energy (1–10^4 eV) to break the adsorbate-surface bonds by collisional energy transfer or by electronic excitation and ionization of the adsorbed species. This last part of the mechanism would then be rather similar to ESD. According to this scheme, both neutral and ionic species could be desorbed. The authors further suggested that this model could also be applicable to metal and insulator materials. The ideas proposed by these authors appear to be quite interesting and may have some relevance for certain chemical systems.

Other factors that should be considered in the electronically excited desorption include the surface-enhanced electromagnetic field effects. As discussed by Gersten and others (Gersten and Nitzan, 1980; Gersten, 1980; McCall and Platzman, 1980), molecules adsorbed on rough surfaces could experience an electric field several orders of magnitude larger than the incident light field. The origins for such local field enhancement can be due to both surface electromagnetic resonances (plasmons) and shape effects. Surface plasmons tend to appear at visible frequencies with silver near the UV, and copper and gold in the red region. The shape-induced field enhancement, e.g., "lightening rod" effect (Gersten and Nitzan, 1980), is produced by large surface curvatures and can appear for all metals at all frequencies. As demonstrated theoretically by Aravind et al. (1982), large laser field enhancement could also be obtained in the medium or far infrared region (~200–1100 cm^{-1}) if proper shapes and materials (e.g., metal spheres on SiC or doped InSb) were chosen. Such surface-enhanced electric fields could influence not only the radiative (optical absorption and emission) and nonradiative properties, but also surface adsorption and desorption behaviors.

3.3.2 Vibrationally Actived Desorption

When molecules are adsorbed on solid surfaces, the system exhibits characteristic vibrational bands due to intramolecular vibrational motions modified by the surface adsorption, and the stretching and bending vibrations of the adsorbate-surface bonds. The vibrational motions (frequency, linewidth, and lineshape) are both molecular and mode specific, and can be studied by electron energy loss spectroscopy (EELS), surface infrared spectroscopy (IRS), and other surface vibrational techniques. Taking CO adsorption on a Cu(100) surface as an example, a rather well-studied system by EELS and IRS, the linear "end-on" adsorption can give rise to two EELS and IRS active vibrational modes, namely, the C-O stretching mode occurs at about 2090 cm^{-1} and the M-C (metal-carbon) stretching mode at about 340 cm^{-1} (Anderson, 1979; Sexton, 1979; Horn and Pritchard, 1976). There are four other modes involving hindered rotation and translation, all generating dipoles oscillating parallel to the metal surface and are EELS and IRS inactive. On this surface, even in a bridged site with C_{2v} bonding symmetry, when the hindered translation and rotation lose their degeneracy, only the same two vibrational modes are observable with EELS and IRS. For CO on Pt(111), most of the other modes become observable (Lehwald et al., 1982). The C-O stretching mode of the chemisorbed CO vibrates at a substantially lower frequency than the frequency (at 2155 cm^{-1}) in the gas phase, which has only one vibrational mode. The downshift in frequency has been attributed to the donation of charge from Cu metal into the CO-$2\pi^*$ antibonding orbitals and to the mechanical coupling with the solid surface. Conceptually, in a given condition, if one could directly excite the M-C stretching vibration with infrared photons in resonance with this mode, the vibrational motion might be activated to such a level that the adsorbate-metal bond could be ruptured and the molecule could be desorbed. Alternatively, one could excite the higher frequency C-O stretching vibration and indirectly excite the M-C bond to induce the breakage of the surface bond. Because of the distinctive vibrational frequencies, one can also imagine that for a mixed isotopic system such as $^{12}C^{16}O$ and $^{12}C^{18}O$ coadsorbed on Cu(100), one might be able to selectively excite and desorb one isotope species. The process that competes against vibrationally activated desorption and other chemical processes is clearly the vibrational energy damping that can occur very efficiently on solid surfaces. This is an interesting subject that has attracted considerable theoretical as well as experimental attention recently. Granted that vibrational energy relaxation on surfaces can be extremely fast, the question to be answered is whether a sufficiently high-energy pumping rate can be achieved with infrared lasers to excite either the adsorbate-surface (M-A) stretching vibration or the intramolecular vibrational modes without causing severe thermal effects. Before the discussion on existing theoretical studies, some general considerations are in order.

First, it is useful to consider the photon absorption steps in the vibrational ladder. Depending on the laser wavelength, energy, and power, the vibrational states can be excited in a number of ways, as shown schematically in Fig. 2.7. The most common excitation is the one-photon absorption for fundamental transitions ($\Delta v = 1$), i.e., process (a) in Fig. 2.7. This activation can provide at most 9–10 kcal/mole of energy for such high-frequency modes as N-H, C-H, and O-H stretching vibrations near $\lambda = 3\ \mu$m, and may be important for desorbing weakly bound species, provided that the adsorbed photon energy can be totally utilized for breaking surface bonds. For most if not all molecules, consecutive $\Delta v = 1$ ladder-climbing-like photoexcitation (process (b) in Fig. 2.7) is impractical, because once the first vibrational level is excited ($v = 0 \rightarrow 1$) by a resonant IR photon, the vibrational anharmonicity will necessitate a lower frequency for the absorption of a second photon for the $v = 1 \rightarrow 2$ transition, a still lower one for the third photon for $v = 2 \rightarrow 3$, and so on. To reach the nth level for desorption to occur, n different photon frequencies would be required. Therefore, for a single frequency excitation, process (b) alone can not be important. However, if the intermolecular energy transfer rate is much faster than vibrational energy relaxation rate, a situation like this might happen:

$$\mathrm{CO}(v=0) \underset{}{\overset{h\nu}{\rightleftharpoons}} \mathrm{CO}(v=1) \tag{2.5}$$

$$\mathrm{CO}(v=1) + \mathrm{CO}(v=1) \rightleftharpoons \mathrm{CO}(v=2) + \mathrm{CO}(v=0) \tag{2.6}$$

$$\mathrm{CO}(v=2) + \mathrm{CO}(v=1) \rightleftharpoons \mathrm{CO}(v=3) + \mathrm{CO}(v=0). \tag{2.7}$$

$$\vdots$$

Namely, a molecule such as CO can be optically pumped *via* the fundamental transition, a one-photon process. Rapid near-resonant

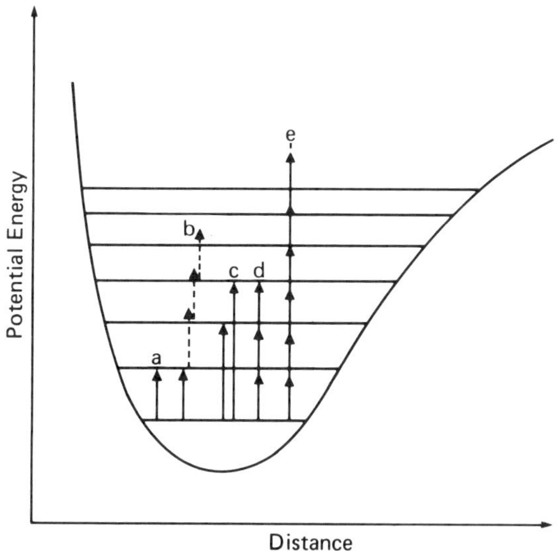

Fig. 2.7 Schemes for photoexcitation of vibrational levels in an anharmonic potential: process (a), $\Delta v = 1$, one-photon absorption; (b) $\Delta v = 1$, consecutive one-photon ladder-climbing-like excitation, an improbable process involving single-photon frequency for an anharmonic potential; (c) $\Delta v \geq 2$, one-photon absorption for overtone transitions; (d) coherent multiphoton absorption; and (e) multiple-photon excitation involves coherent multiphoton absorption in low-lying levels and subsequent incoherent single-photon absorption in upper states.

energy exchange *via* interaction with neighboring excited molecules can effectively promote the molecules into rather high vibrational states. For molecular CO, this phenomenon has been observed in both the gas phase (see, e.g., Weitz and Flynn, 1975) (energy transfer by bimolecular collisions) and the condensed phases (Legay, 1977; Legay-Sommaire and Legay, 1980). In both phases, CO vibrational relaxation rates are known to be slow. For instance, Legay-Sommaire and Legay (1980) were able to excite the molecules in solid CO near 25°K with a cw CO laser in the $\nu = 2070\text{–}2110\ \text{cm}^{-1}$ region and observed infrared emission from such vibrational levels as high as $v = 23$. Similar excitation *via* optical pumping and efficient energy exchange for vibrational ladder climbing might also be possible for molecules adsorbed on solid surfaces. For this reason, CO molecules might be a good candidate for observing single-photon excited desorption.

When high-level excitation is necessary for desorption or other surface chemical processes, as often the case, one-photon overtone absorption ($\Delta v \geq 2$) [process (c) in Fig. 2.7] and multiple-photon absorption [processes (d) and (e)] should also be considered. Direct one-photon high overtone ($\Delta v = 4$–7) excitation with visible and near infrared lasers has been observed for polyatomic molecules such as C_6H_6, and C_6D_6 (Bray and Berry, 1979; Reddy et al., 1982) in the gas phase. By this approach, a unimolecular isomerization reaction was demonstrated and the photochemical process appeared to be nonstatistical in nature (Reddy and Berry, 1979). These highly forbidden transitions have the absorption cross sections typically in the 10^{-24} to 10^{-27} cm^2 range. If $\Delta v = 2$ optical pumping is adequate for surface chemical processes, the excitation can have a much higher cross section in the 10^{-20}–10^{-21} cm^2 region. It should be noted that for molecules adsorbed on surfaces, the overtone absorption strength could sometimes be substantially enhanced because of the increased anharmonicity. However, except for $\Delta v = 2$, very high laser power is usually required for high-level overtone excitation. Under such excitation conditions, it is probably difficult to separate the quantum effects from the surface heating effect. Process (d) in Fig. 2.7 involves coherent *multi*photon absorption. This simultaneous absorption of two or three photons may be effective for excitation to $v = 2$–3 levels. For higher level excitation ($v > 3$), this coherent nonlinear absorption alone would not be efficient because of the rapid decreasing absorption cross sections. Process (e) involves the coherent *multi*photon absorption from the ground state to certain excited levels followed by incoherent stepwise one-photon absorption to even higher levels. This *multiple* photon absorption process has been known for many years (see the review: Ambartsumyan and Letokhov, 1977). Depending on the fluence and the power of the laser beam, it can even dissociate molecules with relatively high-absorption cross sections, i.e., about $\sim 10^{-19}$–10^{-20} cm^2, as compared with 10^{-18}–10^{-20} cm^2 for fundamental transitions. The mechanism of multiple-photon excitation has been elucidated by Yablonovitch (1980) and others (Black et al., 1977). As shown schematically in Fig. 2.8, the vibrational-rotational energy levels in the ground electronic state can be roughly divided into three regions, i.e., the low-lying discrete levels, the *quasi*continuum, and the true continuum regions. The most important feature of this energy diagram is the existence of several vibrational degrees of freedom that are available in polyatomic molecules. Unlike an isolated diatomic molecule, a polyatomic molecule has the important property that the density of states grows rapidly with energy, due to the rapid expansion in the volume of accessible phase space. The number of possible permutations and combinations of vibrational modes increases rapidly with available energy. Using SF_6 as an example, it was shown that the v_3 mode could be initially excited to the $v = 3$ (or higher) level by simultaneous absorption of three (or more) CO_2 laser photons at 10.6 μm

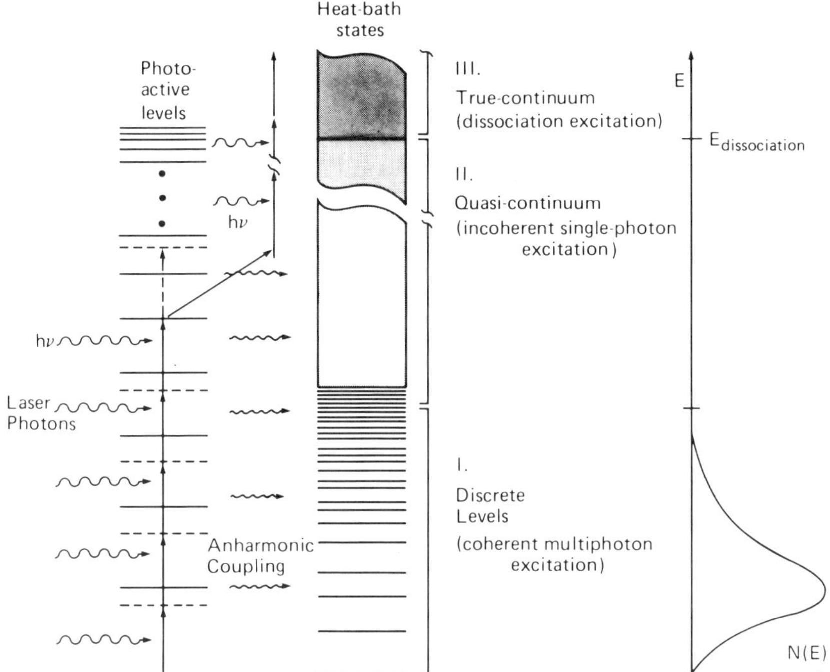

Fig. 2.8 Schematic representation of the discrete levels, the quasicontinuum, and the true-continuum regions of the vibrational energy diagrams of a gaseous polyatomic molcule or a simple molecule adsorbed on a solid surface. The initial population distribution function $N(E)$ is the product of the vibrational density of states and the room-temperature Boltzmann factor. In this intramolecular-heat-bath picture, the energy levels of an infrared-active mode are shown separately from the energy levels of the remaining modes, which form a quasicontinuous heat bath. Multiple-photon excitation is initiated by coherent multiphoton pumping of the photoactive mode and followed by incoherent single-photon interaction in the quasicontinuum states. Anharmonic coupling can cause phase-fluctuation damping (τ_2) and lifetime broadening (τ_1) of the photoactive mode.

and excited further by an incoherent single-photon interaction. Such a multiple-photon excitation process can also be important for polyatomic molecules adsorbed on surfaces. For diatomic molecules, the multiple-photon process is not important in the gas phase, but in the adsorbed state the process may not be neglected. This is because as an isolated molecule it has only one vibrational mode, and in the adsorbed state it can have six vibrational modes, including hindered translational and rotational motions. Taking CO on Pt(111) as an example, tentative assignments gave vibrations ν_1 (Pt-C) at $470\,cm^{-1}$, ν_2 at $1850\,cm^{-1}$, ν_3

(C-O) at 2100 cm^{-1}, ν_{rot} (hindered rotation) at 360 cm^{-1} and ν_{trans} (hindered translation) at 720 cm^{-1} (Lehwald et al., 1982). From the density of state point of view, the *quasi*continuum states of the CO-metal system should also be accessible by multiple photon absorption process. Thus for adsorbed diatomic and polyatomic molecules, multiple-photon excitation should be included in the theoretical consideration of photon-induced desorption and other surface chemical processes, particularly if the activation barrier for desorption or reaction is less than four or five infrared photons. The relative importance of processes (c), (d), and (e) depends on the vibrational level of excitation that is required for promoting a specific surface chemical process. For instance, if $v = 2$ excitation is adequate, then both processes (c) and (d) can contribute to the excitation. If $v = 3$–5 activation is needed, then process (e) is likely to be the most important one to consider. It should be further recognized that for pumping to $v = n (n \geq 2)$ level, process (c) requires the laser frequency precisely at the $v = 0 \rightarrow n$ transition frequency. Yet for processes (d) and (e), the laser frequency has to be near $v = 0 \rightarrow 1$ transition. Thus, depending on the desired excitation process to be investigated, a suitable laser wavelength should be chosen accordingly.

Second, it is usually considered that the most effective means to promote vibrationally activated desorption is to excite the M-A stretching mode directly. Such considerations are derived from the modeling of thermal desorption in which it has been suggested that phonon energy from the thermally heated lattice can excite the vibration of the adsorbate-surface bond and such excitation can induce desorption (De and Landman, 1980; Kreuzer, 1980). Photoexcitation of the M-A vibrational mode is just the reverse of the thermal process, except that the excitation energy would be primarily utilized for breaking the surface bond rather than decaying into the lattice phonons. These surface modes, however, vibrate at rather low frequencies, e.g., H-W at about 1000 cm^{-1} and O-Cu at 250 cm^{-1}. Clearly, excitation of the fundamental band ($\Delta v = 1$) by single-photon absorption would not be adequate to overcome an activation barrier for desorption with a barrier height of 3 kcal/mole or higher. For an activation energy of 9 kcal/mole, a relatively weak surface bond, it would require the absorption of more than 10 photons at 300 cm^{-1} to acquire enough energy to induce bond breakage. Such a high level of excitation, even involving multiple-photon absorption, appears fundamentally difficult to accomplish without causing a severe surface heating effect. To observe the quantum effects, with the exception of a very weak adsorbate-substrate physisorption bond, direct adsorbate-surface vibrational excitation by photon absorption (whether *via* the single-photon or multiple-photon adsorption process) is not likely to be effective for inducing desorption.

In contrast to direct photoexcitation of M-A vibration, excitation of intramolecular modes can provide much higher energy for promoting desorption or other chemical processes. Absorption of a single photon at 3.2 μm can furnish about 9 kcal/mole of energy, which if efficiently utilized may be sufficient for breaking a weak surface bond. For CO on Cu(100), fundamental excitation ($\Delta v = 1$) of the C-O stretching mode at 2090 cm^{-1} alone would not be able to provide enough energy for breaking the CO-Cu bond. However, excitation to the $v = 2$ or 3 level might just be adequate to "kick" the molecule loose from the metal surface. Pumping to the CO $v = 2$ level at about 4200 cm^{-1} (Lehwald et al., 1982) can be accomplished with a one-photon overtone ($\Delta v = 2$) excitation at $\lambda = 2.38$ μm or possibly by multiple-photon excitation at λ near 4.76 μm. As discussed above, multiple-photon absorption may be possible even for diatomic molecules such as CO adsorbed on solid surfaces. As illustrated in Eqs. 2.5 through 2.7, another possibility of exciting CO would be *via* single-photon absorption at $v = 2090$ cm^{-1}, coupled with efficient intermolecular energy exchange to pump the molecule to higher vibrational levels. For CO on Cu(100), intermolecular energy coupling through a dipole-dipole interaction has been shown to be very effective (Andersson and Persson, 1980; Persson and Ryberg, 1980). Persson and Ryberg (1981) showed that the strong surface dipole-dipole interaction could result in vibrational frequency shift, phase shift, amplitude modification, and rapid energy transfer and sharing between adsorbed molecules.

If an excitation is initiated in an intramolecular mode not directly connected to the M-A stretching mode, it might take a finite period of time for that excitation energy to be transferred to excite the M-A mode before A could depart from M. This intramolecular energy transfer usually occurs very rapidly (see Fig. 2.9), but the rate may or may not be faster than the vibrational energy relaxation rate. For diatomic adspecies, M-(A-B), where A-B represents the diatomic molecule, excitation of A-B stretching is very likely to result in instantaneous excitation of M-A stretching because the two modes appear so closely connected. Obviously, many time scales involving excitation and relaxation are crucial for the surface chemical process. Direct one-photon absorption (whether $\Delta v = 1$ or $\Delta v = n$) and coherent *multi*photon absorption can take place instantaneously in less than 10^{-14} sec. The incoherent stepwise excitation in the part of the *multiple*-photon absorption process may not be so fast, and the rate of pumping depends on the fluence and power of the laser

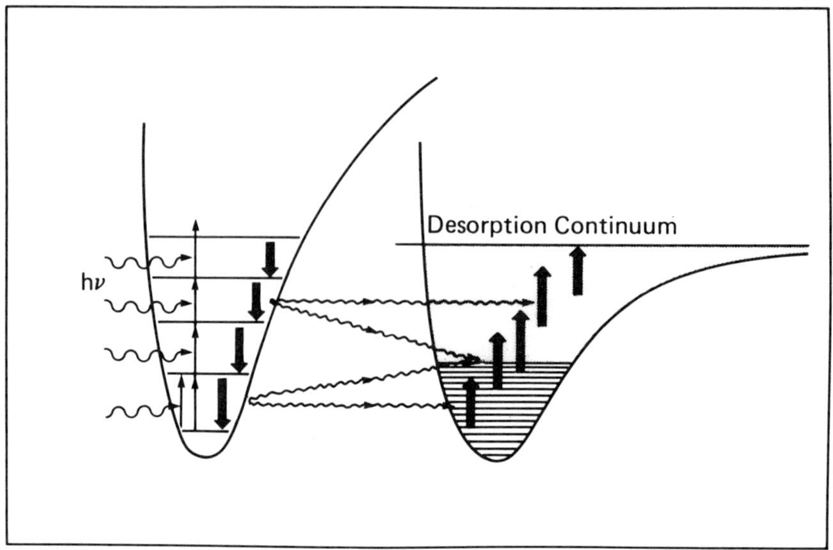

Fig. 2.9 Intramolecular energy transfer from the excited high-frequency photoactive mode to the low-frequency adsorbate-surface stretching mode may cause desorption of the adsorbate species.

beam. The time dependence of the desorption step may also be complex. If the desorption may be considered to be similar to the direct photodissociation of a diatomic molecule, the desorption rate may be as fast as the optical excitation rate. On the other hand, if the adsorbate does not acquire enough kinetic energy to move away from the surface at the moment it is excited, then there is a good chance that the particle may be recaptured into the attractive potential, and the particle may not desorb. These various possibilities have to be properly taken into account.

Third, both the *direct* substrate heating due to direct photon absorption by substrate atoms and the *indirect* heating due to energy relaxation of excited adsorbates to lattice phonons should be considered for photodesorption. This aspect tends to be neglected in existing theoretical treatments. The extent of direct substrate heating can be estimated according to the formula developed by Ready (1971). For experiments with a pulsed laser beam of 10^{-8} sec pulse duration or longer, the ordinary bulk thermal diffusion laws are adequate. The light absorption does not take place just at the surface of the solid, but extends into the bulk, with an optical absorption depth determined by the extinction coefficient of the material at a given photon wavelength. For many metals in the visible and near-IR region, this optical absorption depth is a few hundred angstroms, and about 4000 Å for silicon at $\lambda = 0.5\ \mu m$ (see Fig. 2.4). The optical absorption depth is usually much shorter than the thermal diffusion length ($\Delta\chi$) determined by thermal conductivity (κ) and heat capacity (c) *via*

$$\Delta\chi = \left(\frac{\kappa\,\Delta t}{c}\right)^{1/2} \tag{2.8}$$

where Δt is the laser pulse duration. The temperature increase on the solid surface due to the optical absorption of a light pulse can be readily shown to be (Ready, 1971)

$$\Delta T(t) = \frac{\varepsilon}{(kc\pi)^{1/2}} \int_0^t \frac{p(t-t')}{t'^{1/2}}\, dt', \tag{2.9}$$

where ε is the optical absorptivity and $p(t)$ is the laser pulse shape. For a laser beam with a triangular or a Gaussian pulse shaped irradiated on Si at $\lambda = 10\ \mu m$, $I = 5 \times 10^6\ W/cm^2$ and $\Delta = 5 \times 10^{-8}$ sec, the peak temperature rise (ΔT) is calculated to be about 100°K, using $\varepsilon = 0.15$, $\kappa = 14\ W/cm\ K$ and $c = 1.66\ J/cm^3\ K$. Under the same excitation condition on a silver or copper surface, ΔT is about 20°K with $\varepsilon = 0.015$ and the known values of κ and c. For a cw laser beam, ΔT can be calculated by

$$\Delta T = \frac{\varepsilon I d\sqrt{\pi}}{2\kappa} \tag{2.10}$$

where d is the radius of the laser beam. The indirect substrate heating from energy relaxation of the excited adsorbate could be quite localized on the surface on the time scales involved in desorption. The combination of these direct and indirect laser surface heating can affect the activation barrier for desorption and may not be negligible in many cases.

In short, in trying to formulate and solve the rate equations for vibrationally excited desorption, all these complex excitation mechanisms including the direct and indirect surface heating effects, the energy transfer, and the damping processes have to be considered. In the following, we will discuss a few existing theoretical studies on this subject and outline their primary approaches, major features, and general conclusions, if available.

Lin and George (1980) were among the first to theoretically study light-enhanced desorption involving vibrational excitation. Their earlier efforts were concentrated on studying models, with the purpose of mapping out possible desorption behavior. Attempts have also been made to generate qualitative predictions concerning the efficiency of the photon-stimulated desorption. Both classical and quantum mechanical approaches were adopted with various assumptions and simplifications. In their series of studies, photon excitation of both adsorbate-surface bonds and intramolecular vibrations were considered. The most prominent features of these studies appeared to be the inclusion of multiphoton excitation, multiphonon coupling, and intramolecular vibrational relaxation processes. Specifically, they divided a multilevel adsorbate-surface system into the optical active mode (A) with a high vibrational frequency (1000–3000 cm^{-1}), the intermediate frequency modes (B) (200–400 cm^{-1}), and all the other low frequency modes (C) ($\leq$200 cm^{-1}), including the lattice phonons, as shown schematically in Fig. 2.10. The

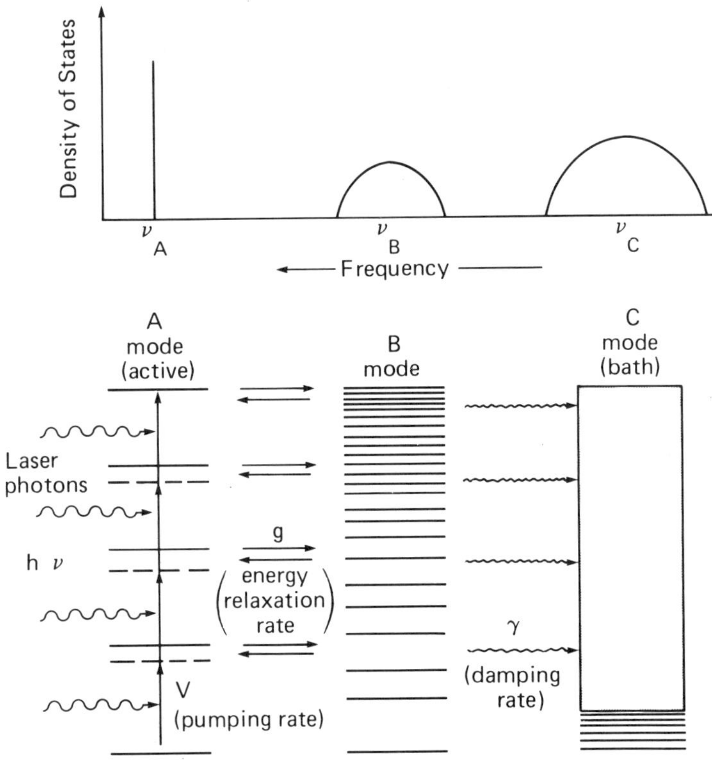

Fig. 2.10 Schematic representation of vibrational excitation and energy transfer processes of a model system considered by Lin and George (1980). The adsorbate-substrate system consists of the molecular high-frequency photoactive mode (A), the intermediate frequency modes (B), and the low-frequency bath modes (C). The optical pumping rate of the A mode, the intramolecular energy transfer rate between A and B modes, and the energy damping rate from the B to C modes are represented by v, g, and γ, respectively.

active mode was an internal molecular vibrational mode and was pumped by the laser field. Optical excitation was achieved by single-photon and multiphoton absorption. The excited A mode would relax its energy *via* intramolecular energy transfer to B modes, which usually were the optically inactive stretching and bending modes of the system. The transfer of energy was characterized by a coupling factor. Finally, the photon energy stored in the B modes would leak to the C modes with a damping rate. For a diatomic adsorbate such as CO on copper, the C-O stretching mode at 2090 cm^{-1} might be considered as the A mode in the proposed scheme. The B modes would consist of the other five vibrational modes, including the C-Cu stretching motion, and the lattice phonons would be the C mode. The level populations of A, B, and C modes were then calculated for various sets of the optical pumping rate (v), the coupling factor (g), and the damping rate (γ). The results showed that for a high pumping rate ($v > g \gg \gamma$), the A mode might be selectively excited with the lifetime of the vibrational excited level in the 10^{-6} sec range if $v \simeq 10^8$/sec. They estimated that such a rate of excitation could be achieved with a cw laser of 10–30 W/cm^2 intensity. (This estimated laser intensity seems to be unusually low. Perhaps, the authors had used an unrealistically high photon absorption cross section.) Also, not surprisingly, for large coupling and damping factors ($g \simeq \gamma \simeq v$), the steady-state populations of A and B modes were very low, and most of the photon energy would relax rapidly into the heat bath (C modes). If $g \simeq v > \gamma$, selective excitation of both A and B modes could be obtained with optimum efficiency, and the excitation could be utilized for nonthermal surface chemical processes such as desorption. While the theory remains so general that direct applications appear rather cumbersome, the proposed approaches have brought forth some very useful concepts such as the excitation of molecular high-frequency modes, multiphoton absorption, and intramolecular vibrational relaxation in the context of laser-stimulated surface processes. In this proposed model, the thermal contribution to the surface process and substrate heating effects were completely ignored.

A rather different theoretical approach has been advanced by Jedrzejek, Korzeniewski, and their coworkers (Jedrzejek et al., 1981; Korzeniewski et al., 1982a). They constructed a one-dimensional model for describing the states of the adspecies in the presence of the average potential of the lattice. An anharmonic oscillator (i.e., Morse potential) was used to describe the chemisorption bond, and only one vibrational mode, namely, the adsorbate-surface vibration was considered. The particle was adsorbed if it occupied a bound state and was free to desorb if its energy was located in the continuum of the potential. The rate of desorption was determined by the rate with which the particle was promoted into the continuum states. Both the phonon (thermal) and laser contribution to the transition rate were included in the evaluation of the desorption rate. To compute the photon-induced transition rate, the authors used the golden rule formula, including the consideration of local electric field, its interaction with the oscillating dipole, and absorption linewidths. The results were numerically illustrated with CO on Cu(100) as an example. The gas-metal system had the vibrational mode along C-Cu axis at $340\,cm^{-1}$, a heat of desorption of about 16 kcal/mole. The total desorption rate, caused by the concerted action of laser and lattice phonons, was then calculated as a function of laser power. It was found that the laser power required to enhance photodesorption was very large, i.e., much greater than $5 \times 10^{10}\,W/cm^2$. One should realize that at this laser intensity, physical ablation and other effects would occur on copper surfaces. Thus according to this model, the quantum effect in photodesorption could not be observed. The authors further pointed out the limitations of their model and other important factors that were not included in their model calculation such as the substrate heating effects by the laser radiation and the possibility of electromagnetic field enhancement on rough surfaces. However, above all, the most important reason for such low probability of photon-enhanced desorption, as pointed out in our general consideration given above, is the fundamental difficulty involved in the excitation of such a low-frequency mode as the adsorbate-metal stretching vibration. For the CO-Cu system, it would require the accumulation of more than 16 photons of energy at $340\,cm^{-1}$ (0.97 kcal/mole) in order to have enough energy for breaking a bond with 16 kcal/mole of bond strength. Since only single-photon transition was considered by these authors, high-level excitation could not be achieved simply because the anharmonicity would prevent the absorption of photons between any two adjacent levels except the two specific levels in resonance with the photon frequency, i.e., $v = 0 \rightarrow 1$ transition at $340\,cm^{-1}$ in this case. For this and other reasons discussed before, direct photoexcitation of the adsorbate-surface modes, even including multiple-photon absorption, is not likely to be an efficient means for observing photodesorption, except perhaps for very weakly physisorbed species. From this point of view, the conclusion obtained by Korzeniewski et al. (1982a) is not surprising.

In studies by Kreuzer, Gortel, and their coworkers (Kreuzer and Lowy, 1981; Gortel et al., 1983), it was proposed that photodesorption of a diatomic adsorbate could be induced by exciting the internal vibrations of the molecule. Unlike Lin and George (1980), who adopted multiphoton absorption, Kreuzer et al. considered the molecular vibration as a harmonic oscillator in which only one-photon absorption was involved. According to these authors, desorption was initiated by the photon-induced rapid excitation of many low-lying molecular vibrational states. Such excitation was in turn coupled *via* the surface potential to the thermal motion of the surface. If the released vibrational energy was sufficiently large, the molecule might in the process be excited up into the continuum of the surface potential and desorb from the surface. The transition from the vibrational excitation in the molecular potential (harmonic) to the surface potential (anharmonic) could take place *via* a phonon-mediated bound state-to-bound state transition, a phonon-mediated bound state-to-continuum transition, and elastic tunneling into the continuum. The interaction of the two potentials is similar to that shown in Fig. 2.9. The harmonic assumption for the molecular potential facilitated the photoexcitation to upper vibrational levels by consecutive one-photon absorption ($\Delta v = 1$), very similar but not identical to process (b) in Fig. 2.7. High-level excitation was required because the $v = 0 \rightarrow 1$ transition alone could not provide sufficient energy for desorption. Whereas the harmonic assumption has drastically simplified the theoretical treatment, it is obviously unrealistic. When the model calculation on CH_3F desorption from NaCl surfaces was compared with the experimental results obtained by Heidberg et al. (1980, 1982), it was found that the computed photodesorption rate was several orders of magnitude higher than the experimentally observed values. The authors pointed out several possible reasons for such a large discrepancy, namely, the dipole-dipole coupling for energy relaxation was not included in the model calculation, the assumed transition dipole moment might be too large, or the dipole might not lie perpendicular to the surface. It appears, however, that the major problem is likely to be due to the harmonic assumption, which overestimates the excitation of high vibrational levels. More practically, an anharmonic oscillator, as treated recently by Jedrzejek (1985), and multiple-photon excitation should be considered.

The elastic tunneling process considered by Gortel et al. (1983) could take place if a molecule was excited into a high vibrational level so that its total energy would be degenerate with some momentum continuum state. Under such a condition, elastic tunneling into the continuum momentum states would desorb the molecule. This process was also proposed by Lucas and Ewing (1981) in their model of vibrational predissociation of an adsorbed molecule. In the theoretical approach, they perceived the desorption problem as one conceptually similar to the case of vibrational predissociation of a van der Waals molecule, with one partner of the van der Waals complex being replaced by a surface. As numerically demonstrated by the authors, such treatment produced rather long vibrational lifetimes for excited surface species, e.g., 6×10^{-8} sec for H_2^* on Al_2O_3 and 1×10^{-6} sec for D_2^* on the same substrate. For the adsorbed CH_4^*, the vibrational lifetime could even approach 0.1 sec. The reason for such a long lifetime and thus a high vibrationally excited desorption probability was due to the assumption that ignored energy relaxation into the solid substrate. In a similar approach, by considering vibrational predissociation Casassa and Celii et al. (1982, 1984) had taken the rapid energy decay channels into account. They, nevertheless, concluded that even if the energy damping rates were faster than 10^{13} sec, vibrationally excited desorption could still be observed by using an infrared laser with the laser power as low as 10 mW, provided that the desorption rates were in the 10^{12}/sec range. This is, of course, an unknown factor that urgently requires clear-cut experimental determinations.

Insofar as relating the theoretical considerations and calculations to the experimentally observed quantities, the work by Wu and Fain et al. (1984, 1985) also appears to be quite interesting. They used a macroscopic master equation approach to treat the photodesorption without having to follow the exact paths of the energy flow. They were able to show that for a system with multiple vibrational levels successively absorbing multiple (n) IR photons, the desorption yield would be proportional to the nth power of the laser intensity. Furthermore, under the multiple-photon excitation conditions, the bandwidth of the desorption spectrum would be substantially narrower than the linear IR absorption spectrum. Using this approach, they were able to obtain the desorption spectrum and the desorption yield dependence on the laser intensity for the C_5H_5N/Ag system, in good agreement with those observed experimentally by Chuang and Seki (1982d, 1983a). With a similar treatment, Fain and Lin (1985) showed that the desorption rate could depend on the number of monolayers of the adsorbate if intermolecular vibrational energy exchange resulting in the excitation of higher vibrational levels was taken into account. Based on this mechanism, as discussed in the previous section, a $v = 2$ or higher states could be produced due to rapid energy transfer between the neighboring $v = 1$ molecules *via* the dipole-dipole coupling. In other words, the vibrational coupling could be very strong between molecules in the adlayers. It is indeed interesting that the results of such calculation agrees rather well with the experimental data for the C_5H_5N/KCl system (Chuang, 1983a). Both the experimental and the theoretical results demonstrate that intermolecular vibrational energy transfer between the same or different isotopic species can play an important role in infrared-induced photodesorption, particularly in multilayer systems.

3.3.3 Isotope Effects

Isotope effects in gas-phase electronic and vibrational photochemistry have been known for a long time (see the review by Letokhov and Moore, 1977). The effects have been very useful for mechanistic understanding of gaseous photochemical processes. Isotope separation by lasers has also raised firm expectation of new and less costly methods for chemical processing of various isotopes. Similar isotope effects are expected to take place for chemical species adsorbed on solid surfaces. Indeed, the effect has been clearly observed in electron-stimulated desorption (ESD) (Madey et al., 1970; Jelend and Menzel, 1973; Leung et al., 1977). Madey et al. (1970) studied oxygen desorption induced by surface bombardment with 100-eV electrons. They found that the probability of $^{16}O^+$ ion desorption was about a factor of 1.5 greater than that of $^{18}O^+$ ion from tungsten surfaces. The observation was shown to be consistent with the Menzel–Gomer–Redhead (MGR) model (Menzel and Gomer, 1964; Redhead, 1964) and inconsistent with the mechanism proposed by Zingerman and Ishchuk (1968), who suggested that ESD was due to the interaction of primary electrons with surface atoms of tungsten and not with the adsorbed oxygen. According to the MGR model, as described briefly earlier, ESD consists of a two-step process involving the electron impact excitation of the adsorbate-surface complex into a nonbonding state followed by the reformation of the adsorbate bond for a substantial fraction of the excited species during their course of departure from the surface. The total recapture probability in the second step for a particle desorbing from the surface would depend on its velocity. Two isotope species with the same potential curves and excitation cross sections, but different mass and therefore different velocity, should thus have different recapture probabilities and desorption cross sections. The MGR theory predicted an isotope effect for both ionic and neutral desorption. In the oxygen desorption experiments performed by Madey et al. (1970), it was found that the ionic desorption behavior followed the theory, but the isotope effects were too small for neutral desorption. Jeland and Menzel (1973) also observed large isotope effects in ESD of hydrogen and deuterium ions from tungsten but were unable to detect the effect for neutral desorption. This apparent discrepancy between the MGR theory and experimental observations was elucidated by Leung et al. (1977) in their ESD study of CO and O_2 from tungsten surfaces. They found that in all cases examined, the isotope effects for neutral desorption were much smaller than those for ionic production. For instance, the desorption cross section for neutral $^{12}C^{16}O$ was only about 10% greater than $^{12}C^{18}O$, whereas the difference was greater than 40% for ionic desorption. They attributed this difference to the fact that the escape probability for a neutral species was much larger than for an ionic species and that the lifetime for ion neutralization and recapture was shorter than the lifetime for transition of an excited neutral antibonding state to the ground state. By taking the lifetimes into consideration, they concluded that the MGR model was basically correct. The foregoing argument should be quite relevant to photodesorption by UV light. If the photodesorption mechanisms follow the MGR model, then there should be isotope effects in photon-induced neutral desorption, i.e., the desorption probability (D_1) should follow

$$D_1 = C_1 \exp(-C_2 M^{1/2}), \tag{2.11}$$

where C_1 and C_2 are constants and M is the mass of the desorbing particle. In Antoniewiez's model of desorption (Antoniewiez, 1980), the probability for desorption would also depend on the mass of the desorbed ion or neutral. For neutral desorption, the desorption rate dependence on the mass was the same as the MGR model, i.e., following Eq. 2.11. However, for ionic desorption, the desorption probability (D_2) could be

$$D_2 = C_1 M \exp(-C_2 M^{1/2}). \tag{2.12}$$

This different mass dependence was due to the involvement of the two-electron tunneling process proposed by the author. As pointed out by Antoniewiez, this model would predict larger isotope dependences for ions than neutrals in both ESD and photon-stimulated desorption, as experimentally observed in ESD.

In electronically excited desorption, according to the MGR or Antoniewiez model, the isotope effect comes into play not in the initial excitation step, but in the de-excitation and desorption step. As shown in ESD experiments, the isotope effect for neutral desorption is small. In vibrationally excited desorption, only a specific molecular vibrational mode is initially excited by the laser beam. Conceptually, the isotope effects could be very large if only or mainly the photon-excited isotope species could desorb. This possibility has inspired some researchers to speculate that perhaps infrared photon-stimulated desorption could be utilized for efficient separation of isotopes (Gangwer and Goldstein, 1976). Thus far, to our knowledge, isotope effects have not been clearly demonstrated in desorption experiments with infrared radiation (further discussed in Section 4). In methane dissociative chemisorption on rhodium and tungsten surfaces studied by Steward and Ehrlich (1975), and Winters (1975, 1976), respectively, isotope effects were clearly observed. As mentioned earlier, both studies suggested that vibrational excitation was involved in CH_4 chemisorption processes on these metal surfaces. There are certain factors that can make isotope effects difficult to observe in infrared photon-stimulated desorption. The effect of substrate heating by both direct photon absorption by substrate atoms and indirect energy relaxation from excited adsorbate to lattice phonons is not easy to avoid, even for relatively optically transparent substrates and highly IR-reflective metals. Such a substrate heating effect, which would enhance the total desorption yield, would reduce the isotope effects. Perhaps the most important factor that is difficult to overcome is the effective intermolecular interactions that can occur on solid surfaces. As elucidated by Andersson, Persson, and Ryberg (Andersson and Persson, 1980; Persson and Ryberg, 1980, 1981) for CO adsorbed on Cu(100), intermolecular dipole-dipole interactions could cause collective excitation not only among identical adsorbate molecules, but also among different isotopes. Therefore, very rapid energy transfer could take place from one molecule to another. In high-resolution electron energy loss (EELS) experiments, Andersson and Persson (1980) showed that the inelastic electron scattering was predominantly determined by CO dipole-dipole coupling. In the infrared spectroscopy study of mixed $^{12}C^{16}O$ and $^{12}C^{18}O$ adsorption on copper, Persson and Ryberg (1980, 1981) showed that the two isotope species were also strongly coupled in their vibrational motions with the high-frequency mode from $^{12}C^{16}O$ dominant for all surface compositions. For a one-component system, the authors suggested that the vibrational eigenmodes would not be localized vibrations but rather like phonon modes characterized by a two-dimensional wave vector. The stationary vibrational states of a completely uniform monolayer of identical molecules would behave like plane waves. In the presence of isotope species, this behavior would be altered. For adsorbed $^{12}C^{16}O$ and $^{12}C^{18}O$, the difference in the stretching vibration is about $47\,cm^{-1}$, and their dynamic dipoles are rather large (Persson and Ryberg, 1980, 1981). The strong dipole interactions and the relatively small difference in vibrational frequency would make selective vibrational excitation difficult to accomplish for molecules such as CO. For molecules with weaker intermolecular interactions and a larger difference in vibrational frequency, it might be possible to observe isotope effects in vibrationally activated desorption. Following this reasoning, H_2 and D_2 appear to be a good system to investigate the isotope effects.

3.4. *Energy Transfer and Relaxation*

If photon energy is to be effectively utilized for promoting surface chemistry, the rates of energy transfer and relaxations should be slow in comparison with chemical reaction rates. Therefore, it is important to understand the mechanisms and rates of energy transfer, storage, and damping at gas-solid interfaces. Such information, although incomplete, is much easier to obtain for gas-phase photochemistry because of extensive gas-phase studies in the past two decades. Comparable information for gas-surface systems is rather limited. Fortunately studies on these dynamic surface processes have begun to emerge in the last few years. Experimental methods for investigating energy-relaxation processes include measurements of absorption and scattering lineshapes, luminescence quantum yields, and direct determinations of excited-state decay rates by luminescence, absorption, or scattering techniques. Lineshape analyses can give interesting structure information, yet extraction of the population lifetime (τ_1) is complicated by a variety of effects such as site inhomogeneity, coupling with electron-hole pairs and phonons, dipole-dipole interactions, and, importantly, the dephasing time (τ_2) of this system. The dephasing effects can be caused by the dephasing of the adspecies-radiation field interaction because of the conformational fluctuation of the excited species and time-dependent changes in effective dipole moment. It can also be induced by excitation-induced surface migration, which leads to elastic collisions with other adspecies, and other broadening mechanisms involving phonon bandwidth and the anharmonic coupling of the surface potential (George et al., 1980). τ_2 is generally very short in condensed phases, typically in the 10^{-12} sec region, and can be dominant in the overall broadening of linewidths for many gas-solid systems. The lineshape technique works well if $\tau_1 \gg \tau_2$. Luminescence quantum yields can be converted to lifetime provided that the quantum yields are determined by the simple relationship between the radiative and the nonradiative rates. Direct measurements of the excited-state population decay can be obtained by absorption or Raman scattering with short laser pulses, such as picosecond light pulses. Luminescence decay, if feasible, can also yield reliable measurement of τ_1. In the following, we divide our discussion into electronic and vibrational relaxation processes.

3.4.1 Electronic Relaxation

In early studies by Drexhage et al. (1970), the fluorescence lifetimes of various dye molecules attached to a fatty acid monolayer of adjustable lengths on evaporated metal films (Au, Ag, and Cu) were investigated. The dye-fatty acid-metal system was prepared in aqueous solution by the Langmuir-Blodgett dipping technique (Kuhn, 1970). It was found that for large distances between the light-emitting dye and the metal surface, the fluorescence lifetime oscillated as a function of distance, whereas for small distances the lifetime decreased monotonically toward zero. The minimum distance between the dye and the surface was about 50 Å. The phenomenon was theoretically dealt with by Morawitz (1969, 1974), Philpott (1975), and Chance et al. (1978). In these studies, it was shown that these optical effects could be quantitatively explained by a classical electromagnetic theory. In these models, the electronically excited molecule was treated as a point dipole located above a metal surface with a local dielectric constant that was separated from the dielectric ambient at an infinitely sharp boundary. For molecules located far from the surface but with a distance still much shorter than the wavelength of the incident light, the emitted photon would interfere with itself upon reflection from the metal. This "image effect" would cause the radiative decay of the emitting species to fluctuate depending on the phase relationship between the reflected field and the oscillating molecular dipole. At small molecule-surface separations, energy loss mechanisms such as energy transfer to surface plasmons became important. The total decay rate of the excited species could then be separated into radiative and nonradiative components, with the latter representing the rate of energy transfer to the metal. In spite of the classical macroscopic treatment, these models have appeared to stand well against the tests of a variety of experiments ranging from atmospheric to UHV ambient and for molecule-metal separations as small as 7 Å (Rossetti and Brus, 1980, 1982; Campion et al., 1980; Whitmore et al., 1982). In the experiment performed in a UHV environment by Whitmore et al. (1982), the phosphorescence lifetime of pyrazine above an Ag(111) surface was measured as a function of molecule-metal separation (with Ar as spacer) between 10 and 420 Å. They observed a monotonic decrease in the lifetime as the dipole-surface separation was reduced in quantitative agreement with the theory by Chance et al. (1978). Because of the diminished emission intensity, the authors were not able to measure the lifetime for distances shorter than 10 Å. Model analyses further showed that surface plasmons played an important role in the energy transfer of the gas-metal system excited by near-UV photons. For pyrazine-Ag(111), a different experimental measurement was carried out by Demuth and Avouris (1981), who determined the excited-state lifetimes for the molecule directly adsorbed on the metal surface. They used high-resolution electron-energy-loss spectroscopy (EELS) to study vibronic broadening of the pyrazine excited by low-energy electron scattering. By analyzing the EEL spectra of the $^1B_{2u}$ state of the monolayer and subsequent adsorbed layers, they were able to determine the lifetime broadening of about 100 meV for the first (directly) adsorbed pyrazine layer on Ag surfaces, and about 20 meV for the second layer, corresponding to about 6×10^{-15} and 3×10^{-14} sec, respectively. The classical model would predict the corresponding lifetime to be about 1×10^{-15} sec and 1×10^{-13} sec, about a factor of 3–6 longer than the vibronic broadening measurements. The discrepancy suggested that a more rigorous theoretical treatment might be needed for species directly adsorbed on solid surfaces. Korzeniewski et al. (1982b) recently advanced an electromagnetic theory that introduced two important effects: that of nonlocality and the continuous variation of the dielectric response across the interface, which were assumed to be local and discontinuous in the model used by Chance et al. (1978). In addition, the effect of surface roughness on the fluorescence lifetime was considered by Arias et al. (1982). In a model calculation of an emitting dipole near a surface with small and random roughness, they showed that the rate of nonradiative energy transfer to the solid was increased by surface roughness and was strongly influenced by the surface plasmon. The dependence of the energy transfer on dipole-surface distance could deviate from the inverse cube law obtained by Chance et al. Similar lifetime modifications for a molecule emitting from the neighbourhood of a sphere were also considered by Nitzan, Brus, and Gersten (1981).

Very fast electronic relaxation appears to be a quite general phenomenon. The very weak chemiluminescence in the UV-visible region observed in a variety of exothermic gas-surface reactions, e.g., O_2, NO, and CO on W (McCarroll, 1969); O_2 on Si (Brus and Comas, 1971); O_2 on Al and Mg (Kasemo, 1974); and halogens on Na (Kasemo and Wallden, 1975), was suggested to be related to the very short excited-state lifetimes. The observed photon emission probability per reactive adsorption is typically in the 10^{-5}–10^{-8} range. Since the radiative lifetime for free molecules are around 10^{-7}–10^{-8} sec, the low photon emission probability would indicate that the excited state lifetime might be in the 10^{-12}–10^{-15} sec region. Accurate lifetime measurements from chemiluminescence are difficult to achieve, mainly because the identities of the emitting surface species and their radiative lifetimes as free molecules are largely unknown. Such measurements can only provide estimates in orders of magnitude.

3.4.2 Vibrational Relaxation

Vibrational lifetimes in the gas phase and condensed phases can range from picoseconds to seconds. For a number of molecules trapped in matrices at low temperatures, the lifetimes of vibrational states excited by infrared photons can be long enough for photochemical reactions to take place. On metal and semiconductor surfaces, vibrational relaxation rates are generally expected to be very fast because of the existence of efficient de-excitation channels: free electrons and phonons. Yet there is substantial experimental evidence to show that the vibrational lifetimes of adsorbed species may also exhibit a fairly large range and that gas-surface collisions are not necessarily always effective in quenching excited molecular vibrations. In the exothermic XeF_2-silicon surface reaction studied by Chuang (1979, 1980a), it was found that light emission occurred in both near-UV-visible and infrared regions. In the 400 to 700-nm region, the emission was broad and structureless, whereas in the IR region, the chemiluminescence exhibited fine structures ranging from 2.5 to 14.5 μm for the Si temperature of less than 50°C. The proximity of these fine structures to the known SiF_2, SiF_3, and SiF_4 vibrational spectra suggested that the emission might be related to these excited species produced from the surface reaction. It was further observed that the emitted IR radiation was polarized and had strong angular dependence. Namely, for the p-polarized light (parallel to the plane of incidence), the maximum emission occurred at near 60° from the surface normal. At near grazing incidence, the emission intensity was greatly reduced. For s-polarized photons (perpendicular to plane of incidence), the emission intensity decreased monotonically from 0° to 90°. These polarization and angular dependences (Greenler, 1977) provided evidence to show that at least a substantial part of the IR emission was due to excited molecular dipoles at or near the Si surfaces. In a subsequent study of the same chemical system with XPS, the Si sample was in fact covered with chemisorbed fluorine species (Chuang, 1980c). Thus, the light emission originated from the XeF_2 reaction with fluorinated Si surfaces. The experiment showed that vibrationally excited species could be produced from surface reaction and desorbed from the surface without being completely quenched by the solid surfaces. In a similar study, it was also found that XeF_2 could react with fluorinated Ta surfaces and produce vibrationally excited tantalum fluoride species (Chuang, 1980a). The probability of infrared photon emission per chemisorption reaction in these systems was in the 10^{-7}–10^{-8} range. Since the vibrational radiative lifetimes are longer than 10^{-3} sec, it is likely that the vibrational lifetimes, dominated by the nonradiative processes of the species produced from the surface reaction, are in the 10^{-10} sec region. More recently, infrared chemiluminescence has also been detected in other gas-surface interactions. Mantell et al. (1981) studied CO oxidation reaction on Pt surfaces and observed infrared emission from a vibrationally excited CO_2 reaction product in the gas phase. When the Pt surface was held at 775°K, the emission spectra showed that the rotational levels could have two

Boltzmann distributions corresponding to 400°K and 1150°K. The vibrational temperature could be as high as 2000°K. The experiment clearly showed that a surface reaction product could have much more internal excitation than would be the case if it were in equilibrium with the surface temperature. Bernasak and Leone (1981) also studied the same gas-surface system and obtained the same conclusion that the product CO_2 molecules were vibrationally much hotter than the temperature of the Pt metal surface. By a different experimental approach, Thorman et al. (1980) observed that the vibrational temperature of N_2 molecules desorbed from a polycrystalline iron surface was substantially higher than the temperature of iron. They used electron-beam-induced fluorescence to measure the vibrational energy distribution of N_2 following atomic permeation and recombination on Fe surfaces. The results showed that part of the chemical N-atom recombination energy remained with the desorbing N_2 as internal excitation, in agreement with the work by Halpern and Rosner (1978), who investigated N-atom recombination reactions on a variety of metal surfaces.

Vibrational deactivation due to gaseous collisions with solid surfaces was investigated by Misewich et al. (1983) using an IR-laser-excited fluorescence quenching technique. It was observed that the vibrational deactivation probabilities per surface collision were very high, e.g., 0.16, 0.20, and 0.22 for an excited $CO_2(001)$ colliding with a silver, nickel, and stainless-steel surface, respectively; namely, it required only five gas-surface collisions to damp the excited state. Similar high deactivation efficiency was found for excited $CO(v=1)$ and $CO_2(101)$ molecules colliding with an Ag film (Apkarian et al. 1984). Surface trapping followed by electron-hole pair formation was proposed to be the dominant mechanism for the effective decay of the vibrational energy. The vibrational relaxation mechanism was theoretically considered by Bawagan et al. (1981) and Gerber et al. (1981). It was suggested that an efficient vibrational to rotational energy transfer could be induced by the short-range repulsive part of the molecule-surface potential during the collision process, and thus it should play a significant role in the vibrational damping process. In a separate study by Zacharias et al. (1982), an IR laser was used to excite a NO molecular beam before it was scattered from a LiF surface. The NO molecules were excited into $v=1$ and $J=3/2$ state, direct-inelastically scattered from the surface, and its internal state distribution was analyzed by resonantly enhanced two-photon ionization. In the initial report, they found an extremely low vibrational survival probability (i.e., $<2\times10^{-4}$) for the scattered NO ($v=1$) state. Also, there was insignificant vibrational to rotational energy exchange. The finding was in stark contrast to the result of a stochastic trajectory calculation performed by Lucchese and Tully (1984) on the same system. It was shown in the calculation that when trapping of NO on the LiF surface was negligible (i.e., scattering dominated by the direct-inelastic process), the vibrational energy accommodation coefficient to the solid surface should be less than 1%. Furthermore, the NO vibrational mode was most strongly coupled to the surface phonons with a much weaker coupling to the rotational and translational modes. In a later experimental study by Misewich et al. (1985) using a vacuum-cleaved LiF crystal, the research group determined that indeed the vibrational survival probability for the scattered NO($v = 1$) molecules could be very high, and highly rotationally excited states could be produced in the surface scattering process.

A direct measurement of the vibrational energy relaxation rate for molecules adsorbed on dielectric surfaces was carried out by Heilweil et al. (1985) using the picosecond IR laser spectroscopic method. Here, a picosecond IR pulse was used to excite some of OH groups adsorbed on silica particles, and a probe pulse after some delay was employed to determine the time required for the system to return to its original state. The relaxation time for the excited O-H stretching vibration near 3000 cm^{-1} was about 204 psec for the system in vacuum. The time decreased to 140 psec when SiO_2 particles were surrounded by a solvent such as CCl_4. This relaxation time corresponds to about 10^4 vibrational periods. It was suggested that the relative slow relaxation rate in vacuum was controlled by the vibrational coupling to lower frequency modes of the SiO_2-OH system. With large differences in vibrational frequencies, such vibrational to vibrational (V-V) energy transfer rates were expected to be slow. These V-V transfer rates were enhanced by CCl_4-OH collisions when CCl_4 molecules were present. Further experiments showed comparable relaxation rates for the OD groups, but a substantially shorter relaxation time for the surface NH_2 group (Heilweil, 1985). These results clearly indicated that the observed IR linewidth of 8 cm^{-1} for the OH stretching mode was not due to lifetime broadening but rather to the vibrational dephasing, inhomogeneous broadening, or both. This study also suggested that one should exercise extreme caution in trying to correlate the IR absorption linewidth with the vibrational lifetime for adsorbed species. Naturally, the relative long vibrational relaxation time has important implications for surface photochemistry.

In an interesting theoretical study performed by Tully (1980) on carbon oxidation reaction on a Pt(111) surface, it was shown that the product CO molecules desorbed from the metal surface with considerable internal excitation. A stochastic classical trajectory calculation of a gaseous oxygen atom reacting with carbon adsorbed on a Pt surface was carried out. The flow of energy, including phonons, between the reaction zone and the crystal was taken into account by adopting an empirical interaction potential. The computation showed that the oxidation reaction occurred with a high probability, and, surprisingly, for the 6 eV of chemical energy released from the exothermic reaction only 10% or less was deposited on the metal. The remaining (>90%) energy was carried by the gaseous CO product in the form of excited vibration (~2.8 eV), rotation (~1 eV), and translation (1.8 eV). The high vibrational energy meant that the CO molecules were excited into very high vibrational levels, with the average vibrational quantum number approaching $v = 11$. The author suggested that infrared emission from such high CO vibrational states might be experimentally observable. The variations of the interaction potential within reasonable limits did not substantially affect the calculated results. The author also considered possible mechanisms for CO vibrational energy relaxation. Because of the large difference between the CO stretching frequency (4×10^{14} sec) and the surface Debye frequency of Pt (1.5×10^{13} sec), the phonon relaxation mechanism was concluded to be unimportant. The fraction of vibrational energy dissipated by conduction electrons of the metal was calculated to be no more than 20%, even if the energy relaxation rate was assumed to be as high as 1.6×10^{12}/sec for the CO-Pt distance $z < 3$ Å and to decrease as a function of z^{-3} for $z > 3$ Å. The high energy relaxation rate ($\tau_1 = 0.63 \times 10^{-12}$ sec) was obtained if one assumed that the 10 cm^{-1} linewidth of the infrared absorption band of CO on Pt(111) was due entirely to the energy relaxation effect, which the author considered highly unlikely. In any event, the author concluded that substantial deactivation of the vibrationally excited CO molecule after it was formed from the surface oxidation reaction did not occur, either by the electronic or phonon damping mechanisms. It was suggested that the damping contribution from the phonon mechanism was negligible and that from the electronic mechanism was far less than the computed 22%.

The mechanisms of vibrational relaxation on surfaces have also been theoretically considered by several other researchers. For the low-frequency modes such as the adsorbate-substrate vibrations, which are not very far from the Debye frequency of the solid, the phonon mechanisms, including multiphonon effects, appear to be the dominant factor for damping surface vibrations. Korzeniewski et al. (1982) and Jedrzejek et al. (1981) showed that in this low-frequency regime, for almost any lattice and oscillator model, the rates of energy transfer between the surface vibration and the electron-hole excitations of the metal were substantially smaller than those induced by phonons. The relative importance of one-phonon and multiphonon coupling mechansims would depend on the phonon lifetime and the difference between the surface vibrational frequency (ν) and the Debye frequency of the solid (ν_D). For $\nu < \nu_D$, one phonon would clearly be the most important damping mechanism. For $\nu > \nu_D$, multiphonon would be important but the one-phonon interaction might not be completely neglected if the phonon had a finite lifetime. In the high vibrational frequency regime, it is generally considered that energy relaxation to electronic motions in the solid should be very efficient on metal surfaces (Persson, 1978; Kozhushner et al., 1979; Persson and Persson, 1980). Surface plasmons are important in the near UV-visible region but should be unimportant for damping surface vibrational motions because of the large difference in resonant frequencies. In the electronic considerations, Persson (1978) suggested that electron-hole pair excitation could be the dominant channel for the relaxation of excited adsorbate molecules. In the model calculation, the metal was treated as a semi-infinite electron gas confined to one half-space by an infinite potential barrier, and the excited molecule was treated as an oscillating dipole. The damping efficiency was evaluated as a function of the distance (z) between the dipole and metal surface, and it was found to be proportional to z^{-3} for small z, the same as that obtained by Brus (1980). In a later study, Persson and Persson (1980) refined this model and numerically calculated the lifetime of CO($\nu = 1$) adsorbed on a Cu(100) surface. With a dynamic dipole moment of 0.1 D, a vibrational frequency at 2090 cm^{-1}, and a distance of 2.5 Å between the center of the mass of the CO molecule and the first surface of the Cu lattice plane, they estimated z to be about 0.7 Å according to their infinite barrier model and the vibrational lifetime τ_1 to be about 1×10^{-10} sec. When the CO dynamic dipole was adjusted to a high value (Persson and Persson, 1980), τ_1 was shortened to 2×10^{-12} sec. These values could be compared with that obtained by Kozhushner et al. (1979), who made a calculation based on the same mechanism but with a different description of the metal. According to this calculation, τ_1 should be about 2×10^{-11} sec. It, therefore, appears that the theoretically computed lifetime of the CO($\nu = 1$) state on Cu surfaces can be between 1×10^{-10} sec and 2×10^{-12} sec based on the electron-hole pair damping mechanism. Experimentally, Ryberg (1982) had performed infrared absorption studies and found that the linewidth of the IR absorption band of $^{12}C^{16}O$ on Cu(100) in the ordered C(2×2) structure was 4.5 cm^{-1} and that for $^{12}C^{18}O$ was 4.0 cm^{-1}. The author also found that the structure or isotopic disorder would cause the linewidth to be broadened by only about 1 cm^{-1}. It was, therefore, suggested that the 3 cm^{-1} linewidth was due entirely to lifetime broadening, i.e., $\tau_1 \simeq 2 \times 10^{-12}$ sec. In Ryberg's estimate, the possibility of a short vibrational dephasing time (τ_2) contributing to the observed IR absorption linewidth was completely ignored. In view of the model calculations by Tully (1980) and Persson and Ryberg (1981), it seems possible that the adsorbed CO molecules on metal surfaces could strongly interact with each other due to dipole-dipole coupling. In the presence of the IR radiation field, vibrational dephasing could occur extremely rapidly, which could make a major contribution to the IR absorption linewidth. Once the molecule is vibrationally excited, whether by IR photon absorption or from an exothermic chemical reaction, the excitation energy could be very quickly transferred by a near-resonant mechanism to other molecules in the adsorbed CO layer, even to the adsorbed isotopic species. It appears likely that this intermolecular energy transfer process can take place before the energy is eventually relaxed to lattice phonons *via* the electronic damping mechanism.

From these various experimental observations and theoretical considerations, it appears certain that in the high-frequency regime, vibrational energy relaxation *via* substrate electronic interaction is important. It is, however, equally certain that although vibrational energy decay rates are generally very fast whether it is due to electronic, phonon, or vibrational-to-rotational energy transfer mechanisms, many interesting chemical processes can still take place on solid surfaces.

4. Experimental Studies on Laser-Stimulated Desorption

4.1. Laser-Induced Thermal Desorption (LITD)

Laser-induced thermal desorption is accomplished by optical heating of the substrate. The unique features of the technique, as pointed out by Cowin et al. (1978), include the vast range of heating rates ranging from 10^5 to 10^{11} K/sec, the easy control of heating period from 1 msec or longer to 1 nsec or shorter, and such technical advantages as no need for a chopper for desorption velocity measurements, no special requirements for thermal or electrical properties of the sample or its shape, and the usually high signal-to-noise ratio, allowing angular resolved measurements to be performed with high precision. The wide range of heating rates may allow one to examine desorption kinetics over a wide dynamic range and help to clarify the mechanisms for multistep reactions. It may also be possible to desorb chemical intermediates not detectable by other methods. Therefore, the laser thermal desorption technique should be very useful for surface studies and should also be a powerful supplement to conventional thermal-desorption and flash-desorption methods. In the case of metals, the absorption of laser radiation by the conduction electrons of the substrates is rapidly transformed into lattice heating. For a laser pulse width of 10^{-9} sec or longer, the ordinary bulk thermal conduction laws are applicable. Solution of the heat conduction equations for application to laser experiments were given by Ready (1971) and Bechtel (1975). One such solution is shown in Eq. 2.9, which was also used by Cowin et al. (1978) for their study on desorption kinetics of D_2 from tungsten. Specifically, Cowin et al. investigated deuterium atom-recombination desorption with a Q-switched Nd:glass laser at 1.06 μm with a pulse width of 30 nsec irradiated on W surfaces covered with D atoms. D_2 desorption time-of-flight was measured with a differentially pumped mass spectrometer. The time-of-flight distributions were converted into velocity distributions and thermal distributions. The desorption flux was also determined as a function of angle from the surface normal. They detected a noncosine distribution in the pulsed laser-induced desorption. However, after careful model analyses, they attributed this observation to the effect of collisions between molecules after they left the metal surface. The authors concluded that the desorption of D_2 from tungsten at rates of 5×10^7 monolayers/sec by the laser technique was governed by the same kinetics as obtained by extrapolating previous thermal desorption meaurements made at a rate about 10^5 or more slowly. Wedler and Ruhmann (1982) in a LITD study of CO from Fe(110) surfaces adopted essentially the same experimental approach. The time-dependent CO desorption signal was recorded by a mass spectrometer following a 30-nsec Nd:glass laser pulse and compared with the calculated desorption signal due to the temperature rise induced by the laser irradiation. The maximum amplitudes of the desorption signals were examined rather thoroughly as a function of the laser intensity and the CO coverage for both single pulses and sequences of laser pulses. For a monolayer of CO adsorbed on Fe, the threshold for laser-induced desorption was about 8 MW/cm^2, and at about 30 MW/cm^2 a saturation value was reached, corresponding to all the CO being desorbed by one pulse. There was also evidence to show that in the low temperature range, the temperature of desorbed molecules (T_d) followed a Maxwell–Boltzmann distribution identical to the maximum surface temperature (T_s). Above $T_s = 600°K$, T_d was found to be smaller than T_s. The experimental observations were analyzed successfully with the first-order rate equation for desorption. A similar LITD behavior was observed in CO/Cu(100) studied by Burgess et al. (1983). It was again found that Boltzmann distributions were obeyed for all coverages and for all incident laser power levels. T_d was always lower than the temperature predicted from temperature-programmed thermal desorption or the calculated T_s. Other TOF experiments (Schäfer and Hess, 1985) involving ablation of thick adsorbate layers by laser pulses showed that Boltzmann distributions could still be obtained for very large desorption fluxes, namely, many monolayers of molecules desorbed by a single pulse.

Questions then arise from these studies: a) Could the velocity distributions be influenced, and how are they affected by large desorption fluxes in LITD in which the heating rate and the desorption rate are usually high? b) Could T_d be different from T_s, namely, the question of equilibration? If so, are there reasons other than the fact that in many cases, the molecule can desorb at temperatures lower than the maximum surface temperature reached by the laser pulse? c) What are the internal energy distributions of the desorbed molecules? Are they in thermal equilibrium with T_s? d) Is LITD governed by the same kinetics as obtained by extrapolation of the conventional thermal desorption carried out at much lower heating rates? e) What are the actual surface temperatures raised by laser pulses? There are no clear answers to these questions from the experimental data presently available. There were, however, theoretical studies and model calculations by Tully (1981) and Lucchese and Tully (1984b). It was predicted that the mean energies of the translational and the internal degrees of freedom of desorbing molecules were significantly lower than those corresponding to the surface temperatures at the instant of desorption. Obviously, we need systematic studies to determine the velocity and the angular as well as the internal state distributions and to measure the surface temperature in the time-resolved manner in order to fully characterize the LITD process.

As an initial attempt to correlate T_d and T_s, Hussla et al. (1986) recently performed a TOF experiment using an excimer laser to desorb Xe atoms adsorbed on a Cu film that is deposited on top of a pyroelectric detector. The thin-film calorimetric detector can monitor the surface temperature with a rapid time response (a few nanoseconds). Some results of this study are shown in Figs. 2.11 and 2.12. Some slight but noticeable deviations from the Boltzmann velocity distributions are detected. Furthermore, T_d's are apparently lower than the measured T_s's, particularly in the high laser intensity regime. This initial experiment is, however, complicated by the fact that many layers of Xe atoms are desorbed by a single UV pulse. As discussed above, gaseous collisions above the Cu surface can affect the measured velocity distribution as well as the translational temperature. The study nevertheless demonstrates that the thin-film calorimetry is a sensitive surface-temperature sensing technique and it can be applied to probe LITD properties.

In spite of the fundamental problems that remain to be solved, LITD has been successfully utilized to study surface diffusion of CO adsorbed on Cu(100) (Viswanathan et al., 1982) and H_2 adsorbed on Ni(100)

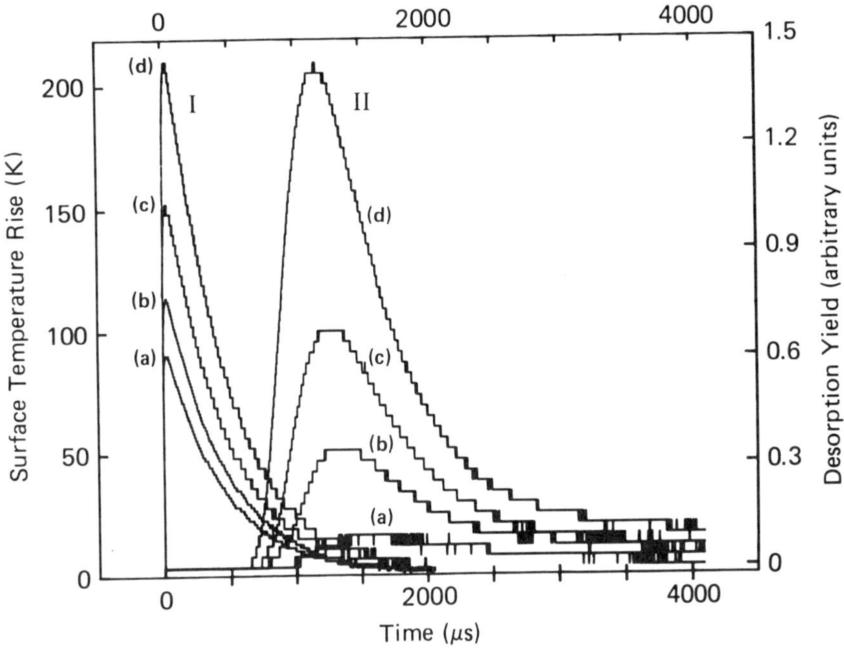

Fig. 2.11 Time-resolved signals of (I) surface temperature rise and (II) laser-induced thermal desorption yield of Xe from a Cu film on top of a pyroelectric thin film calorimeter with Xe surface coverage of about 100 monolayers at the substrate temperature of 20°K: (a) laser fluence $1.05 \times 10^{-2}\ \mathrm{J/cm^2}$; (b) $1.35 \times 10^{-2}\ \mathrm{J/cm^2}$; (c) $1.75 \times 10^{-2}\ \mathrm{J/cm^2}$; and (d) $2.35 \times 10^{-2}\ \mathrm{J/cm^2}$. Data according to Hussla et al. (1986b).

(George et al., 1985). It was used to measure desorption energies of H_2 and CO from stainless-steel surfaces (Tagle and Pospieszczyk, 1983). The technique in conjunction with short laser pulses was also applied to determine the reaction intermediates present on surfaces. For example, the determination of the $SiF_x (x < 4)$ products in the Si etching reaction with XeF_2 (Chuang et al., 1984b) and the product of methanol decomposition on Ni surfaces (Hall et al., 1984, 1985) was obtained with the LITD method. In a different approach, the role of surface microstructure in enhancing molecular desorption by laser radiation was examined (Fletcher et al., 1984).

4.2. *Infrared Laser-Induced Photodesorption (IRPD)*

IRPD due to resonant absorption of IR photons by adsorbed molecules performed under a UHV condition was first reported by Heidberg et al. (Heidberg et al., 1980, 1982, 1985). In these studies, a monolayer or a

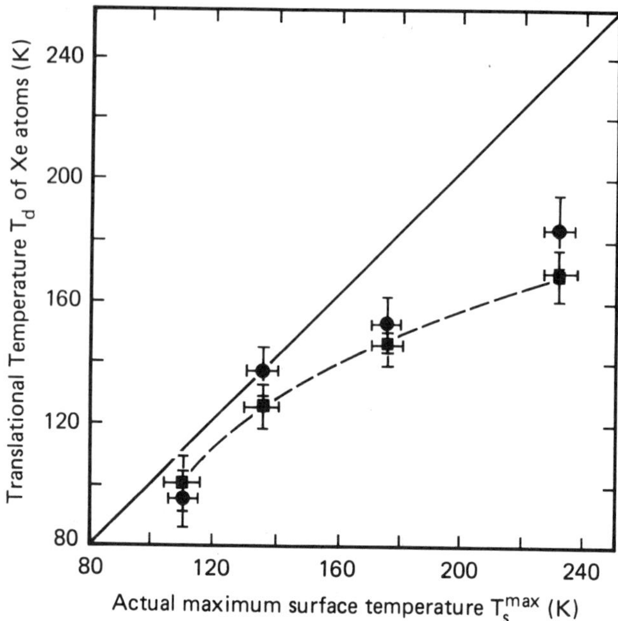

Fig. 2.12 Translational temperatures (T_d) of Xe desorbed by LITD *versus* maximum surface temperatures (T_s^{max}) determined by the thin-film calorimeter: T_d's obtained from both the fitting of the time-of-flight spectra to Boltzmann velocity distributions (solid squares) and the calculation based on the peaks of the TOF spectra (solid circles) are displayed. The solid line indicates the linear relationship if $T_d = T_s^{max}$. Data according to Hussla et al. (1986b).

multilayer of CH_3F molecules was adsorbed on a NaCl film at 70°K and the adsorbate was excited by a line-tunable pulsed CO_2 laser. Desorption of neutral molecules was observed when the stretching mode of the molecule absorbed the IR photons in the 970–990 cm^{-1} region. The observed bandwidth in the photodesorption spectrum was about 10–15 cm^{-1}, substantially narrower than the IR absorption spectrum. The desorption yield (Y) dependence on the laser fluence (I) was found to be $Y \propto I^{2.8}$, suggesting that three photons might be involved in inducing the desorption. The photodesorption cross section was estimated to be about $2 \times 10^{-19}\ cm^2$ (Heidberg et al., 1982), only slightly smaller than the IR absorption cross section, indicating a very high desorption quantum yield (i.e., number of molecules desorbed per absorbed IR photon). The yield, however, was not directly determined from the *in-situ* analyses of the surface coverages before and after the laser radiation.

The phenomenon of IRPD was also investigated under UHV conditions by Chuang, Seki, and Hussla (1982b, 1982d, 1983a, 1984a, 1984c,

1985) with pyridine and deuterated-pyridine molecules adsorbed on KCl, Ag film, and Ag(110) surfaces, and with NH_3 and ND_3 molecules on NaCl, Ag film, and Cu(100) surfaces. The pyridine systems involved the excitation of the symmetric and the antisymmetric ring modes of adsorbed molecules with a pulsed CO_2 laser in the 9–11 μm range. For the ammonia systems, the high-frequency N-H and N-D stretching modes were excited with a tunable pulsed IR laser in the 2.5–4.2 μm region, while the low-frequency bending mode was excited with a pulsed CO_2 laser. The systems were characterized with XPS, both the conventional and the laser-induced thermal desorption, and in the case of C_5H_5N/Ag film, also with the surface-enhanced Raman scattering. Time-of-flight (TOF) mass spectrometry was employed in ammonia photodesorption studies. Some typical results of NH_3 photodesorption from Cu(100) are shown in Figs. 2.13 and 2.14.

The IR photodesorption behavior for the pyridine and the ammonia systems is quite similar. The major results of these experiments can be summarized as follows: a) Desorption due to resonant absorption of IR photons by adsorbates could occur from both metallic and dielectric substrates. The quantum yields were generally very low, possibly less than 10^{-3}, with the desorption cross section estimated to be 10^{-22} cm^2 or less, excluding the direct laser substrate heating effect. b) No clear vibrational mode (i.e. chemical bond) selectivity or significant isotope enhancement in desorption was detected. For instance, when an isotopic species was photoexcited in an isotopic mixture, different (nonphoto-excited) isotopic molecules could be desorbed with almost the same efficiency. The desorption yield appeared to depend on the IR absorption cross section and not on the particular vibrational motion (mode) with respect to the surface. c) The desorption could be induced by both single and multiple photon absorption. There was only a narrow range of laser fluences in which the resonantly excited desorption could be observed. Below a certain laser threshold, desorption was not detectable, whereas if the laser intensity was too high, the direct laser substrate heating effect overtook and washed out the resonant adsorbate excitation effect. Direct laser substrate heating was an important factor in enhancing the desorption yield, even for such an IR-transparent substrate as KCl and NaCl, or for highly IR-reflective metals such as a polished Ag or Cu crystal. d) Desorption yields increased with the thickness of the adsorbed molecular layers. This strongly indicated that accumulation of adsorbed energy by neighboring molecules was important in enhancing desorption. e) For a surface coverage of one to two monolayers, large desorption signals were detected in the first few laser pulses. Then the signal diminished or even disappeared, although the vast majority of adsorbed

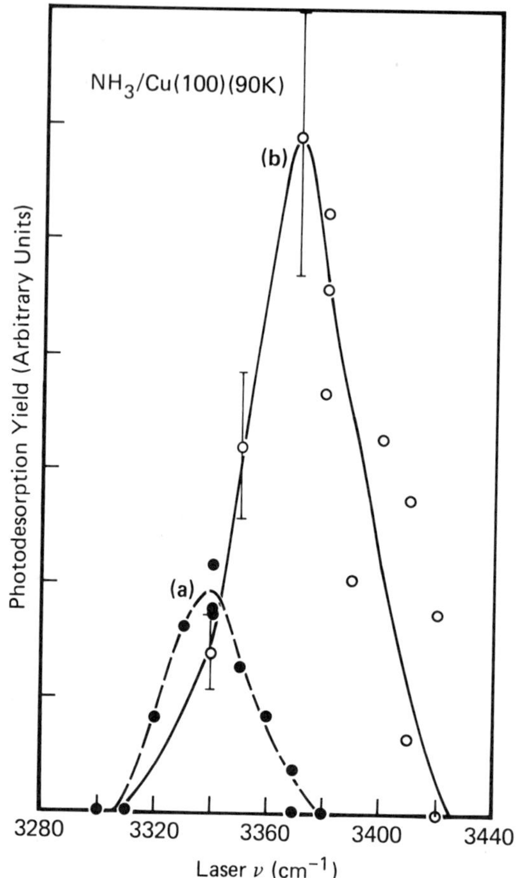

Fig. 2.13 IR photodesorption yields of NH_3 on Cu(100) at 90°K as a function of laser frequency: (a) surface coverage about a monolayer, $\theta = 1$ and laser intensity, $I =$ 10 mJ/cm^2; (b) multilayer coverage, $\theta = 3.4$ and $I = 10$ mJ/cm^2. The laser is p-polarized and at a 75° angle of incidence. Each data point is an average of mass spectrometer signals due to 20 laser pulses. Data according to Chuang et al. (1985).

molecules still remained intact on the surface. This suggested the desorption of some weakly bound species (the minority species). When the laser fluence was increased, some more strongly bound molecules could also be desorbed. f) The translational temperature (T_d) of desorbed molecules was rather low and in some cases about the same as the original substrate temperature (T_s). This T_d was usually much lower than the peak desorption temperature observed in the conventional thermal desorption. For example, the observed T_d for desorbed NH_3 was around

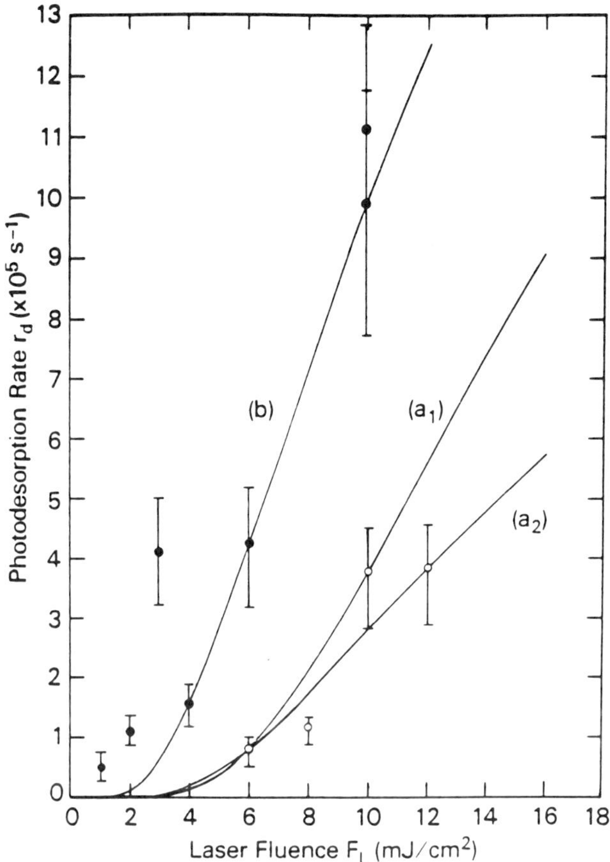

Fig. 2.14 IR photodesorption yields of NH_3/Cu(100) at 90°K as a function of laser fluence: (a) surface coverage $\theta = 1$, $v = 3340\ cm^{-1}$; (b) $\theta = 3.4$, $v = 3370\ cm^{-1}$, sampling average of 20 laser pulses. The dots and circles are experimental data, and the solid curves are the results of a model calculation according to Hussla et al. (1985) based on the phonon and electronic damping mechanisms.

90°K, while the thermal peak desorption temperatures were about 140°K for the physisorbed NH_3 and 190–260°K for the chemisorbed species. Many of the observed photodesorption behaviors were in line with the concept of "indirect" or "resonant" heating that Chuang (1983a) and Gortel et al. (1983b) had suggested. Basically, it was proposed that the photon energy absorbed by the adsorbed molecules decayed rapidly to create local heating of the molecular layer and the substrate surface so that during the laser pulse width there was substantial thermal excitation of the molecule-surface potential. This resulted in the desorption of some weakly bound molecules. Based on the experimental observation, it was further suggested that the direct laser substrate heating always played an important role in reducing the apparent activation barrier for desorption. These thermally assisted processes, including the direct and the indirect heating effects could explain the lack of mode and isotope selectivities in IRPD and the desorption yield dependence on the overlay thickness. These thermally assisted effects alone, however, could not explain the low translational temperatures observed for the ammonia system. In a recent theoretical calculation performed for the NH_3/Cu(100) system by Hussla et al. (1985) involving phonon-mediated and electron-hole pair damping mechanisms, a reasonably good agreement between the calculated desorption yield as a function of laser fluence with the experimentally measured yields was obtained when a reasonable set of parameters, including the energy damping rates in the 10^{11}/sec range, were used. The results of this calculation are also shown in Fig. 2.14. The same calculations, however, showed that the resultant desorption temperature could reach 200°K or higher, in contradiction with the observed $T_d = 90$°K. Clearly, if the local temperatures due to the energy decay from the excited vibrational modes were so high, then the desorption yields due to the thermal effects should be much higher than those observed experimentally.

This discrepancy had led the authors to suggest (Chuang et al., 1985; Hussla et al., 1985) that in IRPD the desorbed molecules might still be internally excited when they desorbed into the gas phase. The excited molecules or the excited internal modes (vibrational and/or rotational modes) might not be the original molecules or the original vibrational modes that absorbed the IR photons. This internal excitation might be acquired *via* the ultrafast intermolecular and intramolecular energy transfer processes from the initially photoexcited molecules. In any event, the detailed dynamics that lead to IR photodesorption are not yet completely clear. Further experiments to determine the angular and the internal energy distributions of the desorbed particles are needed in order to better understand the photodesorption phenomenon.

The IRPD behavior in multilayer systems was also investigated by Schäfer and Hess (1984, 1985) using relatively thick (10–100 μm) molecular films condensed on Ge and metal substrates at low temperatures. The systems were excited by a pulsed CO_2 laser. The time-of-flight (TOF) measurements showed that the velocities of the photodesorbed molecules such as CH_3F, CH_3OH, CCl_4, etc., could be fitted into Maxwell–Boltzmann distributions, in spite of the fact that many monolayers of molecules were ablated per laser pulse and there were certainly lots of molecular collisions in the gas phase above the surface. The observed TOF translational temperature (T_d) and the desorption yield dependences upon the laser wavelength exhibited the same spectral features as the IR absorption spectra. In other words, a stronger absorption of IR laser photons near the center of the IR absorption band would result in a larger desorption yield and a higher T_d than those produced by the weaker absorption near the wing of the band. Furthermore, there existed an approximately Arrhenius relationship between the desorption yields and the TOF temperatures. This strongly suggested a thermally activated process due to the "indirect" or "resonant" heating discussed above. The detected T_d's were in the 200–1200°K range depending on the IR absorption strengths and the heats of vaporization of the condensed layers. Thus, for very thick condensed molecular films or solids, thermally assisted processes seem to be dominating in the IRPD. Other IR laser-induced desorption studies included the work by Abbate et al. (1981) who showed that dehydroxylation on silica, alumina, and zinc oxide surfaces could be induced by a cw CO_2 laser. The effect was, however, attributed to purely thermal desorption, in contrast to an earlier study of a similar system by Djidjoev et al. (1976), in which selective IR photochemical reactions were suggested. Surface desorption and cleaning by IR lasers were also performed by Allen et al. (1984), who used pulsed HF and DF lasers at 2.8 μm and 3.7 μm, respectively, to irradiate contaminated solid surfaces. It was found that both H_2O and hydrocarbons could be desorbed by the IR pulses.

4.3. *Ultraviolet and Visible Laser-Induced Photodesorption*

Photodesorption studies involving electronic excitation of adsorbate-surface complexes by UV and visible lasers on well-characterized surfaces are still relatively few. In a recent experiment by Bourdon et al. (1984), it was observed that CH_3Br molecules adsorbed on a LiF crystal could be photodesorbed by an excimer laser at 222 nm or 308 nm. The result was interpreted in terms of the UV absorption by the color centers of the substrate causing a shock wave to pass through the crystal and thereby inducing the ejection of the adsorbed molecules. It was further found that adsorbed CH_3Br could be photodissociated by the single-photon absorption of the 222-nm light to produce CH_3 radicals that subsequently desorbed from the substrate. In contrast to the photodesorbed CH_3Br molecules, which possessed rather low kinetic energies, the photogenerated CH_3 radicals desorbed with relatively high velocities. Apparently, in the antibonding state reached by the photoexcitation, part of the electronic energy after breaking of the chemical bond was transformed into the translational energy of the desorbing radicals. Laser irradiation at 308 nm that was not absorbed by CH_3Br molecules did not produce the same results. Studies of the CH_3 radical yields as a function of the 222-nm laser intensity and the surface coverage confirmed the photochemical effects on the LiF surface. Photodesorption of neutrals and ions from surfaces of molecular solids or molecular films deposited on metal substrates irradiated by UV or visible lasers had been investigated by Letokhov et al. (1981), Antonov et al. (1981), and Nishi et al. (1984). Quite often, nonthermal photodesorption characteristics were observed. There were also reports of CO_2 and H_2O desorption from oxide semiconductors induced by band-gap excitation with the UV-visible light (Kawai and Sakata, 1980; Van Hieu and Lichtman, 1981). Other UVPD studies have been related mainly to photochemical etching of solid surfaces (see, e.g., Chuang, 1986). Clearly, more systematic and more detailed spectroscopic investigations in this spectral region are needed in view of the fact that the UV-visible light possesses relatively high photon energy comparable to chemical bond energies, and it should be quite efficient in promoting desorption and surface photochemistry. There are also many interesting theories ready to be tested concerning electronically excited desorption. This includes the MGR valence excitation model, the Knotek–Feibelman (KF) core-level excitation mechanism, and the photon-sputtering effect, as discussed in Section 3.3. Although the MGR and KF desorption models were originally intended for dealing with excitations by high-energy photons (>10 eV) and electrons, they could be quite relevant to laser-induced photodesorption in certain spectral and laser intensity regimes. The photon-sputtering effect can be particularly important on dielectric substrates and in the very high laser fluence region.

5. Summary

In this chapter, the basic processes that are important in laser-induced gas-surface interactions have been outlined. Many current experimental and theoretical studies on the subject are reviewed and discussed with the objective of clarifying the relative importance of the interaction steps involved in the photon-stimulated surface processes. Attempts are also made to relate the resulting concepts to the experimentally observed phenomena. Specifically, the photon-enhanced adsorption, adsorbate-adsorbate and adsorbate-absorbent reactions, product formation, and desorption processes are examined in length with available data. The dynamic processes involved in photoexcitation of the electronic and vibrational states, and their energy transfer and relaxation in competition with surface chemical interactions are considered in detail. These include both single and multiple photon absorption; fundamental and overtone transitions in the excitation process; intermolecular and intramolecular energy transfer; and coupling with phonons, electron-hole pairs, and surface plasmons in the energy relaxation process. Throughout the discussion, there has been extensive use of CO adsorbed on metals and SF_6 interaction with silicon as examples to illustrate the many facets of the electronically and vibrationally activated surface processes. SF_6 gas is inert to silicon in the absence of radiation, and its level of excitation can be easily controlled. One can activate the molecules *via* vibrational excitation without dissociation or photodecompose the gas to produce F atoms and investigate their different surface chemical behavior. The gas can also be condensed on solid surfaces at low temperatures. Resonant excitation in the adsorbed state reveals further details of the reaction dynamics. The observation of SiF_4 formation and its desorption from the cold Si substrate in reaction with excited SF_6, along with studies on IR chemiluminescence and IR laser-stimulated desorption in similar systems has provided strong evidence to suggest that some surface reaction products are formed in vibrationally excited states and that these excited species have higher desorption rates than the unexcited molecules. This latter aspect has been more thoroughly studied on the infrared laser-stimulated desorption of pyridine and deuterated pyridine, and ammonia and deuterated ammonia from various solids such as KCl, Ag(110), and Cu(100). The observation of resonant absorption of photon energy by molecular vibrations on surfaces has raised the expectation that perhaps molecular—or even bond-selective—desorption may be possible with infrared photons. Although that possibility exists, the results show that the rapid intermolecular energy transfer, very likely *via* dipole-dipole coupling mechanisms, the very fast vibrational relaxation, and the direct and indirect substrate heating effects not only can cause low selectivity between isotope species, but also can cause rather small photodesorption cross sections that can be attributed to quantum effects. For electronic excitation on metal surfaces, with the exception of direct photodissociation of gaseous or adsorbed species, the photon-stimulated nonthermal process, whether it is enhanced adsorption, reaction, or desorption, appears to be rather difficult to accomplish. However, band-gap excitation on semiconductor surfaces to promote photoadsorption, catalytic reactions, and desorption are known to be efficient. Quantum effects associated with electronically excited photodissociation and photodesorption on insulator surfaces can also be more easily observed. In addition, some examples are given, along with recent studies on oxidation and etching reactions on semiconductor surfaces enhanced by photogeneration of electron and hole pairs. In short, laser applications to gas-surface studies can provide fundamental insight into the mechanisms involved in hetergeneous interactions, in particular, the molecular dynamics on surfaces. Such studies also offer the exciting possibility for technical innovations in practical applications such as material processing for microelectronics.

References

Abbate, A.D., Kawai, T., Moore, C.B., and Chin, C.-T. (1984). Activation and cleaning of oxide surfaces by a cw CO_2 laser. *Surf. Sci.* **136,** L19-L24.

Allen, S.D., Porteus, J.O., and Faith, W.N. (1982). Infrared laser-induced desorption of H_2O and hydrocarbon from optical surfaces. *Appl. Phys. Lett.* **41,** 416–418.

Ambartsumyan, R.V., and Letokhov, V.S. (1977). Multiple photon infrared laser photochemistry. In *Chemical and Biochemical Applications of Lasers* (Moore, C.B., ed.). Academic Press, New York, Vol. 3, 167–316.

Andersson, S. (1979). Vibrational excitations and structure of CO chemisorbed on Cu(100). *Surf. Sci.* **89,** 477–485.

Andersson, S., and Persson, B.N.J. (1980). Inelastic electron scattering by a collective vibrational mode of adsorbed CO. *Phys. Rev. Lett.* **45,** 1421–1424.

Antoniewicz, P.R. (1980). Model for electron- and photon-stimulated desorption. *Phys. Rev.* **B21,** 3811–3815.

Antonov, V.S., Letokhov, V.S., and Shibanov, A.N. (1981). Nonthermal desorption of molecular ions of polyatomic molecules induced by UV laser radiation. *Appl. Phys.* **25,** 71–76.

Apkarian, V.A., Hamers, R., Houston, P.L., and Misewich, J. (1984). Laser studies of vibrational energy exchange in gas-solid collisions. In *Dynamics on Surfaces* (Pullman, B., Jortner, J., Nitzan, A., and Gerber, B., eds.). D. Reidel, Dordrecht, 135–148.

Aravind, P.K., Rendell, R.W., and Metiu, H. (1982). A new geometry for field enhancement in surface-enhanced spectroscopy. *Chem. Phys. Lett.* **85,** 396–403.

Arias, J., Aravind, P.K., and Metiu, H. (1982). The fluorescence lifetime of a molecule emitting near a surface with small, random roughness. *Chem. Phys. Lett.* **85,** 404–408.

Baddour, R.F., and Modell, M. (1970). The effect of visible and ultraviolet light on the palladium-catalyzed oxidation of carbon monoxide. *J. Phys. Chem.* **74,** 1392–1394.

Baidyaroy, S., Bottoms, W.R., and Mark, P. (1971). Photodesorption from CdS. *Surf. Sci.* **28,** 517–524.

Baklanov, M.R., Beterov, I.M., Repinskii, S.M., Rzhanov, A.V., Chebotaev, V.P., and Yurshina, N.I. (1974). Initiation of a surface chemical reaction between single-crystal germanium and bromine gas by using a powerful argon laser. *Sov. Phys. Dokl.* **19,** 312–314.

Basov, N.F., Beknov, E.M., Isakov, V.A., Leonov, Yu.S., Markin, E.P., Oraevskii, A.N., Romanenko, V.I., and Ferapontov, N.B. (1975). Condensation of vibrationally excited gas. *JETP Lett.* **22,** 102–104.

Bass, H.E., and Fanchi, J.R. (1976). The effect of N_2O laser irradiation on the nitrous oxide-copper reaction. *J. Chem. Phys.* **64,** 4417–4421.

Bawagan, A.O., Beard, L.H., Gerber, R.B., and Kouri, D.J. (1981). Vibrational-rotational-translational energy exchange in molecule-surface collisions. *Chem. Phys. Lett.* **84,** 339–342.

Bechtel, J.H. (1975). Heating of solid targets with laser pulses. *J. Appl. Phys.* **46,** 1585–1593.

Bedair, S.M., and Smith, H.P., Jr. (1969). Atomically clean surface by pulsed laser bombardment. *J. Appl. Phys.* **40,** 4776–4781.

Bernasek, S.L., and Leone, S.R. (1981). Direct detection of vibrational excitation in the CO_2 product of the oxidation of CO on a platinum surface. *Chem. Phys. Lett.* **84,** 401–404.

Beterov, I.M., Chebotaev, V.P., Yurshina, N.I., and Yurshin, B.Ya. (1978). Effect of the laser radiation intensity on the kinetic of the heterogeneous photochemical reaction between single crystal germanium and bromine gas. *Sov. J. Quantum Electron.* **8,** 1310–1312.

Bickley, R.I., and Jayanty, R.K.M. (1974). Photo-adsorption and photo-catalysis on titanium oxide surfaces. *Disc. Faraday Soc.* **58,** 194–204.

Black, J.G., Yablonovitch, E., Bloembergen, N., and Mukamel, S. (1977). Collisionless multiphoton dissociation of SF_6: A statistical thermodynamic process. *Phys. Rev. Lett.* **38,** 1131–1134.

Bourdon, E.B.D., Cowin, J.P., Harrison, I., Polanyi, J.C., Segner, J., Stanners, C.D., and Young, P.A. (1984). UV photodissociation and photodesorption of adsorbed molecules I. CH_3Br on LiF(001). *J. Phys. Chem.* **88,** 6100–6103.

Brass, S.G., Reed, D.A., and Ehrlich, G. (1979). Vibrational excitation and surface reactivity: An examination of the ν_3 and $2\nu_3$ modes of CH_4. *J. Chem. Phys.* **70,** 5244–5250.

Bray, R.G., and Berry, M.J. (1979). Intramolecular rate processes in highly vibrationally-excited benzene. *J. Chem. Phys.* **71,** 4909–4922.

Brown, W.L. (1980). Transient laser-induced processes in semiconductors. In *Laser and Electron Beam Processing of Materials* (White, C.W., and Peercy, P.S., eds.). Academic Press, New York, 20–36.

Brueck, S.R.J., Deutsch, T.F., and Osgood, R.M., Jr. (1979). Vibrational kinetics of SF_6 dissolved in simple cryogenic liquids. *Chem. Phys. Lett.* **60,** 242–246.

Brus, L.E., and Comas, J. (1971). Chemisorptive luminescence: Oxygen on Si(111) surfaces. *J. Chem. Phys.* **54,** 2771–2776.

Brus, L.E. (1980). Application of classical electromagnetic theory to an understanding of molecular vibrational energy transfer into metal surfaces. *J. Chem. Phys.* **73,** 940–945.

Brzhazovskii, Y.V., Kusner, Y.S., Rebrov, A.K., Troshin, B.I., and Chebotaev, V.P. (1976). *JETP Lett.* **23,** 260–262.

Burgess, D.R., Jr., Viswanathan, R., Hussla, I., Stair, P.C., and Weitz, E. (1983). Pulsed laser-induced thermal desorption of CO from copper surfaces. *J. Chem. Phys.* **79,** 5200–5202.

Campion, A., Gallo, A.R., Harris, C.B., Robota, H.J., and Whitmore, P.M. (1980). Electronic energy transfer to metal surfaces: A test of classical image dipole theory at short distances. *Chem. Phys. Lett.* **73,** 447–450.

Casassa, M.P., Celii, F.G., and Janda, K.C. (1982). Photodissociation and photodesorption line shapes. *J. Chem. Phys.* **76,** 5295–5302.

Celii, F.G., Casassa, M.P., and Janda, K.C. (1984). Photodesorption of weakly bound molecules. *Surf. Sci.* **141,** 169–190.

Chance, R.R., Prock, A., and Silbey, R. (1978). Molecular fluorescence and energy transfer near interfaces. *Adv. Chem. Phys.* **37,** 1–65.

Chen, B.-H., Close, J.S., and White, J.M. (1977). The role of ultraviolet radiation in promoting the palladium-catalyzed oxidation of carbon monoxide. *J. Catal.* **46,** 253–258.

Christmann, K., Shobes, O., Ertl, G., and Neumann, M. (1974). Adsorption of hydrogen on nickel single crystal surfaces. *J. Chem. Phys.* **60,** 4528–4540.

Chuang, T.J. (1979). Infrared chemiluminescence from XeF_2-silicon surface reactions. *Phys. Rev. Lett.* **42,** 815–817.

Chuang, T.J., and Winters, H.F. (1979). Unpublished.

Chuang, T.J. (1980a). Infrared chemiluminescence from reactions at metal and semiconductor surfaces. In *Proceedings of the 4th International Conference on Solid Surfaces and the 3rd European Conference on Surface Science* (Degras, D.A., and Costa, M., eds.). Sociéte Francaise du Vide, Paris, France, Vol. 1, 486–489.

Chuang, T.J. (1980b). Infrared laser-induced reaction of SF_6 with silicon surfaces. *J. Chem. Phys.* **72,** 6303–6304.

Chuang, T.J. (1980c). Electron spectroscopy study of silicon surfaces exposed to XeF_2 and the chemisorption of SiF_4 on silicon. *J. Appl. Phys.* **51,** 2614–2619.

Chuang, T.J. (1981a). Multiple photon-excited SF_6 interaction with silicon surfaces. *J. Chem. Phys.* **74,** 1453–1460.

Chuang, T.J. (1981b). Infrared laser radiation effects on XeF_2 interaction with silicon. *J. Chem. Phys.* **74,** 1461–1466.

Chuang, T.J. (1982a). Vibrational activation and surface reactivity: SF_6 interaction with silicon induced by infrared laser radiation. In *Vibrations at Surfaces* (Caudano, R., Gilles, J.-M. and Lucas, A.A.). Plenum Press, New York, 573–577.

Chuang, T.J. (1982b). Infrared laser stimulated desorption of pyridine from KCl surfaces. *J. Chem. Phys.* **76,** 3828–3829.

Chuang, T.J. (1982c). Laser-enhanced chemical etching of solid surfaces. *IBM J. Res. Develop.* **26,** 145–150.

Chuang, T.J., and Seki, H. (1982d). Resonantly stimulated desorption of pyridine from silver surfaces by polarized infrared laser radiation. *Phys. Rev. Lett.* **49,** 382–385.

Chuang, T.J. (1983a). Infrared laser-stimulated surface processes. *J. Electro. Spectr. Relat. Phenom.* **29,** 125–138.

Chuang, T.J. (1983b). Laser-induced gas-surface interactions. *Surf. Sci. Reports* **3,** 1–105.

Chuang, T.J., Coufal, H., and Trager, F. (1983c). Infrared laser photoacoustic spectroscopy of adsorbed species. *J. Vac. Sci. Technol.* **A1,** 1236–1239.

Chuang, T.J., and Hussla, I. (1984a). Time-resolved mass spectrometric study on infrared laser photodesorption of ammonia from Cu(100). *Phys. Rev. Lett.* **52,** 2045–2048.

Chuang, T.J., Hussla, I., and Sesselman, W. (1984b). Laser-assisted chemical etching of inorganic materials. In *Laser Processing and Diagnostics* (Bäuerle, D., ed.). Springer, Heidelberg, 300–314.

Chuang, T.J., and Hussla, I. (1984c). Molecule-surface interactions stimulated by laser radiation. In *Dynamics on Surfaces* (Pullman, B., Jortner, J., Nitzan, A., and Gerber, B., eds.). D. Reidel, Dordrecht, 313–327.

Chuang, T.J., Seki, H., and Hussla, I. (1985). Infrared photodesorption: Vibrational excitation and energy transfer processes on surfaces. *Surf. Sci.* **158,** 525–552.

Chuang, T.J. (1986). Laser-induced molecular processes on surfaces. *Surf. Sci.* **178,** 763–786.

Cottrell, T.L. (1958). *The Strengths of Chemical Bonds,* Butterworths, London, 2nd edition, p. 252.

Cowin, J.P., Auerbach, D.J., Becker, C., and Wharton, L. (1978). Measurement of fast desorption kinetics of D_2 from tungsten by laser-induced thermal desorption. *Surf. Sci.* **78,** 545–564.

De, G.S., and Landman, U. (1980). Microscopic theory of thermal desorption and dissociation processes catalyzed by a solid surface. *Phys. Rev.* **B21,** 3256–3268.

Demuth, J.E., and Avouris, Ph. (1981). Lifetime broadening of excited pyrazine adsorbed on Ag(111). *Phys. Rev. Lett.* **47,** 61–63.

Deutsch, T.F., Ehrlich, D.J., and Osgood, R.M., Jr. (1979). Laser photodeposition of metal films with microscopic features. *Appl. Phys. Lett.* **35,** 175–177.

Djidjoev, M.S., Khoklov, R.V., Kiselev, A.V., Lygin, V.I., Namiot, V.A., Osipov, A.I., Panchenko, V.I., and Provotorov, B.I. (1976). Isotope separation and laser driven chemical reactions. In *Tunable Lasers and Applications* (Mooradian, A., Jaeger, T., and Stokseth, P., eds.). Springer, Berlin, 100–107.

Drexhage, K.H. (1970). Influence of dielectric interface on fluorescence decay time. *J. Lumin.* **1/2,** 693–701.

Ehrlich, D.J., and Osgood, R.M., Jr. (1981). UV photolysis of van der Waals molecular films. *Chem. Phys. Lett.* **79,** 381–388.

Ehrlich, D.J., Osgood, R.M., Jr., and Deutsch, T.F. (1981). Spatially delineated growth of metal films via photochemical prenucleation. *Appl. Phys. Lett.* **38,** 946–948.

Fain, B., and Lin, S.H. (1985). Effect of vibrational energy transfer on laser-induced desorption. *Chem. Phys. Lett.* **114,** 497–502.

Farneth, W.E., Zimmermann, P.G., Hogenkamp, D.J., and Kennedy, S.D. (1983). Infrared laser induced heterogeneous reactions: 2-propanol with cupric oxide. *J. Am. Chem. Soc.* **105,** 1126–1129.

Feibelman, P.J., and Knotek, M.L. (1978). Reinterpretation of electron-stimulated desorption data from chemisorption systems. *Phys. Rev.* **B18,** 6531–6539.

Fletcher, R.A., Chabay, I., Weitz, D.A., and Chung, J.C. (1984). Laser desorption mass spectrometry of surface-adsorbed molecules. *Chem. Phys. Lett.* **104,** 615–619.

Gangwer, T.E., and Goldstein, M.K. (1976). The production of heavy water by photodesorption. In *SPIE Proceedings of Industrial Applications of High Power Laser Technology,* Vol. 86, 154–159.

Gauthier, R., and Guittard, C. (1976). Mechanism investigations of a pulsed laser light induced desorption. *Phys. Status Solidi* **A38,** 447–486.

George, S.M., DeSantolo, A.M., and Hall, R.B. (1985). Surface diffusion of hydrogen on Ni(100) studied using laser-induced thermal desorption. *Surf. Sci.* **159,** L425-L432.

George, T.F., Lin, J., Lam, K.-S., and Chang, C. (1980). Theory of the interaction of laser radiation with molecular dynamical processes occurring at a solid surface. *Opt. Engin.* **19,** 100–112.

Gerber, R.B., Beard, L.H., and Kouri, D.J. (1981). Vibrational deactivation of diatomic molecules by collisions with solid surfaces. *J. Chem. Phys.* **74,** 4709–4725.

Gersten, J. (1980). The effect of surface roughness on surface enhanced Raman scattering, *J. Chem. Phys.* **72,** 5779–5780.

Gersten, J., and Nitzan, A. (1980). Electromagnetic theory of enhanced Raman scattering by molecules adsorbed on rough surfaces. *J. Chem. Phys.* **73,** 3023–3037.

Gersten, J., and Nitzan, A. (1981). Spectroscopic properties of molecules interacting with small dielectric particles. *J. Chem. Phys.* **75,** 1139–1152.

Gochelashvili, K.S., Karlov, N.V., Ovchenkov, A.I., Orlov, A.N., Petrov, R.P., Petrov, Yu.N., and Prokhorov, A.M. (1976). Methods for selective heterogeneous separation of vibrationally excited molecules. *Sov. Phys. JETP* **43,** 274–277.

Goodman, A.M. (1966). Photoemission of electrons from n-type degenerate silicon into silicon dioxide. *Phys. Rev.* **152,** 785–787.

Gortel, Z.W., Kreuzer, H.J., Piercy, P., and Teshima, R. (1983a). Theory of photodesorption of molecules by resonant laser-molecular vibrational coupling. *Phys. Rev.* **B27,** 5066–5083.

Gortel, Z.W., Kreuzer, H.J., Piercy, P., and Teshima, R. (1983b). Resonant heating in photodesorption via laser-adsorbate coupling. *Phys. Rev.* **B28,** 2119–2124.

Gray, D.E. (1972), ed. *American Institute of Physics Handbook,* 3rd ed. McGraw-Hill, New York, Chap. 6.

Greenler, R.G. (1977). Light emitted from molecules adsorbed on a metal surface. *Surf. Sci.* **69,** 647–652.

Hall, R.B., and DeSantolo, A.M. (1984). Pulsed laser induced excitation of metal surfaces: Application as a probe of surface reaction kinetics of methanol on Ni. *Surf. Sci.* **137,** 421–441.

Hall, R.B., DeSantolo, A.M., and Bares, S.J. (1985). Time-resolved measurements of methanol decomposition on Ni(100) utilizing laser induced desorption. *Surf. Sci.* **161,** L533–L542.

Halpern, B., and Rosner, D.E. (1978). Chemical energy accommodation at catalyst surfaces: Flow reactor studies of the association of nitrogen atoms on metals at high temperatures. *J. Chem. Soc. Faraday Trans.* **I74,** 1883–1912.

Hanabusa, M., Namiki, A., and Yoshihara, K. (1979). Laser-induced vapor deposition of silicon. *Appl. Phys. Lett.* **35,** 626–627.

Hanabusa, M., Suzuki, M., and Nishigaki, S. (1981). Dynamic of laser-induced vaporization for ultrafast deposition of amorphous silicon films. *Appl. Phys. Lett.* **38,** 385–387.

Heidberg, J., Stein, H., Riehl, E., and Nestmann, A., (1980). Evaporation and desorption by resonant excitation of molecular normal vibrations with laser infrared. *Z. Physik. Chem. N.F.* **121,** 145–164.

Heidberg, J., Stein, H., and Riehl, E. (1982). Resonance, rate, and quantum yield of infrared-laser-induced desorption by multiquantum vibrational excitation of the adsorbate CH_3F on NaCl. *Phys. Rev. Lett.* **49,** 666–669.

Heidberg, J., Stein, H., Riehl, E., Szilagyi, Z., and Weiss, H. (1985). Vibration predesorption. *Surf. Sci.* **158,** 553–578.

Heilweil, E.J., Casassa, M.P., Cavanagh, R.R., and Stephenson, J.C. (1985a). Vibrational deactivation of surface OH chemisorbed on SiO_2: Solvent effects. *J. Chem. Phys.* **82,** 5216–5231.

Heilweil, E.J., Casassa, M.P., Cavanagh, R.R., and Stephenson, J.C. (1985b). Time-resolved vibrational energy relaxation of surface adsorbates. *J. Vac. Sci. Technol.* **B3,** 1471–1473.

Hemminger, J.C., Carr, R., Lo, W.J., and Somorjai, G.A. (1980). The adsorption and reactions of gaseous CO_2 and H_2O on Pt-$SrTiO_3$ single-crystal sandwiches. In *Interfacial Photoprocesses: Energy Conversion and Synthesis* (Wrighton, M.S., ed.). Adv. Chem. Series, No. 184, American Chemical Society, Washington, D.C., 233–252.

Herzberg, G. (1950). *Spectra of Diatomic Molecules,* 2nd ed., D. Van Nostrand, Princeton, New Jersey.

Horn, K., and Pritchard, J. (1976). Infrared spectrum of CO chemisorbed on Cu(100). *Surf. Sci.* **55,** 701–704.

Houle, F.A. (1983). Non-thermal effects in laser-enhanced etching of silicon by XeF_2. *Chem. Phys. Lett.* **95,** 5–8.

Hussla, I., and Chuang, T.J. (1985). CO_2 laser induced photodesorption of physisorbed ammonia from Cu(100) single crystal. *Ber. Bunsenges. Phys. Chem.* **89,** 294–297.

Hussla, I., Seki, H., Chuang, T.J., Gortel, Z.W., Kreuzer, H.J., and Piercy, P. (1985). Infrared laser-induced photodesorption of NH_3 and ND_3 adsorbed on Cu(100) and Ag (film). *Phys. Rev.* **B32,** 3489–3501.

Hussla, I., Coufal, H., Träger, F., and Chuang, T.J. (1986a). Surface temperature measurement during pulsed laser-induced thermal desorption of xenon from copper film. *Can. J. Phys.* **64,** 1070–1073.

Hussla, I., Coufal, H., Träger, F., and Chuang, T.J. (1986b). Pulsed laser-induced thermal desorption of xenon. *Ber. Bunsenges. Phys. Chem.* **90,** 240–245.

Itoh, N. (1982). Mechanism of electron-excitation-induced defect creation in alkali halides. *Radiation Effects* **64,** 161–169.

Itoh, N., and Nakayama, T. (1982). Mechanism of neutral particle emission from electron-hole plasma near solid surface. *Phys. Lett.* **92A,** 471–475.

Jedrzejek, C. (1985). Selective laser-stimulated desorption of molecules by internal vibrational excitation. *J. Vac. Sci. Technol.* **B3,** 1431–1435.

Jedrzejek, C., Freed, K.F., Efrima, S., and Metiu, H. (1981). A one-dimensional microscope quantum mechanical theory of light-enhanced desorption. *Surf. Sci.* **109,** 191–206.

Jelend, W., and Menzel, D. (1973). Deuterium isotope effect in electron impact desorption of hydrogen on tungsten. *Chem. Phys. Lett.* **21,** 178–180.

Kasemo, B. (1974). Photon emission during chemisorption of oxygen on Al and Mg surfaces. *Phys. Rev. Lett.* **32,** 1114–1117.

Kasemo, B., and Wallden, L. (1975). Photon and electron emission during halogen adsorption on sodium. *Surf. Sci.* **53,** 393–407.

Kawai, T., and Sakata, T. (1980). Dynamics of photo-induced surface reactions on semiconductors studied by a pulsed-laser-dynamics-mass-spectrometer technique. *Chem. Phys. Lett.* **69,** 33–36.

Khmelev, A.V., Apollonov, V.V., Borman, V.D., Nikolaev, B.I., Sazykin, A.A., Troyan, U.I., Firsov, K.N., and Frolov, B.A. (1977). Stimulation of a heterogeneous reaction of decomposition of ammonia on the surface of platinum by CO_2 laser radiation. *Sov. J. Quantum Electron.* **7,** 1302–1305.

Knebe, M., and Wolfrum, J. (1980). Biomolecular reactions of vibrationally excited molecules. *Ann. Rev. Phys. Chem.* **31,** 47–79.

Knotek, M.L., and Feibelman, P.J. (1978). Ion desorption by core-hole Auger decay. *Phys. Rev. Lett.* **40,** 964–967.

Koel, B.E., White, J.M., Erskine, J.L., and Antoniewicz, P.R. (1980). Photoeffects on reactions over transition metals. In *Interfacial Photoprocesses: Energy Conversion and Synthesis* (Wrighton, M.S., ed.). Adv. Chem. Series, No. 184 American Chemical Society, Washington D.C., 27–45.

Korzeniewski, G.E., Hood, E., and Metiu, H. (1982a). A one-dimensional model for the study of the influence of heat, sound, photons and electron-hole pairs on the rate of desorption. *J. Vac. Sci. Technol.* **20,** 594–599.

Korzeniewski, G.E., Maniv, T., and Metiu, H. (1982b). Electrodynamics at metal surfaces. IV. The electric fields caused by the polarization of a metal surface by an oscillating dipole. *J. Chem. Phys.* **76,** 1564–1573.

Kozhushner, M.A., Kustarev, V.G., and Shub, B.R. (1979). Heterogeneous relaxation of molecule vibrational energy on metals. *Surf. Sci.* **81,** 261–272.

Kreuzer, H.J. (1980). Quantum statistical theory of adsorption and desorption of gas at a solid surface. *Surf. Sci.* **100,** 178–198.

Kreuzer, H.J., and Gortel, Z.W. (1984). Time-of-flight spectra in photodesorption via laser-adsorbate coupling. *Phys. Rev.* **B29,** 6926–6931.

Kreuzer, H.J., and Lowy, D.N. (1981). Photodesorption of diatomic molecules by laser-molecular vibrational coupling. *Chem. Phys. Lett.* **78,** 50–53.

Kronauer, P., and Menzel, D. (1972). Photodesorption of carbon monoxide from tungsten.

In *Adsorption-Desorption Phenomena* (Ricca, F., ed.). Academic Press, New York, 313–328.

Kuhn, H. (1970). Classical aspects of energy transfer in molecular systems. *J. Chem. Phys.* **53,** 101–108.

Lee, R.N., and Farnsworth, H.E. (1965). Leed studies of adsorption on clean (100) copper surfaces. *Surf. Sci.* **3,** 461–479.

Legay, F. (1977). Vibrational relaxation in matrices. In *Chemical and Biochemical Applications of Lasers* (Moore, C.B., ed.). Academic Press, New York, Vol. 2, 43–86.

Legay-Sommaire, N., and Legay, F. (1980). Observation of a strong vibrational population inversion by CO laser excitation of pure solid carbon monoxide. *IEEE J. Quantum Electr.* **QE-16,** 308–314.

Lehwald, S., Ibach, H., and Steininger, H. (1982). Overtones and multiphonon processes in vibration spectra of adsorbed molecules. *Surf. Sci.* **117,** 342–351.

Letokhov, V.S., and Moore, C.B. (1977). Laser isotope separation. In *Chemical and Biochemical Applictions of Lasers* (Moore, C.B., ed.). Academic Press, New York, Vol. 3, 1-165.

Letokhov, V.S., Movshev, V.G., and Chekalin, S.V. (1981). Nonthermal desorption of molecular ions of polyatomic molecules induced by UV laser radiation. *Sov. Phys. JETP* **54,** 257–260.

Leung, C., Steinbruchel, Ch., and Gomer, R. (1977). Isotopic effects in electron impact desorption of CO and O_2 adsorbed on the (110) plane of tungsten. *Appl. Phys.* **14,** 79–87.

Levine, L.P., Ready, J.F., and Bernal, E. (1967). Gas desorption produced by a giant pulsed laser. *J. Appl. Phys.* **38,** 331–336.

Lichtman, D. (1979). Mechanisms of desorption due to electrons or photons. *Surf. Sci.* **90,** 579–587.

Lichtman, D., and Shapira, Y. (1978). Photodesorption: A critical review. *CRC Crit. Rev. Solid State Mater. Sci.* **8,** 93–118.

Lin, C.T., Atvars, T.D.Z., and Pessine, F.B.T. (1977). Laser isotopic enrichment of boron using catalysis. *J. Appl. Phys.* **48,** 1720–1721.

Lin, C.T., and Atvars, T.D.Z. (1978). The role of a catalyst in the isotopically excited laser photochemistry. *J. Chem. Phys.* **68,** 4233–247.

Lin, J., and George, T.F. (1980a). Quantum-stochastic approach to laser-stimulated desorption dynamics and population distribution of chemisorbed species on solid surfaces. *J. Chem. Phys.* **72,** 2554–2569.

Lin, J., and George, T.F. (1980b). Dynamical model of selective versus nonselective laser-stimulated surface processes. *Surf. Sci.* **100,** 381–387.

Lin, J., and George, T.F. (1980c). Dynamical model of selective versus nonselective laser-stimulated surface processes. *J. Phys. Chem.* **84,** 2957–2968.

Lin, J., and George, T.F. (1982). On the synergistic effects of laser/phonon-stimulated processes: A master equation approach. *Surf. Sci.* **115,** 569–575.

Lucas, D., and Ewing, G.E. (1981). Spontaneous desorption of vibrationally excited molecules physically-adsorbed on surfaces. *Chem. Phys.* **58,** 385–393.

Lucchese, R.R., and Tully, J.C. (1984a). Trajectory studies of vibrational energy transfer in gas-surface collisions. *J. Chem. Phys.* **80,** 3451–3462.

Lucchese, R.R., and Tully, J.C. (1984b). Laser-induced thermal desorption from surfaces. *J. Chem. Phys.* **81,** 6313–6319.

Lundquist, B.I., Gunnarsson, O., Hjelmberg, H., and Norskov, J.K. (1979). Theoretical description of molecules-metal interaction and surface reactions. *Surf. Sci.* **89,** 196–225.

Madey, T.E. (1981). The use of angle-resolved electron and photon stimulated desorption for surface structural studies. In *Inelastic Particle-Surface Collisions* (Taglauer, E., and Heiland, W., eds.). Springer, Berlin, 80-103.

Madey, T.E., Yates, J.T., Jr., King, D.A., and Uhlanger, C.J. (1970). Isotopic effect in electron stimulated desorption: Oxygen chemisorbed on tungsten. *J. Chem. Phys.* **52,** 5215–5220.

Mantell, D.A., Ryali, S.B., Halpern, B.L., Haller, G.L., and Fenn, J.B. (1981). The exciting oxidation of CO on Pt. *Chem. Phys. Lett.* **81,** 185–187.

McCall, S.L., and Platzman, P.M. (1980). Surface enhanced Raman scattering. *Phys. Lett.* **77A,** 381–383.

McCarrol, B. (1969). Chemisorptive luminescence. *J. Chem. Phys.* **50,** 4758–4765.

Menzel, D., and Gomer, R. (1964). Desorption from metal surfaces by low-energy electrons. *J. Chem. Phys.* **41,** 3311–3328.

Menzel, D. (1982). Recent developments in electron and photon stimulated desorption. *J. Vac. Sci. Technol.* **20,** 538–543.

Mimeault, V.J., and Hansen, R.S. (1966). Nitrogen adsorption on iridium and rhodium. *J. Phys. Chem.* **70,** 3001–3003.

Misewich, J., Plum, C.N., Blyholder, G., Houston, P.L., and Merrill, R.P. (1983). Vibrational relaxation during gas-surface collisions. *J. Chem. Phys.* **78,** 4245–4249.

Misewich, J., Zacharias, H., and Loy, M.M.T. (1985). State-to-state molecular beam scattering of vibrationally excited NO from cleaved LiF(100) surfaces. *Phys. Rev. Lett.* **55,** 1919–1922.

Morawitz, H. (1969). Self-coupling by a two-level system by a mirror. *Phys. Rev.* **187,** 1792–1796.

Morawitz, H., and Philpott, M.R. (1974). Coupling of an excited molecule to surface plasmons. *Phys. Rev.* **B10,** 4863–4868.

Morrison, S.R. (1977). *The Chemical Physics of Surfaces.* Plenum Press, New York.

Nakayama, T., Okigawa, M., and Itoh, N. (1984). Laser-induced sputtering of oxides and compound semiconductors. *Nucl. Instr. Meth. Phys. Res.* **B1,** 301–306.

Nishi, N., Shinohara, H., and Okuyama, T. (1984). Photodetachment, photodissociation, and photochemistry of surface molecules of icy solids containing NH_3 and pure H_2O ices. *J. Chem. Phys.* **80,** 3898–3910.

Nitzan, A., and Brus, L.E. (1981a). Can photochemistry be enhanced on rough surfaces? *J. Chem. Phys.* **74,** 5321–5322.

Nitzan, A., and Brus, L.E. (1981b). Theoretical model for enhanced photochemistry on rough surfaces. *J. Chem. Phys.* **75,** 2205–2214.

Norskov, J.K., Newns, D.M., and Lundquist, B.I. (1979). Molecular orbital description of surface chemiluminescence. *Surf. Sci.* **80,** 179–188.

Persson, B.N.J. (1978). Theory of the damping of excited molecules located above a metal surface. *J. Phys.* **C11,** 4251–4269.

Persson, B.N.J., and Persson, M. (1980a). Vibrational lifetime for CO adsorbed on Cu(100). *Solid State Commun.* **36,** 175–179.

Persson, B.N.J., and Persson, M. (1980b). Damping of vibrations in molecules adsorbed on a metal surface. *Surf. Sci.* **97,** 609–624.

Persson, B.N.J., and Ryberg, R. (1980). Collective vibrational modes in isotopic mixtures of CO adsorbed on Cu(100). *Solid State Commun.* **36,** 613–617.

Persson, B.N.J., and Ryberg, R. (1981). Vibrational interaction between molecules adsorbed on a metal surface: The dipole-dipole interaction. *Phys. Rev.* **B24,** 6954–6970.

Petro, W.G., Hino, I., Eglash, S., Lindau, I., Su, C.Y., and Spicer, W.E. (1982). Effect of low-intensity laser radiation during oxidation of the GaAs(110) surface. *J. Vac. Sci. Technol.* **21,** 405–408.

Philpott, M.R. (1975). Effect of surface plasmons on transitions in molecules. *J. Chem. Phys.* **62,** 1812–1817.

Ready, J.F. (1971). *Effects of High-Power Laser Radiation.* Academic Press, New York.

Reddy, K.V., and Berry, M.J. (1979). A nonstatistical unimolecular chemical reaction: Isomerization of state-selected allyl isocyanide. *Chem. Phys. Lett.* **66,** 223–229.

Reddy, K.V., Heller, D.F., and Berry, M.J. (1982). Highly vibrationally-excited benzene: Overtone spectroscopy and intramolecular dynamics of C_6H_6, C_6D_6 and partially deuterated or substituted benzenes. *J. Chem. Phys.* **76,** 2814–2837.

Redhead, P.A. (1964). Interaction of slow electrons with chemisorbed oxygen. *Can. J. Phys.* **42,** 886–905.

Rossetti, R., and Brus, L.E. (1980). Time resolved molecular electronic energy transfer into a silver surface. *J. Chem. Phys.* **73,** 572–577.

Ryberg, R. (1982). Intrinsic linewidth of CO adsorbed on Cu(100). In *Vibrations at Surfaces* (Caudano, R., Gilles, J.-M., and Lucas, A.A., eds.). Plenum Press, New York, 309–313.

Schäfer, B., and Hess, P. (1984). Measurement of time-of-flight distributions for wavelength-dependent IR laser-stimulated desorption. *Chem. Phys. Lett.* **105,** 563–566.

Schäfer, B., and Hess, P. (1985). Time-of-flight diagnostics of wavelength-dependent CO_2 laser-induced desorption from condensed layers. *Appl. Phys.* **B37,** 197–204.

Schafer, S.A., and Lyon, S.A. (1981). Optically enhanced oxidation of semiconductors. *J. Vac. Sci. Technol.* **19,** 494–497.

Schafer, S.A., and Lyon, S.A. (1982). Wavelength dependence of laser-enhanced oxidation of silicon. *J. Vac. Sci. Technol.* **21,** 422–425.

Seki, H., and Chuang, T.J. (1982). The detection by SERS of resonantly excited desorption of pyridine from silver island films by IR laser radiation. *Solid State Commun.* **44,** 473–475.

Sexton, B.A. (1979). Vibrational spectrum of carbon monoxide chemisorbed on a copper(100) surface. *Chem. Phys. Lett.* **63,** 451–454.

Shapira, Y., Cox, S.M., and Lichtman, D. (1975). Photodesorption from powdered zinc oxide. *Surf. Sci.* **50,** 503–514.

Shapira, Y., McQuistan, R.B., and Lichtman, D. (1977). Relationship between photodesorption and surface conductivity in ZnO. *Phys. Rev.* **B15,** 2163–2169.

Slutsky, M.S., and George, T.F. (1978). Quantum theory of laser-stimulated desorption. *Chem. Phys. Lett.* **57,** 474–476.

Slutsky, M.S., and George, T.F. (1979). Laser-stimulated migration of adsorbed atoms on solid surface, *J. Chem. Phys.* **70,** 1231–1235.

Steinfeld, J.I. (1981), ed., *Laser-Induced Chemical Processes.* Plenum Press, New York.

Stewart, C.N., and Ehrlich, G. (1975). Dynamics of activated chemisorption: Methane on rhodium. *J. Chem. Phys.* **62,** 4672–4682.

Tamir, M., and Levine, R.D. (1977). The multiphoton collisionless dissociation of polyatomic molecules: An intramolecular amplification mechanism. *Chem. Phys. Lett.* **46,** 208–214.

Tagle, J.A., and Pospieszczyk, A. (1983). Measurement of desorption energies of H_2 and CO from SS-304 LN and Inconel 625 surfaces by laser-induced thermal desorption. *Appl. Surf. Sci.* **17,** 189–206.

Thorman, R.P., Anderson, D., and Bernasek, S.L. (1980). Internal energy of heterogeneous reaction products: Nitrogen-atom recombination on iron. *Phys. Rev. Lett.* **44,** 743–746.

Tibbetts, G.G. (1979). Electronically activated chemisorption of nitrogen on a copper(100) surface. *J. Chem. Phys.* **70,** 3600–3603.

Townsend, P.D. (1979). Photon-induced sputtering. *Surf. Sci.* **90,** 256–264.

Träger, F., Coufal, H., and Chuang, T.J. (1982). Infrared laser photoacoustic spectroscopy for surface studies: SF_6 adsorption on silver surfaces. *Phys. Rev. Lett.* **49,** 1720–1723.

Tsong, T.T. (1979). Field penetration and band bending near semiconductor surfaces in high electric fields. *Surf. Sci.* **81,** 28–42.

Tu, Y.Y., Chuang, T.J., and Winters, H.F. (1981). The chemical sputtering of fluorinated silicon. *Phys. Rev.* **B23,** 823–835.

Tully, J.C. (1980). Dynamics of gas-surface interactions: Reaction of atomic oxygen with adsorbed carbon on platinum. *J. Chem. Phys.* **73,** 6333–6342.

Tully, J.C. (1981). Dynamics of gas-surface interactions: Thermal desorption of Ar and Xe from platinum. *Surf. Sci.* **111,** 461–478.

Umstead, M.E., and Lin, M.C. (1978). Effect of laser radiation on the catalytic decomposition of formic acid on platinum. *J. Phys. Chem.* **82,** 2047–2048.

Umstead, M.E., Talley, L.D., Tevault, D.E., and Lin, M.C. (1980). Laser applications to heterogeneous catalysis: Reactant excitation and product diagnostics. *Opt. Engin.* **19,** 94–99.

Van Hieu, N., and Lichtman, D. (1981a). Band gap radiation-induced photodesorption from titanium oxide surfaces. *Surf. Sci.* **103,** 535–541.

Van Hieu, N., and Lichtman, D. (1981b). Band gap radiation-induced photodesorption from V_2O_5 powder and vanadium oxide surface. *J. Vac. Sci. Technol.* **18,** 49–53.

Van Vechten, J.A. (1980). Evidence for and nature of a nonthermal mechanism of pulsed laser annealing of Si. In *Laser and Electron Beam Processing of Materials* (White, C.W., and Peercy, P.S., eds.). Academic Press, New York, 53–58.

Viswanathan, R., Burgess, D.R., Jr., Stair, P.C., and Weitz, E. (1982). Summary abstract: Laser flash desorption of CO from clean copper surface. *J. Vac. Sci. Technol.* **20,** 605–606.

Volkenshtein, F.F., and Nagaev, V.B. (1973). Electron theory of photocatalytic reactions on semiconductors. *Kinet. Catal.* **14,** 1291–1296.

Volkenshtein, F.F., and Nagaev, V.B. (1975). The electron theory of photocatalytic reactions on semiconductors. *Kinet. Catal.* **16,** 320–325.

Von Allmen, M. (1980). Coupling of beam energy to solids. In *Laser and Electron Beam Processing of Materials,* (White, C.W., and Peercy, P.S., eds.). Academic Press, New York, 6–19.

Wautelet, M. (1980). Solid surfaces under laser irradiation. *Surf. Sci.* **95,** 299–308.

Wedler, G., and Ruhmann, H. (1982). Laser-induced thermal desorption of carbon monoxide from Fe(110) surfaces. *Surf. Sci.* **121,** 464–486.

Weitz, E., and Flynn, G. (1975). Laser studies of vibrational and rotational relaxation in small molecules. *Ann. Rev. Phys. Chem.* **25,** 275–315.

Whitmore, P.M., Robota, H.J., and Harris, C.B. (1982). Mechanisms for electronic energy transfer between molecules and metal surfaces: A comparison of silver and nickel. *J. Chem. Phys.* **77,** 1560–1568.

Winters, H.F. (1975). The activated dissociative chemisorption of methane on tungsten. *J. Chem. Phys.* **62,** 2454–2460.

Winters, H.F. (1976). The kinetic isotope effect in the dissociative chemisorption of methane. *J. Chem. Phys.* **64,** 3495–3500.

Winters, H.F., and Coburn, J.W. (1979). The etching of silicon with XeF_2 vapor. *Appl. Phys. Lett.* **34,** 70–73.

Winters, H.F., and Houle, F.A. (1983). Gaseous products from the reaction of XeF_2 with silicon. *J. Appl. Phys.* **54,** 1218–1223.

Winters, H.F., Coburn, J.W., and Chuang, T.J. (1983). Surface processes in plasma-assisted etching environment. *J. Vac. Sci. Technol.* **B1,** 469–480.

Wu, G.S., Fain, B., Ziv, A.R., and Lin, S.H. (1984). Theoretical studies of laser-stimulated surface processes. *Surf. Sci.* **147,** 537–554.

Yablonovitch, E. (1980). Infrared laser chemistry. In *Relaxation of Elementary Excitations* (Kubo, R., and Hanamura, E., eds.). Springer, New York, 197–205.

Yates, J.T., Jr., Zinck, J.J., Sheard, S., and Weinberg, W.H. (1979). Search for vibrational activation in the chemisorption of methane. *J. Chem. Phys.* **70,** 2266–2272.

Young, E.M., and Tiller, W.A. (1983). Photon-enhanced oxidation of silicon. *Appl. Phys. Lett.* **42,** 63–65.

Zacharias, H., Loy, M.M.T., and Roland, P.A. (1982). Scattering of vibrationally excited NO off LiF and CaF_2 surfaces. *Phys. Rev. Lett.* **49,** 1790–1794.

Zingerman, Ya.P., and Ishchuck, V.A. (1968). Mechanism of electron-stimulated desorption of oxygen from the surface of tungsten. *Sov. Phys. Solid State* **9,** 2638–2640.

CHAPTER 3

Spectroscopy and Photochemistry of Gases, Adsorbates, and Liquids*

M. ROTHSCHILD
Lincoln Laboratory
Massachusetts Institute of Technology
Lexington, Massachusetts

1. Introduction

The photochemical properties of ambients in their gaseous, adsorbed, or liquid phases play a prominent role in any laser-induced thin film microfabrication process. The effectiveness of the laser in initiating and sustaining deposition or etching reactions, for instance, is frequently dominated by the interaction of the ambient molecules with the incident photons. The design of a successful laser-induced process will therefore hinge on a judicious selection process that matches the appropriate ambient molcule with the right laser. This match is frequently limited by

* This work was sponsored by the Defense Advanced Research Projects Agency and by the Deparment of the Air Force, in part under a specific program supported by the Air Force Office of Scientific Research.

0-12-233430-2

the second element in this match, namely, the laser. There is a large variety of commercially available lasers, having a wide range of properties, such as mode of operation (cw or pulsed), wavelength, power, and beam size. Unfortunately, the largest selection is in the visible range of the spectrum where few molecules—with some notable exceptions—exhibit a significant degree of photochemical activity. At shorter wavelengths a larger number of molecules have strong photoabsorption, but at the same time there are fewer available photon sources in the UV. The most notable of these are the pulsed excimer lasers operating most powerfully at 193, 248, 308, and 351 nm; and harmonics of visible lasers, including the second harmonic of the 514.5-nm Ar-ion laser line (257 nm) and the fourth harmonic of the Nd:YAG laser (266 nm). We therefore make explicit comments when possible about the known photochemistry at these wavelengths.

It is obvious that the choice of the system laser/chemical ambient is influenced by the absorption cross section at the laser wavelength. Moreover, even when the absorption cross section is nonnegligible, the details of the photon-molecule interaction may and may not lead to useful chemistry activity. In particular, the nature of the photon-induced electronic excitation determines whether absorption leads to molecular dissociation, and if so determines what the fragmentation pattern is. Frequently, several dissociation channels may coexist, their relative quantum yields being wavelength dependent. Furthermore, even when the primary photofragments are chemically identical, the degree and distribution of their internal excitation may vary with wavelength, and this in turn may affect their interactions at the surface. Finally, multiphoton absorption by the parent molecule, or absorption by the primary photofragments, may become important at high laser flux. Such effects lead to new photoprocesses that are not encountered in low-power, lamp-induced "classical" photochemistry.

It is the purpose of this chapter to serve as an aid to researchers and technologists interested in laser-induced thin film processes by providing basic spectroscopic and photochemical information on a large number of molecules. In view of the foregoing comments, the emphasis here is on absorption spectra; primary photoreactions including fragmentation patterns; and secondary reactions, including photoabsorption by fragments and collisional processes involving the photoproducts. For further details, including experimental and analytical techniques, the reader is urged to consult the original publications listed in the references. There are also several excellent reviews that cover some of the topics discussed here. These include Herzberg (1966), Okabe (1978) for up to five-atom non-metal-containing molecules, and Geoffroy and Wrighton (1979) for organometallics (mostly in solutions and condensed phases).

A brief comment is warranted on the boundaries that had to be put on this chapter, mainly because of space limitations. The emphasis here is on electronic spectra and photochemistry, covering the visible and UV, down to ~180 nm in most cases. There are relatively few applications at shorter wavelengths, both because of difficulties in beam propagation (such as the need to purge the lines of absorbing O_2) and because of the very limited selection of optically transmissive materials that could serve as windows in processing chambers. The list of molecules included in this compendium was limited to those compounds that a) have usable vapor pressure (>10 mtorr) at room temperature and b) are moderately small (up to ethane derivatives). The limitation on size excluded certain organic and organometallic molecules, such as metal acetylacetonates and their derivatives, which have otherwise been found to have applications in photochemical thin-film processes.

The structure of the chapter is as follows. First, the three phases—gas, adsorbate, and liquid— are considered separately. The ordering of the molecules within each phase is based on a division between molecules containing only nonmetallic atoms and those containing metal atoms. Within the subgroup of nonmetals, the listing is by column in the periodic table (from left to right) and within each column by increasing atomic number. Thus, the first are boron-containing molecules (nonmetal, column IIIa), then carbon-, silicon-, and germanium-containing molecules (nonmetals, column IVa, in order of increasing atomic number), and so on. The only exceptions are several molecules containing nitrogen and oxygen, which are discussed separately at the end of the section on gas phase molecules. The metal containing molecules are subdivided into inorganics (halides and oxyhalides), carbonyls, and alkyls. Within each of these categories, the ordering is as with nonmetals, i.e., by column in the periodic table and then by increasing atomic number. For instance, in the carbonyls the first to be discussed are those of column VIb (Cr, Mo, W), then column VIIb (Mn, Re), VIII (Co, Fe, Ni).

The absorption spectra are presented for gas-phase molecules as cross sections, in units of cm^2 $molec^{-1}$ base e, and for adsorbates and liquids as fractional absorption and transmission. Conversion factors to other units can be found in many tables, for instance, in Okabe (1978), Table A-3.

2. Gases

2.1. *Molecules Containing Boron*

2.1.1 Boron Trichloride, BCl_3

The absorption spectrum of BCl_3 at wavelengths above 200 nm consists of a structureless feature, centered at 207.6 nm and with 5.0 nm full width at

Table 3.1 The 193-nm absorption cross section of molecules containing boron.

Molecule	$\sigma_{193\,nm}$ (cm^2)	Reference
BCl_3	5.4×10^{-20}	Deutsch et al., 1981
B_2H_6	2.2×10^{-19}	Clark and Anderson, 1982
	4.3×10^{-20}	Irion and Kompa, 1982
$B(C_2H_5)_3$	4.4×10^{-19}	Ibbs and Lloyd, 1983

half maximum (Rockwood and Rabideau, 1974). The peak absorption coefficient is 4.0×10^{-19} cm^2 (see also Table 3.1).

Absorption at 207.6 nm leads to rapid photochemistry in mixtures of BCl_3 with O_2 or C_2H_4 (Rockwood and Rabideau, 1974). At 193 nm, absorption induces molecular photodissociation (Deutsch et al., 1981).

2.1.2 Diborane, B_2H_6

The absorption spectrum of B_2H_6 is a continuum, which starts at ~210 nm and increases to 105 nm, with secondary maxima at 183 and 135 nm (Irion and Kompa, 1982). The values of the absorption cross section in the 190- to 200-nm range are tabulated by Clark and Anderson (1978). They are about five times higher than those of Irion and Kompa (1982) (see also Table 3.1).

Absorption at these wavelengths leads to photodissociation. Clark and Anderson (1978) suggest that the primary photoprocess is

$$B_2H_6 + h\nu \rightarrow B_2H_5 + H \tag{3.1}$$

Irion and Kompa (1982), however, based on time-of-flight mass-spectrometric data, conclude that the primary photoprocess at 193 nm is

$$B_2H_6 + h\nu \rightarrow BH_3 + BH_3. \tag{3.2}$$

The photofragments are vibrationally and translationally excited, since the photon energy exceeds the dissociation energy by 4.7 eV. Secondary processes are recombination,

$$BH_3 + BH_3 \rightarrow B_2H_6, \tag{3.3}$$

and reaction with diborane,

$$BH_3 + B_2H_6 \rightarrow B_3H_9. \tag{3.4}$$

The recombination reaction is slightly endothermic ($E_a = -1$ kcal/mol). In the high pressure limit the rate constant for recombination is

$7 \times 10^{-11}\,cm^3\,molec^{-1}\,s^{-1}$; the rate constant for the formation of the metastable species B_3H_9 is $5 \times 10^{-14}\,cm^3\,molec^{-1}\,s^{-1}$ (Irion and Kompa, 1982).

2.1.3 Triethylboron, $B(C_2H_5)_3$

Triethylboron exhibits continuous absorption at wavelengths shorter than ~245 nm (Ibbs and Lloyd, 1983; see also Table 3.1).

2.2. Molecules Containing Carbon, Silicon, and Germanium

2.2.1 Methane, CH_4

The absorption spectrum of CH_4 is a continuum that starts at 160 nm. The dissociation energy $D(CH_3\text{-}H)$ is 4.48 eV (Okabe, 1978).

2.2.2 Acetylene, C_2H_2

Absorption of C_2H_2 starts at ~240 nm. The spectrum consists of numerous bands throughout the region of 110 to 240 nm, superimposed on a continuum that becomes prominent below 200 nm (Nakayama and Watanabe, 1964). The bands in the region of 210–240 nm show rotational structure and are assigned to the $\tilde{A}\,^1A_u \leftarrow \tilde{X}\,^1\Sigma_g^+$ transition (Herzberg, 1966), whereas those in the 150- to 200-nm region are more diffuse and belong to the $\tilde{B}\,^1B_u \leftarrow \tilde{X}\,^1\Sigma_g^+$ transition (Foo and Innes, 1973). Detailed spectroscopic measurements were performed by Okabe (1983) in the 180- to 190-nm region. At 193 nm, the absorption is assigned by McDonald et al. (1978) to the $\tilde{A} \leftarrow \tilde{X}$ transition. See also Table 3.2.

At 184.9 nm the primary photoprocess is molecular excitation,

$$C_2H_2(\tilde{X}\,^1\Sigma_g^+) + h\nu \rightarrow C_2H_2(\tilde{B}\,^1B_u), \tag{3.5}$$

followed by three unimolecular processes (Okabe, 1983):

$$C_2H_2(\tilde{B}) \rightarrow C_2H + H \tag{3.6}$$

$$C_2H_2(\tilde{B}) \rightarrow C_2(\tilde{X}\,^1\Sigma_g^+) \tag{3.7}$$

$$C_2H_2(\tilde{B}) \rightarrow C_2H_2 \text{ (metastable triplet state).} \tag{3.8}$$

The quantum yields of processes 3.6–3.8 are 0.06, 0.10, and 0.84, respectively. The predominant photoproduct, metastable C_2H_2, has a radiative lifetime of ~50 μs (Okabe, 1983). Secondary reactions with ground state C_2H_2 are

$$C_2H_2 \text{ (metastable)} + C_2H_2 \rightarrow C_2H + C_2H_3 \tag{3.9}$$

$$C_2H_2 \text{ (metastable)} + 2C_2H_2 \rightarrow C_6H_6. \tag{3.10}$$

The first process, Eq. 3.9, is dominant at low pressures, and its rate constant is $1.1 \times 10^{-13}\ cm^3\ molec^{-1}\ s^{-1}$. Formation of benzene (Eq. 3.10) becomes significant above ~10 torr. A secondary reaction involving C_2H is

$$C_2H + C_2H_2 \rightarrow C_4H_2 + H, \tag{3.11}$$

with a rate constant of $3.1 \times 10^{-11}\ cm^3\ molec^{-1}\ s^{-1}$ (Laufer and Bass, 1979).

At 193 nm the primary photoprocess is dissociation into $C_2H + H$, with most of the C_2H fragments in their electronic ground state and with their vibrational bending mode highly excited (Irion and Kompa, 1982a; Wodtke and Lee, 1985). Dissociation into $C_2 + H_2$ has a yield of less than ~0.1 (Wodtke and Lee, 1985). A secondary photoprocess, which is important at fluences of $\geq 1\ J\ cm^{-2}$, is photodissociation of the nascent C_2H (Wodtke and Lee, 1985):

$$C_2H(X\,^2\Sigma) + h\nu \rightarrow C_2(^1\Sigma_g^+, {}^3\Pi_u, {}^1\Pi_u) + H. \tag{3.12}$$

The dissociation energy $D(C_2H\text{-}H)$ is 5.6 eV (Wodtke and Lee, 1985).

2.2.3 Ethylene, C_2H_4

The absorption spectrum of C_2H_4 starts at 210 nm. From there to 175 nm it is a sequence of diffuse vibrational bands superimposed on an increasingly intense continuum (Wilkinson and Mulliken, 1955; see also Table 3.2). Absorption in this wavelength region probably corresponds to excitation into a single state (Mulliken, 1977), designated as the *V* state,

Table 3.2 The 193-nm absorption cross section of molecules containing carbon, silicon, and germanium.

Molecule	$\sigma_{193\ nm}$ (cm^2)	reference
CH_4	$<10^{-21}$	Okabe, 1978
C_2H_4	1.9×10^{-19}	McDonald et al., 1978
C_2H_4	1.5×10^{-20}	Wilkinson and Mulliken, 1955
	1.4×10^{-20}	Rothschild and Ehrlich, 1986
SiH_4	1.1×10^{-21}	Clark and Anderson, 1978
$Si(CH_3)_4$	$<10^{-21}$	Harada et al., 1968
$SiH_3(C_6H_5)$	3.0×10^{-16}	Inoue and Suzuki, 1984
	3.8×10^{-17}	Baggott et al., 1986
$Si_2(CH_3)_6$	4.6×10^{-17}	Harada et al., 1968
GeH_4	2×10^{-20}	Osmundsen et al., 1985

although it has been suggested that two states, $1\,^1B_{1u}$ and $2\,^1B_{1u}$, are involved (Buenker and Peyerimhoff, 1975). At shorter wavelengths there is a series of intense absorption lines into Rydberg states (Wilkinson and Mulliken, 1955).

The primary photoprocess at 147 nm (Sauer and Dorfman, 1961), and also with a flashlamp in the 155- to 190-nm region (Back and Griffiths, 1967), is molecular dissociation along the two pathways with almost equal yields

$$C_2H_4 + h\nu \rightarrow C_2H_2 + H_2 \tag{3.13}$$

$$C_2H_4 + h\nu \rightarrow C_2H_2 + 2H. \tag{3.14}$$

The quantum yield of H_2 was shown to be nearly constant, 0.42 ± 0.05, for excitation at 147, 163, and 185 nm, and for all pressures in the range 0.1–100 torr (Glasgow and Potzinger, 1972).

2.2.4 Silane, SiH_4

The absorption spectrum of SiH_4 is a continuum that starts at about 200 nm (Clark and Anderson, 1978), grows rapidly below 170 nm, and peaks at ~115 nm (Harada et al., 1968; Itoh et al., 1986; see also Table 3.2). The two-photon absorption cross section at 193 nm is $\sim 6 \times 10^{-44}$ cm^4 s (Fuchs et al., 1988).

2.2.5 Dichlorosilane, SiH_2Cl_2

The absorption spectrum of SiH_2Cl_2 is a continuum that starts at 180 nm. The first peak is at 151 nm, with absorption cross section 1.8×10^{-17} cm^2 (Causley and Russell, 1976).

2.2.6 Trichlorosilane, $SiHCl_3$

The absorption spectrum of $SiHCl_3$ is a continuum that starts at 170 nm and has several secondary peaks in the 110- to 140-nm range (Washida et al., 1985). Following excitation below 170 nm, fluorescence from $SiCl_2$ ($\tilde{A}\,^1B_1$) fragments is observed in the 300- to 400-nm region (Washida et al., 1985).

2.2.7 Phenylsilane, $SiH_3(C_6H_5)$

The absorption cross section of $SiH_3(C_6H_5)$ at 193 nm was determined by Inoue and Suzuki (1984) and by Baggott et al. (1986). See Table 3.2.

Two primary photodissociative channels are operative at 193 nm

(Baggott et al., 1986)

$$SiH_3(C_6H_5) + h\nu \rightarrow SiH(C_6H_5) + H_2 \tag{3.15}$$

$$SiH_3(C_6H_5) + h\nu \rightarrow SiH_2 + C_6H_6. \tag{3.16}$$

Since the relative yield of H_2 is five times higher than that of C_6H_6, Baggott et al. (1986) conclude that the first dissociative pathway is the major one. Secondary reactions of the photoproducts are:

$$SiH(C_6H_5) + SiH_3(C_6H_5) \rightarrow Si_2H_4(C_6H_5)_2 \tag{3.17}$$

$$SiH_2 + SiH_3(C_6H_5) \rightarrow Si_2H_5(C_6H_5). \tag{3.18}$$

2.2.8 Hexamethyldisilane, $Si_2(CH_3)_6$

The absorption spectrum of $Si_2(CH_3)_6$ is a continuum that starts at ~220 nm and has successively stronger peaks at 191, 163, and 138 nm (Harada et al., 1968; see also Table 3.2).

At 193 nm, a primary photoprocess is

$$Si_2(CH_3)_6 + h\nu \rightarrow Si(CH_3)_3 + Si(CH_3)_3. \tag{3.19}$$

The reaction quantum yield of this process is 0.5 (Shimo et al., 1986). This low yield indicates the existence of other yet undetermined photoprocesses. The bimolecular recombination rate constant of $Si(CH_3)_3$ is 2.5×10^{-11} cm^3 molec^{-1} s^{-1} (Shimo et al., 1986).

2.2.9 Germane, GeH_4

The absorption spectrum of GeH_4 is a continuum that starts at 250 nm, increases rapidly below 180 nm, and peaks at ~120 nm (Osmundsen et al., 1985; Itoh et al., 1986; see also Table 3.2).

At 248 nm, the linear absorption cross section (6×10^{-23} cm^2) is smaller than the two-photon absorption cross section for fluences higher than 20 mJ cm^{-2}. At these higher fluences the primary photoprocess is (Osmundsen et al., 1985)

$$GeH_4 + 2h\nu \rightarrow GeH_2 + 2H. \tag{3.20}$$

Secondary reactions are formation of digermane,

$$GeH_2 + GeH_4 + Ar \rightarrow Ge_2H_6 + Ar, \tag{3.21}$$

with an estimated rate constant of 8×10^{-12} cm^3 molec^{-1} s^{-1}, and formation of germyl radicals,

$$H + GeH_4 \rightarrow GeH_3 + 2H. \tag{3.22}$$

2.3. Molecules Containing Phosphorus, Arsenic and Antimony

2.3.1 Phosphine, PH_3

The absorption spectrum of PH_3 is a continuum in the range 210–160 nm, with a peak at 183 nm (Humphries et al., 1963; Di Stefano et al., 1977; see also Table 3.3).

The primary photoprocess is molecular dissociation (Clark and Anderson, 1978; Di Stefano et al., 1977; Sam and Yardley, 1978),

$$PH_3 + h\nu \rightarrow PH_2 + H. \quad (3.23)$$

A fraction of the PH_2 fragments are in the electronically excited $\tilde{A}\ (^2A_1)$ state, which may decay radiatively (400- to 600-nm fluorescence) to the ground $\tilde{X}\ (^2B_1)$ state. From luminescence yields, Sam and Yardley (1979) determined that this fraction is rather small at 193 nm, ~0.014, with most of the nascent PH_2 molecules being in the electronic ground state. The radiative lifetime of the $\tilde{A}$ state is ~2.2 μs, and the rate constant for collisional quenching by PH_3 is $\sim 1.6 \times 10^{-10}\ cm^3\ molec^{-1}\ s^{-1}$ (Sam and Yardley, 1979).

2.3.2 Trimethylphosphine, $P(CH_3)_3$

The absorption spectrum of $P(CH_3)_3$ is a continuum that starts at 220 nm and peaks at 202 nm (Karlicek et al., 1984; see also Table 3.3). The primary photoprocess at 193 nm is elimination of one methyl group (Donnelly et al., 1984)

$$P(CH_3)_3 + h\nu \rightarrow P(CH_3)_2 + CH_3. \quad (3.24)$$

Neither photofragment exhibits strong absorption at 193 nm (Donnelly et al., 1984).

Table 3.3 The 193-nm absorption cross section of molecules containing phosphorus and arsenic.

Molecule	$\sigma_{193\ nm}$ (cm^2)	Reference
PH_3	4.0×10^{-17}	Di Stefano et al., 1977
	1.3×10^{-17}	Clark and Anderson, 1978
$P(CH_3)_3$	3.4×10^{-17}	Karlicek et al., 1984
$P(C_2H_5)_3$	8.5×10^{-18}	Karlicek et al., 1984
AsH_3	1.8×10^{-17}	Clark and Anderson, 1978
$As(CH_3)_3$	4.5×10^{-17}	McCrary and Donnelly, 1987
$As(C_2H_5)_3$	1.8×10^{-17}	McCrary and Donnelly, 1987

2.3.3 Triethylphosphine, $P(C_2H_5)_3$

The absorption spectrum of $P(C_2H_5)_3$ is a continuum that starts at 250 nm and peaks at 210 nm (Karlicek et al., 1984; see also Table 3.3).

2.3.4 Arsine, AsH_3

The absorption spectrum of AsH_3 is a continuum that starts at 210 nm, and peaks at 180 nm (Humphries et al., 1963; Clark and Anderson, 1978; see also Table 3.3).

The primary photoprocess is molecular dissociation (Dixon and Yee, 1967; Clark and Anderson, 1978; Ni et al., 1986),

$$AsH_3 + h\nu \rightarrow AsH_2 + H. \tag{3.25}$$

A fraction of the AsH_2 fragments are in the electronically excited $\tilde{A}\,^2A_1$ state, which may decay radiatively to the ground $\tilde{X}\,^2B_1$ state (400- to 650-nm radiation). The radiative lifetime of the $\tilde{A}$ state is 0.13 μs (Ni et al., 1986).

2.3.5 Trimethylarsine, $As(CH_3)_3$

The absorption spectrum of $As(CH_3)_3$ is a continuum that starts at 240 nm and peaks at 200 nm (McCrary and Donnelly, 1987; see also Table 3.3). The primary photoprocesses at 193 nm is (Donnelly et al., 1987)

$$As(CH_3)_3 + h\nu \rightarrow AsCH_3 + 2CH_3. \tag{3.26}$$

A secondary photoprocess is (Donnelly et al., 1987)

$$AsCH_3 + h\nu \rightarrow As + CH_3, \qquad \sigma \approx 4 \times 10^{-17}\ \text{cm}^2. \tag{3.27}$$

2.3.6 Triethylarsine, $As(C_2H_5)_3$

The absorption spectrum of $As(C_2H_5)_3$ is a continuum that starts at 250 nm and peaks at 210 nm (McCrary and Donnelly, 1987; see also Table 3.3).

2.3.7 Stibine, SbH_3

The first absorption band of SbH_3 consists of a broad continuum with a maximum at 197 nm (Humphries et al., 1963). A primary photoprocess at 193 nm is the formation of electronically excited SbH_2 (Ni et al., 1986a),

$$SbH_3 + h\nu \rightarrow SbH_2(\tilde{A}\,^2A_1) + H. \tag{3.28}$$

The nascent $SbH_2(\tilde{A})$ is also vibrationally excited, and its radiative decay to $SbH_2(\tilde{X}^2B_1)$ is observed in the 400- to 700-nm region. The radiative lifetime is 70 ± 20 ns (Ni et al., 1986a).

2.4. Molecules Containing Sulfur, Selenium, and Tellurium

2.4.1 Hydrogen Sulfide, H_2S

The absorption spectrum of H_2S above 165 nm is a weakly structured continuum that starts at 270 nm and peaks at 193 nm (Goodeve and Stein, 1931; Watanabe and Jursa, 1964; see also Table 3.4).

Photoexcitation in the UV induces a transition from the $^1A_1$ ground state to a bound $^1B_1$ state, which is vibrationally predissociation by the $^1A_2$ valence state (Van Veen et al., 1983). This valence state correlates to the ground-state fragments $H(^2S) + SH(X\,^2\Pi)$ (Roberge and Salahud, 1979).

Since the bond energy D(SH-H) is 3.9 eV, considerable excess energy is imparted to the photofragments, most of it as kinetic energy. At 185 nm, 25% to 40% of the excess energy is converted into internal excitation of SH (Compton et al., 1969; Sturm and White, 1969). At 193 nm there is no vibrational excitation of the photofragments, and the rotational temperature is 220–375°K (Hawkins and Houston, 1980; Heaven et al., 1981). According to Van Veen et al. (1982), the fraction of vibrationally excited SH fragments is 5.5%, 18%, and 10.5% at 248, 222, and 193 nm, respectively; at 222 and 193 nm the vibrational excitation has a bimodal distribution.

2.4.2 Sulfur Dioxide, SO_2

The absorption spectrum of SO_2 exhibits several complex systems (see Okabe, 1978, for a detailed review, and also Table 3.4). There is a weak,

Table 3.4 Absorption cross sections of molecules containing sulfur and tellurium.

Molecule	$\sigma_{193\,nm}$ (cm^2)	$\sigma_{248\,nm}$ (cm^2)	$\sigma_{257\,nm}$ (cm^2)	$\sigma_{266\,nm}$ (cm^2)	$\sigma_{308\,nm}$ (cm^2)	Reference
H_2S	7.0×10^{-18}	7.1×10^{-20}				Goodeve and Stein, 1931
	6.4×10^{-18}					Watanabe and Jursa, 1964
	5.7×10^{-18}					Deutsch et al., 1982
		2.5×10^{-20}				Wight and Leone, 1983
SO_2	9.3×10^{-18}					Golomb et al., 1962
		7.5×10^{-20}	1.9×10^{-19}	4.5×10^{-19}	6.3×10^{-19}	Warneck et al., 1964
$Te(CH_3)_2$	3.8×10^{-18}					Morris, 1986
	6.6×10^{-18}	1.0×10^{-17}	2.3×10^{-18}			Brewer et al., 1988
$Te(C_2H_5)_2$		6×10^{-18}	2×10^{-18}	1×10^{-18}		Irvine et al., 1984
	1.5×10^{-17}	1.0×10^{-17}	1.1×10^{-18}			Brewer et al., 1988

structured system in the range 340–400 nm, which corresponds to a transition from the ground (singlet) state $\tilde{X}\,^1A_{1u}$ to the first excited (triplet) state $\tilde{a}\,^3B_1$ (Brand et al., 1971). The peak absorption cross section of this band, at 370 nm, is $3.5 \times 10^{-22}\,cm^2$. The radiative lifetime of the $\tilde{a}$ state is 8.1 ms (Su et al., 1977). The more intense absorption spectrum in the region 235–340 nm (Warneck et al., 1964) is attributed to transitions to two singlet states (Hamada and Merer, 1975): $\tilde{A}\,^1A_2$, starting at 340 nm, and $\tilde{B}\,^1B_1$, starting at 310 nm. The former has a radiative lifetime of 50 μs, and the latter, 80–530 μs (Brus and McDonald, 1974). The intense absorption bands in the 170- to 235-nm region (Golomb et al., 1962) correspond to a transition to the $\tilde{C}\,^1B_2$ state. The $\tilde{C}$ state is predissociation at excitation wavelengths shorter than 219 nm, as evidenced by a dramatic drop in fluorescence yield (Okabe, 1971), shortening of the radiative lifetime (Hui and Rice, 1972), and an isotropic angular distribution of the photofragments in time-of-flight mass spectrometric studies (Kawasaki et al., 1982).

The primary photoprocess at 193 nm is predissociative dissociation,

$$SO_2 + h\nu \rightarrow SO + O. \qquad (3.29)$$

The fragments are in their respective electronic ground states, but the SO molecule is vibrationally excited (Freedman et al., 1979; Kanamori et al., 1985; Kolbe and Leskovar, 1986). The vibrational distribution is bimodal, with 70% of the molecules in $v = 2$, 20% in $v = 1$, and some in $v = 5$ (Kanamori et al., 1985; Kolbe and Leskovar, 1986). The rotational distribution in each vibrational level is nonthermal (Kanamori et al., 1985).

At excitation wavelengths longer than the dissociation limit, photodissociation may still be accomplished via a two-photon absorption mechanism. Two dissociation channels are operative simultaneously

$$SO_2 + 2h\nu \rightarrow SO + O \qquad (3.30)$$

$$SO_2 + 2h\nu \rightarrow S + O_2. \qquad (3.31)$$

Venkitachalam and Bersohn (1984), using excitation at 285–311 nm, detected SO in its ground electronic state and S atoms in ground (^3P) and excited (^1D) states. At 248 nm, with excitation fluences of 0.1 J/cm^2, Wildt et al. (1983) observed generation of excited SO(b $^1\Sigma^+$). At higher fluences (above 0.4 J/cm^2), Wilson et al. (1982) observed the simultaneous photogeneration of SO(X $^3\Sigma^-$) and S(^3P). Fotakis et al. (1983) also observed the formation of SO(X $^3\Sigma^-$). A secondary photoprocess is excitation of the SO molecule to the B ($v = 2$) state (Wilson et al., 1982; Fotakis et al., 1983).

2.4.3 Hydrogen Selenide, H_2Se

The absorption spectrum of H_2Se is diffuse, with a maximum at 197 nm (Herzberg, 1966).

2.4.4 Dimethyltellurium, $Te(CH_3)_2$

The absorption spectrum of $Te(CH_3)_2$ in the UV consists of two systems (Connor et al., 1969; Stuke, 1984; Brewer et al., 1988): One has a maximum at 250 nm, with secondary peaks at 243 and 258 nm; the other is a structured continuum with a peak at 200 nm. See also Table 3.4.

A photoprocess in UV flash photolysis is the formation Te atoms and possibly also of the $TeCH_3$ intermediate (Connor et al., 1969). A secondary reaction is

$$Te + Te(CH_3)_2 \rightarrow Te_2 + 2CH_3 \text{ or } C_2H_6, \qquad (3.32)$$

with a rate constant of 2.8×10^{-10} cm^3 molec^{-1} s^{-1} (Connor et al., 1969). A primary photoprocess at 248 nm is the production of Te atoms in their electronic ground state (Brewer et al., 1988).

2.4.5 Diethyltellurium, $Te(C_2H_5)_2$

The absorption spectrum of $Te(C_2H_5)_2$ is a structured continuum that starts at 270 nm and peaks at 248, 220, and 198 nm (Irvine et al., 1984; Brewer et al., 1988; see also Table 3.4). A primary photoprocess at 248 and 193 nm is the production of Te atoms in their ^3P electronic states (Brewer et al., 1988).

2.5. Halogen-Containing Molecules

2.5.1 Hydrogen Halides

The absorption spectra of the hydrogen halides have similar features: a broad continuum with a threshold, a peak, and a decreasing absorption coefficient at shorter wavelengths, followed by sharp transitions in the vacuum UV. The location of the threshold and peak are red-shifted, and the maximum value of the absorption coefficient is reduced with increasing mass of the halogen atom (see Fig. 3.1). The threshold for HF is 160 nm (Safary et al., 1951), for HCl it is 210 nm with the peak at 155 nm (Myer and Samson, 1970), for HBr it is 230 nm with the peak at 180 nm, and for HI it is 310 nm with the peak at 220 nm (Huebert and Martin, 1968; see also Table 3.5).

The primary photoprocess induced by absorption in these continua is

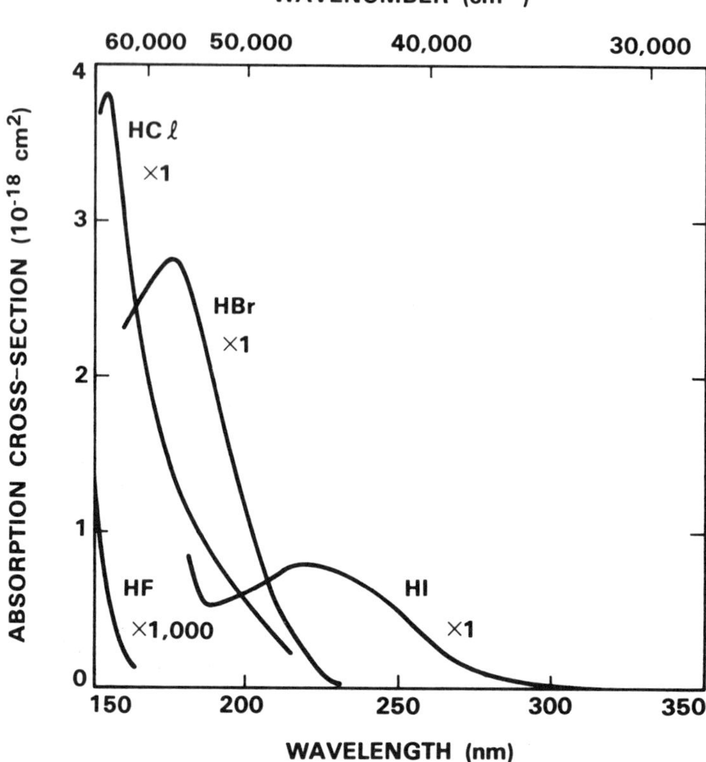

Fig. 3.1 The absorption cross sections of the hydrogen halides in the region 150–350 nm. The HF spectrum was adapted with permission from Safary et al. (1951), and the HCl spectrum was adapted with permission from Myer and Samson (1970). The HBr and HI spectra were adapted with permission from Huebert and Martin, *J. Phys. Chem.* **72,** 3046; copyright 1968, American Chemical Society.

molecular dissociation with quantum yield of unity

$$HX + h\nu \rightarrow H(^2S) + X(^2P_{3/2\,,\,1/2}), \qquad X = F, Cl, Br, I. \tag{3.33}$$

The halogen atoms are produced predominantly in their ground electronic state $^2P_{3/2}$. In the case of HI, the yield of the excited state $^2P_{1/2}$ was found to be wavelength dependent (Compton and Martin, 1969) up to 0.36 at 266 nm (Clear et al., 1975). At 193 nm it is 0.11 (Wight and Leone, 1983). In the photolysis of HBr at 193 nm, the fraction of excited $Br(^2P_{1/2})$ is ~0.15 (Magnotta et al., 1981).

Table 3.5 Absorption cross sections of inorganic halogen-containing molecules.

Molecule	$\sigma_{193\,\mathrm{nm}}$ (cm^2)	$\sigma_{248\,\mathrm{nm}}$ (cm^2)	$\sigma_{266\,\mathrm{nm}}$ (cm^2)	$\sigma_{308\,\mathrm{nm}}$ (cm^2)	$\sigma_{351\,\mathrm{nm}}$ (cm^2)	Reference
HF	$<10^{-21}$					Safary et al., 1951
HCl	4×10^{-19}					Myer and Samson, 1970
	8.9×10^{-19}					Rothschild and Ehrlich, 1986
HBr	1.7×10^{-18}					Huebert and Martin, 1968
HI	5.4×10^{-19}	5.1×10^{-19}	1.8×10^{-19}			Huebert and Martin, 1968
	5.8×10^{-19}					Wight and Leone, 1983
F_2	$<10^{-21}$	1.3×10^{-20}	1.5×10^{-20}	1.9×10^{-20}	5.7×10^{-21}	Steunenberg and Vogel, 1956
Cl_2			1.0×10^{-20}	1.9×10^{-19}	1.7×10^{-19}	Gibson and Bayliss, 1933
	2.5×10^{-21}	$<10^{-21}$			1.2×10^{-19}	Rothschild and Ehrlich, 1986
Br_2	$<10^{-22}$	$<10^{-22}$	$<10^{-22}$	3.8×10^{-22}	3.4×10^{-20}	Passchier et al., 1967
I_2		2.7×10^{-20}	5.4×10^{-20}			Passchier and Gregory, 1968
	6.1×10^{-18}					Myer and Samson, 1970a
NF_3	5.3×10^{-21}					Hirose et al., 1985
	6.1×10^{-21}					Rothschild and Ehrlich, 1986
SF_6	6.8×10^{-22}					Rothschild and Ehrlich, 1986

Secondary reactions are

$$H + HX \rightarrow H_2 + X \tag{3.34}$$

$$X + X + M \rightarrow X_2 + M, \tag{3.35}$$

where X = F, Cl, Br, or I, and M is a third body in the collision, which is necessary for the conservation of energy and momentum.

2.5.2 Halogen Diatomics

The absorption spectra of the halogen molecules X_2 show a clear regularity (see Fig. 3.2 and Table 3.5). They all have a peaked continuum, and at longer wavelengths there is a sequence of sharp bands. As the mass of the halogen atom increases, so do the strengths of both the continuum and the discrete lines (in fact, no discrete lines are observed in the lightest molecule, F_2). In addition with increasing mass the continuum and the discrete bands shift toward longer wavelengths, but at different rates. Thus, in Cl_2 the discrete spectrum is in the 470- to 600-nm region (Douglas et al., 1963), superimposed on the very tail of the continuum centered at 330 nm, while in Br_2 and I_2 they overlap the long-wavelength portion of the respective continua (above approximately 500 nm). At shorter wavelengths, Br_2 and I_2 have weak continua, whose absorption cross section is temperature and pressure dependent because of the formation of (and absorption by) the dimers Br_4 and I_4,

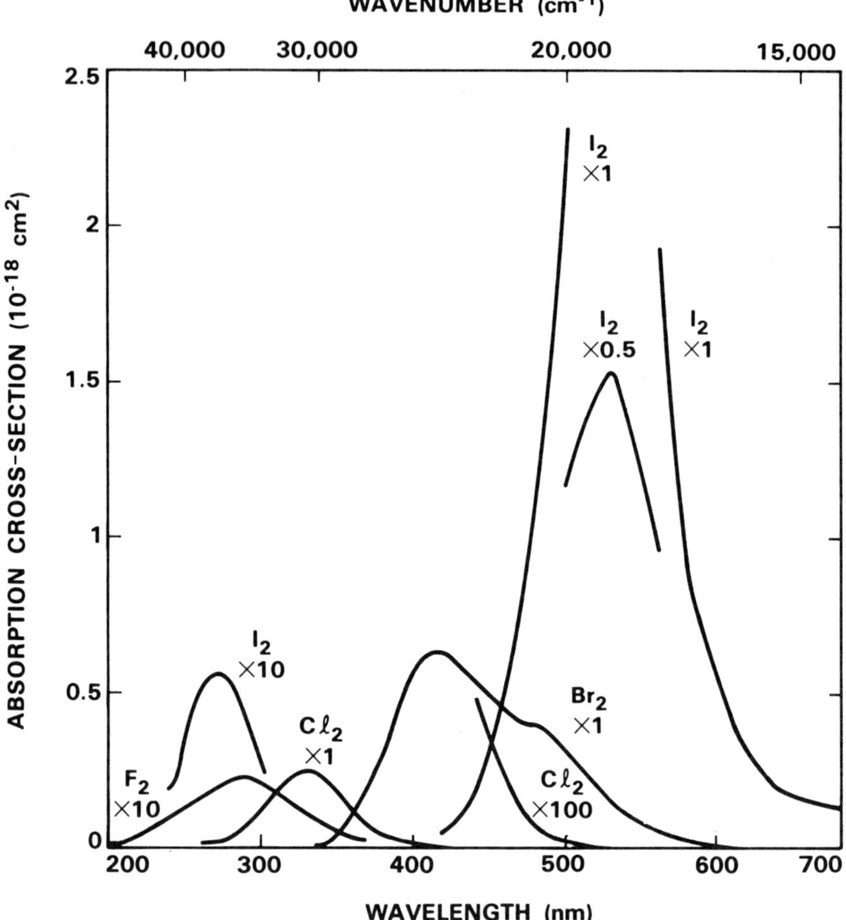

Fig. 3.2 The absorption cross sections of the halogen diatomics in the region 200–700 nm. The F_2 spectrum was adapted with permission from Steunenberg and Vogel, *J. Am. Chem. Soc.* **78,** 901; copyright 1956, American Chemical Society. The Cl_2 spectrum below 420 nm was adapted with permission from Gibson and Bayliss (1933) and above 420 nm from Rothschild and Ehrlich (1986). The Br_2 spectrum was adapted with permission from Passchier et al., *J. Phys. Chem.* **71,** 937; copyright 1967, American Chemical Society. The I_2 spectrum below 320 nm was adapted with permission from Passchier and Gregory, *J. Phys. Chem.* **72,** 2697; copyright 1968, American Chemical Society. The I_2 spectrum above 420 nm was adapted with permission from Tellinghuisen (1973).

respectively. These effects were analyzed by Passchier et al. (1967) for Br_2 in the region 200–230 nm and by Tamres et al. (1968) and Passchier and Gregory (1968) for I_2 in region 230–300 nm.

The primary photoprocess at all wavelengths with energy higher than the respective bond energy is molecular dissociation. Three separate upper states are involved in these transitions, one of them repulsive, $^1\Pi$ (1_u), and the other two weakly bound, $^3\Pi$ (1_u) and $^3\Pi$ (0_u^+). The $^1\Pi$ and the $^3\Pi$ (1_u) states correlate with two-ground-state $^2P_{3/2}$ halogen atoms, while $^3\Pi$ (0_u^+) correlates with one ground-state and one excited $^2P_{1/2}$ atom (Mulliken, 1936). The interactions between these states lead to various unimolecular processes, such as curve crossing and predissociation. As a result, the relative yield of ground and excited halogen atoms varies strongly with excitation wavelength, and so does the absolute atomic quantum yield.

Absorption in the 200- to 400-nm continuum of F_2 results in a quantum yield of unity for dissociation into two $F(^2P_{3/2})$ atoms. Absorption in the 250- to 450-nm region of Cl_2 likewise results in a quantum yield of unity for dissociation into $Cl(^2P_{3/2})$ atoms (Busch et al., 1969). At longer excitation wavelengths, $Cl(^2P_{3/2})$ and $Cl(^2P_{1/2})$ are produced in approximately equal amounts.

Absorption by Br_2 at 532 nm and at 347 nm leads to dissociation into ground-state atoms, while at 466 nm the main dissociation channel is into ground- and excited-state Br atoms (Oldman et al., 1975).

Absorption by I_2 in the 500- to 620-nm region leads to wavelength-dependent dissociative quantum yield (Brewer and Tellinghuisen, 1972). The quantum yield for the production of excited state $I(^2P_{1/2})$ is also wavelength dependent in the region 440- to 530-nm, with a peak of ~0.65 at 495–505 nm (Hunter and Leong, 1987).

2.5.3 Other Halogen-Containing Inorganic Molecules

The absorption spectra of NF_3 and SF_6 start at ~200 nm. See also Table 3.5.

2.5.4 Halogenated Methane and Ethane

2.5.4.1 Methyl Chloride, CH_3Cl

The first absorption band of CH_3Cl is a continuum that starts at 220 nm (Robbins, 1976; Hubrich and Stuhl, 1980; see also Table 3.6). Absorption in this band corresponds to a $n \rightarrow \sigma^*$ transition (lone-pair p orbital on Cl to antibonding orbital localized on the C-Cl bond). The primary photoprocess at 193 nm is direct photodissociation into Cl and CH_3 (Kawasaki et al., 1984).

Table 3.6 Absorption cross sections of chlorine-containing methanes and ethanes.

Molecule	$\sigma_{193\ nm}$ (cm^2)	$\sigma_{248\ nm}$ (cm^2)	Reference
CH_3Cl	$5{\cdot}5\times10^{-20}$		Robbins, 1976; Hubrich and Stuhl, 1980
	2×10^{-21}		Brewer et al., 1984
	6.1×10^{-20}		Rothschild and Ehrlich, 1986
	7×10^{-20}		Lee and Suto, 1987
CH_2Cl_2	4.0×10^{-19}		Hubrich and Stuhl, 1980
	3.5×10^{-19}		Lee and Suto, 1987
$CHCl_3$	8.7×10^{-19}		Hubrich and Stuhl, 1980
	9.0×10^{-19}		Lee and Suto, 1987
CCl_4	1.0×10^{-18}	2.3×10^{-21}	Hubrich and Stuhl, 1980
	4×10^{-19}		Tachibana et al., 1988
CH_2FCl	1.0×10^{-20}		Hubrich and Stuhl, 1980
CHF_2Cl	1.2×10^{-21}		Hubrich and Stuhl, 1980
$CHFCl_2$	2.0×10^{-19}		Hubrich and Stuhl, 1980
CF_2Cl_2	3.2×10^{-19}		Hubrich and Stuhl, 1980; Rothschild and Ehrlich, 1986
CF_3Cl	8.4×10^{-22}		Hubrich and Stuhl, 1980
	2×10^{-21}		Brewer et al., 1984
	1.4×10^{-21}		Rothschild and Ehrlich, 1986
C_2H_5Cl	3.5×10^{-20}		Hubrich and Stuhl, 1980
C_2F_5Cl	1.9×10^{-21}		Hubrich and Stuhl, 1980
1,2-$C_2F_4Cl_2$	4.4×10^{-20}		Hubrich and Stuhl, 1980

2.5.4.2 Methylene chloride, CH_2Cl_2

The first absorption band of CH_2Cl_2 is a continuum that starts at 230 nm (Hubrich and Stuhl, 1980; see also Table 3.6). The primary photoprocess is molecular dissociation into CH_2Cl and Cl. A very small quantum yield was detected at 193 nm for the molecular formation of Cl_2 and HCl (Kenner et al., 1986).

2.5.4.3 Chloroform, $CHCl_3$

The first absorption band of $CHCl_3$ is a continuum that starts at 240 nm (Hubrich and Stuhl, 1980; see also Table 3.6). The primary photoprocess is molecular dissociation into $CHCl_2$ and Cl. At 193 nm, Kenner et al. (1986) observed molecular elimination of HCl and Cl_2, the latter in its electronic ground state and also in its first excited triplet state. The quantum yield for this process was estimated to be very low.

2.5.4.4 *Carbon tetrachloride, CCl_4*

The first absorption band of CCl_4 is a continuum that starts at 250 nm (Hubrich and Stuhl, 1980; see also Table 3.6). At wavelengths longer than ~200 nm, the primary photoprocess is molecular dissociation into CCl_3 and Cl, whereas at shorter wavelengths photodissociation into CCl_2 and 2Cl (or Cl_2) becomes increasingly the dominant process (Okabe, 1978). At 193 nm, CCl_2 was detected with an undetermined quantum yield (Tiee et al., 1979). The formation at 193 nm of Cl_2 molecules is probably a minor process (Kenner et al., 1986).

2.5.4.5 *Chlorofluoromethane, CH_2FCl*

The first absorption band of CH_2FCl is a continuum that starts at 200 nm (Hubrich and Stuhl, 1980; see also Table 3.6). The primary photoprocess is molecular dissociation into CH_2F and Cl.

2.5.4.6 *Chlorodifluoromethane, CHF_2Cl*

The first absorption band of CHF_2Cl is a continuum that starts at 200 nm (Hubrich and Stuhl, 1980; see also Table 3.6). The primary photoprocess is molecular dissociation into CHF_2 and Cl. At high intensities, irradiation at 248 nm generates also CF_2 (Hack and Langel, 1983).

2.5.4.7 *Dichlorofluoromethane, $CHFCl_2$*

The first absorption band of $CHFCl_2$ is a continuum that starts at 220 nm (Hubrich and Stuhl, 1980; see also Table 3.6). The primary photoprocess at 214 nm is molecular dissociation into CHFCl and Cl (Rebbert et al., 1978).

2.5.4.8 *Dichlorodifluoromethane, CF_2Cl_2*

The first absorption band of CF_2Cl_2 is a continuum that starts at 220 nm (Hubrich and Stuhl, 1980; see also Table 3.6). The primary photoprocess at 214 nm is molecular photodissociation into CF_2Cl and Cl; at shorter wavelengths (185 nm) dissociation into CF_2 and 2Cl represents ~1/3 of the photoprocess (Rebbert and Ausloos, 1975). High-intensity irradiation at 193 nm and at 248 nm yields electronically excited CF and CF_2 (Tiee et al., 1979a; Hack and Langel, 1983).

2.5.4.9 *Chlorotrifluoromethane, CF_3Cl*

The first absorption band of CF_3Cl is a continuum that starts at 200 nm (Hubrich and Stuhl, 1980; see also Table 3.6). The primary photoprocess is molecular dissociation into CF_3 and Cl (Brewer et al., 1984).

2.5.4.10 Chloroethanes

The first absorption bands of the chlorethanes are continua that start in the region 200–240 nm; the continua of molecules with more chlorine atoms start at longer wavelengths (Hubrich and Stuhl, 1980; see also Table 3.6). Time-of-flight mass spectrometric studies of C_2H_5Cl indicate that the primary photoprocess at 193 nm is molecular dissociation into C_2H_5 and Cl, with only 53% of the excess energy in translational energy of the fragments and the rest in vibrational/rotational excitation of C_2H_5 (Kawasaki et al., 1984).

2.5.4.11 Methyl bromide, CH_3Br

The first absorption band of CH_3Br is a continuum that starts at 260 nm and peaks at 200 nm (Robbins, 1976; Molina et al., 1982; see also Table 3.7. The absorption spectrum in this region is the sum of three continua, each with its own center, width, and strength, (Van Veen et al., 1985). These correspond to transitions from the electronic ground state to three different dissociative states, $^3Q_1$, ^1Q, and $^3Q_0$. Of these, $^3Q_1$ and ^1Q correlate with ground-state $Br(^2P_{3/2})$, and $^3Q_0$ correlates with the excited-state $Br(^2P_{1/2})$. At 222 nm the absorption is to $^3Q_1$ and $^3Q_0$ in equal amounts, and at 193 nm the excitation is to ^1Q (84%) and $^3Q_0$ (16%) (Van Veen et al., 1985).

The primary photoprocess is molecular dissociation into CH_3 and Br.

Table 3.7. Absorption cross sections of bromine-containing methanes and ethanes.

Molecule	$\sigma_{193\text{ nm}}$ (cm^2)	$\sigma_{248\text{ nm}}$ (cm^2)	$\sigma_{257\text{ nm}}$ (cm^2)	Reference
CH_3Br	5.2×10^{-19}	1.9×10^{-20}	4.4×10^{-21}	Robbins, 1976
	5.7×10^{-19}	1.3×10^{-20}	3.1×10^{-21}	Molina et al., 1982
	5.2×10^{-20}			Brewer et al., 1984
CF_3Br	8.0×10^{-20}			Pence et al., 1981
	8.6×10^{-20}	2.0×10^{-21}	4.5×10^{-22}	Molina et al., 1982
	1×10^{-20}			Brewer et al., 1984
	6.8×10^{-20}	7.9×10^{-21}		Rothschild and Ehrlich, 1986
CH_2Br_2		3.6×10^{-19}	1.1×10^{-19}	Molina et al., 1982
CF_2Br_2	1.1×10^{-18}	6.5×10^{-19}	2.4×10^{-19}	Molina et al., 1982
		6×10^{-19}		Krajnovich et al., 1984
C_2F_5Br	1.1×10^{-19}			Pence et al., 1981
	1.8×10^{-19}	4.2×10^{-21}	1.0×10^{-21}	Molina et al., 1982
1,2-$C_2F_4Br_2$	1.1×10^{-18}	5.1×10^{-20}	1.6×10^{-20}	Molina et al., 1982

From time-of-flight mass spectrometric studies, Van Veen et al. (1985) conclude that at 222 nm Br is produced in its ground and excited states in about equal amounts, and at 193 nm the quantum yield of $Br(^2P_{1/2})$ is only 17%. The CH_3 fragment is in all cases vibrationally excited in its umbrella mode.

2.5.4.12 Bromotrifluoromethane, CF_3Br

The first absorption band of CF_3Br is a continuum that starts at 260 nm and peaks at 205 nm (Pence et al., 1981; Molina et al., 1982; see also Table 3.7). The primary photoprocess is molecular dissociation into CF_3 and Br. The quantum yield for production of excited $Br(^2P_{1/2})$ at 193 nm is 0.56 (Pence et al., 1981).

2.5.4.13 Methylene bromide, CH_2Br_2

The first absorption band of CH_2Br_2 is a continuum that starts at 280 nm and peaks at 220 nm (Molina et al., 1982; see also Table 3.7).

2.5.4.14 Dibromodifluoromethane, CF_2Br_2

The first absorption band of CF_2Br_2 is a continuum that starts at 290 nm and peaks at 225 nm; the second absorption band starts at ~200 nm (Molina et al., 1982; see also Table 3.7).

The primary photoprocess at 265 nm is molecular dissociation into CF_2Br and Br (Walton, 1972). Excitation at 248 and 193 nm, at fluences of 0.1 J/cm^2 and up, was reported to cause dissociation into CF_2 and Br_2, followed at 248 nm by resonant photoexcitation of the CF_2 fragment (Sam and Yardley, 1979; Wampler et al., 1979). However, in crossed laser-molecular beam experiments at 248 nm and fluences of 2 J/cm^2, Krajnovich et al. (1984) found no evidence for the molecular photo-elimination of Br_2. Brannon (1986) suggested that both photodissociation channels may be operative with a branching ratio that is wavelength-dependent.

2.5.4.15 Bromoethanes

The first absorption band of C_2F_5Br is a continuum that starts at 260 nm and peaks at 195 nm (Pence et al., 1981; Molina et al., 1982). The absorption of 1,2-$C_2F_4Br_2$ starts at 270 nm and peaks at 200 nm (Molina et al., 1982; see also Table 3.7).

The primary photoprocess in C_2F_5Br at 193 nm is molecular dissociation into C_2F_5 and Br (Krajnovich et al., 1984a). The quantum yield of excited state $Br(^2P_{1/2})$ is 0.16 at 193 nm (Pence et al., 1981).

2.5.4.16 *Methyl iodide, CH_3I*

The first absorption band of CH_3I is a continuum in the region 220–300 nm, with a peak at 258 nm (Baughcum and Leone, 1980; Pence et al., 1981; see also Table 3.8). Below 220 nm there are several sharper features that form vibrational progressions (Felps et al., 1976).

Absorption in the continuum induces a $n \rightarrow \sigma^*$ transition, promoting an electron from a nonbonding orbital localized on the I atom to an antibonding σ^* orbital localized on the C-I bond. The absorption spectrum is the sum of three continua, each with its own center, width, and intensity (Gedanken and Rowe, 1975). These continua correspond to photoexcitation to three states, two of which correlate with CH_3 and ground-state $I(^2P_{3/2})$, and one with CH_3 and excited-state $I(^2P_{1/2})$. Absorption below 220 nm is assigned to $p \rightarrow s$ transitions into Rydberg states, in which an electron is promoted from a nonbonding p orbital to a nonbonding s orbital, localized on the I atom (Felps et al., 1976).

The primary photoprocess following excitation in the UV is dissociation into CH_3 and I. Because of the complex nature of the upper states, the distribution of the excess energy above the 2.4 eV of the CH_3-I bond strength is distributed among the fragments in a manner that is strongly wavelength dependent. The quantum yield of excited-state $I(^2P_{1/2})$ was found to be as follows: at 266 nm, 0.73–0.78 (Riley and Wilson, 1972; Hess et al., 1986); at 248 nm, 0.58–0.81 (Hunter and Kristjansson, 1978; Baughcum and Leone, 1980; Van Veen et al., 1984; Barry and Gorry, 1984); and at 193 nm, 1.0 (Van Veen et al., 1985a). The CH_3 photofragment is vibrationally excited in its umbrella mode. At 248 and 266 nm, the vibrational distribution is peaked at $v = 2$ for the channel leading to $I(^2P_{1/2})$ and is broader with a peak at $v = 2$–4 for the other channel (Sparks et al., 1981; Van Veen et al., 1984). At 193 nm, the vibrational distribution is bimodal, with peaks at $v = 2$ and at $v = 5,6$ (Van Veen et al., 1985a).

2.5.4.17 *Trifluoroiodomethane, CF_3I*

The first absorption band of CF_3I is a continuum in the region 248–282 nm, with a peak at 265 nm (Herzberg, 1966). According to Gedanken (in Van Veen et al., 1985b) the continuum extends from 210 to 300 nm. See also Table 3.8.

Absorption in this region induces a $n \rightarrow \sigma^*$ transition, promoting an electron from a nonbonding orbital localized on the I atom to an antibonding orbital localized on the C-I bond. The absorption spectrum is the sum of three separate continua, two of which correlate with CF_3 and

Table 3.8 Absorption cross sections of iodine-containing methanes and ethanes.

Molecule	$\sigma_{193\ nm}$ (cm^2)	$\sigma_{248\ nm}$ (cm^2)	$\sigma_{257\ nm}$ (cm^2)	$\sigma_{266\ nm}$ (cm^2)	$\sigma_{308\ nm}$ (cm^2)	$\sigma_{351\ nm}$ (cm^2)	Reference
CH_3I			1.1×10^{-18}	9.5×10^{-19}			Baughcum and Leone, 1980
		8.2×10^{-19}			8.4×10^{-21}		Pence et al., 1981
CF_3I		2.6×10^{-19}			3.4×10^{-20}		Gerck, 1983
	3.8×10^{-21}	2.7×10^{-19}				$<10^{-21}$	Rothschild and Ehrlich, 1986
CH_2I_2				2×10^{-18}			Kroger et al., 1976
		1.6×10^{-18}	1.3×10^{-18}	1.4×10^{-18}	3.3×10^{-18}	2.5×10^{-19}	Baughcum and Leone, 1980
	2.9×10^{-17}	1.6×10^{-18}			3.3×10^{-18}		Pence et al., 1981
C_2F_5I	$<10^{-21}$			6.4×10^{-19} (268 nm)			Pence et al., 1981
		3.3×10^{-19}			5.8×10^{-20}		Gerck, 1983
1,2-$C_2F_4I_2$		9.7×10^{-19}	1.5×10^{-18}	2.1×10^{-18}	1.0×10^{-19}		Gerck, 1983

ground-state $I(^2P_{3/2})$, and one correlating with CF_3 and excited-state $I(^2P_{1/2})$ (Van Veen et al., 1985b).

The primary photoprocess is molecular dissociation into CF_3 and I. The CF_3-I bond energy is 2.3 eV. The excess energy is partitioned among the fragments. The quantum yield of excited-state $I(^2P_{1/2})$ is 0.83 at 308 nm (Gerck, 1983), 0.91 at 266 nm (Ershov et al., 1978), and 0.75–0.92 at 248 nm (Gerck, 1983; Van Veen et al., 1985b). At 248 nm the CF_3 fragment is vibrationally excited in its umbrella mode, with a broad distribution extending to $v = 17$ and a peak at $v = 5$–7; the symmetric stretching mode may also be excited (Van Veen et al., 1985b).

2.5.4.18 Methylene iodide, CH_2I_2

The absorption spectrum of CH_2I_2 starts at 380 nm, and at wavelengths above 200 nm it is a complex continuum that can be deconvoluted into the sum of several Gaussian-shaped broad bands (Baughcum and Leone, 1980; see also Table 3.8).

The primary photoprocess in the region 200–380 nm is molecular dissociation into CH_2I and I (Kawasaki et al., 1975; Kroger et al., 1976; Baughcum and Leone, 1980; Koffend and Leone, 1981). Five different upper states can be excited in this spectral region (Kawasaki et al., 1975). Because of their varying energies, excitation strengths, and possible curve crossings, partitioning of the excess energy into the fragments is strongly wavelength dependent. The quantum yield for production of excited-state $I(^2P_{1/2})$ becomes appreciable at wavelengths below 330 nm, and it increases with decreasing wavelength, at least to 250 nm (Koffend and Leone, 1981; Hunter and Kristjansson, 1982). At 248 nm it is 0.46 (Pence et al., 1981). The CH_2I photofragment is highly vibrationally and rotationally excited following absorption at 266 nm (Kroger et al., 1976) and 248 nm (Baughcum and Leone, 1980).

A secondary photoprocess at 266 nm is absorption by the CH_2I fragment ($\sigma \approx 1.1 \times 10^{-19}$ cm^2), followed by molecular dissociation into CH_2 and I (Kroger et al., 1976). Collisional deactivation by CH_2I_2 of the vibrational excitation of the CH_2I fragment is rapid, with rate constants in the range $(1.6$–$4.6) \times 10^{-11}$ cm^3 molec^{-1} s^{-1}.

At wavelengths below 200 nm a primary photoprocess is molecular dissociation into CH_2 and I_2, where I_2 may be in an electronically excited state (Dyne and Style, 1952; Style and Ward, 1952). Dissociation at 193 nm into CH_2I and $I(^2P_{1/2})$ is a minor channel, with a quantum yield of 0.05 (Pence et al., 1981).

2.5.4.19 Iodoethanes

The first absorption band of C_2F_5I is a continuum in the region 330–220 nm, with a peak at 268 nm (Pence et al., 1981). The first absorption band of 1,2-$C_2F_4I_2$ is similarly a continuum in the range 350–220 nm, with a peak at 280 nm; a second intense continuum is seen below 220 nm (Gerck, 1983; see also Table 3.8).

The primary photoprocess in the first continua of C_2F_5I and $C_2F_4I_2$ is molecular dissociation into C_2F_5 or C_2F_4I, respectively, and I. The quantum yield for producing excited-state $I(^2P_{1/2})$ from C_2F_5I is near 1.0 at 248, 268, and 308 nm (Pence et al., 1981). For 1,2-$C_2F_4I_2$ the quantum yield of $I(^2P_{1/2})$ is 0.6 at 248 nm and 0.9 at 308 nm (Gerck, 1983).

2.6. Metal-Containing Molecules

2.6.1 Metal Halides and Oxyhalides

2.6.1.1 Titanium tetrachloride, $TiCl_4$

The absorption spectrum of $TiCl_4$ is a structured continuum that starts at 400 nm and has two local maxima in the near-UV, at 280 and 232 nm (Alderdice, 1965; Becker et al., 1969; see also Fig. 3.3 and Table 3.9). These absorption bands correspond to ligand-to-metal charge transfer, whereby an electron is promoted from a molecular π orbital localized mainly on the ligands (chlorine atoms) to a molecular orbital that is essentially a metal (titanium) 3d orbital (Alderdice, 1965).

2.6.1.2 Titanium tetrabromide, $TiBr_4$

The absorption spectrum of $TiBr_4$ is a structured continuum that starts at 460 nm and has two peaks in the near UV, at 340 and at 278 nm (Alderdice, 1965; see also Fig. 3.3 and Table 3.9). These absorption bands correspond to ligand-to-metal charge transfer, whereby an electron is promoted from a π orbital localized mainly on the ligands (bromine atoms) to a molecular orbital that is essentially a metal (titanium) 3d orbital (Alderdice, 1965).

2.6.1.3 Vanadium tetrachloride, VCl_4

The absorption spectrum of VCl_4 is a structured continuum that starts at 480 nm and has peaks at 400 and at 293 nm (Pennella and Taylor, 1963; see also Fig. 3.4 and Table 3.9). These absorption bands correspond to ligand-to-metal charge transfer, whereby an electron is promoted from an orbital that is localized mainly on the ligands (chlorine atoms) to a

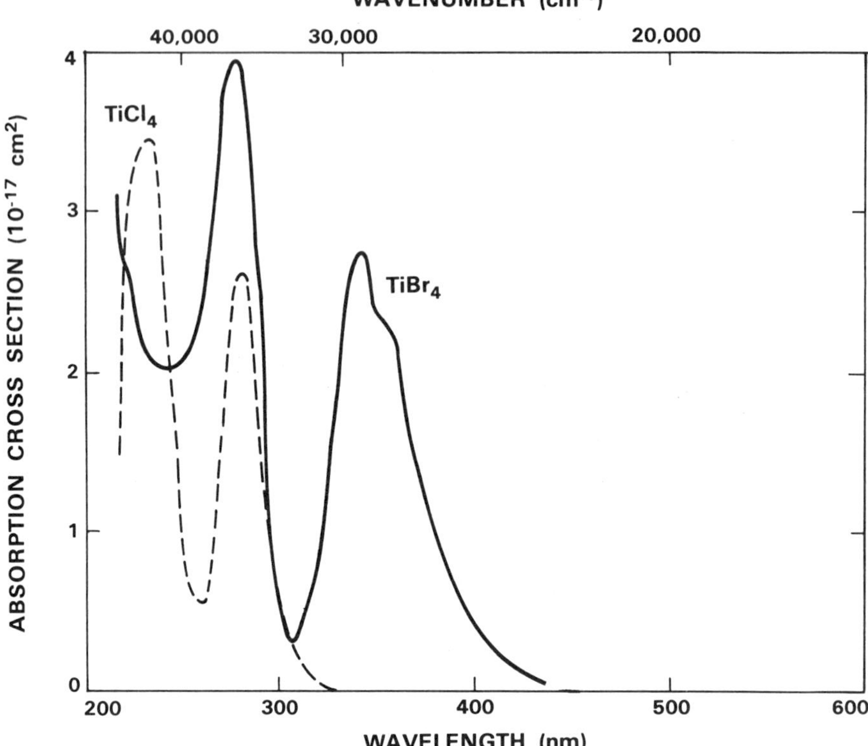

Fig. 3.3 The absorption cross sections of $TiCl_4$ and $TiBr_4$ in the region 200–600 nm. Adapted with permission from Alderdice (1965).

Table 3.9 Absorption cross sections of metal halides and oxyhalides.

Molecule	$\sigma_{193\,nm}$ (cm^2)	$\sigma_{248\,nm}$ (cm^2)	$\sigma_{257\,nm}$ (cm^2)	$\sigma_{266\,nm}$ (cm^2)	$\sigma_{308\,nm}$ (cm^2)	Reference
$TiCl_4$		1.1×10^{-17}	5.5×10^{-18}	1.0×10^{-17}	2.3×10^{-18}	Alderdice, 1965
	3×10^{-17}					Becker et al., 1969
$TiBr_4$		2.0×10^{-17}	2.3×10^{-17}	3.0×10^{-17}	1.2×10^{-18}	Alderdice, 1965
VCl_4		1.0×10^{-17}	9.6×10^{-18}	8.9×10^{-18}	1.1×10^{-17}	Pennella and Taylor, 1963
$VOCl_3$		1.8×10^{-17}	1.3×10^{-17}	8.9×10^{-18}	5.0×10^{-18}	Miller and White, 1957
CrO_2Cl_2		3.0×10^{-18}	3.1×10^{-18}	5.5×10^{-18}	5.1×10^{-18}	Halonbrenner et al., 1968
MoF_6		2.0×10^{-19}	1.4×10^{-19}	9.0×10^{-20}	$<10^{-21}$	Tanner and Duncan, 1951
	4.9×10^{-18}					McDiarmid, 1974
WF_6	3.5×10^{-19}					Tanner and Duncan, 1951
ReF_6	7.3×10^{-18}					McDiarmid, 1971
$SnCl_4$	3.8×10^{-17}					Fernandez et al., 1986
		8.3×10^{-18}	5.2×10^{-18}	1.3×10^{-18}		Tabuchi et al., 1987

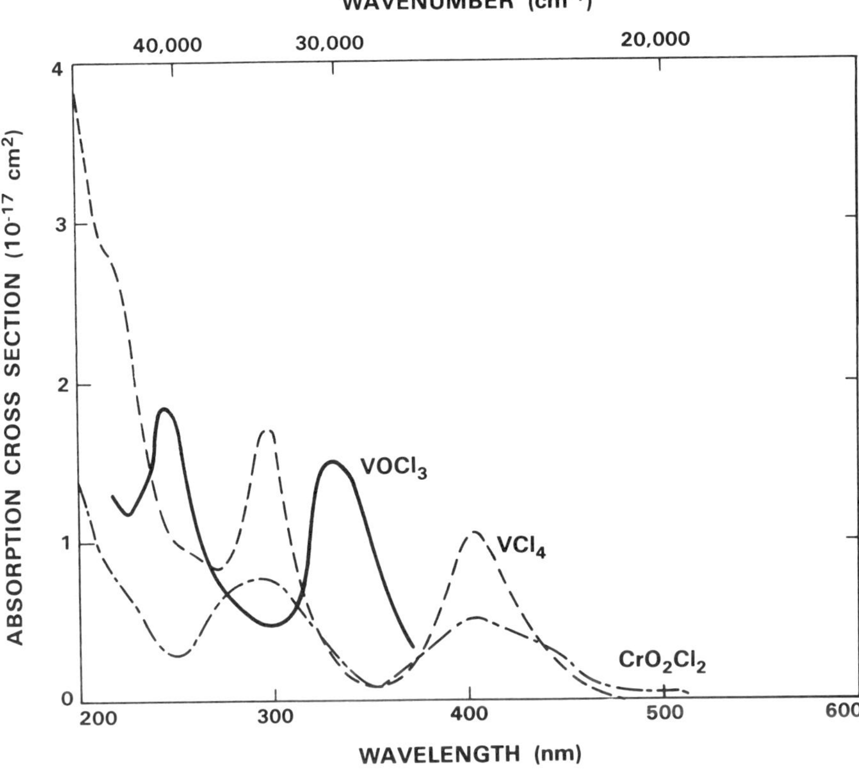

Fig. 3.4 The absorption cross sections of VCl_4, $VOCl_3$, and CrO_2Cl_2 in the region 200–600 nm. The VCl_4 spectrum was adapted with permission from Pennella and Taylor (1963). The $VOCl_3$ spectrum was adapted with permission from Miller and White, *Spectrochim. Acta* **9,** 98, copyright 1957, Pergamon Journals, Ltd. The CrO_2Cl_2 spectrum was adapted with permission from Halonbrenner et al., *J. Phys. Chem.* **72,** 3929; copyright 1968, American Chemical Society.

molecular orbital that is essentially a metal (vanadium) 3d orbital (Pennella and Taylor, 1963; Alderdice, 1965).

2.6.1.4 Vanadium oxychloride, $VOCl_3$

The absorption spectrum of $VOCl_3$ is a structured continuum that starts at 380 nm and has peaks at 333 and 244 nm (Miller and White, 1957; see also Fig. 3.4 and Table 3.9).

2.6.1.5 *Chromyl chloride, CrO_2Cl_2*

The absorption spectrum of CrO_2Cl_2 consists of sharp ro-vibronic lines in the wavelength region 565–600 nm and an increasingly prominent continuum below 570 nm (McDonald, 1975). The continuum extends to below 200 nm, and it has peaks at 400 nm and at 290 nm (Halonbrenner et al., 1968; see also Fig. 3.4 and Table 3.9).

Absorption at wavelengths below 515 nm leads to molecular photodecomposition by sequential elimination of the two chlorine atoms and formation of CrO_2 (Halonbrenner et al., 1968; Arnone et al., 1986). At longer wavelengths there is a complex interplay between decomposition and internal electronic-to-vibrational energy conversion (McDonald, 1975). Irradiation at 193 nm may also lead to elimination of a Cl_2 molecule (Arnone et al., 1986).

2.6.1.6 *Molybdenum hexafluoride, MoF_6*

The absorption spectrum of MoF_6 starts at 280 nm; it is an increasingly intense continuum down to 200 nm, and below 200 nm there are several sharp bands that represent a vibrational progression (Tanner and Duncan, 1951; McDiarmid, 1974; see also Fig. 3.5 and Table 3.9). Absorption at wavelengths above 180 nm corresponds to a ligand-to-metal charge transfer, whereby an electron is promoted from a molecular π orbital that is localized mainly on the ligands (the fluorine atoms) to a molecular orbital that is essentially a metal (molybdenum) d orbital (McDiarmid, 1974).

2.6.1.7 *Tungsten hexafluoride, WF_6*

The absorption of WF_6 starts at 225 nm and is a structureless continuum down to 170 nm (Tanner and Duncan, 1951; McDiarmid, 1974; see also Fig. 3.5 and Table 3.9). Absorption at these wavelengths corresponds to a ligand-to-metal charge transfer, whereby an electron is promoted from a molecular π orbital that is localized mainly on the ligands (the fluorine atoms) to a molecular orbital that is essentially a metal (tungsten) d orbital (McDiarmid, 1974).

2.6.1.8 *Rhenium hexafluoride, ReF_6*

The absorption spectrum of ReF_6 consists of a series of weak and sharp bands in the region 1.4–2.1 μm and of a more intense, structured continuum at wavelengths below 245 nm (McDiarmid, 1971; see also Fig. 3.5 and Table 3.9). The first UV feature is a broad and diffuse continuum, which peaks at 210 nm. The second feature, in the region

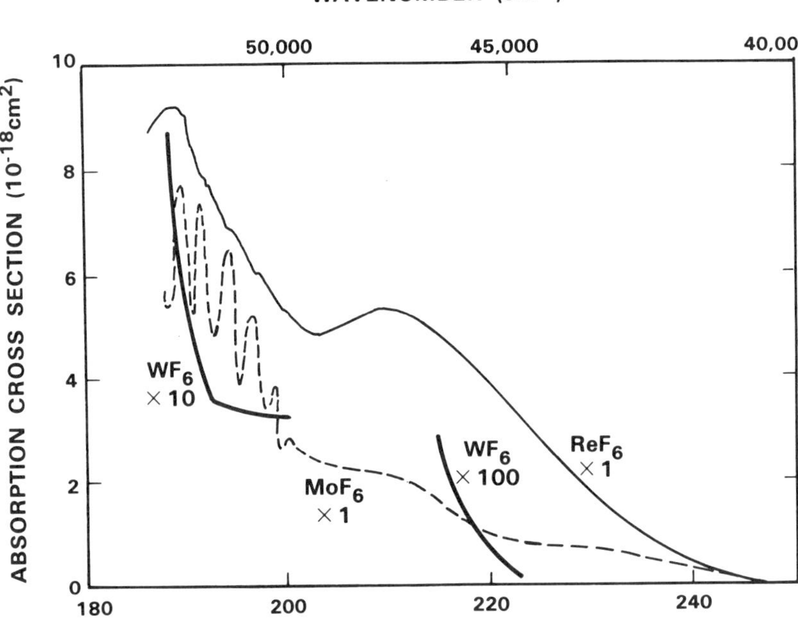

Fig. 3.5 The absorption cross sections of MoF_6, WF_6, and ReF_6 in the region 180–250 nm. The MoF_6 spectrum was adapted with permission from McDiarmid (1974). The WF_6 spectrum was adapted with permission from Tanner and Duncan, *J. Am. Chem. Soc.* **73,** 1164; copyright 1951, American Chemical Society. The ReF_6 spectrum was adapted with permission from McDiarmid (1971).

180–200 nm, is a vibrational progression superimposed on an increasingly intense continuum. Absorption in the UV corresponds to electronic transitions of the charge transfer type (McDiarmid, 1974).

2.6.1.9 Tin tetrachloride, $SnCl_4$

The absorption spectrum of $SnCl_4$ starts at ~280 nm and is a broad band with a peak at ~210 nm (Fernandez et al., 1986; Tabuchi et al., 1987; see also Table 3.9). Absorption in this band corresponds to the transition of a nonbonding electron localized on the Cl atoms to an antibonding σ^* orbital localized on the Sn-Cl bond (Fernandez et al., 1986). The primary photoprocess is thus expected to be molecular dissociation into $SnCl_3$ and Cl.

2.6.2 Metal Carbonyls

2.6.2.1 Chromium hexacarbonyl, $Cr(CO)_6$

The absorption spectrum of $Cr(CO)_6$ consists of several broad overlapping bands, starting at 340 nm (Mayer et al., 1982): two weak shoulders centered at 319 and 195 nm, and two intense lines at 280 and 225 nm (Gray and Beach, 1963; see also Fig. 3.6 and Table 3.10). The 319-nm shoulder is a metal d-d transition where the initial and final molecular orbitals are localized mainly on the metal (chromium) atom and involve 3d atomic levels; the weakness of this transition stems from its forbiddenness in the electric dipole approximation. The two intense lines correspond to metal-to-ligand charge transfer, whereby an electron is

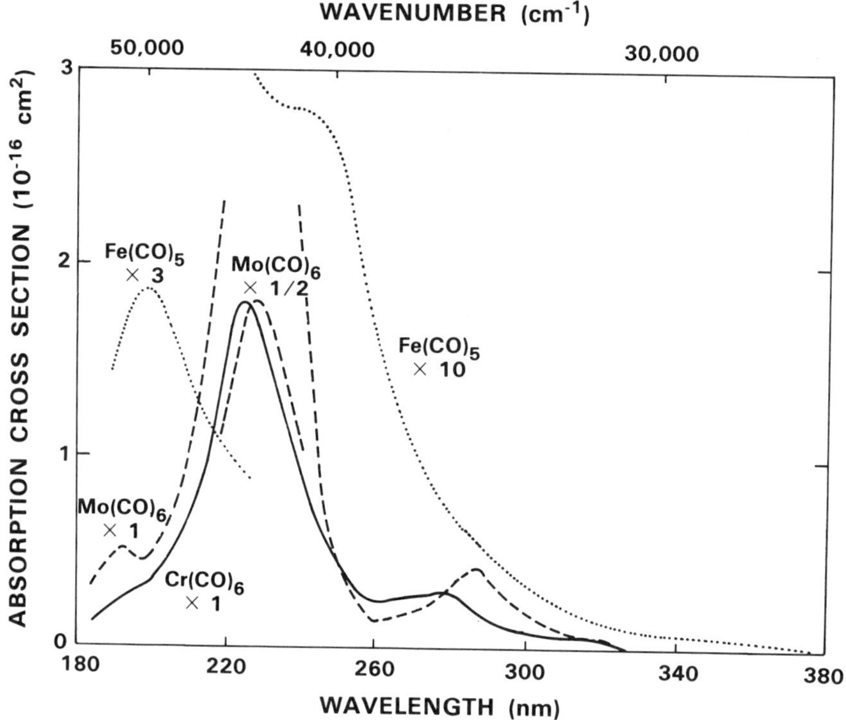

Fig. 3.6 The absorption cross sections of $Cr(CO)_6$, $Mo(CO)_6$, and $Fe(CO)_5$ in the region 180–380 nm. The $Cr(CO)_6$ spectrum was adapted with permission from Mayer et al. (1982). The $Mo(CO)_6$ spectrum was adapted with permission from Gray and Beach, *J. Am. Chem. Soc.* **85,** 2922; copyright 1963, American Chemical Society. The $Fe(CO)_5$ spectrum was adapted with permission from Yardley et al. (1981).

Table 3.10 Absorption cross sections of metal carbonyls.

Molecule	$\sigma_{193\,nm}$ (cm^2)	$\sigma_{248\,nm}$ (cm^2)	$\sigma_{257\,nm}$ (cm^2)	$\sigma_{266\,nm}$ (cm^2)	$\sigma_{308\,nm}$ (cm^2)	Reference
$Cr(CO)_6$	2.6×10^{-17}	5.0×10^{-17}	2.5×10^{-17}	2.5×10^{-17}	5.5×10^{-18}	Mayer et al., 1982
		5.3×10^{-17}				Tumas et al., 1982
	1.2×10^{-17}	3.3×10^{-17}			5.2×10^{-18}	Flynn et al., 1986
$Mo(CO)_6$	5.1×10^{-17}	5.8×10^{-17}	2.0×10^{-17}	1.7×10^{-17}	1.3×10^{-17}	Gray and Beach, 1963
	6.0×10^{-17}	4.4×10^{-17}			1.1×10^{-17}	Flynn et al., 1986
$W(CO)_6$	1.2×10^{-17}	4.5×10^{-18}			2.4×10^{-18}	Flynn et al., 1986
$Fe(CO)_5$	5.7×10^{-17}	2.7×10^{-17}	1.9×10^{-17}	1.3×10^{-17}	2.4×10^{-18}	Nathanson et al., 1981; Yardley et al., 1981
			8.2×10^{-18}			Ehrlich et al., 1981
$Ni(CO)_4$					2.4×10^{-18}	Rösch et al., 1986

promoted from a molecular orbital localized on the metal (chromium) to an antibonding π^* molecular orbital localized mainly on the carbonyl ligands (Gray and Beach, 1963). The origin of the 195-nm feature is uncertain (Beach and Gray, 1968).

The energy required to remove one CO ligand is 1.6 eV, and that required to remove two CO groups is 3.3 eV (Lewis et al., 1984). These values correspond to long-wavelength thresholds of 775 nm and 376 nm, respectively. Nevertheless, at 355 nm the only primary photoprocess is (Breckenridge and Sinai, 1981; Breckenridge and Stewart, 1986)

$$Cr(CO)_6 + h\nu \rightarrow Cr(CO)_5 + CO. \tag{3.36}$$

At 351 nm there is still only one predominant primary photoprocess, namely, the cleavage of a single Cr-CO bond (Seder et al., 1985)

$$Cr(CO)_6 + h\nu \rightarrow Cr(CO)_5 + CO. \tag{3.37}$$

At 248 nm the $Cr(CO)_5$ photoproduct has sufficient internal energy to undergo further rapid (<100 nsec) unimolecular dissociation (Tumas et al., 1982; Fletcher and Rosenfeld, 1985)

$$Cr(CO)_6 + h\nu \rightarrow [Cr(CO)_5]^* + CO \tag{3.38}$$

$$[Cr(CO)_5]^* \rightarrow Cr(CO)_4 + CO. \tag{3.39}$$

Thus, the primary photoproduct at 248 nm is $Cr(CO)_4$, although small amounts of $Cr(CO)_3$ and $Cr(CO)_2$ have also been observed in chemical trapping experiments (Tumas et al., 1982).

At fluences of 3–40 mJ cm^{-2}/pulse, photoexcitation of $Cr(CO)_6$ results also in the formation of excited Cr atoms, apparently via two- or three-photon absorption in the parent molecule (Tyndall and Jackson, 1987).

Secondary reactions involving the various photofragments are:

$$Cr(CO)_5 + CO \rightarrow Cr(CO)_6, \qquad (3.40)$$
$$k \approx 3 \times 10^{-11}\ cm^3\ molec^{-1}\ s^{-1}$$

(Seder et al., 1985; Fletcher and Rosenfeld, 1985);

$$Cr(CO)_4 + CO \rightarrow Cr(CO)_5, \qquad (3.41)$$
$$k \approx 2.3 \times 10^{-10}\ cm^3\ molec^{-1}\ s^{-1}$$

in the high-pressure limit (Fletcher and Rosenfeld, 1986);

$$Cr(CO)_5 + Cr(CO)_6 \rightarrow Cr(CO)_5 \cdot Cr(CO)_6, \qquad (3.42)$$
$$k \approx 4.1 \times 10^{-11}\ cm^{-3}\ molec^{-1}\ s^{-1}$$

(Breckenridge and Stewart, 1986). The complex $Cr(CO)_5 \cdot Cr(CO)_6$ may decompose, and it also further reacts with CO (Breckenridge and Stewart, 1986):

$$Cr(CO)_5 \cdot Cr(CO)_6 \rightarrow Cr(CO)_5 + Cr(CO)_6, \qquad (3.43)$$
$$k = 5 \times 10^5\ s^{-1},$$

$$Cr(CO)_5 \cdot Cr(CO)_6 + CO \rightarrow 2Cr(CO)_6, \qquad (3.44)$$
$$k = 1.2 \times 10^{-11}\ cm^3\ molec^{-1}\ s^{-1}.$$

Also,

$$Cr(CO)_4 + Cr(CO)_6 \rightarrow Cr_2(CO)_{10}, \qquad (3.45)$$
$$k = 5.6 \times 10^{-10}\ cm^3\ molec^{-1}\ s^{-1}$$

(Fletcher and Rosenfeld, 1985).

2.6.2.2 *Molybdenum hexacarbonyl, $Mo(CO)_6$*

The absorption spectrum of $Mo(CO)_6$ starts at 340 nm, and in the range 185–340 nm it has a prominent peak at 228 nm, a weaker one at 287 nm, and three weak shoulders at 193, 265, and 319 nm (Gray and Beach, 1963; Iverson and Russell, 1970; see also Fig. 3.6 and Table 3.10).

The 228-nm and 287-nm features correspond to metal-to-ligand charge transfer, whereby an electron is promoted from a molecular orbital localized mainly on the metal (molybdenum) to an antibonding π^* molecular orbital localized on the ligand (CO groups). The weaker 265-nm and 319-nm structures correlate to d-d electronic transitions, whereby the initial and final molecular orbitals are localized on the molybdenum atom, and the transitions are essentially atomic excitations within the d-electron manifold. Their low intensity is due to their forbidenness in the electric dipole approximation (Gray and Beach, 1963). The origin of the 193-nm line is uncertain (Beach and Gray, 1968).

The dissociation energy of the $Mo(CO)_5$-CO bond is 1.7 eV (Lewis et al., 1984).

2.6.2.3 Tungsten hexacarbonyl, $W(CO)_6$

The absorption spectrum of $W(CO)_6$ exhibits an intense peak at 224 nm, a weaker one at 287 nm, and secondary weak features at 307, 269, and 195 nm (Gray and Beach, 1963; Iverson and Russell, 1970; see also Table 3.10). The first two bands correspond to metal-to-ligand charge transfer, whereby an electron is promoted from a molecular orbital localized on the metal (tungsten) atom to antibonding π^* molecular orbital localized mainly on the ligand (CO groups). The 307-nm and 269-nm absorption features correlate to d-d electronic transitions, whereby the initial and final molecular orbitals are localized on the tungsten atom and are essentially spin-allowed, electric dipole-forbidden excitations within the atomic d-electron energy manifold (Gray and Beach, 1963). The origin of the 195-nm line is uncertain (Beach and Gray, 1968).

The dissociation energy of the $W(CO)_5$-CO bond is 2.0 eV (Lewis et al., 1984).

2.6.2.4 Manganese decacarbonyl, $Mn_2(CO)_{10}$

The absorption of $Mn_2(CO)_{10}$ starts at 350-nm or longer wavelengths, since photochemical studies have been performed in the region 320–351 nm (Freedman and Bersohn, 1978; Leopold and Vaida, 1984; Seder et al., 1986, 1986a; Bray et al., 1986; Prinslow and Vaida, 1987). Absorption in this spectral region corresponds to an electronic σ-σ^* transition, whereby an electron is promoted from a bonding σ molecular orbital localized on the two metal atoms to an antibonding σ^* molecular orbital localized mainly on the same atoms (Seder et al., 1986; Prinslow and Vaida, 1987). Absorption at wavelengths below 270 nm corresponds to the promotion of an electron localized on the metal atoms to an antibonding π^* molecular orbital localized mainly on the ligands (CO groups).

The primary photoprocess in the region 320–351 nm is molecular dissociation, involving homolytic cleavage of the Mn-Mn band (Freedman and Bersohn, 1978; Seder et al., 1986, 1986a; Bray et al., 1986; Prinslow and Vaida, 1987):

$$Mn_2(CO)_{10} + h\nu \rightarrow Mn(CO)_5 + Mn(CO)_5. \qquad (3.46)$$

There is however, evidence that a significant fraction (up to one half) of the $Mn_2(CO)_{10}$ molecules undergo dissociative loss of (apparently two) CO groups (Seder et al., 1986, 1986a; Prinslow and Vaida, 1987):

$$Mn_2(CO)_{10} + h\nu \rightarrow Mn_2(CO)_8 + 2CO. \quad (3.47)$$

The nascent $Mn(CO)_5$ are vibrationally excited, and the rate constant for collisional deactivation by $Mn_2(CO)_{10}$ is 2.7×10^{-10} cm^3 $molec^{-1}$ s^{-1} (Bray et al., 1986).

The primary photoprocess at 248 and 193 nm is molecular dissociation by loss of a CO ligand (Seder et al., 1986a; Prinslow and Vaida, 1987):

$$Mn_2(CO)_{10} + h\nu \rightarrow Mn_2(CO)_9 + CO. \quad (3.48)$$

At 193 nm the $Mn_2(CO)_9$ is formed with high internal excitation, and it apparently further decomposes into $Mn_2(CO)_5$ and four CO groups (Seder et al., 1986a; Prinslow and Vaida, 1987).

Secondary reactions are (Seder et al., 1986a):

$$2Mn(CO)_5 \rightarrow Mn_2(CO)_{10}, \quad (3.49)$$

$$k \approx 7.5 \times 10^{-11}\ cm^3\ molec^{-1}\ s^{-1}$$

$$Mn_2(CO)_9 + CO \rightarrow Mn_2(CO)_{10}, \quad (3.50)$$

$$k \approx 4.0 \times 10^{-15}\ cm^3\ molec^{-1}\ s^{-1}$$

2.6.2.5 Rhenium decacarbonyl, $Re_2(CO)_{10}$

The first absorption band of $Re_2(CO)_{10}$ peaks at 300 nm (Freedman and Bersohn, 1978). It corresponds to a σ-σ^* transition, whereby an electron is promoted from a bonding σ molecular orbital localized on the rhenium atoms to an antibonding σ^* orbital localized on these atoms.

The primary photoprocess at 300 nm is homolytic cleavage of the Re-Re bond,

$$Re_2(CO)_{10} + h\nu \rightarrow 2Re(CO)_5, \quad (3.51)$$

with two thirds of the available energy as internal excitation of the nascent fragments and one third as kinetic energy (Freedman and Bersohn, 1978).

2.6.2.6 Iron pentacarbonyl, $Fe(CO)_5$

The absorption spectrum of $Fe(CO)_5$ is a continuum that starts at 380 nm, has a shoulder at 250 nm, and has a broad peak at 200 nm (Nathanson et al., 1981; Yardley et al., 1981; see also Fig. 3.6 and Table 3.10).

The long-wavelength absorption corresponds to ligand field d-d electronic transition, whereby the initial and final molecular orbitals are localized on the Fe atom and the photoexcitation takes place within the atomic d-electron manifold. At shorter wavelengths, the absorption spectrum is a superposition of several closely spaced bands (Dick et al., 1982; Daniel et al., 1984). Most of them correspond to metal-to-ligand charge transfer, whereby an electron is promoted from a molecular orbital localized on the Fe atom to an antibonding π^* molecular orbital localized mainly on the ligands (CO groups). The excited states corresponding to these transitions are further split into singlets and triplets, and a total of 18 states exist between 4.2 eV (295 nm) and 6.3 eV (196 nm) (Daniel et al., 1984).

The primary photoprocess at all wavelengths is molecular dissociation. The dissociative lifetime of the $Fe(CO)_5$ excited state depends on the excitation wavelength: at 310–300 nm it is 2 psec and at 280–275 nm it is 0.6 psec (Whetten et al., 1983). The distribution of the primary photoproducts is also wavelength dependent (Yardley et al., 1981): at 352 nm, the ratio $Fe(CO)_4$:$Fe(CO)_3$:$Fe(CO)_2$ is 0.23:0.46:0.31; at 248 it is 0.10:0.35:0.55; and at 193 nm it is 0.09:0.09:0.81, with a trace of Fe(CO). Thus, shorter lifetimes and increased decarbonylation are observed with increasing photon energy. Formation of $Fe(CO)_x$ (x = 2, 3, 4) following single-photon excitation is apparently a sequential elimination process, resulting from the high internal energy retention of successive fragments, followed by unimolecular decomposition (Yardley et al., 1981; Seder et al., 1986b; Waller et al., 1987). The energy partitioning among the various degrees of freedom was modeled by Yardley et al. (1981) to be nonstatistical. However, the experimentally determined rotational, vibrational, and translational energy distribution in the CO fragments is consistent with an entirely statistical process (Waller et al., 1987).

At 351 nm the spin-forbidden $^3E_1$ state is populated directly; at 248 nm it is populated by internal conversion from the higher-lying spin-allowed ^1E state (Daniel et al., 1984). The ^3E state then decomposes into $Fe(CO)_x$ ($x = 2, 3, 4$) whose ground states are also all triplets (Daniel et al., 1984; Seder et al., 1986b). At 193 nm an additional dissociation channel may open up, namely, the excitation of a high-lying ^1E state in $Fe(CO)_5$, which then dissociates into an excited state (^1E) of $Fe(CO)_3$ via the ^1A state of $Fe(CO)_4$ (Seder et al., 1986b).

Secondary reactions are (Seder et al., 1986b):

$$Fe(CO)_4 + CO \rightarrow Fe(CO)_5, \qquad k \approx 5.8 \times 10^{-14}\ cm^3\ molec^{-1}\ s^{-1} \tag{3.52}$$

$$Fe(CO)_3 + CO \rightarrow Fe(CO)_4, \qquad k \approx 2.2 \times 10^{-11}\ cm^3\ molec^{-1}\ s^{-1} \tag{3.53}$$

$$Fe(CO)_2 + CO \rightarrow Fe(CO)_3, \qquad k \approx 3.0 \times 10^{-11}\ cm^3\ molec^{-1}\ s^{-1}. \tag{3.54}$$

2.6.2.7 Nickel tetracarbonyl, $Ni(CO)_4$

The absorption spectrum of $Ni(CO)_4$ is a continuum that starts at ~380 nm, has shoulders at ~250 and 220 nm, and peaks at ~205 nm (Schreiner and Brown, 1968; Preston and Zink, 1987; see also Table 3.10).

The spectrum corresponds to metal-to-ligand charge transfer transitions, whereby an electron is promoted from a molecular orbital localized mainly on the nickel atom to an antibonding π^* orbital localized mainly on the CO ligands (Schreiner and Brown, 1968). However, ligand-field $d \rightarrow s$ and $d \rightarrow p$ transitions of an electron localized on the nickel atom may also contribute to the absorption spectrum (Dick et al., 1982).

The primary photoprocess in UV flash photolysis is molecular decomposition by sequential two-photon absorption (Callear, 1961):

$$Ni(CO)_4 + h\nu \rightarrow Ni(CO)_3 + CO \tag{3.55}$$

$$Ni(CO)_3 + h\nu \rightarrow Ni(CO)_2 + CO. \tag{3.56}$$

The dissociative lifetime of excited-state $Ni(CO)_4$ is 7 ns (Callear, 1961).

At 308 nm, a primary photoprocess following single-photon excitation is the formation of electronically excited $Ni(CO)_3$, which decays within >10 μs via visible emission peaked at ~650 nm. The quantum yield of this dissociation channel is ~0.1 (Rösch et al., 1986). Excitation at 315 or 364 nm is followed by visible fluorescence, with a maximum at 730 nm and a lifetime of ~200 ns; this fluorescence may originate from excited $Ni(CO)_3$ (Preston and Zink, 1987).

2.6.3 Metal Alkyls

2.6.3.1 Dimethylzinc, $Zn(CH_3)_2$

The absorption spectrum of $Zn(CH_3)_2$ is a weakly structured continuum that starts at 250 nm and peaks at 200 nm (Chen and Osgood, 1984; see also Fig. 3.7 and Table 3.11). The spectrum can be resolved into two features: a long-wavelength continuum that peaks at 210 nm and a series of vibrational bands in the region 215–195 nm (Chen and Osgood, 1984). These features correspond to photoexcitation of the ground-state $X\,^1\Sigma_g^+$ into the first ($A\,^1A_1$) and second ($B\,^1\Pi$) electronically excited states, respectively; the A state has a bent geometry and correlates to ground-state $ZnCH_3$, while the B state is linear and correlates to electronically excited $ZnCH_3$ (Chen and Osgood, 1984).

The primary photoprocess at all wavelengths is molecular dissociation into Zn and $2CH_3$, possibly by sequential elimination of the two methyl

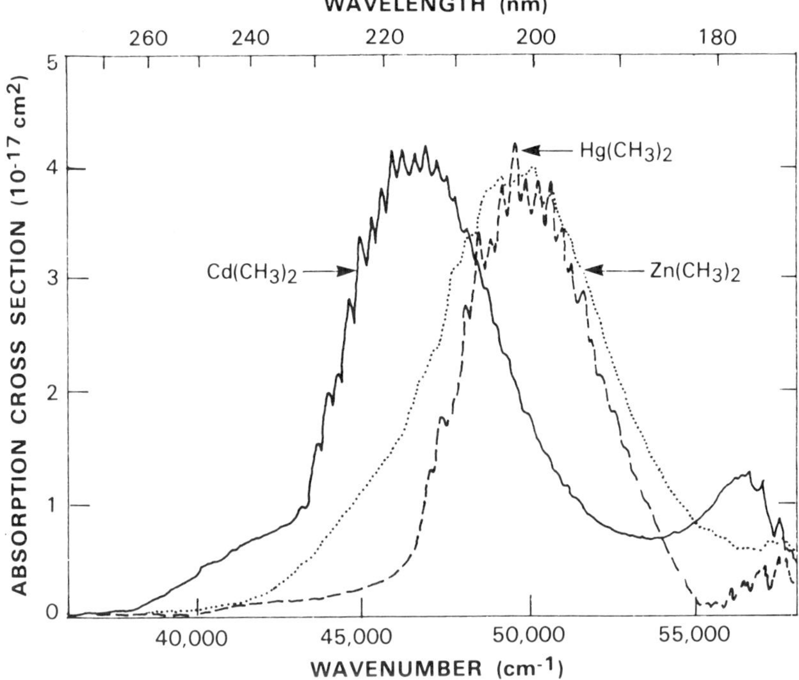

Fig. 3.7 The absorption cross sections of $Zn(CH_3)_2$, $Cd(CH_3)_2$, and $Hg(CH_3)_2$ in the region 170–280 nm. Adapted with permission from Chen and Osgood (1984).

groups (Chen and Osgood, 1984). The CH_3 fragments at 248 and 193 nm possess a high degree of vibrational excitation, the out-of-plane umbrella mode being more populated than the C-H stretching mode (Chu et al., 1985). The fraction of $Zn(CH_3)_2$ that at 193 nm is excited to the B $^1\Pi_u$ state dissociates first into electronically excited $ZnCH_3$ and CH_3; radiative decay of $ZnCH_3$ to its electronic ground state is observed at 410–440 nm with a lifetime of ~65 ns (Yu et al., 1986). The average translational energy of the CH_3 photofragments at 193 nm is 0.2 eV (Yu et al., 1986).

2.6.3.2 Diethylzinc, $Zn(C_2H_5)_2$

The absorption spectrum of $Zn(C_2H_5)_2$ above 200 nm is a continuum that starts at 280 nm and peaks at 220 nm; several diffuse bands are superimposed on the continuum in the region 240–220 nm (Thompson, 1934; Krchnavek et al., 1987; see also Fig. 3.8).

Table 3.11 Absorption cross sections of metal alkyls.

Molecule	$\sigma_{193\,nm}$ (cm^2)	$\sigma_{248\,nm}$ (cm^2)	$\sigma_{257\,nm}$ (cm^2)	$\sigma_{266\,nm}$ (cm^2)	Reference
$Zn(CH_3)_2$	1.5×10^{-17}				Chen and Osgood, 1984
$Cd(CH_3)_2$			1.4×10^{-18}		Ehrlich et al., 1982
	4×10^{-18}	2.1×10^{-18}	8.9×10^{-19}	4.5×10^{-19}	Irvine et al., 1984
	9.1×10^{-18}	4.0×10^{-18}	1.7×10^{-18}		Chen and Osgood, 1984
$Hg(CH_3)_2$	2.5×10^{-17}	3.2×10^{-19}			Baughcum and Leone, 1982
		7.1×10^{-20}	5×10^{-20}		Irvine et al., 1984
	2.7×10^{-17}	2×10^{-19}			Chen and Osgood, 1984
$Al(CH_3)_3$		1.1×10^{-20}	3×10^{-21}		Ehrlich et al., 1982
	2.3×10^{-17}				Gilgen et al., 1984
	1.8×10^{-17}				Suzuki et al., 1986
$Al(C_2H_5)_3$	3.3×10^{-18}				Haigh, 1983
	6.1×10^{-18}				Tokumitsu et al., 1988
$Ga(CH_3)_3$	4.2×10^{-18}				Haigh, 1983
	5.4×10^{-18}	2.4×10^{-19}	8.7×10^{-20}		Rytz-Froidevaux et al., 1983
	2.6×10^{-17}	7.7×10^{-19}			Gilgen et al., 1984
	2.6×10^{-17}	2.1×10^{-18}	1.8×10^{-18}		McCrary and Donnelly, 1987
	1.3×10^{-17}				Tokumitso et al., 1988
$Ga(C_2H_5)_3$	6.6×10^{-18}				Haigh, 1983
	5.7×10^{-18}				Tokumitso et al., 1988
	9×10^{-18}				McCrary and Donnelly, 1987
$In(CH_3)_3$	1.5×10^{-17}	6.9×10^{-18}			Karlicek et al., 1984
	1.2×10^{-17}	1.5×10^{-18}	1.5×10^{-18}	1.5×10^{-18}	Gilgen et al., 1984
	1.0×10^{-17}	1.2×10^{-18}			Zuhoski et al., 1988
$Sn(CH_3)_4$	4.0×10^{-17}				Fernandez et al., 1988
$Pb(CH_3)_4$			3.7×10^{-19}	6.2×10^{-20}	Leighton and Mortensen, 1936

2.6.3.3 Dimethylcadmium, $Cd(CH_3)_2$

The absorption spectrum of $Cd(CH_3)_2$ above 190 nm consists of a weak continuum that starts at 265 nm and a more intense broad band starting at 230 nm and peaking at 215 nm; below 230 well-resolved vibrational bands are superimposed on the continuum (Irvine et al., 1984; Chen and Osgood, 1984; see also Fig. 3.7 and Table 3.11).

The spectrum can be resolved into two features (Chen and Osgood, 1984): a long-wavelength continuum that peaks at ~240 nm, and that corresponds to photoexcitation from the linear $X\,^1\Sigma_g^+$ ground state to the (bent) first excited state $A\,^1A_1$; and a series of vibrational bands in the region 230–190 nm, which corresponds to photoexcitation from $X\,^1\Sigma_g^+$ to the second excited state, the linear $B\,^1\Pi_u$. The A state correlates to ground-state $CdCH_3$, and the B state correlates to electronically excited $CdCH_3$ (Chen and Osgood, 1984).

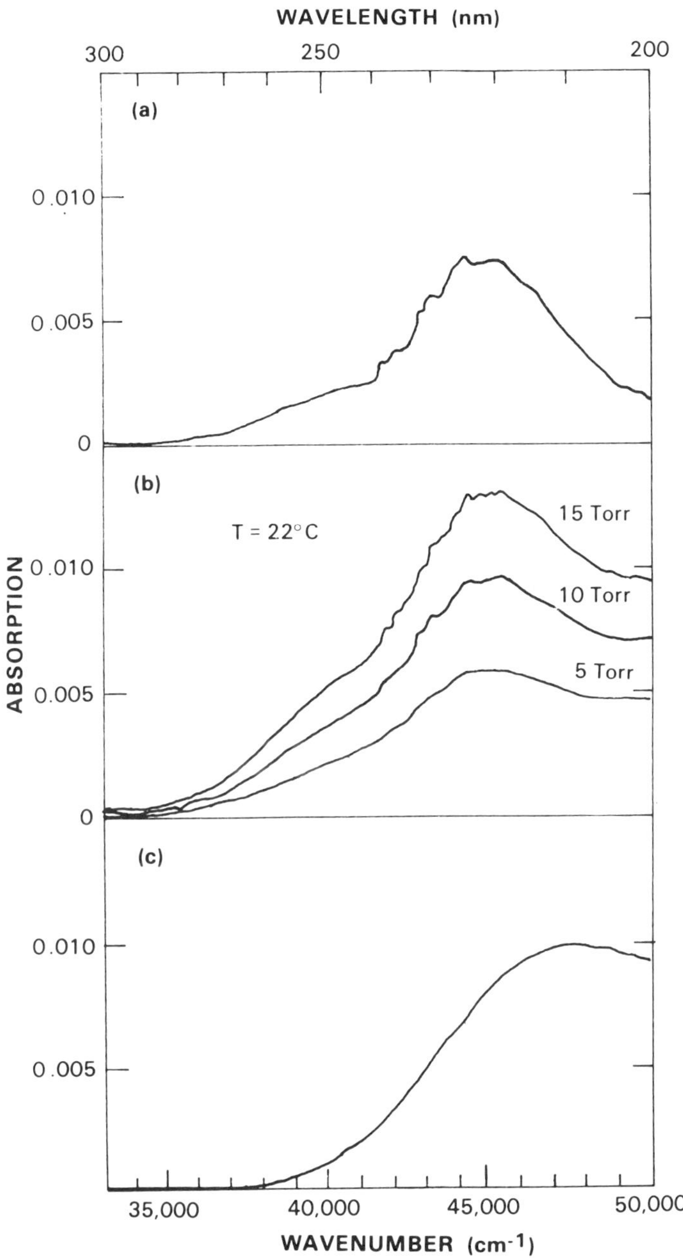

Fig. 3.8 The fractional absorption spectrum of gaseous and adsorbed $Zn(C_2H_5)_2$ in the region 200–300 nm. (a) The absorption of an equivalent monolayer of gas-phase $Zn(C_2H_5)_2$. (b) The absorption of physisorbed $Zn(C_2H_5)_2$ on fused silica at room temperature for three values of $Zn(C_2H_5)_2$ pressure. (c) The absorption of $Zn(C_2H_5)_2$ chemisorbed on fused silica. Adapted with permission from Krchnavek et al. (1987).

The primary photoprocess at all wavelengths is molecular dissociation into a Cd atom and two CH_3 radicals, possibly by sequential elimination of the methyl groups (Jonah et al., 1971; Irvine et al., 1984; Chen and Osgood, 1984). The CH_3 fragments at 248 and 193 nm possess a high degree of vibrational excitation, the out-of-plane umbrella mode being more populated than the C-H stretching mode (Chu et al., 1984). At 193 nm practically all the excitation is into the B state, which dissociates into electronically excited $CdCH_3$ and CH_3; radiative decay of $CdCH_3$ to its electronic ground state is observed at 420–460 nm, with a lifetime of ~70 ns (Yu et al., 1986). The average translational energy of the CH_3 photofragments at 193 nm is 0.25 eV (Yu et al., 1986).

2.6.3.4 *Dimethyl mercury, $Hg(CH_3)_2$*

The absorption spectrum of $Hg(CH_3)_2$ starts at 250 nm, and it consists of a weakly modulated continuum to 220 nm and a strongly structured intense band in the region 220–190 nm (Irvine et al., 1984; Chen and Osgood, 1984; see Fig. 3.7 and Table 3.11).

The two spectral features correspond to photoexcitation of the $X\,^1\Sigma_g^+$ ground state to the first $(A\,^1A_1)$ and second $(B^1\,\Pi_u)$ excited electronic states, respectively. The A state has bent geometry and correlates to ground-state $HgCH_3$ and CH_3, while the B state is linear, and correlates to electronically excited $HgCH_3$ and CH_3 (Chen and Osgood, 1984).

The primary photoprocess at all wavelengths is molecular dissociation into a Hg atom and two CH_3 radicals, possibly by sequential elimination of the methyl groups (Irvine et al., 1984; Chen and Osgood, 1984). The methyl fragments at 248 and 193 nm are characterized by a high degree of vibrational excitation in the out-of-plane umbrella mode and in the C-H stretching mode; a rotational temperature of 1200–1500°K was determined for the stretching mode at 248 nm (Baughcum and Leone, 1982).

2.6.3.5 *Trimethylaluminum, $Al(CH_3)_3$*

The absorption spectrum of $Al(CH_3)_3$ above 200 nm is a continuum that starts at 260 nm and peaks at ~190 nm (Ehrlich et al., 1982; Gilgen et al., 1984; Suzuki et al., 1986; see also Table 3.11). The main primary photoprocess in the UV is elimination of one or more methyl groups (Ehrlich et al., 1982). At 248 nm most molecules undergo complete photodissociation into Al and three CH_3 groups, while at 193 nm complete dissociation occurs in only ~80% of the molecules, the remaining 20% retaining one methyl to form $AlCH_3$ (Stuke et al., 1988). At room temperature $Al(CH_3)_3$ is mostly dimerized as $Al_2(CH_3)_6$ (Laubengayer and Gilliam, 1947).

2.6.3.6 *Triethylaluminum, $Al(C_2H_5)_3$*

The absorption spectrum of $Al(C_2H_5)_3$ is a continuum that starts at ~370 nm and has a broad peak in the range 190–210 nm (Haigh, 1983; see also Table 3.11). The primary photoprocesses at 193 and 248 nm are dissociation into Al and $3C_2H_5$, and into AlC_2H_5 and $2C_2H_5$, with relative yields of 2:1, respectively; the AlC_2H_5 photofragment decomposes by β elimination to form AlH and C_2H_6 with unit efficiency (Stuke et al., 1988).

2.6.3.7 *Trimethylgallium, $Ga(CH_3)_3$*

The absorption spectrum of $Ga(CH_3)_3$ is a continuum that starts at 260 nm and peaks at 200 nm (Haigh, 1983; Rytz-Froidevaux et al., 1983; Gilgen et al., 1984; McCrary and Donnelly, 1987; see also Table 3.11).

The primary photoprocess at 193 nm is (Donnelly et al., 1987)

$$Ga(CH_3)_3 + h\nu \rightarrow GaCH_3 + 2CH_3. \tag{3.57}$$

A secondary photoprocess is (Donnelly et al., 1987)

$$GaCH_3 + h\nu \rightarrow Ga + CH_3, \qquad \sigma \approx 4 \times 10^{-17}\ \text{cm}^2. \tag{3.58}$$

2.6.3.8 *Triethylgallium, $Ga(C_2H_5)_3$*

The absorption spectrum of $Ga(C_2H_5)_3$ is a continuum that starts at ~310 nm, has a shoulder at 270 nm, and has a broad peak in the region 190–210 nm (Haigh, 1983; McCrary and Donnelly, 1987; see also Table 3.11).

2.6.3.9 *Trimethylindium, $In(CH_3)_3$*

The absorption spectrum of $In(CH_3)_3$ is a broad band starting at 300 nm, with a shoulder at 250 nm, and peaking at 210 nm (Haigh, 1983; Karlicek et al., 1984; Gilgen et al., 1984; Zuhoski et al., 1988; see also Table 3.11).

2.6.3.10 *Triethylindium, $In(C_2H_5)_3$*

The absorption spectrum of $In(C_2H_5)_3$ is a broad band starting at ~350 nm, and peaking at 230 nm, followed by a second band below 190 nm (Haigh, 1983).

2.6.3.11 *Tetramethyltin, $Sn(CH_3)_4$*

The absorption spectrum at wavelengths above 180 nm is a broad band that starts at 220 nm and peaks at 186 nm (Fernandez et al., 1986; see also Table 3.11). This band is attributed to a pure Rydberg-type transition (Fernandez et al., 1986). Two primary photoprocesses are observed at 193 nm: molecular dissociation into $Sn(CH_3)_3$ and CH_3, and dissociation into $Sn(CH_3)_2$ and $2CH_3$, with relative yields of 1 and 0.6, respectively (Kawasaki et al., 1987).

2.6.3.12 Tetramethyl lead, $Pb(CH_3)_4$

The absorption spectrum of $Pb(CH_3)_4$ above 255 nm is a continuum that starts at ~280 nm (Leighton and Mortensen, 1936; see also Table 3.11). At 240 nm and 295°K, the absorption cross section is $6 \times 10^{-18}\,cm^2$ (Homer and Hurle, 1972). Photolysis at 254 nm results in molecular decomposition into a Pb atom and organic fragments, with a quantum yield slightly higher than 1.0 (Leighton and Mortensen, 1936).

2.6.3.13 Tetraethyl lead, $Pb(C_2H_5)_4$

The absorption of $Pb(C_2H_5)_4$ has been reported to be continuous starting at 255 nm (Duncan and Murray, 1934) or at 350 nm (Leighton and Mortensen, 1936), or to consist of sharp lines in the region 270–225 nm, with only a weak continuum below 220 nm (Thompson, 1934). At 254 nm the absorption cross section is $3.8 \times 10^{-18}\,cm^2$ (Rigby, 1969). The primary photoprocess in the UV is (Rigby, 1969)

$$Pb(C_2H_5)_4 + h\nu \rightarrow Pb(C_2H_5)_3 + C_2H_5. \qquad (3.59)$$

2.7. Miscellaneous Molecules Containing Nitrogen and Oxygen

2.7.1 Ammonia, NH_3

The absorption spectrum of NH_3 starts at 220 nm, and in the range 170–220 nm it consists of a vibrational sequence superimposed on a continuum that peaks at ~190 nm (Watanabe, 1954; Douglas, 1963; see also Table 3.12). Absorption in this wavelength region corresponds to a transition from the ground $\tilde{X}\,^1A_1$ state to the predissociative $\tilde{A}\,^1A_2''$ excited state (Ashfold et al., 1986).

The major primary photoprocess in the region 185–214 nm is molecular dissociation into NH_2 and H. Partitioning of the excess energy is wavelength dependent but as a rule less than one half is kinetic energy of the H atom (Back and Koda, 1977; Koplitz et al., 1987). The remainder is internal excitation of NH_2, mostly as ro-vibrational excitation of the NH_2 ($\tilde{X}\,^2B_1$) fragment (Di Stefano et al., 1977a). At 193 nm, the quantum yield for the production of electronically excited $NH_2(\tilde{A}\,^2A_1)$ is only ~2.5% (Donnelly et al., 1979), even though 70% of

Table 3.12 Absorption cross sections of molecules containing nitrogen and oxygen.

Molecule	$\sigma_{193\,nm}$ (cm^2)	$\sigma_{248\,nm}$ (cm^2)	$\sigma_{257\,nm}$ (cm^2)	$\sigma_{266\,nm}$ (cm^2)	$\sigma_{308\,nm}$ (cm^2)	Reference
NH_3	5.2×10^{-18}					Donnelly et al., 1979
	1.2×10^{-17}					Kenner et al., 1985
H_2O_2	$\sim6\times10^{-19}$	7.7×10^{-20}	6.5×10^{-20}	4.3×10^{-20}	4.2×10^{-21}	Lin et al., 1978
	6.0×10^{-19}	8.2×10^{-20}	5.7×10^{-20}	3.7×10^{-20}	4.2×10^{-21}	Molina and Molina, 1981
HNO_3	1.1×10^{-17}	2.0×10^{-20}	1.9×10^{-20}	1.8×10^{-20}	1.2×10^{-21}	Johnston and Graham, 1973
	1.3×10^{-17}	2.0×10^{-20}	1.9×10^{-20}	1.7×10^{-20}	1.0×10^{-21}	Molina and Molina, 1981
NO	2.0×10^{-20}					Rothschild and Ehrlich, 1986
N_2O	9.3×10^{-20}					Zelikoff et al., 1953
	7.9×10^{-20}					Hubrich and Stuhl, 1980
NO_2	6.4×10^{-19}	3.3×10^{-20}	4.8×10^{-20}	5.6×10^{-20}		Nakayama et al., 1959
	2.6×10^{-19}	1.4×10^{-20}	1.8×10^{-20}	2.4×10^{-20}	1.6×10^{-19}	Bass et al., 1976
O_2	1.4×10^{-21}					Rothschild and Ehrlich, 1986

the NH_2 molecules have energies in excess of the $\tilde{A}\,^2A_1$ origin (Koplitz et al., 1987). A minor primary photoprocess at 193 nm is molecular dissociation into NH (a $^1\Delta$) and H_2, with a quantum yield of less than 1% (Kerner et al., 1987).

Secondary reactions are

$$NH_2(\tilde{X}) + NH_2(\tilde{X}) \rightarrow N_2H_4, \quad (3.60)$$

$$k = 3.9 \times 10^{-12}\ cm^3\ molec^{-1}\ s^{-1}$$

(Hanes and Bair, 1963) and

$$NH_2(\tilde{A}) + NH_3(\tilde{X}) \rightarrow NH_2(\tilde{X}) + NH_3, \quad (3.61)$$

$$k = 6.1 \times 10^{-10}\ cm^3\ molec^{-1}\ s^{-1}$$

(Donnelly et al., 1979). The radiative decay of $NH_2(\tilde{A})$ in the 660- to 1100-nm region has a lifetime of 31 μs (Donnelly et al., 1979). A secondary photoreaction at 193 nm (20–100 mJ cm^{-2}) is absorption by the primary photofragment NH_2 followed by dissociation into NH(A $^3\Pi$) and H, and by subsequent radiative decay of NH(A $^3\Pi \rightarrow X\,^3\Sigma^-$) at 336 nm (Donnelly et al., 1979; Kenner et al., 1985; Ni et al., 1986b).

2.7.2 Water, H_2O

The absorption spectrum of H_2O is a continuum that starts at ~190 nm and peaks at 167 nm (Watanabe and Zelikoff, 1953; Wang et al., 1977).

The primary photoprocess at 193 nm is molecular dissociation into electronic ground-state fragments (Grunewald et al., 1985)

$$H_2O(\tilde{X}\,^1A_1) + h\nu \rightarrow OH(X\,^2\Pi) + H(^2S). \quad (3.62)$$

The OH fragment is vibrationally cold, with less than 3% in $v = 1$, and the rotational temperature is ~400°K; approximately 97% of the excess energy is found as kinetic energy of the photogragments (Grunewald et al., 1987).

2.7.3 Hydrogen Peroxide, H_2O_2

The absorption spectrum of H_2O_2 is a continuum that starts at ~330 nm and increases continuously to 180 nm (Lin et al., 1978; Molina and Molina, 1981; Suto and Lee, 1983; see also Table 3.12).

The primary photoprocess in this wavelength region is molecular photodissociation into electronic ground-state fragments

$$H_2O_2 + h\nu \rightarrow 2OH(X\,^2\Pi). \tag{3.63}$$

The excess energy is found predominantly as kinetic energy of the photofragments. At 266 nm there is no vibrational excitation and the rotational temperature is 1530°K (Klee et al., 1986; Gericke et al., 1986); at 248 nm there is no vibrational excitation, and up to 11% of the excess energy is rotational excitation (Ondrey et al., 1983; Docker et al., 1986); at 193 nm up to 16% of the excess energy is converted into rotational excitation of the OH fragments, but still with no vibrational excitation (Ondrey et al., 1983; Jacobs et al., 1983; Grunewald et al., 1986; Jacobs et al., 1987).

Although the thermodynamic threshold for formation of electronically excited $OH(A\,^2\Sigma^+)$ is ~200 nm, the quantum yield for this dissociation channel is negligible above 175 nm (Suto and Lee, 1983).

At 193 nm (1–30 mJ cm^{-2}) a primary photoprocess is also another molecular dissociation channel,

$$H_2O_2 + h\nu \rightarrow H + HO_2, \tag{3.64}$$

which has a quntum yield of 12% (Gerlach-Meyer et al., 1987).

2.7.4 Nitric acid, HNO_3

The absorption spectrum of HNO_3 is a continuum that starts at ~310 nm, levels off in the range of 250–270 nm, and then increases to at least 190 nm (Johnston and Graham, 1973; Molina and Molina, 1981; see also Table 3.12).

The primary photoprocess in the UV is molecular photodissociation with unit quantum yield (Johnston et al., 1974)

$$HNO_3 + h\nu \rightarrow OH + NO_2. \tag{3.65}$$

The nascent OH radicals at 193 nm have no vibrational excitation and little, non-Boltzmannian rotational excitation (Jacobs et al., 1983).

A secondary reaction is (Johnston et al., 1974)

$$OH + HNO_3 \rightarrow H_2O + NO_3, \quad (3.66)$$

$$k = 1.5 \times 10^{-13}\ cm^3\ molec^{-1}\ s^{-1}.$$

2.7.5 Nitric Oxide, NO

The absorption spectrum of NO below 230 nm consists of a series of sharp lines superimposed on a weak continuum that starts at ~200 nm (Okabe, 1978 and references therein; see also Table 3.12). The lines are assigned to several ro-vibrational progressions: the γ bands ($A\,^2\Sigma^+ \leftarrow X\,^2\Pi$) starting at ~230 nm, the β bands ($B\,^2\Pi \leftarrow X\,^2\Pi$) starting at ~210 nm, and the δ bands ($C\,^2\Pi \leftarrow X\,^2\Pi$) starting at 190 nm; at shorter wavelengths there are the ε, β', and γ' bands, corresponding to transitions to the $D\,^2\Sigma^+$, $B'\,^2\Delta$, and $E\,^2\Sigma^+$ states, respectively.

Photoexcitation at wavelengths longer than ~195 nm results in mainly nondissociative processes (fluorescence, collisional deactivation, intramolecular energy conversion), since the N-O bond energy is 6.5 eV, corresponding to the 191-nm wavelength; below 191 nm predissociation is the dominant primary photoprocess (Okabe, 1978 and references therein). At 193 nm, rotationally excited molecules can be promoted to the $B\,^2\Pi$ ($v' = 7$) state, from which radiative decay has been observed in the region 200–300 nm (Shibuya and Stuhl, 1982).

2.7.6 Nitrous Oxide, N_2O

The absorption spectrum of N_2O is a weakly structured continuum that starts at 210 nm and peaks at 183 nm (Zelikoff et al., 1953; Hubrich and Stuhl, 1980; see also Table 3.12).

The primary photoprocess at wavelengths longer than 185 nm is molecular dissociation into N_2 and metastable $O(^1D)$ (Okabe, 1978; Zavelovich et al., 1981)

$$N_2O + h\nu \rightarrow N_2 + O(^1D). \quad (3.67)$$

Photodissociation into $N_2 + O(^3P)$ or $NO + N(^4S)$, although energetically possible, is spin forbidden and has not been observed (Okabe, 1978). Secondary processes are the reaction of $O(^1D)$ with N_2O (Schofield, 1978; Amimoto et al., 1979; Davidson et al., 1979)

$$O(^1D) + N_2O \rightarrow N_2 + O_2, \quad (3.68)$$

$$k = 6.7 \times 10^{-11}\ cm^3\ molec^{-1}\ s^{-1},$$

$$O(^1D) + N_2O \rightarrow NO + NO, \quad (3.69)$$

$$k = 5.3 \times 10^{-11}\ cm^3\ molec^{-1}\ s^{-1}.$$

Some of the NO molecules thus formed are vibrationally excited, up to at least $v = 7$ (Chamberlain and Simons, 1975; Zavelovich et al., 1981). A secondary photoprocess at 193 nm is the excitation of these NO molecules to high-lying electronic states (Zavelovich et al., 1981).

2.7.7 Nitrogren Dioxide, NO_2

The absorption spectrum of NO_2 is a complex, structured continuum throughout the visible and the UV (Hall and Blacet, 1952; Nakayama et al., 1959; Bass et al., 1976; Okabe, 1978; Merer and Hallin, 1978; see also Table 3.12). Below ~400 nm the dimer N_2O_4 exhibits intense absorption (Bass et al., 1976).

The primary photoprocess at wavelengths below 398 nm is molecular dissociation (Okabe, 1978 and references therein)

$$NO_2 + h\nu \rightarrow NO + O. \tag{3.70}$$

The oxygen atom is in its ground ^3P state above 244 nm; below this wavelength the quantum yield for the formation of metastable $O(^1D)$ increases to $\sim 0.4 \pm 0.1$ in the range 242–241 nm (Uselman and Lee, 1976). Although energetically possible below 275 nm, there is no evidence of photodissociation into $N(^4S) + O_2(^3\Sigma_g^-)$ (Preston and Cvetanovič, 1966). The nascent NO photofragments at wavelengths longer than 244 nm possess considerable internal excitation: The vibrational population is inverted, peaking at values as high as $v = 6$–8 at 248 nm, and the rotational distribution exhibits a low temperature for low J values and a high temperature for higher J values (Zacharias et al., 1981; McKendrick et al., 1982; Morrison and Grant, 1982; Slanger et al., 1983). The vibrational deactivation of NO ($v = 8$) by NO_2 has a rate constant of $1.1 \times 10^{-11}\ cm^3\ molec^{-1}\ s^{-1}$ (Slanger et al., 1983).

The primary photoprocess at wavelengths longer than 398 nm is electronic excitation to one or more upper states, followed by radiative decay in the visible, intramolecular energy conversion, and collisional deactivation (Okabe, 1978, and references therein).

2.7.8 Oxygen, O_2

The absorption spectrum of O_2 is a series of rotationally resolvable vibrational bands starting at 205 nm, which are superimposed on an increasingly intense continuum below 190 nm (Ackerman et al., 1970; Hudson and Mahle, 1972; see also Table 3.12). Absorption in these bands (the so-called Schumann-Runge system) corresponds to the electronic transition $B\,^3\Sigma_u^- \leftarrow X\,^3\Sigma_g^-$. The B state is perturbed by a repulsive

$^5\Pi_u$ state, which induces predissociation with an efficiency that peaks at $v = 4$ (Julienne and Krauss, 1975).

The primary photoprocess at wavelengths shorter than ~194 nm is molecular dissociation into two ground-state oxygen atoms within 10^{-11}–10^{-12} s (Ackerman et al., 1970; Hudson and Mahle, 1972). Radiative decay (210–570 nm) to the electronic ground state has also been observed, even from the most strongly predissociated $v = 4$ level (Creek and Nichols, 1975; Shibuya and Stuhl, 1982).

3. Adsorbates

3.1. General Considerations

The electronic spectra and photochemical reactions of adsorbed molecules often differ significantly from their respective counterparts in the gas phase. The differences arise from perturbations of the molecular wave function, caused by interactions with the surface as well as with other adsorbed molecules. The nature and magnitude of these perturbations vary not only with the adsorbate, but also with the chemical and physical properties of the adsorbent. This, of course, adds a considerable degree of complexity to the fundamental study of adsorbate photoreactions. On the experimental side, it also requires carefully controlled surface preparation prior to and during dosing, in order to ensure reproducible and meaningful data.

The perturbations caused by surface interactions are the largest for chemisorbed species. In fact, in chemisorption the surface-adsorbate bond strength may be comparable to the strength of intramolecular bonds, and the resultant adsorbate may bear little chemical or structural resemblance to the gas-phase parent molecule. Consequently, the spectra of chemisorbed molecules typically differ significantly (in overall shape as well as in the magnitude of the absorption cross section) from the respective gas-phase spectra.

For physisorbed molecules the perturbations are small compared to the intramolecular bond strengths, and the spectrum of the adsorbate correlates well with that of the gas-phase molecule. A notable change commonly encountered in electronic spectra of monolayer physisorbed molecules is a spectral shift, whereby the gas-phase spectrum is shifted to shorter or longer wavelengths because of the adsorbate-adsorbent interaction. The direction and magnitude of the shift depend on the shapes of the potential curves representing the interactions between the surface and ground-state molecule, and between the surface and the electronically excited molecule (Terenin, 1964 and references therein). In the case of multilayer physisorption, the perturbations of the free-molecule wave function are caused by the high-density environment in the adsorbed phase; this environment may be more closely related to the liquid or solid phases than to the gas phase (Ehrlich and Osgood, 1981).

In addition to affecting the absorption spectrum, the above-mentioned perturbations may alter other aspects of the photoprocess as well. In particular, one can expect changes in the quantum yield and energy distribution of the primary photoproducts and also the existence of different secondary processes.

While there exists a significant body of work on the spectroscopy of adsorbates (see, for instance, the review by Terenin, 1964), most of it is concerned, mainly for experimental reasons, with adsorbents with only limited use in thin-film processes, such as porous glasses and gels. The present survey will include only those results that can be more directly related to thin-film applications. As will be noted, the majority of the pertinent studies is quite recent, and their number and breadth are still extremely limited. Clearly, and in spite of its great technological significance, the field of adsorbate photoreactions is still in its infancy. Much more exciting work remains to be done in order to bring it to the level of maturity of gas-phase spectroscopy and photochemistry.

3.2. Molecules Containing Nonmetals

3.2.1 Hydrogen Sulfide, H_2S

The photoprocesses of H_2S physisorbed on LiF(001) were studied by Bourdon et al. (1986). Prior to dosing, the LiF crystal was annealed at 300°C for 12 hours; the studies were performed at a surface temperature of 110°K.

At 222 nm, a primary photoprocess is molecular photodissociation

$$H_2S(ads) + h\nu \rightarrow H(g) + HS. \tag{3.71}$$

The kinetic energy distribution of H is bimodal, with peaks at 1.0 and 1.53 eV, corresponding to vibrationally excited and unexcited HS photofragments, respectively. The ratio of these two components is ~0.5 at submonolayer H_2S coverage, and it decreases with increasing coverage.

Secondary reactions involve the abstraction by the ejected H of another H atom from a nearby adsorbed H_2S:

$$H + H_2S(ads) \rightarrow H_2(g) + HS. \tag{3.72}$$

The relative abundance $[H_2(g)]/[H(g)]$ increases with increasing coverage and is near unity for a monolayer of H_2S(ads). The H_2 molecule has a kinetic energy with bimodal distribution, with peaks at 0.07 and 0.6 eV. This distribution is atttributed to the existence of several encounters with the two-dimensional H_2S(ads) "cage" for H_2 formed by ejection of H nearly parallel to the surface and to the lack of such encounters for H_2 formed by ejection of H nearly perpendicular to the surface.

Molecular desorption of H_2S is also observed. In contrast with the photodissociation channel, its yield is nearly constant with wavelength (193–308 nm), and the velocity distribution is Maxwellian. This desorption is attributed to an optoacoustic shock generated by photoabsorption by color centers in LiF.

3.2.2 Methyl bromide, CH_3Br

Photoreactions of CH_3Br physisorbed on LiF(001) have been studied at cryogenic temperatures (Bourdon et al., 1984, 1986; Tabares et al., 1987). Both photofragmentation and molecular desorption were observed at 222 and 193 nm

$$CH_3Br(ads) + h\nu \rightarrow CH_3(g) + Br(g), \tag{3.73}$$

$$CH_3Br(ads) + h\nu \rightarrow CH_3Br(g). \tag{3.74}$$

The energy distribution of the primary photofragments CH_3 and Br apparently depends on experimental conditions, including laser wavelength, surface temperature, and surface treatment prior to dosing. If the LiF was prebaked in vacuum at 300°C for 12 hours, and dosing is at 110°K, then photolysis at 222 nm (3–30 mJ cm^{-2}/pulse) generates CH_3 fragments with a narrow kinetic energy distribution (FWHM $\approx$ 0.5 eV) whose peak shifts from 1.7 eV at submonolayer coverage to 1.4 eV at multilayer coverage (Bourdon et al., 1984, 1986). When the LiF was prebaked at 150°C in vacuum for 12 hours, the CH_3 energy distribution was much broader (at 222 nm, 110°K, Bourdon et al., 1986; at 193 nm, 30°K, Tabares et al., 1987).

The Br photofragments have at 222 nm a kinetic energy peak at 0.22 eV, but some Br have energies as high as 1 eV, which is 0.6 eV above the gas-phase thermodynamically allowed maximum value (Bourdon et al., 1984, 1986). This last effect, corrected for the difference in photon energy, is seen also at 193 nm (Tabares et al., 1987). In this case the maximum Br energy is 1.9 eV. These results are interpreted in terms of inelastic collisions between the CH_3 and Br photofragments as they leave the surface.

Photoinduced molecular desorption is apparently a secondary reaction. It is caused by an optoacoustic shock originating from photoabsorbing color centers in the crystal (Bourdon et al., 1986) or by inelastic encounters of high-energy photofragments with neighboring CH_3Br(ads) (Tabares et al., 1987).

3.3. Molecules Containing Metals

3.3.1 Metal Halides

3.3.1.1 Titanium tetrachloride, $TiCl_4$

$TiCl_4$ adsorbed on fused silica or $LiNbO_3$, and in equilibrium at room temperature with its gas, decomposes at 257 nm to precipitate Ti atoms (unlike gas-phase photolysis, where the photoreaction terminates after the removal of only one Cl); the photoinduced surface dechlorination is probably autocatalytic, and the cross section of the rate-limiting step is $\sim 5 \times 10^{-21}$ cm^2, reflecting the decomposition of a partially chlorinated intermediate (Tsao et al., 1983).

In the presence of coadsorbed trimethylaluminum, a chemical chain reaction is induced by the partial photodechlorination of $TiCl_4$ (possibly involving $TiCl_3$) to form a Ti:Cl:Al:CH_3 deposit (Tsao and Ehrlich, 1984), which acts a a Ziegler–Natta catalyst (Ehrlich and Tsao, 1985).

3.3.2 Metal Carbonyls

3.3.2.1 Molybdenum carbonyl, $Mo(CO)_6$

The photoreactions of physisorbed $Mo(CO)_6$ on Si were studied at cryogenic temperatures (Creighton, 1986; Bartosch et al., 1986). The absorption cross section at 248 nm is $(5 \pm 3) \times 10^{-17}$ cm^2, close to the gas-phase value; the primary photoprocess at 248 nm ($Mo(CO)_6$ on Si(100) at 150°K) is partial decarbonylation, the stoichiometry of the newly formed film being $Mo_1C_1O_{0.3}$ (Creighton, 1986). Low power, 257-nm irradiation of $Mo(Co)_6$ on Si(111) at 90°K was also shown to induce photochemical partial decarbonylation (Bartosch et al., 1986; Gluck et al., 1987).

3.3.2.2 Tungsten carbonyl, $W(CO)_6$

When $W(CO)_6$, physisorbed on Si(111)-(7 × 7) at 120°K is irradiated at 248 nm, a primary photoprocess is partial decarbonylation, resulting in surface-stable $W(CO)_x$ ($x < 6$) fragments (Swanson et al., 1988). Similar effects are seen upon cw irradiation at 257 nm of $W(CO)_6$ on Si(111) at 90°K (Gluck et al., 1987).

3.3.2.3 Iron pentacarbonyl, $Fe(CO)_5$

The UV photoreactions of multilayer $Fe(CO)_5$ physisorbed on Si have been the subject of several studies at cryogenic temperatures. Broad-band UV illumination (Hg-lamp) of $Fe(CO)_5$ on Si(100) at 77°K leads to nearly complete decarbonylation (Foord and Jackman, 1984). Similarly, pulsed irradiation at 248, 308, and 337 nm of $Fe(CO)_5$ adsorbed on Si(111) at 120°K causes the removal of all CO ligands in a one-photon process (Swanson et al., 1987, 1988). On the other hand, Bartosch et al. (1985) and Gluck et al. (1987) observe sequential and only partial (up to three CO groups removed per photoreacted molecule) decarbonylation of $Fe(CO)_5$ on Si(100) at 80°K, and on Si(111) at 90°K using a 257-nm cw source.

3.3.3 Metal Alkyls

3.3.3.1 Diethylzinc, $Zn(C_2H_5)_2$

The spectrum of $Zn(C_2H_5)_2$ adsorbed at room temperature on fused silica has been reported by Krchnavek et al., (1987). There is a chemisorbed layer that is stable in vacuum, and it exhibits a continuous absorption spectrum that is blue shifted with respect to the gas-phase spectrum: It starts at ~270 nm (compared to ~290 nm in the gas phase) and it peaks at ~215 nm (compared to ~225 nm). Physisorbed layers exist only under equilibrium with the gas phase. Their spectrum is similar to the gas-phase spectrum, except for some broadening of the main lines and loss of superimposed fine structure (see Fig. 3.8).

3.3.3.2 Dimethylcadmium, $Cd(CH_3)_2$

$Cd(CH_3)_2$ has near room temperature a chemisorbed layer on fused silica, which is stable in vacuum, and physisorbed layers that exist only in equilibrium with the gas phase (Ehrlich and Osgood, 1981). The chemisorbed layer is probably formed by interaction with surface hydroxyl groups, forming a monomethyl bound to the surface through the Cd atom (Chen and Osgood, 1983). Its spectrum is a continuum that starts at ~280 nm and peaks at ~205 nm (Chen and Osgood, 1983). The spectrum of the physisorbed layers, when compared to that of the gas phase, exhibits broadening of the main 215-nm peak, red shift of this peak, and a smoothing of the ro-vibronic spectrum superimposed on it; a long-wavelength tail in the apparent absorption is attributed to Rayleigh scattering due to nonuniform molecular coverage (Chen and Osgood, 1983; see also Fig. 3.9).

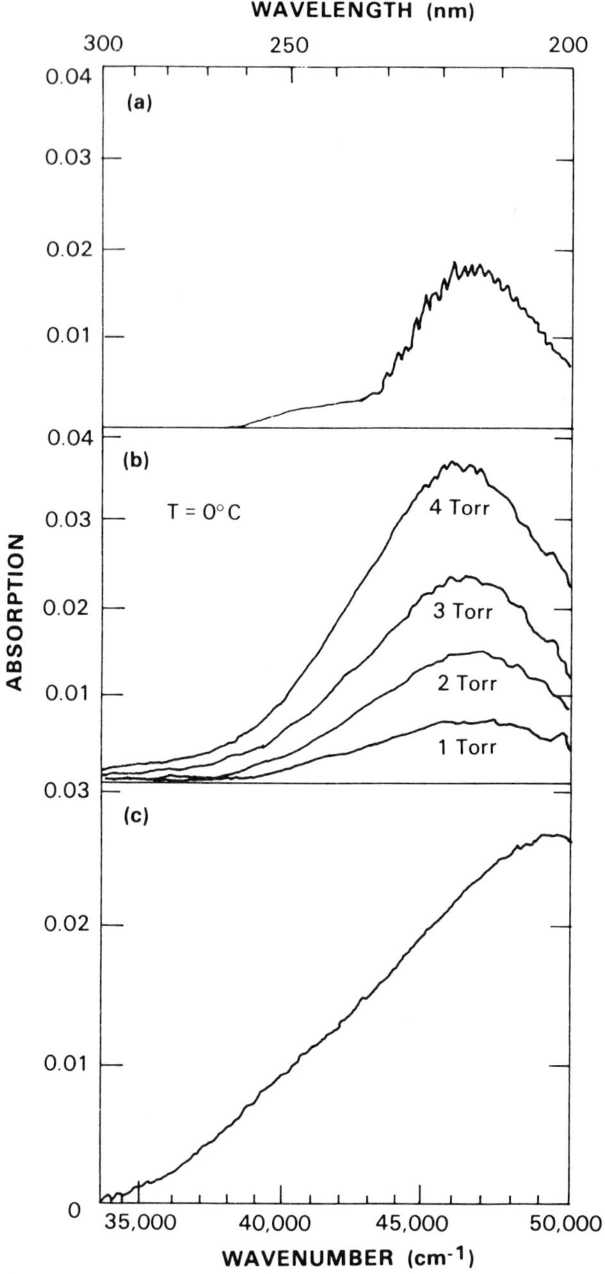

Fig. 3.9 The fractional absorption spectrum of gaseous and adsorbed $Cd(CH_3)_2$ in the region 200–300 nm. (a) The absorption of an equivalent monolayer of gas-phase $Cd(CH_3)_2$. (b) The absorption of physisorbed $Cd(CH_3)_2$ on fused silica at 0°C for four values of $Cd(CH_3)_2$ pressure. (c) The absorption of $Cd(CH_3)_2$ chemisorbed on fused silica. Adapted with permission from Chen and Osgood, *Chem. Phys. Lett.* **98,** 363 (1983).

3.3.3.3 Trimethylaluminum, $Al(CH_3)_3$

When adsorbed on fused silica near room temperature, $Al(CH_3)_3$ has a chemisorbed layer that is stable in vacuum, and physisorbed layers that exist only in equilibrium with the gas phase (Ehrlich and Osgood, 1981). The chemisorbed layer exhibits new absorption bands in the region 240–300 nm, to the red of the main 185-nm gas-phase peak (Ehrlich and Osgood, 1981). This layer is probably formed by interaction with surface hydroxyl groups, and it contains CH_3 groups bound to aluminum (Higashi et al., 1985). The effects of irradiation in vacuum of the chemisorbed layer are strongly wavelength dependent (Higashi and Rothberg, 1985): at 248 nm (1–20 mJ cm^{-2}/pulse) no noticeable changes in chemical composition are detected; at 193 nm, however, there is efficient nonthermal photodesorption of CH_3 groups and enhancement of the metallic properties of the remaining film. The cross section for this process may vary by orders of magnitude, depending on experimental conditions (Higashi, 1988). The methyl groups desorb with only 0.025 eV kinetic energy, compared to the available excess energy of ~3.5 eV (Higashi, 1988). At cryogenic temperatures (100°K), trimethylaluminum is adsorbed on Si(100) as dimers, $Al_2(CH_3)_6$; irradiation in vacuum at 248 nm monomerizes the adsorbate and induces desorption of CH_3 groups (Lubben et al., 1986).

4. Liquids

The spectra and photochemical processes of molecules in a high-density liquid environment may differ significantly from their isolated-molecule counterparts, the latter of which are often quite similar to those at low-pressure gas-phase conditions. Deviations from gas-phase spectra may exist even if the compound under study is a solute at low concentrations, because solvent-solute interactions must be taken into account, and so must any photon-solvent interactions. The latter may become noticeable at shorter wavelengths, where few solvents are truly transparent. Figures 3.10–3.13 are the transmission spectra through a 1-cm pathlength of several commonly used solvents (Rothschild and Ehrlich, 1986). It is seen that even the more highly transmissive solvents exhibit small amounts of residual absorption in the UV. This photoabsorption by the solvent may effect the photochemistry of the solute, even if the latter is photochemically more reactive.

The solvent-solute interaction may take several forms. In the case of polar solvents, a significant fraction of the solute is in ionic form, and the resultant spectrum and photochemistry may be radically different from

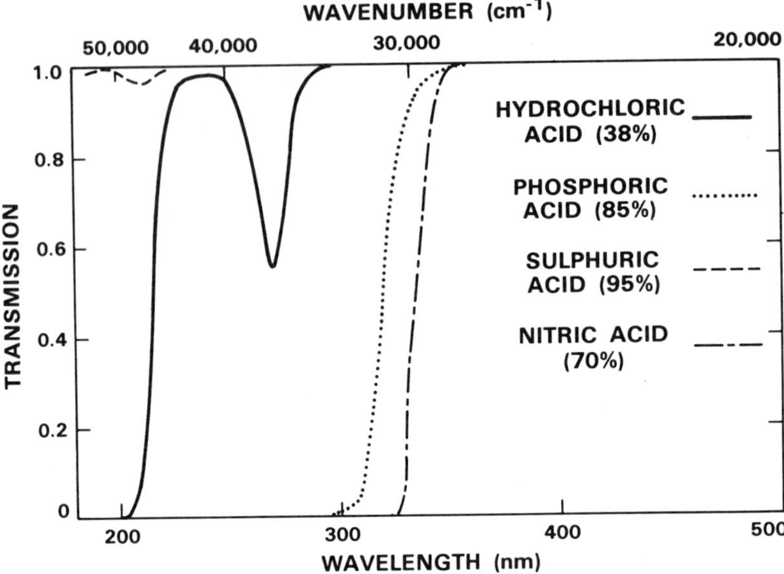

Fig. 3.10 The transmission spectra of 1 cm of aqueous solutions of HCl, H_3PO_4, H_2SO_4, and HNO_3 in the region 190 to 500 nm. From Rothschild and Ehrlich (1986).

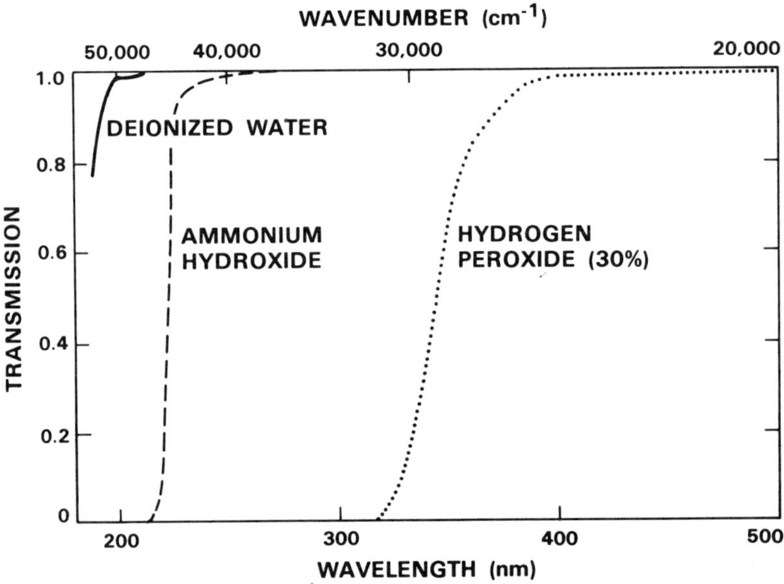

Fig. 3.11 The transmission spectra of 1-cm liquid H_2O and aqueous solutions of H_2O_2 and NH_4OH in the region 190–500 nm. From Rothschild and Ehrlich (1986).

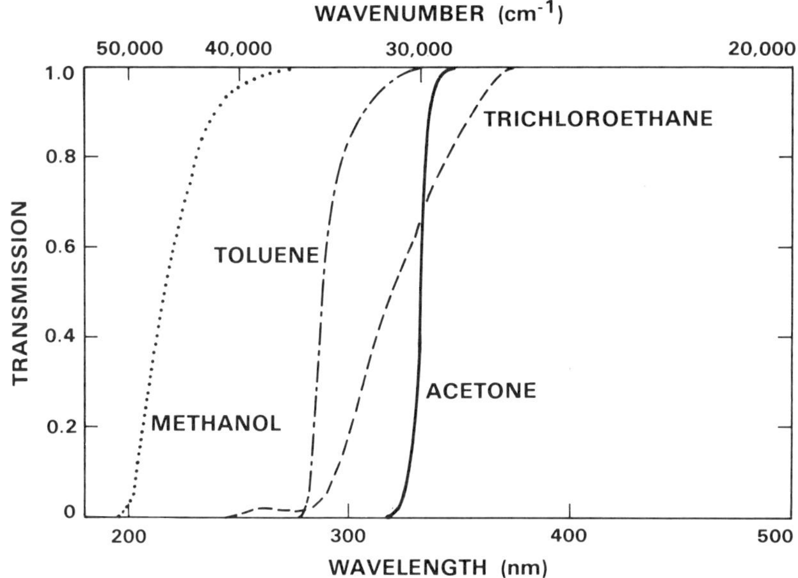

Fig. 3.12 The transmission spectra of 1 cm of liquid methanol (spectroscopic grade), toluene, acetone, and 1,1,1-trichloroethane in the region 190–500 nm. From Rothschild and Ehrlich (1986).

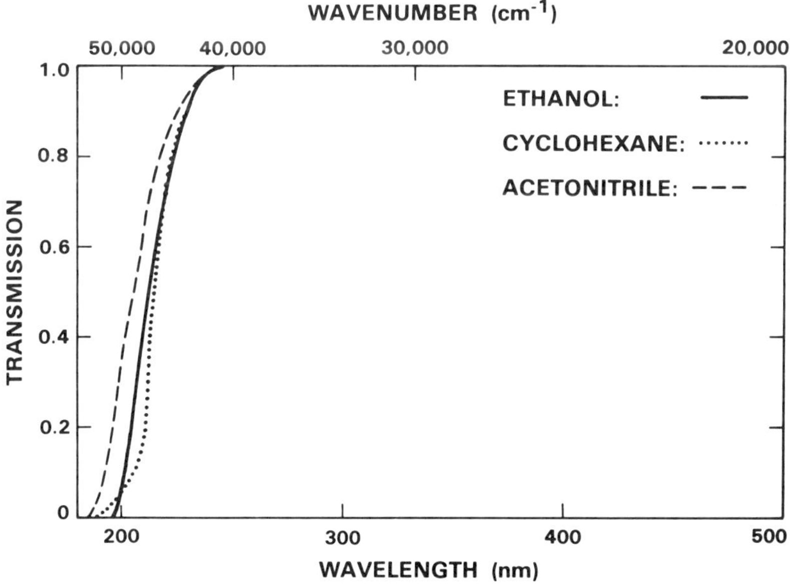

Fig. 3.13 The transmission spectra of 1 cm of liquid ethanol, cyclohexane, and acetonitrile in the region 190–500 nm. From Rothschild and Ehrlich (1986).

those of the electrically neutral solute molecule. If the solvent is nonpolar the effects of its interaction with the solute may be more subtle. Nevertheless, they frequently constitute a severe enough perturbation to modify the solute's spectroscopic and photochemical properties. Solvents can cause broadening and shifting of absorption bands, in a manner somewhat analogous to those observed in the adsorbed phase. Solvents may also modify photochemical reactions by promoting recombinative back-reactions of photodissociating molecules ("cage" effect) or by solvent-induced relaxation of the photoexcited solute molecule. Following is a partial listing of publications on the spectroscopy and photochemical reactions of solutions: Calvert and Pitts, 1966; Turro, 1967; Geoffroy and Wrighton, 1979; Berg et al., 1985.

References

Ackerman, M., Biaumé, F., and Kockarts, G. (1970). Absorption cross sections of the Schumann–Runge bands of molecular oxygen. *Planet. Space Sci.* **18,** 1639.

Alderdice, D.S. (1965). Electronic spectra of transition metal complexes. Part I. *J. Mol. Spectrosc.* **15,** 509.

Amimoto, S.T., Force, A.P., Gulotty, R.G., and Wiesenfeld, J.R. (1979). Collisional deactivation of $O(2\,^1D_2)$ by the atmospheric gases. *J. Chem. Phys.* **71,** 3640.

Arnone, C., Rothschild, M., Black, J.G., and Ehrlich, D.J. (1986). Visible-laser photodeposition of chromium oxide films and single crystals. *Appl. Phys. Lett.* **48,** 1018.

Ashfold, M.N.R., Bennett, C.L., and Dixon, R.N. (1986). Dissociation dynamics of $NH_3(\tilde{A}\,^1A_2'')$. Experiment and theory. *Faraday Discuss. Chem. Soc.* **82,** 163.

Back, R.A., and Griffiths, D.W.L. (1967). Flash photolysis of ethylene. *J. Chem. Phys.* **46,** 4839.

Back, R.A., and Koda, S. (1977). The photodissociation of ammonia in the $\tilde{A} \leftarrow \tilde{X}$ absorption system. Part II. Translational excitation of the hydrogen atoms produced, and the mechanism of the predissociation. *Can. J. Chem.* **55,** 1387.

Bagott, J.E., Frey, H.M., Lightfoot, P.D., and Walsh, R. (1986). The photodissociation of phenylsilane at 193 nm. *Chem. Phys. Lett.* **125,** 22.

Barry, M.D., and Gorry, P.A. (1984). Photofragmentation dynamics of CH_3I at 248 nm. *Mol. Phys.* **52,** 461.

Bartosch, C.E., Stroscio, J.A., and Ho, W. (1985). Thermal and laser induced decomposition of $Fe(CO)_5$ on Si(100). In *Beam-Induced Chemical Processes,* Extended Abstracts, (von Gutfeld, R.J., Greene, J.E., and Schlossberg, H., eds.). Materials Research Society, Pittsburgh, 67.

Bartosch, C.E., Gluck, N.S., Ho, W., and Ying, Z. (1986). Laser-surface-adsorbate interactions: Thermal versus photoelectronic excitation of $Mo(CO)_6$ on Si(111). *Phys. Rev. Lett.* **57,** 1425.

Basco, N., and Yee, K.K. (1967). Formation of metastable atoms of phosphorus, arsenic and antimony by flash photolysis. *Nature* **216,** 998.

Bass, A.M., Ledford, A.E., Jr., and Laufer, A.H. (1976). Extinction coefficients of NO_2 and N_2O_4. *J. Res. Nat. Bur. Stand.* **80A,** 143.

Baughcum, S.L., and Leone, S.R. (1980). Photofragmentation infrared emission studies of vibrationally excited free radicals CH_3 and CH_2I. *J. Chem. Phys.* **72,** 6531.

Baughcum, S.L., and Leone, S.R. (1982). Laser photodissociation of $Hg(CH_3)_2$: Infrared emission studies of vibrational and rotational excitation in the CH_3 fragments. *Chem. Phys. Lett.* **89,** 183.

Beach, N.A., and Gray, H.B. (1968). Electronic structures of metal hexacarbonyls. *J. Am. Chem. Soc.* **90,** 5713.

Becker, C.A.L., Ballhausen, C.J., and Trabjerg, I. (1969). Investigation of the electronic structure of $TiCl_4$. *Theoret. Chim. Acta* (Berl.) **13,** 355.

Berg, M., Harris, A.L., and Harris, C.B. (1985). Rapid solvent-induced recombination and slow energy relaxation in a simple chemical reaction: Picosecond studies of iodine photodissociation in CCl_4. *Phys. Rev. Lett.* **54,** 951.

Bourdon, E.B.D., Cowin, J.P., Harrison, I., Polanyi, J.C., Segner, J., Stanners, C.D., and Young, P.A. (1984). UV photodissociation and photodesorption of adsorbed molecules. 1. CH_3Br on LiF(001). *J. Phys. Chem.* **88,** 6100.

Bourdon, E.B.D., Das, P., Harrison, I., Polanyi, J.C., Segner, J., Stanners, C.D., Williams, R.J., and Young, P.A. (1986). Photodissociation, photoreaction and photodesorption of adsorbed species. Part 2. CH_3Br and H_2S on LiF(001). *Faraday Discuss. Chem. Soc.* **82,** 343.

Brand, J.C.D., Jones, V.T., and di Lauro C. (1971). The $^3B_1$-$^1A_1$ 3880 Å band system of sulfur dioxide: Rotational analysis of the 0-0 band. *J. Mol. Spectrosc.* **40,** 616.

Brannon, J.H. (1986). Glass etching initiated by excimer laser photolysis of CF_2Br_2. *J. Phys. Chem.* **90,** 1784.

Bray, R.G., Seidler, P.F., Gethner, J.S., and Woodin, R.L. (1986). Photofragment infrared fluorescence in the photodissociation of gas-phase $Mn_2(CO)_{10}$. *J. Am. Chem. Soc.* **108,** 1312.

Breckenridge, W.H., and Sinai, N. (1981). Pulsed laser photolysis of chromium hexacarbonyl in the gas phase. *J. Phys. Chem.* **85,** 3557.

Breckenridge, W.H., and Stewart, G.M. (1986). Pulsed laser photolysis of chromium hexacarbonyl in the gas phase. *J. Am. Chem. Soc.* **108,** 364.

Brewer, L., and Tellinghuisen, J. (1972). The quantum yield for unimolecular dissociation of I_2 in visible absorption. *J. Chem. Phys.* **56,** 3929.

Brewer, P., Halle, S., and Osgood, R.M. (1984). Excimer-laser-initiated dry etching of single crystal GaAs. *Mat. Res. Soc. Symp. Proc.* **29,** 179.

Brewer, P.D., Jensen, J.E., Olson, G.L., Tutt, L.W., and Zinck, J.J. (1988). Photodissociation dynamics of alkyltellurides. *Mat. Res. Soc. Symp. Proc.* **101,** 327.

Brus, L.E., and McDonald, J.R. (1974). Time-resolved fluorescence kinetics and $^1B_1(^1\Delta_g)$ vibronic structure in tunable ultraviolet laser excited SO_2 vapor. *J. Chem. Phys.* **61,** 97.

Buenker, R.J., and Peyerimhoff, S. (1975). All-valence-electron configuration mixing calculations for the characterization of the $^1(\pi, \pi^*)$ states of ethylene. *Chem. Phys.* **9,** 75.

Busch, G.E., Mahoney, R.T., Morse, R.I., and Wilson, K.R. (1969). Translational spectroscopy: Cl_2 photodissociation. *J. Chem. Phys.* **51,** 449.

Callear, A.B. (1961). The decomposition of nickel carbonyl studied by flash photolysis. *Proc. R. Soc. London* **A265,** 71.

Calvert, J.C., and Pitts, J.N., Jr. (1966). *Photochemistry.* Wiley, New York.

Causley, G.C., and Russell, B.R. (1976). Vacuum ultraviolet absorption spectra of dichlorosilane, dichloromethylsilane and dichlorodimethylsilane. *J. Electron Spectrosc. Rel. Phenom.* **8,** 71.

Chamberlain, G.A., and Simons, J.P. (1975). Vacuum ultraviolet photolysis of nitrous oxide. Energy disposal in the reaction $O(2\,^1D) + N_2O \rightarrow 2\,NO$. *J. Chem. Soc. Faraday II* **71,** 402.

Chen, C.J., and Osgood, R.M. (1983). Measurement of the electronic spectra of physisorbed molecular layers. *Chem. Phys. Lett.* **98,** 363.

Chen, C.J., and Osgood, R.M. (1984). A spectroscopic study of the excited states of dimethylzinc, dimethylcadmium, and dimethylmercury. *J. Chem. Phys.* **81,** 327.

Chu, J.O., Flynn, G.W., Chen, C.J., and Osgood, R.M., Jr. (1985). Infrared emission studies of vibrational excitation in CH_3 fragments produced from ArF and KrF laser photolysis of $Cd(CH_3)_2$ and $Zn(CH_3)_2$. *Chem. Phys. Lett.* **119,** 206.

Clark, J.G., and Anderson, R.G. (1978). Silane purification via laser-induced chemistry. *Appl. Phys. Lett.* **32,** 46.

Clear, R.D., Riley, S.J., and Wilson, K.R. (1975). Energy partitioning and assignment of excited states in the ultraviolet photolysis of HI and DI. *J. Chem. Phys.* **63,** 1340.

Compton, L.E., Gole, J.L., and Martin, R.M. (1969). Kinetics of hot hydrogen atoms from H_2S photodissociation at 1850 Å. *J. Phys. Chem.* **73,** 1168.

Compton, L.E., and Martin, R.M. (1969). Photodissociation dynamics. Production of $I^*(^2P_{1/2})$ atoms in the photolysis of hydrogen iodide. *J. Phys. Chem.* **73,** 3474.

Connor, J., Greig, G., and Strausz, O.P. (1969). The reactions of tellurium atoms. I. *J. Am. Chem. Soc.* **91,** 5695.

Creek, D.M., and Nicholls, R.W. (1975). A comprehensive re-analysis of the $O_2(B^3\Sigma_u^- - X\,^3\Sigma_g^-)$ Schumann–Runge band system. *Proc. R. Soc. London* **A341,** 517.

Creighton, J.R. (1986). Photodecomposition of $Mo(CO)_6$ adsorbed on Si(100). *J. Appl. Phys.* **59,** 410.

Daniel, C., Bénard, M., Dedieu, A., Wiest, R., and Veillard, A. (1984). Theoretical aspects of the photochemistry of organometallics. 3. Potential energy curves for the photodissociation of $Fe(CO)_5$. *J. Phys. Chem.* **88,** 4805.

Davidson, J.A., Howard, C.J., Schiff, H.I., and Fehsenfeld, F.C. (1979). Measurements of the branching ratios of the reaction of $O(^1D_2)$ with N_2O. *J. Chem. Phys.* **70,** 1697.

Deutsch, T.F., Ehrlich, D.J., Rathman, D.D., Silversmith, D.J., and Osgood, R.M. (1981). Electrical properties of laser chemically doped silicon. *Appl. Phys. Lett.* **39,** 825.

Deutsch, T.F., Fan, J.C.C., Ehrlich, D.J., Turner, G.W., Chapman, R.L., and Gale, R.P. (1982). Efficient GaAs solar cells formed by UV laser chemical doping. *Appl. Phys. Lett.* **40,** 722.

Dick, B., Freund, H.-J., and Hohlneicher, G. (1982). Calculation of transition metal compounds using an extension of the CNDO formalism. IV. CNDO-CI calculations on $Ni(CO)_4$ and $Fe(CO)_5$; electronic spectra and photochemical implications. *Mol. Phys.* **45,** 427.

Di Stefano, G., Lenzi, M., Margani, A., Mele, A., and Xuan, C.N. (1977). Fluorescence excitation spectroscopy of photofragments of PH_3. *J. Photochem.* **7,** 335.

Di Stefano, G., Lenzi, M., Margani, A., and Xuan, C.N. (1977a). $NH_2(\tilde{A}\,^2A_1)$ from the predissociation of $NH_3(\tilde{A}\,^1A_2'')$. *J. Chem. Phys.* **67,** 3832.

Docker, M.P., Hodgson, A., and Simons, J.P. (1986). Photodissociation of H_2O_2 at 248 nm: Translational anisotropy and OH product state distributions. *Chem. Phys. Lett.* **128,** 264.

Donnelly, V.M., Baronavsky, A.P., and McDonald, J.R. (1979). ArF laser photodissociation of NH_3 at 193 nm: Internal energy distributions in $NH_2\,\tilde{X}\,^2B_1$ and $\tilde{A}\,^2A_1$, and two photon generation of NH $A\,^3\Pi$ and $b\,^1\Sigma^+$. *Chem. Phys.* **43,** 271.

Donnelly, V.M., Geva, M., Long, J., and Karlicek, R.F. (1984). Excimer-laser induced deposition of InP and indium-oxide films. *Mat. Res. Soc. Symp. Proc.* **29,** 73.

Donnelly, V.M., McCrary, V.R., Appelbaum, A., Brasen, D., and Lowe, W.P. (1987). ArF excimer-laser-stimulated growth of polycrystalline GaAs thin films. *J. Appl. Phys.* **61,** 1410.

Douglas, A.E. (1963). Electronically excited states of NH_3. *Discuss. Faraday Soc.* **35,** 158.

Douglas, A.E., Møller, C.K., and Stoicheff, B.P. (1963). The absorption spectrum of $^{35}Cl_2$ from 4780 to 6000Å. *Can. J. Phys.* **41,** 1174.

Duncan, A.B.F., and Murray, J.W. (1934). The Raman and ultraviolet absorption spectra of some metal carbonyls and alkyls. *J. Chem. Phys.* **2,** 636.

Dyne, P.J., and Style, D.W.G. (1952). 'Fluorescence obtained from formic acid, carbonyl chloride, and methylene iodide. *J. Chem. Soc.,* 2122.

Ehrlich, D.J., and Osgood, R.M. (1981). UV photolysis of van der Waals molecular films. *Chem. Phys. Lett.* **79,** 381.

Ehrlich, D.J., Osgood, R.M., and Deutsch, T.F. (1981). Direct writing of refractory metal thin film structures by laser photodeposition. *J. Electrochem. Soc.* **128,** 2039.

Ehrlich, D.J., Osgood, R.M., and Deutsch, T.F. (1982). Photodeposition of metal films with ultraviolet laser light. *J. Vac. Sci. Technol.* **21,** 23.

Ehrlich, D.J., and Tsao, J.Y. (1985). UV laser photodeposition of patterned catalyst films from adsorbate mixtures. *Appl. Phys. Lett.* **46,** 198.

Ershov, L.S., Zaleskii, V.Yu., and Sokolov, V.N. (1978). Laser photolysis of perfluoroalkyl iodides. *Soviet J. Quantum Electron.* **8,** 494.

Felps, S., Hochmann, P., Brint, P., and McGlynn, S.P. (1976). Molecular Rydberg transitions. The lowest-energy Rydberg transitions of s-type in CH_3X and CD_3X, X = Cl, Br, and I. *J. Mol. Spectrosc.* **59,** 355.

Fernandez, J., Lespes, G., and Dargelos, A. (1986). Theoretical and experimental study of the vacuum ultraviolet spectrum of tetrasubstituted tin derivatives of $SnCl_4$ and $Sn(CH_3)_4$. *Chem. Phys.* **111,** 97.

Fletcher, T.R., and Rosenfeld, R.N. (1985). Studies on the photochemistry of chromium hexacarbonyl in the gas phase: Primary and secondary processes. *J. Am. Chem. Soc.* **107,** 2203.

Fletcher, T.R., and Rosenfeld, R.N. (1986). Reactivity of $Cr(CO)_4$ in the gas phase. *J. Am. Chem. Soc.* **108,** 1686.

Flynn, D.K., Steinfeld, J.I., and Sethi, D.S. (1986). Deposition of refractory metal films by rare-gas halide laser photodissociation of metal carbonyls. *J. Appl. Phys.* **59,** 3914.

Foo, P.D. and Innes, K.K. (1973). Spectrum of acetylene: 1650–1950 Å. *Chem. Phys. Lett.* **22,** 439.

Foord, J.S., and Jackman, R.B. (1984). Chemical vapour deposition of silicon: In situ surface studies. *Chem. Phys. Lett.* **112,** 190.

Fotakis, C., Torre, A., and Donovan, R.J. (1983). Two-photon UV excitation and laser-induced fluorescence from SO. *J. Photochem.* **23,** 97.

Freedman, A., and Bersohn, R. (1978). Photodissociation of molecular beams. Cleavage of metal-metal bonds in rhenium and manganese decacarbonyl. *J. Am. Chem. Soc.* **100,** 4116.

Freedman, A., Yang, S.C., and Bersohn, R. (1979). Translational energy distribution of the photofragments of SO_2 at 193 nm. *J. Chem. Phys.* **70,** 5313.

Fuchs, C., Boch, E., Fogarassy, E., Aka, B., and Siffert, P. (1988). Two-photon absorption cross section for silane under pulsed ArF (193 nm) excimer laser irradiation. *Mat. Res. Soc. Symp. Proc.* **101,** 361.

Gedanken, A., and Rowe, M.D. (1975). Magnetic circular dichroism spectra of the methyl halides. Resolution of the $n \rightarrow \sigma^*$ continuum. *Chem. Phys. Lett.* **34,** 39.

Geoffroy, G.L., and Wrighton, M.S. (1979). *Organometallic Photochemistry* Academic Press, New York.

Gerck, E. (1983). Quantum yields of $I(^2P_{1/2})$ for CF_3I, C_2F_5I, i-C_3F_7I, n-C_3F_7I, n-$C_6F_{13}I$, and 1,2-$C_2F_4I_2$ at 308 and 248 nm. *J. Chem. Phys.* **79,** 311.

Gericke, K-H., Klee, S., Comes, F.J., and Dixon, R.N. (1986). Dynamics of H_2O_2 photodissociation: OH product state and momentum distribution characterized by sub-Doppler and polarization spectroscopy. *J. Chem. Phys.* **85,** 4463.

Gerlach-Meyer, U., Linnebach, E., Kleinermanns, K., and Wolfrum, J. (1987). H-atom photofragments from H_2O_2 dissociated at 193 nm. *Chem. Phys. Lett.* **133,** 113.

Gibson, G.E., and Bayliss, N.S. (1933). Variation with temperature of the continuous absorption spectrum of diatomic molecules: Part I. Experimental, the absorption spectrum of chlorine. *Phys. Rev.* **44,** 188.

Gilgen, H.H., Chen, C.J., Krchnavek, R., and Osgood, R.M. (1984). The physics of ultraviolet photodeposition. In *Laser Processing and Diagnostics,* (Bäuerle, D., ed.). Springer, Berlin, 225.

Glasgow, L.C., and Potzinger, P. (1972). Ethylene as an actinometer in the wavelength region 147–185 nanometers. *J. Phys. Chem.* **76,** 138.

Gluck, N.S., Ying, Z., Bartosch, C.E., and Ho, W. (1987). Mechanisms of laser interaction with metal carbonyls adsorbed on Si(111) 7×7: Thermal vs. photoelectronic effects. *J. Chem. Phys.* **86,** 4957.

Golomb, D., Watanabe, K., and Marmo, F.F. (1962). Absorption coefficients of sulfur dioxide in the vacuum ultraviolet. *J. Chem. Phys.* **36,** 958.

Goodeve, C.F., and Stein, N.O. (1931). Absorption spectra and the optical dissociation of the hydrides of the oxygen group. *Trans. Faraday Soc.* **27,** 393.

Gray, H.B., and Beach, N.A. (1963). The electronic structures of octahedral metal complexes. I. Metal hexacarbonyls and hexacyanides. *J. Am. Chem. Soc.* **85,** 2922.

Grunewald, A.U., Gericke, K.-H., and Comes, F.J. (1986). Photofragmentation dynamics of H_2O_2 at 193 nm. *Chem. Phys. Lett.* **132,** 121.

Grunewald, A.U., Gericke, K.-H., and Comes, F.J. (1987). Photodissociation of room-temperature and jet-cooled water at 193 nm. *Chem. Phys. Lett.* **133,** 501.

Hack, W., and Langel, W. (1983). Photophysics of CF_2 ($\tilde{A}\,^1B_1(0,6,0)$) excited by a KrF laser at 248 nm. I: Quenching of CF_2 ($\tilde{A}$) by its common precursors. *J. Photochem.* **21,** 105.

Haigh, J. (1983). The vapor-phase ultraviolet spectra of metallorganic precursors of III-V compounds. *J. Mat. Sci.* **18,** 1072.

Hall, T.C., Jr., and Blacet, F.E. (1952). Separation of the absorption spectra of NO_2 and N_2O_4 in the range 2400–5000 Å. *J. Chem. Phys.* **20,** 1745.

Halonbrenner, R., Huber, J.R., Wild, U., and Gunthard, J.J. (1968). A flash-photolysis study of chromyl chloride. *J. Phys. Chem.* **73,** 3929.

Hamada, Y., and Merer, A.J. (1975). Rotational structure in the absorption spectrum of SO_2 between 3000 Å and 3300 Å. *Can. J. Phys.* **53,** 2555.

Hanes, M.H., and Bair, E.J. (1963). Reactions of nitrogen-hydrogen radicals. I. NH_2 recombination in the decomposition of ammonia. *J. Chem. Phys.* **38,** 672.

Harada, Y., Murrell, J.N., and Sheena, H.H. (1968). The far ultraviolet spectra of methylsilanes. *Chem. Phys. Lett.* **1,** 595.

Hawkins, W.G., and Houston, P.L. (1980). 193 nm photodissociation of H_2S: The SH internal energy distribution. *J. Chem. Phys.* **73,** 297.

Heaven, M., Miller, T.A., and Bondybey, V.E. (1981). Production and characterization of temperature-controlled free radicals in a free jet expansion. *Chem. Phys. Lett.* **84,** 1.

Herzberg, G. (1966). *Molecular Spectra and Molecular Structure III, Electronic Spectra and Electronic Structure of Polyatomic Molecules.* Van Nostrand, Princeton, New Jersey.

Hess, W.P., Kohler, S.J., Haugen, H.K., and Leone, S.R. (1986). Application of an InGaAsP diode laser to probe photodissociation dynamics: I* quantum yields from n- and i-C_3F_7I and CH_3I by laser gain vs. absorption spectroscopy. *J. Chem. Phys.* **84,** 2143.

Higashi, G.S., and Rothberg, L.J. (1985). Investigation of the surface photochemical basis for metal film nucleation in laser chemical vapor deposition. *Appl. Phys. Lett.* **47,** 1288.

Higashi, G.S., Rothberg, L.J., and Fleming, C.G. (1985). Vibrational spectroscopy of growth surfaces during photochemical deposition from trimethylaluminum vapor. *Chem. Phys. Lett.* **115,** 167.

Higashi, G.S. (1988). Sub-thermal photoproduct surface desorption distribution arising from excimer laser excitation of adsorbed trimethylaluminum. *J. Chem. Phys.* **88,** 422.

Hirose, M., Yokoyama, S., and Yamakage, Y. (1985). Characterization of photochemical processing. *J. Vac. Sci. Technol.* **B3,** 1445.

Homer, J.B., and Hurle, I.R. (1972). Shock-tube studies on the decomposition of tetramethyl-lead and the formation of lead oxide particles. *Proc. R. Soc. London* **A327,** 61.

Hubrich, C. and Stuhl, F. (1980). The ultraviolet absorption of some halogenated methanes and ethanes of atmospheric interest. *J. Photochem.* **12,** 93.

Hudson, R.D., and Mahle, S.H. (1972). Photodissociation rates of molecular oxygen in the mesosphere and lower thermosphere. *J. Geophys. Res.* **77,** 2902.

Huebert, B.J., and Martin, R.M. (1968). Gas-phase far-ultraviolet absorption spectrum of hydrogen bromide and hydrogen iodide. *J. Phys. Chem.* **72,** 3046.

Hui, M.-H., and Rice, S.A. (1972). Decay of fluorescence from single vibronic states of SO_2. *Chem. Phys. Lett.* **17,** 474.

Humphries, C.M., Walsh, A.D., and Warsop, P.A. (1963). Absorption spectra of the hydrides, deuterides and halides of group 5 elements. *Discuss. Faraday Soc.* **35,** 148.

Hunter, T.F., and Kristjansson, K.S. (1978). Photoacoustic measurements of photofragmentation in CH_3I. *Chem. Phys. Lett.* **58,** 291.

Hunter, T.F., and Kristjansson, K.S. (1982). Yield of $I(^2P_{1/2})$ in the photodissociation of CH_2I_2. *Chem. Phys. Lett.* **90,** 35.

Hunter, T.F., and Leong, C.M. (1987). Absolute yields of $I(^2P_{1/2})$ in I_2 photodissociation using a laser optoacoustic technique. *Chem. Phys.* **111,** 145.

Ibbs, K.G., and Lloyd, M.L. (1983). Ultra-violet laser doping of silicon. *Opt. Laser Technol.* **15,** 35.

Irion, M.P., and Kompa, K.L. (1982). UV-laser photochemistry of diborane at 193.3 nm: The exchange reaction with deuterium. *J. Chem. Phys.* **76,** 2338.

Irion, M.P., and Kompa, K.L. (1982a). UV laser photochemistry of acetylene at 193 nm. *Appl. Phys.* **B27,** 183.

Irvine, S.J.C., Mullin, J.B., Robbins, D.J., and Glasper, J.L. (1984). UV absorption spectra and photolysis of some group II and group VI alkyls. *Mat. Res. Soc. Symp. Proc.* **29,** 253.

Itoh, U., Toyoshima, Y., Onuki, H., Washida, N., and Ibuki, T. (1986). Vacuum ultraviolet absorption cross sections of SiH_4, GeH_4, Si_2H_6, and Si_3H_8. *J. Chem. Phys.* **85,** 4867.

Iverson, A., and Russell, B.R. (1970). The vacuum ultraviolet spectra of three transition metal hexacarbonyls. *Chem. Phys. Lett.* **6,** 307.

Jacobs, A., Kleinermanns, K., Kuge, H., and Wolfrum, J. (1983). $OH(X\,^2\Pi)$ state distribution from HNO_3 and H_2O_2 photodissociation at 193 nm. *J. Chem. Phys.* **79,** 3162.

Jacobs, A., Wahl, M., Weller, R., and Wolfrum, J. (1987). Rotational distribution of nascent OH radicals after H_2O_2 photolysis at 193 nm. *Appl. Phys.* **B42,** 173.

Johnston, H., and Graham, R. (1973). Gas-phase ultraviolet absorption spectrum of nitric acid vapor. *J. Phys. Chem.* **77,** 62.

Johnston, H.S., Chang, S.-G., and Whitten, G. (1974). Photolysis of nitric acid vapor. *J. Phys. Chem.* **78,** 1.

Jonah, C., Chandra, P., and Bersohn, R. (1971). Anisotropic photodissociation of cadmium dimethyl. *J. Chem. Phys.* **55,** 1903.

Julienne, P.S., and Krauss, M. (1975). Predissociation of the Schumann–Runge bands of O_2. *J. Mol. Spectrosc.* **56,** 270.

Kanamori, H., Butler, J.E., Kawaguchi, K., Yamada, C., and Hirota, E. (1985). Spin polarization in SO photochemically generated from SO_2. *J. Chem. Phys.* **83,** 611.

Karlicek, R., Long, J.A., and Donnelly, V.M. (1984). Thermal decomposition of metalorganic compounds used in the MOCVD of InP. *J. Cryst. Growth* **68,** 123.

Kawasaki, M., Lee, S.J., and Bersohn, R. (1975). Photodissociation of molecular beams of methylene iodide and iodoform. *J. Chem. Phys.* **63,** 809.

Kawasaki, M., Kasatani, K., Sato, H., Shinohara, H., and Nishi, N. (1982). Photodissociation of molecular beams of SO_2 at 193 nm. *Chem. Phys.* **73,** 377.

Kawasaki, M., Kasatani, K., Sato, H., Shinohara, H., and Nishi, N. (1984). Photodissociation of molecular beams of halogenated hydrocarbons at 193 nm. *Chem. Phys.* **88,** 135.

Kawasaki, M., Sato, H., Shinohara, H., and Nishi, N. (1987). Photodissociation of tetramethyltin at 193 nm. *Laser Chem.* **7,** 109.

Kenner, R.D., Rohrer, F., and Stuhl, F. (1985). Determination of the excitation mechanism for photofragment emission in the ArF laser photolysis of NH_3, N_2H_4, HNO_3 and CH_3NH_2. *Chem. Phys. Lett.* **116,** 374.

Kenner, R.D., Haak, H.K., and Stuhl, F. (1986). Cl_2 and HCl emissions in the ArF-laser photolyses of chlorinated compounds: Identification and mechanism of generation. *J. Chem. Phys.* **85,** 1915.

Kenner, R.D., Rohrer, F., and Stuhl, F. (1987). Generation of NH($a\,^1\Delta$) in the 193 nm photolysis of ammonia. *J. Chem. Phys.* **86,** 2036.

Klee, S., Gericke, K.-H., and Comes, F.J. (1986). Doppler spectroscopy of OH in the photodissociation of hydrogen peroxide. *J. Chem. Phys.* **85,** 40.

Koffend, J.B., and Leone, S.R. (1981). Tunable laser photodissociation: Quantum yield of $I^*(^2P_{1/2})$ from CH_2I_2. *Chem. Phys. Lett.* **81,** 136.

Kolbe, W.F., and Leskovar, B. (1986). Millimeter-wave observations of vibrationally excited SO and CS produced by laser photolysis. *J. Chem. Phys.* **85,** 7117.

Koplitz, B., Xu, A., and Wittig, C. (1987). Product energy distributions from the 193 nm photodissociation of NH_3, *Chem. Phys. Lett.* **137,** 505.

Krajnovich, D., Zhang, Z., Butler, L., and Lee, Y.T. (1984). Photodissociation of CF_2Br_2 at 248 nm by the molecular beam method. *J. Phys. Chem.* **88,** 4561.

Krajnovich, D., Butler, L., and Lee, Y.T. (1984a). UV photodissociation of C_2F_5Br, C_2F_5I, and 1,2-C_2F_4BrI. *J. Chem. Phys.* **81,** 3031.

Krchnavek, R.R., Gilgen, H.H., Chen, J.C., Shaw, P.S., Licata, T.J., and Osgood, R.M. (1987). Photodeposition rates of metal from metal alkyls. *J. Vac. Sci. Technol.* **B5,** 20.

Kroger, P.M., Demou, P.C., and Riley, S.J. (1976). Polyhalide photofragment spectra. I. Two-photon two-step photodissociation of methylene iodide. *J. Chem. Phys.* **65,** 1823.

Laubengayer, A.W., and Gilliam, W.F. (1941). The alkyls of the third group elements. I. Vapor phase studies of the alkyls of aluminum, gallium and indium. *J. Am. Chem. Soc.* **63,** 477.

Laufer, A.H., and Bass, A. (1979). Photochemistry of acetylene. Bimolecular rate constant for the formation of butadiyne and reactions of ethynyl radicals. *J. Phys. Chem.* **83,** 310.

Lee, L.C., and Suto, M. (1987). Flourescence yields from photodissociative excitation of chloromethanes by vacuum ultraviolet radiation. *Chem. Phys.* **114,** 423.

Leighton, P.A., and Mortensen, R.A. (1936). The photolysis of lead tetramethyl and lead tetraphenyl. *J. Am. Chem. Soc.* **58,** 448.

Leopold, D.G., and Vaida, V. (1984). Photochemistry of gas-phase $Mn_2(CO)_{10}$ and $Re_2(CO)_{10}$: Mass spectrometric evidence for a dinuclear primary photoproduct. *J. Am. Chem. Soc.* **106,** 3720.

Lewis, K.E., Golden, D.M., and Smith, G.P. (1984). Organometallic bond dissociation energies: Laser pyrolysis of $Fe(CO)_5$, $Cr(CO)_6$, $Mo(CO)_6$, and $W(CO)_6$. *J. Am. Chem. Soc.* **106,** 3905.

Lin, C.L., Rohatgi, N.K., and DeMore, W.B. (1978). Ultraviolet absorption cross sections of hydrogen peroxide. *Geophys. Res. Lett.* **5,** 113.

Lubben, D., Motooka, T., Gorbatkin, S., and Green, J.E. (1986). Summary abstract: Photostimulated reactions in $Al_2(CH_3)_6$: Gas phase and adsorbed layer photolysis. *J. Vac. Sci. Technol.* **A4,** 668.

Magnotta, F., Nesbitt, D.J., and Leone, S.R. (1981). Excimer laser photolysis studies of translational to vibrational energy transfer. *Chem. Phys. Lett.* **83,** 21.

Mayer, T.M., Fisanik, G.J., and Eichelberger, T.S. (1982). Deposition of chromium films by multiphoton dissociation of chromium hexacarbonyl. *J. Appl. Phys.* **53,** 8462.

McCrary, V.R., and Donnelly, V.M. (1987). The ultraviolet absorption spectra of selected organometallic compounds used in the chemical vapor deposition of gallium arsenide. *J. Cryst. Growth* **84,** 253.

McDiarmid, R. (1971). Higher electronic states of ReF_6. *J. Mol. Spectrosc.* **39,** 332.

McDiarmid, R. (1974). Assignments in the ultraviolet spectra of MoF_6 and WF_6. *J. Chem. Phys.* **61,** 3333.

McDonald, J.R. (1975). Emission spectroscopy and excited-state dynamics of chromyl-chloride vapor. *Chem. Phys.* **9,** 423.

McDonald, J.R., Baronavsky, A.P., and Donnelly, V.M. (1978). Multiphoton-vacuum-ultraviolet laser photodissociation of acetylene: Emission from electronically excited fragments. *Chem. Phys.* **33,** 161.

McKendrick, G.B., Fotakis, C., and Donovan, R.J. (1982). Laser photodissociation of NO_2 at 248 nm and production of $NO(A\,^2\Sigma^+ \rightarrow X\,^2\Pi)$ fluorescence. *J. Photochem.* **20,** 175.

Merer, A.J., and Hallin, K.-E.J. (1978). The temperature dependence of the 8920 Å band of NO_2 and the location of the 000–000 band of the $\tilde{A}\,^2B$-$\tilde{X}^2A_1$ transition. *Can. J. Phys.* **56,** 838.

Miller, F.A., and White, W.B. (1957). Electronic spectrum and apparent thermochromism of $VOCl_3$. *Spectrochim. Acta* **9,** 98.

Molina, L.T., and Molina, M.J. (1981). UV absorption cross sections of HO_2NO_2 vapor. *J. Photochem.* **15,** 97.

Molina, L.T., Molina, M.J., and Rowland, F.S. (1982). Ultraviolet absorption cross sections of several brominated methanes and ethanes of atmospheric interest. *J. Phys. Chem.* **86,** 2672.

Morris, B.J. (1986). Photochemical organometallic vapor phase epitaxy of mercury cadmium telluride. *Appl. Phys. Lett.* **48,** 867.

Morrison, R.J.S., and Grant, E.R. (1982). Dynamics of the two-photon photodissociation of NO_2: A molecular beam multiphoton ionization study of NO photofragment internal energy distributions. *J. Chem. Phys.* **77,** 5995.

Mulliken, R.S. (1936). The low electronic states of simple heteropolar diatomic molecules. I. General survey. *Phys. Rev.* **50,** 1017.

Mulliken, R.S. (1977). The excited states of ethylene. *J. Chem. Phys.* **66,** 2448.

Myer, J.A., and Samson, J.A.R. (1970). Vacuum-ultraviolet absorption cross sections of CO, HCl, and ICN between 1050 and 2100 Å. *J. Chem. Phys.* **52,** 266.

Myer, J.A., and Samson, J.A.R. (1970a). Absorption cross section and photoionization yield of I_2 between 1050 and 2200 Å. *J. Chem. Phys.* **52,** 716.

Nakayama, T., Kitamura, M.Y., and Watanabe, K. (1959). Ionization potential and absorption coefficients of nitrogen dioxide. *J. Chem. Phys.* **30,** 1180.

Nakayama, T., and Watanabe, K. (1964). Absorption and photoionization coefficients of acetylene, propyne, and 1-butyne. *J. Chem. Phys.* **40,** 558.

Nathanson, G., Gitlin, B., Rosan, A.M., and Yardley, J.T. (1981). Ultraviolet laser photolysis: Primary photochemistry of $Fe(CO)_5$ in PF_3. *J. Chem. Phys.* **74,** 361.

Ni, T., Lu, Q., Ma, X., Yu, S., and Kong, F. (1986). The emission spectrum of AsH_2 ($\tilde{A}\,^2A_1 \rightarrow \tilde{X}\,^2B_1$) via UV photolysis of AsH_3. *Chem. Phys. Lett.* **126,** 417.

Ni, T., Yu, S., Ma, X., and Kong, F. (1986a). The UV laser photolysis of stibine. *Chem. Phys. Lett.* **128,** 270.

Ni, T., Yu, S., Ma, X., and Kong, F. (1986b). $NH(A^3\Pi \rightarrow X^3\Sigma^-)$ emission from 193 nm two-photon photolysis of NH_3. *Chem. Phys. Lett.* **126,** 413.

Okabe, H. (1971). Fluorescence and predissociation of sulfur dioxide. *J. Am. Chem. Soc.* **93,** 7095.

Okabe, H. (1978). *Photochemistry of Small Molecules*. Wiley, New York.

Okabe, H. (1983). Photochemistry of acetylene at 1849 Å. *J. Chem. Phys.* **78,** 1312.

Oldman, R.J., Sander, R.K., and Wilson, K.R. (1975). Photofragment spectrum of bromine. *J. Chem. Phys.* **63,** 4252.

Ondrey, G., Van Veen, N., and Bersohn, R. (1983). The state distribution of OH radicals photodissociated from H_2O_2 at 193 and 248 nm. *J. Chem. Phys.* **78,** 3732.

Osmundsen, J.F., Abele, C.C., and Eden, J.G. (1985). Activation energy and spectroscopy of the growth of germanium films by ultraviolet laser-assisted chemical vapor deposition. *J. Appl. Phys.* **57,** 2921.

Passchier, A.A., Christian, J.D., and Gregory, N.W. (1967). The ultraviolet-visible absorption spectrum of bromine between room temperature and 440°. *J. Phys. Chem.* **71,** 937.

Passchier, A.A., and Gregory, N.W. (1968). Evidence of molecular dimers, $(I_2)_2$, in iodine vapor. *J. Phys. Chem.* **72,** 2697.

Pence, W.H., Baughcum, S.L., and Leone, S.R. (1981). Laser UV photofragmentation of halogenated molecules. Selective bond dissociation and wavelength-specific quantum yields for excited $I(^2P_{1/2})$ and $Br(^2P_{1/2})$ atoms. *J. Phys. Chem.* **85,** 3844.

Pennella, F., and Taylor, W.J. (1963). Ultraviolet spectrum of VCl_4 vapor. *J. Mol. Spectrosc.* **11,** 321.

Preston, D.M., and Zink, J.I. (1987). Multiple luminescence spectra from cw excitation of room temperature gas-phase $Ni(CO)_4$. *J. Phys. Chem.* **91,** 5003.

Preston, K.F., and Cvetanović, R.J. (1966). Collisional deactivation of excited oxygen atoms in the photolysis of NO_2 at 2288 Å. *J. Chem. Phys.* **45,** 2888.

Prinslow, D.A., and Vaida, V. (1987). Wavelength-dependent photofragmentation of gas-phase dimanganese decacarbonyl. *J. Am. Chem. Soc.* **109,** 5097.

Rebbert, R.E., and Ausloos, P. (1975). Photodecomposition of trichlorofluoromethane and dichlorodifluoromethane. *J. Photochem.* **4,** 419.

Rebbert, R.E., Lias, S.G., and Ausloos, P. (1978). The gas phase photolysis of $CHFCl_2$. *J. Photochem.* **8,** 17.

Rigby, L.J. (1969). Photodeposition from tetra ethyl lead. *Trans. Faraday Soc.* **65,** 2421.

Riley, S.J., and Wilson, K.R. (1972). Excited fragments from excited molecules. Energy partitioning in the photodissociation of alkyl iodides. *Faraday Discuss. Chem. Soc.* **53,** 132.

Robbins, D.E. (1976). Photodissociation of methyl chloride and methyl bromide in the atmosphere. *Geophys. Res. Lett.* **3,** 213.

Roberge, R., and Salahud, D.R. (1979). Valence and Rydberg excited states of H_2S: An SCF-Xα-SW molecular orbital study. *J. Chem. Phys.* **70,** 1177.

Rockwood, S., and Rabideau, S.W. (1974). Boron isotope separation by laser-induced photochemistry. *IEEE J. Quant. Electron.* **QE-10,**, 789.

Rösch, N., Kotzian, M., Jörg, H., Schröder, H., Rager, B., and Metev, S. (1986). On visible transients in gas phase UV photolysis of transition metal compounds: Experimental and theoretical results for $Ni(CO)_4$. *J. Am. Chem. Soc.* **108,** 4238.

Rothschild, M., and Ehrlich, D.J. (1986). Unpublished.

Rytz-Froidevaux, Y., Salathé, R.P., and Gilgen, H.H. (1983). Laser-initiated Ga-deposition on quartz substrates. *Mat. Res. Soc. Symp. Proc.* **17,** 29.

Safary, E., Romand, J., and Vodar, B. (1951). Ultraviolet absorption spectrum of gaseous hydrogen fluoride. *J. Chem. Phys.* **19,** 379.

Sam, C.L., and Yardley, J.T. (1978). Laser induced production of excited states of PH and PH_2 from phosphine. *J. Chem. Phys.* **69,** 4621.

Sam, C.L., and Yardley, J.T. (1979). Excimer laser photolysis in CBr_4 and CF_2Br_2: Luminescence from Br_2, Br, CF_2, and C_2. *Chem. Phys. Lett.* **61,** 509.

Sauer, M.C., Jr. and Dorfman, L.M. (1961). Molecular detachment processes in the vacuum UV photolysis of gaseous hydrocarbons. I. Ethylene. II. Butane. *J. Chem. Phys.* **35,** 497.

Schofield, K. (1978). Rate constants for the gaseous interactions of $O(2\,^1D_2)$ and $O(2\,^1S_0)$—a critical evaluation. *J. Photochem.* **9,** 55.

Schreiner, A.F., and Brown, T.L. (1968). A semiempirical molecular orbital model for $Cr(CO)_6$, $Fe(CO)_5$, and $Ni(CO)_4$. *J. Am. Chem. Soc.* **90,** 3366.

Seder, T.A., Church, S.P., Oudekirk, A.J., and Weitz, E. (1985). Gas-phase photofragmentation of $Cr(CO)_6$: Time-resolved infrared spectrum and decay kinetics of "naked" $Cr(CO)_5$. *J. Am. Chem. Soc.* **107,** 1432.

Seder, T.A., Church, S.P., and Weitz, E. (1986). Gas-phase infrared spectroscopy and recombination kinetics for $Mn(CO)_5$ generated via XeF laser photolysis of $Mn_2(CO)_{10}$. *J. Am. Chem. Soc.* **108,** 1084.

Seder, T.A., Church, S.P., and Weitz, E. (1986a). Photodissociation pathways and recombination kinetics for gas-phase $Mn_2(CO)_{10}$. *J. Am. Chem. Soc.* **108,** 7518.

Seder, T.A., Oudekirk, A.J., and Weitz, E. (1986b). The wavelength dependence of excimer laser photolysis of $Fe(CO)_5$ in the gas phase. Transient infrared spectroscopy and kinetics of the $Fe(CO)_x$ ($x = 4, 3, 2$) photofragments. *J. Chem. Phys.* **85,** 1977.

Shimo, N., Nakashima, N., and Yoshihara, K. (1986). The UV absorption spectrum of trimethylsilyl radical in the gas phase. *Chem. Phys. Lett.* **125,** 303.

Shibuya, K., and Stuhl, F. (1982). Single vibronic emissions from NO B $^2\Pi$ ($v' = 7$) and O_2 B $^3\Sigma_u^-$ ($v' = 4$) excited by 193 nm ArF laser. *J. Chem. Phys.* **76,** 1184.

Slanger, T.G., Bischel, W.K., and Dyer, M.J. (1983). Nascent NO vibrational distribution from 2485 Å NO_2 photodissociation. *Chem. Phys.* **79,** 2231.

Sparks, R.K., Shobatake, K., Carlson, L.R., and Lee, Y.T. (1981). Photofragmentation of CH_3I: Vibrational distribution of the CH_3 fragment. *J. Chem. Phys.* **75,** 3838.

Steunenberg, R.K., and Vogel, R.C. (1956). The absorption spectrum of fluorine. *J. Am. Chem. Soc.* **78,** 901.

Stuke, M. (1984). Ultrasensitive fingerprint detection of organometallic compounds by laser multiphoton ionization mass spectroscopy. *Appl. Phys. Lett.* **45,** 1175.

Stuke, M., Zhang, Y., and Kuper, S. (1988). Fundamentals of laser photochemistry for surface modification. *Mat. Res. Soc. Symp. Proc.* **101,** 139.

Sturm, G.P., and White, J.M. (1969). Photodissociation of hydrogen sulfide and methanethiol. Wavelength dependence of the distribution of energy in the primary products. *J. Chem. Phys.* **50,** 5035.

Style, D.W.G., and Ward, J.C. (1952). Origins of the fluorescence obtained from formic acid and methylene iodide. *J. Chem. Soc.*, 2125.

Su, F., Bottenheim, J.W., Thorsell, D. L., Calvert, J.G., and Damon, E.K. (1977). The efficiency of the phosphoresence decay of the isolated $SO_2(^3B_1)$ molecule. *Chem. Phys. Lett.* **49,** 305.

Suto, M., and Lee, L.C. (1983). OH $(A^2\Sigma^+ \rightarrow X^2\Pi)$ yield from photodissociation of H_2O_2 at 106–193 nm. *Chem. Phys. Lett.* **98,** 152.

Suzuki, N., Anayama, C., Masu, K., Tsubouchi, K., and Mikoshiba, N. (1986). Pyrolysis and photolysis of trimethylaluminum. *Jpn. J. Appl. Phys.* **25,** 1236.

Swanson, J.R., Friend, C.M., and Chabal, Y.J. (1987). Deposition of iron on Si(111)-(7×7): Photo- and electron-assisted decomposition of $Fe(CO)_5$. *Mat. Res. Soc. Symp. Proc.* **75,** 559.

Swanson, J.R., Friend, C.M., and Chabal, Y.J. (1988). Laser-assisted deposition of Fe and W: Photodecomposition of $Fe(CO)_5$ and $W(CO)_6$ on Si(111)-(7×7), *Mat. Res. Soc. Symp. Proc.* **101,** 201.

Tabares, F.L., Marsh, E.P., Bach, G.A., and Cowin, J.P. (1987). Laser photofragmentation and photodesorption of physisorbed CH_3Br on lithium fluoride. *J. Chem. Phys.* **86,** 738.

Tabuchi, T., Yamagishi, K., and Tarui, Y. (1987). Low-temperature growth of transparent and conducting tin oxide film by photo-chemical vapor deposition. *Jpn. J. Appl. Phys.* **26,** L186.

Tachibana, H., Nakaue, A., and Kawate, Y. (1988). Deposition of amorphous carbon films by laser induced CVD. *Mat. Res. Soc. Symp. Proc.* **101,** 367.

Tamres, M., Duerksen, W.K., and Goodenow, J.M. (1968). Vapor-phase charge-transfer complexes. II. The $2I_2 \rightarrow I_4$ system. *J. Phys. Chem.* **72,** 966.

Tanner, K.N., and Duncan, A.B.F. (1951). Raman effect and ultraviolet absorption spectra of molybdenum and tungsten hexaflourides. *J. Am. Chem. Soc.* **73,** 1164.

Tellinghuisen, J. (1973). Resolution of the visible-infrared absorption spectrum of I_2 into three contributing transitions. *J. Chem. Phys.* **58,** 2821.

Terenin, A. (1964). Electronic spectroscopy of adsorbed gas molecules. *Adv. Catalysis.* **15,** 227.

Thompson, H.W. (1934). The absorption spectra of some polyatomic molecules containing methyl and ethyl radicals. *J. Chem. Soc.*, 790.

Tiee, J.J., Wampler, F.B., and Rice, W.W. (1979). Laser-induced fluorescence excitation spectra of CCl_2 and CFCl radicals in the gas phase. *Chem. Phys. Lett.* **65,** 425.

Tiee, J.J., Wampler, F.B., and Rice, W.W. (1979a). UV laser photochemistry of CF_2Cl_2. *Chem. Phys. Lett.* **68,** 403.

Tokumitsu, E., Yamada, T., Konagai, M., and Takahashi, K. (1988). Excimer laser irradiation effect on metalorganic molecular beam deposition of Al and AlAs: Prospects for laser-assisted epitaxial growth of GaAlAs. *Mat. Res. Soc. Symp. Proc.* **101,** 307.

Tsao, J.Y., Becker, R.A., Ehrlich, D.J., and Leonberger, F.J. (1983). Photodeposition of Ti and application to direct writing of Ti:$LiNbO_3$ waveguides. *Appl. Phys. Lett.* **42,** 559.

Tsao, J.Y., and Ehrlich, D.J. (1984). UV-laser photodeposition from surface-adsorbed mixtures of trimethylaluminum and titanium tetrachloride, *J. Chem. Phys.* **81,** 4620.

Tumas, W., Gitlin, B., Rosan, A.M., and Yardley, J.T. (1982). Olefin rearrangement resulting from the gas-phase KrF laser photolysis of $Cr(CO)_6$. *J. Am. Chem. Soc.* **104,** 55.

Turro, N.J. (1967). *Molecular Photochemistry*. W.A. Benjamin, New York.

Tyndall, G.W., and Jackson, R.L. (1987). UV multiple-photon dissociation of $Cr(CO)_6$ to Cr* and CO: Evidence for direct and sequential dissociation processes. *J. Am. Chem. Soc.* **109,** 582.

Uselman, W.M., and Lee, E.K.C. (1976). A study of nitrogen dioxide ($2\,^2B_2$) photodecomposition to $O(^1D)$ and $NO(^2\Pi)$ in its second predissociation region 2500–2139 Å. *J. Chem. Phys.* **65,** 1948.

Van Veen, G.N.A., Mohamed, K.A., Baller, T., and de Vries, A.E. (1983). Photofragmentation of H_2S in the first continuum. *Chem. Phys.* **74,** 261.

Van Veen, G.N.A., Baller, T., de Vries, A.E., and Van Veen, N.J.A. (1984). The excitation of the umbrella mode of CH_3 and CD_3 formed from photodissociation of CH_3I and CD_3I at 248 nm. *Chem. Phys.* **87,** 405.

Van Veen, G.N.A., Baller, T., and de Vries, A.E. (1985). Photofragmentation of CH_3Br in the A band, *Chem. Phys.* **92,** 59.

Van Veen, G.N.A., Baller, T., and de Vries, A.E. (1985a). Predissociation of specific vibrational states of CH_3I upon excitation around 193.3 nm. *Chem. Phys.* **97,** 179.

Van Veen, G.N.A., Baller, T., de Vries, A.E., and Shapiro, M. (1985b). Photofragmentation of CF_3I in the A band. *Chem. Phys.* **93,** 277.

Venkitachalam, T. and Bersohn, R. (1984). Two-photon dissociation of SO_2 in the region 285–311 nm. *J. Photochem.* **26,** 65.

Waller, I.M., Davis, H.F., and Hepburn, J.W. (1987). Photofragment spectroscopy of metal carbonyls: A molecular beam study of $Fe(CO)_5$ photolysis at 193 nm. *J. Phys. Chem.* **91,** 506.

Walton, J.C. (1972). Photolysis of dibromodifluoromethane at 265 nm. *J. Chem. Soc. Faraday Tran.* 1 **68,** 1559.

Wampler, F.B., Tiee, J.J., Rice, W.W., and Oldenborg, R.C. (1979). UV laser photochemistry of CF_2Br_2. *J. Chem. Phys.* **71,** 3926.

Wang, H.-T., Felps, W.S., and McGlynn, S.P. (1977). Molecular Rydberg states. VII. Water. *J. Chem. Phys.* **67,** 2614.

Warneck, P., Marmo, F.F., and Sullivan, J.O. (1964). Ultraviolet absorption of SO_2: Dissociation energies of SO_2 and SO. *J. Chem. Phys.* **40,** 1132.

Washida, N., Matsumi, Y., Hayashi, T., Ibuki, T., Hiraya, I., and Shobatake, K. (1985). Emission spectra of $SiH(A\,^2\Delta \rightarrow X^2\Pi)$ and $SiCl_2(\tilde{A}\,^1B_1 \rightarrow \tilde{X}\,^1A_1)$ in the VUV photolyses of silane and chlorinated silanes. *J. Chem. Phys.* **83,** 2769.

Watanabe, K., and Zelikoff, M. (1953). Absorption coefficients of water vapor in the vacuum ultraviolet. *J. Opt. Soc. Am.* **43,** 753.

Watanabe, K. (1954). Photoionization and total absorption cross section of gases. I. Ionization potentials of several molecules. Cross sections of NH_3 and NO. *J. Chem. Phys.* **22,** 1564.

Watanabe, K., and Jursa, A.S. (1964). Absorption and photoionization cross sections of H_2O and H_2S. *J. Chem. Phys.* **41,** 1650.

Whetten, R.L., Fu, K.-J., and Grant, E.R. (1983). Photodissociation dynamics of $Fe(CO)_5$: Excited state lifetimes and energy disposal. *J. Chem. Phys.* **79,** 4899.

Wight, C.A., and Leone, S.R. (1983). Vibrational state distributions and absolute excitation efficiencies for T-V transfer collisions of NO and CO with H atoms produced by excimer laser photolysis. *J. Chem. Phys.* **79,** 4823.

Wildt, J., Fink, E.H., Winter, R., and Zabel, F. (1983). Radiative lifetime and quenching of $SO(b\,^1\Sigma^+, v' = 0)$. *Chem. Phys.* **80,** 167.

Wilkinson, P.G., and Mulliken, R.S. (1955). Far ultraviolet absorption spectra of ethylene and ethylene-d_4. *J. Chem. Phys.* **23,** 1895.

Wilson, M.W., Rothschild, M., Muller, D.F., and Rhodes, C.K. (1982). Multiphoton photofragmentation of SO_2 at 248 nm. *J. Chem. Phys.* **77,** 1837.

Wodtke, A.M., and Lee, Y.T. (1985). Photodissociation of acetylene at 193.3 nm. *J. Phys. Chem.* **89,** 4744.

Yardley, J.T., Gitlin, B., Nathanson, G., and Rosan, A.M. (1981). Fragmentation and molecular dynamics in the laser photodissociation of iron pentacarbonyl. *J. Chem. Phys.* **74,** 370.

Yu, C.F., Youngs, F., Tsukiyama, K., Bersohn, R., and Preses. J. (1986). Photodissociation dynamics of cadmium and zinc dimethyl, *J. Chem. Phys.* **85,** 1382.

Zacharias, H., Geilhaupt, M., and Welge, K.H. (1981). Laser photofragment spectroscopy of the NO_2 dissociation at 337 nm. A nonstatistical decay process. *J. Chem. Phys.* **74,** 218.

Zavelovich, J., Rothschild, M., Gornik, W., and Rhodes, C.K. (1981). VUV fluroescence following photodissociation of N_2O at 193 nm. *J. Chem. Phys.* **74,** 6787.

Zelikoff, M., Watanabe, K., and Inn, E.C.Y. (1953). Absorption coefficients of gases in the vacuum ultraviolet. Part II. Nitrous oxide. *J. Chem. Phys.* **21,** 1643.

Zuhoski, S.P., Killeen, K.P., and Biefeld, R.M. (1988). Photolytic deposition of InSb films. *Mat. Res. Soc. Symp. Proc.* **101,** 313.

CHAPTER 4

Photophysics and Thermophysics of Light Absorption and Energy Transport in Solids

C.I.H. ASHBY AND J.Y. TSAO

Sandia National Laboratories
Albuquerque, New Mexico

1. Introduction

Surface physical or chemical processes may be influenced by laser light in a variety of ways. As discussed in Chapter 3, these influences may be

0-12-233430-2

categorized according to where the light is absorbed: in the surrounding external phase, in adsorbates bound to the surface, or in the solid substrate itself. It is the third case, primary photoexcitation of the solid substrate, that is the focus of this chapter. In particular, we discuss the following sequence of events: a) light absorption by the solid leading to "hot" carriers with possibly enhanced photochemical activity, b) thermal relaxation of carriers, leading both to "cold" carriers, still with possibly enhanced photochemical reactivity, and to a "hot" lattice with enhanced thermal reactivity, c) carrier recombination and carrier diffusion into the bulk and d) heat diffusion into the bulk.

There are two extremes of behavior, depending on the partitioning of absorbed energy into the available electronic or vibrational degrees of freedom. At one extreme are processes in which the laser is merely a spatially localized heat source enhancing the rate of a thermally activated chemical reaction. Laser-driven processes for metals and insulators generally fall into this category, as do processes for semiconductors illuminated at very high power densities. Photon absorption may occur either a) via band-to-band electronic transitions in semiconductors and insulators, in which case photons must possess energies greater than the band-gap energy, or b) via vibrational transitions or free carrier absorption in all three classes of materials, in which case sub-band-gap energies are sufficient. Following absorption, relaxation and recombination thermalize the absorbed energy; the resultant temperature rise accelerates the rate of a thermally activated chemical reaction. The spatial extent of the region that undergoes enhanced reaction is determined by the spatial and temporal properties of the laser beam and by the optical absorption and thermal diffusion properties of the material, since these properties determine the local temperature rise.

At the other extreme are processes in which photogenerated electrons and holes participate directly in chemical reactions. These processes generally involve semiconductors and are most prominent at lower power densities. Carriers may participate while in the "hot" excited state in which they are initially created, but more typically do so after fast intraband relaxation to a "cold," but nevertheless excited, state. If the "hot" carriers have enough energy to produce additional free carriers by impact ionization, then quantum yields may be enhanced. However, loss of surface carriers through recombination and diffusion will ultimately limit the quantum yield, as well as determine the spatial extent of the reactive region.

The degree to which processes fall at one or the other extreme depends both on the partitioning of the absorbed energy into the available electronic and vibrational degrees of freedom and on the processes that transport energy away from the surface. In this chapter, we discuss the photophysics and thermophysics of these partitioning and energy transport questions from a laser microfabrication point of view. The assumption is made that the important determinants of light-induced reaction enhancements or suppressions, and hence the quantities we wish to understand, are the surface carrier density for semiconductors at lower illumination levels and the temperature distributions for metals, insulators, and semiconductors at higher illumination levels.

It should be emphasized that this chapter is concerned only with the *physical* events that follow light absorption. We defer to Chapter 2 a discussion of the important and complex second half of the problem, which has to do with the *chemistry* of how carriers and thermal energy affect surface reactions. Here, we only point out that photochemical reactions are typically linear in light intensity, whereas thermal reactions are super-linear (Arrhenian). Therefore, for a given average power density incident on a surface, photochemical effects will tend to be favored for cw (continuous wave) irradiation at relatively low power densities, whereas thermal effects will tend to be favored for pulsed irradiation. In the extreme case, peak power densities may be so high in the pulsed case as to cause substrate damage, while the duty cycle may be so low that photochemical effects are negligible.

The rest of this chapter is organized in the following manner. In the first half of the chapter, we discuss semiconductor photoexcitation from the point of view of laser-assisted photochemistry. For such photochemistry, it is the free-carrier density at the surface that is crucial. This density is determined by a balance between free-carrier creation by photon absorption and loss through recombination, drift, and diffusion away from the reacting surface. In general, determining this balance is complex and sample specific, and numerical techniques are required to model a particular case. For example, it will depend in a complex way on laser-wavelength-dependent optical absorption in the near surface, on environment-dependent surface recombination velocities, on space-charge-field-dependent drift in the near-surface region, and on laser-spot-size-dependent three-dimensional carrier diffusion into the bulk. In this chapter, therefore, we give only a qualitative discussion of these phenomena, from a laser-assisted processing viewpoint.

In the second half of the chapter, we discuss laser heating of surfaces from the point of view of laser-assisted thermal chemistry. For such chemistry, it is the temperature at the surface due to patterned laser irradiation that is crucial. This temperature is determined by a balance between heating through photon absorption and loss through heat conduction into the substrate. Determining this temperature accurately generally also requires numerical techniques. However, the problem is not as complex as that of determining the carrier densities described above. Therefore, we will outline semiquantitative approaches from which order-of-magnitude estimates may be obtained. We also introduce a simple classification scheme to describe different heat-flow regimes. Simple analytic treatments are summarized for the spatiotemporal temperature profiles in the different regimes. A set of tables of thermophysical properties of solids is also given that is sufficient for semiquantitative estimates of heat flow and of the resulting temperature profiles.

2. Electronic Properties of Semiconductors Important for Carrier-Driven Photochemistry

This section reviews some of the important electronic properties of semiconductors relevant to laser-assisted reactions requiring direct participation of carriers. These properties can be divided into two general categories: a) the inherent properties that characterize a particular semiconductor material or sample and b) the properties that may be modified by laser irradiation or other process parameters. The following discussion assumes that reaction rates for carrier-dependent photochemical reactions are proportional to the free carrier density at the surface. This assumption will hold in the most common situation, in which the photochemical surface reaction itself, rather than mass transport of reactants or products to or from the surface, is rate limiting. It is generally the minority carrier density that determines the reaction rate since both types of carriers are required to drive a balanced oxidation-reduction reaction, and the minority carrier supply will usually be rate limiting.

At the outset, two general comments may be useful concerning laser-induced photochemical reactions. First, a range of chemical reactions may be enhanced by photogenerated carriers. A recent introductory review may be found in Plaskov and Gurevich (1986); current research in the field may be found in Schiavello (1985). However, virtually all cases studied thus far for microfabrication purposes have been etching reactions. Consequently, the following discussion address mainly photoenhanced etching reactions. Nevertheless, the discussion is not meant to exclude the possibility of other photochemical reaction types, such as photostimulated deposition, in which the deposited material is itself photosensitive.

Second, carrier-driven etching reactions may be divided into two general types: photoelectrochemical and photochemical. Many specific examples of both types will be discussed in Chapter 6 of this book and in Ashby (1987). Photoelectrochemical (or photoelectrolytic) processes require ohmic contacts to the semiconductor and an external power supply. The semiconductor is immersed in an electrolyte; within limits, the specific chemical properties of the electrolyte are often of secondary importance (Ashby, 1987). Etching is controlled by applying a bias voltage while photons create electron/hole pairs in the near surface. The application of an external voltage can alter the complex interplay between internal electric fields, carrier concentrations, and carrier motion. For example, the field-assisted movement of carriers may be controlled by the externally imposed field rather than by the characteristic field in the space-charge region.

In contrast, photochemical processes require neither ohmic contacts nor an external power supply. With aqueous etchants, the illuminated and unilluminated regions of the semiconductor surface serve as anode and cathode, respectively, for a galvanic reaction. With gaseous etchants, both oxidation and reduction occur at the etching surface, but the analogy to a galvanic reaction cell does not strictly apply. For these processes, the chemical properties of the liquid or gas and the characteristic electronic properties of a particular material are very important in determining whether etching will occur and at what rate.

2.1. Inherent Optical and Electronic Properties

The inherent bulk and surface properties of a semiconductor material are determined by stoichiometry for compound semiconductors, width and type of band gap, equilibrium majority carrier concentration as determined by impurity doping, and crystallographic orientation of the surface. The first three affect bulk optical and electronic properties, while the last affects primarily surface electronic properties.

2.1.1 Band Gap and Absorption Coefficient

The characteristics of the band gap place the most severe restrictions on semiconductor photoexcitation. To produce free carriers, photons must have energies greater than the width of the band gap. Band-gap energies and types for most of the technologically important crystalline semiconductors are listed in Table 4.1 (Pankove, 1975). In the corresponding amorphous materials band gaps are narrower and are sensitive to the method of preparation (Alimoussa et al., 1981; Matsumoto and Kumabe, 1980; Piller, 1985).

Semiconductors can be divided into two major categories: direct gap and indirect gap. For direct gap semiconductors, the lowest energy

Table 4.1 Band Gaps of Semiconductors (after Pankove, 1975).

Material	Band Gap (eV, 0°K)	Band gap (eV, 300K)	Type	$dE_g/dT \times 10^4$ (eV/K)
Si	1.166	1.11	i	−2.3
Ge	0.74	0.67	i	−3.7
AlP	2.5	2.43	i	−3.5
AlAs	2.24	2.16	i	—
AlSb	—	1.6	d	−4
GaP	2.4	2.25	i	−5.5
GaAs	1.52	1.43	d	−4.5
GaSb	0.81	0.69	d	−4.1
InP	1.42	1.28	d	−4.6
InAs	0.43	0.36	d	−3.3
InSb	0.235	0.17	d	−2.9
ZnS	—	3.8	d	−4 to −5
ZnSe	2.8	2.58	d	−7.2
ZnTe	2.39	2.28	d	−5
CdS	2.58	2.53	d	−5
CdSe	1.85	1.74	d	−4.6
CdTe	1.60	1.50	d	−4.1
HgS	—	2.5	d	—
HgSe	−0.24	−0.15	d	—
HgTe	−0.28	−0.15	d	+5.6

optical transition that generates free carriers occurs at the Γ-point in reciprocal space, where momentum transfer to the carrier is nearly zero. The absorption cross section for such a direct transition is generally large and rises rapidly with increasing photon energy near the band edge. For indirect gap semiconductors, the lower-energy carrier-generating transition requires a nonzero momentum transfer. Since photons supply negligible momentum, such a transition must be lattice-vibration (phonon) assisted. The absorption cross section for such phonon-assisted, indirect processes is typically two to four orders of magnitude smaller than for direct processes and increases much more slowly with increasing photon energy above the gap. The dependence of the absorption coefficient, α, on photon energy is illustrated for the direct gap semiconductor GaAs in Fig. 4.1 and for the indirect gap semiconductor GaP in Fig. 4.2.

The absorption coefficient at the wavelength of the light source determines how near the surface the carriers are created. When the lowest energy gap is indirect, the absorption coefficient for photons with energies near the gap is on the order of 1–10 cm^{-1}. A large fraction of

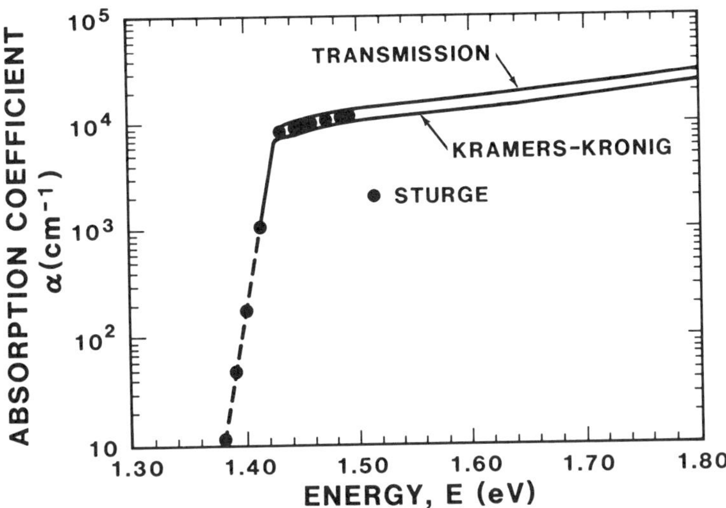

Fig. 4.1 Absorption coefficient of the direct-gap semiconductor GaAs (Casey et al., 1975).

the photons with these energies are absorbed relatively deep within the substrate. In contrast, when the lowest energy gap is direct, the absorption coefficient near the band gap exceeds $10^4\ cm^{-1}$. In this case, many of the carriers are created within a diffusion length of the surface. The probability of photogenerated carriers reaching the reacting surface

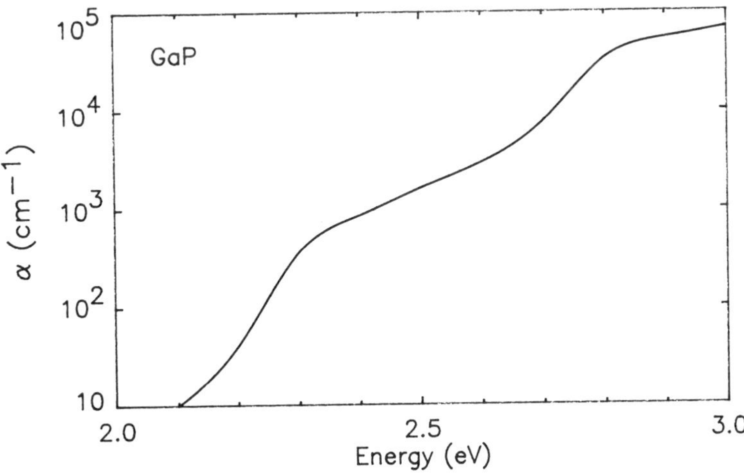

Fig. 4.2 Absorption coefficient of the indirect-gap semiconductor GaP (Aspnes and Studna, 1983).

is higher when they are generated closer to the surface, especially if bulk recombination rates are high. Therefore, a high absorption coefficient at the excitation wavelength is desirable to achieve high quantum yields.

The absorption coefficient, α, can be calculated from the extinction coefficient, k, using

$$\alpha = \frac{4\pi k}{\lambda}, \tag{4.1}$$

where the complex index of refraction is $n + ik$. Values for the extinction coefficient, k, and refractive index, n, are listed in Table 4.2 for several common laser wavelengths for selected semiconductors and insulators. The optical constants from 1.5 to 6.0 eV (within which lies the wavelengths readily accessible with common cw ion or dye lasers) have been recently been reported for Si, Ge, GaP, GaAs. GaSb, InP, InAs, and InSb (Aspnes and Studna, 1983). The values of n and k over this energy range are presented graphically for Si, Ge, GaAs, and InP in Figs. 4.5 through 4.8. Optical constants for the preceding materials and for CdTe, PbSe, PbS, PbTe, ZnS, SiC, and a-Si are tabulated in Palik (1985). The optical constants for some important insulators such as SiO_2, SiO, Si_3N_4, and TiO_2 are also tabulated there.

For some materials, such as Ga(As, P) (Tietjen and Amick, 1966) and (Al, Ga)As (Keuch et al., 1987, Oelgart et al., 1987, and Kaneko et al., 1976), band-gap energies not only change with alloy stoichiometry, but the nature of their gaps changes from direct to indirect. This is illustrated in Figs. 4.3 and 4.4 for the $GaAs_{1-x}P_x$ and $Al_xGa_{1-x}As$ alloy systems.

2.1.2 Impurity Concentration

Equilibrium majority and minority carrier concentrations depend to some extent on the band gap and temperature of a semiconductor. However, except for highly intrinsic material, they usually depend, in the absence of photoexcitation, primarily on the type and concentration of impurities. Such impurities are also important in determining the relaxation and recombination behavior, and hence the carrier diffusion length, in photoexcited materials. Carriers tend to diffuse a shorter distance before recombination in low-purity than in high-purity samples. A short carrier diffusion length can have a pronounced effect on the ultimate lateral resolution that can be achieved: In such applications as holographic etching of gratings (Lum et al., 1985), heavier doping has been found to increase resolution. A short carrier diffusion length also implies that only photons absorbed very near the surface produce carriers that ultimately reach the surface reaction zone. For example, the quantum yield for

Table 4.2 Room-temperature Optical Constants (n, k) of Selected Semiconductors and Insulators at Various Wavelengths (in μm) (after Palik, 1985).

λ (μm)	0.193	0.248	0.488	0.514	0.632	1.06	10.6
Group IV Elements and Compounds							
d-C	(2.9, 1.0E-3)	(2.7, 2E-6)	(2.4, 3E-7)	(2.4, 3E-7)	(2.4, 3E-7)	(2.4, 1E-6)	(2.4,)
c-Si	(0.88, 2.8)	(1.7, 3.6)	(4.4, 0.080)	(4.2, 0.060)	(3.9, 0.019)	(3.5, 1.0E-3)	(3.4, 1.3E-4)
a-Si	(1.0, 2.1)	(1.7, 2.8)	(4.5, 1.2)	(4.5, 1.0)	(4.2, 0.4)	(3.6, 5E-2)	(3.7, 0.0)
Ge	(1.1, 2.1)	(1.4, 3.2)	(4.3, 2.4)	(4.6, 2.4)	(5.5, 0.7)	(4.4, 0.1)	(4.0, 1E-5)
SiC	(4.0, 1.4)	(3.2, 0.26)	(2.7, 1.2E-5)	(2.7, ~1E-5)	(2.6, ~1E-5)	(2.6, ~5E-5)	(0.06, 1E-2)
III–V Compounds							
GaP	(1.2, 1.9)	(3.7, 3.6)	(3.7, 3.4E-3)	(3.5, 1.2E-3)	(3.3,)	(3.1,)	(3.0, 2E-3)
GaAs	(1.4, 2.0)	(2.3, 4.1)	(4.4, 0.48)	(4.2, 0.38)	(3.8, 0.19)	(3.5,)	(3.3, 7.6E-7)
GaSb							(3.8, 1.0E-2)
InP	(1.5, 2.0)	(2.1, 3.5)	(3.9, 0.53)	(3.7, 0.46)	(3.5, 0.30)	(3.3,)	(3.0,)
InAs	(1.6, 2.1)	(1.5, 2.9)	(4.2, 1.9)	(4.5, 1.3)	(4.0, 0.6)	(3.6, 0.4)	(3.4,)
InSb	(1.1, 1.7)	(1.3, 2.4)	(3.5, 2.2)	(3.8, 2.3)	(4.2, 1.8)	(4.2, 0.3)	(3.9,)
II–VI and Chalcogenide Compounds							
c-ZnS	(2.2, 1.6)	(3.0, 0.7)	(2.4, ~6E-6)	(2.4, ~4E-6)	(2.4, 3.4E-6)	(2.3, 3.0E-6)	(2.2, 2E-5)
h-ZnS	(2.0, 1.3)	(2.6, 0.91)	(2.4, 0.1)	(2.4, 0.1)	(2.4, 6E-2)	(2.3,)	(2.3,)
CdTe		(2.5, 1.9)	(3.1, 0.5)	(3.0, 0.4)	(2.9, 0.2)	(2.8, 0.0)	(2.7, 5.1E-8)
PbS	(0.95, 1.5)	(1.5, 2.1)	(4.3, 2.0)	(4.3, 1.7)	(4.3, 1.4)	(4.4, 0.54)	(4.0, 0.0)
PbSe	(0.67, 0.80)	(0.54, 1.2)	(~3.5, ~3.0)	(~3.5, ~3.0)	(3.7, 2.8)	(4.7, 1.4)	(4.7, 1.3E-3)
PbTe	(0.9, 0.93)	(0.72, 1.0)	(1.2, 2.7)	(~1.7, 2.9)	(~2.6, 3.0)	(4.2, 2.6)	(5.5, 6.0)
a-As_2S_3		(2.5, 1.2)	(2.9, 0.026)	(2.8, 5.5E-3)	(2.6, 4E-6)	(2.5, 2E-7)	(2.4, 9E-5)
a-As_2Se_3						(2.9,)	(2.8, 1.1E-6)

Table 4.2 *(continued)*

λ (μm)	0.193	0.248	0.488	0.514	0.632	1.06	10.6
I/II–VII Compounds							
LiF	(1.4, E-7)	(1.4, E-7)	(1.4, 1E-8)	(1.4, 1E-8)	(1.4, E-8)	(1.4, E-8)	(1.0, 1E-2)
NaCl	(1.8, <E-7)	(1.6,)	(1.6, 1.1E-10)	(1.5, 4.9E-11)	(1.5,)	(1.5,)	(1.5, 8E-8)
KCl	(1.8, <E-7)	(1.6,)	(1.5, 7E-11)	(1.5, E-11)	(1.5, E-11)	(1.5,)	(1.5, 2.5E-8)
CaF		(1.46,)	(1.43,)	(1.43,)	(1.43,)	(1.43,)	(1.3,)
Oxides and Nitrides							
a-SiO_2	(1.6, <E-6)	(1.5, <E-6)	(1.5,)	(1.5,)	(1.5,)	(1.4,)	(2.2, 0.09)
a-SiO	(1.8, 0.72)	(2.0, 0.60)	(2.0, 0.05)	(2.0, 0.04)	(2.0, 0.005)	(1.9,)	(2.9, 0.8)
a-Si_3N_4	(2.6, 0.24)	(2.3, 4.9E-3)	(2.0,)	(2.0,)	(2.0,)	(2.0,)	
TiO_2	(1.5, 1.7)	(2.1, 1.4)	(2.9,)	(2.9,)	(2.7,)	(2.6,)	(1.1, 0.07)
$LiNbO_3$	(1.9, 1.3)	(3.4, 0.8)	(2.3,)	(2.3,)	(2.2,)	(2.2,)	(1.2, ~1)

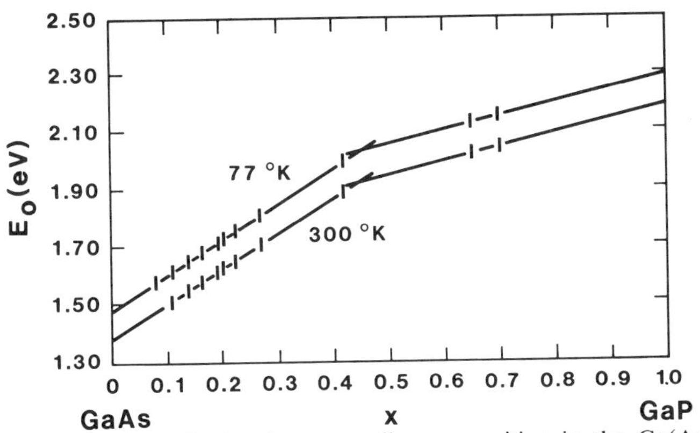

Fig. 4.3 Dependence of the band gap on alloy composition in the Ga(As, P) system (Tietjen and Amick, 1966). GaAs has a direct gap; GaP has an indirect gap.

etching GaAs or Ga(As, P) is reduced by more than an order of magnitude when impurity doping is high enough to produce degenerate material (Ashby and Biefeld, 1988; Ashby and Myers, 1988).

In heavily doped materials, free-carrier absorption of sub-band-gap radiation can be comparable in strength to indirect band-to-band

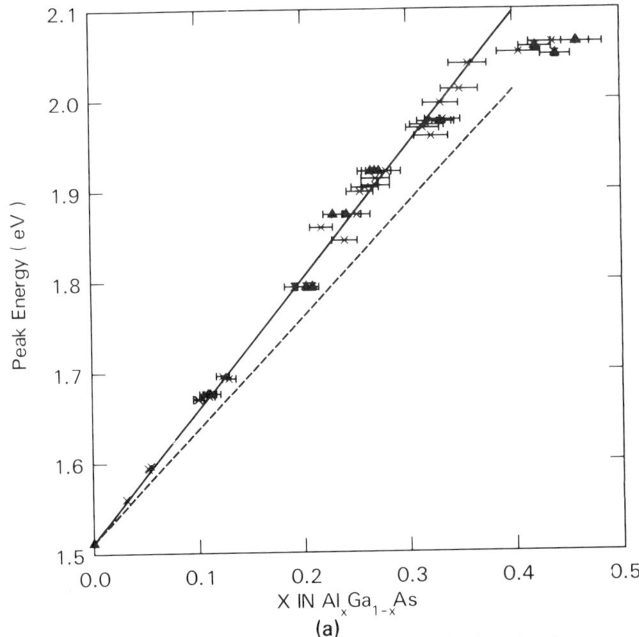

Fig. 4.4 (a) Band gap of $Al_xGa_{1-x}As$ as measured by peak photoluminescence intensity at 2°K (Keuch et al., 1987). (b) Energy diagram of $Al_xGa_{1-x}As$ (Kaneko et al., 1976).

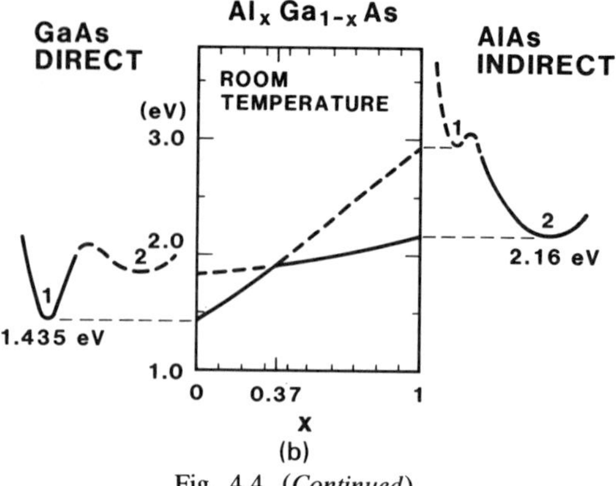

Fig. 4.4 (*Continued*)

absorption, as shown in Fig. 4.9 for GaAs (Casey et al., 1975). This process of raising a preexistent carrier to a higher energy level in the same band does not create new minority carriers and, consequently, will not contribute to carrier-dependent surface photochemical reactions. However, the subsequent relaxation will contribute to heating of the material and could enhance thermally activated etching.

2.1.3 Crystallographic Orientation

The crystallographic orientation of the semiconductor surface determines the number and types of surface states that will be present in a given

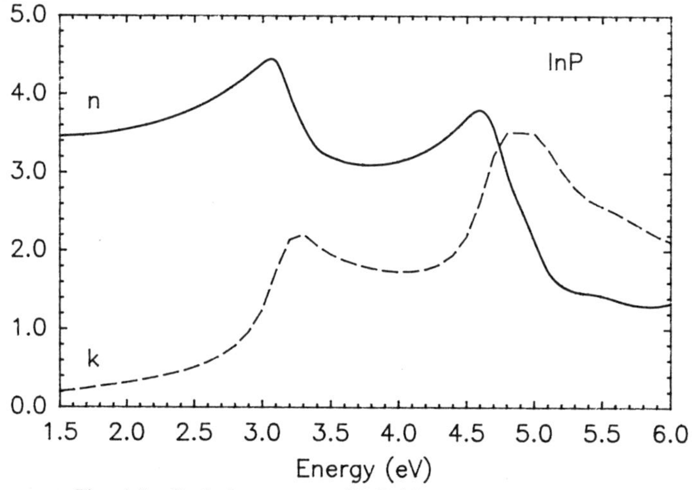

Fig. 4.5 Optical constants of InP (Aspnes and Studna, 1983).

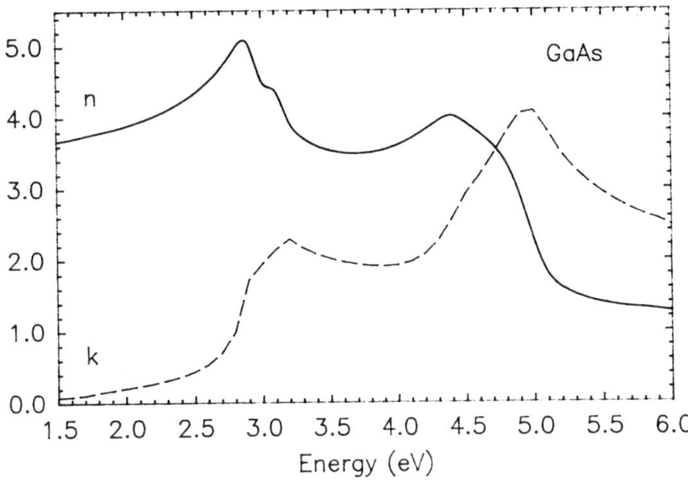

Fig. 4.6 Optical constants of GaAs (Aspnes and Studna, 1983).

ambient. To the extent that surface states contribute to such phenomena as surface band bending, Fermi-level pinning, and surface recombination, crystallographic orientation may influence surface photochemistry. The possibility of anisotropic carrier-dependent etching based on these phenomena has not yet been explored. However, for non-carrier-dependent processes resulting from absorption-induced heating, one would expect the same crystallographic etch-rate dependence seen in ordinary thermal

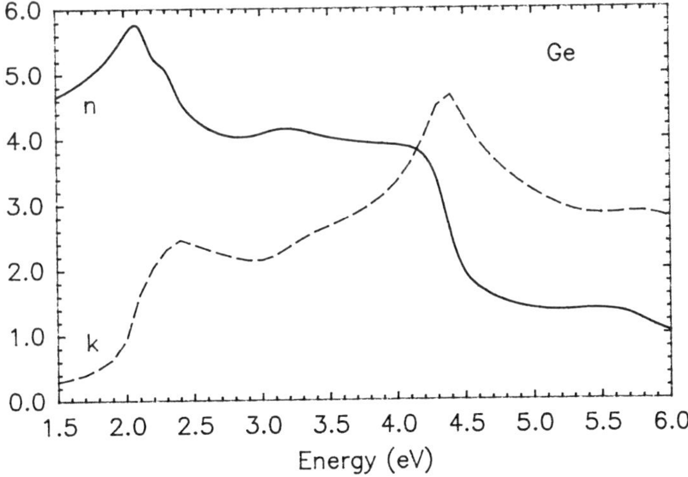

Fig. 4.7 Optical constants of Ge (Aspnes and Studna, 1983).

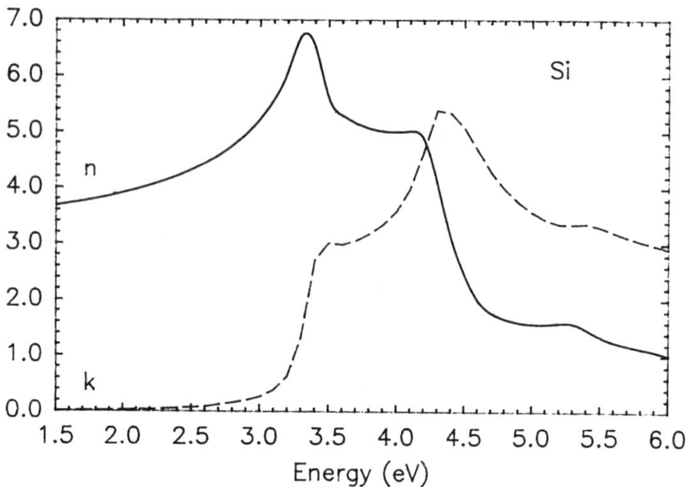

Fig. 4.8 Optical constants of Si (Aspnes and Studna, 1983).

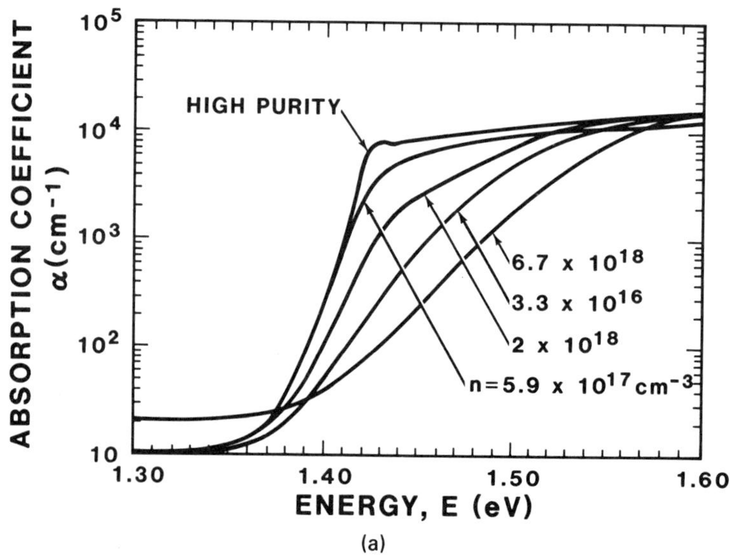

(a)

Fig. 4.9 (a) Absorption coefficient of heavily doped n-GaAs at 297°K (Casey et al., 1975). (b) Absorption coefficient of heavily doped p-GaAs at 297°K (Casey et al., 1975).

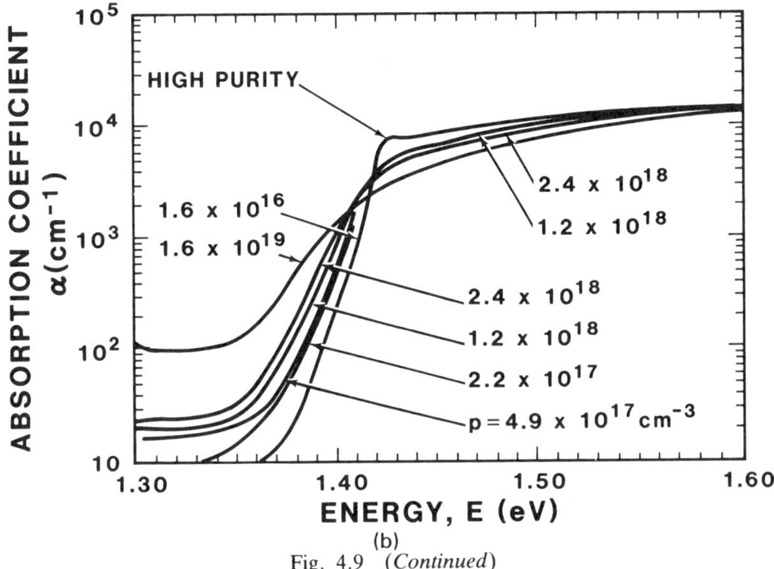

Fig. 4.9 (*Continued*)

reactions. For example, the ordering of the reactivities of the various crystallographic orientations of (111) As $\geq$ (100) $\geq$ (110) > (111) Ga observed for dry (Ibbotson et al., 1983) and wet thermal (Tarui et al., 1971) processes can be expected.

2.2. Process-Dependent Optical and Electronic Properties

During laser processing, many optical and electronic properties are modified either by the radiation itself or by the particular process operating conditions. These include such properties as the number and type of surface states, surface band bending, and nonequilibrium carrier concentrations. These modifications can have profound influences on carrier-driven photochemistry.

2.2.1 Temperature-Dependent Properties

One important effect of an increase in temperature is to shift the relative rates of thermal (nonphotochemical) and carrier-driven (photochemical) reactions. Since a carrier-dependent reaction may exhibit little dependence on substrate temperature (Ashby, 1984), a competing temperature-dependent dark reaction may become dominant as the substrate temperature is increased. In that event, the utility of a photochemical etching process will be diminished or destroyed. In some cases, the competition between carrier-driven and thermal etching may be seen even near room temperature, as observed by Houle (1983) for the etching of Si by F.

Substrate temperature can also affect photochemical etching through its effect on the band-gap energy. Band-gap energies usually decrease as the temperature is increased due to thermal expansion and electron–phonon interactions (Tietjan and Amick, 1966; Weakliem and Redfield, 1979). The dependence of band-gap energy on temperature is listed in Table 4.1 for some representative semiconductors. The temperature dependence of the band gap may vary with stoichiometry, as observed in the (Al, Ga)As system (Oelgart et al., 1987).

Temperature-dependent band-gap energies are of considerable concern for selective photoenhanced etching of ternary and higher order compound semiconductors (Ashby and Biefeld, 1985). They complicate the selection of laser wavelength for composition-selective etching, since excessive variation in band-gap energies due to variation in surface temperature make it more difficult to discriminate between ternary and higher order compound semiconductors differing only slightly in composition (Ashby and Biefeld, 1985).

Temperature can also affect the critical fraction in (Al, Ga)As alloys that corresponds to the direct-to-indirect gap crossover. This fraction has been found to increase with temperature due to the different temperature dependences of the two gaps (Oelgart et al., 1987) and may be important for applications in which photoetching selectivity is controlled by wavelength selection.

Finally, substrate temperature can affect photochemical etching through its effect on carrier recombination processes. A given sample will have a particular concentration of defects and impurity-associated traps. Since carrier recombination at traps is thermally activated, an increase in substrate temperature can decrease carrier densities through trap-mediated recombination.

2.2.2 Surface States

Surface states exist as a consequence of the termination of the solid and the resultant loss of three-dimensional translational symmetry. Crystallographic orientation has a pronounced effect on the number and types of such intrinsic states. On GaAs, for example, there is a wide range of possible behavior ranging from GaAs (110), which has virtually no in-gap states, to GaAs (100), which has such a high density of in-gap states that the Fermi level is usually pinned midgap both in air and in vacuum (Chiang et al., 1983).

However, it is not these intrinsic surface states that are the principal determinant of the electronic behavior at the surface under reaction conditions. One would expect these intrinsic surface states to be altered by adsorption in a chemically reactive environment. This is the case for GaAs (100) in the presence of Cl atoms, which unpin the normally pinned Fermi level (Ashby, 1986), and is probably the case for many other systems undergoing a chemical reaction. The density of surface states and their state of occupation by electrons or holes play an important role in determining another important processing-dependent property, the surface band bending.

2.2.3 Surface Band Bending

Surface band bending is one of the most important process-dependent properties, since the associated near-surface field either increases or decreases the flow of carriers toward the surface. When the Fermi level is unpinned, the surface potential (Φ) and associated electric field are determined, assuming weak photoexcitation, by the dopant concentration (N_d for donor impurities or N_a for acceptor impurities) and the number of occupied surface states below the Fermi level (N_s)

$$\Phi = \frac{eN_s^2}{(2\varepsilon N_{d,a})}. \tag{4.2}$$

Here, e is the electronic charge and ε is the dielectric constant (Smith, 1978). The concentration terms, N_s and $N_{d,a}$ have units of cm^{-2} and cm^{-3}, respectively. The depth of the space-charge region is approximately given by $N_s/N_{d,a}$. This depth can vary from a few crystal layers in very heavily doped materials to 10 micrometers in high-purity intrinsic materials. Therefore, field-driven carrier migration will be important over a greater depth in materials with lower dopant concentrations.

When a semiconductor is immersed in an electrolyte (as in photoelectrochemical etching), or when the gas in contact with the surface is partially ionized (as in dry plasma-based etching), application of a bias voltage can alter the surface band bending by shifting the Fermi level. This in turn can alter the depth of the space-charge region. Since this depth often determines the region within which the near-surface field sweeps carriers toward the surface, changes in surface band bending are important for the control of etching which discriminates on the basis of dopant concentration or type (Ashby, 1985; Ostermayer and Kohl, 1981).

2.2.4 Nonequilibrium Carrier Concentrations

The property most readily altered by the selection of process parameters is the nonequilibrium carrier concentration resulting from photoexcitation. Except at very high intensities, the free carrier density, and hence the resulting etching rate, will be linearly proportional to the number of photons absorbed. At very high intensities, saturation effects can occur that lead to a sublinear relationship between intensity and carrier concentration.

The depth below the surface within which most of the carrier generation occurs is determined by the absorption coefficient. However, the most important excitation events occur in or near the space-charge region, since carrier separation is enhanced if a field is present at or within a diffusion length of the point of carrier generation. Therefore, the shallower penetration of shorter wavelength light can lead to a higher photochemical etching rate than that achieved with longer wavelength light (Reksten et al., 1986; Podlesnik et al., 1984; Tisone and Johnson, 1983). One might also expect a slightly stronger wavelength dependence in heavily doped materials, in which both the space-charge region and the carrier diffusion length are shorter than in intrinsic or lightly doped materials.

As discussed below, nonequiibrium carrier densities can be reduced by increasing the density of lattice defects or other traps that promote carrier recombination. This has been achieved using ion implantation for both dry processes, such as the reaction of Cl with GaAs (Ashby and Myers, 1988) and wet photoelectrochemical processes (Yamamoto and Yano, 1975; Cummings et al., 1986).

3. Creation and Thermal Relaxation of Free Carriers

When an etching reaction requires direct participation of carriers, the highest material-removal rates will be achieved when the free-carrier density at the surface is maximized. In this section, we describe briefly the processes by which light is absorbed in semiconducors. Some of these processes contribute to carrier-dependent etching due to the creation of free carriers. Others contribute mainly to substrate heating through post-absorption relaxation but do not produce a significant increase in the total number of free carriers; these can enhance a thermal reaction rate but will not promote carrier-driven photochemical reactions.

3.1. Band-to-Band Absorption

The absorption process most likely to contribute to etching is simple band-to-band absorption, leading to the production of an electron in the conduction band and a hole in the valence band. It is the extra minority carriers resulting from this excitation that drives photochemistry.

Since carriers produced near the surface have a greater chance of reaching the surface than do those generated deep within the sample, one might expect higher etching quantum yields, i.e., atoms removed per incident photon, for direct-gap materials than for indirect-gap materials. This is because the high absorption coefficient in direct-gap materials results in most of the carriers being generated within one micrometer from the surface, i.e., within a diffusion length of the space-charge field region.

However, the critical quantity is not necessarily the initial number of electron-hole pairs that are generated near the surface, but rather the number that survive long enough to participate in a chemical reaction at the surface. This quantity is determined not only by the initial location of carrier creation, but also by the opportunities for carrier loss through recombination, both within the bulk and at the surface. The critical role of recombination processes in determining quantum yield is discussed in Section 4.

3.2. Free-Carrier, Exciton, and Impurity Absorption

For completeness, we mention here three other absorption processes that generally have little effect on carrier-driven reactions. Free-carrier absorption involves heating a preexistent carrier within a band. Unless the carrier gains sufficient energy, i.e., is sufficiently "hot," to produce an additional electron and hole by impact ionization, no new carriers are generated by this process. Consequently, little if any effect on carrier-driven etching should be observed. The rapid (<1 psec) thermalization of the excited carriers within the same band can heat the semiconductor lattice, however, and this might lead to an increase in the rate of thermal etching.

Exciton absorption involves interactions of light with bound electron-hole pairs. At typical etching temperatures of 290°K or higher, however, equilibrium exciton concentrations should be quite low. For those excitons that may be present, excitation to higher level bound exciton states would be expected to contribute primarily to sample heating by relaxation, although sufficiently high excitation will produce photodissociation to form free carriers.

Finally, impurity absorption involves ionization of either shallow or deep impurity levels. At typical etching temperatures (room temperature or higher), however, most shallow impurities present in a semiconductor are already thermally ionized, and photon absorption is unlikely to produce a significant increase in total carrier concentration. Absorption leading to ionization of deep-level impurities might, however, contribute to the free carrier density.

3.3. Intraband Thermalization and Impact Ionization

If the photon energy exceeds the band gap, then carriers having excess kinetic energy will be created. These "hot" carriers can themselves participate in surface chemistry if they are created near enough to the surface. However, such hot carriers typically thermalize within the band on a subpicosecond or picosecond time scale (Liu, 1982), due to strong electron-phonon scattering. Therefore, except for photochemistry induced by extremely short laser pulses, the dominant reactive carriers will be cold, rather than hot. If the photon energy exceeds the band-gap energy by a sufficently large amount, additional free carriers may be produced by impact ionization or avalanche multiplication (Smith, 1978).

4. Carrier Recombination and Diffusion

The competition between carrier loss through recombination, carrier diffusion away from the reaction zone, and carrier consumption through reaction, determines the reaction quantum yield. When recombination is fast, the equilibrium free-carrier density at the surface will remain low, as will the rate of reaction by carrier-driven photochemistry. If recombination is nonradiative, heat may be generated, enhancing competing thermal reactions. As the temperature increases, recombination rates further increase, making photochemical reactions even less likely. Also, as the temperature increases, carrier diffusion lengths decrease; the resulting spatial localization of the recombination can lead to a further increase in the temperature. These effects have been studied extensively in the context of laser annealing of semiconductors (see, e.g., von Allmen, 1982).

For photochemical reactions, however, it should be noted that recombination is a two-edged sword. While it reduces the quantum yield for material removal, it also increases the resolution of a carrier-driven process by reducing lateral diffusion of carriers (Lum et al., 1985).

In general, recombination can occur by a one-step process, as in direct or indirect band-to-band recombination and multicarrier (Auger) recombination, or by a series of steps, as in surface recombination or recombination at trap sites. A detailed discussion of carrier transport and recombination processes may be found in Smith (1978) and Sze (1981).

4.1 One-Step Recombination Processes

There are two main types of one-step recombination: two-carrier and multicarrier Auger processes. Examples of two-carrier recombination are recombination either of a free electron in the conduction band with a free hole in the valence band or of an electron-hole pair bound as an exciton. In either case, the energy released by the recombination may be emitted as light, which can be used as a process diagnostic. These processes may or may not involve the additional emission or absorption of a phonon. In multicarrier Auger processes, the energy released by the recombination of an electron-hole pair is not emitted as light, but rather excites another carrier to a higher energy level.

4.2. *Two-Step Recombination Processes*

Two-step recombination involves the sequential trapping of an electron and a hole. The trap may be an impurity atom, a bulk lattice defect, or an unpassivated surface site. Such recombination may be radiative or nonradiative, but nonradiative processes, which usually proceed by the sequential emission of phonons ($\simeq$0.1 eV/phonon), generally dominate (Smith, 1978).

Surface recombination is an especially important type of trap-mediated recombination. Its rate is characterized by a surface recombination velocity, S. When S is large, carriers are lost rapidly through surface recombination and the equilibrium concentration of carriers at the surface is greatly reduced. The chemical environment at the surface can influence the surface recombination velocity if, by changing surface stoichiometry or adsorption chemistry, it also changes the nature of the surface sites at which recombination occurs (Offsey et al., 1986).

The importance of recombination in determining the quantum yield of a carrier-driven etching process is seen in the comparison between the etching of GaAs by Cl (Ashby, 1984, 1985) and Si by F (Reksten et al., 1986). Direct-gap GaAs has an absorption coefficient greater than 10^4/cm and typical surface recombination velocities of 10^5–10^6 cm/s. The quantum yield for the carrier-driven reaction is 10^{-7}–10^{-5} atoms removed/incident photon. Indirect-gap high-purity Si has a much lower absorption coefficient of 1–10/cm; however, it also has a much slower surface recombination velocity of 10–10^3 cm/s. The reduced loss of carriers through surface recombination and possible differences in bulk recombination rates largely cancel the effect of fewer carriers initially generated near the surface in Si; it, too, has a quantum yield of 10^{-7}–10^{-6} atoms removed/incident photon.

4.3. *Shifting the Balance Between Reaction and Recombination*

The quantum yield of a solid-excitation-based reaction can vary greatly for a given semiconductor material depending on its dopant and defect concentrations. This effect is predominantly due to enhanced recombination in the bulk when dopant or defect concentrations are high. For example, the quantum yield for etching degenerate, mid-$10^{18}/cm^3$ Si-doped n-GaAs is one to two orders lower in magnitude than that for etching nondegenerate, low-$10^{17}/cm^3$ nGaAs under the same conditions (Ashby and Myers, 1988). Similar behavior is seen with indirect-gap Ga(As, P) (Ashby and Biefeld, 1988).

The defect concentration in a semiconductor can be spatially patterned by ion implantation. This permits the etching of patterns into a semiconductor through a suppression of photochemical etching in the implanted region relative to that in adjacent, unimplanted regions (Yamamoto and Yano, 1975, Cummings et al., 1986; Ashby and Myers, 1988). Selective suppression can be achieved by this approach, even in degenerate materials, and an optimum wavelength range exists that maximizes the interaction between implantation-induced defects and photogenerated carriers, while producing good quantum yields in unimplanted regions of the surface (Ashby and Myers, 1988).

Finally, recombination effects can be especially important in ternary or higher order compound semiconductors. For example, in the $GaAs_{1-x}P_x$ system, surface and bulk recombination rates do not vary much with stoichiometry. However, for stoichiometries for which the band gap is direct gap, the alloy is etched at rates comparable to those for GaAs (Ashby and Biefeld, 1985). For slightly different stoichiometries for which the band gap is indirect, the alloy does not etch detectably if the exciting photons have energies intermediate between the indirect- and direct-gap energies. This occurs because the free carriers are generated over a much deeper range when the absorption coefficient is less than 1000/cm, as is typical of indirect-gap excitation. A large reduction in etching quantum yield occurs because fewer carriers reach the surface before undergoing recombination in the bulk. Excitation of these indirect materials at energies in excess of their direct gap with associated absorption coefficients greater than $10^4/cm$ produces free carriers much nearer the surface and leads to etching comparable to that obtained for direct-gap materials (Ashby and Biefeld, 1988).

4.4. Carrier-Diffusion Effects on Resolution

The size of features, i.e., resolution, that can be produced using solid-excitation-based photochemistry is ultimately determined by the pattern-formation system (Ashby, 1987). However, the ultimate resolution can only be achieved when sample or process characteristics are selected to minimize lateral diffusion of photoexcited carriers. This can be achieved by increasing the reaction rate at the surface, which results in reactive consumption of minority carriers before they diffuse a significant distance. It can also be achieved by increasing the rate of carrier recombination by increasing impurity doping concentrations, although this can have the undesirable side effect of reducing the overall reaction rate.

These effects are clearly illustrated in the formation of holographic gratings in InP using photoelectrochemistry (Lum et al., 1985). The improvement in resolution resulting from increasing the reaction rate by increasing reactant concentration can be seen in Fig. 4.10, in which grating modulation amplitude is a measure of process resolution. The improvement in resolution resulting from faster carrier recombination and consequently decreased lateral carrier diffusion in more heavily doped n^+-InP is illustrated in Fig. 4.11. It is worth noting, however, that although one can vary reactant concentrations with any substrate, the dopant concentration is usually predetermined by the characteristics of the device that is being fabricated.

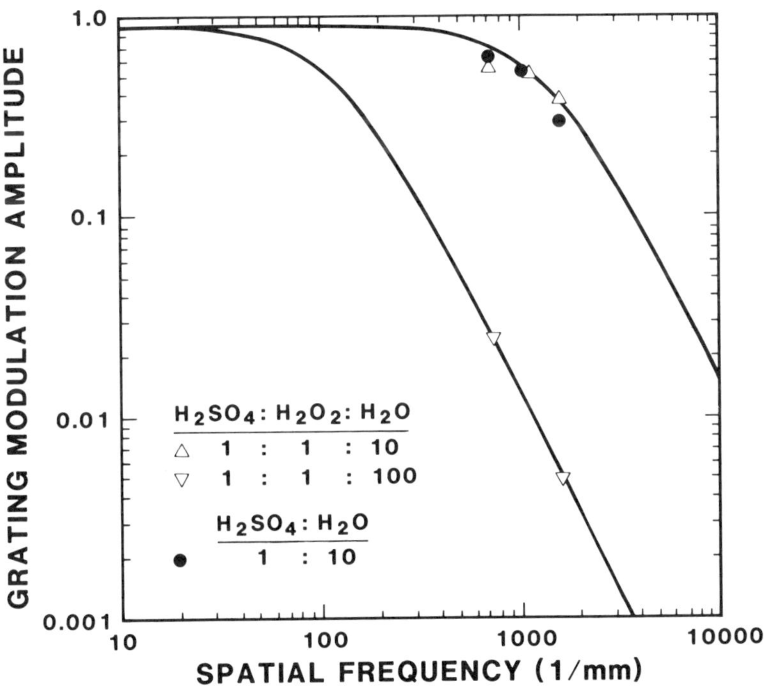

Fig. 4.10 Improvement of photoelectrochemical etching resolution by increased reactant concentration (Lum et al., 1985).

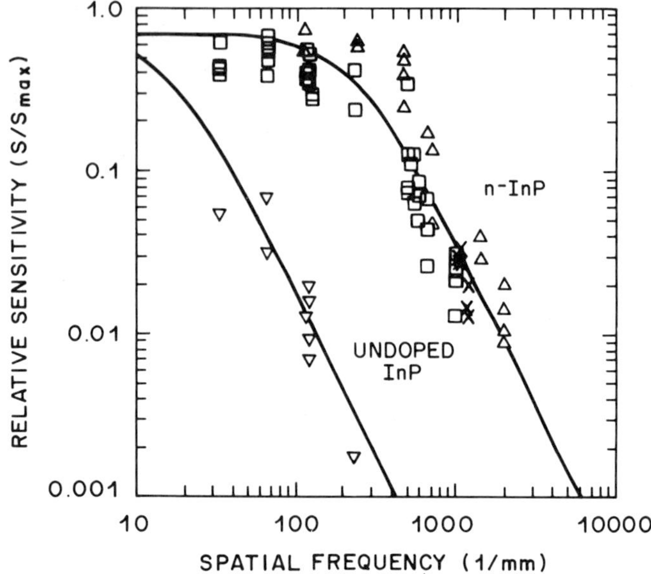

Fig. 4.11 Improvement of photoelectrochemical etching resolution by increased impurity doping concentration (Lum et al., 1985).

5. Interaction of Laser Beams with Metals

In the case of metals, true photochemical reaction enhancements due to light absorption by the solid are rare. This is because light absorption in metals, unlike in semiconductors, typically *conserves* carrier density. Under some conditions, though, photoemission of electrons from the surface may occur, both into liquid and vapor ambients, and may give rise to chemistry (Gurevich et al., 1980). The quantum yields for photoemission, however, are not high and have not been explored to date for laser microfabrication. Hence, laser microfabrication techniques based on light absorption by metals generally rely on simple heating of the surface. Note, however, that quantum yields for secondary electron emission due to electron impact can be quite high and have been recently found to stimulate surface chemistry (Kunz and Mayer, 1987).

The interaction of high-power cw laser beams with metals has been studied extensively for nonchemical materials processing applications. Fairly comprehensive introductions to this area have been written by Ready (1978), and, more recently, by Duley (1983). A useful recent bibliography of that field has been compiled by Gomersall (1986). The interaction of high-fluence pulsed laser beams with metals has also been studied extensively, as reported in the proceedings of the annual meetings of the Materials Research Society (e.g., Donnelly et al., 1987; Picraux et al., 1987; Kurz et al., 1986). Recent edited volumes devoted to the materials science aspects of the interaction of pulsed laser light with surfaces are Poate and Mayer (1982) and Poate et al. (1983). Here, we outline briefly only the most important characteristics of the interaction of laser beams with metals.

5.1. Absorption and Reflectance

The optical properties of metals are determined to a large extent by the strong interaction of electromagnetic radiation with free electrons (Born and Wolf, 1970). Because these free electrons are generally dense, optical absorption lengths are short, typically on the order of 100 Å. This length is shorter than both the spatial scale of even the highest resolution laser processing, as well as the thermal diffusion length of all but the shortest (sub-picosecond) laser pulses. Hence, for all practical purposes in calculating subsequent spatial and temporal temperature profiles, light may be considered to be absorbed at the surface, rather than in the bulk. Furthermore, as for semiconductors, hot electrons undergo intraband relaxation on a picosecond time scale to come to thermal equilibrium with the lattice. This relaxation is generally sufficiently fast, and electron mean free paths sufficiently short, that heat can be considered to be deposited where the absorption occurs.

In an ideal free-electron gas, the reflectance at frequencies below the plasma resonance is high. This is the case for most metals in the infrared and for a few metals (e.g., Ag and Al) even throughout much of the visible. In these cases laser processing is complicated by the need for very high powers. Processing that relies on melting is further complicated by the decrease in reflectance that generally occurs upon melting. This can result in a very sudden, large increase in the rate of light absorption and, hence, melting that is difficult to control.

In Fig. 4.12a is shown time-resolved measurements of the grazing-angle reflectance of TM (transverse magnetic)-polarized 488-nm light from Al during pulsed laser annealing (Tsao et al. 1986a,b). As the surface melts, the reflectance decreases, due to the lower optical reflectance of liquid Al than of solid Al. As the surface freezes, the reflectance recovers. Note that in this case the well-defined shape of the reflectance trace indicates a well-controlled, planar liquid/solid interface. Although the reflectance of Al is high in the visible, as can be seen from Table 4.3, it is not so high that a small change in reflectance results in an uncontrolled increase in light absorption upon melting.

Most metals, however, are not ideal free-electron gases, especially in the visible or UV. Bound, inner-shell electronic transitions contribute additional oscillator strength for absorption. The main practical result is a significant decrease in the optical reflectance, as can be seen from Table 4.3. This both increases the net light absorption and decreases the change in the net light absorption due to reflectance changes upon melting. Therefore, laser processing of most metals presents no particular difficulties in the visible/UV spectral regions.

This is illustrated in Fig. 4.12b, in which time-resolved measurements similar to those in Fig. 4.12a are shown for the grazing-incidence reflectance of TM-polarized 488-nm light from Ni during pulsed laser annealing (Atwater et al., 1988). As in the case of Al, the well-defined steps in the reflectance trace indicate a well-controlled planar liquid/solid interface. In this case, however, the initial downward step in the reflectance has two well-defined contributions: a reflectance change due to heating the room temperature solid to its melting temperature and an additional reflectance change due to melting. This is also seen clearly in the recovery of the reflectance, in which solidification of the 100 to 200-Å

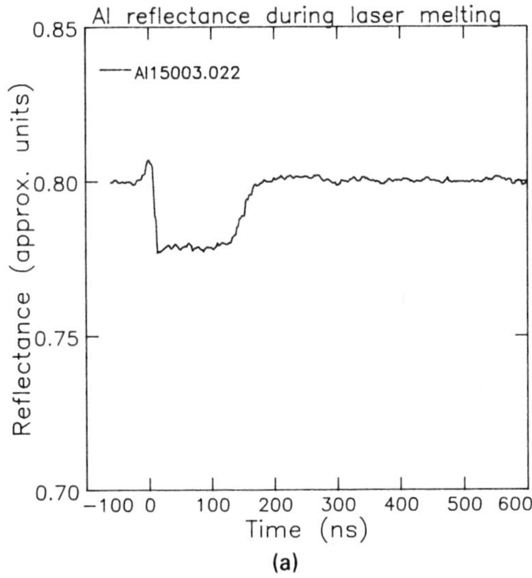

Fig. 4.12 Measured grazing-incidence-angle reflectivities during pulsed laser annealing with 30-nsec FWHM 694.3-nm ruby laser irradiation of (a) Al at 488 nm (85° incidence angle, annealed at 1.14 J/cm^2), (b) Ni at 676 nm (≈80° incidence angle, annealed at 0.449 J/cm^2), and (c) Si at 488 nm (≈80° incidence angle, annealed at 1.16 J/cm^2).

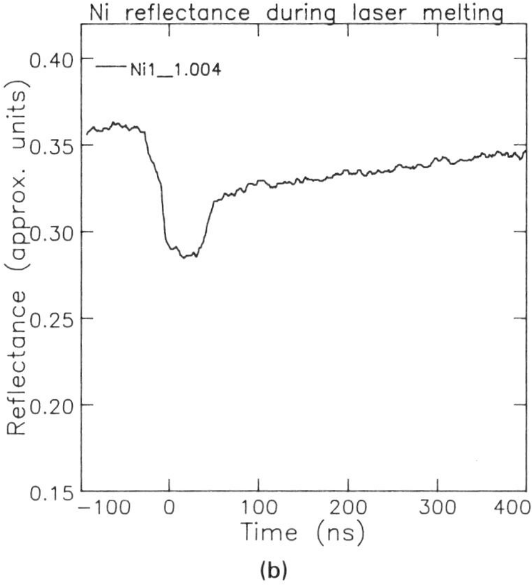

(b)

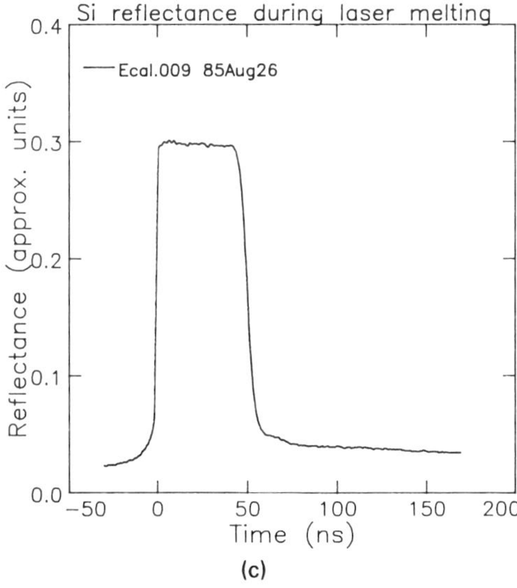

(c)

Fig. 4.12 (*Continued*)

near-surface layer causes an initial step recovery, followed by the much slower cooling of the hot solid.

For comparison, Fig. 4.12c shows time-resolved measurements similar to those in Fig. 4.12a for the grazing-incidence reflectance of TM-polarized 488-nm light from Si during pulsed laser annealing. As in the

Table 4.3 Normal Incidence Optical Reflectances of Selected Metals at Various Wavelengths (in μm) (after Palik, 1985).

	$\lambda(\mu m)$	0.193	0.248	0.488	0.514	0.632	1.06	10.6
Z	Element							
13	Al	0.93	0.92	0.92	0.92	0.91	0.96	0.99
28	Ni	0.34	0.45	0.57	0.59	0.65	0.72	0.98
29	Cu	0.34	0.36	0.58	0.60	0.92	0.97	0.98
42	Mo	0.64	0.69	0.59	0.59	0.57	0.68	0.98
45	Rh	0.50	0.64	0.75	0.76	0.79	0.83	0.98
47	Ag	0.25	0.27	0.94	0.95	0.97	0.98	0.98
74	W	0.64	0.50	0.49	0.49	0.52	0.60	0.98
76	Os	0.55	0.60	0.54	0.51	0.42	0.61	0.99
77	Ir	0.4	0.56	0.67	0.69	0.70	0.80	0.97
78	Pt	0.25	0.37	0.61	0.62	0.67	0.75	0.97
79	Au	0.21	0.33	0.44	0.64	0.93	0.98	0.98

previous two cases, the well-defined steps in the reflectance trace indicate a well-controlled planar liquid/solid interface. In this case, however, semiconducting Si becomes metallic upon melting, and its reflectance *increases*. Note that for these measurements the incidence angles were near Brewster's angle for the solid; the reflectance changes will not be as pronounced at normal incidence.

5.2. Temperature-Dependent Optical and Electrical Properties

Laser processing of metals generally relies on heating. Consequently, the temperature dependences of the optical properties of metals are important. For the wavelength regimes in which a metal behaves as a nearly ideal free-electron metal, its dielectric properties can be approximated using a Drude model, in which the complex dielectric constant is written as

$$\varepsilon = 1 - \frac{\omega_p^2}{\omega^2 + \gamma^2} + i\frac{\gamma\omega_p^2}{\omega(\omega^2 + \gamma^2)}, \tag{4.3}$$

where $\omega_p = (4\pi n_e e^2/m^*)^{1/2}$ is the plasma frequency, $\gamma = \omega_p^2/4\pi\sigma$ is the damping rate, $\omega = 2\pi c/\lambda$ is the light frequency, λ is the wavelength, m^* is the effective mass of the carrier, n_e is the carrier density, and σ is the electrical conductivity. The temperature dependence of the optical properties is contained essentially in the temperature dependence of the electrical conductivity. The absorption coefficient, α, and incident-angle-dependent reflectivity, $R(\Theta_{inc}) = r \cdot r^*$, can then be deduced readily from the complex index of refraction, $N = n + ik = \varepsilon^{1/2}$.

$$\alpha = \frac{4\pi k}{\lambda} \tag{4.4}$$

$$r_{TM} = \frac{\cos(\Theta_{inc}) - \cos(\Theta_{inc})/N}{\cos(\Theta_{inc}) + \cos(\Theta_{inc})/N} \tag{4.5}$$

$$r_{TE} = \frac{\cos(\Theta_{inc}) - N\cos(\Theta_{inc})}{\cos(\Theta_{inc}) + N\cos(\Theta_{inc})} \tag{4.6}$$

for the transverse magnetic and transverse electric polarizations, respectively. Here, Θ_{inc} is the incidence angle relative to the surface normal, and we have assumed the external medium to have unity index of refraction. Note that $1/\alpha$ is the $1/e$ decay length for the electric field intensity, rather than for its amplitude.

The incident angle dependences of the TE- and TM-polarized reflectances of Al and Ni are shown in Fig. 4.13. Usually, it is the normal incidence reflectivity that is important,

$$R(\Theta_{inc} = 0) = \frac{(n-1)^2 + (k)^2}{(n+1)^2 + (k)^2}, \tag{4.7}$$

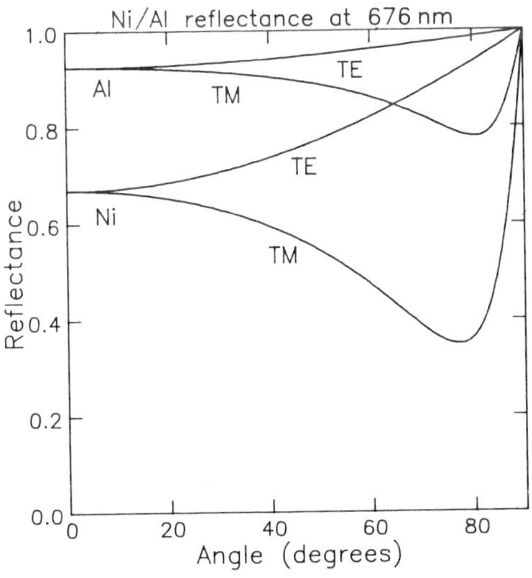

Fig. 4.13 Calculated dependence of room-temperature TE (transverse electric) and TM (transverse magnetic) reflectivities on incidence angle of Al and Ni. Room-temperature complex refractive indices $N_{Al} = 0.73 + 5.95(i)$ and $N_{Ni} = 1.67 + 2.89(i)$ were taken from Palik (1985).

although for diagnostic purposes a near-Brewster's-angle (TM-polarized) probe laser is often used, as illustrated by the measurements shown in Fig. 4.12.

The temperature-dependent conductivity σ is generally known; values for the temperature-dependent resistivity, ρ $(=1/\sigma)$, for the common pure metals and compounds are listed in Table 4.4 and 4.5, respectively. To convert from σ in units of $\Omega^{-1}\,\mathrm{cm}^{-1}$ to σ in units of s^{-1}, multiply by 9.0×10^{11}. Within the accuracy of the Drude approximation, the plasma frequency ω_p can be estimated from knowledge of the normal-incidence reflectance in the infrared, i.e., below the cutoff for inner-shell absorption.

Some materials, such as silicon, become nearly ideal free-electron metals upon melting, and their optical properties may be deduced in a similar manner. As an illustration, the properties of liquid silicon near its melting temperature (1685°K) are reasonably well represented by a plasma frequency of $2.50 \times 10^{16}\,\mathrm{s}^{-1}$ and a temperature-dependent electrical conductivity $\sigma_{1685\mathrm{K}} + [d\sigma/dT]_{1685\mathrm{K}}(\mathrm{T}\text{-}1685\ \mathrm{K})$, where $\sigma_{1685\mathrm{K}} \approx 1.172 \times 10^4\,\Omega^{-1}\,\mathrm{cm}^{-1} = 1.054 \times 10^{16}\,\mathrm{s}^{-1}$ and $[d\sigma/dT]_{1685\mathrm{K}} \approx -8.13\,\Omega^{-1}\,\mathrm{cm}^{-1}\,\mathrm{K}^{-1} = -7.3 \times 10^{12}\,\mathrm{s}^{-1}\,\mathrm{K}^{-1}$. This choice of values reproduces the measured value of the dielectric constant of liquid Si at 632.8 nm of $\varepsilon(632.8\ \mathrm{nm}) = -19.2 + 32.0(\mathrm{i})$ (Li and Fauchet, 1987), as well as

Table 4.4 Selected Thermal and Electrical Properties of Metals. [Values compiled from Gray (1972), Weast (1974), Kubaschewski (1979), and Ubbelohde (1978)].

Z	Elem	T_m (°C)	T_b (°C)	ΔH_f (J/cm^3)	ΔH_v $\left(\frac{\mathrm{kJ}}{\mathrm{cm}^3}\right)$	$c_p(\infty)$ $\left(\frac{\mathrm{J}}{\mathrm{cm}^3\mathrm{K}}\right)$	Θ (K)	ρ_{RT} ($\mu\Omega$-cm)	$d\rho/\rho_{RT}$ (K^{-1})	$\beta\left(\frac{\rho_{l,MT}}{\rho_{S,MT}}\right)$
3	Li	180.5	1330					8.55		1.64
4	Be	1278	2970	2508		5.1	1440	4.0	0.025	
11	Na	97.8	892					4.2		1.45
12	Mg	649	1107					4.45	0.00545	1.78
13	Al	660	3200	1073	28.4	2.5	428	2.65	0.00429	2.20
19	K	63.7	760					6.15		1.56
21	Sc	1539	2730					61.0	0.00282	
20	Ca	839	1440					3.91	0.00416	
22	Ti	1668	3260	1887	39.8	2.3	420	42.0		2.06
23	V	1900	3450	2501		3.0	380	25.4		
24	Cr	1875	2665	2377	42.2	3.4	630	12.9	0.003	
25	Mnα	1245	2150	1990	30.3	3.4	410	185.0		0.061
26	Fe	1536	3000	2138	49.9	3.5	467	9.71	0.00651	1.01
27	Co	1495	2900	2458	58.8	3.8	445	6.24	0.00604	1.09

Table 4.4 *(continued)*

Z	Elem	T_m (°C)	T_b (°C)	ΔH_f (J/cm^3)	ΔH_v $\left(\frac{kJ}{cm^3}\right)$	$c_p(\infty)$ $\left(\frac{J}{cm^3 K}\right)$	Θ (K)	ρ_{RT} ($\mu\Omega$-cm)	$d\rho/\rho_{RT}$ (K^{-1})	$\beta\left(\frac{\rho_{l,MT}}{\rho_{S,MT}}\right)$
28	Ni	1453	2730	2644	57.4	3.8	450	6.84	0.0069	1.33
29	Cu	1083	2595	1837	43.0	3.5	343	1.673	0.0068	2.04
30	Zn	419	906	807	12.5	2.7	327	5.916	0.00419	2.24
31	Ga	30	2237			2.1	320	17.4		
34	Se	217	685					12.0		1.00
37	Rb	38.9	688					12.5		1.60
38	Sr	768	1380					23.0		
39	Y	1509	2927					57.0	0.0027	
40	Zr	1852	3580	100		1.8	291	40.0	0.0044	
41	Nb	2468	3300	2390		2.3	275	12.5		
42	Mo	2610	5560	2945		2.7	450	5.2		1.23
44	Ru	2500	4900			3.0	600	7.6		
45	Rh	1966	4500	2594	64.0	3.0	480	4.51	0.0042	
46	Pd	1552	3980		42.0	2.8	274	10.8	0.00377	
47	Ag	961	2210	1164	24.7	2.4	225	1.59	0.0041	2.09
48	Cd	321	765	478	7.7	1.9	209	6.83	0.0042	1.97
49	In	156	2000	54		1.6	108	8.37		2.18
50	Sn	232	2270	431	14.1	1.5	205	11.0	0.0047	2.10
51	Sb	631	1750	1066	10.6	1.4	211	39.0		0.61
52	Te	450	990		2.4	1.2	153			
57	La	920	3470					5.7	0.00218	1.08
58	Ce	798	3468					75.0	0.00087	
59	Pr	935	3127					68.0	0.00171	
60	Nd	1024	3027					64.0	0.00164	
62	Sm	1072	1900					88.0	0.00184	
64	Gd	1312	3000					140.5	0.00176	
66	Dy	1407	2600					57.0	0.00119	
67	Ho	1461	260					87.0	0.00171	
68	Er	1497	2900					107.0	0.00201	
69	Tm	1545	1727					79.0	0.00195	
70	Yb	824	1427					29.0	0.0013	
71	Lu	1652	3327					79.0	0.00240	
72	Hf	2222	5400			1.8	252	35.1	0.0038	
73	Ta	2996	5425	2848		2.3	240	12.45	0.00383	
74	W	3410	5930	3715		2.6	400	5.65		1.08
75	Re	3180	5900			2.8	430	19.3	0.00395	
76	Os	3000	5500	3215		3.0	500	9.5	0.0042	
77	Ir	2454	5300	3107		2.9	420	5.3	0.003925	
78	Pt	1769	4530	2149	56.0	2.7	240	10.6	0.003927	1.40
79	Au	1063	2970	1211	30.4	2.4	165	2.35	0.004	2.08
80	Hg	−38.4	357	154	4.0	1.7	72	98.4		3.74
81	Tl	303	1457		9.4	1.4	78	18		2.06
82	Pb	327.4	1725	262	9.8	1.4	105	20.648	0.00336	1.92
83	Bi	271	1560	508	8.1	1.2	119			
90	Th	1750	3850					13.0	0.0038	
92	U	1132	3818					30.0		
94	Pu	640	3235					141.4		

Table 4.4 (*continued*)

Z	Elem	κ_T (W/cmK) (273°K)	κ_T (W/cmK) (973°K)	Density $\left(\frac{g}{cm^3}\right)$	Mass $\left(\frac{g}{mole}\right)$	ΔH_f $\left(\frac{cal}{g}\right)$	ΔH_v $\left(\frac{kcal}{mole}\right)$
4	Be	2.2	0.96	1.85	9	324	
13	Al	2.35		2.70	27	95	67.9
22	Ti	0.22	0.20	4.51	47.9	100	101
23	V	0.30		6.1	51	98	
24	Cr	0.95	0.66	7.19	52	79	73
25	Mn	0.077		7.43	55	64	53.7
26	Fe	0.83	0.34	7.86	55.8	65	84.6
27	Co	1.0		8.9	58.9	66	93
28	Ni	0.91	0.71	8.9	58.7	71	90.5
29	Cu	4.0	3.6	8.96	63.5	49	72.8
30	Zn	1.2		7.14	65.4	27	27.4
31	Ga	0.85		5.91	69.7		
40	Zr	0.22	0.21	6.49	91.2		
41	Nb	0.51	0.60	8.4	92.9	68	
42	Mo	1.35	1.13	10.2	96	69	
44	Ru	1.17		12.2	101		
45	Rh	1.51		12.4	102.9	50	127
46	Pd	0.76		12.0	106.4		89
47	Ag	4.28	3.76	10.5	107.9	26.5	60.7
48	Cd	0.98		8.65	112.4	13.2	23.9
49	In	0.87		7.31	114.8		
50	Sn	0.67		7.30	118.7	14.1	55
51	Sb	0.26		6.62	121.7	38.5	46.7
52	Te	0.04		6.24	127.6		11.9
72	Hf	0.22		13.1	178.5		
73	Ta	0.57	0.61	16.6	180.9	41	
74	W	1.70	1.22	19.3	183.8	46	
75	Re	0.49	0.44	21.0	186.2		
76	Os	0.88		22.6	190.2	34	
77	Ir	1.60		22.5	192.2	33	
78	Pt	0.73	0.74	21.4	195.1	24	122
79	Au	3.18	2.79	19.3	197	15	74.2
80	Hg			13.6	200.6	2.7	14.0
81	Tl	0.47		11.85	204.4		38.8
82	Pb	0.35		11.4	207.2	5.5	42.5
83	Bi	0.11		9.8	209	12.4	41.1

the measured temperature dependences of the reflectivity (Lampert, 1981) and electrical conductivity (Glazov, 1969). Note, however, that these values are somewhat different from those measured by Jellison (1985). The wavelength dependences of the complex refractive index and dielectric constant for liquid Si are shown in Fig. 4.14, and the

Table 4.5 Selected Thermal and Electrical Properties of Compounds [Values taken from Gray (1972), Weast (1974), Kubaschewski (1979), and Harper (1970)].

Compound	ρ ($\mu\Omega$-cm)	κ_T (W/cmK) (300°K)	T_m C	Density (g/cm^3)	ΔH_f $\left(\frac{\text{kcal}}{\text{mole}}\right)$	Mass $\left(\frac{\text{g}}{\text{mole}}\right)$	ΔH_f $\frac{\text{J}}{\text{cm}^3}$	Mass $\frac{\text{g}}{\text{atom-mole}}$	$c_p(\infty)$ J/cm^{3}K	E_g eV
Gp IV										
d-C		20	>3550	3.51	25	12	30595	12	7.3	5.47
g-C		20	>3550	2.25		12		12	4.7	0
c-Si		1.4	1415	2.33	12.1	28.1	4198	28	2.08	1.12
a-Si			~1185			28.1	~2900	28		
Ge		0.6	937	5.33	8.8	72.6	2703	72.6	1.8	0.66
III–V										
BP		3.5	>1500	2.85		41.8		20.9		6.0
AlP		1.3	>1600	3.81		57.9		29.0		3.0
AlAs			1050	4.22		101.9		50.9		2.16
AlSb		0.57	s800	6.1	19.6	148.7		74.4		1.58
GaN		1.7	1465	4.09		83.7		41.8	3.6	3.36
GaP		1.0	1237	5.31		100.7		50.3		2.26
GaAs		0.46	712	5.62		144.6		72.3	1.8	1.424
GaSb		0.390	1055	4.79		191.5		95.7		0.72
InP		0.68	942	5.7	15.0	145.8		72.9		1.29
InAs		0.273	523	5.78	18.4	189.7		94.9		0.36
InSb		0.17			11.8	236.5		118.3		0.17

Table 4.5 (*continued*)

Compound	ρ ($\mu\Omega$-cm)	κ_T (W/cmK) (300°K)	T_m C	Density (g/cm^3)	ΔH_f $\left(\frac{\text{kcal}}{\text{mole}}\right)$	Mass $\left(\frac{\text{g}}{\text{mole}}\right)$	ΔH_f $\frac{\text{J}}{\text{cm}^3}$	Mass $\frac{\text{g}}{\text{atom-mole}}$	$c_p(\infty)$ J/cm^{3}K	E_g eV
II/IV–VI										
ZnS			1700	4.1		97.4		48.7	2.0	3.58
ZnSe			1515	5.42		144.3		72.2	1.9	2.67
CdS		0.16	1750	4.82		144.4		72.2	1.7	2.59
CdSe			>1350	5.81		191.4		95.7	1.5	1.70
CdTe			1041	6.1		240.0		120	1.3	1.50
PbS		0.024	1119	7.5	8.7	239.2	1141	119.6	1.6	0.37
PbSe		0.017	1076	8.1	11.8	286.2	1397	143.1	1.4	0.26
PbTe		0.022	920	8.16	13.7	334.8	1397	167.4	1.2	0.29
As_2S_3			312	3.43	6.85	246.0	400	49.2	1.7	
As_2Se_3			377	4.75	9.75	386.7	501	77.4	1.5	1.6
HgSe			800	8.25		279.6		139.8	1.5	0.6
HgTe			670	8.42		328.2		164.1	1.3	0.025
ZnTe			1238	5.72		193.0		96.5	1.5	2.26
HgS			1450			2327.6		116.3		2.0
I/II–VII										
LiF			848	2.60	6.4	25.9	2688	13.0	5.0	~12
NaCl			801	2.16	6.7	58.4	1037	29.2	1.8	8.6
KCl			772	1.99	6.35	74.6	709	37.3	1.3	8.5
CaF_2			1418	3.18	7.1	78.0	1211	29.5	2.7	~10
Borides										
TiB_2	28.4	0.24	2980	4.52				23.2	4.9	
VB_2	6.45	0.17	2100	5.1				24.2	5.3	
ZrB_2	9.2	0.24	3040	6.1				37.6	4.0	
HfB_2	10		3100	10.5				66.7	3.9	

Carbides										
SiC		4.9	~2800	3.22				20	4.0	2.35
TiC	180	0.24	3150	4.9				30.0	4.1	
VC	150	0.25	2830	5.8				31.5	4.6	
ZrC	70	0.20	3530	6.4				51.6	3.1	
NbC	74	0.14	3760	7.85				52.4	3.7	
Mo_2C	97	0.07	2570	9.2				40	5.7	
HfC	109	0.06	3890	12.7				95.2	3.3	
TaC	30	0.22	3880	14.5				96.5	3.7	
WC	53	0.29	2870	15.8				98	4.0	
Nitrides										
BN		1.80	>3000	2.25				12.0	4.5	~7.5
AlN		2.0	2400							
Si_3N_4			1900	3.1				20.0	3.9	5.0
TiN	21.7	0.2	3205	5.43				31.0	4.4	
VN	200	0.18	2360	6.1				32.5	4.7	
ZrN	13.6	0.17	2980	7.35				52.6	3.5	
NbN	200	0.03	2300	7.28				53.5	3.4	
Ta2N	135	0.05	3090	14.1				125.3	2.8	
Oxides										
Al_2O_3		0.30	2051	4.0	25.7	102	4217	20.4	4.9	8.3
BeO		3.1	2580	3.05	19.3	25	9852	12.5	6.1	
MgO		0.36	2825	3.65	18.5	40.3	7011	20.2	4.5	7.3
SiO_2		0.14	1722	2.2	2.6	60.0	399	20.0	2.74	
SiO						44		22		
TiO_2		0.065	1870	4.28	16.0	79.9	3586	26.6	4.0	3.05
ZnO						81.4				3.4
GeO_2			1116	4.7	10.5	104.6	1974	34.9	3.4	
In_2O_3						277.6				3.75
ZrO_2		0.020	2677	5.82	20.8	123.2	4111	41.0	3.54	
$LiNbO_3$						147.8		29.6		4.0
$SrTiO_3$			1910	5.11		183.5		36.7	3.5	3.4

Table 4.5 *(continued)*

Compound	ρ ($\mu\Omega$-cm)	κ_T (W/cmK) (300°K)	T_m C	Density (g/cm^3)	ΔH_f $\left(\frac{\text{kcal}}{\text{mole}}\right)$	Mass $\left(\frac{\text{g}}{\text{mole}}\right)$	ΔH_f $\frac{\text{J}}{\text{cm}^3}$	Mass $\frac{\text{g}}{\text{atom-mole}}$	$c_p(\infty)$ J/cm^{3}K	E_g eV
Silicides (Murarka, 1983; Filyand and Semenova, 1970)										
$TiSi_2$	13		1540	4.0				34.6	2.9	
VSi_2	14.4		1670	4.6				35.7	3.2	
CrSi2	600		1550	5.0				36.0	3.5	
FeSi2	>1000		1212	4.9				37.3	3.3	
CoSi2	18		1277	4.9				38.3	3.2	
NiSi2	50		~990	4.8				38.2	3.1	
ZrSi2	35		1700	4.9				49.0	2.5	
NbSi2	6.3		1950	5.7				49.6	2.9	
MoSi2	21.5		1980	6.2				50.7	3.1	
Ru2Si3				7.0				57.2	3.1	
RhSi				8.5				65.5	3.2	
Pd2Si			1398	9.6				80.3	3.0	
HfSi2	45		1800	8.0				78.1	2.6	
TaSi2	8.5		2200	9.1				79.0	2.9	
WSi2	33		2165	9.9				80.0	3.1	
ReSi2			1980							
OsSi2										
IrSi3										
PtSi	28		1773	12.4				111.6	2.8	

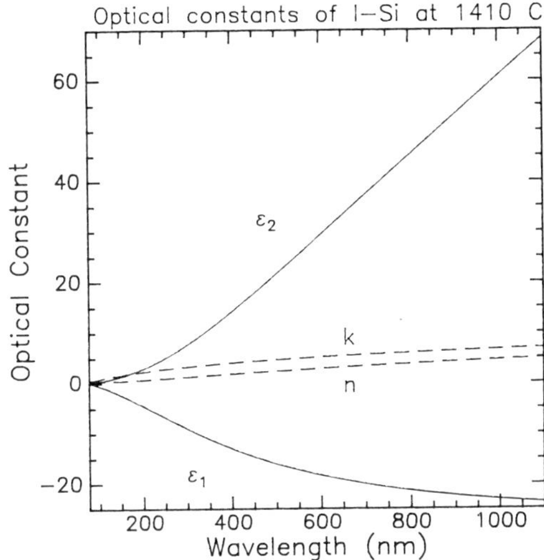

Fig. 4.14 Real and imaginary parts to the dielectric constant (ε_1, ε_2) and refractive index (n, k) for liquid Si as a function of wavelength, at the melting temperature. We use the simple Drude model and the parameters discussed in the text.

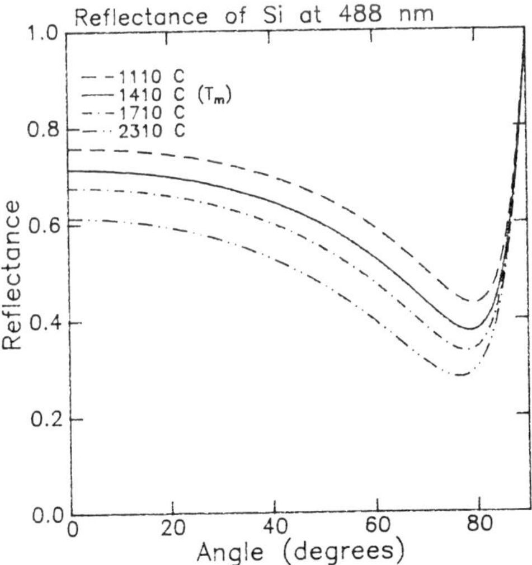

Fig. 4.15 Calculated dependence of reflectivity at 488 nm of liquid Si on incidence angle for various temperatures. We use the simple Drude model and the parameters discussed in the text.

incident-angle-dependent reflectivity at 488 nm is shown in Fig. 4.15 for various temperatures above and below the melting temperature.

For most materials, however, the optical properties have significant contributions from transitions originating from bound, inner-shell electronic states. When this is so, the temperature dependence is difficult to predict and cannot be determined solely from knowledge of the temperature-dependent electrical conductivity. At high temperatures, and particularly at phase transitions, the electronic band structure changes and can influence optical properties significantly. However, as noted above, since optical reflectivities are already low (and net light absorption already high) at room temperature for these metals, temperature-dependent changes in optical reflectivity lead to frictional changes in net light absorption small enough to disregard in many cases.

6. Lattice Heating, Thermal Diffusion, and Spatial-Temporal Temperature Profiles

The problem of calculating the spatial-temporal temperature profiles due to laser irradiation has been examined extensively with regard to non-chemical laser processing of materials. As the considerations are essentially the same, the reader is referred to the references by Ready (1978) and Duley (1983). Other references containing more detailed treatments of the heat-flow problem are the edited volumes by Bass (1983) and Poate (1982). Here, we discuss only general features of the heat-flow problem, with special emphasis on heat flow resulting from very tightly focused or finitely patterned laser beams.

6.1. Heat Flow Regimes

Once heat is deposited at the surface of a solid it will diffuse into the bulk. The balance between the rate at which heat is deposited and the rate at which it is removed determines the resulting temperature distribution in space and time. Typically, it is the skin temperature that matters in laser chemical processing; however, in some cases, e.g., in laser doping, the temperature profile in the bulk matters as well. In both cases, the full heat generation and transport equations must be solved. In general, since thermal properties such as thermal conductivities, heat capacities, etc., are temperature dependent, very accurate solutions for the temperature distributions must be obtained numerically (Gibbons and Sigmon, 1982; Thompson, 1984; Wood, 1986; Baeri and Campisano, 1982). Here, we describe the qualitative behavior analytically by assuming temperature-independent thermal properties.

We start by noting that the nature of the heat-flow problem can be classified according to whether the feature size w_o of the laser irradiation is greater than or less than the thermal diffusion length $l_T \approx \sqrt{4D_T\tau}$ during the exposure time τ, where $D_T = \sqrt{\kappa_T/c_p}$ is the thermal diffusivity, κ_T is the thermal conductivity, and c_p is the specific heat. With the caveats described below, projection patterning is usually the $w_o \gg l_T$ limit; direct writing ("cw" exposure) is usually the $w_o \ll l_T$ limit. Because steady-state heat flow begins when the thermal diffusion length increases beyond the feature size (assuming a substrate much thicker than either length), projection patterning can also be considered the transient heat-flow limit, while direct writing can be considered the steady-state heat-flow limit. These two limits are illustrated schematically in Fig. 4.16.

In most cases, this nomenclature accords with conventional terminology. For example, for high-resolution excimer laser projection patterning of low-thermal-diffusivity substances like glasses or polymers, typically $w_o \approx 1.0\ \mu m$, $\tau \approx 20$ nsec, and $D_T \cong 0.01\ cm^2/s$. Then, $l_T \approx 0.30\ \mu m$ and

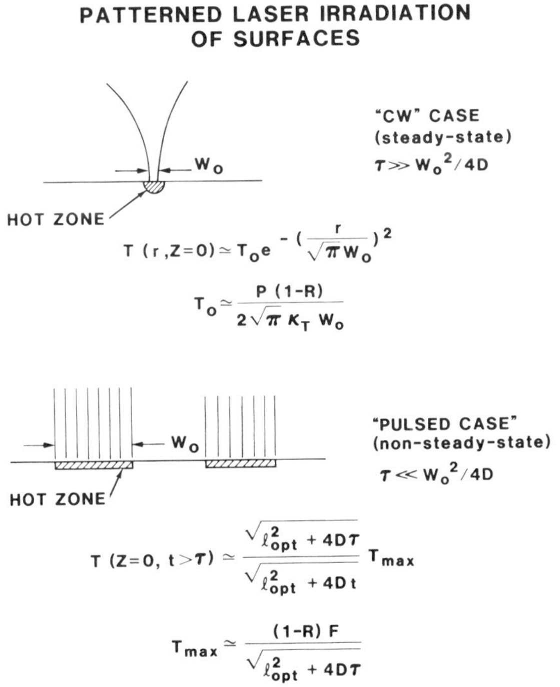

Fig. 4.16 Schematic of temperature distributions in the (a) cw (steady-state) and (b) pulsed (non-steady-state) heat-flow regimes. The symbols are defined in the text.

$w_o > l_T$. For low-resolution ($w_o > 5\ \mu m$) excimer-laser projection patterning of even high-thermal diffusivity ($D_T \approx 1\ cm^2/s$) substances like metals, $l_T \approx 3\ \mu m$ and $w_o > l_T$. However, for high-resolution ($w_o \approx 1\ \mu m$) excimer-laser projection patterning of medium-to-high-thermal-diffusivity substances, $D_T \approx 0.1\ cm^2/s$, $l_T \approx 0.9\ \mu m$, and $w_o > l_T$ is only marginally fulfilled. In such cases, the onset of steady-state heat flow must be considered.

Also note that direct writing with cw laser beams may effectively take place in the pulsed regime if the scan velocity $v = w_o/\tau$ exceeds $4D_T/w_o$. However, this will only occur under extreme conditions of high scan velocities, low thermal diffusivities, and large spot sizes, as illustrated in Fig. 4.17. For example, for a laser spot size of 10 μm and a substrate thermal diffusivity of 0.03 cm^2/s, the critical scan velocity is 120 cm/s.

The physical difference between the pulsed and cw heat flow regimes can be seen as follows. For a surface heat source of characteristic lateral dimension w_o, a quasi-steady-state temperature distribution due to three-dimensional heat flow into the substrate is set up on the time scale $\tau_0 \approx w_o^2/4D_T$. For times much shorter than this, steady-state heat flow has not yet been established. Heat flows mainly perpendicularly into the substrate rather than radially along the substrate surface, simplifying the

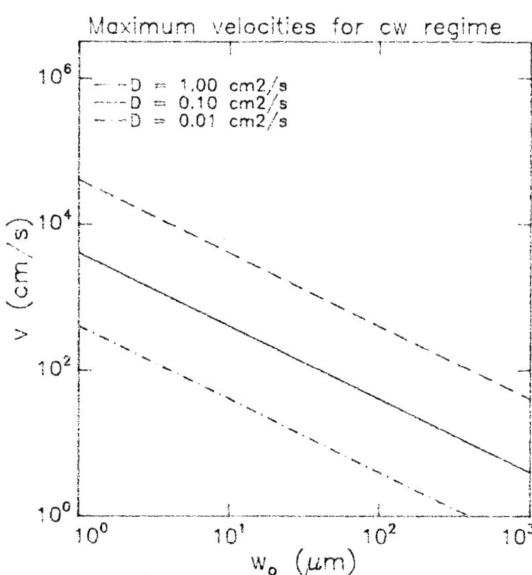

Fig. 4.17 Maximum cw-laser scan velocity for steady-state heat flow for various thermal diffusivities, D_T. For velocities greater than these, non-steady-state heat flow must be considered.

problem considerably, as only *one*-dimensional heat flow needs to be considered. This is the pulsed-exposure limit. For times much longer than this, *three*-dimensional heat diffusion determines the eventual quasi-steady-state temperature distribution. As described later, for a circular heat source, the resulting hot zone is a hemisphere of approximate diameter $\sqrt{\pi w_o}$. This is the cw-exposure limit.

6.2. Projection Patterning: Heat Diffusion in One Dimension

Very accurate solutions for the temperature distributions must be obtained numerically. However, for the purposes of making simple estimates and determining general trends, the qualitative behavior may be determined by assuming temperature-independent thermal properties. In the following, we make use of much of the simple, but reasonably accurate, analytic formulas developed by Thompson (1984).

6.2.1 Bulk Substrates

In this case, the feature size is large compared to the thermal diffusion length; alternatively, the time time scale of interest is short compared to the time for lateral diffusion across the feature size of interest. Under these conditions heat, first deposited in the near surface, is transported into the bulk largely by one-dimensional heat conduction according to

$$J(z, t) = -\kappa_T \nabla T(z, t) \tag{4.8}$$

$$(c_p)\frac{dT}{dt} = \frac{-dJ}{dz}, \tag{4.9}$$

where the z axis points perpendicularly into the substrate. Here, J is the heat flux, T is the temperature, and κ_T and c_p are the thermal conductivity and heat capacity, respectively.

The thickness l_0 of the near-surface region that is initially heated up by a short laser pulse (typically 5–100 nsec) is equal to the optical absorption length l_{opt}, or the thermal diffusion length $l_T \approx \sqrt{4D_T\tau_{pulse}}$ during the pulsewidth τ_{pulse}, whichever is greater: $l_0 = \text{Max}\{l_{opt}, l_T\}$. For the case of metals (strong absorbers), the optical absorption depth is much shorter than the thermal diffusion length during the pulse, so that $l_0 \rightarrow l_T$, except for exceptionally short (sub-psec) laser pulses. For semiconductors and insulators in the infrared (weak absorbers), however, the optical absorption depth can be greater, so that $l_0 \rightarrow l_{opt}$. This is the case for silicon excited by ruby-laser (694.3 nm) radiation of pulsewidth < 25 nsec. For semiconducors above their band gaps (intermediate absorbers), the two thicknesses may be comparable. Unless melting occurs, the maximum temperature rise ΔT_{max} in this layer is determined approximately by

$$c_p l_o \, \Delta T_{max} = F(1 - R_{sol}), \tag{4.10}$$

where F is the incident laser fluence (in J/cm^2), $F(1 - R_{sol})$ is the *absorbed* laser fluence and R_{sol} is the optical reflectance of the solid.

Thermally enhanced chemistry will, of course, only occur while the surface is hot. Except for the shortest laser pulses and the highest thermal conductivity materials, the surface is hot for a duration determined not by the temperature rise time (approximately the laser pulse width itself), but rather by the temperature fall time. This fall time is in turn determined by heat diffusion into the bulk *after* the pulse. The dependence of the temperature distribution on depth, z, and time, t, assuming the pulse to be centered at $t = 0$, is characterized approximately by the (half) Gaussian

$$T(z, t) = T_{room} + \Delta T(z = 0, t) \exp\left\{ -\left(\frac{z}{l_0 + \sqrt{[4D_T t]}} \right)^2 \right\}, \tag{4.11}$$

with the time-dependent skin temperature change

$$\Delta T(z = 0, t) = \frac{l_0}{l_0 + \sqrt{[4D_T t]}} \Delta T_{max}. \tag{4.12}$$

The temperature fall time is seen to be approximately

$$\tau_{fall} \approx \frac{l_0^2}{(4D_T)}, \tag{4.13}$$

typically much longer than the temperature rise time. This fall time, rather than the laser pulsewidth itself, is the main determinant of the duration of the reaction enhancement due to laser irradiation.

Often, in the event of a phase change (e.g., solid to liquid transformation) that may occur at high fluences, surface chemistry is enhanced considerably. Such an enhancement has been observed in the etching of Si by chlorine vapor during laser irradiation (Ehrlich, 1981; von Gutfeld, 1982). In this event, the important quantities are the fluence threshold for melting, the maximum molten layer thickness, and the duration of the surface melt. In particular, it is the duration of the surface melt that should be taken to be the duration of the reaction enhancement due to laser irradiation. All of these quantities have been modeled and measured extensively by Thompson (1984) for pulsed laser melting of silicon; the simple estimates derived in that work (and summarized here) should be applicable to other materials, however.

The fluence threshold for melting is basically given by Eq. 4.10 above, by substituting $T_m - T_{room}$ for ΔT_{max}, where T_m is the melting temperature

$$F_{thresh} = c_p l_o \frac{(T_m - T_{room})}{(1 - R_{sol})}. \tag{4.14}$$

Beyond this threshold fluence, the maximum molten layer thickness d_{max} increases with fluence with a slope approximately equal to

$$\frac{\partial(d_{max})}{\partial F} = \frac{(1 - R_{liq})}{\Delta H + \int c_p \, dT}, \tag{4.15}$$

where ΔH is the latent heat of melting, R_{liq} is the optical reflectance of the liquid, and $\int c_p \, dT$ is the integral of the energy density required to heat room-temperature solid to the melting temperature.

After the pulse has gone, the liquid solidifies as latent heat is conducted away into the solid substrate. The time-dependent solidification velocity, $v_{sol}(t)$, is determined principally by heat flow,

$$v_{sol}(t) \, \Delta H = \kappa_s \frac{\partial T}{\partial z}\bigg|_{d^+} - \kappa_l \frac{\partial T}{\partial z}\bigg|_{d^-}, \tag{4.16}$$

where the temperature gradients are evaluated on the solid (d^+) and the liquid (d^-) sides of the liquid-solid interface, and κ_s and κ_l are the solid and liquid thermal conductivities, respectively. For the case of Si, the thermal gradient in the liquid is negligible due to its greater thermal conductivity. The heat flow on the solid side of the interface can be estimated by treating the molten layer as a constant temperature bath. Then,

$$\frac{\partial T}{\partial z}\bigg|_{d^-} \approx \frac{T_m - T_{room}}{\sqrt{\pi l/2}}, \tag{4.17}$$

where $l = [l_o^2 + 4D_T t]^{1/2}$. Since the solidification velocity is generally much slower than the melting velocity, the duration of surface melting is dominated by the solidification velocity. Furthermore, at long times $l \rightarrow \sqrt{4D_T t}$ so that

$$v_{sol}(t) \, \Delta H = \kappa_s \frac{T_m - T_{room}}{\sqrt{\pi D_T t}}. \tag{4.18}$$

An approximate expression for the duration, $\tau_{regrowth}$, of the solidification phase is found by integrating this time-dependent solidification velocity

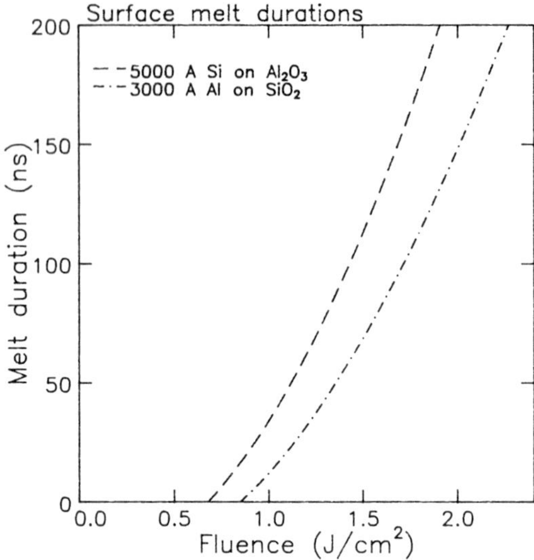

Fig. 4.18 Duration of surface melts as a function of fluence for Si on sapphire and thin Al films on SiO_2 for irradiation by 30-nsec FWHM ruby laser pulses. Fluence thresholds, taken from measurements, are 0.68 J/cm^2 for Si on sapphire (Thompson, 1986) and 0.85 J/cm^2 for Al on SiO_2 (Tsao et al., 1986).

up to the maximum melt depth d_{max}

$$d_{\text{max}} = \int_{\tau_{\text{pulse}}}^{\tau_{\text{pulse}}+\tau_{\text{regrowth}}} v_{\text{sol}}(t)\,dt = \int_{\tau_{\text{pulse}}}^{\tau_{\text{pulse}}+\tau_{\text{regrowth}}} \kappa_{\text{s}} \frac{T_{\text{m}} - T_{\text{room}}}{\sqrt{\pi D_{\text{T}} t}}. \qquad (4.19)$$

This gives

$$\tau_{\text{regrowth}} \approx \left[\frac{\sqrt{\pi D_{\text{T}}} d_{\text{max}}\, \Delta H}{2\kappa_{\text{T}}(T_{\text{m}} - T_{\text{room}})} + \sqrt{\tau_{\text{pulse}}} \right]^2 + \tau_{\text{pulse}}, \qquad (4.20)$$

where d_{max} can be estimated from Eqs. (4.14) and (4.15).

The total melt duration can then be estimated by adding to τ_{regrowth} the pulsewidth, to take into account in an approximate way the duration of the melt-in phase. The melt duration for Si thin films irradiated by 30-nsec pulses is shown in Fig. 4.18, and has been found to be in reasonable agreement with experiment (Thompson, 1986).

6.2.2 Thin Films on Thick Substrates

For multilayer films, the heat-flow problem is far more complex, and numerical techniques are required for accurate results. However, estimates may be obtained readily in the simple and common case of a single film of high thermal conductivity supported by a substrate (or insulating film) of much poorer thermal conductivity (Tsao, et al., 1986a; 1986b). This would include, for example, the case of thin Al films supported by an oxidized (>0.5 μm thick) Si wafer. In this case, for simple heating of the solid, the estimates for the maximum temperature rise ΔT_{max} (Eq. 4.10) and the temperature fall time (Eq. 4.13) are valid provided the thickness of the heated film, l_0, is taken to be the film thickness itself, and the thermal diffusivity is taken to be that of the substrate. Furthermore, for melting, the estimates for the melt threshold (Eq. 4.14), maximum melt depth (Eq. 4.15), and melt duration (Eq. 4.18) are also valid provided the same substitutions are made.

The dependence of melt durations for an Al film supported by oxidized Si is shown in Fig. 4.18. Note that for these time scales oxides thicker than ~0.5 μm are essentially infinitely thick.

6.3 Direct Writing: Heat Diffusion in Three Dimensions

For focused radiation from a cw laser beam, the thermal diffusion problem is three-, rather than one-dimensional. In that case, it has been shown by Lax (1977) that for cw irradiation by a Gaussian beam of waist-size w_o (radical intensity distribution $I_o \exp(-r^2/w_o^2)$), the steady-state increase in the surface temperature distribution is

$$\Delta T(r, t=\infty) = \frac{1}{\sqrt{\pi}} \int_0^{\pi} \tfrac{1}{2} J_o\left(\frac{\lambda r}{w_o}\right) e^{-\lambda^2/4}\, d\lambda, \tag{4.21}$$

where J_o is the Bessel function of order 0. For most practical purposes, this may be approximated by a Gaussian with a lateral spatial extent larger by $\sqrt{\pi}$ than the original intensity distribution, or

$$\Delta T(r, t=\infty) \approx \Delta T_o \exp\left(\frac{-r^2}{\pi w_o^2}\right). \tag{4.22}$$

Here, ΔT_o, the peak temperature increase at spot center, is given by

$$\Delta T_o = \Delta T(r=0, t=\infty) = \frac{P(1-R)}{2\sqrt{\pi}\,\kappa_T w_o}, \tag{4.23}$$

where P is the incident power and R is the optical reflectance. The peak temperature may be understood physically by assuming temperature gradients on the order of T_o/w_o, and then equating, in steady state, the heat conducted out through a hemispherical shell of radius w_o (approximately $\kappa_T[2\pi w_o^2\, \Delta T_o/(\sqrt{\pi}\, w_o)]$) with the adsorbed power $P(1-R)$.

These equations are valid in the steady state, which is reached on a time scale approximately equal to the diffusion time over one spot size, $\tau_o = w_o^2/4D_T$. For cw irradiation scanned at velocities v such that the duration of the irradiation at any point, $\tau = w_o/v$, is less than τ_o, steady-state conditions can no longer be assumed. This case must in general be treated numerically (Cline and Anthony, 1977); the resulting temperature distributions have been reviewed comprehensively by Gibbons and Sigmon (1982). In many cases, however, it is sufficient to know only the peak temperature at spot center; this temperature has been shown to be (Abraham and Halley, 1987)

$$\Delta T_o(\tau) = \frac{P(1-R)}{\pi^{3/2} w_o \kappa_T} \arctan\left(\frac{\tau}{\tau_o}\right). \tag{4.24}$$

Note that Eq. 4.24 approaches Eq. 4.23 in the steady-state limit. This dependence is shown in Fig. 4.19. At early times ($\tau \ll \tau_o$), the lateral spatial scale of the temperature distribution is not characterized by the length scale $\sqrt{\pi} w_o$, as in the steady state, but approaches that of the Gaussian laser intensity profile (w_o) itself (Abraham and Halley, 1987).

An alternative approach to the early-time (non-steady-state) behavior, as discussed earlier, is to approximate the early-time heat-flow behavior as essentially one dimensional. The analysis described above for pulsed irradiation then holds, with an effective pulsewidth of $\tau \approx w_o/v$ and an

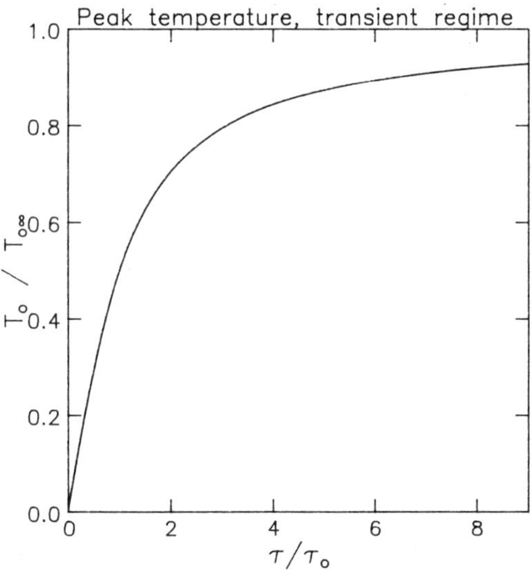

Fig. 4.19 Peak surface temperature, normalized to the long-time (steady-state) peak surface temperature, as a function of laser pulse width, normalized to the time, $\tau_0 = w_0^2/4D$, for thermal diffusion across an irradiated zone of dimension w_0. The temperature rises linearly in the pulsed regime, then saturates in the cw regime.

effective fluence of $I\tau$, where I is the laser intensity incident on the surface.

The efficiency of direct-write laser processing can be very different depending on whether it occurs in the cw or pulsed limits. This can be illustrated for the case of processing that relies on melting. In the cw limit, the surface is essentially in equilibrium with the intensity distribution of the laser beam, and so the melt duration is equal to the laser-beam dwell time, w_o/v. In the pulsed limit, however, the surface is not in equilibrium with the intensity distribution of the laser beam, and the melt duration can be much longer than w_o/v. In this limit, the beam leaves in its wake a trail of molten material whose length increases with increasing scan velocity (for constant absorbed fluence, $P\tau$). Note that a constant $P\tau$ product implies that power is increased as scan rate is increased.

For multilayered structures, Eqs. 4.21–4.23 will hold, provided a) the films are thinner than the laser-beam spot size, b) the thermal conductivity of the substrate is used, and c) the thermal conductivity of the film is not much greater than that of the substrate (Lin, 1967). In the transient regime, Eq. 4.24 will also hold, unless the effective dwell time, τ, is so short that the thermal diffusion length in that time is less than the film thickness. Then, the thermal properties of the film must be considered, and Eq. 4.24 will not be valid.

Also note that a number of techniques have been developed for monitoring the surface temperature rise due to cw laser irradiation. These include infrared emission techniques (Salathe, Gilgen, Rytz-Froidevaux, 1981); photoluminescence from semiconductors (Salathe, Gilgen, and Rytz-Froidevaux, 1981); the use of films that undergo a phase transition (Shaapur and Allen, 1987); and thin-film thermocouples (Kodas, Baum, and Comita, 1987).

7. Summary

When light is absorbed by a solid, the resulting change in temperature or minority carrier density can produce highly localized regions of enhanced chemical reaction. In this chapter, we have tried to outline the physical processes that determine the spatial and temporal extent of these regions. As mentioned at the outset, we have avoided discussing the detailed chemical mechanisms by which such enhancements occur. These mechanisms are complex and will differ with material type and laser-irradiation conditions. However, the rapidity with which few processes have been discovered and refined during these past few years attest to the versatility of laser microfabrication based on laser-enhanced chemistry. Further development of these laser-based processes can be expected to expand process options for a wide range of microfabrication applications.

8. Notes on Tables of Optical and Thermophysical Properties of Solids

Tables 4.1, 4.2, and 4.3 list important optical properties for selected semiconductors, metals, and insulators. Tables 4.4 and 4.5 list those thermophysical properties of materials that are important in the semi-quantitative heat-flow analyses described in Section 6. To the degree of accuracy of these analyses, it is reasonable to consider these properties approximately temperature independent. Very accurate analyses require both a consideration of the temperature dependence of these quantities, which can be found in the comprehensive tabulation by Touloukian, et al. (1970), as well as numerical integration of the heat-flow equations.

The specific heats in the tables are the Dulong-Petit infinite-temperature values ($c_p(\infty) = 3R$, where $R = 8.3143$ J/mole-K is the gas constant), which can be expected to be good estimates for temperatures on the order of or above the Debye temperature (Θ_D). For temperatures much less than the Debye temperature, the Debye formula

$$c_p(T) = c_p(\infty) \cdot D(T/\Theta_D) = c_p(\infty) \cdot 3(T/\Theta_D)^3 \cdot \int_0^{\Theta_D/T} \frac{x^4 e^x}{(e^x - 1)^2} dx \quad (4.25)$$

may be used, where the Debye function, $D(T/\Theta_D)$, is drawn in Fig. 4.20.

Except in unusual situations, T/Θ_D will be on the order of or greater than unity, for which $D(T/\Theta_D) \geq 0.95$, so that $c_p(\infty)$ should give a reasonable estimate. For convenience, the Debye temperatures for the pure metals are also listed. For the compounds, the Debye temperatures are not listed; a useful rule of thumb, however, is that the Debye temperature is generally of the order of one-fifth to one-half the melting point in degrees Kelvin (Kingery, 1976). The Debye temperature may also be estimated from the velocity of sound, c_s, via $k\Theta_D = \hbar\omega_D$, where $\omega_D = c_s[6\pi^2 N/V]^{1/3}$, and N/V is the atomic density (Reif, 1965).

Thermal conductivities κ_T for the pure metals are listed at 273°K and at 973°K. For the compounds, they are listed only at room temperature. Note that for the good conductors, the Wiedemann-Franz law may be used to estimate the thermal conductivity

$$\kappa_T = \frac{T}{\rho} \cdot \left[\left(\frac{\pi^2}{3}\right)\left(\frac{k_B}{e}\right)^2\right] = \frac{T}{\rho} \cdot [2.45 \cdot 10^{-8}\ \mathrm{W\Omega/K^2}], \quad (4.26)$$

where ρ is the electrical resistivity in Ω-cm, T is the temperature in degrees Kelvin, k_B is Boltzman's constant and e is the electron charge. In

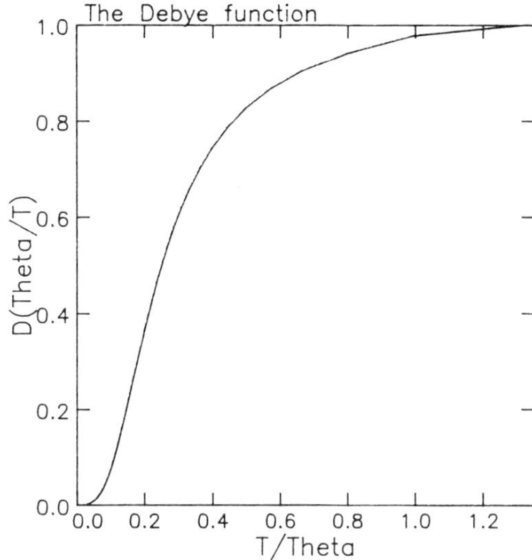

Fig. 4.20 The Debye function $D(\Theta_D/T)$, as a function of temperature, normalized to the Debye temperature, Θ_D.

all cases, even for poor conductors, this formula may be used to obtain a lower bound to the thermal conductivity. Once the specific heat and thermal conductivities are known, thermal diffusivities may be calculated using $D_T = \kappa_T/c_p$.

In all the tables, our notation is T_m for the melting temperature, T_b for the boiling temperature, ΔH_f for the latent heat of fusion, ΔH_v for the latent heat of vaporization, E_g for the room-temperature band gap, and ρ_{RT} for the room-temperature resisitivity.

References

Abraham, E., and Halley, J.M. (1987). Some calculations of temperature profiles in thin films with laser heating. *Appl. Phys.* **A42,** 279.

Alimoussa, L., Carchano, H., and Thomas, J.P. (1981). Influence of hydrogen partial pressure on deposition and properties of sputtered amorphous gallium arsenide. *J. Phys. Colloq.* **C4,** 683.

Ashby, C.I.H. (1984). Photochemical dry etching of GaAs. *Appl. Phys. Lett.* **45,** 892.

Ashby, C.I.H. (1985). Doping level selective photochemical dry etching of GaAs. *Appl. Phys. Lett.* **46,** 752.

Ashby, C.I.H. (1986). Photochemical dry etching of semiconductors and its relationship to semiconductor electronic properties. *Proc. Symp. Photon, Beam and Plasma Stimulated Processes at Surfaces,* 1986 Fall Mtg. of Materials Research Society, Boston, Dec. 1–6, 1986, 369.

Ashby, C.I.H. (1987). In *Physics of Thin Films,* Vol. 13, (Francombe, M.H., and Vosen, J.L., eds.). Academic Press, Orlando, FL, 151.

Ashby, C.I.H., and Biefeld, R.M. (1985). Composition-selective photochemical etching of compound semiconductors. *Appl. Phys. Lett.* **47,** 62.

Ashby, C.I.H., and Biefeld, R.M. (1989). Effect of bandgap characteristics on laser-induced dry etching of semiconductors. *J. Electrochem. Soc.,* to be published.

Ashby, C.I.H., and Myers, D.R. (1987). Control of solid-excitation-based photochemical dry etching of semiconductors by ion-bombardment-induced damage. *Proc. Symp. Laser and Particle Beam Chemical Processing for Microelectronics,* 1987 Fall Mtg. of Materials Research Society, Boston Dec. 1–3, 1987, 429.

Ashby, C.I.H., Myers, D.R., and Vook, F. L. (1989). Selective laser-induced photochemical dry etching of semiconductors controlled by ion-bombardment-induced damage. *J. Electrochem. Soc.,* to be published.

Aspnes, D.E., and Studna, A.A. (1983). Dielectric functions and optical parameters of Si, Ge, GaP, GaAs, GaSb, InP, InAs and InSb from 1.5 to 6.0 eV. *Phys. Rev.* **B27,** 985.

Atwater, H., et al. (1988). Manuscript in preparation.

Baeri, P., and Campisano, S.U. (1982). Heat flow calculations. In *Laser Annealing of Semiconductors* (Poote, J.M., and Mayer, J.W., eds.). Academic Press, New York, Chap. 4.

Bass, M., ed. (1983). *Laser Materials Processing,* Materials Processing, Materials Processing—Theory and Practices Vol. 3, North-Holland, Amsterdam.

Bauerle, D. (1986). *Chemical Processing with Lasers.* Springer Series in Material Science, Volume 1. Springer-Verlag, Berlin.

Born, M., and Wolf, E. (1970). *Principles of Optics,* 4th edition. Pergamon, New York.

Calder, I.D., and Sue, R. (1982). Modeling of CW Laser Annealing of Multilayer Structures. *J. Appl. Phys.* **53,** 7545.

Carslaw, H.S., and Jaeger, T.C. (1959). *Conduction of Heat in Solids.* Oxford U. Press, Oxford.

Casey, H.C. Jr., Sell, D.D., and Wecht, K.W. (1975). Concentration dependence of the absorption coefficient for n- and p-type GaAs between 1.3 and 1.6 eV. *J. Appl. Phys.* **46,** 250.

Chiang, T.-C., Ludeke, R., Aono, M., Landgren, G., Himpsel, F.J., and Eastman, D.E. (1983). Angle-resolved photoemission studies of GaAs(100) surfaces grown by molecular-beam epitaxy. *Phys. Rev.* **B27,** 4770.

Cline, H.E., and Anthony, T.R. (1977). Heat treating and melting material with a scanning laser or electron beam. *J. Appl. Phys.* **48,** 3895.

Cummings, K.D., Harriott, L.R., Chi, G.C., and Ostermayer, F.W., Jr, (1986). Using focused ion beam damage patterns to photoelectrochemically etch features in III–V materials. *Appl. Phys. Lett.* **48,** 659.

Donnelly, V.M., Herman, I.P. and Hirose, M., eds. (1987). *Photon, Beam, and Plasma Stimulated Chemical Processes at Surfaces.* Materials Research Society Symposia Proceedings Vol. 75, Materials Research Society, Pittsburgh.

Driscoll, W.G., and Vaughan, W., eds. (1978). *Handbook of Optics.* McGraw-Hill, New York.

Duley, W.W. (1983). *Laser Processing and Analysis of Materials.* Plenum Press, New York.

Ehrlich, D.J., Osgood, R.M., and Deutsch, T.F. (1981). Laser chemical technique for rapid direct writing of surface relief in silicon. *Appl. Phys. Lett.* **38,** 1018.

Ehrlich, D.J., and Tsao, J.Y. (1983). A review of laser-microchemical processing. *J. Vac. Sci. Technol.* **B1,** 969.

Filyand, M.A. and Semenova, E.I. (1970). *Handbook of the Rare Elements Vol. II:*

Refractory Elements. Boston Technical Publishers, Cambridge, MA.

Gibbons, J.F., and Sigmon, T.W. (1982). Solid phase regrowth. In *Laser Annealing of Semiconductors* (Poote, J.M., and Mayer, J.W., eds.). Academic Press, New York, Chap. 4.

Glazov, V.M., Chizhevskaya, S.N., and Glagoleva, N.N. (1969). *Liquid Semiconductors*. Plenum Press, New York.

Gomersall, A., ed. (1986). *Lasers in Materials Processing, A Bibliography of a Developing Technology*. Springer-Verlag, Berlin.

Gray, D.E., ed. (1972). *American Institute of Physics Handbook, 3rd ed.* McGraw-Hill, New York.

Gurevich, Y.Y., Pleskov, Y.V., and Rotenberg, Z.A. (1980). *Photoelectrochemistry*. Consultants Bureau, New York.

Harper, C.A., ed. (1970). *Handbook of Materials and Processes for Electronics*. McGraw-Hill, New York, Chap. 7.

Houle, F.A. (1983). Non-thermal effects in laser-enhanced etching of Si by XeF_2. *Chem. Phys. Lett.* **95,** 5.

Ibbotson, D.E., Flamm, D.L., and Donnelly, V.M. (1983). Crystallographic etching of GaAs with bromine and chlorine plasmas. *J. Appl. Phys.* **54,** 5974.

Jellison, G.E., Jr. and Lowndes, D.H. (1985). Time-resolved ellipsometry measurements of the optical properties of silicon during pulsed excimer laser irradiation. *Appl. Phys. Lett.* **47,** 718.

Kaneko, K., Ayabe, M., and Watanabe, N. (1976). Electrical properties of n-$Al_xGa_{1-x}As$. In *Gallium Arsenide and Related Compounds* (Hilsum, C., ed.). Proceedings of the Sixth International Symposium on Gallium Arsenide and Related Compounds, Edinburgh, 20-22 September 1976. The Institute of Physics Conference Series Number 33a, Bristol and London.

Kingery, W.D. (1976). *Introduction to Ceramics*. Wiley & Sons, New York.

Kodas, T.T., Baum, T.H., and Comita, P.B. (1987). Surface temperature rise in multilayer solids induced by a focus beam. *J. Appl. Phys.* **61,** 2749.

Kubaschewski, O., and Alcock, C.B. (1979). *Metallurgical Thermochemistry*, 5th ed., Pergamon Press, Oxford.

Kuech, T.F., Wolford, D.J., Potemski, R., Bradley, J.A., Kelleher, K.H., Yan, D., Farrell, J.P., Lesser, P.M.S., and Pollack, .F.H. (1987). Dependence of the $Al_xGa_{1-x}As$ band edge on alloy composition based on the absolute measurement of x. *Appl. Phys. Lett.* **51,** 505.

Kunz, R.R., and Mayer, T.M. (1987). Surface reaction enhancement via low energy electron bombardment and secondary electron emission. *J. Vac. Sci. Techn.* **B5,** 427.

Kurz, H., Olson, G.L., and Poate, J.M. (1986). *Beam-Solid Interactions and Phase Transformations*. Materials Research Society Symposia Proceedings, Vol. 51, Materials Research Society, Pittsburgh.

Lampert, M.O., Koebel, J.M., and Siffert, P. (1981). Temperature dependence of the reflectance of solid and liquid silicon. *J. Appl. Phys.* **52,** 4975.

Lax, M. (1977). Temperature rise induced by a laser beam. *J. Appl. Phys.* **48,** 3919.

Lax, M. (1978). Temperature rise induced by a laser beam II. The nonlinear case. *Appl. Phys. Lett.* **33,** 786.

Li, K.D., and Fauchet, P.M. (1987). Drude parameters of liquid silicon at the melting temperature. *Appl. Phys. Lett.* **51,** 1747.

Lin, T.P. (1967). Estimate of temperature rise in electron beam heating of thin films. *IBM J. Res. Develop.* **11,** 527.

Liu, J.M. (1982). *Thermal Model of Picosecond Laser Interactions with Silicon.* Ph.D thesis, Harvard University, Cambridge.

Lum, R.M., Glass, A.M., Ostermayer, F.W. Jr., Kohl, P.A., Ballman, A.A., and Logan, R.A. (1984). Holographic photoelectrochemical etching of diffraction gratings in n-InP and n-GaInAsP for distributed feedback lasers. *J. Appl. Phys.* **57,** 39.

Matsumoto, N., and Kumabe, K. (1980). Amorphous GaAs films by molecular beam deposition. *Jpn. J. Appl. Phys.* **19,** 1583.

Metzbower, E.A., ed. (1979). *Applications of Lasers in Materials Processing.* American Society for Metals, Metals Park, Ohio.

Mukherjee, K., and Mazumder, J., eds. (1985). *Laser Processing of Materials.* The Metallurgical Society of AIME, Warrendale, Pennsylvania.

Murarka, S.P. (1983). *Silicides for VLSI Applications.* Academic Press, New York.

Nicolet, M.-A., and Lau, S.S. (1983). Formation and characterization of transition-metal silicides. In *VLSI Electronics, Microstructure Science,* Vol. 6. (Einspruch, N.G., ed.). Academic Press, New York, Chap. 6.

Oelgart, G., Schwabe, R., Heider, M., and Jacobs, B. (1987). Photoluminescence of $Al_xGa_{1-x}As$ near the Γ-X crossover. *Semicond. Sci. Technol.* **2,** 468.

Offsey, S.D., Woodall, J.M., Warren, A.C., Kirchner, P.D., Chappell, T.I., and Pettit, G.D. (1986). Unpinned (100) GaAs surfaces in air using photochemistry. *Appl. Phys. Lett.* **48,** 475.

Ostermayer, F.W., Jr., and Kohl, P.A. (1981). Photoelectrochemical etching of p-GaAs. *Appl. Phys. Lett.* **39,** 76.

Palik, E.D., ed. (1985). *Handbook of Optical Constants of Solids.* Academic Press, Orlando.

Pankove, J.I. (1975). *Optical Processes in Semiconductors.* Dover Publications, New York, 412.

Picraux, S.T., Thompson, M.O., and Williams, J.S., eds. (1987). *Beam-Solid Interactions and Transient Processes.* Materials Research Society Symposia Proceedings, Vol. 74. Materials Research Society, Pittsburg.

Piller, H. (1985). In *Handbook of Optical Constants of Solids* (Palik, E.D., ed.), Academic Press, Orlando.

Pleskov, Y.V., and Gurevich, Y.Y. (1986). *Semiconductor Photoelectrochemistry.* Consultants Bureau, New York.

Poate, J.M., and Mayer, J.W., eds. (1982). *Laser Annealing of Semiconductors.* Academic Press, New York.

Poate, J.M., Foti, G., and Jacobson, D.C. (1983). *Surface Modification and Alloying by Laser, Ion and Electron Beams.* Proceedings of a NATO Advanced Study Institute Surface Modification and Alloying, August 24–28, 1981, Trevi, Italy. Plenum Press, New York.

Podlesnik, D.V., Gilgen, H.H., and Osgood, R.M., Jr. (1984). Deep-ultraviolet induced wet etching of GaAs. *Appl. Phys. Lett.* **45,** 563.

Ready, J.F. (1978). *Industrial Applications of Lasers,* Academic Press, New York.

Ready, J.F. (1977). *Effects of High-Power Laser Radiation.* Academic, New York.

Reif, F. (1965). *Fundamentals of Statistical and Thermal Physics.* McGraw-Hill, New York, Sec. 10-2.

Reksten, G.M., Holber, W., and Osgood, R.M., Jr. (1986). Wavelength dependence of laser enhanced plasma etching of semiconductors. *Appl. Phys. Lett.* **48,** 551.

Salathe, R.P., Gilgen, H.H. and Rytz–Froidevaux, Y. (1981). Efficient luminescence band created in (Al, Ga)As multilayers by athermal laser processing. *IEEE J. Quant. Electronics* **17,** 1989.

Schiavello, M., ed. (1985). *Photoelectrochemistry, Photocatalysis and Photoreactors: Fundamentals and Developments.* NATO Advanced Science Institutes, Series C, Vol. 146. Proceedings of the NATO Advanced Study Institute on Fundamentals and Developments of Photocatalytic and Photochemical Processes, Erice, Trapani, Italy, May 20-June 2, 1984. D. Reidel, Dordrecht, Holland.

Shaapur, F., and Allen, S.D. (1987). Experimental determination of laser heated surface temperature distributions. *Appl. Phys. Lett.* **50,** 723.

Smith, R.A. (1978). *Semiconductors,* 2nd ed. Cambridge University Press, New York.

Sze, S.M. (1981). *Physics of Semiconductor Devices, 2nd Ed.* John Wiley & Sons. New York.

Tarui, Y., Komiya, Y., and Harada, Y. (1971). Preferential Etching and Etched Profile of GaAs. *J. Electrochem. Soc.* **118,** 118.

Thompson, M.O. (1984). *Liquid-Solid Interface Dynamics during Pulsed Laser Melting of Silicon-on-Sapphire.* Ph.D thesis, Cornell University, Ithaca, NY.

Tietjen, J.J., and Amick, J.A. (1966). The preparation and properties of vapor-deposited epitaxial $GaAs_{1-x}P_x$ using arsine and phosphine. *J. Electrochemical Soc.* **113,** 724.

Tisone, G.C., and Johnson, A.W. (1983). Laser-controlled etching of chromium-doped $\langle 100 \rangle$ GaAs. *Appl. Phys. Lett.* **42,** 530.

Tsao, J.Y., Picraux, S.T., Peercy, P.S., and Thompson, M.O. (1986a). Direct measurements of liquid/solid interface kinetics during pulsed-laser-induced melting of aluminum. *Appl. Phys. Lett.* **48,** 278–280.

Tsao, J.Y., Picraux, S.T., Peercy, P.S., and Thompson, M.O. (1986b). Transient conductance measurements and heat-flow-analysis of pulsed-laser-induced melting of aluminum thin films. *Materials Research Society Symposia Proceedings* **51,** 283–288.

Touloukian, Y.S., Powell, R.W., Ho, C.Y., and Klemens, P.G. (1970). *Thermophysical Properties of Matter,* Vols. 1–10. IFI/Plenum, New York.

Ubbelohde, A.R. (1978). *The Molten State of Matter: Melting and Crystal Structure.* John Wiley, Chichester, 1978.

von Allmen, M.F. (1982). Fundamentals of energy deposition. In *Laser Annealing of Semiconductors* (Poate, J.M., and Mayer, J.W., eds.). Academic Press, New York, Chap. 3.

von Gutfeld, R.J., and Hodgson, R.T. (1982). Laser enhanced etching in KOH. *Appl. Phys. Let.* **40,** 352–354.

Weakliem, H.A., and Redfield, D. (1979). Temperature dependence of the optical properties of silicon. *J. Appl. Phys.* **50,** 1491.

Weast, R.C., ed. (1974). *CRC Handbook of Chemistry and Physics, 55th ed.* CRC Press, Cleveland.

Wood, R.F., and Geist, G.A. (1986). Modeling of nonequilibrium melting and solidification in laser-irradiated materials. *Phys. Rev.* **B34,** 2606–2620.

Yamamoto, A., and Yano, S. (1975). Anodic dissolution of n-type gallium arsenide under illumination. *J. Electrochem. Soc.* **122,** 260.

CHAPTER 5

Transport and Kinetics

H.J. ZEIGER and D.J. EHRLICH
Lincoln Laboratory
Massachusetts Institute of Technology
Lexington, Massachusetts

J.Y. TSAO
Sandia National Laboratories
Albuquerque, New Mexico

1. Introduction

It is widely recognized that an important and unique feature of laser-beam processing is the shortness of the time scale of the laser-surface interaction. Because of this, the nature and the kinetics of the surface processes can be very different from those that govern ordinary surface processes. For example, with picosecond laser pulses, re-crystallization of laser-melted surfaces can be suppressed in favor of resolidification into amorphous phases. Even with nanosecond laser

ISBN 0-12-233430-2

pulses, greatly enhanced solute trapping can occur during resolidification of laser-melted surfaces.

Perhaps the most striking feature of patterned laser microfabrication, however, is that the spatial scale of the laser-surface interaction can be made very small. In this new regime of small-area reactions, a general observation has been that reaction rates are often orders of magnitude larger than those of the corresponding large-area reactions. This observation has been explained by numerous workers in terms of enhanced mass transport of chemical reactants and products to and from very small reaction zones.

In this chapter, we discuss the fundamental mass-transport processes involved in laser-microchemical reactions, with an emphasis on those strongly dependent on the size of the reaction zone. We classify, as is customary, the various reactions according to whether they are initiated by light absorption a) in the external (gas- or liquid-phase) medium contacting the surface, b) in molecular layers adsorbed on the surface, or c) in the (solid) substrate itself.

An example of (a) is photodeposition in the regime of moderate pressures. In this case, a UV-laser beam is focused onto a surface, photodissociating molecules in an adjacent vapor. The free atoms and radicals produced can then further react or simply condense on the surface. Because of nucleation constraints, condensation can often be arranged to occur predominantly in a localized area as small as the laser beam spot itself. Such reactions have been studied for a wide variety of metals deposited from their alkyl or carbonyl vapors, e.g., Cd from $Cd(CH_3)_2$, Zn from $Zn(CH_3)_2$, Al form $Al_2(CH_3)_6$, Fe from $Fe(CO)_5$, W from $W(CO)_6$, and Cr from $Cr(CO)_6$ (Krchnavek et al., 1987).

An example of (b) is photopolymerization. In this case, a focused UV laser beam initiates a chain-polymerization reaction in an adsorbed surface layer of monomer. This reaction has been studied for the methyl methacrylate system, in which poly (methyl methacrylate) (PMMA), a high-quality resist material is produced. Another example of (b) is photodeposition in the regime of low pressures (usually less than a tenth of the vapor pressure of the molecule used). Clear examples of these reactions are Ti from $TiCl_4$, Cd from $Cd(CH_3)_2$, Zn from $Zn(CH_3)_2$, and Al from $Al_2(CH_3)_6$.

Examples of (c) are thermal reactions in which a laser beam locally heats a substrate to activate a deposition or etching reaction. Systems that have been studied include the laser etching of Si in a Cl_2 ambient and the laser deposition of Si in a SiH_xCl_{4-x} ambient.

The rest of this chapter is organized as follows: Section 2 briefly discusses some of the techniques used to measure rates of laser-induced surface chemistry. Sections 3, 4, and 5 form the core of the chapter. Section 3 describes the transport of molecules from the localized column of light above the substrate to the surface (where they react); this is the case of photodeposition due to molecular photoexcitation in the gas phase. Section 4 describes the transport of molecules by volume diffusion to a localized reactive spot on the surface defined by the spatial extent of the laser beam; this is the case of localized laser-thermal chemistry replenished by diffusion from the vapor (or liquid) ambient. Section 5 describes the transport of molecules by surface diffusion to a localized reaactive spot on the surface defined by the spatial extent of the laser beam; this is the case of localized photodeposition due to molecular photoexcitation in the adsorbed layer, under conditions where replenishment is by diffusion from the surrounding adlayer.

We note here at the outset, however, some striking differences between the three mass-transport problems. Both the first and second problems are three dimensional, and hence admit nonzero steady-state solutions to the reaction rate. The third problem is two dimensional and has no nonzero steady-state reaction rate. Therefore transient effects are inherently important in the third case but may or may not be in the first two cases.

In general, the onset of steady-state behavior for three-dimensional diffusion problems occurs for times longer than that required for diffusion across the feature size of interest. Therefore, laser processing with very small feature sizes and dilute ambients (with high diffusivities for mass transport) will tend to be steady state. For example, even for feature sizes as large as $10\,\mu$m and ambient gas pressures as high as 1 atmosphere, steady-state conditions are reached in times less than $1\,\mu$s.

For larger feature sizes, denser ambients, or shorter laser dwell times (high scan velocities or short pulse widths), transient effects must be considered. In the limit of very short laser dwell times, or pulsewidths, there may be no time for mass transport during either the laser pulse, or in the case of laser-thermal processing, during the time over which the surface is hot. In that case, laser-chemical processing must rely on the formation of adsorbed layers *in between* pulses and hence will become dependent on the laser repetition rate.

We also note that the mass-transport limitations discussed in this chapter are not the only bounds to chemical processing rates. In practice, other effects may set in that limit rates, such as substrate damage in the case of laser-thermal processing or homogeneous nucleation of gas-phase particles in the case of laser photodeposition due to gas-phase photoexcitation. Note further that other factors may set in to enhance rates. In this chapter we deal exclusively with diffusive mass transport. In dense external media (such as liquids) and for large thermal gradients, convective mass transport becomes an important additional mechanism for the replenishment of molecules reacting at a surface (Von Gutfeld et al., 1979).

We also consider here only the most simplified geometries (which are already difficult), namely, hemispherical or planar reaction surfaces. The kinetics of laser processing in cases where aspect ratios become very large is more difficult to treat exactly, although in some cases approximations may be used. For example, in the etching of through-wafer via holes, the problem simplifies, under some conditions, to one of simple one-dimensional diffusion down the via hole. In the deposition of rods, the problem simplifies, under some conditions, to one of diffusion to a spherical, rather than hemispherical, reacting surface.

Finally, for the case of three-dimensional hemispherical diffusion through an ambient gas, we restrict ourselves to pressures high enough that mean-free paths are less than the spatial scale of the surface reaction zone. Then, it is possible to treat mass transport diffusively, rather than ballistically. For many high-throughput laser microfabrication processes, this is not a severe restriction. For example, at atmospheric pressure mean-free paths are on the order of $0.4\,\mu$m. However, at low pressures and small laser spot sizes, the onset of ballistic transport must be considered.

2. Experimental and Measurement Techniques

There are a variety of methods for in-situ monitoring of the interaction of laser beams with materials, ranging from techniques based on photo-acoustic spectroscopy (see, e.g., Melcher, 1984) to time-resolved electrical conductance (Thompson, 1984) and optical reflectance (Auston, 1978). It is much more difficult to measure directly, in situ, the time-resolved rate of laser-induced chemical reactions.

A particularly convenient, in-situ measurement of the rate of laser deposition of a material can be made by monitoring the self-transmitted light through the deposit of the laser beam itself. Although there will be nonlinearities associated with the initial nucleation phase of deposition, for intermediate deposit-thicknesses the thickness profile $d(r)$ of the deposit can be approximated by the Gaussian intensity profile $I_{inc}(r) = I_0 e^{-2r^2/\omega_0^2}$ of the focused laser beam itself. Then we can write $d(r) = De^{-2r^2/\omega_0^2}$, where D is the center thickness of the deposit.

For a transparent deposit, such as PMMA, the primary effect of the highly curved Gaussian deposit will be to diffract and defocus the beam via a transmission function of the form

$$T(r) = \exp\left[i\frac{2\pi(n-1)D}{\lambda}e^{-2r^2/\omega_0^2}\right], \tag{5.1}$$

where n is the refractive index of the deposit. The far-field, on-axis undiffracted intensity can be calculated by evaluating the Fourier transform of the transmitted electric field profile at $k_r = 0$:

$$\frac{I_{\text{undiff}}}{I_0} = \frac{1}{\lambda^2 z^2}\left|\mathscr{F}\left\{\frac{E_{\text{inc}}(r)}{E_0}T(r)\right\}_{k_r=0}\right|^2. \tag{5.2}$$

In practice, the lensing behavior of the system is dominated by the form of the transmission function $T(r)$, so for convenience of integration, we take the incident electric field profile to be $E_{\text{inc}}(r) = E_0 e^{-2r^2}/\omega_0^2$. Then, it can be shown that

$$\begin{aligned}\frac{I_{\text{undiff}}}{I_0} &= \frac{1}{\lambda^2 z^2}\left|\sum_{q=0}^{\infty}\frac{\left[i\frac{(2\pi(n-1)D}{\lambda}\right]^q}{q!}\mathscr{F}\{e^{-2(q+1)r^2/\omega_0^2}\}_{k_r=0}\right|^2 \\ &= \frac{z_0^2}{z^2}\left|\sum_{q=0}^{\infty}\frac{\left[i\frac{2\pi(n-1)D}{\lambda}\right]^q}{q!\,(q+1)}\right|^2 \\ &= \frac{z_0^2}{z^2}4\sin^2\left[\frac{\pi(n-1)D}{\lambda}\right],\end{aligned} \tag{5.3}$$

where $z_0 = \pi\omega_0^2/\lambda$ is the confocal laser-beam parameter. The transmitted on-axis intensity, measured through an aperture, undergoes periodic modulations with which the thickness of the deposit can be calibrated. For an incident electric field amplitude equal to $E_{\text{inc}} = E_0 e^{-r^2/\omega_0^2}$, the behavior of the undiffracted intensity will be similar but with a reduced modulation depth, as shown in Fig. 5.1.

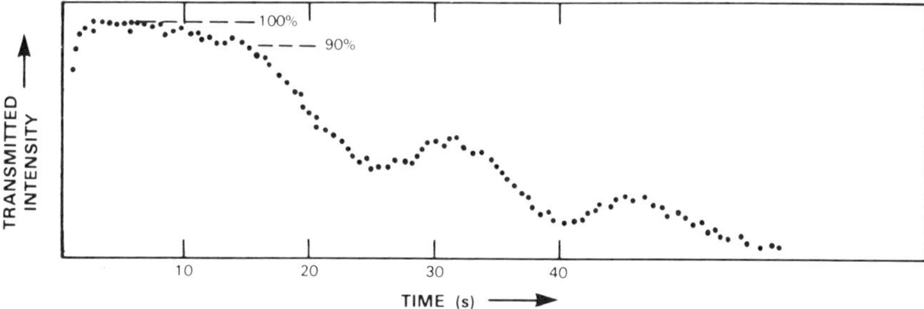

Fig. 5.1 On-axis transmitted-intensity oscillation as measured experimentally during in-situ PMMA polymerization. After Tsao and Ehrlich, 1983.

For a nontransparent deposit, such as a metal, the relationship between transmitted intensity and deposit thickness is more complicated, due to the thickness dependence of thin-film reflectivity. If the transmitted intensity is re-imaged, and measured on-axis, then as a function of center thickness D it will be (Heavens, 1955)

$$T = \frac{8(n^2+K^2)}{\alpha \cosh(4\pi KD/\lambda)+\beta \sinh(4\pi KD/\lambda)+\varepsilon \cos(4\pi nD/\lambda)+\gamma \sin(4\pi nD/\lambda)}, \tag{5.4}$$

where $\alpha = 4n^2 + (1+n^2+K^2)^2$, $\beta = -4n(1+n^2+K^2)$, $\varepsilon = 4K^2 - (1-n^2-K^2)^2$, and $\gamma = -4K(1-n^2-K^2)$.

If the total transmitted intensity is measured, then this expression for T must be multiplied by the incident intensity profile and then integrated. A phenomenological and more easily integrated expression for T can be approximated by $1-R-A$, where the reflectivity $R = R_0(1-e^{-\alpha d})$ is taken to increase exponentially to its bulk value R_0 with film thickness, and the absorption is $A = (1-R)(1-e^{-\alpha d})$. Then $T \simeq 1-R-A \simeq (1-R_0)e^{-\alpha d} + R_0 e^{-2\alpha d}$, and the total transmitted intensity is

$$\frac{I_{\text{trans}}}{I_0} = \int_0^\infty \{(1-R_0)\exp[-\alpha D e^{-2r^2/\omega_0^2}] + R_0 \exp[-2\alpha D e^{-2r^2/\omega_0^2}]\} \cdot e^{-2r^2/\omega_0^2}(2\pi r)\, dr \simeq \frac{1-e^{-\alpha(1+R_0)D}}{\alpha(1+R_0)D} \simeq e^{-\alpha/2(1+R_0)D}, \tag{5.5}$$

where $\alpha = 2\pi K/\lambda$ is the film absorption, $R_0 = [(1+n)^2+K^2]/[(1-n)^2+K^2]$ is the bulk film reflectivity, and n and K are the optical constants of the film.

These formulas are not exact but are generally accurate enough for relative measurements of deposition rates. Similar measurements of etch rates through thin films can also be made conveniently using optical transmission.

The methods described above, though convenient, can only be considered approximate, principally because of uncertainties in the optical constants of deposited material, as well as because of their simplified derivation. Perhaps the most accurate methods are those based on piezoelectric microbalances (Chuang, 1981). They are ideal for measuring large-area laser-induced processes, for which sensitivities of less than one monolayer are possible. Such microbalances, however, are also very sensitive to small temperature changes, either in the ambient or due to laser heating, which can be a disadvantage in some cases.

If the deposited feature is large enough, deposition rates can be measured directly by optical microscopy, as has been elegantly demonstrated by Bäuerle (1986). Ultimately, postmortem analysis by stylus profilometry is a standard way of quickly measuring absolute deposition/etch thicknesses. Cross-section scanning electron microscopy is necessary for very high-resolution features and for obtaining cross-sectional profile information.

Finally, laser-induced thermal desorption has also been shown to be a powerful technique for monitoring surface chemistry in small zones. In this technique, a laser pulse causes transient heating of the substrate, desorbing molecules that can then be measured by mass spectrometry. Often, the desorption is fast enough to bypass chemistry during the heat pulse (Hall and DeSantolo, 1984). Then, the technique is effectively a "passive" probe of species undergoing reaction due to some other source of thermal or photoexcitation.

More recently, George and coworkers (Mak et al., 1986) have used laser-induced thermal desorption extensively to measure surface diffusion rates on fairly small (240-μm spot sizes) spatial scales.

3. Reactions Initiated by External-Phase Absorption

In this class of laser microchemistry, as mentioned above, the sequence of events is as follows. First, a column of gas- (or liquid-) phase molecules in the laser-beam path is excited, either vibrationally, using infrared radiation, or electronically, using UV radiation. Either form of excitation can lead to direct fragmentation or to secondary photochemical reactions through collisions. Second, the excited species molecules diffuse from the column to the surface. On the way, they may undergo collisions, leading to quenching through energy transfer or radical recombination, or to possible chain reactions. Finally, the excited species reacts with or on the surface, either to form a volatile (etching) or nonvolatile (deposition) product. The reaction, if it leads to etching, can be other than first order in the surface concentration of reactants; here, however, we assume first-order kinetics. If the reaction leads to deposition, there will be both nucleation and growth phases, in which the surface-sticking coefficients can be expected to be quite different. In what follows we first assume that the surface reaction is in the linear growth regime and that, in particular, sticking coefficients are unity; deviations due to nucleation phenomena are not considered.

3.1. Approximate Solution

The geometry for the diffusion problem is fairly simple. For UV laser beams, a practical focal spot diameter is $\omega_0 \simeq 1\ \mu$m, which implies a confocal beam parameter of $z_0 = \pi\omega_0^2/\lambda = 12\ \mu\text{m}$. As this is large compared to the size of the focal spot itself, we can, without losing any qualitative features of the model, take the laser beam to be an infinite cylindrical column with a uniform intensity I_0 within its diameter $\sqrt{2}\omega_0$. This column is a source of excited molecules whose flux can be estimated by equating the rate of creation per unit surface area of the column,

$$j_{\text{creation}} = I\sigma\left(\frac{\pi\omega_0^2/2}{\sqrt{2}\,\pi\omega_0}\right)(n_0 - n), \tag{5.6}$$

with the rate of diffusion out of the column,

$$j_{\text{diff}} = \left(\frac{D}{\omega_0^2}\right)\frac{(\pi\omega_0^2/2)}{\sqrt{2}\,\pi\omega_0}\,n. \tag{5.7}$$

The result is that the steady-state creation rate, including depletion of excitable species, is

$$j_{\text{creation}} = \left(\frac{Dn_0}{\sqrt{2}\,\omega_0}\right)\frac{I\sigma\omega_0^2/D}{1 + I\sigma\omega_0^2/D}. \tag{5.8}$$

The flux produced by this column of excited molecules to the surface, where deposition or etching with unit sticking probability is assumed to occur, can be calculated by considering the solid angle that this column subtends at a point on the surface. This approach automatically compensates for geometrical divergence effects, since solid angles subtended by various parts of the column decrease in proportion of their flux contribution. Therefore, the surface flux is (Tsao and Ehrlich, 1984)

$$\begin{aligned} j_{\text{surf}}(r) &= j_{\text{creation}} \cdot \frac{1}{2\pi}\int_0^{\pi/2} \tan^{-1}\left(\frac{\sqrt{2}\,\omega_0}{r}\right)\cos\phi\, d\phi \\ &= j_{\text{creation}} \cdot \frac{1}{2\pi}\tan^{-1}\left(\frac{\sqrt{2}\,\omega_0}{r}\right). \end{aligned} \tag{5.9}$$

As shown in Fig. 5.2, this is a mildly peaked function of radius along the surface from the beam axis, with a long tail due to film contributions from excited molecules created a distance $\gg\omega_0$ away from the surface.

Since the geometric factor in brackets in Eq. 5.9 is normalized to unity, the on-axis growth rate in the linear regime is proportional simply to j_{creation}. As with all reactions in which mass transfer to or from a localized site plays a role, there are two clear kinetic regimes. At very low light intensities,

$$j_{\text{creation}} \to I\sigma\left(\frac{\omega_0}{\sqrt{2}}\right)n_0, \tag{5.10}$$

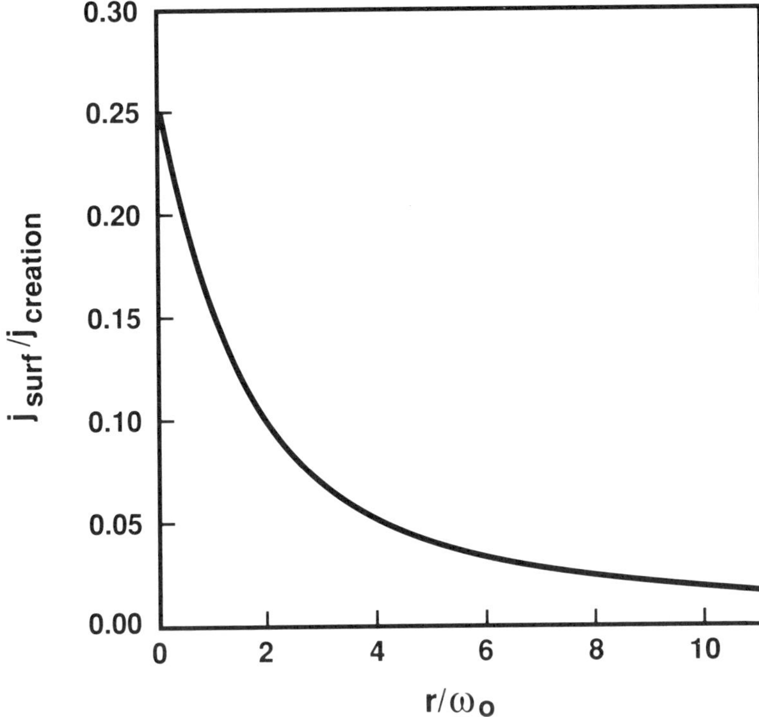

Fig. 5.2 Approximate calculation (Eq. 5.9) of radial dependence of rate of photodeposition from the gas phase.

so that the flux of excited molecules is linear in laser intensity and in beam diameter. This is the reaction-rate limited regime. The linear dependence on beam diameter can be understood intuitively by considering that the contributions to the on-axis growth rate are dominated by the bottom portion of the column, a pillbox of height ω_0. The contributions from the rest of the column decrease as $1/z^2$ and rapidly become insignificant. In this reaction-rate-limited regime, therefore, larger beam spot sizes increase the height of the contributing pillbox and thereby increase growth rate.

At very high light intensities, however, $j_{\text{creation}} \rightarrow (D/\sqrt{2}\,\omega_0)n_0$ so that the flux of excited molecules is independent of laser intensity, but depends linearly on the diffusivity and inversely on beam spot size. This is the diffusion-limited regime. The inverse dependence on beam spot size can be understood by the greater ease of excited-molecule diffusion out of a narrower column and the lessened degree of saturation in the rate of excited-molecule creation. Although in most practical cases, other effects limit rates at high intensities, as discussed below, the potential rate enhancement due to very small reaction zones is clear from this expression.

As mentioned above, the geometrical factor governing the laser-induced surface reaction will have a linewidth comparable to the laser beam spot size ω_0; the long tail is generally detrimental to many applications. As discussed previously, the long tail can be largely eliminated, however, if the excited species undergoes quenching or recombination as it diffuses to the surface. This can be made semiquantitative by multiplying the solid angle integrand by $e^{-r \sec \phi}/l_{qu}$ before integrating. This gives

$$j_{surf}(r) = j_{creation} \frac{1}{2\pi} \tan^{-1}\left(\frac{\sqrt{2}\omega_0}{r}\right) \int_0^{\pi/2} e^{-(r/l_{qu}) \sec \phi} \cos \phi \, d\phi, \quad (5.11)$$

where l_{qu} is an effective quenching or recombination length.

3.2. *Nucleation Effects*

As discussed in the above section, the linear growth regime leads to only moderately confined thin-film growth, unless quenching effects become significant. In most situations, however, linear growth must be preceded by nucleation, at least in the case of deposition. This introduces a surface nonlinearity into the overall process that can confine the growth even more.

To a first approximation, we can model this as a heterogeneous nucleation problem in which surface clusters of atoms must attain a certain critical size before they become stable. The details of this problem are complicated; however, for our purposes a qualitative treatment is adequate. The nucleation rate I_N of critical nuclei of N atoms can be modeled as

$$I_N \sim j_{surf}\left[\frac{j_{surf}}{\nu_{des}}\right]^N, \quad (5.12)$$

where j_{surf} is the surface flux in units of atoms/(surface-lattice-site) and ν_{des} is the desorption rate of atoms. When $j_{surf} \gg \nu_{des}$, then nucleation is rapid, whereas when $j_{surf} \ll \nu_{des}$, nucleation is slow. Therefore, if the on-axis j_{surf} is arranged to be only somewhat larger than ν_{des} then nucleation for $r > \omega_0$ can be suppressed, leading to sharply reduced linewidths.

In practice, however, the desorption frequency for radicals that may be fairly tightly bound to the surface can be rather low ($<1\ \text{sec}^{-1}$) so that the surface flux must also be low. This imposes severe limitations on the overall growth rate and can make diffusion-limited reactions incompatible with high-resolution processes.

For maximum process control, in fact, it has been found best to maintain j_{surf} below the critical flux even on-axis, relying on photochemistry in adsorbed molecular layers to create nuclei. Subsequent growth can then be due to either external-phase or adsorbed-phase photolysis. The kinetics of adsorbed-phase processes are discussed in the following section.

Another, less severe, limitation on the overall growth rate is imposed by gas-phase nucleation. At radical concentrations well below the onset of saturation effects, the uncontrolled agglomeration of gas-phase nuclei will occur. In the case where photochemical chain reactions may occur, the problem is compounded.

3.3. *Exact Numerical Solution*

Chen (1987) has presented a more rigorous kinetic theory model of laser photochemical reactions leading to deposition on a substrate. In particular, we will describe his treatment of deposition produced by gas-phase decomposition. Chen considers gas-phase photochemical deposition of a product species in two limiting cases: ballistic deposition, where the mean-free path of product molecules, $\bar{\lambda}$, is greater than the laser beam width w_0 ($\bar{\lambda} > w_0$); and diffusive deposition, ($\bar{\lambda} < w_0$).

For ballistic deposition the rate per unit volume $R(\vec{\mathbf{r}}')$ at which product molecules are produced in the gas phase at point $\vec{\mathbf{r}}'$ is given by

$$R(\vec{\mathbf{r}}') = \Phi(\vec{\mathbf{r}}')\mathrm{n}(\vec{\mathbf{r}}')\sigma_{\mathrm{d}}(\lambda), \tag{5.13}$$

where $\Phi(\vec{\mathbf{r}}')$ is the photon flux at point $\vec{\mathbf{r}}'$, $n(\vec{\mathbf{r}}')$ is the density of parent molecules, and $\sigma_d(\lambda)$ is the cross section for decomposition. Assuming an isotropic distribution of product molecules, and steady state, the deposition rate at point $(x, y, 0)$ on the substrate surface is

$$R(x, y) = \frac{\alpha_{\mathrm{p}} n \sigma_{\mathrm{d}}(\lambda)}{4\pi} \cdot \lim_{z \to 0} \left[\int_{z'>0} dx'\, dy'\, dz'\, \Phi(\vec{\mathbf{r}}') \cdot \frac{\cos\phi}{|\vec{\mathbf{r}}' - \vec{\mathbf{r}}|^2} \right], \tag{5.14}$$

where α_p is the surface sticking coefficient, ϕ is the angle between the vector $(\vec{\mathbf{r}}' - \vec{\mathbf{r}})$ and the normal to the surface, and n is assumed constant

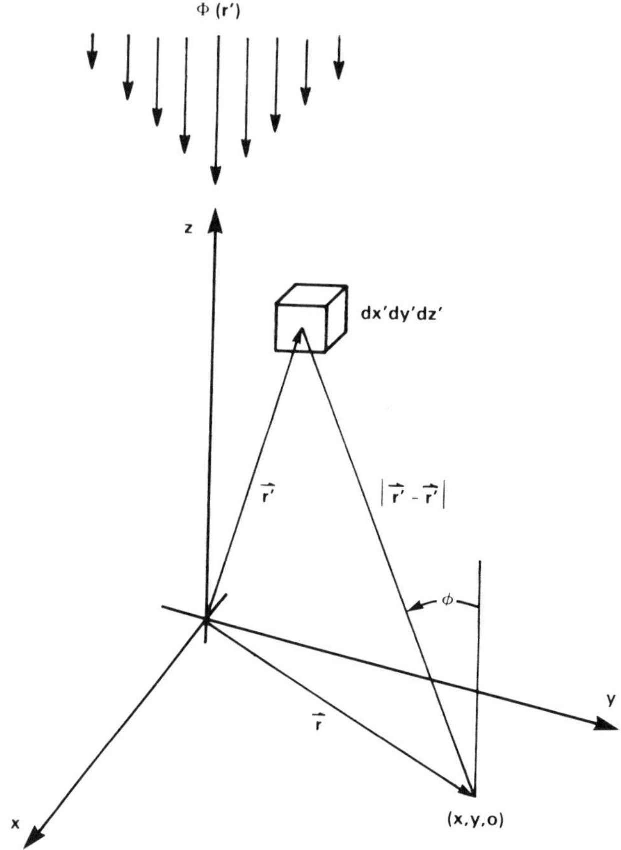

Fig. 5.3 Geometry for the calculation of deposition due to photochemical reaction in the gas phase. After Chen, 1987.

(see Fig. 5.3). Noting that

$$\frac{\cos\phi}{|\vec{\mathbf{r}}' - \vec{\mathbf{r}}|^2} = \frac{\partial}{\partial z}\frac{1}{|\vec{\mathbf{r}}' - \vec{\mathbf{r}}|} = -\frac{\partial}{\partial z'}\frac{1}{|\vec{\mathbf{r}}' - \vec{\mathbf{r}}|}, \tag{5.15}$$

and assuming $\Phi(\vec{\mathbf{r}}')$ is independent of z' near the surface, Eq. (5.14) can be rewritten

$$\begin{aligned} R(x, y) &= \alpha_{\mathrm{p}} \cdot \left[\frac{n\sigma_{\mathrm{d}}(\lambda)}{4\pi}\int_{z'=0}^{\infty} dx'\,dy'\,dz' \cdot \frac{\partial}{\partial z}\frac{\Phi(x', y')}{|\vec{\mathbf{r}}' - \vec{\mathbf{r}}|}\right]_{z=0} \\ &= \alpha_{\mathrm{p}} \cdot \left[\frac{-n\sigma_{\mathrm{d}}(\lambda)}{4\pi}\int_{z'=0}^{\infty} dx'\,dy'\,dz' \cdot \frac{\partial}{\partial z'}\frac{\Phi(x', y')}{|\vec{\mathbf{r}}' - \vec{\mathbf{r}}|}\right]_{z=0} \end{aligned}$$

or

$$R(x, y) = \alpha_p \cdot \frac{n\sigma_d(\lambda)}{4\pi} \int dx'\, dy' \cdot \frac{\Phi(x', y')}{|\vec{\mathbf{r}}' - \vec{\mathbf{r}}|}, \tag{5.16}$$

where $\vec{\mathbf{r}}$ and $\vec{\mathbf{r}}'$ are both now surface vectors. For a Gaussian beam of the form

$$\Phi(x, y) = \frac{2I}{hw_0^2} \exp\left\{ -\frac{2(x^2 + y^2)}{w_0^2} \right\}, \tag{5.17}$$

Chen has shown that the exact solution of Eq. (5.16) is fit well by the simple analytic expression,

$$R(r) = \frac{\alpha_p n I \sigma_d(\lambda)}{2\sqrt{2\pi}(w_0^2 + 2r^2)} \frac{2w_0^4 + r^4}{2w_0^4 + \sqrt{\pi}\, r^4}. \tag{5.18}$$

For diffusive deposition ($\bar{\lambda} < w_0$), the transport of product molecules to the surface by diffusion is governed by the diffusion equation,

$$\frac{\partial \rho}{\partial t} = -D\nabla^2 \rho + S(\vec{\mathbf{r}}, t), \tag{5.19}$$

where ρ is the concentration of product molecules in the gas phase, D is the diffusivity, and $S(\vec{\mathbf{r}}, t)$ is a source function. In the present case, the source function is of the form given by Eq. (5.13).

The boundary condition on ρ at the surface where deposition occurs is

$$D \frac{\partial \rho}{\partial z} = \tfrac{1}{4} \rho \bar{v} \alpha_p, \tag{5.20}$$

where $\bar{v}$ is the average velocity of product atoms.

Near the center of the beam Chen argues that

$$\frac{\partial \bar{\rho}}{\partial z} \sim \frac{\bar{\rho}}{w_0}, \tag{5.21}$$

where $\bar{\rho}$ is the average density. The elementary expression for the diffusion coefficient is

$$D = \tfrac{1}{3} \bar{v} \bar{\lambda}. \tag{5.22}$$

Combining Eqs. (5.20), (5.21), and (5.22) then gives the approximate value of ρ at the surface,

$$[\rho]_{z=0} = \frac{4\bar{\lambda}}{3\alpha_p w_0} \bar{\rho}. \tag{5.23}$$

For $\alpha_p \sim 1$, and $\bar{\lambda}/w_0 \ll 1$, then $[\rho]_{z=0} \ll \bar{\rho}$, and the boundary condition

Eq. (5.20) can be replaced by the approximate boundary condition on ρ,

$$[\rho]_{z=0} \sim 0. \tag{5.24}$$

With this boundary condition, the concentration of product molecules $\rho(\vec{\mathbf{r}})$ at any point $\vec{\mathbf{r}}$ in the gas phase above the surface, due to photodecomposition at the point $\vec{\mathbf{r}}'$ is

$$\rho(\vec{\mathbf{r}}) = \int_{z'>0} S(\vec{\mathbf{r}}')G(\vec{\mathbf{r}}, \vec{\mathbf{r}}')\, dr', \tag{5.25}$$

Where $S(\vec{\mathbf{r}})$ is given by Eq. (5.13), and $G(\vec{\mathbf{r}}, \vec{\mathbf{r}}')$ is the Green's function,

$$G(\vec{\mathbf{r}}, \vec{\mathbf{r}}') = \frac{1}{4\pi D} \cdot \frac{1}{[(x-x')^2 + (y-y')^2 + (z-z')^2]^{1/2}} - \frac{1}{[(x-x')^2 + (y-y')^2 + (z+z')^2]^{1/2}}. \tag{5.26}$$

Equation (5.26) is the Green's function for diffusion that satisfies the boundary condition, Eq. (5.24). The deposition rate on the surface is

$$\begin{aligned} R(x, y) &= \left[D \frac{\partial}{\partial z} \rho(\vec{\mathbf{r}}) \right]_{z=0} \\ &= Dn\sigma_{\mathrm{d}}(\lambda) \int_{z'=0}^{\infty} \left[\frac{\partial}{\partial z} G(\vec{\mathbf{r}} - \vec{\mathbf{r}}') \right]_{z=0} \Phi(x', y')\, dx'\, dy'\, dz' \\ &= \frac{n\sigma_{\mathrm{d}}(\lambda)}{4\pi} \int_{z'=0}^{\infty} \left[\frac{-\partial}{\partial z'} \left\{ \frac{1}{[(x-x')^2 + (y-y')^2 + (z-z')^2]^{1/2}} \right.\right. \\ &\quad \left.\left. + \frac{1}{[(x-x')^2 + (y-y')^2 + (z+z')^2]^{1/2}} \right\} \right]_{z=0} \\ &\quad \cdot \Phi(x', y')\, dx'\, dy'\, dz', \end{aligned} \tag{5.27}$$

or

$$R(x, y) = \frac{n\sigma_{\mathrm{d}}(\lambda)}{2\pi} \int \frac{\Phi(x', y')\, dx'\, dy'}{|\vec{\mathbf{r}} - \vec{\mathbf{r}}'|}, \tag{5.28}$$

where $\vec{\mathbf{r}}$ and $\vec{\mathbf{r}}'$ are both surface vectors. Comparing Eqs. (5.16) and (5.28), we see that the shape of the deposition profile is the same in the ballistic regime and in the diffusion-controlled regime so that Eq. (5.18) gives the shape of the deposition profile in both cases. Chen notes that when $\alpha_p = 1$, the overall deposition in the diffusion-controlled case is exactly 1/2 that in the ballistic regime.

4. Reactions Initiated by Surface-Phase or Substrate Absorption: Mass Transport by Gas-Phase Diffusion

4.1. Hemispheric Model

4.1.1 Simple Analytical Treatment

Tsao and Ehrlich (1984) have introduced a model to describe the effect of mass transport in the vapor phase on surface reactions induced by localized laser irradiation. The real problem is that of the chemistry excited by the laser on a plane surface, with mass transport of reactant species to the reaction zone and of products away from the reaction zone. The model problem consists of a small, hemispheric surface of radius ω_0 at which a reaction is taking place, protruding from an inactive plane surface of infinite extent. Tsao and Ehrlich consider three-dimensional diffusion of vapor-phase reactant molecules only, to the surface, which is a sink for a fraction of these molecules. It is assumed in the following discussion that the ambient vapor consists largely of an inert molecular species so that pressure equilibrium is readily maintained during the reaction. The diffusion of the reactant molecules is governed by the diffusion equation

$$D\,\nabla^2 n - \frac{\partial n}{\partial t} = 0. \tag{5.29}$$

The molecules diffuse to the surface subject to the boundary conditions

$$D\left(\frac{\partial n}{\partial r}\right)_{r=\omega_0} = \left(\tfrac{1}{2}\eta\bar{v}n\right)_{r=\omega_0}, \tag{5.30}$$

$$n(r, t=0) = n_\infty,$$

$$n(r=\infty, t) = n_\infty, \qquad \text{and}$$

$$\left(\frac{\partial n}{\partial z}\right)_{\substack{z=0\\ x^2+y^2>\omega_0^2}} = 0. \tag{5.30a}$$

Here, $(\frac{1}{2})$ $(\eta\bar{v}n)_{r=\omega_0}$ is a surface reaction current density, where $\bar{v}$ is the rms velocity of reactant molecules and η is the temperature-dependent fraction of surface collisions that result in a reaction. The hemispheric surface represents a reaction zone area in which a focused laser spot excites a photolytic or a thermally activated surface reaction. The surface boundary condition, Eq. (5.30), states that the molecular diffusion current density into the surface is equal to the current density of molecules that are removed from the vapor by the surface reaction.

The solution to this problem is equivalent to that for a spherical surface of radius ω_0 with the same boundary conditions (the last satisfied automatically for the solution, which has spherical symmetry). The result is an exactly soluble problem analogous to a heat flow problem (Carslaw and Jaeger, 1959, p. 248) with solution

$$\frac{n(r,t)}{n_\infty} = 1 - \frac{\omega_0^2/r}{r_0+\omega_0}\left\{\operatorname{erfc}\left(\frac{r-\omega_0}{2\sqrt{Dt}}\right) - \exp\left[\left(\frac{1}{\omega_0}+\frac{1}{r_0}\right)(r-\omega_0) + \left(\frac{1}{\omega_0}+\frac{1}{r_0}\right)^2 Dt\right] \cdot \operatorname{erfc}\left[\left(\frac{r-\omega_0}{2\sqrt{Dt}}\right) + \left(\frac{1}{\omega_0}+\frac{1}{r_0}\right)\sqrt{Dt}\right]\right\}, \tag{5.31}$$

Where $r_0 \equiv 2D/\eta\bar{v}$ is a length scale on the order of 3 mm torr for typical conditions. Evaluated at the surface $r=\omega_0$, this gives

$$j_{\text{surf}}(t) = \frac{Dn_\infty}{r_0+\omega_0} \cdot \left\{1 + \frac{\omega_0}{r_0} e^{(1/r_0+1/\omega_0)^2 Dt} \operatorname{erfc}\left[\left(\frac{1}{r_0}+\frac{1}{\omega_0}\right)\sqrt{Dt}\right]\right\}, \tag{5.32}$$

which is plotted as a function of time for various spot sizes in Fig. 5.4. We note that under most conditions, and particularly for smaller reaction zones, steady-state conditions are reached very rapidly. In addition, the time it takes to approach halfway the steady-state reaction rate is shorter for smaller hemispheres.

In the steady state the surface concentration is given by $n(d,\infty)/n_\infty = r_0/(r_0+\omega_0)$, and the surface reaction rate is

$$j_{\text{surf}}(t\to\infty) = \frac{Dn_\infty}{(r_0+\omega_0)}. \tag{5.33}$$

There are two kinetic regimes. If $r_0 \gg \omega_0$, the reaction rate $j_{\text{surf}}(t\to\infty)$ approaches $Dn_\infty/r_0 = (\frac{1}{2})\eta\bar{v}n_\infty$, is limited by interfacial kinetics, is nearly proportional to pressure, and is independent of the hemisphere radius ω_0. If $r_0 \ll \omega_0$, then the reaction rate $j_{\text{surf}}(t\to\infty)$ approaches Dn_∞/ω_0, is limited by diffusion kinetics, is independent of pressure (since $D \sim P^{-1}$), and is inversely proportional to ω_0. The pressure dependence of the reaction rate is plotted in Fig. 5.5 for various hemisphere radii. The enhancement at smaller hemispherical radii is due to the higher pressure of reactant molecule that can be sustained before the reaction becomes diffusion limited.

Since most uniform planar reactions are diffusion limited, these same

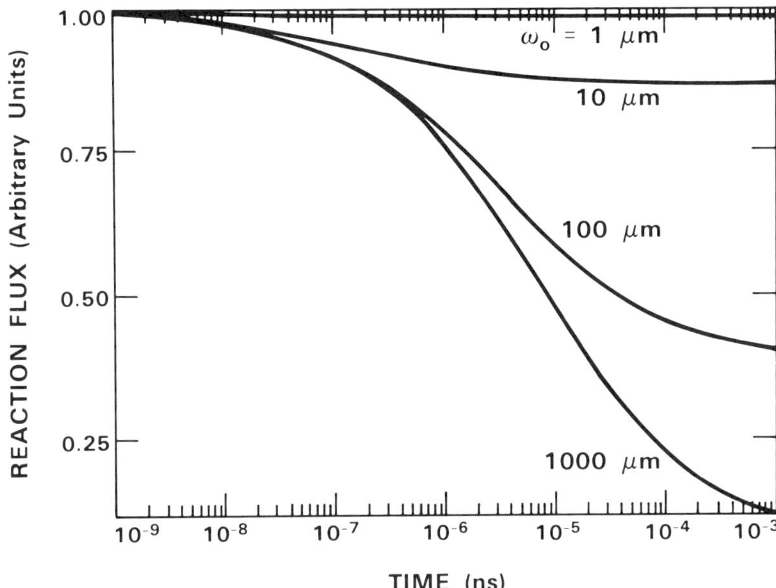

Fig. 5.4 Temporal evolution of the molecular reaction flux into or out of a laser microreaction zone of radius ω_0. The plots are for a pressure of 100 torr, a temperature of 1000°C, a diffusion coefficient D of 2.8 $\mathrm{cm^2\,s^{-1}}$, and a reaction efficiency η of 0.1. After Ehrlich and Tsao, 1983.

reactions, confined to small localized regions, have much faster reaction rates. In the high-pressure limit, the ratio of the reaction rate on the localized hemisphere to that on a uniformly reactive planar surface approaches infinity. In practice, of course, reactions on planar surfaces are always carried out in flowing reactors, in which diffusive transport need only occur across a thin boundary layer of width $\delta = 5(vx/u)^{1/2}$ where v is the kinematic viscosity, x is a characteristic size of the surface, and u is the velocity of the flow far from the surface (Rosenberger, 1979). The small-zone enhancement then becomes δ/ω_0. For typical values (in an atmospheric pressure reactor) of $\delta \sim 3$ mm and $\omega_0 \sim 3\ \mu$m, the enhancement is three orders of magnitude.

4.1.2 Generalization for Forward and Reverse Reactions

The hemispheric model yields additional important insights into the kinetics of chemical reactions involving localized laser irradiation. To explore the effect of reverse chemical reactions on the net reaction rates

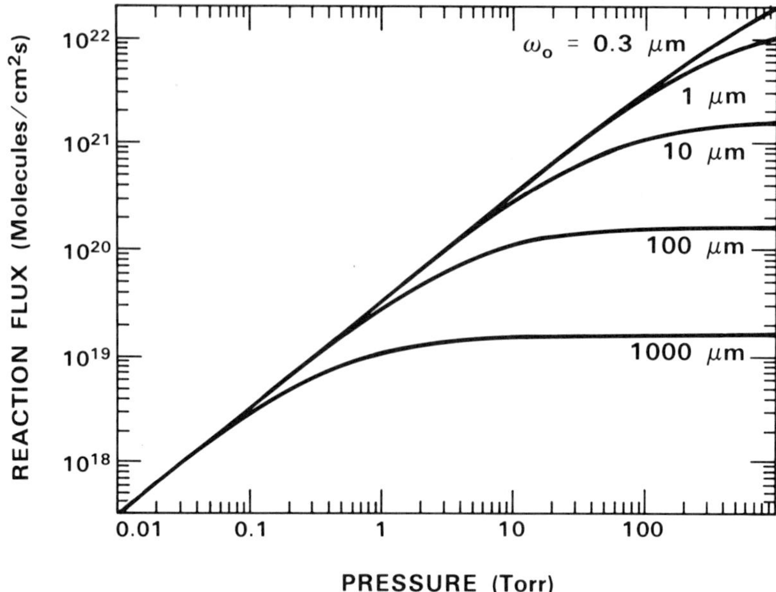

Fig. 5.5 Steady-state values for the molecular reaction flux as a function of pressure for various values of the spot radius ω_0. Curves are plotted for a diffusivity $D = a/P$, $a = 2 \times 10^2$ torr cm^2 s^{-1}, and length scale factor $l_0 = \beta/P$, $\beta = 5 \times 10^2$ torr cm, under the conditions $T = 1000°$C and $\eta = 0.1$. These values are typical of thermally activated laser etching. After Ehrlich and Tsao, 1983.

at surfaces, we consider as an example a reaction of the type

$$A + B \underset{k_C}{\overset{k_B}{\rightleftarrows}} \rho C, \tag{5.34}$$

where A is a surface species undergoing deposition or etching, B is a reactant vapor-phase species of diffusivity D_B reacting on the surface, and C is a product vapor-phase species of diffusivity D_C. When driven to the right, Eq. (5.34) is an etching reaction. We will include in our treatment both the kinetics of the reacting species and the back reaction of the product species.

The hemispheric model is again replaced by an equivalent spherical model. To emphasize the nature of the surface chemistry, we use slightly different notation from that used in Tsao and Ehrlich (1984). The sphere, of radius d, provides a surface at which the forward and reverse reactions of Eq. (5.34) are occurring. The diffusion of species B and C are governed, respectively, by the diffusion equations

$$D_B \nabla^2 n_B - \frac{\partial n_B}{\partial t} = 0$$

$$D_C \nabla^2 n_C - \frac{\partial n_C}{\partial t} = 0. \tag{5.35}$$

The boundary condition, Eq. 5.30, for the case of a single reacting species when the laser is switched on is replaced by the boundary conditions

$$\begin{aligned} D_B\left(\frac{\partial n_B}{\partial r}\right)_{r=d} &= \left(k_B n_B - \frac{1}{\rho} k_C n_C\right)_{r=d} \\ D_C\left(\frac{\partial n_C}{\partial r}\right)_{r=d} &= -\rho\left(k_B n_B - \frac{1}{\rho} k_C n_C\right)_{r=d}, \end{aligned} \tag{5.36}$$

where $k_B n_B$ is the rate per unit area at which molecules of species B are eliminated at the surface, and $k_C n_C$ is the rate per unit area at which molecules of species C are eliminated at the surface. We have assumed, in writing Eq. (5.36), that for every molecule of species B eliminated, ρ molecules of species C are generated, and vice versa, as indicated by Eq. (5.34). We have also assumed in Eq. (5.36) that when the laser is switched on, the reaction of Eq. (5.34) is driven to the right.

The equations (5.35) with boundary conditions (5.36) are difficult to solve unless $D_B \simeq D_C \simeq D$. For this case, the coupled Eqs. (5.36) can be decoupled by introducing new time-dependent variables n_1 and n_2, which are linear combinations of n_B and n_C:

$$\begin{aligned} n_1 &= \left(n_B + \frac{1}{\rho} n_C\right) \\ n_2 &= \left(n_B - \frac{1}{\rho}\frac{k_C}{k_B} n_C\right). \end{aligned} \tag{5.37}$$

In terms of these new variables, with $D_B \simeq D_C \simeq D$, Eqs. 5.35 and 5.36 become

$$\begin{cases} D\nabla^2 n_1 - \dfrac{\partial n_1}{\partial t} = 0 & \text{(5.38a)} \\ D\left(\dfrac{\partial n_1}{\partial r}\right)_{r=d} = 0 & \text{(5.38b)} \end{cases}$$

$$\begin{cases} D\nabla^2 n_2 - \dfrac{\partial n_2}{\partial t} = 0 & (5.39a) \\ D\left(\dfrac{\partial n_2}{\partial r}\right)_{r=d} = (k_2 n_2)_{r=d}, & (5.39b) \end{cases}$$

where $k_2 \equiv (k_B + k_C)$.

Now, both Eqs. 5.38 and 5.39 conform to the conditions of Section 4.1.1, and the solutions obtained for n_1 and n_2, analogous to Eq. 5.31, are

$$\frac{n_1(r, t)}{n_1(0)} = 1, \tag{5.40}$$

$$\frac{n_2(r, t)}{n_2(0)} = 1 - \frac{d^2/r}{r_2 + d}\left\{\operatorname{erfc}\left(\frac{r-d}{2\sqrt{Dt}}\right) - \exp\left[\left(\frac{1}{d}+\frac{1}{r_2}\right)(r-d) + \left(\frac{1}{d}+\frac{1}{r_2}\right)^2 Dt\right]\right.$$
$$\left.\cdot \operatorname{erfc}\left[\left(\frac{r-d}{2\sqrt{Dt}}\right) + \left(\frac{1}{d}+\frac{1}{r_2}\right)\sqrt{Dt}\right]\right\}, \tag{5.41}$$

where $r_2 \equiv D/k_2 = D/(k_B - k_C)$, and $n_1(0)$ and $n_2(0)$ are the initial, uniform values of n_1 and n_2 when the reaction is started by switching the laser on at $t = 0$. Equation 5.40 follows because the value of $r_1 \equiv (D/k_1) \to \infty$, where k_1, defined as the factor of n_1 on the right-hand side of Eq. 5.38, is zero.

From Eq. 5.41, $n_2(r, t)$ evaluated at the surface is

$$\frac{n_2(d, t)}{n_2(0)} = \frac{r_2}{r_2 + d}\left[1 + \frac{d}{r_2}\exp\left\{\left(\frac{1}{d}+\frac{1}{r_2}\right)^2 Dt\right\}\right.$$
$$\left.\cdot \operatorname{erfc}\left\{\left(\frac{1}{d}+\frac{1}{r_2}\right)\sqrt{Dt}\right\}\right], \tag{5.42}$$

and the limiting value of $n_2(d, t)$ as $t \to \infty$ is

$$\frac{n_{2\,\mathrm{lim}}}{n_2(0)} = \frac{r_2}{r_2 + d}. \tag{5.43}$$

Combining Eq. 5.42 with the limiting value of n_1, $n_{1\,\mathrm{lim}}/n_1(0) = 1$, and using Eq. 5.37 to solve for $n_{B\,\mathrm{lim}}$ and $n_{c\,\mathrm{lim}}$, we obtain

$$\begin{aligned} (n_{B\,\mathrm{lim}} - n_B(0)) &= -\left(\frac{d}{r_2 + d}\right)\frac{k_B}{k_B + k_c}\left[n_B(0) - \frac{1}{\rho}\frac{k_C}{k_B} n_C(0)\right] \\ (n_{C\,\mathrm{lim}} - n_C(0)) &= \rho\left(\frac{d}{r_2 + d}\right)\frac{k_B}{k_B + k_C}\left[n_B(0) - \frac{1}{\rho}\frac{k_C}{k_B} n_C(0)\right], \end{aligned} \tag{5.44}$$

where $r_2 = D/k_2 = D/(k_B + k_C)$. The net limiting value of the surface reaction rate of species B, taking into account the effect of the back reaction, is

$$
\begin{aligned}
j_{\text{surf}}^{\text{lim}} &= \left[k_B n_{B\,\text{lim}} - \frac{1}{\rho} k_C n_{C\,\text{lim}}\right] \\
&= \left(\frac{r_2}{r_2 + d}\right)\left[k_B n_B(0) - \frac{1}{\rho} k_C n_C(0)\right], \\
&= \left(\frac{D n_B(0)}{r_2 + d}\right)\left[\frac{k_B}{k_2} - \frac{1}{\rho}\frac{k_C}{k_2}\frac{n_C(0)}{n_B(0)}\right],
\end{aligned}
\tag{5.45}
$$

where $r_2 = D/k_2$ and $k_2 = k_B + k_C$.

This is the generalization of Eq. 5.33 taking into account the effect of the back reaction. For $k_B \gg k_C$, the result reduces to the equivalent of Eq. 5.33.

For $k_C \gg k_B$, and $n_C(0)$ negligibly small, the result is equivalent to Eq. 5.33 multiplied by the reduction factor k_B/k_C, with $r_2 = D/k_C$. In this case, for $d \gg r_2$, the reaction rate is limited by diffusion and is inversely proportional to k_C. This indicates that in this limit it is the back reaction due to the pileup of product species C that depresses the net forward reaction rate.

4.2. *Detailed Model*

4.2.1 Green's Function Equation

In Section 4.1 a simple hemispheric model of the effect of mass transport in a laser-driven localized surface reaction was discussed. In this section, we consider a more detailed model of a thermally driven localized surface reaction occurring on a plane surface (Zeiger and Ehrlich, 1989). The treatment makes use of a three-dimensional Green's function defined in the half-space above the surface to describe vapor-phase diffusion. It is assumed that the major constituent of the vapor above the surface is initially the reactant species B. To explore the possible effects of the back reaction, and yet minimize the complexity of the calculation, it is further assumed that even at equilibrium on the laser-heated surface the concentration n_C of the product species C in the vapor remains small compared to that of the reactant species. Since pressure equilibrium is maintained during the reaction, if the temperature variation of the gas is neglected, the concentration of the reactant species n_B can then be taken as roughly constant. This leads to a single Green's function equation for the concentration of product species. The calculation shows in more detail some of the effects of product pileup and back reaction discussed already in Section 4.1.2.

The reaction treated is an etching-deposition reaction given by

$$A + B \underset{k_2}{\overset{k_1}{\rightleftharpoons}} 2C. \tag{5.46}$$

In this model, Zeiger and Ehrlich assume the back-reaction rate is second order in concentration of species C. the rates at which molecules of species B and species C are eliminated by the surface reaction Eq. 5.46 per unit area are taken respectively as

$$\begin{aligned} R_B &= -\frac{dN_B}{dA\,dt} = k_1 n_B \\ R_C &= -\frac{dN_C}{dA\,dt} = k_2 n_C^2. \end{aligned} \tag{5.47}$$

However, for a laser-heated surface, k_1 and k_2, which are temperature dependent, will be functions of position on the surface. The equation governing the vapor-phase diffusion of species C is

$$D\nabla^2 n_C(\vec{\mathbf{r}}, t) - \frac{\partial n_C(\vec{\mathbf{r}}, t)}{\partial t} = 0, \tag{5.48}$$

subject to the boundary condition at the surface,

$$\frac{\partial n_C(\vec{\boldsymbol{\beta}}, t)}{\partial t} = [k_1(\vec{\boldsymbol{\beta}}) n_B - \tfrac{1}{2} k_2(\vec{\boldsymbol{\beta}}) n_C(\vec{\boldsymbol{\beta}}, t)], \tag{5.49}$$

where $\vec{\mathbf{r}}$ labels a point in the vapor above the surface and $\vec{\boldsymbol{\beta}}$ is a surface coordinate measured from the center of the focused laser spot. The laser is switched on at $t = 0$, when the initial surface temperature is T_0. After the laser is switched on, the surface temperature distribution has the form, $T = T_0 + T_L e^{-\beta^2/d^2}$. The temperature dependences of k_1 and k_2 are taken to be of the Arrhenius form,

$$\begin{aligned} k_1 &= A_1 e^{-E_1/kT} \\ k_2 &= A_2 e^{-E_2/kT}. \end{aligned} \tag{5.50}$$

The Green's function that satisfies Eq. (5.48) and describes the evolution of a source of unit strength emitted at position $\vec{\mathbf{r}}'$ at time t' to any point $\vec{\mathbf{r}}$ at a later time t is

$$G(\vec{\mathbf{r}} - \vec{\mathbf{r}}'; t - t') = \frac{2}{[4\pi D(t - t')]^{3/2}} e^{-(\vec{\mathbf{r}} - \vec{\mathbf{r}}')^2/4D(t - t')}. \tag{5.51}$$

The Green's function is defined so that its integral over $\vec{\mathbf{r}}$ in the half-space above the surface is equal to one. It satisfies the boundary conditions that

$$\left(\frac{\partial G}{\partial z}\right)_{\vec{\mathbf{r}}'=\vec{\mathbf{r}}} = 0,$$

so that molecules introduced or removed by this source at the surface contribute only to lateral diffusion currents at the surface.

Making use of this Green's function, the equation describing the concentration of species C at points $\vec{\boldsymbol{\beta}}$ just above the surface is

$$n_C(\vec{\boldsymbol{\beta}}, t) = n_C(0) + 2\int_0^t \int G(\vec{\boldsymbol{\beta}} - \vec{\boldsymbol{\beta}}'; t - t') \cdot [k_1(\vec{\boldsymbol{\beta}}')n_B - \tfrac{1}{2}k_2(\vec{\boldsymbol{\beta}}')n_C^2(\vec{\boldsymbol{\beta}}', t')]\, d\vec{\boldsymbol{\beta}}'\, dt', \tag{5.52}$$

where $n_C(0)$ is the initial value of n_C, assumed to be in equilibrium at $t = 0$. Once $n_C(\vec{\boldsymbol{\beta}}, t)$ is known over a range of values of t, the total number of molecules of species A per unit area at point $\vec{\boldsymbol{\beta}}$ and time t, $\mathcal{N}_A(\vec{\boldsymbol{\beta}}, t)$, can be found from the relation

$$\mathcal{N}_A(\vec{\boldsymbol{\beta}}, t) = -\int_0^t [k_1(\vec{\boldsymbol{\beta}})n_B - \tfrac{1}{2}k_2(\vec{\boldsymbol{\beta}})n_C^2(\vec{\boldsymbol{\beta}}, t')]\, dt'. \tag{5.53}$$

It is convenient to introduce normalized variables $\vec{\boldsymbol{\rho}} \equiv \vec{\boldsymbol{\beta}}/\beta_0$, and $\tau \equiv (4D/\beta_0^2)t$, where β_0 is a distance of the order of the radius of the focused laser spot and will be defined more precisely in a moment. In terms of these variables, Eq. 5.52 becomes

$$n_C(\vec{\boldsymbol{\rho}}, \tau) = n_C(0) + \frac{\beta_0}{\pi^{3/2}D}\int_0^\tau \int \frac{e^{-(\vec{\boldsymbol{\rho}}-\vec{\boldsymbol{\rho}}')^2/(\tau-\tau')}}{(\tau - \tau')^{3/2}} \cdot [k_1(\vec{\boldsymbol{\rho}}')n_B - \tfrac{1}{2}k_2(\vec{\boldsymbol{\rho}}')n_C^2(\vec{\boldsymbol{\rho}}', \tau')]\, d\vec{\boldsymbol{\rho}}'\, d\tau'. \tag{5.54}$$

If the focused laser spot has cylindrical symmetry, then the angular integral in Eq. 5.54 can be performed (Carslaw and Jaeger, 1959), yielding

$$n_C(\rho, \tau) = n_C(0) + \frac{2\beta_0}{\pi^{1/2}D}\int_0^\tau \int_0^\infty \frac{e^{-(\rho^2+\rho'^2)/(\tau-t')}}{(\tau - \tau')^{3/2}} I_0\left(\frac{2\rho\rho'}{\tau - \tau'}\right) \cdot [k_1(\rho')n_B - \tfrac{1}{2}k_2(\rho')n_C^2(\rho', \tau')]\rho'\, d\rho'\, d\tau', \tag{5.55}$$

where $I_0(x)$ is a modified Bessel function. Introducing additional normalized parameters, Eq. (5.55) becomes

$$\nu_C(\rho, \tau) = 1 + w\int_0^\tau \int_0^\infty \frac{e^{-(\rho^2+\rho'^2)/(\tau-\tau')}}{(\tau - \tau')^{3/2}} I_0\left(\frac{2\rho\rho'}{\tau - \tau'}\right) e^{-\varepsilon\xi[f(\rho')-f(0)]} \cdot [1 - \exp\{-\varepsilon(\xi - 1)[1 - f(\rho')]\}\nu_C^2(\rho', \tau')]\rho'\, d\rho'\, d\tau', \tag{5.56}$$

where

$$\nu_C(\rho, t) \equiv \frac{n_C(\rho, \tau)}{n_C(0)},$$

$$\varepsilon \equiv \frac{E_2}{kT_0}, \qquad \varepsilon\xi \equiv \frac{E_1}{kT_0}$$

$$f(\rho) \equiv [1 + Ue^{-\sigma\rho^2}]^{-1}, \qquad U = \frac{T_L}{T_0},$$

$$w = \frac{2\beta_0}{\pi^{1/2} D n_C(0)} R_1(T_0 + T_L),$$

$$R_1(T_0 + T_L) \equiv n_B A_1 e^{-\varepsilon\xi f(0)}. \tag{5.57}$$

In Eq. 5.56, ρ is a normalized surface radial distance and τ is a normalized time. $\nu_C(\rho, \tau)$ is a normalized vapor-phase concentration just above the surface at (ρ, τ), and ε and $\varepsilon\xi$ are normalized Arrhenius activation energies. $Ue^{-\sigma\rho^2}$ is the normalized surface-temperature-profile contribution due to the focused laser spot, with Gaussian width $d = \beta_0\sigma^{1/2}$. The parameter σ is a numerical factor introduced for convenience in numerical computation. None of the real physical quantities, in the end, depend upon the value chosen for σ. Finally, $R_1(T_0 + T_L)$ is the reaction rate per unit area at which molecules of species B are removed at temperature $(T_0 + T_L)$. The back reaction rate does not appear in Eq. 5.56, since it has been eliminated by using the initial condition that at $T = T_0$, $[k_1(T_0)n_B - (\frac{1}{2})k_2(T_0)n_C^2(0)] = 0$.

Equation 5.56 is in general analytically intractable. It must be solved by numerical integration in ρ' and τ', with $(w, \varepsilon, \xi, U, \sigma)$ chosen for particular cases. Once $\nu_C(\rho, \tau)$ has been obtained for a range of values of τ, the number of molecules of species A per unit area at normalized radius ρ and normalized time τ, $\mathcal{N}_A(\rho, t)$, can be obtained from the normalized equivalent of Eq. 5.53. The rate of deposition of species A, (the negative of the etching rate), at the center of rhe focused laser spot is then found to be

$$R(\tau) = -R_1(T_0 + T_L)R'(\tau),$$

where

$$R'(\tau) = [1 - \exp\{-\varepsilon(\xi - 1)[1 - f(0)\}\nu_C^2(0, \tau)]. \tag{5.58}$$

$R'(\tau)$ will be referred to as the reduced reaction rate. It includes the effect of the back reaction in reducing the net forward reaction rate.

4.2.2 Limiting Reaction Rates

Figure 5.6 shows the results of a numerical calculation of $|\mathcal{N}_A(0, \tau)|$ versus τ obtained (Zeiger and Ehrlich, 1989) for the set of parameters ($w = 1$, $\varepsilon = 30$, $\xi = 1.2$, $U = 3$, $\sigma = 4$) and is typical of the results obtained generally. After a brief, steep rise, $|\mathcal{N}_A(0, \tau)|$ increases with a constant slope, indicative of the onset of a limiting steady-state behavior in $R(\tau)$. Figure 5.7 shows the behavior of $\nu_C(\rho, \tau)$, and Fig. 5.8 the shape of $\mathcal{N}_A(\rho, \tau)$ plotted versus ρ for the same set of parameters. The Gaussian half-width of the laser-generated temperature profile is ~0.42, and it can be seen that the width of ν_C and $\mathcal{N}_A$ are considerably smaller, because of the large activation energies for the forward and reverse reactions compared to kT_0. This leads to a reaction zone that is smaller than the laser-heated temperature zone.

Figure 5.9 is a plot of the reduced reaction rate (at $\rho = 0$), $R'(\tau)$ versus τ for a number of different combinations of parameters. These plots indicate that the approach to a limiting value is quite general, as would be expected on physical grounds.

Figure 5.10 shows a plot of the limiting values of reduced reaction rate,

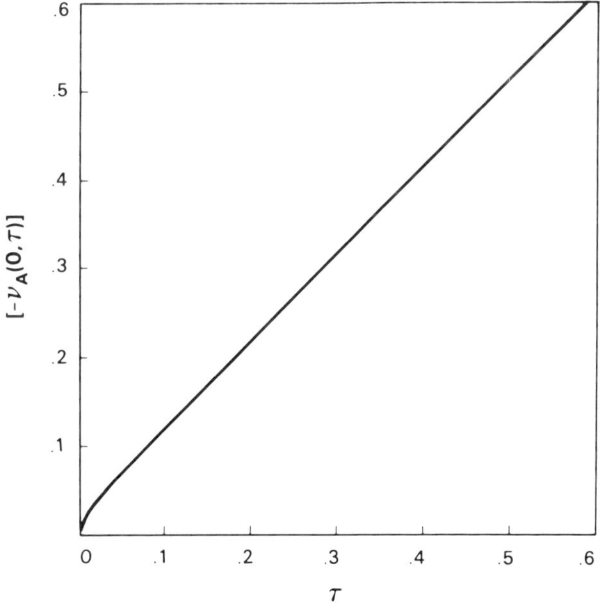

Fig. 5.6 $[-\nu_A(0, \tau)]$ versus τ for the parameters ($w = 1$, $\varepsilon = 30$, $\xi = 1.2$, $U = 3$, $\sigma = 4$). After Zeiger and Ehrlich, 1989.

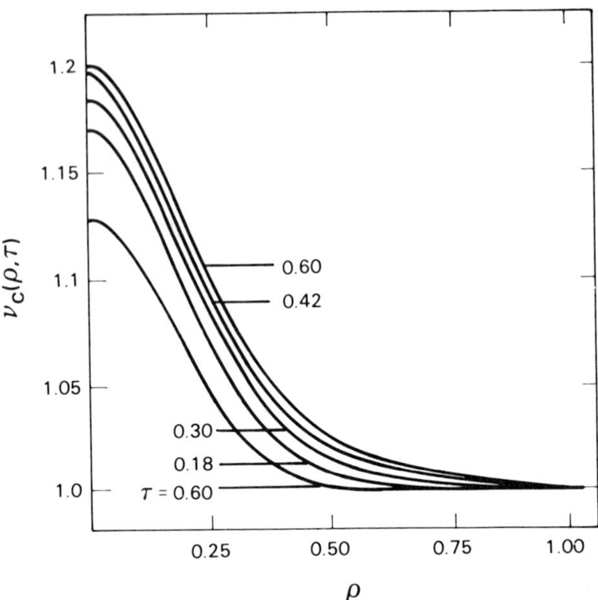

Fig. 5.7 Plots of normalized concentration of product species C, $\nu_C(\rho, \tau)$ versus normalized radial distance ρ for several values of normalized time τ for the parameters of Fig. 5.6 After Zeiger and Ehrlich, 1989.

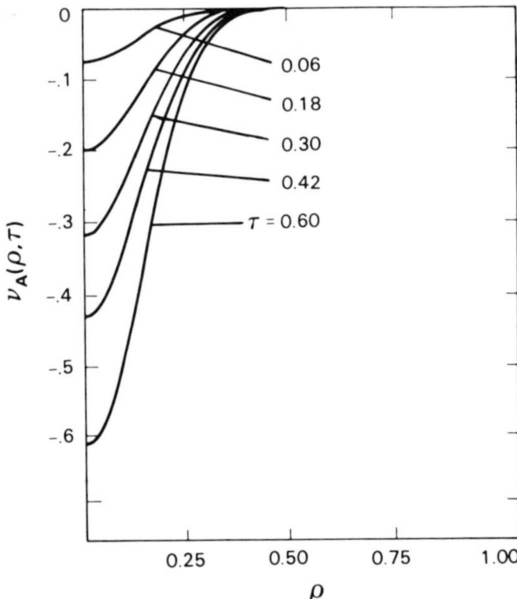

Fig. 5.8 Plots of normalized surface concentration of species A, $\nu_A(\rho, \tau)$ versus normalized distance ρ for several values of normalized time τ for the parameters of Fig. 5.6. After Zeiger and Ehrlich, 1989.

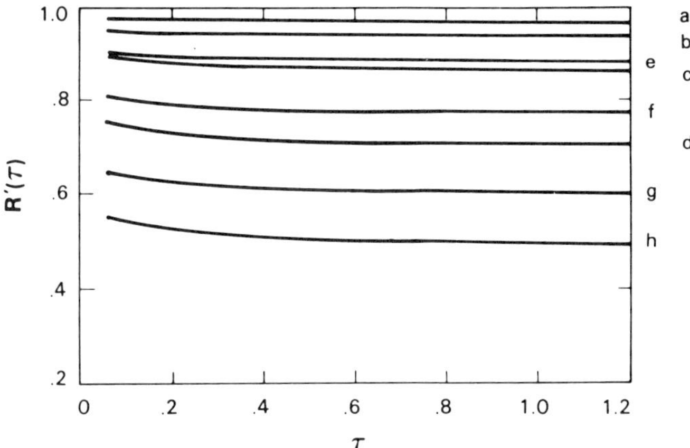

Fig. 5.9 Plots of reduced reaction rate $R'(\tau)$ versus τ. In all cases $U = 3$, $\sigma = 16$. Curves a, b, c, and d are for the cases $w = 4$, 10, 20, 40, respectively, with $\varepsilon = 30$, $\varepsilon\xi = 36$. Curves e, f, g, and h are for the cases $w = 4$, 10, 20, 40, respectively, with $\varepsilon = 32$, $\varepsilon\xi = 36$. After Zeiger and Ehrlich, 1989.

R'_{lim} versus

$$\beta_0^* \equiv w = \frac{2\beta_0}{\pi^{1/2} D n_C(0)} R_1(T_0 + T_L)$$

for a number of sets of parameters. β_0^* represents a normalized laser spot radius so that Fig. 5.10 gives the dependence of R'_{lim} on the spot size. The value of R'_{lim} continues to decrease indefinitely as spot size increases. This

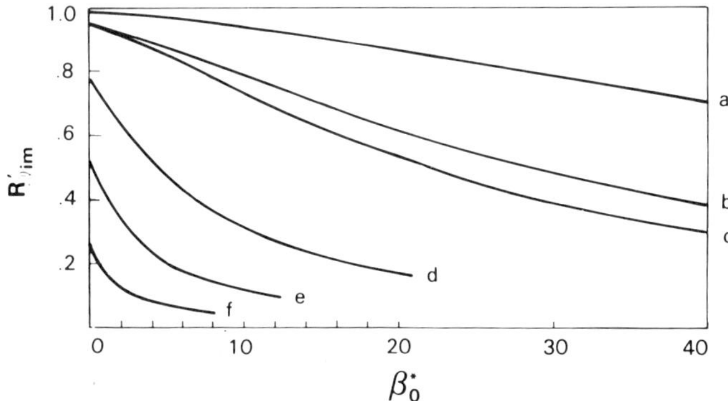

Fig. 5.10 Plots of limiting reduced reaction rate R'_{lim} versus normalized laser spot radius β_0^*. In all cases, $U = 3$, $\sigma = 16$. Curves a and b are for the cases ($\varepsilon = 30$, $\xi = 1.2$) and ($\varepsilon = 32$, $\xi = 1.125$), respectively, both with $\varepsilon\xi = 36$. Curves c, d, e, and f are for the cases $\varepsilon = 20$, 10, 5, and 2, respectively, all with $\xi = 1.2$. After Zeiger and Ehrlich, 1989.

is indicative of the effect of a pileup of reaction product because of limitations of diffusion, causing an increase in back reaction. This pile-up for a uniformly heated surface, ($\beta_0^* \to \infty$), can in principle eventually choke off the forward reaction entirely. Of course, this conclusion must be treated with caution, since the flow of gases has been neglected in the present calculation, and the time of duration of an experiment (or the dwell time) can also cut off the effect of product pile-up.

Another way of presenting the same data shown in Fig. 5.10 is to plot R'_{lim} versus

$$D^* \equiv \frac{1}{w} = \frac{\pi^{1/2} n_C(0) D}{2\beta_0 R_1(T_0 + T_L)},$$

where D^* represents a normalized diffusivity. Plots of R'_{lim} versus D^* are shown in Fig. 5.11.

4.2.3 Contour Reversal and Competing Reactions

Thus far, the cases discussed have had large values of normalized activation energies $\varepsilon\xi$ and ε. Figure 5.12 is a plot of the shape of $n_A(\rho, \tau)$ versus ρ for the parameters ($w = 4$, $\varepsilon = 2$, $\xi = 1.2$, $U = 3$, $\sigma = 16$) for several times τ. For this case, with the relatively small activation energies $\varepsilon\xi = 2.4$ and $\varepsilon = 2$, etching of species A occurs in the center of the pattern, but a deposit of species A is found in the wings of the laser heating pattern. (in Zeiger and Ehrlich, 1989, this is referred to as *contour reversal.*) The etching occurs because heating of the surface near the laser spot center unbalances the reaction rate constants so that the reaction of Eq. 5.46 is shifted toward the right-hand side. On the other hand, the molecules of species C generated by the etching reaction

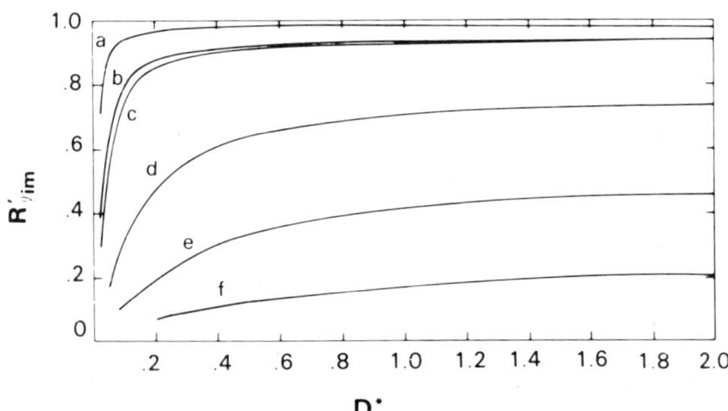

Fig. 5.11 Plots of limiting reduced reaction rate R'_{lim} versus normalized diffusivity D^* for the cases a–f of Fig. 5.10. After Zeiger and Ehrlich, 1989.

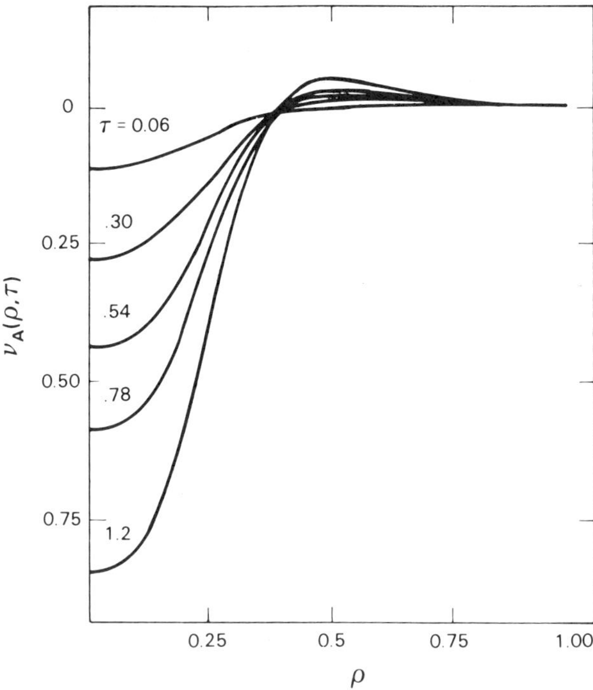

Fig. 5.12 Plots of normalized surface concentration of species A, $\nu_A(\rho, \tau)$ versus normalized radial distance ρ for several values of normalized time τ for the parameters ($w = 4$, $\varepsilon = 2$, $\xi = 1.2$, $U = 3$, $\sigma = 16$). After Zeiger and Ehrlich, 1989.

diffuse toward the cooler wings of the laser-heated region, producing an excess of species C over its equilibrium concentration, and shifting the net reaction rate toward the left, or deposition direction, in Eq. 5.46. Such a pattern of etching and deposition of Si has been observed when a Si surface is heated with a focused laser beam in a silane atmosphere.

Unfortrunately, a model such as the present one is inapplicable to many reactions, because the normalized activation energy $\varepsilon\xi = 2.4$ implies an activation energy of 0.06 eV, whereas activation energies for surface reactions can be 0.5 eV or greater. For both forward and reverse activation energies that large, reaction rates drop off so rapidly away from $\rho = 0$ that the excess of diffusing species C is unable to induce significant deposition of species A. If the reverse reaction has a much smaller activation energy, the prefactor in the reverse reaction must be very small in order to produce equilibrium at T_0, and the deposition rate in the wings would still be small.

Other examples of contour reversal have been observed, and a number of other possible explanations have been offered. Bäuerle (1986) has observed a similar effect during the laser pyrolysis of silane and has attributed it to melting of Si near the center of the laser spot, and the pulling away of the molten Si by temperature-dependent surface tension. Similar patterns have been observed by Allen (1981) in the deposition of Ni from nickel carbonyl and by Moylan et al. (1986) in the deposition of Cu. Skouby and Jensen (1988) have found such "volcanolike" contours in their highly detailed numerical modeling of laser deposition, taking into account temperature-dependent thermal properties of gas, substrate, and metal deposit, as well as time-dependent laser radiation absorption of the metal deposit.

In any case, an etching reaction of the type given by Eq. 5.46 alone cannot account for the contour reversal observed in the Si/silane reaction produced by a tightly focused laser beam. This is because, with less localized heating, pyrolysis of silane with the *deposition* of Si is the dominant reaction observed. However, it is possible to have competition between an etching reaction of the type given by Eq. 5.46 and a deposition reaction of the form

$$B \underset{k_2'}{\overset{k_1'}{\rightleftarrows}} A + 2E. \tag{5.59}$$

In both these equations, species B represents silane, and it is possible, at the center of the laser spot, for the etching reaction to dominate at tight focusing, while the deposition reaction is dominant for more uniform heating. A model calculation to exemplify this situation is given in Zeiger and Ehrlich, 1989, and the results of the calculation are summarized in Fig. 5.13. The curve labeled $r^{(e)}(\rho)$ is a canonical shape (essentially independent of w) for the limiting etching reaction rate as a function of normalized radial distance ρ from the laser-heated spot center, computed for parameters ($\varepsilon = 32$, $\xi = 1.125$, $U = 3$, $\sigma = 16$). $r^{(e)}(\rho)$ is normalized so that $r^{(e)}(0) = 1$. The curve labeled $r^{(d)}(\rho)$ is the canonical limiting deposition rate curve, also independent of w for the set of parameters ($\varepsilon = 20$, $\xi = 1.4$, $U = 3$, $\sigma = 16$), and also normalized so that $r^{(d)}(0) = 1$. The figure also shows plots of

$$r(\rho) \equiv r^{(d)}(\rho) - fr^{(e)}(\rho), \tag{5.60}$$

which depict the net shape of the competing deposition-etching curves for a number of values of the parameter f. Because the etching reaction has been taken to have a larger normalized activation energy $\varepsilon\xi$ than the deposition reaction, the etching reaction rate dies off more rapidly in the

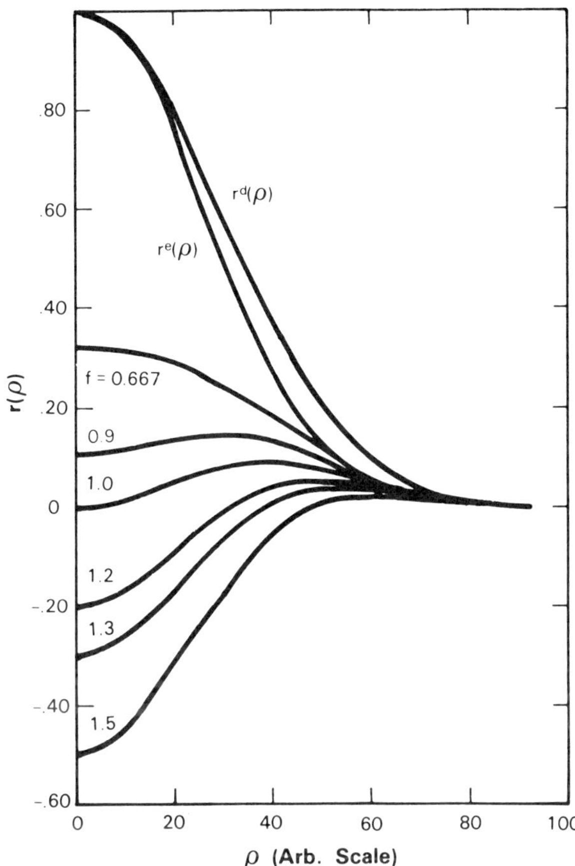

Fig. 5.13 Plots of the canonical deposition and etching curves $r^d(\rho)$ and $r^e(\rho)$ versus ρ for a model competing reaction. Also plotted is $r(\rho) \equiv r^d(\rho) - fr^e(\rho)$ versus ρ for several values of f. After Zeiger and Ehrlich, 1989.

wings than the deposition reaction rate. For f between ~1.2–1.5, contour reversal is seen to occur, much as is observed in Figs. 5.14 and 5.15 for the etching of Si in a silane atmosphere.

To study the laser spot-size dependence of the deposition-etching rate, a further extension of the model is discussed, on the simplifying assumption that for a given spot size, w given by Eqs. (5.57) is approximately the same for the etching reaction and the deposition reaction. Then, the limiting reduced reaction rates at laser spot center, $R'^{(e)}_{\text{lim}}$ and $R'^{(d)}_{\text{lim}}$, can be computed as functions of a common parameter

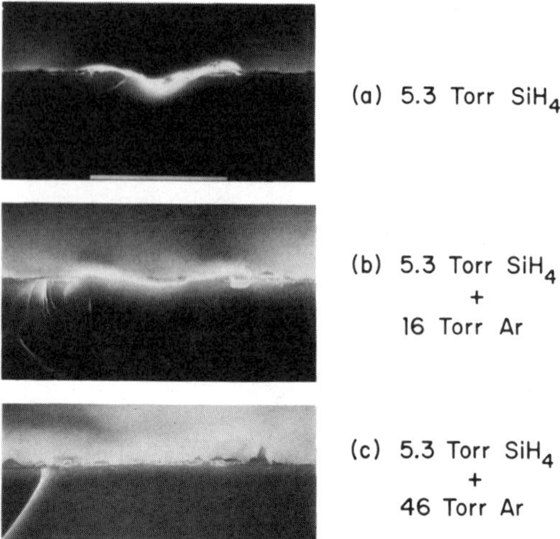

Fig. 5.14 SEM pictures of cleaved profiles of Si after treatment by a scanned, focused, Ar-ion laser beam in an SiH_4 atmosphere and in the presence of three different pressures of Ar buffer gas. After Zeiger and Ehrlich, 1989.

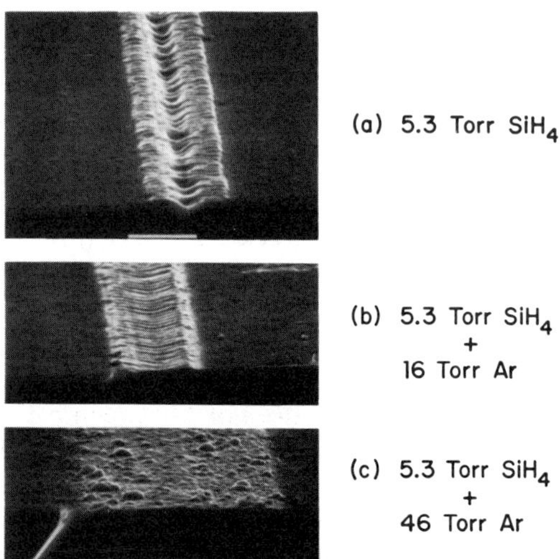

Fig. 5.15 SEM pictures of a Si surface after treatment by a scanned, focused, Ar-ion laser beam in an SiH_4 atmosphere, and in the presence of three different pressures of Ar buffer gas. After Zeiger and Ehrlich, 1989.

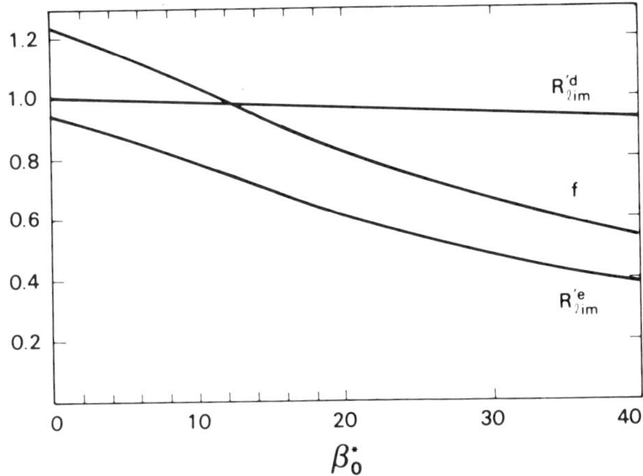

Fig. 5.16 Plots of limiting reduced reaction rates $R_{\text{lim}}^{\prime d}$, $R_{\text{lim}}^{\prime e}$; and f versus normalized laser spot radius β_0^* calculated for a model competing reaction. After Zeiger and Ehrlich, 1989.

$\beta_0^* \equiv w$. These are shown plotted as functions of β_0^* in Fig. 5.16. The overall reaction rate as a function of ρ is found to be

$$\begin{aligned} R(\rho) &= R^{(d)}(0)r^{(d)}(\rho) - R^{(e)}(0)r^{(e)}(\rho) \\ &= R^{(d)}(0)[r^{(d)}(\rho) - \{(R^{(e)}(0)/R^{(d)}(0))r^{e}(\rho)], \end{aligned} \tag{5.61}$$

where

$$R^{(d)}(0) = R_1(T_0 + T_{\text{L}})R_{\text{lim}}^{\prime(d)},$$
$$R^{(e)}(0) = R_1(T_0 + T_{\text{L}})R_{\text{lim}}^{\prime(e)}.$$

Comparing Eq. (5.61) with the definition of $r(\rho)$, the factor f is identified as

$$f = \left\{\left(\frac{R^{(e)}(0)}{R^{(d)}(0)}\right)\right\}. \tag{5.62}$$

For the case at hand, f was taken to be 1.2 at $\beta_0^* = 2$. Figure 5.16 shows the resulting plot of f versus β^*. For the present example, f is 1.23 at $\beta_0^* = 0$ and decreases to $f \sim 1$ at $\beta_0^* \sim 12$. Above $\beta_0^* \sim 12$, there is no etching observed at all.

Although the example considered was deliberately rigged to explain the observations of the Si/silane reaction, it nevertheless does serve to show that the masking of one competing reaction by another can occur, except for small values of focused-laser spot size. It also does indicate a possible explanation of contour reversal. The model could be applicable to the Si/silane reaction if the deposition reaction is the familiar pyrolysis reaction,

$$SiH_4 \underset{k_2}{\overset{k_1}{\rightleftharpoons}} Si + 2H_2,$$

while the etching reaction could be, for example,

$$SiH_4 + Si \underset{k'_2}{\overset{k'_1}{\rightleftharpoons}} 2SiH_2.$$

5. Reactions Initiated by Surface-Phase or Substrate Absorption: Mass Transport by Surface-Phase Diffusion

5.1. Circular ring Model

The importance of vapor-phase diffusion of reactant and product molecules in determining net reaction rates was stressed in Sections 4.1 and 4.2. The transport of surface-phase molecules can be equally significant in determining net reaction rates. This is the basis for the sophisticated measurements of surface diffusion rates using laser-induced thermal desorption (Mak et al., 1986). The importance of surface diffusion in laser-stimulated surface chemistry has been discussed by Tsao et al., (1985) and Zeiger et al., (1985).

In Section 4.1 it was shown that the considerable insight could be gained in understanding the vapor-phase diffusion problem by solving the hemispheric model. The solution of this problem is in turn identical to the solution of the problem of a reaction occurring on a spherical surface with three-dimensional vapor-phase diffusion of both reactant and product molecules. For the case of surface diffusion, the analogous model problem would be that of a reaction occurring at the circumference of a circular ring embedded in a two-dimensional plane surface, with diffusion of reactant and product molecules occurring on the surface.

In analogy with the spherical surface model of Section 4.1, then, we consider the problem of surface diffusion to a circular reactive ring of radius d. Surface diffusion is governed by the diffusion equation,

$$D\nabla^2 n - \frac{\partial n}{\partial t} = 0,$$

where ∇^2 is the two-dimensional Laplacian, D is the surface diffusivity, and n is the number of surface molecules per unit area. As is Section 4.1.1, for simplicity only diffusion of reactant species is considered, and the boundary conditions are taken to be

$$D\left(\frac{\partial n}{\partial r}\right)_{r=d} = (kn)_{r=d}, \tag{5.63}$$

$$n(r, t=0) = n(r=\infty, t) = n_0, \tag{5.64}$$

where k is a reaction rate constant. The reaction is assumed to begin at $t=0$.

The solution to this model problem for the concentration of reactant molecules as a function of time can be obtained from the analogous heat flow problem (Carslaw and Jaeger, 1959, p. 337). The concentration of molecules at the circular ring boundary as a function of time is given by the integral

$$n(d, t) = 2\eta \frac{n_0}{\pi} \int_0^\infty \exp\left\{-\left(\frac{D}{d^2}\right)x^2 t\right\} \cdot \frac{[Y_0(x)J_1(x) - J_0(x)Y_1(x)]\,dx}{[xJ_1(x) + \eta J_0(x)]^2 + [xY_1(x) + \eta Y_0(x)]^2}, \tag{5.65}$$

where $\eta = kd/D$, and $Y_0(x)$, $J_0(x)$, $Y_1(x)$, and $J_1(x)$ are Bessel functions. Figure 5.17 shows plots of $n(d, t)/n_0$ versus $(D/d^2)t$ for various values of η (Carslaw and Jaeger, 1959, p. 338). All curves start at $n(d, t)/n_0 = 1$ at $t=0$. After a rapid initial drop, they all continue to decrease more slowly. In the limit $t \to \infty$, all the curves approach zero, a result immediately apparent from Eq. 5.65.

The approach of $n(d, t)$ to zero as $t \to \infty$ for two-dimensional diffusion in the circular ring model is distinctly different from the behavior of $n(d, t)$ for three-dimensional diffusion in the hemispheric model discussed in Section 4.1.1. In the limit $t \to \infty$, $n(d, t)$ for the hemispheric model approaches a finite value that depends on the relative values of r_0 and ω_0. The difference in behavior is due to the two spatial degrees of freedom available for lateral diffusion of reactant molecules in the three-dimensional case, while there is only one lateral degree of freedom for diffusion in the two-dmensional case. In Section 5.2, a more detailed Green's function model of surface diffusion in the presence of localized laser-driven photolysis will be presented. A qualitative comparison of the results of that calculation and those obtained from the circular ring problem will be discussed.

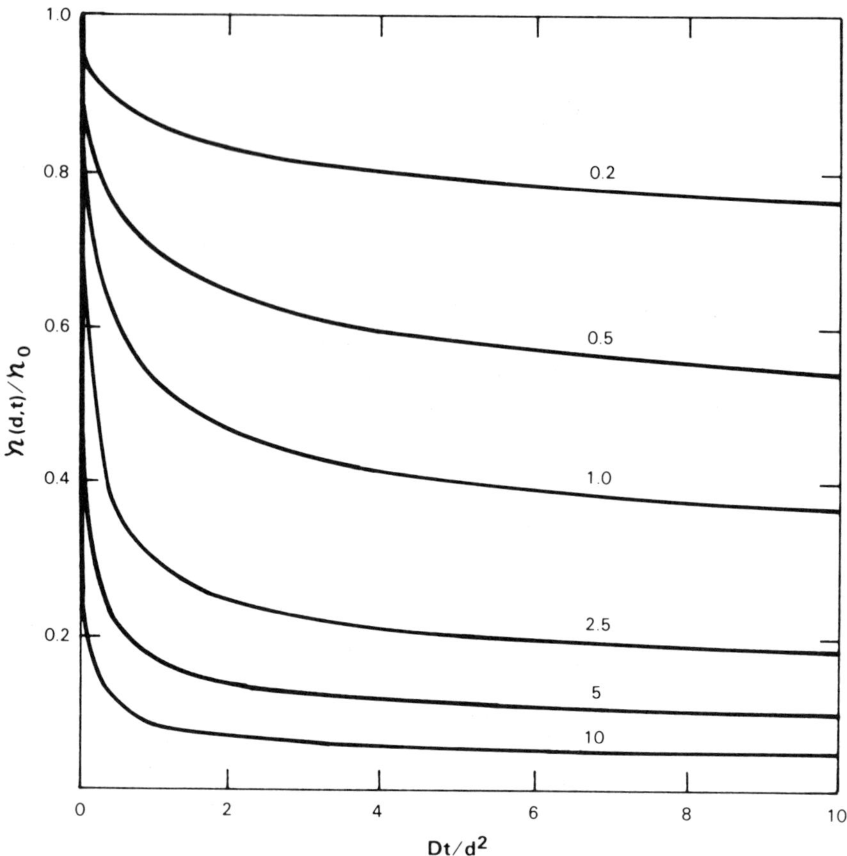

Fig. 5.17 Solution of Eq. 5.65 for $n(d, t)/n(0)$ versus $(D/d^2)t$ for values of η indicated. After Carslaw and Jaeger, 1959.

5.2. *Detailed Model*

5.2.1 Green's Function Equation

In Section 5.1 a brief discussion of the "circular ring model" of localized surface reaction kinetics in the presence of surface diffusion was presented. That development was based on the two-dimensional analogue of the spherical model, which involved vapor-phase diffusion in three dimensions. Tsao et al., (1985) have considered a more detailed model of a surface reaction driven by a laser, with surface diffusion, for which a Green's function solution can be found. Their treatment is outlined in this section.

Included in the model problem is surface diffusion of a reactant species only and photolysis by a focused laser beam resulting in irreversible deposition of an immobile surface species with the release of a vapor-phase species. The calculation also includes the possibility of replenishment of the surface reactant species from the vapor phase. The equation governing the concentration of the reactant surface phase species is

$$\frac{\partial n(\vec{\mathbf{r}},t)}{\partial t} = D\nabla^2 n(\vec{\mathbf{r}},t) - An(\vec{\mathbf{r}},t) + BP[n_{\max} - n(\vec{\mathbf{r}},t)] - cI(\vec{\mathbf{r}},t)n(\vec{\mathbf{r}},t), \tag{5.66}$$

where $n(\vec{\mathbf{r}},t)$, D and A represent the surface density, the surface diffusivity, and the desorption rate of adsorbed parent molecules, respectively; $BP[n_{\max} - n(\vec{\mathbf{r}},t)]$ represents a Langmuir-type adsorption term, where P is the ambient pressure (assumed time independent) and $n_{\max}$ corresponds to monolayer coverage; $cI(\vec{\mathbf{r}},t)n(\vec{\mathbf{r}},\mathrm{t})$ is a photodeposition reaction rate per unit area proportional to the light intensity $I(\vec{\mathbf{r}},t)$; and ∇^2 is the two-dimensional Laplacian. In this equation, it is assumed for tractability that D is everywhere constant, including the region of deposition. Equation 5.66 can be rewritten as

$$\frac{\partial n}{\partial t} = D\nabla^2 n - \gamma n - J(\vec{\mathbf{r}},t), \tag{5.67}$$

where

$$\gamma = A + BP, \tag{5.68}$$

$$J(\vec{\mathbf{r}},t) = cIn - Bpn_{\max}. \tag{5.69}$$

The parameter γ is the equilibration rate between surface and gas-phase molecules. Even though the restrictive Langmuir adsorption assumption was made in obtaining Eq. 5.66, the form of Eq. 5.67 has greater generality.

To solve Eq. 5.66 the two-dimensional Green's function that satisfies the equation

$$\frac{\partial G(\vec{\mathbf{r}},\vec{\mathbf{r}}';t-t')}{\partial t} = D\nabla^2 G(\vec{\mathbf{r}},\vec{\mathbf{r}}';t-t') - \gamma G(\vec{\mathbf{r}},\vec{\mathbf{r}}';t-t') \tag{5.70}$$

is used. The Green's function can be shown to have the form

$$G(\vec{\mathbf{r}},\vec{\mathbf{r}};t-t') = \frac{1}{4\pi D(t-t')} e^{-(\vec{\mathbf{r}}-\vec{\mathbf{r}}')^2/4D(t-t')} e^{-\gamma(t-t')} \quad (t > t'). \tag{5.71}$$

For light switched on at $t' = 0$ and localized spatially on the surface, the solution for $n(\vec{\mathbf{r}}, t)$ is given in terms of that Green's function by

$$n(\vec{\mathbf{r}}, t) = \int_s [G(\vec{\mathbf{r}}, \vec{\mathbf{r}}'; t - t')n(\vec{\mathbf{r}}', t')]_{t'=0}\, d\vec{\mathbf{r}}'$$

$$- \int_s \int_0^t J(\vec{\mathbf{r}}', t')G(\vec{\mathbf{r}}, \vec{\mathbf{r}}'; t - t')\, d\vec{\mathbf{r}}'\, dt', \tag{5.72}$$

where the surface integral extends to infinity. The first integral and the portion of the second integral due to the constant term $BPn_{\max}$ in Eq. (5.69) can be evaluated, and the result is

$$n(\vec{\mathbf{r}}, t) = n_0 e^{-\gamma t} + n_{\text{eq}}(1 - e^{-\gamma t})$$

$$- \int_s \int_0^t cI(\vec{\mathbf{r}}', t')n(\vec{\mathbf{r}}', t')G(\vec{\mathbf{r}}, \vec{\mathbf{r}}'; t - t')\, d\vec{\mathbf{r}}'\, dt', \tag{5.73}$$

where n_0 is the initial uniform value of n when the light is first switched on and is not necessarily the equilibrium value, $n_{\text{eq}} = BPn_{\max}/\gamma$, at the ambient pressure. In the absence of the photolytic reaction, Eq. 5.73 correctly gives the time dependence of $n(t)$ when the surface coverage does not initially have its equilibrium value.

If $I(r, t)$ has cylindrical symmetry, then the angular integral in Eq. 5.73 can be evaluated by inserting the Green's function (Eq. 5.71) and using a familiar integral from the theory of heat flow. The result is

$$n(r, t) = n_0 e^{-\gamma t} + n_{\text{eq}}(1 - e^{-\gamma t})$$

$$- c \int \int_0^t I(r')n(r', t') \frac{r'}{2D(t - t')} e^{-(r^2 + r'^2)/4D(t - t')}$$

$$\cdot\, \mathscr{I}_0\left(\frac{rr'}{2D(t - t')}\right) e^{-\gamma(t - t')}\, dr'\, dt', \tag{5.74}$$

where $\mathscr{I}_0(x)$ is the modified Bessel function and $I(r')$ has been assumed constant in time once the light is switched on.

It is convenient to introduce normalized quantities, and, assuming the spatial light distribution is Gaussian,

$$\mu(\rho, \tau) = e^{-\Gamma\tau} + \eta(1 - e^{-\Gamma\tau}) - \alpha \int \int_0^\tau \mu(\rho', \tau')\left(\frac{\rho'}{\tau - \tau'}\right) e^{-(\rho^2 + \rho'^2)/(\tau - \tau')}$$

$$\cdot\, \mathscr{I}_0\left(\frac{2\rho\rho'}{\tau - \tau'}\right) e^{-\Gamma(\tau - \tau')} e^{-\beta\rho'^2}\, d\rho'\, d\tau', \tag{5.75}$$

where

$$\mu = \frac{n}{n_0}, \qquad \rho = \frac{r}{r_0}, \qquad \tau = \frac{4Dt}{r_0^2},$$

$$I(r) = I_0 \exp\left[-\beta\left(\frac{r}{r_0}\right)^2\right] = I_0 \exp(-\beta\rho^2),$$

$$\Gamma = \gamma r_0^2 4D = [A + BP] r_0^2 4D, \qquad \eta = \frac{n_{eq}}{n_0}, \qquad \alpha = \frac{cI_0 r_0^2}{2D}. \tag{5.76}$$

the parameters Γ and α are, respectively, the ambient-pressure-dependent rate at which surface-gas equilibration occurs and the consumption rate by photolysis, both normalized to the surface-diffusion replenishment rate into a spot of Gaussian half-width r_0. The parameter η is the ratio between the equilibrium coverage at pressure P and the actual (possibly nonequilibrium) coverage. For computational convenience the geometrical scaling parameter β has been introduced in Eq. 5.75 so that the Gaussian half-width of the beam is $r_0\beta^{-1/2}$. β plays the same role in this calculation as the parameter σ played in Section 4.2. In all the calculations described, the choice $\beta = 4$ has been made. In the end, the parameter values extracted by comparing theory and experiment are independent of the choice of β. The computational advantage of the integral in Eq. 5.75 over the differential equation (5.66) is that the surface density at any time and location depends only on the surface density at all earlier times within a radius $\rho' < m/(\beta^{1/2})$, $(m \sim 2)$, rather than on the surface density at earlier times and at all radii.

5.2.2 Calculation of Reactant and Deposited Molecule Profiles

In order to make a connection with the experimentally measured rates, Eq. 5.75 was solved in the following manner. For given values of η, α, and Γ, the equation was solved for a discrete grid of ρ, $\rho' < m/(\beta^{1/2})$ for incrementally increasing values of τ. Having computed $\mu(\rho, \tau)$ by numerically solving Eq. 5.75, the next step is to calculate the number of deposited molecules per unit area, $n_{dep}(\rho, \tau)$ due to photolysis produced by the laser beam.

Assuming that one molecular unit is deposited for each molecule of the original surface species photolyzed,*

$$n_{dep}(\rho, t) = \int_0^\tau cI_0 e^{-\beta\rho^2} n(\rho, \tau')\, dt'$$

$$= \frac{cI_0 r_0^2}{4D} r_0^2 \int_0^\tau e^{-\beta\rho^2} \mu(\rho, \tau')\, d\tau', \qquad \text{or}$$

$$\frac{n_{dep}(\rho, \tau)}{n_0} \equiv A(\rho, \tau) = \frac{\alpha}{2} \int_0^\tau e^{-\beta\rho^2} \mu(\rho, \tau')\, d\tau'. \tag{5.77}$$

* The factor $e^{-\beta\rho^2}$ in Eq. 5.77 was inadvertently omitted in the definition of $A(\rho, \tau)$ given in Tsao et al. (1985). However, $A(\rho, \tau)$ was correctly calculated as given by Eq. 5.77.

Figures 5.18 and 5.19 show $\mu(\rho, \tau)$ and $A(\rho, \tau)$ computed for small values of Γ (the low-pressure regime, with negligible replenishment of mobile surface species from the vapor) for $\alpha = 10$ (moderate illumination) and $\alpha = 20$ (strong illumination). The value of Γ was small enough that $\Gamma\tau \ll 1$ for the largest value of τ shown, $\tau = 30$. The rapid photolysis of molecules near the rim of the laser-illuminated region for $\alpha = 20$ leads to a dip in the deposition near the center of the laser spot.

It is interesting to note the rapid initial drop of $\mu(\rho, \tau)$ followed by a slower decay for increasing τ near the boundary of the laser deposition profiles shown in Figs. 5.18a and 5.19a. This is consistent with the results of the circular ring model calculation summarized in Fig. 5.17.

5.2.3 Photodepeosition Rate

The results summarized in Eqs. 5.75 and 5.77 were obtained for application to an experiment on the photolysis of tetraethyl lead (TEL) with a focused Ar^+ laser beam. The TEL molecules were adsorbed on a sapphire wafer, in an atmosphere of TEL vapor at various pressures. As

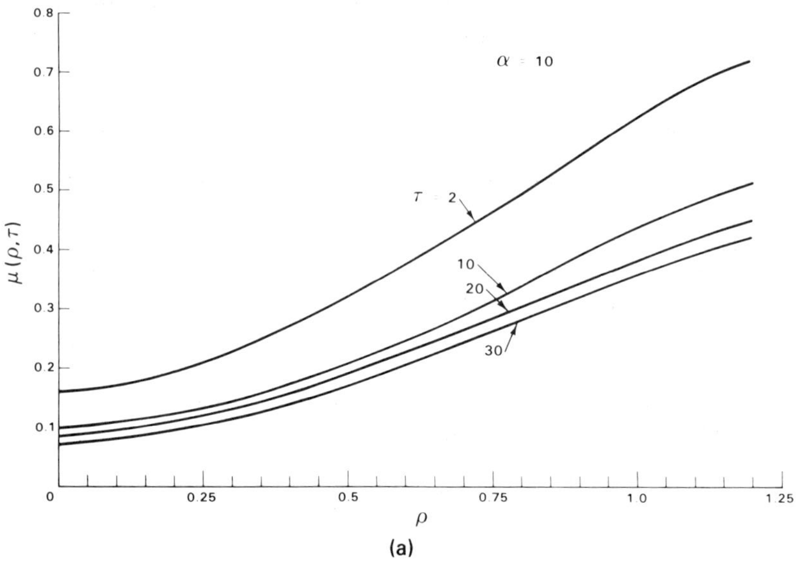

Fig. 5.18 Numerically calculated spatial profiles for a photoreacting adsorbed layer at a number of normalized times τ after a laser is switched on. ρ is the normalized radial distance from beam center, and the Gaussian half-width of the laser beam is at $\rho = 0.5$. The system is under moderate UV illumination and in the moderately diffusion limited regime ($\alpha = 10$). (a) Normalized parent molecule density $\mu(\rho, \tau)$. (b) Normalized photodeposit thickness $A(\rho, \tau)$. After Tsao et al., 1985.

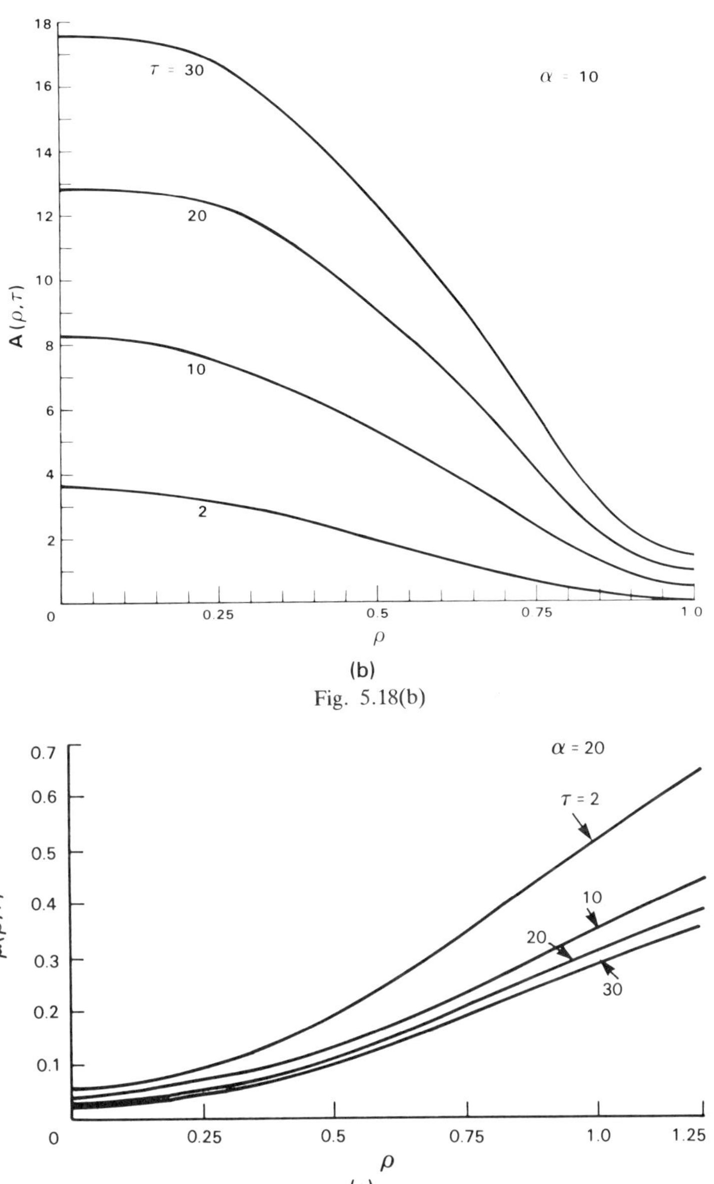

Fig. 5.19 Numerically calculated spatial profiles for a photoreacting adsorbed layer at a number of normalized times τ after a laser is switched on. The system is under strong UV illumination and in the strongly diffusion limited regime ($\alpha = 20$). (a) Normalized parent molecule density $\mu(\rho, \tau)$. (b) Normalized photodeposit thickness $A(\rho, \tau)$. After Tsao et al., 1985.

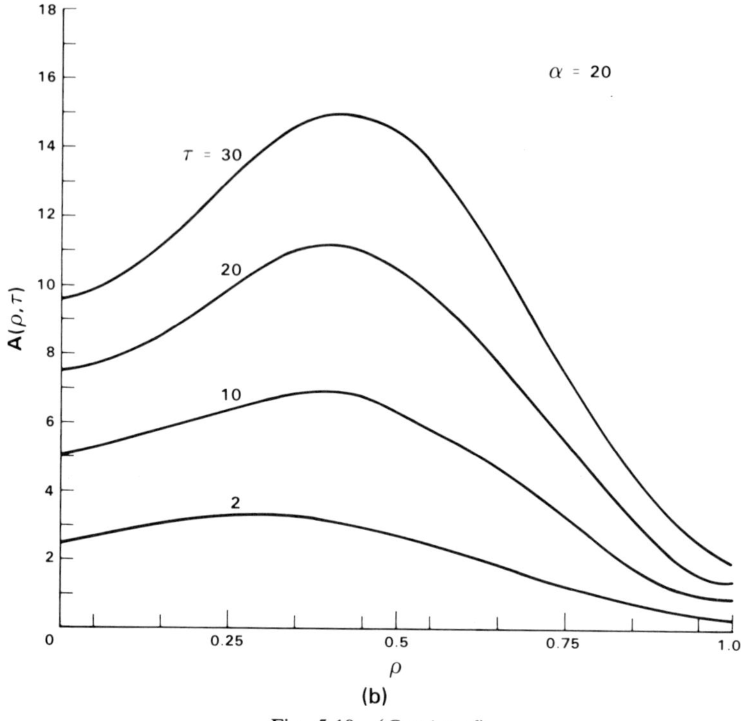

Fig. 5.19 (*Continued*)

surface and vapor molecules of TEL diffused into the region of laser-beam irradiation, a deposit of mainly lead atoms formed, which reduced the transmission of laser radiation. The experimental quantity measured was the inverse time, t_0^{-1} (defined as the photodeposition rate) for the transmission of the laser beam to be reduced to 70% of its initial value when the laser was first switched on. To compare the model calculation with experiment, the transmission $T(\tau)$ was therefore calculated in the following manner:

$$\int e^{-\beta\rho^2} 2\pi\rho \exp\left[-K\left(\frac{n_0}{n_{\max}}\right)A(\rho, \tau)\right] d\rho = T(\tau) \int e^{-\beta\rho^2} 2\pi\rho \, d\rho, \quad (5.78)$$

where K is a dimensionless absorption constant per monolayer of reactant molecules deposited. Performing the integration on the right-hand side,

$$T(\tau) = 2\beta \int e^{-\beta\rho^2} \rho \exp\left[-K\left(\frac{n_0}{n_{\max}}\right)A(\rho, \tau)\right] d\rho.$$

Now, 70% transmission is known to be produced by the uniform deposition of 10 monolayers of TEL for the experiment to be analyzed, so that $\exp(-10K) = 0.7$, or $K = 0.036$. A transmission index was therefore defined as

$$\text{Index}(\tau) \equiv \frac{2\beta}{0.7}\int e^{-\beta\rho^2}\rho \exp\left[-0.036\left(\frac{n_0}{n_{\max}}\right)A(\rho,\tau)\right]d\rho. \quad (5.79)$$

Index(τ) was numerically calculated for each value of τ, and τ_0 was determined from the condition Index(τ_0) = 1.

Accordingly, for each value of τ, Eqs. 5.75 and 5.77 were calculated numerically, as well as Index(τ). When Index(τ_0) = 1 was reached, τ_0^{-1} could be calculated, corresponding to the experimentally determined t_0^{-1} for a given set of parameters.

In Fig. 5.20 are plotted values of $1/\tau_0$ as a function of the parameter α (proportional to the normalized laser intensity) for $\eta = 1$ and for several values of Γ. Two distinct pressure regimes are observed. At high pressures ($\Gamma \gg 1$), the intensity dependence is linear. This indicates that replenishment by adsorption from the gas phase is much faster than consumption by absorbed-phase photoreaction so that the overall photodeposition rate is reaction-rate limited. In this case the coverage, even while the laser is on, never deviates from its starting value.

As the pressure is decreased ($\Gamma < 1$), the overall intensity dependence becomes more and more nonlinear. This indicates that replenishment by adsorption from the gas phase is becoming as slow as or slower than replenishment by surface migration. In the extreme case where adsorption from the gas becomes very slow relative to surface migration ($\Gamma \ll 1$), two distinct intensity regimes are observed. At low intensities ($\alpha \ll 1$), the photolysis rate is low compared to the surface diffusion rate. In this regime depletion is unimportant, the coverage never deviates from its starting value, and the intensity dependence is linear. At high intensities ($\alpha \gg 1$), surface diffusion cannot keep up with photolysis, depeletion occurs, and the photodeposition rate shows a sharp saturation, consistent with the data obtained in the experiments. From a comparison between the saturated value of $1/t_0$ and the saturated value of $1/\tau_0$ shown in Fig. 5.20 for the $\Gamma \ll 1$ simulations, the magnitude of the surface diffusivity was estimated to be $3 \times 10^{-7}\ \text{cm}^2/\text{s}$.

At sufficiently high intensities and low pressures, the simulation curves actually roll off slightly, although this was not apparent from the experimental data. Physically, this roll off occurs because photoreaction rates are so large that every parent molecule that migrates into the laser beam spot reacts almost immediately. Since molecules are diffusing in from the rim of the spot, they are preferentially consumed in the vicinity

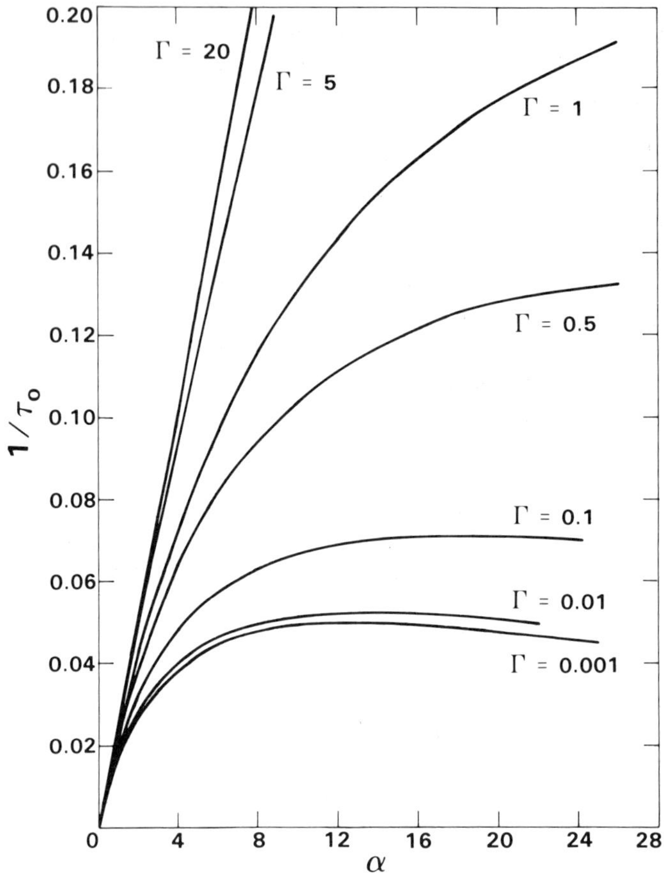

Fig. 5.20 Numerical simulations of normalized photodeposition rate $1/\tau_0$ versus normalized incident UV laser intensity α for $\eta = 1$ and for various ambient pressures (i.e., values of Γ). After Tsao et al., 1985.

of the rim, where the laser intensity itself is weakest. This means that longer reaction times are necessary to achieve the same attenuation of the laser beam. This effect is consistent with the spatial profiles shown in Fig. 5.19b.

6. Reactions Initiated by Surface-Phase or Substrate Absorption: Mass Transport by Bulk Diffusion

Finally, we note in closing that for some processes, such as photodoping, one of the reaction products is driven into the substrate by solid-state diffusion. In most covalently bonded solids, this solid-state diffusion is extremely slow and will be the rate-limiting step, although diffusion into liquified semiconductors can be extremely rapid ($D \simeq 10^{-4}\,\text{cm}^2/\text{s}$).

For the cases where solid-state diffusion is rate limiting, the driving forces for diffusion, due to the very small laser-beam diameters that may be used, need not be restricted to concentration gradients (Tsao & Ehrlich, 1983). For example, temperature gradients can drive Soret diffusion characterized by a flux

$$J_{\nabla T} = -DNS_T \nabla T, \tag{5.80}$$

where the Soret coefficient is $S_T = Q^*/RT^2$, and Q^* is the "heat of transport" for impurity motion.

Another example is electromigration, which can occur if the diffusing species is charged. Then, for the case of doping in semiconductors, even in the absence of externally imposed electric fields, the space-charge field $\nabla\Psi$ created by the dopant concentration gradient ∇N can drive diffusion. This can be modeled as

$$J_{\nabla\Psi} = \frac{-eND\nabla\Psi}{kT} = \frac{-D}{\sqrt{1 + 4n_i^2/N^2}} \nabla N, \tag{5.81}$$

where n_i is the intrinsic carrier density at the temperature of the diffusion.

In addition to Soret diffusion, temperature gradients can lead to stress gradients $\nabla\sigma$, which in turn can interact with the excess volume V^* associated with the impurity atoms and cause a net stress-gradient-driven diffusion flux (Tsao and Ehrlich, 1983):

$$J_{\nabla\sigma} = \frac{DN}{RT} V^* \nabla\sigma. \tag{5.82}$$

References

Allen, S.D. (1981). Laser chemical vapor deposition: a technique for selective area deposition. *J. Appl. Phys.* **52,** 6501.

Auston, D.H., Surko, C.M., Venkatesan, T.N.C., Slusher, R.E., and Golovchenko, J.A. (1978). Time-resolved reflectivity of ion-implanted silicon during laser anealing. *Appl. Phys. Lett.* **33,** 437.

Bäuerle, D. (1984). *Laser Processing and Diagnostics.* Proceedings of an International Conference, University of Linz, Austria, July 15–19, 1984. Springer Series in Chemical Physics, Vol. 39. Springer-Verlag, Berlin.

Bäuerle, D. (1986). *Chemical Processing with Lasers.* Springer-Verlag, Berlin.

Carslaw, H.S., and Jaeger, J.C. (1959). *Conduction of Heat in Solids.* 2nd ed. Oxford Press, Oxford, Chap. XIV.

Chen, C.J. (1987). Kinetic theory of laser photochemical deposition. *J. Vac. Sci. Technol. A* **5,** 3386–3398.

Chuang, T.J. (1981). Multiple photon-excited SF_6 interaction with silicon surfaces. *J. Chem. Phys.* **74,** 1453–1460.

Hall, R.B., and DeSantolo, A.M. (1984). Pulsed laser induced excitation of metal surfaces: Application as a probe of surface reaction kinetics of methanol on Ni. *Surf. Sci.* **137,** 421–441.

Heavens, O.S. (1955). *Optical Properties of Thin Solid Films.* Dover Publications, New York.

Krchnavek, R.R., Gilgen, H.H., Chen, J.C., Shaw, P.S., Licata, T.J., and Osgood, R.M. Jr. (1987). Photodeposition rates of metal from metal alkyls. *J. Vac. Sci. Technol.* **B5,** 20–26.

Mak, C.H., Brand, J.L., Deckert, A.A., and George, S.M. (1986). Surface diffusion of hydrogen on Ru(001) studied using laser-induced thermal desorption. *J. Chem. Phys.* **85,** 1676.

Melcher, R.L. (1984). Thermal and acoustic techniques for monitoring pulsed laser processing. In: *Laser Processing and Diagnostics* (Bäuerle, D., ed.). Springer-Verlag, Berlin, 418–424.

Moylan, C.R., Baum, T.H., and Jones, C.R. (1986). LCVD of copper: deposition rates and deposit shapes. *Appl. Phys.* **A40,** 1.

Rosenberger, F. (1979). *Fundamentals of Crystal Growth.* Springer-Verlag Series in Solid-State Sciences, Vol. 5. Springer, Berlin, 264.

Skouby, D.C., and Jensen, K.F. (1988). Modeling of pyrolytic laser-assisted vapor deposition: mass transfer and kinetic effects influencing the shape of the deposit. *J. App. Phys.* **63,** 198.

Thompson, M.O. (1984). *Liquid-Solid Interface Dynamics during Pulsed Laser Melting of Silicon-on-Sapphire.* Ph.D. thesis, Cornell University.

Tsao, J.Y., and Ehrlich, D.J. (1983). Submicrometer-linewidth laser doping. *Mat. Res. Soc. Symp. Proc.* **17,** 235–241.

Tsao, J.Y., and Ehrlich, D.J. (1984). Surface and gas processes in photodeposition in small zones. *SPIE Proceedings,* Vol. 459, Laser Assisted Deposition, Etching and Doping. SPIE, The International Society for Optical Engineering, Bellingham, Washington.

Tsao, J.Y., Zeiger, H.J., and Ehrlich, D.J. (1985). Measurement of surface diffusion by laser-beam-localized surface photochemistry. *Surf. Sci.* **160,** 419–442.

von Gutfeld, R.J., Tynan, E.E., Melcher, R.L., and Blum, S.E. (1979). Laser-enhanced electroplating and maskless pattern generation. *Appl. Phys. Lett.* **35,** 651.

Zeiger, H.J., Tsao, J.Y., and Ehrlich, D.J. (1985). Technique for measuring surface diffusion by laser-beam-localized surface photochemistry. *J. Vac. Sci. Technol.* **B3,** 1436.

Zeiger, H.J., and Ehrlich, D.J., Lateral confinement of microchemical surface reaction: effects on mass diffusion and kinetics. To be published, *J. Vac. Sci. Tech.* **B,** 1989.

PART III

Reactions

CHAPTER 6

Laser Etching

JOHN J. RITSKO

IBM T.J. Watson Research Center
Yorktown Heights, New York

1. Introduction

There are a wide variety of laser-driven processes that can be used to directly and controllably etch fine features in semiconductors, insulators, and metals. These processes show promise in a number of microelectronic applications. Laser etching has been applied to semiconductor laser and device fabrication, including patterning of semiconductors and metallization, trench isolation, integrated optoelectronic coupling, and grating formation. Interconnection packages may be fabricated using metal and polymer etching techniques, and a variety of circuit repair, customization, and device trimming applications are feasible. Many of these techniques offer considerable advantages over conventional lithographic techniques in process simplicity, flexibility, or speed. This chapter provides an overview of significant recent work on the laser etching of materials commonly used in microelectronic fabrication. The

0-12-233430-2

emphasis is on processes and parameters that control etch rates, resolution, and application. Tables are included that summarize the etch rates obtainable for different materials under a variety of processing conditions.

2. Semiconductors

2.1. *Silicon and Germanium*

Focused laser light has been shown by several authors to produce very rapid local etching of silicon in gas or liquid ambients by purely photothermal processes. Ehrlich et al. (1981a) have shown that silicon can be etched very rapidly and cleanly in Cl_2 or HCl gas using a focused argon-ion laser. At power levels above ~5 MW/cm^2, small holes (40 μm, large diameter; 15 μm, small diameter) can be etched through 250-μm thick wafers in 35–45 seconds (etch rate ~7 μm/sec). Alternatively, 3.9-μm deep grooves are etched when scanning a focused 7-W beam at

Table 6.1 Semiconductor Etching.

Semi-conductor	Ambient	Laser	Intensity/ Energy	Etch Rate	Reference
Si	Cl_2, HCl	Ar^+	≳5 MW/cm^2	7 μm/sec	Ehrlich, 1981a
Si	KOH	Ar^+	~10^7 W/cm^2	15 μm/sec	von Gutfeld, 1982
Si	NaOH	Nd:YAG	NA	4 μm/min	Bunkin, 1985
Si	NaOH	CO_2	NA	2 μm/min	Bunkin, 1985
α-Si	KOH	Ruby	0.5 J/cm^2	500 Å/pulse	Krimmel, 1985
Si	Cl_2(0.1 torr)	N_2	0.12 J/cm^2	~1 Å/pulse	Sesselmann, 1985
Ge	Br_2	Ar^+	0.1 KW/cm^2	36 μm/sec	Sullivan, 1968
Ge	Br_2(10^{-4} torr)	DYE	0.1–0.3 J/cm^2	0.2 Å/pulse	Davis, 1984
Ge	Br_2(2.5 torr)	Ar^+	40 W	860 Å/sec	Baklanov, 1974
GaAs	H_2SO_4	Ar^+	0.2 MW/cm^2	30 μm/min	Osgood, 1982
GaAs	HNO_3	Ar^+	60 MW/cm^2	2 μm/sec	Tisone, 1983a
GaAs	HF	Kr^+	60 MW/cm^2	0.7 μm/min	Tsukada, 1984
GaAs	HNO_3	Ar^+ (257 nm)	1 W/cm^2	12 μm/min	Podlesnik, 1986
GaAs	Cl	Ar^+	1.4 KW/cm^2	5 Å/sec	Ashby, 1984
GaAs	CH_3Br (750 torr)	Ar^+ (257 nm)	1 KW/cm^2	60 Å/sec	Ehrlich, 1980
GaAs	HBr	ArF (193 nm)	32 mJ/cm^2	8.2 μm/min	Brewer, 1985
GaAs	CCl_4	Ar^+	0.1 MW/cm^2	6 μm/sec	Takai, 1985
GaAs	Cl_2	Ar^+ (488 nm)	2 MW/cm^2	33 μm/sec	Tucker, 1984
InP	CH_3Br (750 torr)	Ar^+ (257 nm)	100 W/cm^2	9.4 Å/sec	Ehrlich, 1980
InP	H_3PO_4	Ar^+	1.6 MW/cm^2	140 μm/sec	Bjorkholm, 1983
GaP	KOH	Ar^+ (351 nm)	3.5 KW/cm^2	600 Å/sec	Johnson, 1984

440 μm/sec in an atmosphere of 200 torr of Cl_2. These rapid etch rates are only achieved when the silicon is melted (4 W focused to ~5-μm diameter spot), and hence there is little dependence on crystal orientation or doping. Once melting is achieved, the etch rate saturates so that higher powers lead to little additional etching. The etching mechanism is a chemical reaction between silicon and chlorine to form volatile silicon chlorides, thus no significant debris appears at the edges of the etched feature.

The use of a focused argon-ion laser to melt Si and produce rapid etching can also be done in solutions (von Gutfeld, 1982). KOH is a well-known etchant for Si (Bean, 1978). It etches the (111) plane 600 times slower than the (110) plane at 80°C. However, with a 15-W argon ion laser focused to a small spot (laser power ~10^7 W/cm^2) etch rates for through holes in Si (111) can be 15 μm/sec in a concentrated KOH solution. Apparently both melting and vaporization contribute to this rapid etching by direct material removal, increase in effective surface area of material in contact with the etchant, and a local temperature increase that promotes the thermally activated etching kinetics.

Thermally activated etching of Si in an aqueous NaOH solution was studied by Bunkin et al. (1985). A comparison was made between CO_2 ($\lambda = 10.6$ μm) radiation, which was absorbed in the solution, and Nd:YAG ($\lambda = 1.06$ μm) radiation, which was absorbed only in the Si. Both lasers etched the silicon well (~2 μm/min), and the difference in etch rates was ascribed to photovoltaic phenomena in the case of Nd:YAG radiation. In either case the cw laser beam was chopped and the optimum modulation period was found to approximately coincide with the cooling time of the system.

With silicon wafers immersed in concentrated KOH and subjected to pulsed ruby radiation, Krimmel et al. observed simultaneous etching and annealing of amorphous regions (Krimmel, 1985). Single crystal silicon wafers had been amorphized by high-dose phosphorous ion implantation. When subjected to a single pulse of ruby laser radiation (0.5 J/cm^2 pulse length 20 nsec) the etch rate in KOH solution is enhanced so that 500 Å of the substrate is removed in the irradiated area. The implanted phosphorus is activated, and the radiation damage from the ion implantation is annealed as well (Krimmel, 1985). The annealing of defects is a very important application of lasers to semiconductor processing (Swenson, 1983; Young, 1983). Once annealed however, the amorphous silicon recrystallizes and cannot be etched by subsequent pulses (Krimmel, 1985). The difference in etch rate between crystalline and amorphous silicon is attributable to the higher optical absorption coefficient of the amorphous material at the ruby wavelength and the lower heat of melting of the amorphous Si. Hence while both crystalline and amorphous regions melt, the amorphous region remains melted for a longer time and, as has been pointed out in several earlier articles (Ehrlich, 1981; von Gutfeld, 1982), molten Si is easily etched rapidly.

Laser etching processes involving melting Si have extremely high local etch rates but cannot easily produce intricate fine patterns over large areas. Such fine structures may be achievable by laser-assisted chemical etching techniques in which a surface chemical reaction is accelerated by energy absorbed in the sample through thermal or photochemical processes. These lower temperature processes enable larger areas to be illuminated by a particular laser so that intricate patterns can be created in reasonable times. In a series of papers, T. Chuang (1982a, 1982b, 1984) has described many laser enhanced gas-surface chemical reactions. These basic mechanisms are described in detail in Chapter 2. Here we will describe laser enhanced etching of Si by fluorine and chlorine and the contributions of thermal and photochemical effects.

Molecular fluorine in the form of XeF_2 gas reacts with and spontaneously etches Si (Winters, 1979, 1983). F. Houle has studied how weakly focused or unfocused light from a cw argon ion laser can greatly enhance the spontaneous etch rate (Houle, 1983a, 1983b, 1984a, 1984b). In these experiments, a molecular beam of XeF_2 (flux $\sim 0.3 \times 10^{16}$ mol/cm^2 sec) is incident on single-crystal silicon surfaces. Unfocused light from an argon ion laser is also incident on the surface raising the temperature by at most 40°C (Houle, 1983a). Under these circumstances, the etch rate at first decreases as the laser power increases, and then above 0.75 W, the etch rate increases linearly. For n-type Si at 4 W, the etch rate has been enhanced by a factor of four over the spontaneous etch rate, while for p-type Si the enhancement is a factor of three under similar conditions. The maximum enhancement attributable to thermal effects is less than 10%. The nonthermal enhanced etching is attributed to electron-hole pair excitation, as shown in Fig. 6.1.

Figure 6.1 shows typical electronic energy bands near the surfaces of n- and p-type semiconductors. These bands are such that radiation exceeding the band gap produces electron-hole pairs. At low intensity, holes are swept to the surface by the fields in an n-type semiconductor and electrons are swept into the bulk of the sample. In p-type semiconductors the opposite is true. There can be surface traps that keep both electrons and holes near the surface. Nevertheless, the differences between n- and p-type semiconductors can account for differences in etch rates as a function of doping if the etching mechanism requires electrons or holes to result in increased chemical etching. At high intensities, both electrons and holes can be present in sufficient quantity to take part in the photochemistry.

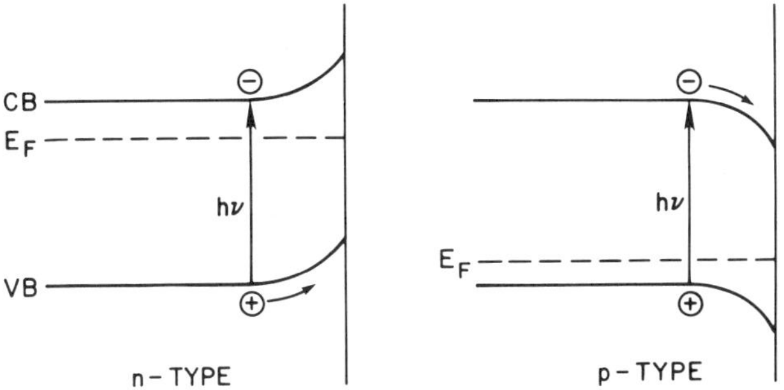

Fig. 6.1 Band bending and photoexcitation in semiconductors.

The photochemistry associated with the laser-enhanced etching of Si in fluorine (XeF_2) can be explained by increased chemical pathways involving electrons and holes that lead to volatile products (Houle, 1984b). One such scheme involves holes (p) associated with SiF_x centers while free fluorine atoms capture electrons (n). Coulombic attraction then increases the reaction cross section

$$SiF_3 + p^+ \rightarrow SiF_{3p}^+$$

$$F + n^- \rightarrow F_n^-$$

$$SiF_{3p}^+ + F_n^- \rightarrow SiF_4 \text{ (volatile gas).}$$

The detailed mechanisms of photochemical etching are quite complex. Experiments show that the reduction in spontaneous etching of Si in XeF_2 at low light intensity (<0.75 W) is due to light-enhanced diffusion of fluorine below the surface leading to reduction in surface F concentration and lower etch rate (Houle, 1983b). The laser light also alters the distribution of volatile reaction products, as shown in Fig. 6.2. At low light intensities SiF_4 is the dominant product, with some SiF^+ and SiF_2^+. At high intensities SiF_4 still predominates, but much SiF_3^+ is observed and SiF^+ and SiF_2^+ are much weaker. Under spontaneous etching, even at elevated temperatures, SiF_3 is only formed in small amounts. Thus the effect of heat and light induce very different chemical pathways in the etching of Si by XeF_2.

When Si is exposed to Cl_2 gas in the absence of light, spontaneous etching does not occur. There is a surface reaction that stops after a thin passivation layer is formed (Sesselmann, 1985). At least part of this layer (formed in Cl_2 at 0.1 torr at room temp) can be removed with a pulsed N_2 laser at 0.12 J/cm^2, resulting in relatively low etch rates ~1 Å/pulse for Si films (Sesselmann 1985). Under relatively low temperature and pressure conditions, surface halogenation of silicon is similar to surface oxidation

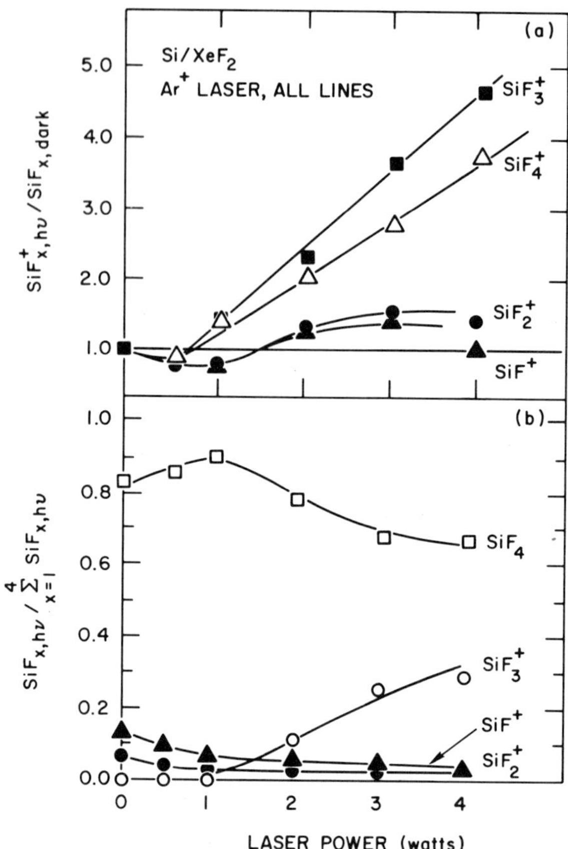

Fig. 6.2 Effect of light on the product mass distribution of Si etched in XeF_2.

of metals as described by the Mott–Cabrera theory (Winters, 1983b; Chuang, 1984; Cabrera, 1949; Fehner, 1970). The surface halide grows by electron tunnelling from Si through a very thin halide layer to the surface where new negatively charged halogen ions are formed following dissociation of halogen molecules. The high fields that result in the surface region strongly assist the diffusion of ions and the formation of additional halide. However, the probability of electron tunnelling falls off exponentially with increasing halide thickness and the electric field decreases as well. The result is that the Mott–Cabrera mechanism results in very thin halide layers.

While Si is not spontaneously etched by Cl_2 gas, it is etched by chlorine atoms (Ehrlich, 1981a). The etch rates are orders of magnitude lower than for the purely thermal processes described earlier. The reactive chlorine radicals can be created by photolysis of Cl_2 gas, which has a broad dissociative continuum centered at about 330 nm (Aickin, 1937). Thus the etching of Si in chlorine gas with a pulsed N_2 laser described above (Sesselmann, 1985) is due partly to the ablation of the surface halide that forms in the dark and partly to the etching caused by photoproduced chlorine atoms.

In the presence of UV light, heavily doped n-type poly-Si was shown to be spontaneously etched by Cl radicals generated in the gas phase by photodissociation of Cl_2 (Arikado, 1984). In this work it was shown that two conditions are needed for a substantial etch rate: a large number of electrons in the conduction band that can form halides by the Mott–Cabrera mechanism and Cl radicals. These conditions apparently result in significant amounts of $SiCl_4$, (a volatile liquid) being formed on the surface. When the number of conduction-band electrons is reduced as in undoped or p-type Si, no etching can occur, even in the presence of gas-phase Cl radicals. However, if UV light is incident normally on the undoped or p-type Si surface, then electrons are available to form halides and etching occurs (Arikado, 1984). This later condition was explored by Ehrlich, Osgood, and Deutsch for a focused argon-ion laser normally incident on Si (100) and (111) in a chlorine-gas atmosphere (Ehrlich, 1981a).

A detailed study of the etching of single-crystal silicon in 100 torr of Cl_2 using an unfocused pulsed excimer laser at 308 nm was also carried out (Arikado, 1984). The etch rate dependence as a function of conduction type, sheet resistance, and beam orientation (parallel or perpendicular to the surface) was reported. The etched depth, normalized to a total dose of 1 J/cm^2 (this required several laser pulses), is given as a function of silicon sheet resistance in Fig. 6.3 for a laser beam at normal incidence to the surface. In n-type Si (phosphorus doped) the etch depth decreases dramatically with increasing sheet resistance. Since the laser parameters are not changing, this can only be due to the changing reactivity of Si and Cl_2 to form volatile halides or halides that are ablated by the laser. In highly doped n-type Si many electrons are available and the highest etch rates are obtained. As the number of electrons decreases, the etch rate drops. The least number of available electrons occurs in heavily doped p-type Si and here the etch rates are lowest.

The difference in etch rates on the (100) and (111) surfaces, as shown in Fig. 6.3, can also be explained by reaction rate limitations. In p-type and lightly doped n-type Si the etch rate is higher on the (100) face than on the (111) face. This difference has been ascribed to the more open

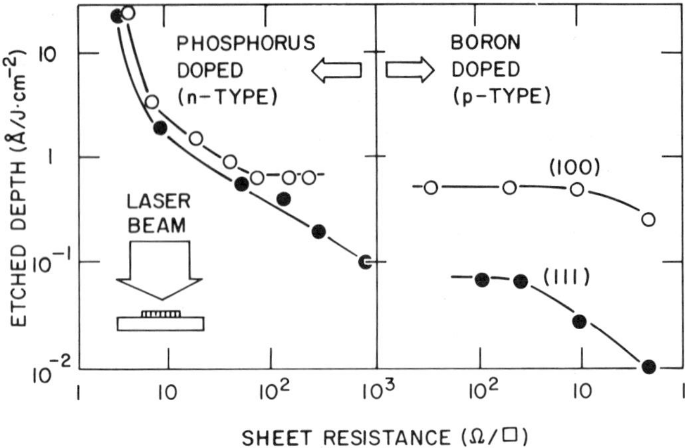

Fig. 6.3 Etched depth per unit energy versus sheet resistance for n- and p-type Si etched in Cl_2 by XeCl laser.

nature of the (100) face, which allows easier penetration of the chlorine atoms than the close-packed (111) face (Arikado, 1984). The ease in forming the surface chloride then controls the etch rate.

The dependence of the etch rate of Si in Cl_2 under 308-nm excimer laser irradiation parallel or perpendicular to the surface of (111) and (100) Si is shown in Fig. 6.4. In the case of parallel irradiation the beam passed about 1 mm above the surface. Since p-type Si was not etched without direct irradiation, Fig. 6.4 shows data for n-type Si. In highly

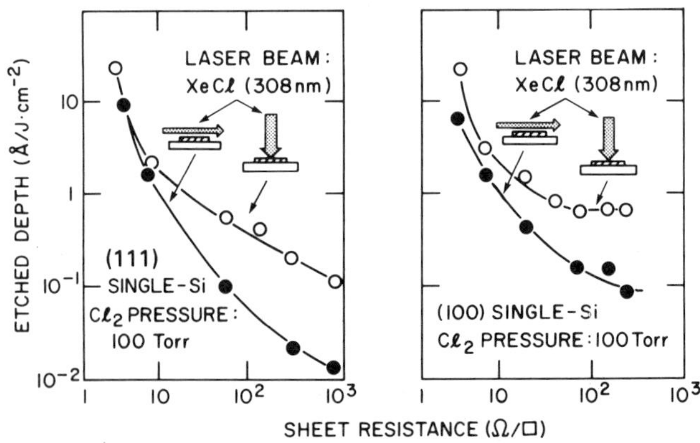

Fig. 6.4 Etched depth per unit energy versus sheet resistance for (111) and (100) Si etched in Cl_2 by XeCl laser with parallel and perpendicular beam orientation.

doped samples, direct irradiation of the surface does not greatly increase the electron concentration at the surface of n-type Si. Hence the etch rate is almost entirely due to spontaneous chemical reaction with Cl radicals generated in the gas phase by dissociation of Cl_2. Under these conditions, the etch rates for parallel and perpendicular radiation are nearly the same. In lightly doped samples, on the other hand, photoelectrons can be a major portion of the total electrons in the conduction band, and hence direct irradiation greatly enhances the etch rate.

Using the etching mechanisms described above, Arikado et al. have created very fine VLSI patterns (64K DRAM) in Si by direct irradiation with an excimer laser of the Si surface in Cl_2 gas. The pattern in silicon was created by placing the sample in near proximity to an Al pattern on a quartz mask plate. High-resolution etching was also obtained using patterned SiO_2 on the Si surface. In this case the more isotropic nature of radical etching of heavily doped n-type samples was apparent. For p-type or lightly doped n-type samples the etching is more anisotropic, since only the illuminated regions have a substantial electron population.

The etching of silicon by photogenerated fluorine radicals was studied by Loper and Tabat (1985a, 1985b). They used a pulsed ArF excimer laser (193 nm) to photolyze carbonyl difluoride (COF_2) and nitrogen trifluoride (NF_3) in the gas phase to create fluorine atoms that reacted with silicon to form volatile SiF_4. One particular experiment involved etching of a 4000-Å polysilicon layer deposited on silicon dioxide in an atmosphere of 260 torr COF_2, 500 torr He. The output of the ArF laser at 10 Hz was weakly focused onto the surface with a fluence of 85 mJ/cm^2 for various exposure times. This fluence was too low to ablate the film in the absence of the gas. The etched depth versus total exposure dose is shown in Fig. 6.5. No etching occurs until a dose of 1.5×10^2 photons/cm^2. This may be due to the removal of a thin oxide layer, as observed for the etching of Si in Cl_2 with an argon-ion laser (Ehrlich, 1981a). Above the threshold dose the etch depth varies linearly with dose until all the polysilicon has been removed. This shows that thin films of polysilicon on SiO_2 can be quite controllably etched by photogenerated radicals. The etch rate, however, is quite low (0.13 Å/laser pulse). Loper and Tabat point out that this low rate can still be quite useful with the advent of commercially available high-power excimer lasers. Assuming a high-pulse-rate, 200-W excimer laser under the conditions of their experiment, they estimate that 4000 Å of polysilion could be removed from a 9-cm^2 area of a wafer at the rate of ~2000 Å/min. This rate is comparable to rates obtained by current RF discharge plasma etching (Loper, 1985a).

Silicon can also be etched by fluorine atoms photogenerated by

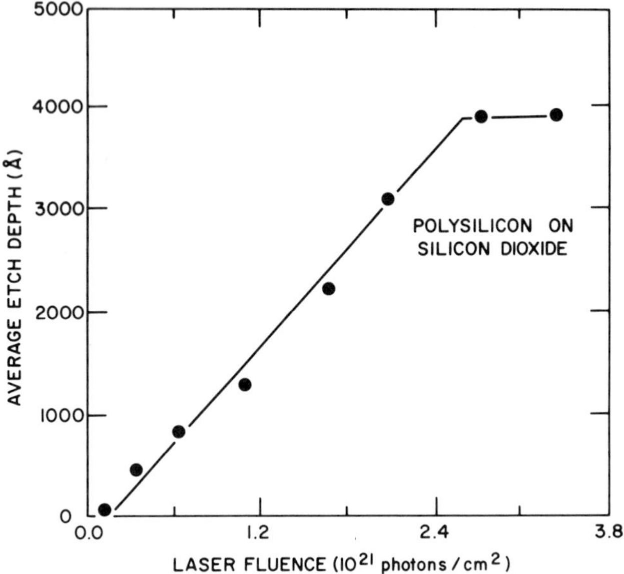

Fig. 6.5 Etched depth versus laser fluence for laser-induced radical etching of polycrystalline Si on SiO_2.

multiphoton dissociation of SF_6 by CO_2 radiation (Chuang, 1980, 1981; Ambartsumyan, 1982). Typical experiments use a TEA CO_2 laser at normal incidence with an energy of up to 5 J/pulse and a pulse rate of up to 5 Hz (Ambartsumyan 1982). The etch rate was determined by measuring the amount of SiF_4 that formed in the sample cell during irradiation. The calculated rates were 15–20 Å/pulse, which indicates that large amounts of fluorine were created with each pulse since each pulse effectively removed several silicon layers.

Germanium has also been shown to be etchable by both laser-assisted thermal and photochemical processes. Some very early work (Sullivan 1968) involved photodissociation of halogens (Br_2 and Cl_2) by a focused cw argon-ion laser. Thin evaporated Ge films (3600 Å thick) on a glass substrate were etched and patterned with a resolution of 5 μm. A combination of photochemical and photothermal effects resulted in etch rates approaching 36 μm/sec.

Germanium etching by a mechanism involving surface chemical reactions with Br_2 gas at 5×10^{-5} torr followed by laser thermal ablation of the reaction products has been studied in detail by workers at A.T.T. Bell Laboratories (Davis, 1984). In this work an excimer-pumped dye laser was tuned to 488 nm and a time-of-flight mass spectrometer was used to monitor the masses and kinetic energy of ablated species. At fluences above 0.5 J/cm^2, atomic Ge was observed in the mass spectrometer due to simple thermal ablation independent of bromine ambient. Without Br_2 no products were observed below 0.5 J/cm^2. However, in the presence of bromine gas at fluences of 120–300 mJ/cm^2, significant amounts of $GeBr_2^+$ and $GeBr^+$ were observed. The time-of-flight distribution could be fit reasonably well by a Maxwell–Boltzmann distribution with an effective temperature of ~2000°K. Small deviations from the Maxwell–Boltzmann distribution were attributed to collisions in the desorption plume or potential energy barriers along the reaction coordinate. In any case, these measurements, as well as some calculations, indicate that the germanium surface is melted by the laser pulse. Under these conditions an etch rate of only 0.1–0.2 Å per pulse is obtained. This low value is due to the relative unreactivity of Ge and Br_2 at room temperature.

When Br_2 is photodissociated to form atomic bromine, there is a spontaneous reaction with germanium, which etches it quite rapidly (Baklanov, 1974; Beterov, 1978). In these studies, a 40-W argon-ion laser tuned to either $\lambda = 488$ nm or 514.5 nm was incident at glancing angles to the surface of thin germanium plates in an atmosphere of bromine gas. At 488 nm Br_2 is photodissociated with a quantum efficiency close to 1, and at 514 nm it is effectively pumped to a predissociation state. Although pressures were varied from 0.01 to 200 torr, a maximum etch rate was observed at 2–3 torr, which corresponded to about 860 Å/sec ($\lambda = 488$ nm). During these experiments the sample was held at about 200°C in the bromine gas, under which conditions there is a small amount of thermally driven reaction. However, the effect of the argon laser is to accelerate the reaction rate by three or four orders of magnitude. The reaction mechanism involves several steps, including photodissociation in the gas phase, adsorption and the formation of various bromides, and finally desorption of $GeBr_4$ gas. Clearly the creation of reactive bromine atoms by photodissociation is the key to high etch rates.

While photodissociation of Br_2 at $\lambda = 488$ or 514 nm can rapidly etch germanium, it is also possible to generate reactive species in the gas phase by thermal dissociation of molecules containing halogens (Karlov 1985). Karlov et al. used a 25-W CO_2 laser to heat a gas containing SF_6 as well as Br_2 or CF_3I. The SF_6 strongly absorbs the radiation, heating the gas to temperatures of ~1000°C, which effectively dissociates the halogen-containing molecules. Surface-localized etching is achieved by making the germanium wafer one window of the gas cell and irradiating the wafer from the back surface. Since germanium does not absorb the

10.6 μm radiation, a hot gas zone is formed at the germanium surface exposed to the gas mixture. Although no quantitative etch rates were reported, it was shown that Ge, Si, GaAs, and ZnSe could be etched by this technique using CF_3I or Br_2.

2.2. III–V Compounds

Gallium arsenide and other III–V semiconductors are important for a variety of applications. Controlled anisotropic etching is necessary for the fabrication of integrated optical components and microwave devices, for example. Using visible laser wavelengths and specific liquid etching solutions, Osgood et al. (1982) have demonstrated the fabrication of high-resolution diffraction gratings and through wafer vias in Cr-doped (semi-insulating) GaAs, n-type GaAs, and InP. For semi-insulating, SI, GaAs low lasers powers produced gratings with periods less than 1 μm. An acidic solution involving H_2SO_4, H_2O_2, and water that etches the SI GaAs at 0.7 μm/min in the dark was used. The laser enhances the etch rate locally by producing electron-hole pairs in the semiconductor. The spatial resolution is limited by the lateral spreading of the photogenerated carriers and is estimated to be on the order of 100 nm. The etching mechanism is clealy photochemical since the estimated temperature rise for the laser powers used was less than 0.1°C and wavelengths below the GaAs band produced negligible etching.

Osgood et al. also showed that by focusing the laser to approximately an 8-μm spot, through wafer vias could be etched in roughly 4 minutes (30 μm/min). Under the conditions of these high intensities, 0.2 MW/cm^2, a variety of thermal and optical processes influence etching, and the detailed mechanism is not easy to determine.

Osgood et al. (1982), Tisone and Johnson (1983a, 1983b), and Rauh and LeLievre (1985) have studied the etching of n-type GaAs in a basic (KOH) solution. The results of etching n-type GaAs in KOH show a strong saturation effect, most likely due to production of nonsoluble products (Tisone, 1983a). There is also evidence of the effects of diffuse scattering on the etched features (Rauh, 1985). GaAs (n-type) is more efficiently and cleanly etched in an acidic (HNO_3) solution where no surface residue is observed (Tisone, 1983a).

Tisone and Johnson (Tisone, 1983b) point out that n-type compounds are more easily photoetched than p-type compounds. Figure 6.1 shows that optical excitation of electron-hole pairs results in excess holes being swept to the surface of n-type compounds by the space-charge field in the semiconductor. Hence n-type compounds can decompose oxidatively,

$$GaAs + 6h^+ \rightarrow Ga^{+3} + As^{+3}.$$

Conversely, p-type compounds are photoreduced

$$GaAs + 3e^- \rightarrow Ga + As^{-3}.$$

In the later process Ga metal remains on the surface, whereas in the former ions can go into solution easily.

In the study of photoetching of semi-insulating Cr-doped GaAs in HNO_3, the etch rate was compared with the optical absorption coefficient as a function of wavelength near the band gap (Tisone, 1983b). These results are given in Fig. 6.6. Here it is seen that although the total absorbed intensity does not vary significantly, the measured etch rate follows the optical absorption coefficient closely. Thermal diffusion calculations show that the surface temperature of GaAs depends only on the total laser power absorbed. Thus, if the laser-induced etching were a thermal process, strong wavelength dependence would not occur, contrary to observation. Hence, Tisone and Johnson conclude that photogenerated holes in n-type and semi-insulating GaAs are responsible for the laser-inducing etching.

In addition to the early work of Osgood et al. (1982), several groups have studied the formation of gratings on GaAs surfaces by laser etching techniques (Tsukada, 1983, 1984a; Aoyagi, 1985). A typical experimental setup is shown in Fig. 6.7. Using this setup, Tsukada et al. (1984a) have formed a series of gratings using 568- and 647-nm light and aqueous solutions of HF or H_2SO_4. At high intensities, HF solutions yielded smoother grating surfaces than H_2SO_4 solutions. The measured etch rates

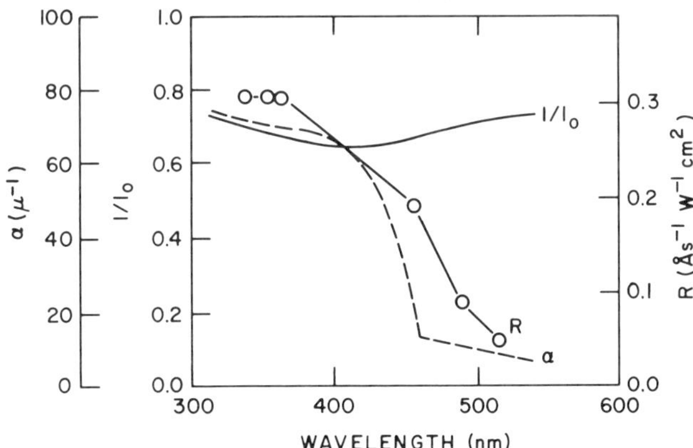

Fig. 6.6 Photochemical etching of Cr-doped GaAs in HNO_3 solution: Wavelength dependence of total absorbed light (I/Io), absorption coefficient (α) and etch rate per unit intensity (R).

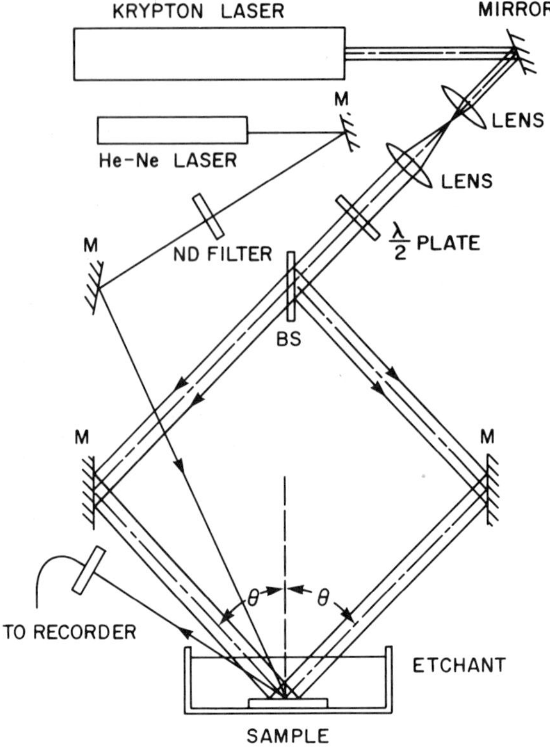

Fig. 6.7 Experimental setup for formation of gratings by holographic methods.

were 0.7, 0.4, and 0.1 μm/min for n-type GaAs, GaAlAs, and Cr-doped SI GaAs. As expected from the arguments of Tisone and Johnson (1983b), p-type GaAs showed negligible etching. Tsukada et al. also measured the grating efficiency. The highest diffraction efficiency was only 2.5% of the incident He-Ne monitoring laser beam. However, taking into account the reflectivity of the various interfaces, the ratio of the diffracted and reflected intensities is about 4.

In addition to the holographic setup shown in Fig. 6.7, Tsukada et al. (1983, 1984a) investigated gratings spontaneously generated by a single laser beam. A single beam creates an interference pattern with a stimulated surface electromagnetic wave. Tsukada et al. point out that in the visible or infrared, a wave such as a surface plasmon or surface polariton that propagates parallel to the surface of GaAs does not satisfy the Maxwell boundary conditions at the interface. Nevertheless, they speculate that an incident photon can be Raman scattered into a state composed of a stimulated surface polariton wave and an energy shifted photon. Another suggestion is that a polarization charge is induced on defect boundaries by the laser field and "results in a sinusoidally varying perturbation of the applied laser field in the vicinity of the defect." Yet, regardless of the mechanism, intriguing periodic structures have been demonstrated. Using a frequency-doubled Nd:YAG laser at 100 mW/cm^2, a 0.4-μm period grating structure was created on Si-doped GaAs. This period corresponds quite well with the laser wavelength in the aqueous solution. The surface ripples created were always well aligned perpendicular to the laser polarization direction. In cross section, the ripples had a triangular shape, with depths more than 2 μm. These very sharp V grooves could not be made by the holographic setup of Fig. 6.7.

The fabrication of high-quality distributed feedback laser gratings by maskless laser etching techniques has been demonstrated by Aoyagi et al. (1985). Samples used were n-GaAs and n-InP (100) single crystals. The GaAs was etched in an aqueous solution of I_2 and KI while the InP was etched in an AZ-303 developer. Etching was accomplished with an argon-ion laser run at 457.9 nm using a conventional holographic technique. Using a rigid covered etching cell, very high quality gratings were obtained. The etched profile and hence diffraction efficiency varied with the groove direction relative to the crystal directions. In InP the diffraction efficiency of the grating with grooves parallel to the $[0\bar{1}1]$ direction is about 4% and the depth of the groove is about 130 nm. Under the same irradiation conditions (4 W/cm^2 for 10 min.) grooves in the [011] direction are shallower and an efficiency of only 0.5% is obtained.

A novel aspect in the laser etching of semiconductors is the ability of the laser to create its own hollow optical waveguide as it etches the semiconductors (Podlesnik, 1984, 1986). Extremely high aspect ratio holes and slits can be created in this fashion. Typical results involve single-crystal GaAs in a dilute HNO_3 aqueous solution that manifests no dark etching and only weakly absorbs the laser light. The laser used was a frequency-doubled argon-ion laser at 257 m, focused to a 3-μm diameter spot at the wafer surface. Typical power densities were 1 W/cm^2, which resulted in negligible temperature rise. Under these conditions a narrow slit 2-3 μm wide and 200 μm long could be etched through the wafer by translating it perpendicular to the laser beam direction. The walls are smooth and vertical, and there is a very little dependence on the crystal face being etched. The measured etch rates for making vias in samples of varying thicknesses are 12 μm/min for a 40-μm thick sample and 7 μm/min for a 250-μm thick sample. These differences are most likely due to the limited mass transport of the etched material out of the narrow high-aspect ratio features.

Thus far the laser etching of GaAs, which has been described, has involved photochemical etching in liquid solutions. Several studies have been made of laser etching of GaAs in a gaseous environment (Ashby 1984, 1985a, 1985b, 1985c). One purpose of these studies was to overcome some of the problems associated with dry reactive ion etching of GaAs. In reactive ion etching, the ions responsible for etching also produce lattice damage in the surface region that must be removed by further annealing. Moreover, reactive ion etching does not exhibit much selectivity between materials with different band gaps or doping levels.

The method employed in these dry photoetching studies (Ashby 1984, 1985a, 1985b, 1985c) employed a focused argon-ion laser at 514.5 nm, normally incident on crystalline GaAs and GaP of various types and doping levels. A plasma source produced an atmosphere of reactive Cl atoms in a gas mixture of 0.2% HCl in He. The sample was held at about 100°C and could be biased electrically. Since the plasma source was upstream of the sample, only neutral Cl atoms were present in the vicinity of the sample. At the concentrations used, there was negligible dark etching. A typical etch profile duplicated the Gaussian shape of the beam. Etch rates were of the order of 5 Å/sec for laser powers of 1.4 KW/cm^2. The etch rates increased linearly with the laser power density, indicative of photochemical rather than photothermal processes.

The necessity of creating a number of electron-hole pairs near the sample surface for efficient photoetching of III–V compound semiconductors was shown by comparing GaAs and GaP (Ashby, 1985c). The direct band gap of GaAs at 400°K is 1.4 eV with an absorption coefficient of 10^4/cm. GaP has an indirect band gap of 2.2 eV with an absorption coefficient of 10/cm and a direct gap of 2.8 eV with an absorption coefficient of 10^4/cm. Laser light at 2.54 eV etches GaAs but not GaP. Thus laser etching exhibits material selectivity by matching the wavelength with strong band-gap absorption.

By voltage biasing the GaAs samples in the Cl atom atmosphere, it is actually possible to selectively etch samples with different doping levels (Ashby, 1985a, 1985c). This is illustrated in Fig. 6.8. Photochemical reactions are often limited by the availability of minority carriers at the surface. As illustrated in Fig. 6.1, for n-type semiconductors, photogenerated holes are swept to the surface by the fields in the space-charge region and can participate in photochemical reactions. However, if an n-type sample is negatively biased, then the band bending is reduced or even inverted, resulting in few photogenerated holes reaching the surface. This is seen in Fig. 6.8a where an n-type sample doped to

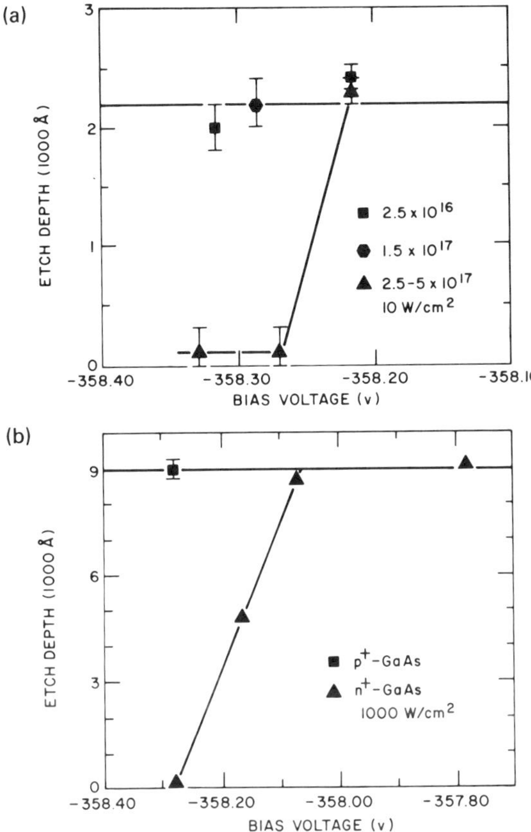

Fig. 6.8 Photochemical dry etching of GaAs—etch depth versus bias voltage and incident intensity: (a) n-GaAs at different doping levels at 10 W/cm^2 (b) p- and n-type GaAs at 1000 W/cm^2.

$2.5–5 \times 10^{17}$ carriers/cm^3 shows very little etching when biased near -358.3 volts. The large voltage is necessary since a large voltage drop across the gas-filled space above the sample surface is necessary to significantly reduce the band bending in the sample. This does not occur for lower doping levels because for these samples the band bending is greater to begin with, and the external voltage cannot reduce it enough to keep photoholes from reaching the surface. Figures 6.8a and 6.8b show the power dependence of these effects on n- and p-doped GaAs.

There are two types of photochemical reactions that can effectively etch compound semiconductors. The first, which has been described above, involves creation of electron-hole pairs in the semiconductor that drive surface chemical reactions. The second involves photochemical reaction in gas or liquid media to produce reactive species that then etch the semiconductor. This latter effect has been studied for gas-phase reactants (Ehrlich, 1980; Brewer, 1984, 1985) and for liquid-phase reactants (Haynes, 1980).

A frequency-doubled argon-ion laser (257 nm) normally incident on the sample surface, was used to photodissociate CH_3Br and CF_3I, which lead to etching of GaAs and InP under conditions where essentially no etching is observed in the dark (Ehrlich, 1980). The etching characteristics for amorphous GaAs films are given in Fig. 6.9. The etch rates for crystalline semiconductors are quite similar. For example, in 750 torr of CH_3Br with an incident power of 100 W/cm^2 at 257 nm, n-GaAs(100) and n-InP(100) etch at 5.2 Å/sec and 9.4 Å/sec, respectively.

The mechanism of photochemical etching involves, first, photodissociation of the halogen-containing methane. The linear relationship between etch rate and laser intensity at low pressure (Fig. 6.9) indicates that this process initiates the etching. Then some bromine atoms chemisorb on the GaAs surface, while others recombine in the gas phase in three-body processes (Ehrlich, 1980). The surface reaction products thermally desorb. At low pressure and laser flux the time for the photogenerated bromine atoms to diffuse to the surface is shorter than the recombination time. Under these conditions, the etch rate is controlled by the bromine atom production rate, which depends linearly on pressure and beam intensity. However, at higher pressures the recombination rate increases and limits the flux of bromine atoms hitting the surface so the etch rate tends to saturate.

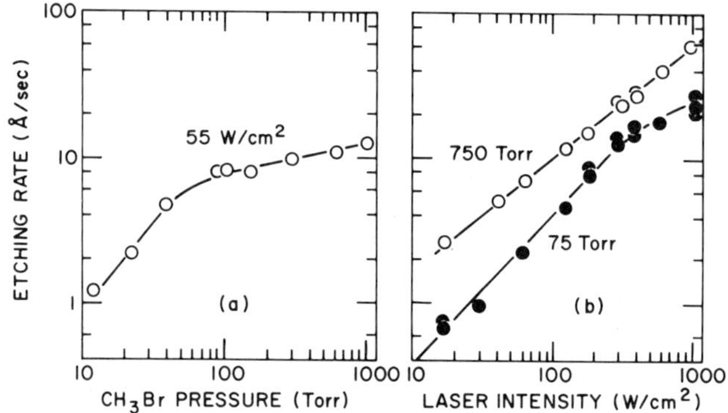

Fig. 6.9 Radical etching of amorphous GaAs in CH_3Br: (a) etch rate versus pressure at constant laser intensity; (b) etch rate versus intensity.

Another interesting aspect of the photochemical etching of compound semiconductors is the high resolution attainable despite the fact that the reactive species are generated in the gas phase. A spatial resolution on the order of 1 μm was demonstrated. This can actually be simply understood from a geometrical model of the atom production (Ehrlich, 1980). Calculations show that relatively localized features result from the portion of the beam within one beam radius of the surface under typical operating conditions. Atoms created at larger distances from the surface contribute only to a diffuse background. Therefore, the ultimate spatial resolution of the etching process is determined to a large extent by the laser intensity profile near the solid surface.

The dry etching of GaAs by photogenerated radicals was studied by Brewer et al. (1984, 1985, 1986) using a pulsed excimer laser operating at 193 nm. Gas-phase photodissociation of several methyl and trifluoromethyl-halogen compounds was used to etch fine steep-walled features in GaAs using an in-situ contact etch mask made of a spun-on liquid organosilicate that was patterned with a focused 350-mW argon-ion laser by a laser-curing mechanism (Brewer, 1984). The chemistry of the photogenerated species is quite complex, with the observed etch products being both Ga and As compounds containing at least one methyl or trifluoromethyl group as well as fully halogenated compounds. These result in a multicomponent liquid as well as gaseous products. The desorption of trifluoromethyl derivatives of Ga and As was confirmed by UV absorption spectra of the cell gases.

A much less complicated product distribution was observed in the case of GaAs etched by photogenerated atomic species created by 193-nm laser light incident on a sample in a flowing HBr gas atmosphere (Brewer, 1985). For laser light incident on the etching cell in such a way that it does not hit the sample, there is a slight but steady increase in etch rate with increasing laser power. For light that hits the surface directly, there is a threshold at about 10 mJ/cm^2 above which the etch rate increases linearly with laser power. This threshold is believed to be due to laser-initiated desorption of the reaction products, either due to electronic excitation or transient substrate heating ($<150°C$). These products, which are primarily $GaBr_3$ and $AsBr_3$, have thermal barriers to desorption of 13.5 and 10.5°K cal/mole, respectively. The highest etch rate measured was 8.2 μm/min at 32 mJ/cm^2 with the substrate heated to 60°C. Above about 40 mJ/cm^2 there is direct ablation of the GaAs at 193 nm.

The importance of surface reaction products is illustrated in Fig. 6.10. Here the variation of the etch rate as a function of temperature for fixed laser power and gas pressure is shown. There are three distinct regions

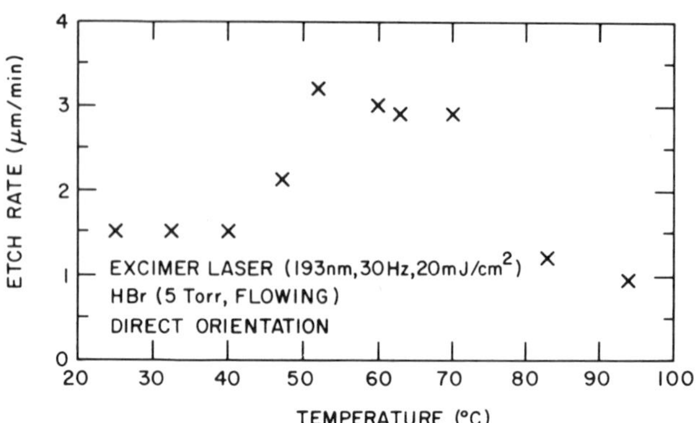

Fig. 6.10 Radical etching of GaAs in HBr as a function of substrate temperature.

with different surface morphology. At low temperatures, 20–42°C, the etch rate is reduced by liquid etching products on the surface. Surface x-ray emission studies showed that these were bromide salts of Ga and As. The surface morphology is smooth with droplet structures. Above 50°C no liquid drops are seen, surfaces are smooth and flat, and the etch rate increases by a factor of two. Above 80°C a blue-brown deposit covers the surface and the surface is grainy and rough. The etch rate is low. X-ray analysis indicates both oxygen and bromine on the surface. The oxide that inhibits etching is apparently a residual impurity in the vacuum system (Brewer, 1985).

When crystals of different orientation were etched, it was observed that different crystal planes etched at different rates (Brewer, 1985). The rates followed the sequence: (111) B > (100) > (110) > (111) A. Since the photoetching experiments described here involve attack of GaAs by Br atoms, the relative etch rates can be compared to low-energy Br_2 plasma etching. In plasma etching, a similar sequence is found (Brewer, 1985).

Brewer et al. (1986) also studied the effect of buffer gases on the etching of GaAs in HBr gas using 193-nm excimer laser radiation. Significant etching is due to photocreated radicals in the gas phase that react with the semiconductor to produce volatile products. Since the radicals are created in the gas, they diffuse to the surface, and if the gas pressure is low, the etched features can be several times as large as the projected pattern linewidth. However, if buffer gases such as argon or hydrogen are introduced, the mean free path of the radicals is limited and much narrower features can be etched. For example, the measured linewidth narrows from 11 μm to 4 μm (the projected optical feature size) when the argon buffer gas pressure is increased from 0 to 20 torr. On the other hand, by restricting the number of radicals that reach the surface, the argon buffer gas reduces the etch rate by a factor of about seven as pressure is increased from zero to 20 torr.

The etching of GaAs by photogenerated reactive species can also take place in a liquid environment (Haynes, 1980). Haynes et al. showed that with a focused Kr^+-ion laser n-type GaAs could be controllably etched in solutions containing dissolved Br_2 or I_2. A typical solution contained 0.1–1% Br_2 and 1–10% of an alkali bromide by volume in distilled water. The solutions were kept cold (5°C), and low laser fluences were used resulting in <1°C temperature rise. The etch rate increased linearly with power density, indicative of photochemical and not photothermal processes. In addition to the optically dissociated halogens, there is evidence of electrolytic (electron-hole pair) effects. GaAs which is n-type etches well because "dissolution proceeds faster if holes are optically injected over the surface." The etching reaction is anodic oxidation. For this reason Cr-doped semi-insulating GaAs did not etch at all, even under high laser power (Haynes, 1980).

While finer features with sharper edges can be created with the photochemical processes described above, coarser features can be etched very rapidly in GaAs by pyrolytic laser-induced chemical processes (Takai, 1983, 1984a, 1984b, 1985). The first of these studies (Takai, 1983) involved an argon-ion laser focused to about a 20-μm spot with laser powers ranging from 0.1 to 0.6 W. The GaAs sample was placed in an atmosphere of CCl_4, Cl_2, or $SiCl_4$. Grooves were etched in the GaAs by scanning the part perpendicular to the laser beam. Typical results are shown in Fig. 6.11. For scan speeds of 6 μm/sec and above, a clean nearly Gaussian profile trench as etched in the GaAs that was narrower than the laser beam profile. The minimum trench width was about 7 μm (a factor of three narrower than the laser beam). These results indicate that etching takes place only in the central portion of the laser beam where the temperature rise is high enough for pyrolytic reactions to take place. In the case of CCl_4 gas, for example, no energy is absorbed in the gas and only substrate heating can occur. The gaseous products of the etching reaction are $AsCl_3$ and $GaCl_3$. In the case of low scanning speed (3 μm/sec), the etched shape was quite different and a deposited layer of carbon filled the etched trench. Apparently, at the temperatures reached at these low scanning speeds both deposition (from pyrolytic decomposition of CCl_4) and etching take place, while at higher scan speeds only etching occurs. In a follow-up study (Takai, 1984a) microscope optics were used to focus the beam to 1.2 μm. By varying scanning speed, CCl_4

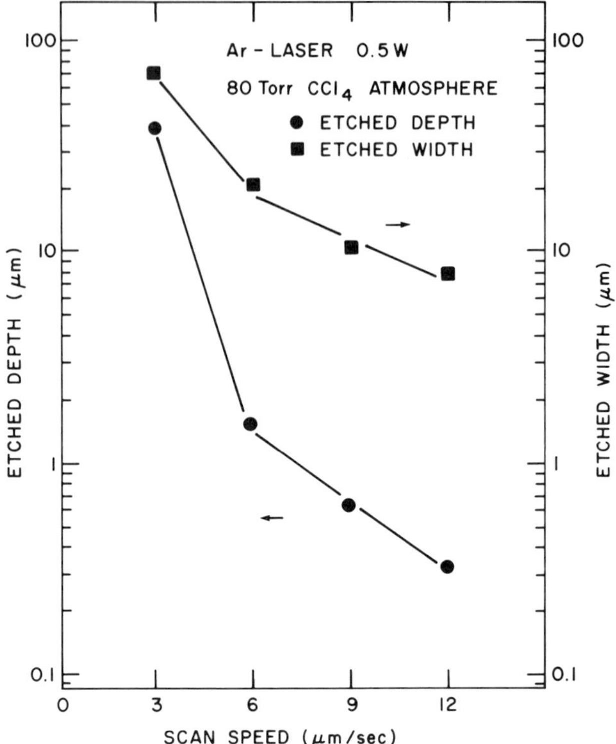

Fig. 6.11 Pyrolytic etching of GaAs in CCl_4 gas as a function of beam scanning speed.

pressure and laser intensity lines as small as 0.6 μm wide could be etched in GaAs. Typical trenches had a rounded cross section and were 0.2 μm deep.

In a more recent study (Takai, 1985), laser-induced pyrolytic etching of GaAs in $SiCl_4$ was studied to determine the minimum laser power and local temperature required for etching. The local temperature rise during etching was measured by the shift in the band-to-band emission peak of the photoluminescence spectrum. Comparisons were made with calculated temperatures by solving the three-dimensional heat equation. The measured temperature as a function of incident laser power was in good agreement with the theoretical model. The threshold power for pyrolytic etching at a $SiCl_4$ pressure of 76 torr was found to be about 0.25 W for a laser spot size of about 13 μm. This corresponds to a temperature of 192°C. This is reasonably close to the boiling point of $GaCl_3$ (201°C at

760 torr), so as to suggest that vaporization of reaction products may be the etch-rate-determining step.

In addition to making fine grooves (Takai, 1983–85), pyrolytic processes were shown to etch through wafer vias extremely rapidly (Tucker, 1983). Using the multiline output of an argon-ion laser at up to 4 W, etch rates on the order of 20 μm/sec were measured in an atmosphere of 5 torr of Cl_2. The etch rate saturates once the melting point of GaAs is reached. Even higher etch rates (33 μm/sec) were observed for monochromatic light at 488 nm. This difference may be attributable to Cl_2 absorption and dissociation that occur along with pyrolytic reactions at the molten GaAs surface.

Photoelectrochemical etching of InP and GaInAsP has been demonstrated to be important for creating several structures needed for semiconductor lasers (Bowers, 1985; Lum, 1985). Mirrors of a semiconductor laser are made by a cleaving procedure normally initiated by manual scribing. This is difficult where precise alignment is necessary. Bowers et al. have shown that photoelectrochemical etching can be used to make deep narrow grooves into the laser substrate to weaken the crystal along specific lines and induce cleaves along these lines when the wafer is flexed (Bowers, 1985). Additionally, as mentioned earlier in the discussion on GaAs etching, a maskless technique for producing gratings on distributed feedback lasers is expected to have many advantages over present multistep photoresist processes. This has been accomplished for InP and GaInAsP using photoelectrochemical techniques by Lum et al. (1985).

A schematic diagram of a typical photoelectrochemical etching apparatus is shown in Fig. 6.12. A sample of InP is immersed in a conducting solution with a low dark etch rate such as 20:1 H_2O:HCl. A bias voltage is applied, as shown in Fig. 6.12, and the current measured. An n-type substrate is positively biased (typically 0.4–0.6 V) and, using a potentiostat, applied voltages can be measured with respect to a saturated calomel reference electrode. Using a laser with a wavelength such that band-gap excitations are created (for example an argon-ion laser), a pattern can be etched at rates of 50 μm/hr over a large area. After one calibration run with a given mask pattern, the etch rate could be accurately predicted from measurements of the etching current and sample area. Using this method narrow, high aspect ratio grooves were etched that successfully induced cleaves for laser facets, resulting in a high yield of low-threshold lasers (Bowers, 1985).

Using a schematically similar apparatus, Lum et al. etched InP and GaInAsP photoelectrochemically in a 2-M HF/0.5-M KOH solution with the sample controlled between 0.2 and 0.5 volts relative to a saturated

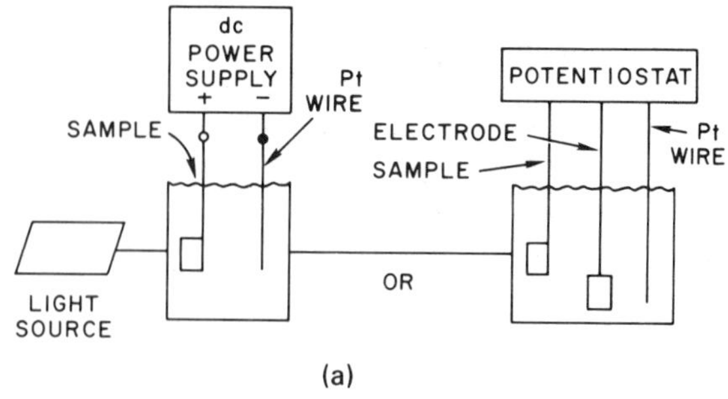

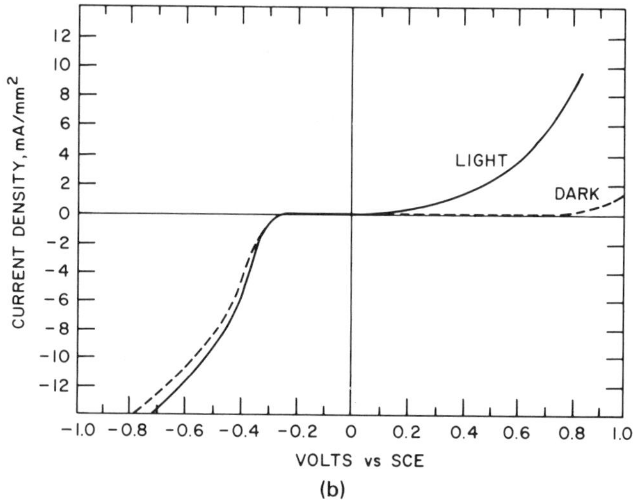

Fig. 6.12 Photoelectrochemical etching of InP: (a) Schematic diagram of etching configuration; (b) dependence of etching current on bias voltage for dark and illuminated n-InP.

calomel electrode (Lum, 1985). A laser interferometer was used to create a sinusoidal spatial intensity variation across the sample surface. Several lasers were used, including HeNe (632.8 nm) argon ion (488 nm), and HeCd(441.6 nm). Etching occurs by oxidation induced by photogenerated holes, and the etch rate ratio between light and dark areas is greater than 100:1. The quantum efficiency for decomposition is 100%. A two-dimensional hole transport model was used to fit the measured sensitivity of these holographic gratings as a function of spatial frequency. There was good agreement between the quantitative predictions of this model and the experimental data. At high spatial frequency the etching is limited by surface-hole diffusion. The measured first-order diffraction efficiency is 20–25% for 8.5-μm gratings and can approach 15–20% for 1.5-μm gratings. Gratings with periods as short as 0.5 μm were produced. Some gratings were of such quality that diffracted beams out to the seventh order could be observed.

Using a photothermal process, Bjorkholm and Ballman showed that laser-driven localized wet-chemical etching of InP could result in very rapid etching of grooves (Bjorkholm, 1983). Using an argon-ion laser at 514.5 nm focused to a 9-μm spot, with the sample in a dilute aqueous solution of H_3PO_4, grooves 7–10 μm deep could be etched while scanning at 100–200 μm/sec. The groove depth did not vary much with scan speed so long as the laser power was increased somewhat at faster scan speeds. For a given scan speed, the groove depth varied nearly exponentially with laser power, as shown in Fig. 6.13. This is exactly what is to be expected for a thermally activated process where the sample temperature is proportional to laser power.

Gallium phosphide is a III–V semiconductor that can be used to

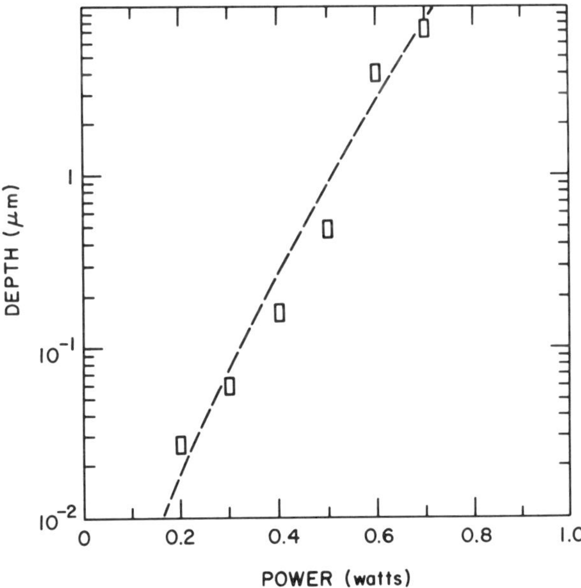

Fig. 6.13 Photothermal etching of InP in H_3PO_4 Solution: Depth of etched groove versus laser power at scan speed 40 μm/sec and spot size 9 μm. The dashed curve shows a fit to an Arrhenius temperature dependence.

fabricate strained layer superlattices and high-speed and high-temperature microelectronics. Generally, etching of this material is carried out with isotropic liquid-etching processes. Laser etching can provide both anisotropic etching and etch-rate selectivity for differently doped materials (Johnson, 1984). Johnson and Tisone have shown that n-type GaP can be efficiently photochemically etched using cw UV light in an aqueous KOH solution (Johnson, 1984). The process is similar to that described above for photochemical etching of GaAs. Laser light capable of exciting electron-hole pairs in n-type GaP results in oxidative dissolution at the liquid interface caused by holes swept to the surface by the space-charge field in the semiconductor (Tisone, 1983b). Due to the large direct band gap of GaP, UV light is needed and the 351-nm line of an argon-ion laser was used. When visible radiation was used the etch rate was essentially zero. With the UV laser the etch rate increased roughly linearly with laser power, reaching about 600 Å/sec at a power of 3.5 KW/cm^2. Both the linearity of the etch rate with laser power and the fact that at the powers used the temperature of the GaP is raised only a few degrees, indicate that the process is predominantly photochemical and not photothermal. In these studies the importance of solution pH was also shown. There is a sharp etch-rate threshold near pH 12, and the etch rate increases steeply with pH above this value. In addition, for etching solutions just above threshold, spatial features with very sharp edges are observed, while for the most basic solutions broad smooth spatial features are observed.

2.3. II–IV Compounds

In the paper dealing primarily with the etching of GaAs in solution, Osgood et al. also showed that undoped CdS could be etched with an argon-ion laser in an etching solution of H_2SO_4, H_2O_2, and H_2O in the ratio of 1:1.3:25 (Osgood, 1982). More recent work by Arnone et al. dealt with CdTe single crystals that were etched by photosublimation with an argon-ion laser (Arnone, 1986).

The dry laser-etching mechanism of CdTe is a complex one (Arnone, 1986). Both thermal and nonthermal photoinduced carrier effects are present. Neutral Cd and Te interstitial complexes generated thermally and optically in the bulk crystal migrate to the surface and then desorb. Both Cd and Te_2 vapors are evolved, although near threshold a ratio of $(Cd/Te_2) > 10$ is measured (Arnone, 1986). The threshold laser power at which etching of CdTe occurs varies with substrate bias temperature. At room temperature, this threshold is nearly 0.6 W for a 1-mm laser spot, but falls to about 0.25 W between 200 and 300°C. Above 300°C the rate of Cd evolution increases dramatically, even in the absence of light.

If the argon-ion laser is operated at 488 nm and focused to a 2.4-μm spot at the surface of the CdTe crystal, a very narrow deep groove is formed in the semiconductor when the sample is translated relative to the laser beam (Arnone, 1986). The groove is tapered at the top from about 13 μm wide to about 0.4 μm wide at a depth of 20 μm. There is then a vertical section 12 μm deep of uniform width 0.4 μm. These grooves have a similar origin to the deep grooves discussed previously in GaAs (Podlesnik, 1984). The grooves are caused by the formation of a hollow waveguide with widths that are narrower than the laser spot and, in fact, comparable to the laser wavelength.

2.4. *Laser-Enhanced Plasma Etching*

Plasma and reactive ion etching of semiconductors are very important technologies for etching patterns. Very fine precise patterns can be etched by these processes if lithographic techniques are employed. In the mass production of devices it is important to increase the etch rate so as to maximize output. Lasers have been shown to increase the etch rates of these processes by both thermal and nonthermal processes. The laser-enhanced plasma etching of Si in CF_4/O_2 and NF_3 plasmas has been studied at Columbia University (Holber, 1985; Reksten, 1984, 1985). A typical experiment involved an 80% CF_4/20% O_2 gas mixture at 120 mtorr. The applied RF power was 0.1 W/cm^2 at 30 KHz. The dark etch rate was 0.2 μm/min. An argon-ion laser at 514 nm was focused on the sample surface to a 30–100 μm spot and the etch-rate enhancement as a function of laser power density and sample doping was recorded. For laser intensities above 10 KW/cm^2 the etch rate enhancement increased from about a factor of 1.6 at 10 KW/cm^2 to 2.5 at about 125 KW/cm^2 and was independent of either the extent or type of sample doping. For this range of laser power a substantial temperature increase is expected at the sample surface (30–145°C for 10–125 KW/cm^2). Since there is little effect on the etch rate due to sample doping, the increased etch rate in this region was attributed almost entirely to thermal effects. These thermal effects increase the surface reaction rate and/or the rate of desorption of reaction products.

In the low-intensity region, the 514-nm laser enhancement of plasma etching of Si shows a strong doping dependence, as shown in Fig. 6.14 (Holber, 1985). At 1 KW/cm^2 the etch-rate enhancement for p-type Si of 1–2 Ω-cm resistivity is five times greater than that for p-type Si of $6\text{–}9 \times 10^{-4}$ Ω-cm. At this laser intensity the temperature rise is <5°C, so

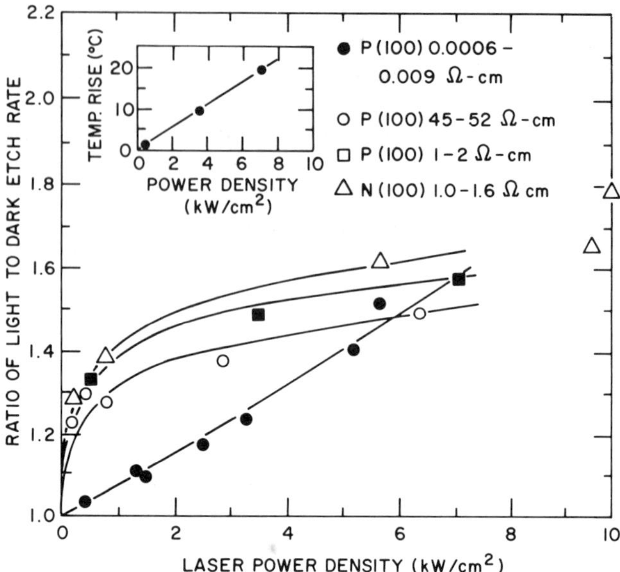

Fig. 6.14 Laser-enhanced plasma etching of Si for a focused laser spot (95 μm) in a CF_4/O_2 plasma.

these differences must be due to nonthermal, i.e., photoinduced carrier effects. To further understand the nonthermal mechanism, the wavelength dependence of the etch rate was measured using a Kr^+ laser (Reksten, 1986). For lightly doped Si (1 Ω-cm) the same laser-enhanced etch rate was observed for visible (514-nm and 647-nm) light as for UV (350-nm) light. However, in more heavily doped samples (10^{-3} Ω-cm) the photoenhancement for the visible lines was much smaller than in the UV. These differences could be explained by considering the flux of photogenerated carriers (which are responsible for the enhanced etching) at the surface. For high doping levels, the carrier lifetime is so short that most of the carriers recombine before they can reach the surface. For example, with 10^{-3} Ω-cm Si, the bulk lifetime is 10^{-9} sec, the space-charge region is only 0.01 μm, and the diffusion length is only 0.2–0.5 μm. For 514-nm light, the absorption depth is 1.0 μm so that most of the carriers are produced and recombine in the bulk. On the other hand, for UV light the absorption depth is about 0.01 μm so that the carriers are produced within the space-charge region and some will arrive at the surface where they will enhance the reactions taking place there. In lightly doped Si (1 Ω-cm) the carrier diffusion length can be hundreds of microns so that all wavelengths produce carriers capable of reaching the surface. Thus in lightly doped Si all of the measured wavelengths showed essentially the same etch rate enhancement, while in heavily doped Si only the UV light showed a similar large enhancement. The visible lines were less effective.

The laser-enhanced reactive ion etching (RIE) of GaAs was studied by Tsukada et al. (Tsukada, 1984b). They set up an etching chamber with flowing CCl_4 and H_2, which produced a dark etch rate of 0.3 μm/min at a total gas pressure of 1.5 Pa and 0.8 μm/min at 15 Pa when operated with RF power of 200 W (13.56 MHz). A frequency-doubled Q-switched Nd:YAG laser ($\lambda = 532$ nm) was weakly focused to an 80–100 μm spot size at normal incidence to a Si-doped GaAs wafer. In general the laser increased the etch rate by a factor of 10. This enhancement results primarily from laser-assisted vaporization of reaction products by localized surface heating. Nevertheless, products such as gallium melt, gallium chlorides, and carbon-chlorine compounds still tend to limit the etching process. One of the prime indications of a simple thermal process is that the diameters of the etched holes (30–40 μm) were significantly smaller than the laser beam diameter (80–100 μm). If the surface temperature that is necessary to activate an etching reaction is near the peak temperature on the sample, the diameter of the etched hole can be less than that of the beam.

2.5. Defects and Defect Detection

Several techniques are used to characterize GaAs crystal perfection and the uniformity of the doping impurity concentration. Preferential etching is one of the most widely used methods since it is quick and easy. Commonly used etchants require several microns of etched depth to delinate crystal imperfections. However, optical illumination and, in particular, low-intensity laser light can greatly shorten the etched depth required to delineate features, and hence even thin epitaxial layers can be studied (Munoz-Yague, 1981).

As described in Section 2.2, at low light intensities, etching of GaAs in solutions is an electrochemical process in which the etch rate at each point on the semiconductor surface depends in particular on the local concentrations of electrons and holes, which in turn are affected by the presence of crystal defects. Illumination of the surface creates electron-hole pairs that recombine at crystal dislocations and defects more rapidly than in the regions without dislocations. In addition, some types of defects or differing impurity concentrations can modify the surface built-in field. Some of the initial work in this area was done by Kuhn-Kuhnenfeld (1972) using intense illumination from a mercury lamp in an oxidizing etching solution (H_2SO_4, H_2O_2, H_2O). He found that illuminated n-type GaAs regions produced a flat-bottomed hole while p-type regions formed a mesa. The height and depth of these structures decreased with increasing doping level. In addition, striations, precipitates, and dislocation lines were observed with excellent resolution, even in undoped or semi-insulating samples.

Munoz-Yague and Bafleur developed a laser-based technique sensitive and selective enough to delineate defects in thin epitaxial GaAs layers (Munoz-Yague, 1981). They used a diluted well-known defect etch consisting of an aqueous solution of $AgNO_3$, CrO_3, and HF that had a dark etch rate of 0.1–0.2 μm/min. With this diluted solution, no defects were revealed without illumination. A He-Ne laser (632.8 nm) was incident on the surface at an intensity of 0.12 W/cm^2, increasing the average etch rate to ~0.5 μm/min. They observed that the etch rate for p-type material was less than for n-type and that for a given impurity (Te) the etch rate increased with doping concentration. Clear features were observed with etched depths of 0.2 μm for n-type and 0.5 μm for p-type and semi-insulating substrates. Both crystal and impurity-related defects were observed. Small hillocks were observed that could be associated with the strain field induced by dopant impurity precipitates. Large diffuse bands on the surface are growth striations associated with impurity segregation during crystal growth. A particular type of hillock was ascribed to dislocations normal to the etched surface. "Boat"-shaped pits indicating the presence of surface damage and long microcavities were thought to arise from dislocation loops or stacking faults. Some improvements on the etching solution used by Munoz-Yague were described by Weyher and van de Ven. They used a simple CrO_3, HF, H_2O system, which ensured high reproducibility without solution aging or precipitation (Weyher, 1982).

A study of laser-induced defects in Al(Ga, As), as well as a means of observing them, was described by Zysset and Salathe (1984). A line of dislocation defects was created in Al(Ga, As) by scanning a focused Kr^+ laser (647 nm) across the surface at power densities in excess of 0.5 MW/cm^2 (typically 1.5 MW/cm^2). The defects were revealed using an unfocused Kr^+ laser at a power density of approximately 1 W/cm^2 in an etching solution of H_3PO_4, H_2O_2, and CH_3OH. In n-type material a decrease in etch rate in the processed zone is observed, leading to etch hillocks, whereas in p-type material, a mesa-like structure was observed. In n-type material photoinduced holes are present at the surface to accelerate etching. However, they recombine rapidly at the defects, giving rise to a reduced etch rate and the observed hillocks. In p-type material, the defect etches at the same rate as the native material but nearby the electron density is reduced and etching is accelerated there, resulting in mesa-like structures.

In their study of laser-induced thermal etching of GaAs in CCl_4 atmospheres, Takai et al. also performed a study of the local strain induced in the semiconductor during the etch process (Takai, 1985b). The etching experiments involved atmospheres of CCl_4, SiC_4, and $GeCl_4$ and a scanned focused argon-ion laser beam. Etch rates of 4–6 μm/sec were obtained for thermally driven processes. The local temperature ranged from 190 to 510°C for laser powers from 0.25 to 0.5 W. The local temperature induces local strains that lead to lattice defects. By using microprobe Raman scattering and ion-channeling measurements, they found that the local strain was 8.6×10^{-3}, corresponding to a tensile stress of 6.4 kbar following laser pyrolysis at 0.5 W incident power in a CCl_4 pressure of 32 torr. At somewhat lower powers (0.4 W) such strain was not induced. These types of measurements are useful for defining the practical limitations of laser etching processes.

3. Insulators

3.1. *SiO₂*

In the microelectronics industry at the present time, silicon dioxide is the most important insulator. Its relative inertness makes it difficult to etch rapidly, but since often only thin layers need to be patterned, slow etch

Table 6.2 Insulator Etching.

Insulator	Ambient	Laser	Intensity/ Energy	Etch Rate	Reference
SiO_2	CF_3Br	CO_2(1 Hz)	0.4 J/pulse	17 Å/min	Steinfeld, 1980
SiO_2	$SF_6:H_2:H_2O$	CO_2	3.5 J/cm^2	100 Å/min	Ambartsumyan, 1982
SiO_2	$NF_3:H_2$	ArF	7.7 mJ/cm^2	0.12 nm/sec	Yokoyama, 1985
SiO_2	SiH_4(Plasma)	KrF	0.3 J/cm^2	40 Å/min	Gee, 1984
Al_2O_3/TiC	KOH	Ar^+	1 MW/cm^2	200 μm/sec	von Gutfeld, 1982
Diamond	Cl_2, O_2, NO_2	ArF	30 J/cm^2	1400 Å/pulse	Rothschild, 1986
Polyimide	KOH	Ar^+	0.03 MW/cm^2	0.3 μm/sec	Moskowitz, 1984
Polyimide	Air	KrF	0.3 J/cm^2	0.15 μm/pulse	Brannon, 1985
Polyimide	Air	CO_2	1.7 J/cm^2	0.8 μm/pulse	Brannon, 1986
Polyethylene terephthalate	Air	ArF	0.37 J/cm^2	0.12 μm/pulse	Srinivasan, 1982
Polyethylene terephthalate	Air	XeCl	1.1 J/cm^2	1 μm/pulse	Andrew, 1983
Polymethyl-methacrylate	Air	ArF	0.25 J/cm^2	0.3 μm/pulse	Srinivasan, 1983

rates may be acceptable. A summary of etch rates for SiO_2 and other insulators is given in Table 6.2. Steinfeld et al. have demonstrated a laser-generated radical etching process that etches SiO_2 without the ion-bombardment damage associated with the more commonly used dry-etching process, namely, reactive ion etching (Steinfeld, 1980). They used a TEA CO_2 laser at 1 Hz with a 0.2–1.5 J/pulse incident parallel to the sample surface about 1 mm from the sample. The sample was placed in a CF_3Br atmosphere at 5.5 torr. Each laser pulse dissociated about 10% of the gas, and the radicals produced attacked the SiO_2, producing products that are volatile at room temperature. The etch rate with the low-power laser used was only 17 Å/min, but with more powerful lasers Steinfeld et al. estimate that rates of 3500 Å/min are achievable. These rates would be comparable to those in present plasma processes.

The radical etching mechanism was also exploited by Ambartsumyan et al. They used a 25-W pulsed TEA CO_2 laser at normal incidence to dissociate SF_6 by multiphoton dissociation (Ambartsumyan 1982). While SF_6 alone was sufficient to etch Si, the products of its dissociation are not able to etch SiO_2. Rather, it was found that etching of SiO_2 occurs when hydrogen is present as well as SF_6 and there is adsorbed water on the SiO_2 surface. The etch rate was monitored by the production of the volatile product, SiF_4, which is actually formed after the laser pulse. The etch rate was estimated as ~100 Å/min for a laser intensity of 3.5 J/cm^2 using gas pressures of 2 torr of SF_6 and 3 torr of H_2.

Yokoyama et al. (1984, 1985) have studied the photochemical etching of thermally grown SiO_2 on Si wafers using an ArF excimer laser (193 nm, 7.7 mJ/cm^2/pulse) in an atmosphere of NF_3 and H_2. The etch rate was studied for various configurations between the incident laser beam and the SiO_2 surface. A two-step method was used. The sample was first irradiated at low pressure ($NF_3 = 480$ Pa, $H_2 = 20$ Pa) for 60 seconds to allow the surface to be "activated" by photocreated fluorine radicals. Then a 2-minute process was carried out at high pressures ($NF_3 = 9.8 \times 10^4$ Pa, $H_2 = 2.6 \times 10^3$ Pa). Without the low pressure preirradiation there is an incubation time at high pressure of more than 2 minutes. With this two-step process etch rates of 0.12 and 0.07 nm/sec were obtained for the normal and parallel incidence of the laser beam to the surface, respectively, in the high-pressure step. The reaction mechanism was studied by in-situ x-ray photoelectron spectroscopy of the etched surface and in-situ infrared absorption measurements of the gas above the sample surface. These studies showed that with the UV irradiation the reaction mechanism was

$$SiO_2 + NF_3 + H_2 \rightarrow SiF_4 + NO_2 + N_2O + HF.$$

While the 193-nm ArF laser can be used at normal incidence to etch SiO_2, it can also produce unwanted radiation damage. This was studied by a group at Toshiba who measured the electron traps generated in MOS capacitors that were irradiated by various excimer laser lines (Sekine, 1985). The power density of XeCl, KrF, and ArF laser beams were 1.5 W/cm^2 at 80 Hz, 5 W/cm^2 at 100 Hz, and 1 W/cm^2 at 100 Hz. The sample configuration was such that to affect the MOS capacitor the light would have to scatter and propagate in the SiO_2 film parallel to the Si substrate. Under these conditions, the damage associated with 193-nm radiation was comparable to that experienced in reactive ion etching, while the XeCl and KrF lasers were essentially damage free. It was concluded that the ArF laser light breaks the Si-OH bonds present in an SiO_2 film and the resulting positively charged Si ions act as electron traps (Sekine, 1985).

Gee and Hargis have described a laser-induced etching process for SiO_2 that takes place in a DC glow discharge in silane gas (Gee, 1984). A silane glow discharge is used to produce reactive chemical species such as Si and H atoms, Si^+, SiH, and H_2. When a pulsed KrF laser is weakly focused onto a sample of CVD SiO_2 (2 μm thick) deposited on a Si wafer, both deposition and etching processes can occur, as shown in Fig. 6.15. After an induction period of about 5 minutes, deposition of

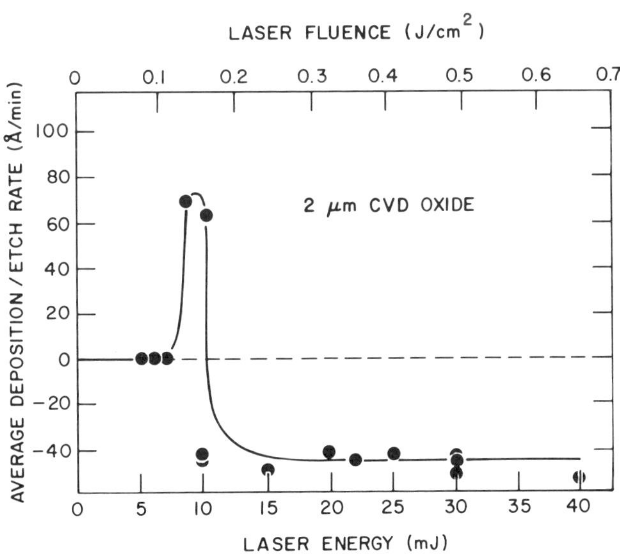

Fig. 6.15 Etch and deposition rates of CVD SiO_2 in a silane plasma under excimer laser irradiation.

polycrystalline Si occurs at fluences of 0.13 and 0.17 J/cm^2. Above this fluence range the SiO_2 is etched at 40–50 Å/min almost independently of laser energy. The mechanism of etching is complex. The induction period appears to be the result of an adsorbed layer of amorphous Si:H deposited by the glow discharge. Once this film builds up, it forms a strongly absorbing layer on the weakly absorbing SiO_2 surface. Pulsed 248-nm radiation causes a temperature rise at the strongly absorbing surface and drives the formation of volatile reaction products. Reaction of SiO_2 with hydrogen to produce volatile silanes and water may be the etching reactions. The saturation of the etch rate is due to the limited material in the adlayer, and the long induction period is the time necessary for the plasma to form the first adlayer. In addition, studies on fused silica showed that the etch rate is independent of wavelength for 248 nm or 351 nm, indicating that photochemical processes are not significant in the etching mechanism.

3.2. Photothermal Patterning of Inorganic Insulators

In addition to demonstrating that holes could be etched in Si very rapidly by photothermal processes in liquid solutions, von Gutfeld and Hodgson have shown that ceramics could also be rapidly etched (von Gutfeld, 1982). They used a concentrated KOH solution and an argon-ion laser focused to a small spot (intensity ~10^6 W/cm^2) to etch holes and slots in a hot-pressed Al_2O_3/TiC ceramic piece. Instantaneous rates of up to 200 μm/sec could be obtained in a rather violent process in which finely dispersed black particles were observed to emanate from the sample during etching. Both melting and vaporization were seen as contributing factors, along with direct material removal.

Diamond is hard to etch. Yet the etching of diamond is required for certain device applications such as backward-wave oscillators operating in excess of 500 GHz (Rothschild, 1986). In addition, hard carbon films with diamondlike bonding could find application as high-resolution photoresists. Using an ArF excimer laser Rothschild et al. have demonstrated that diamond can be successfully laser etched in direct writing or optical projection modes. In the direct writing mode, 20 μm wide lines were etched with etch rates of 1 μm/sec at 20 J/cm^2, 20 Hz. Using a 36 X reduction imaging system, well-defined gratings with periods as small as 0.25 μm were obtained. The etch rate as a function of laser fluence is given in Fig. 6.16. The etching mechanism is quite complex, but in some ways it resembles that described by Gee and Hargis (1984) for SiO_2. The first step in the process is the conversion of a thin surface layer to graphite. This process has a threshold of about 60 mJ/cm^2. The graphite

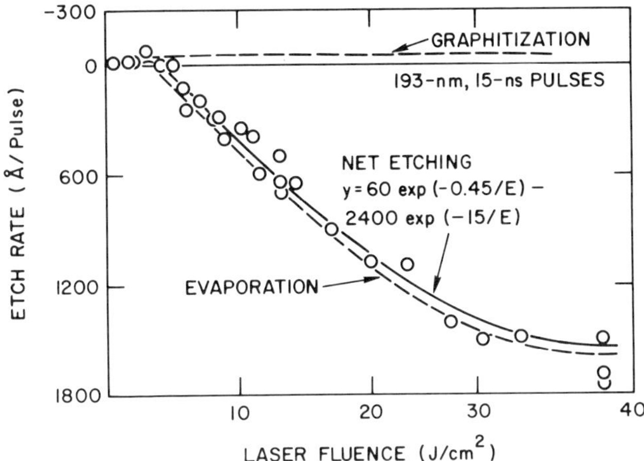

Fig. 6.16 Rate of graphitization and etching as a function of excimer-laser fluence for diamond.

generated at the diamond surface in the early part of the laser pulse has a much higher absorption coefficient than diamond itself at 193 nm. It also has a lower thermal diffusion coefficient. As a result, the energy in the latter part of the laser pulse is deposited and localized in the graphite layer, leading to runaway absorption and vaporization at high fluences. At low fluences (1–3 J/cm^2) only graphite conversion occurs. This is seen in Fig. 6.16 as negative etching since the sample actually becomes thicker. As the fluence is increased beyond this range, sublimation of the graphite layer occurs and net etching results. All of these processes can occur in a vacuum or an inert gas atmosphere. However, since no laser spot is perfectly sharp, there will be a region at the edge of an etched feature that will show a graphite deposit since the laser flux is much reduced there. To eliminate the accumulation of graphite in these areas, reactive gases such as Cl_2, NO_2, and O_2 can be introduced in the etching chamber. These react with the hot graphite and lead to much cleaner etch features.

Iron garnet films, which have been developed for magnetic bubble memories, are also important for waveguiding type magneto-optical devices such as isolators, modulators, circulators, and switches. These are usually prepared as thin films on gadolinium gallium garnet (GGG) transparent substrates (Ando 1982). Laser patterning and flattening techniques for such films were developed by Ando et al. (1982, 1983, 1985). They used a focused argon-ion laser or a dye laser at 581 nm at normal incidence to the sample to etch grooves in the iron garnet film (4 μm thick) on GGG substrates. The sample was in a phosphoric acid solution, and the mechanism of etching was simple thermal heating of the film to drive the thermally activated chemical etching in the acid. Since the substrate is not etched and is a good heat sink, thicker iron garnet films are etched more rapidly than thin films in this thermally driven process. Hence, films of iron garnet can be made flat by repeated laser scanning (Ando, 1985).

3.3. Polymer Etching

In experiments similar to those carried out for laser-enhanced chemical etching of Si and ceramics (von Gutfeld, 1982) holes were etched in polyimide films when subjected to focused argon-ion laser light in alkaline solutions (Moskowitz, 1984). This process is thermally enhanced chemical etching, which takes advantage of the strong absorbance of polyimide for blue light. The laser heating speeds the chemical reaction of polyimide with the alkaline solution, and turbulence in the fluid, caused by the local temperature differentials, constantly brings fresh etchant fluid in contact with the surface. The etch rate decreases with time as the hole develops, but rates of 0.3 μm/sec are typical even at long times when a 0.1-W beam is focused to a 10–20 μm spot.

One mechanism of laser etching that appears to be peculiar to polymers and organic materials in general is direct ablative photodecomposition (Srinivasan, 1982, 1983). In this process photochemical decomposition occurs in such a way as to lead to rapid and clean etching. For example, pulsed excimer radiation from an ArF laser (193 μm) was directed onto polyethylene terephthalate (PET) and polymethylmethacrylate (PMMA) (Srinivasan, 1982, 1983). The high absorption coefficient of these polymers in the near-UV is such that 95% of the radiation is absorbed in the first few thousand angstroms at the surface. At this wavelength the radiation is highly efficient at breaking bonds. The threshold for the etching process is only 10 mJ/cm^2 for PMMA. At lower fluences many of the bonds recombine, although some react in air and oxidative photoetching occurs. Above threshold the bond breaking per unit time exceeds a critical value where ablative decomposition occurs. The driving forces for this phenomenon are the large increase in specific volume of the fragments compared to the polymer chain they replace, and the excess energy absorbed, which excites the fragments vibrationally, rotationally, and translationally. Since this is primarily a photochemical process at 193 nm, there is little dependence of the etch rate on sample temperature. Typical etch rates of PET in air were 1200 Å/pulse at $0.37\,J/cm^2$; and, for PMMA, the etch rate was $0.3\,\mu m$/pulse at $0.25\,J/cm^2$.

The etching of PET using an XeCl excimer laser (308 nm) was studied by Andrew et al. (1983), and their results are plotted in Fig. 6.17. At this wavelength the quantum yield for chain fractures and other photochemical processes in PET is very low and most of the adsorbed energy goes into heating. This was confirmed by thermocouple measurements that showed nearly all absorbed energy appeared as heat when the film was irradiated at below the threshold fluence ($0.17\,J/cm^2$). Above threshold the heat absorbed by the polymer film was less than the deposited laser energy since some energy was carried away with the ejected material. The ablative etching was also carried out at very high pulse repetition rates to investigate fast cutting of PET films (Bishop, 1985). Significant increases in cutting efficiency and etch rate occur at high pulse rates (>100 Hz), apparently due to cumulative heating effects. Cutting rates of 130 cm/sec are obtained at 900 Hz for a 12-μm thick PET film (Bishop, 1985).

The data in Fig. 6.17 can be fit by an equation of the form

$$x = k^{-1} \ln\left(\frac{E}{E_T}\right),$$

where x is the etch depth per pulse, k the absorption coefficient, E the laser fluence, and E_T the threshold fluence. This formula follows from the well-known Beer-Lambert law of optical absorption with the assumption of a threshold fluence. The straight line in Fig. 6.17 corresponds to a k of $2 \times 10^4\,cm^{-1}$. This is a factor of seven higher than the absorption coefficient measured at low light intensity. Andrew et al. explained that this difference is most likely due to a significant increase in k with temperature. This explanation seems reasonable since the above equation applies to other polymers at a variety of wavelengths where the low-intensity absorption coefficient is greater. For other polymers than PET, the measured and calculated k values are in much better agreement.

An example of a good fit to the etch rate equation above is in the work done on polyimide by Brannon et al. (1985). They studied the etching of a polyimide film made by thermal curing of polyamic acid that was made from pyromellitic dianhydride (PMDA) and oxydianiline (ODA). Brannon et al. used different excimer wavelengths of 248, 308, and 351 nm to quantify the etch rates and to further understand the etching mechanism. For each wavelength the etch-rate data fell on a curve that was well described by the above equation. From the slope of the etch rate per

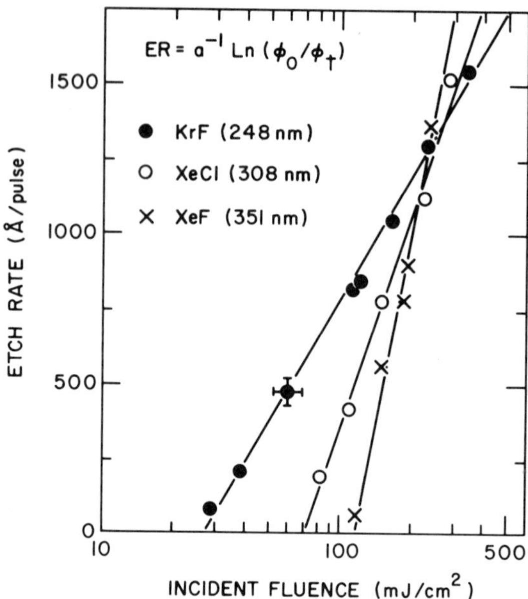

Fig. 6.17 Etch depth per pulse of polyethylene terephthalate exposed to excimer laser radiation at various wavelengths.

pulse versus the log of the fluence, an absorption coefficient was derived that was $1.6 \times 10^5\ \mathrm{cm}^{-1}$ at 248 nm, $0.9 \times 10^5\ \mathrm{cm}^{-1}$ at 308 nm, and $0.47 \times 10^5\ \mathrm{cm}^{-1}$ at 351 nm (Brannon, 1985). These values are all in quite reasonable agreement with the absorption coefficient as a function of wavelength measured at low intensity. The threshold fluence varied from $0.027\ \mathrm{J/cm}^2$ at 248 nm, to $0.07\ \mathrm{J/cm}^2$ at 308 nm, to $0.12\ \mathrm{J/cm}^2$ at 351 nm. Brannon et al. point out that at this threshold the product of the threshold energy and the absorption coefficient is nearly constant independent of wavelength. This product is equal to the absorbed energy density at threshold ($\approx 5 \times 10^3\ \mathrm{J/cm}^3$). That it is the same at all the wavelengths studied suggests that it is a material property that is required for rapid ablation. Since the photochemical ablative photodecomposition mechanism (Srinivasan, 1982, 1983) depends on bond breaking and should be wavelength dependent, Brannon et al. argue that, at the wavelengths they studied, the mechanism of polyimide etching is photothermal. The process is one in which the energy, initially absorbed as electronic excitation, is rapidly converted to vibrational heating of the solid. The intense local heating is confined by the low thermal diffusivity of the polymer and results in an explosive pyrolysis.

The etching of polyimide, PET, and PMMA has also been studied by

Koren and Yeh (1984a, 1984b) using various excimer laser wavelengths. In polyimide they studied the morphology of the etched surface and found that there existed a transition fluence (which depends on wavelength) such that below this fluence the morphology of the etched surface was rough and above this fluence the morphology was very smooth. Although the absorption coefficients varied considerably at the wavelengths studied, the absorbed energy density at the rough-to-smooth transition was independent of wavelength ($\sim 5 \times 10^4$ J/cm^3). This observation suggests that the smoothness may be due to transient melting (Koren, 1984a). Koren and Yeh also characterized the etching process by measuring the optical emission spectra of the bright plume that accompanies the laser ablation. At high fluences they found the emission spectra of several small molecules, while at low fluences only a continuum of emission was observed. In polyimide, they found that the small molecule emission spectra appeared at fluences above the rough-to-smooth transition. Hence this transition may be due to the change in ablated species from large to small molecules.

The laser etching of polyimide films with a pulsed CO_2 laser was studied by Brannon and Lankard (1986). They used two different wavelengths—944 cm^{-1} (10.6 μm) where the absorption is weak, and 1087 cm^{-1} (9.2 μm) where the absorption is much stronger. They found that the threshold for laser ablation in air was much lower at 9.2 μm (0.5 J/cm^2) than at 10.6 μm (2.1 J/cm^2). This is qualitatively consistent with the excimer laser studies (Brannon, 1985), which showed that the threshold decreased as the absorption coefficient increased. The etch rate at a wavelength of 9.2 μm was 0.8 μm pulse for a fluence of 1.7 J/cm^2.

4. Metals

4.1. Aluminum

Because of its general utility in the microelectronics industry, aluminum has been the subject of many laser patterning studies. Thin films of aluminum on substrates with poor thermal conductivity can be patterned with a single pulse of a UV excimer laser (Andrew, 1983b; see Table 6.3). Andrew et al. showed that aluminum films, 0.12 μm thick on glass substrates, could be removed by a single 308-nm excimer laser pulse at 0.2 J/cm^2. With somewhat higher pulse energies, $\lesssim$1 J/cm^2, films up to 1 μm could be removed. The energies needed to remove these films are far below the energy calculated to cause vaporization of the metal. The energy used corresponded more closely to the melting threshold. Hence, the mechanism of ablation was considered to be melting followed by an

Table 6.3 Metal Etching.

Metal	Ambient	Laser	Intensity/ Energy	Etch Rate	Reference
Ag	Cl_2 (0.1 torr)	N_2 (337 nm)	0.12 J/cm^2	500 Å/min	Sesselmann, 1985
Ag	Air	XeCl	0.05 J/cm^2	0.10 μm/pulse	Andrew, 1983b
Al	Cl_2 (0.1 torr)	N_2 (337 nm)	0.12 J/cm^2	330 Å/sec	Sesselmann, 1985
Al	Air	XeCl	0.2 J/cm^2	0.12 μm/pulse	Andrew, 1983b
Al	Cl_2 (2 torr)	XeCl	1.0 J/cm^2	1.4 μm/sec	Koren, 1986a
Au	Air	XeCl	0.03 J/cm^2	0.05 μm/pulse	Andrew, 1983b
Cr	Air	XeCl	0.24 J/cm^2	0.08 μm/pulse	Andrew, 1983b
Cu	Air	XeCl	0.08 J/cm^2	0.10 μm/pulse	Andrew, 1983b
Fe	Cl_2 (0.1 torr)	N_2 (337 nm)	0.12 J/cm^2	850 Å/min	Sesselmann, 1985
Mo	NF_3	ArF	0.057 J/cm^2	0.22 Å/pulse	Loper, 1985a
Mo	Cl_2 (0.1 torr)	N_2 (337 nm)	0.12 J/cm^2	16 Å/sec	Sesselmann, 1985
Mo	Air	Ar^+	$>10^8$ W/cm^2	20 μm/sec	Koren, 1986b
Ni	Air	XeCl	0.24 J/cm^2	0.12 μm/pulse	Andrew, 1983b
Ti	NF_3	ArF	0.115 J/cm^2	0.29 Å/pulse	Loper, 1985a
W	Cl_2 (0.1 torr)	N_2 (337 nm)	0.12 J/cm^2	240 Å/min	Sesselmann, 1985

explosive process due to pressure buildup at the film-substrate interface as a result of decomposition or degassing of the substrate (Andrew, 1983b). In addition to aluminum thin films, Andrew et al. obtained qualitatively similar results for films of Ag, Au, Ni, Cu, and Cr on insulating substrates such as Mylar, Kapton, glass, and fused silica.

A laser-induced chemical technique for localized etching of aluminum films (200–5000 Å thick on glass substrates) was described by Ehrlich et al. (1981b). This method involves photochemical processes to create locally an alloy of aluminum and zinc. This alloy is quite reactive and can be dissolved and washed away in an aqueous acetic-acid solution. The laser process requires irradiation of the sample in a $Zn(CH_3)_2$ vapor by two laser beams, a frequency-doubled Ar-ion beam at 257 nm (1–10 W/cm^2) to dissociate the dimethyl zinc and a beam at 514.5 nm (10^5–10^6 W/cm^2) to heat the film and cause the alloy to form. This technique is capable of writing 5-μm wide lines in films several thousand angstroms deep but may be unable to produce features smaller than several microns and has a relatively narrow range of laser powers over which it is effective (Ehrlich, 1981b).

Laser-enhanced chemical etching of aluminum in chlorine environments has been studied by Koren et al. (1985, 1986a) and by Sesselmann and Chuang (1985). Once the protective aluminum oxide has been removed, aluminum metal films react with chlorine to form aluminum chlorides. These have low boiling points and actually sublime in vacuum, leading to spontaneous etching of aluminum by chlorine. This is illustrated by curve (a) in Fig. 6.18, which is taken from the study by Sesselmann and Chuang. When a clean aluminum film coated on a quartz crystal microbalance is first exposed to chlorine, the balance shows an increase in weight (lower frequency) due to the formation of a chemisorbed chlorine layer. Then the film slowly loses weight as the volatile reaction products desorb. This spontaneous etching occurs at pressures above 0.01 torr. When a N_2 laser at 337 nm irradiates the surface, the etch rate is greatly enhanced. For example, at 0.01 torr of Cl_2 the etch rate is 15 times higher than the dark etch rate under illumination from the N_2 laser at 0.12 J/cm^2 per pulse at a 30-Hz pulse rate. This etching behavior is shown in Fig. 6.18 curve b). Curve (c) shows the time dependence of the weight of the aluminum film when etched in 0.1 torr Cl_2. The fact that the etching at 0.1 torr is much faster than at 0.01 torr clearly indicates that reaction rates can strongly influence the measured laser etch rate.

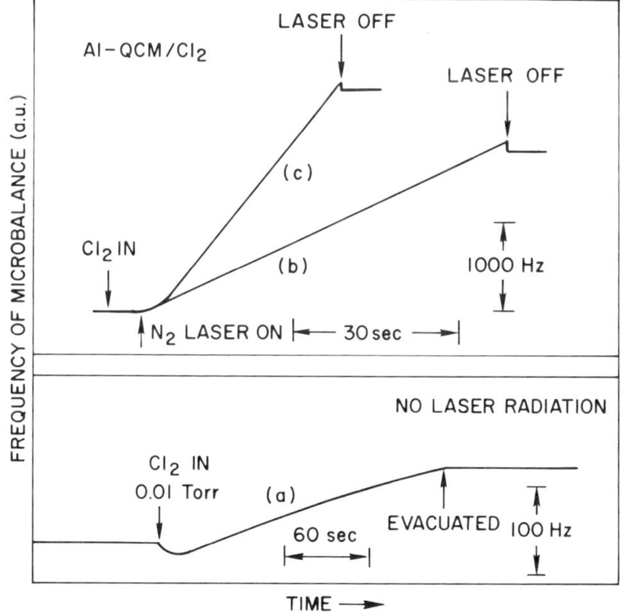

Fig. 6.18 Frequency response of a quartz crystal microbalance coated with aluminum in a Cl_2 Atmosphere (10^{-2} torr): (a) Dark reactions, (b) excited by N_2 laser (0.12 J/cm^2, 30 Hz), (c) excited by the N_2 laser with Cl_2 pressure of 0.1 torr.

Sesselman and Chuang also studied the relative etch rates of a variety of metals under the same operating conditions. When irradiated by a pulsed N_2 laser at a 0.12-J/cm^2, 30-Hz rate in a Cl_2 pressure of 0.1 torr, the etch rates of Al, Ag, Mo, W, Fe, $Ni_{80}Fe_{20}$, Au, and Ni were 20,000, 500, 1000, 240, 850, 120, <10, <10 Å/min, respectively. These differences may be explained by the relative reaction rates of the various metals with chlorine and with the volatility of the reaction products under the particular irradiation conditions used.

A detailed study of the parameters affecting the laser etching of Al in Cl_2 was carried out by Koren and coworkers (1985, 1986a). Figure 6.19 shows the etch rate of aluminum in 0.4 torr Cl_2 as a function of pulse fluence from an XeCl excimer laser (308 nm). In these studies a small rectangular region of the sample was etched, and the depth profile was always deeper at the edges of the rectangular region than in the center. Since the etch rates were determined by measuring the time to penetrate a film of a given thickness (usually 5 μm of Al on Si), two separate etch rates, for the center and edge regions, were measured. These are indicated in Figs. 6.19 and 6.20 by crosses and circles, respectively.

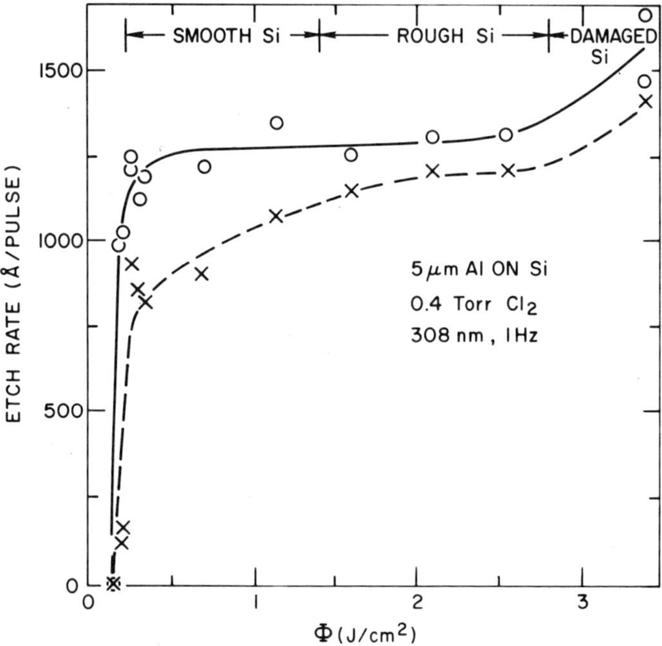

Fig. 6.19 Aluminum etched in Cl_2 gas by 308-nm pulsed-excimer radiation: Etch rate per pulse versus fluence at fixed pressure.

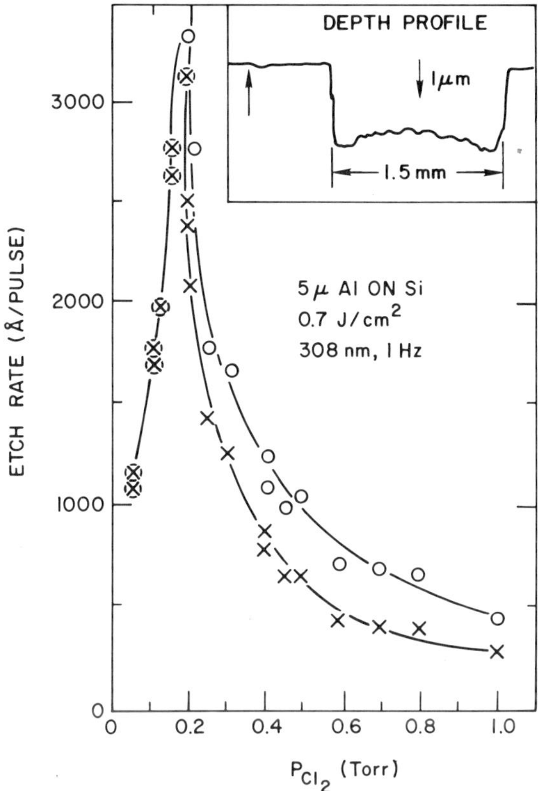

Fig. 6.20 Aluminum etched in Cl_2 gas by 308-nm pulsed excimer radiation: Etch rate per plut versus pressure at fixed fluence.

Figure 6.19 shows that the etch rate (1250 Å/pulse) is essentially independent of the laser fluence between 0.25 and 3.0 J/cm^2 at a Cl_2 pressure of 0.4 torr. At lower fluences, the etch rate drops sharply, with no detectable etching below 0.15 J/cm^2. At fluences above 3 J/cm^2 a very rough surface occurs, along with damage (etching) of the Si substrates. The effect of the chlorine pressure on etch rate is quite dramatic, as shown in Fig. 6.20. Beginning at low pressure, the etch rate at room temperature increases rapidly, reaching 3300 Å/pulse at about 0.18 torr and then rapidly falling off with increasing pressure. While the detailed understanding of this effect is not known, it clearly depends on the complex mechanism of reaction and ablation. That these vary considerably with pressure was shown by Koren et al. (1985, 1986a).

The etching mechanism is mainly reaction between Al and Cl_2 in the dark followed by laser photoablation of the product. Photochemical effects are negligible. This was shown by exposing an Al film for 15 seconds in the dark to chlorine, (0.4 torr) then pumping out the chlorine, and then irradiating the sample with a single pulse at 0.7 J/cm^2. About 1 μm of Al was removed per pulse, and no etching occurred with subsequent pulses until more Cl_2 was admitted to the sample cell. Moreover, at constant low chlorine pressures ($\lesssim$0.2 torr), the etch rate is nearly proportional to the square root of the time between the laser pulses. This is consistent with the notion that reactive diffusion occurs in the time between the laser pulses and that a reaction product is formed, with thickness determined by diffusion limitations. At low pressures (below the peak in Fig. 6.20), the laser removes essentially all of this reaction product and the etch rate is proportional to the square root of the time between pulses. At higher pressures, something happens such that this is not the case. At pressures above about 0.2 torr, the etch rate increases much more slowly than the square root of the time between laser pulses. The higher pressures of Cl_2 in the cell may cause redeposition of the ablated material so that each laser pulse does not fully remove the reaction product as it does at low pressure. A second speculation as to the origin of the peak in the pressure dependence of the etch rate is that the higher Cl_2 pressure inhibits the explosive volatilization of the reaction product by, in effect, inhibiting boiling of the aluminum chloride. This latter concept is partially supported by the fact that the peak in Fig. 6.20 is quite temperature dependent (Koren, 1986a). As temperature is increased, the peak moves to higher pressure and becomes stronger, reaching 1.4 μm/sec at 60°C in 2 torr of Cl_2 when the illumination is 1.0 J/cm^2 at 1 Hz. At higher temperatures the vapor pressure of the aluminum chloride is higher and greater pressures are required to suppress rapid volatilization and the measured etch rate.

4.2. Radical Etching of Metals

While the previous section dealt primarily with laser etching of aluminum, it was pointed out that many other metals can be etched by halogens (in particular chlorine) if they form a thin reactive layer that a pulsed laser can remove. Chemical reactions and chemical etching can also be generated by lasers interacting with gas-phase molecules to create vibrationally excited molecules or photoinduced radicals. These can etch metals. Houle and Chuang (1982) point out that tantalum can be etched by SF_6 when it is vibrationally excited by a CO_2 laser, although the etch rate is very low, 5×10^{14} atoms/pulse at 1 torr of SF_6.

Molybdenum is a metal that forms volatile fluorides. Loper and Tabat (1985a, 1985b) used this fact to etch Mo with an ArF excimer laser and a NF_3 atmosphere. Even at low fluences (57 mJ/cm^2) the 193-nm radiation generates enough fluorine radicals in 752 torr of NF_3 to etch thin films of Mo at the rate of 0.22 Å/pulse. While this rate is very low, Loper and Tabat point out that a 200-W excimer laser could remove 0.5 μm of Mo over a 9 cm^2 area of a wafer at the rate of 6000 Å/min, which is quite respectable. Loper and Tabat achieved very similar results with Ti. Etch rates of 0.29 Å/pulse were obtained at 0.115 J/cm^2 of ArF radiation in 750 torr of NF_3. Although the active species are generated in the gas phase, the gas pressures are high and spatial resolution <1 μm can be achieved (Loper, 1985b).

4.3. Laser-Controlled Oxidation and Electroetching

While many metals can be laser etched in gases containing halogens, several studies have attempted to use laser-driven oxidation to control etching. In what amounts to a negative etching process, Koronkevich et al. (1985) illuminated chromium films 1000–2000 Å thick on glass substrates with a scanned focused laser ($\lambda = 5145$ Å) with power density of $\sim 10^6$ W/cm^2 or about 30% below the threshold for melting the film. This caused an oxide film 50–70 Å thick to grow on the surface. When the films were immersed in an aqueous chromium etching solution of $K_3Fe(CN)_6$ and NaOH, the unexpected regions were rapidly etched, while the laser-exposed regions were etched much more slowly due to the protective oxide coating. This technique has allowed Koronkevich et al. to fabricate optical diffraction elements (Fresnel zone plates) more than 100 mm in diameter with minimum element dimensions of ~ 0.5 μm. Feature edge irregularity is <0.1 μm.

Using a focused argon-ion laser, Koren has shown that holes and slots in thin Mo foils can be etched in air by a simple laser-assisted chemical etching mechanism. Etching occurs due to local heating, which induces Mo oxidation. At high temperatures MoO_3 is volatile and it sublimes, allowing the oxidation to continue. The etch rates for drilling holes in Mo foils are shown in Fig. 6.21 for all lines of the argon-ion laser focused to an area of ~ 1 μm^2. For each sample thickness, a minimum laser power exists below which the sample is not etched. This happens because dissipation of heat in the thicker foils is more effective. Above threshold, the etch rates increased much more than linearly with power, indicative of the thermal nature of the mechanism. At low laser powers, the diameters of the etched holes were found to increase linearly with the laser power. At higher powers, the diameter increased faster than

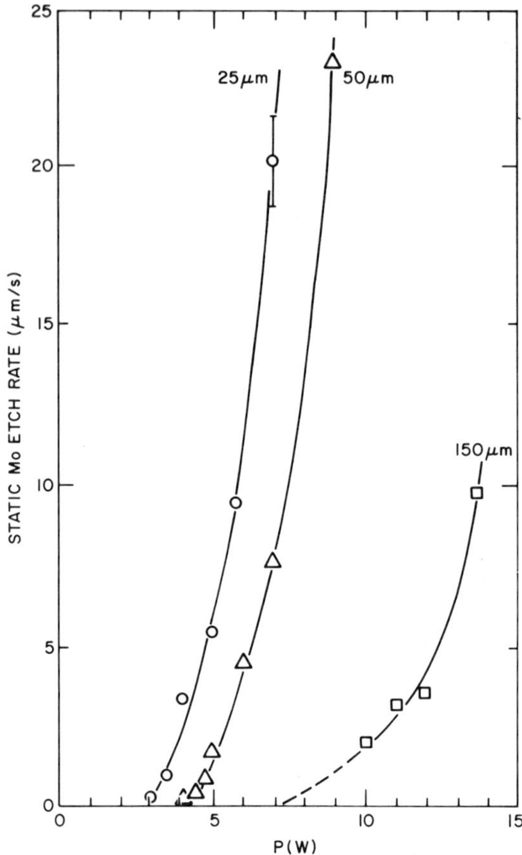

Fig. 6.21 Etch rates for drilling holes in Mo foils in air using an argon-ion laser.

linearly. Slightly tapered holes with straight walls could be etched with diameters less than 10 μm.

Koren showed that by scanning the laser, slots could be etched easily. Using an optical pyrometer, he showed that average etching temperatures were ~1000°C, just below the threshold for etching through a foil, and only increased slowly with laser power. However, once etching through the foil occurred, the temperature increased rapidly. He ascribed this to the heat released by the exothermal oxidation reaction, which is of the same order of magnitude as the direct laser heating and hence greatly affects the etching characteristics.

A final example of laser-driven oxidation for etching does not involve the formation of insoluble oxides, but rather the oxidation of metals in an electrochemical cell to create soluble salts. R. von Gutfeld et al. (1979, 1987) have studied how laser light can enhance metal deposition in an electrochemical cell. Plating solutions typically containing Au, Ni, or Cu ions were placed in an electroplating cell with applied voltages of about 1 volt. For deposition, the substrate is the cathode of the cell. A multimode argon-ion laser or a Kr^+-ion laser at 6471 Å was focused on the sample with power densities of about 10^4 W/cm^2. Under these conditions, plating enhancement rates on the order of 10^3 occur. This enhancement is due to enhanced chemical kinetics with increased temperature, shifts in the local rest potential with temperature, and local agitation of the solution as a result of large temperature gradients. If the applied potential is reversed in the electroplating cell, then electroetching occurs. This effect has been demonstrated on stainless-steel shim stock (50 μm thick) where 50 μm diameter through holes have been drilled using a $NiCl_2$ solution with organic additives as the electrolyte. The holes formed in this fashion are very smooth and regular. They have been used as nozzles for ink jets (von Gutfeld, 1987).

References

Aickin, R.G., and Bayliss, N.S. (1937). The continuous absorption spectrum of chlorine in the region 4000–5000 Å. *Trans. Faraday Soc.* **33,** 1333–1338.

Ambartsumyan, R.V., Gorokhov, Y.A., Gritsenko, A.L., and Lokhman, V.N. (1982). Selectivity of the etching of the basic microelectronics materials through many photon dissociation of SF_6 molecules in an intense CO_2 laser beam. *Sov. Tech. Phys. Lett.* (*USA*) **8,** 276–277.

Ando, K., and Tsukahara, S. (1982). Selective removal of iron garnet film on transparent substrate by laser etching. *Jpn. J. Appl. Phys.* **21,** L347-L348.

Ando, K., Okuda, T., Yokoyama, Y., and Koshizuka, N. (1983). Profiles of laser etched grooves in iron garnet films. *Jpn. J. Appl. Phys. Suppl.* **22-1,** 615.

Ando, K., Takeda, N., and Koshizuka, N. (1985). New flattening technique of iron garnet films by laser etching. *Appl. Phys. Lett.* **46,** 1107–1109.

Andrew, J.E., Dyer, P.E., Forster, D., and Key, P.H. (1983a). Direct etching of polymeric materials using a XeCl laser. *Appl. Phys. Lett.* **43,** 717–719.

Andrew, J.E., Dyer, P.E., Greenough, R.D., and Key, P.H. (1983b). Metal film removal and patterning using a XeCl laser. *Appl. Phys. Lett.* **43,** 1076–1078.

Aoyagi, Y., Masuda, S., Doi, A., and Namba, S. (1985). Maskless fabrication of high quality DFB laser gratings by laser induced chemical etching. *Jpn. J. Appl. Phys.* **24,** L294–L296.

Arikado, T., Sekine, M., Okano, H., and Horiike, Y. (1984). Single-crystal silicon etching charcteristics using excimer laser Cl_2 gas. *Mat. Res. Soc. Symp. Proc.* **29,** 167–172.

Arnone, C., Rothschild, M., and Ehrlich, D.J. (1986). Laser etching of 0.4 μm structures in CdTe by dynamic light guiding. *Appl. Phys. Lett.* **48,** 736–738.

Ashby, C.I.H. (1984). Photochemical dry etching of GaAs. *Appl. Phys. Lett.* **45,** 892–894.

Ashby, C.I.H. (1985a). Doping level selective photochemical dry etching of GaAs. *Appl. Phys. Lett.* **46,** 752–754.

Ashby, C.I.H. (1985b). Composition-selective photochemical etching of compound semiconductors. *Appl. Phys. Lett.* **47,** 62–63.

Ashby, C.I.H. (1985). Laser induced photochemical dry etching of III–V compound semiconductors. *Proc. SPIE Int. Soc. Opt. Eng.* (*USA*) **540,** 467–471.

Baklanov, M.R., Beterov, J.M., Repinskii, S.M., Rzhanov, A.V., Chebotaev, V.P., and Yurshina, N.I. (1974). Initiation of a surface chemical reaction between single-crystal germanium and bromine gas by using a powerful argon laser. *Sov. Phy. Dokl.* **19,** 312–314.

Bean, K.E. (1978). Anisotropic Etching of Silicon. *IEEE Trans. Electron Dev. Ed.* **25,** 1185–1193.

Beterov, I.M., Chebotaev, V.P., Yurshina, N.I., and Yurshin, B.Y. (1978). Effect of the laser radiation intensity on the kinetics of the heterogeneous photochemical reaction between single crystal-germanium and bromine gas, *Sov. J. Quantum Elec.* **8,** 1310–1312.

Bishop, G.J., and Dyer, P.E. (1985). Polymer film cutting and ablative etching using a 1-KHZ XeCl laser. *Appl. Phys. Lett.* **47,** 1229–1331.

Brannon, J.H., Lankard, J.R., Baise, A.I., Burns, F., and Kaufman, J. (1985). Excimer laser etching of polyimide. *J. Appl. Phys.* **58,** 2036–2043.

Brannon, J.H., and Lankard, J.R. (1986). Pulsed CO_2 laser etching of polyimide. *Appl. Phys. Lett.* **48,** 1226–1228.

Bjorkholm, J.E., and Ballman, A.A. (1983). Localized wet-chemical etching of INP induced by laser heating. *Appl. Phys. Lett.* **43,** 574–576.

Bowers, J.E., Hemenway, B.R., and Wilt, D.P. (1985). Etching of deep grooves for the precise positioning of cleaves in semiconductor lasers. *Appl. Phys. Lett.* **46,** 453–455.

Brewer, P., Halle, S., and Osgood, R.M. (1984). Excimer-laser-initiated dry etching of single crystal GaAs. *Mat. Res. Soc. Symp. Proc.* **29,** 179–184.

Brewer, P., McClure, D., and Osgood, R.M. (1985). Dry, laser assisted rapid HBR etching of GaAs. *Appl. Phys. Lett.* **47,** 310–312.

Brewer, P., McClure, D., and Osgood, R. (1986). Excimer laser projection etching of GaAs. *Appl. Phys. Lett.* **49,** 803–805.

Bunkin, F.V., Lukyanchuk, B.S., Shafiev, G.A., Kozlova, E.K., Portniagin, A.I., Yeryomenko, A.A., Mogyorosi, P., and Kiss, J.G. (1985). Si etching affected by IR laser irradiation. *Appl. Phys.* **A37,** 117–119.

Cabrera, N., and Mott, N.F. (1949). Theory of the oxidation of metals. *Rep. Prog. Phys.* **12,** 163–184.

Chuang, T.J. (1982a). Laser enhanced gas-surface chemistry: Basic processes and applications. *J. Vac. Sci. Tech.* **21,** 798–806.

Chuang, T.J. (1982). Ti laser-enhanced chemical etching of solid surfaces. *IBM J. Res. and Dev.* **26,** 145–151.

Chuang, T.J. (1984). Laser-induced chemical etching of solids: Promises and challenges. *Mat. Res. Soc. Symp. Proc.* **29,** 185–194.

Chuang, T.J. (1985). Photodesorption and adsorbate-surface interactions stimulated by laser radiation. *J. Vac. Sci. Tech.* **B3,** 1408–1420.

Davis, G.P., Moore, C.A., and Gottscho, R. (1984). Ti dynamics of laser stimulated etching of germanium by bromine. *J. Appl. Phys.* **56,** 1808–1811.

Ehrlich, D.J., Osgood, R.M., Jr., and Deutsch, T.F. (1980). Ti laser induced microscopic etching of GaAs and InP. *Appl. Phys. Lett.* **36,** 698–700.

Ehrlich, D.J., Osgood, R.M., Jr., and Deutsch, T.F. (1981a). Laser chemical technique for rapid direct writing of surface relief in silicon. *Appl. Phys. Lett.* **38,** 1018–1020.

Ehrlich, D.J., Osgood, R.M., and Deutsch, T.F. (1981b). Laser photochemical microalloying for etching of aluminum thin films. *Appl. Phys. Lett.* **38,** 399–401.

Fehlner, F.P., and Mott, N.F. (1970). Low Temperature Oxidation. *J. Oxidation Metals* **2,** 59–99.

Gee, J.M., and Hargis, P.J. (1984). Laser induced etching of insulators using a DC glow discharge in silane. *SPIE* **459,** 132–137.

Haynes, R.W., Metze, G.M., Kreismanis, V.G., and Eastman, L.F. (1980). Laser photoinduced etching of semiconductors and metals. *Appl. Phys. Lett.* **37,** 344–345.

Holber, W., Reksten, G., and Osgood, R.M. (1985). Laser enhanced plasma etching of silicon. *Appl. Phys. Lett.* **46,** 201–203.

Houle, F.A., and Chuang, T.J. (1982). Laser induced chemical etching of metals and semiconductors. *J. Vac. Sci. Tech.* **20,** 790–791.

Houle, F.A., (1983a). Non-thermal effects in laser enhanced etching of silicon by XEF_2. *Chem. Phys. Lett.* **95,** 5–8.

Houle, F.A. (1983b). Photoeffects on the fluorination of silicon. I. Influence of doping on steady state phenomena. *J. Chem. Phys.* **79,** 4237–4246.

Houle, F.A. (1984a). Photoeffects on the fluorination of silicon. II. Kinetics of the initial response to light. *J. Chem. Phys.* **80,** 4851–4858.

Houle, F.A. (1984b). Mechanism of laser-enhanced etching of silicon. *Mat. Res. Soc. Symp. Proc.* **29,** 203–209.

Johnson, A.W., and Tisone, G.C. (1984). Laser photochemical etching of GaP in KOH aqueous solutions. *Mat. Res. Soc. Symp. Proc.* **29,** 145–150.

Karlov, N.V., Lukyanchuk, B.S., Sisakyan, E.V., and Shafeev, G.A. (1985). Etching of semiconductors by products of laser thermal dissociation of molecular gases. *Sov. J. Quantum Elec.* **15,** 522–526.

Koren, G., and Yeh, J.T.C. (1984a). Emission spectra, surface quality, and mechanism of excimer laser. *Appl. Phys. Lett.* **44,** 1112–1114.

Koren, G., and Yeh, J.T.C. (1984b). Emission spectra and etching of polymers and graphite irradiated by excimer lasers. *J. Appl. Phys.* **56,** 2120–2126.

Koren, G., Ho, F., and Ritsko, J.J. (1985). Excimer laser etching of Al metal films in chlorine environments. *Appl. Phys. Lett.* **46,** 1006–1008.

Koren, G., Ho, F., and Ritsko, J.J. (1986a). XeCl laser controlled chemical etching of aluminum in chlorine gas. *Appl. Phys.* **A40,** 13–23.

Koren, G. (1986b). Ar ion laser-assisted chemical etching of molybdenum foils in air. *J. Appl. Phys.* **59,** 1667–1672.

Koronkevich, V.P., Poleshchuk, A.G., Churin, E.G., and Yurlov, Y.I. (1985). Selective etching of laser exposed chromium thin films. *Sov. Tech. Phys. Lett.* **11,** 57–58.

Kuhn-Kuhnenfeld, F. (1972). Selective photoetching of gallium arsenide. *J. Electrochem. Soc.* **119,** 1063–1068.

Krimmel, E.F., Lutsch, A.G.K., Swanepoel, R., and Brink, J. (1985). Contribution to time resolved enhanced chemical etching and simultaneous annealing of ion implanation amorphized silicon under intense laser irradiation. *Appl. Phys.* **A38,** 109–115.

Loper, G.L., and Tabat, M.D. (1985a). UV laser generated fluorine atom etching of polycrystalline Si, Mo, and Ti. *Appl. Phys. Lett.* **46,** 654–656.

Loper, G.L., and Tabat, M.D. (1985b). Submicrometer-resolution etching of integrated circuit materials with laser-generated atomic fluorine. *J. Appl. Phys.* **58,** 3649–3651.

Lum, R.M., Glass, A.M., Ostermayer, F.W., Jr., Kohl, P.A., Ballman, A.A., and Logan, R.A. (1985). Holographic photoelectrochemical etching of diffraction gratings in n-InP and n-GaInAsP for distributed feedback lasers. *J. Appl. Phys.* **57,** 39–44.

Moskowitz, P.A., Vigliotti, D.R., and von Gutfeld, R.J. (1984). Laser micromachining of polyimide materials. In *Polyimides* (Mital, K., ed.). Plenum, New York, Vol. 1, 365–376.

Munoz-Yague, A., and Bafleur, M. (1981). Shallow defect etching of GaAs using AB solution under laser illumination. *J. Crystal Growth* **53,** 239–248.

Osgood, R.M, Sanchez-Rubio, A., Ehrlich, D.J., and Daneu, V. (1982). Localized laser etching of compound semiconductors in aqueous solution. *Appl. Phys. Lett.* **40,** 391–393.

Podlesnik, D.V., Gilgen, H.H., and Osgood, R.M. (1984). Deep-ultraviolet induced wet etching of GaAs. *Appl. Phys. Lett.* **45,** 563–565.

Podlesnik, D., Gilgen, H.H., and Osgood, R.M. (1986). Waveguiding effects in laser induced aqueous etching of semiconductors. *Appl. Phys. Lett.* **48,** 496–598.

Rauh, R.D., and LeLievre, R.A. (1985). Microphotoelectrochemical etching of N-GaAs using a scanned focused laser. *J. Electrochem. Soc.* **132,** 2811–2812.

Reksten, G., Holber, W., and Osgood, R.M. (1984). Laser controlled plasma etching. *J. Vac. Sci. Tech.* **A2,** 506–507.

Reksten, G., Holber, W., and Osgood, R.M. (1986). Wavelength dependence of laser enhanced plasma etching of semiconductors. *Appl. Phys. Lett.* **48,** 551–553.

Rothschild, M., Arnone, C., and Ehrlich, D.J. (1986). Excimer laser etching of diamond and hard carbon films by direct writing and optical projection. *J. Vac. Sci. Tech.* **B4,** 310–314.

Sekine, M., Okano, H., Yamabe, K., Hayasaka, N., and Horiike, Y. (1985). Radiation damage evaluation in an excimer laser etching. *Sym. VLSI Tech, IEEE Jpn. Soc. Appl. Phys.,* 82–83.

Sesselman, W., and Chuang, T.J. (1985). Chlorine surface interaction and laser induced shift etching reactions. *J. Vac. Sci. Tech.* **B3,** 1507–1512.

Srinivasan, R., and Mayne-Banton, V. (1982). Self-developing photoetching of poly(ethylene terephthalate) films by far ultraviolet excimer laser radiation. *Appl. Phys. Lett* **41,** 576–578.

Srinivasan, R. (1983). Kinetics of ablative photodecomposition of organic polymers in the far ultraviolet (193 nm). *J. Vac. Sci. Tech.* **B1,** 923–926.

Steinfeld, J.I., Anderson, T.G., Reiser, C., Denison, D.R., Hartsough, L.D., and Hollahan, J.R. (1980). Surface etching by laser generated free radicals. *J. Electrochem. Soc.* **127,** 514–515.

Sullivan, M.V., Kolb, G.A. (1986). Direct photoetching of evaporated germanium and its use in mask fabrication. *Electrochemical Technology* **6,** 430.

Swenson, E.J. (1983). Some present and future applications of laser processing—An overview. *Sol. State Technology* **Nov. 1983,** 156–158.

Takai, M., Tokuda, J., Nakai, H., Gamo, K., Namba, S. (1983). Laser induced local etching of gallium arsenide in gas atmosphere. *Jpn. J. Appl. Phys.* **22,** L757–L759.

Takai, M., Tsuchimoto, J., Nakai, H., Gamo, K., Namba, S. (1984a). Maskless dry etching of gallium arsenide with a submicron linewidth by laser pyrolysis in CCl_4 gas atmosphere. *Jpn. J. Appl. Phys.* **23,** L852–L854.

Takai, M., Tokuda, J. Nakai, H., Gamo, K., Namba, S. (1984b). Laser assisted local etching of gallium arsenide. *Mat. Res. Soc. Symp.* **29,** 211–216.

Takai, M., Nakai,H., Tsuchimoto, J., Gamo, K., Namba, S. (1985a). Local temperature rise during laser induced etching of gallium arsenide in $SiCl_4$ atmosphere. *Jpn. J. Appl.Phys.* **24,** L705-L708.

Takai, M., Nakai, H., Nakashima, S., Minamisono, T., Gamo, K., and Namba, S., (1985b). Residual local strain in gallium arsenide induced by laser pyrolytic. *Jpn. J. Appl. Phys.* **24,** L755–L757.

Tisone, G.C., and Johnson, A.W. (1983a). Laser controlled etching of chromium doped and N doped [100] GaAs. *Mat. Res. Soc. Symp. Proc.* **17,** 73.

Tisone, G.C., and Johnson, A.W. (1983b). Laser controlled etching of chromium doped [100] GaAs. *Appl. Phys. Lett.* **42,** 530–532.

Tsao, J.Y., and Ehrlich, D.J. (1983). Laser controlled chemical etching of aluminum. *Appl. Phys. Lett.* **43,** 146–148.

Tsukada, N., Sugata, S., Saitoh, H., and Mita, Y. (1983). Surface ripples in laser photochemical wet etching of gallium arsenide. *Appl. Phys. Lett.* **43,** 189–191.

Tsukada, N., Sugata, S., Saitoh, H., Yamanaka, K., and Mita, Y. (1984a). Grating formation on gallium arsenide by one step laser photochemical etching. In *Second European Conference on Integrated Optics, Florence, Italy,* IEE.

Tsukada, N., Semura, S., Saito, H., Sugata, S., Asakawa, K., and Mita, Y. (1984b). Laser enhanced reactive ion etching of GaAs with CCL_4 and H_2 mixed gas. *J. Appl. Phys.* **55,** 3417–3420.

Tucker, A.W., and Birnbaum, M. (1983). Laser chemical etching method for drilling vias in GaAs. *SPIE* **385,** 131–140.

von Gutfeld, R.J., Tynan, E.E., Melcher, R.L., and Blum, S.E. (1979). Laser enhanced electroplating and maskless pattern generation. *Appl. Phys. Lett.* **35,** 651–653.

von Gutfeld, R.J., and Hodgson, R.T. (1982). Laser enhanced etching in KOH. *Appl. Phys. Lett.* **40,** 352–354.

von Gutfeld, R.J. (1987). Laser enhanced patterning using photothermal effects: Maskless plating and etching. *J. Opt. Soc. Am* **B4,** 272–280.

Weyher, J.L., and van de Ven, J. (1982). Selective etching of N-type GaAs in a CRO_3-HF-H_2O system under laser illumination. *J. De Physique Coll.* **C5, S12,** 313–319.

Winters, H.F., and Coburn, J.W. (1979). The etching of silicon with XeF_2 vapor. *Appl. Phys. Lett.* **34,** 70–73.

Winters, H.F., and Houle, F.A. (1983a). Gaseous products from the reaction of XeF_2 with silicon. *J. Appl. Phys.* **54,** 1218–1223.

Winters, H.F., Coburn, J.W., and Chuang, T.J. (1983b). Surface processes in plasma-assisted etching environments. *J. Vac. Sci. Tech.* **B1,** 469–480.

Young, R.T., Narayan, J., Christie, W.H., van der Leeden, G.A., Levatter, J.I., and Cheng, L.J. (1983). Semiconductor processing with excimer lasers. *Sol. State Technology* **Nov. 1983,** 183–189.

Yokoyama, S., Yamakage, Y., and Hirose, M. (1984). Laser induced chemical dry etching of SiO_2. In *Conference on Solid State Devices, Kobe 1984,* 451.

Yokoyama, S., Yamakage, Y., and Hirose, M. (1985). Laser induced photochemical etching of SiO_2 studied by x-ray photoelectron spectroscopy. *Appl. Phys. Lett.* **47,** 389–391.

Zysset, B., and Salathe, R.P. (1984). Photochemical etching of laser induced defects in (Al, Ga) as heterostructures. *Appl. Phys. Lett.* **45,** 428–430.

CHAPTER 7

Laser Deposition: Energetics and Chemical Kinetics

R.L. JACKSON[a], T.H. BAUM[a], T.T. KODAS[a], D.J. EHRLICH[b], G.W. TYNDALL[a], AND P.B. COMITA[a]

[a]*IBM Almaden Research Center*
San Jose, California

[b]*Lincoln Laboratories*
Massachusetts Institute of Technology
Lexington, Massachusetts

1. Introduction

Laser-induced deposition has been studied vigorously in the last 10 years because of its potential applications in microelectronics and electro-optics. To date, laser deposition has not been widely incorporated into manufacturing processes, but uses in manufacturing are emerging. For example, processes that employ highly localized laser deposition for repair of photolithographic mask defects with a resolution of 1 μm are now relatively common in mask manufacturing (Tison and Cohen, 1987). Another application that shows promise uses highly localized laser deposition of metal and polysilicon lines for discretionary interconnects in integrated circuits (Tsao et al., 1982; Herman, 1984). Continued research promises to yield many other applications.

Despite these early successes, laser deposition technology is very young, with many fundamental aspects yet poorly understood and much

0-12-233430-2

essential material and process characterization only beginning. The next two chapters discuss the underlying fundamentals and early experimental work on laser deposition, respectively. Our intent in this first chapter is to discuss the general aspects of these deposition processes according to method, particularly by strategy of excitation and localization. We cite specific examples only as necessary to clarify more general points. A thorough review of research in specific systems follows in the next chapter of this volume.

The discussion in the first chapter is divided into two major sections, according to the two basic ways in which laser deposition has been carried out. Deposition may be accomplished photochemically, using a laser to induce photochemical reactions of deposition precursors, or thermally, using a laser to locally heat a substrate surface. Photochemical deposition will be discussed first, followed by thermal deposition. We have restricted our discussion to deposition processes involving gas-phase precursors or precursors adsorbed from the gas phase on a solid surface. Laser deposition from condensed phases has also been reported, however, including laser dissociation of metallopolymer films (Fisanick, et al., 1985; Gross, et al., 1985) and laser-driven electroplating and electroless plating (von Gutfeld, 1987). In addition, our review concentrates on laser deposition, but some interesting results have been achieved in ion-beam- (Shedd, et al., 1986; Harriott, et al., 1986) and electron-beam- (Foord and Jackman, 1984; Matsui and Mori, 1987) induced deposition.

The chapter includes references through September 1987. Due to the tremendous number of papers in laser deposition, we have limited our references to journals and conference proceedings published in English or with readily available English translations. When a conference proceedings paper or extended abstract draws heavily on papers published by the same group in a journal, we have chosen to cite only the journal reference. Hopefully, this will avoid references to literature that may not always be readily available.

2. Photochemical Deposition

Laser-induced photochemical deposition of both large-area uniform films and highly localized films has been successfully demonstrated for a wide variety of materials. The goals and mechanisms of blanket and localized deposition processes are clearly very different, but the general term *photodeposition* has been applied to both. For blanket deposition, a large-area laser beam is used to photodissociate a precursor gas, yielding products that condense or react on a nearby substrate to form a film. The laser may be directed at normal incidence onto the substrate or it may be directed parallel to the substrate, passing just above it. The blanket laser deposition process may be referred to as *photochemically-assisted chemical vapor deposition,* by analogy to plasma-assisted or plasma-enhanced chemical vapor deposition (Hess, 1984). For localized deposition, photodissociation of a precursor gas is induced by a focused laser beam impinging on a substrate at normal incidence. A number of factors may operate to restrict deposition to the area of the substrate that is irradiated by the laser beam. For example, deposition may occur solely via photodissociation of an adsorbed precursor (Tsao et al., 1985; Tsao et al., 1983). Alternatively, laser photodissociation of adsorbed species can create nucleation sites so that gas-phase photodissociation processes result in film growth only where the substrate is illuminated by the laser beam (Ehrlich et al., 1980; Ehrlich et al., 1981). Additionally, secondary photochemical or thermal reactions of photoproducts produced in the gas phase and subsequently adsorbed onto the substrate may be driven by the laser to produce a localized deposit (Braichotte and van den Bergh, 1984; Gilgen et al., 1987; Jackson and Tyndall, 1988). Photochemical deposition of lines is accomplished with a focused laser beam by scanning the beam with respect to the substrate during deposition. Following a brief discussion of the types of photochemical reactions encountered in deposition processes, we will discuss both photochemically assisted chemical vapor deposition and localized photochemical deposition.

2.1. *Photochemistry of Deposition Processes*

Photodissociation can be induced in a variety of ways that are appropriate to photochemical deposition processes. Many processes employ single-photon electronic dissociation of a precursor vapor to induce deposition. For example, silicon can be deposited via UV photodissociation of SiH_4 to yield predominantly SiH_2 (Perkins et al., 1979), which can then undergo further reaction on a heated surface to form a silicon film (Andreatta et al., 1982). In some cases, deposition has been driven by gas-phase multiphoton dissociation of a precursor gas with intense IR lasers. An example of this is silicon film deposition initiated by IR-laser dissociation of SiH_4 to form SiH_2 in the gas phase (Deutsch, 1979; Longeway and Lampe, 1981). Extensive photofragmentation of the precursor, requiring a multiphoton electronic dissociation process driven by a pulsed UV or visible laser, may be useful in some cases. One example is deposition of chromium (Solanki et al., 1982; Flynn et al., 1986; Yokayama et al., 1985) via excimer-laser photodissociation of $Cr(CO)_6$ to CO and Cr atoms (Tyndall and Jackson, 1987), which can then deposit on a nearby substrate.

The kinetics of a single-photon dissociation process can be formulated simply in certain limiting cases. Using the example of silane, the rate of formation of SiH_2 in the gas phase is given by

$$\frac{d[SiH_2]}{dt} = \sigma I \phi [SiH_4], \tag{7.1}$$

where $[SiH_2]$ and $[SiH_4]$ denote the gas-phase concentrations of SiH_2 and SiH_4, σ is the gas-phase absorption cross section of SiH_4 at the wavelength of the light source, I is photon flux, and ϕ is the quantum yield for SiH_2 formation, which is the probability than an excited SiH_4 molecule will dissociate to SiH_2. Secondary reactions of SiH_2 are neglected. If I is constant, i.e., if a cw light source uniformly illuminates the sample, and if the quantity $\sigma I l$ (l is the pathlength of the gas sample) is $\ll 1$ (optically thin sample), then the probability of forming SiH_2 is

$$\frac{[SiH_2]}{[SiH_4]_0} = 1 - e^{-\sigma I \phi t}, \tag{7.2}$$

where $[SiH_4]_0$ denotes the SiH_4 concentration at $t = 0$. Furthermore, if $\sigma I \phi t \ll 1$ (i.e., in the absence of saturation), Eq. 7.2 reduces to

$$\frac{[SiH_2]}{[SiH_4]_0} = \sigma I \phi t. \tag{7.3}$$

Thus, the probability of forming a molecule of SiH_2 depends linearly on I and t. For pulsed light sources, a similar solution is obtained by replacing It in Eq. 7.3 with nF, where n is the number of pulses and F is the photon fluence per pulse, provided $\sigma \phi n F \ll 1$ and provided one is not interested in the time evolution of SiH_2 formation during a laser pulse. Equation 7.1 is more difficult to solve if I is time dependent or nonuniform, or if the sample is not optically thin, but the probability of product formation still depends linearly on I or F if the total yield of photoproducts is small.

The kinetics of simple multiphoton dissociation processes can also be readily described. For example, Cr atoms have been formed by two- and three-photon dissociation of $Cr(CO)_6$ with a KrF* laser (Tyndall and Jackson, 1987). Two excitation processes contribute to Cr atom formation: direct multiphoton excitation of $Cr(CO)_6$ to a dissociative continuum and a sequential absorption/fragmentation process involving $Cr(CO)_x$ intermediates (Tyndall and Jackson, 1987). The direct process may involve coherent absorption of multiple photons. In the direct case, it can be shown that the rate of $Cr(CO)_6$ excitation, and thus the rate of Cr atom formation, scales as I^n, where n is the number of photons absorbed, provided saturation does not occur (Steinfeld, 1974). This can be similarly shown for a sequential multiphoton dissociation process. Consider the following example of the sequential dissociation mechanism proposed for production of Cr atoms via two-photon KrF*-laser dissociation of $Cr(CO)_6$ (Tyndall and Jackson, 1987):

$$Cr(CO)_6 + 1\,h\nu \xrightarrow{k_1} Cr(CO)_4^{\dagger} + 2CO \tag{7.4}$$

$$Cr(CO)_4^{\dagger} + 1\,h\nu \xrightarrow{k_2} Cr + 4CO, \tag{7.5}$$

where $Cr(CO)_4^{\dagger}$ denotes vibrationally excited $Cr(CO)_4$. The rate of Cr atom formation is given by the following set of coupled differential equations:

$$\frac{d[Cr(CO)_6]}{dt} = -k_1[Cr(CO)_6] \tag{7.6}$$

$$\frac{d[Cr(CO)_4]}{dt} = k_1[Cr(CO)_6] - k_2[Cr(CO)_4] \tag{7.7}$$

$$\frac{d[Cr]}{dt} = k_2[Cr(CO)_4] \tag{7.8}$$

$$k_1 = \sigma_1 I \phi_1; \qquad k_2 = \sigma_2 I \phi_2, \tag{7.9}$$

where the subscript 1 refers to reaction (7.4) and the subscript 2 refers to reaction (7.5). The probability of forming Cr atoms as a function of time for an optically thin sample irradiated uniformly by a cw beam is

$$\frac{[Cr]}{[Cr(CO)_6]_0} = \frac{k_1 k_2}{k_2 - k_1}\left[\frac{e^{-k_2 t} - 1}{k_2 t} - \frac{e^{-k_1 t} - 1}{k_1 t}\right]. \tag{7.10}$$

Equation 7.10 may be greatly simplified by expanding the exponentials, keeping terms to second order, giving after substitution for k_1 and k_2 from Eq. 7.9

$$\frac{[Cr]}{[Cr(CO)_6]_0} = \frac{\sigma_1 \sigma_2 \phi_1 \phi_2 I^2 t^2}{2}. \tag{7.11}$$

Equation 7.11 shows that the probability of Cr atom formation is proportional to I^2. For pulsed light sources, nF may be substituted for It in Eq. 7.11, subject to the limitations noted earlier. The second-order exponential expansion is accurate to relatively high levels of $Cr(CO)_6$ dissociation, so I^2 behavior will hold well beyond the limit where the first-order expansion in Eq. 7.3 fails It can be similarly shown that I^n behavior holds for n-step sequential multiphoton dissociation processes.

If I is nonuniform or time dependent, the coupled differential equations are more difficult to solve, but the I^n behavior will still hold.

Recent studies of metal carbonyl and metal alkyl photodissociation using a variety of product detection schemes, including multiphoton ionization (Duncan et al., 1979; Whetten et al., 1983; Nagano et al., 1986; Liou et al., 1986; Gerrity et al., 1980; Fisanick et al., 1981; Hossenlopp et al., 1986; Mitchell et al., 1985), luminescence (Karny et al., 1983; Gerrity et al., 1983; Iskizaka et al., 1986; Eres et al., 1987; Tyndall and Jackson, 1987), chemical trapping (Tumas et al., 1982), and kinetic absorption methods (Fletcher and Rosenfeld, 1985; Seder et al., 1986), have shown that dissociation of several ligands can be achieved via single-photon dissociation and that metal atoms can be efficiently produced by multiphoton dissociation. For the most part, these processes take place by sequential loss of the carbonyl or alkyl ligands. Following loss of the first ligand in the primary photochemical step, the metal-containing photofragment retains much of the excitation energy provided by photon absorption and continues to fragment without the need for additional energy input until the remaining energy is insufficient to permit further dissociation. Knowledge of the metal carbonyl and metal alkyl bond-dissociation energies is thus extremely useful in predicting the extent of dissociation that is possible energetically in a particular photodissociation process. These are given in Table 7.1 for the metal carbonyls and metal aklyls that have figured prominently in the laser deposition literature. Note that few of these species can be dissociated to the bare metal atom in the gas phase via a single-photon dissociation process at generally accessible laser wavelengths ($\leq$193 nm, 148 kcal/mole).

The kinetics of IR-laser photodissociation are very complicated, but a few general observations can be made (Sudboe et al., 1986). Excitation begins by absorption into the vibrational band that is resonant with the laser field. The initially excited vibrational mode will no longer be resonant after excitation to $v = 1$–3, due to anharmonicity. For molecules with a large number of low-frequency vibrations, however, the density of rotational/vibrational levels at the energy provided by absorption of only a few photons can be so high that some transition will always be resonant with the laser field. When this quasicontinuum of vibrational levels is reached, absorption of many photons proceeds rapidly. For molecules with few low-frequency vibrations, the quuasicontinuum cannot be reached by direct absorption, but dissociation can still proceed in most cases with the aid of collisions. Vibrational energy transfer from excited molecules will eventually raise the gas vibrational temperature to a high enough level that many molecules will reach the quasicontinuum. Energy

Table 7.1 Gas-phase bond dissociation energies in kcal/mole for metal carbonyls and alkyls. Bond dissociation energies were determined from pyrolysis kinetic data, except where noted. Total dissociation energies were determined from heats of formation measured calorimetrically. Reasonable estimated values (noted by parentheses) are quoted where measurements are not available.

Compound	Metal Carbonyls D_1[a,c]		D_T[b,d]
$Fe(CO)_5$	42		140.2
$Cr(CO)_6$[e]	37		153.6
$Mo(CO)_6$	41		217.8
$W(CO)_6$	46		255.6
$Ni(CO)_4$	22		139.3
	Metal Alkyls		
Compound	D_1[a,f]	D_2[a,g]	D_T[b,d]
$Zn(CH_3)_2$	67		86
$Zn(C_2H_5)_2$	55		69
$Cd(CH_3)_2$	60		69
$Hg(CH_3)_2$	60		60
$Hg(C_2H_5)_2$	48		49
$Al(CH_3)_3$	(65)		201
$Ga(CH_3)_3$	64	39	182
$Ga(C_2H_5)_3$	48		
$In(CH_3)_3$	51	20	119
$Tl(CH_3)_3$	40		
$Sn(CH_3)_4$	69		213
$Sn(C_2H_5)_4$	62		187
$Pb(CH_3)_4$	57		150
$Pb(C_2H_5)_4$	54		125
$P(CH_3)_3$	(72)		200
$P(C_2H_5)_3$	68		165
$As(CH_3)_3$	67		168
$Sb(CH_3)_3$	61		157
$Sb(C_2H_5)_3$	57		129
$Bi(CH_3)_3$	52		105

[a] Gas values of ΔH_{298} for breaking ligand-metal bonds ($\pm \sim 3$ kcal/mole). The subscript denotes first, second, etc. ligand-metal bond. [b] Gas-phase value of ΔH_{298} for forming metal plus ligands ($\pm \sim 1$ kcal/mole), as determined from the gas-phase heat of formation of each compound. [c] Lewis et al., (1984). [d] Cox and Pilcher, (1970). [e] Pilcher et al., (1975). [f] Smith and Patrick, (1983). [g] Price, (1974).

is randomized very rapidly among all vibrational modes of a molecule pumped into the quasicontinuum, even in the absence of collisions, so there are no well-documented cases of IR-laser dissociation breaking a specific bond in preference to the most favorable thermal dissociation channel of the molecule (Sudboe et al., 1986).

2.2. Photochemically-Assisted Chemical Vapor Deposition

Chemical vapor deposition (CVD) has been widely studied in recent years because of its importance in the microelectronics industry (see, for example, Hess et al., 1986). CVD is superior to physical deposition techniques, such as evaporation and sputtering, for covering vertical steps and filling contact holes with vertical sidewalls; it is thus often the method of choice for deposition of metals and insulators on very large scale integrated (VLSI) circuits, where such structural features are common. High deposition temperatures limit some applications, however, since silicon VLSI structures and most compound-semiconductor devices cannot tolerate high temperatures. The use of polymer dielectrics in VLSI circuits (Rothman, 1984) tightens the temperature constraint.

Photochemistry has often been employed in CVD processes as a means of reducing the operating temperature. These methods utilize a light source directed either parallel or perpendicular to the substrate to induce photofragmentation of one or more CVD precursors in the gas phase. The substrate is usually heated, but in some cases, deposition proceeds at room temperature. The light source need not be a laser, and indeed, CVD processes enhanced by lamp radiation, involving both direct excitation of the precursor (Jones et al., 1966; Rigby, 1969; Calloway et al., 1983; Foord and Jackman, 1984; Haigh, 1985; Numasawa et al., 1983; Mishima et al., 1983; Mishima et al., 1984) and mercury vapor sensitization (Saitoh et al., 1983; Tarui et al., 1983; Delahoy, 1985; Inoue et al., 1983; Hamano et al., 1984; Mullin and Irvine, 1986) have been developed. Pulsed lasers offer the advantage of high peak power, which may be required to drive a multiphoton dissociation process.

Photochemically assisted CVD is somewhat analogous to plasma-enhanced CVD (PECVD, Hess, 1984), where precursor fragmentation is induced in the gas phase by electron impact or ion-molecule reactions. Photochemically-assisted reactions, however, do not involve high-energy ions or electrons, which frequently damage the deposited film or delicate substrate structures. Photochemically assisted CVD often requires uniform illumination over the entire substrate area to achieve uniform deposition, and the deposition chamber requires a window that can become fogged with deposition products, but these difficulties can be

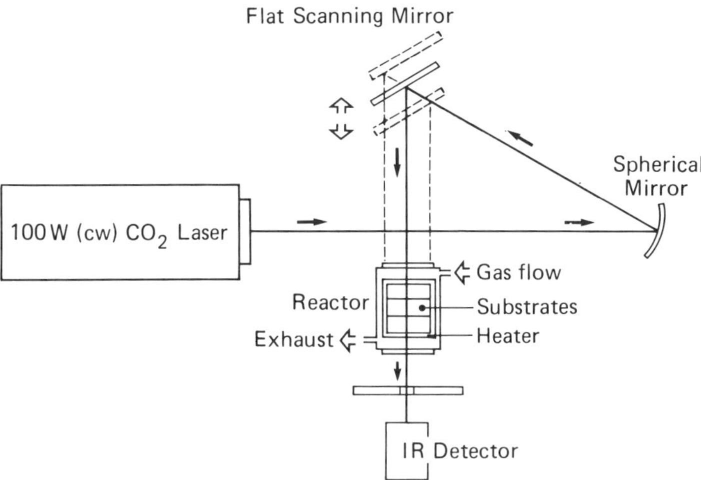

Fig. 7.1 Schematic of the scanning laser beam approach employed to achieve uniform deposition of a-Si: H films using a Gaussian cw CO_2 laser beam. The laser beam is scanned in a plane parallel to the substrate, resulting in time-averaged uniform illumination over the entire substrate area (Bilenchi et al., 1985).

readily overcome. For example, Bilenchi and coworkers (1985) used a scanning mirror, as shown in Fig. 7.1, to move the Gaussian beam of a CO_2 laser rapidly above a heated substrate in a plane parallel to the substrate surface, which results in time-averaged uniform illumination over the entire substrate area. Fogging of the chamber window is commonly reduced or eliminated by purging it with an inert gas, for example as shown in Fig. 7.2.

Below, we will discuss examples where both electronic and vibrational photodissociation of precursor gases have been used to enhance the rate or lower the deposition temperature of CVD processes. These processes frequently achieve results superior to those obtained by purely thermal CVD or PECVD processes employing the same precursors. Additional examples are given in Tables 7.2–7.4.

2.2.1 CVD Processes Assisted by Electronic Photodissociation—Parallel Illumination

Deposition processes performed in the parallel illumination geometry have neither the complications nor the advantages of laser-beam interactions with the substrate surface, so the effect of irradiation can be

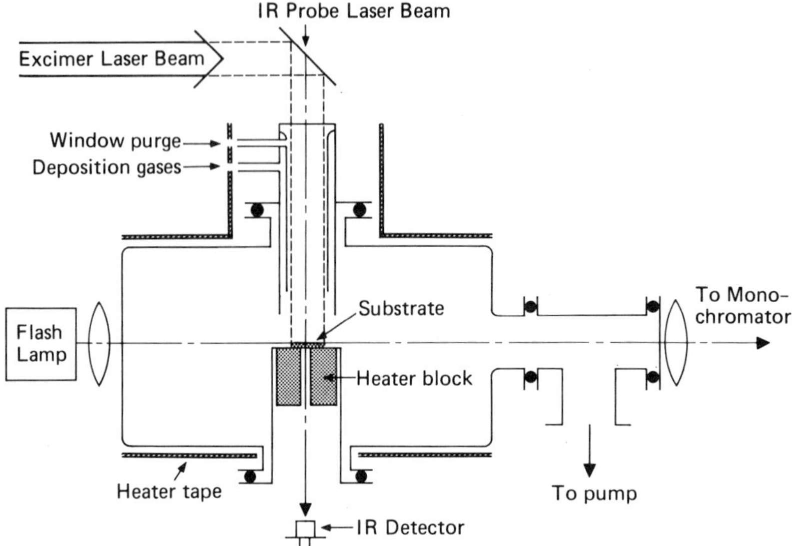

Fig. 7.2 Schematic of the laser deposition reactor used in the excimer-laser-assisted chemical vapor deposition of single-crystal InP films (Donnelly et al. 1985). An IR probe laser beam is used to monitor the film thickness via its IR transmission, and a flashlamp and monochromator are used to record spectra of photoproducts formed via excimer laser irradiation.

attributed exclusively to photodissociation of one of the precursor gases. Particularly interesting results have been achieved for deposition of tungsten and SiO_2, as discussed below. High-quality material can be deposited by the laser process in each case at temperatures where material of similar quality cannot be deposited by traditional CVD or PECVD. Additional examples are summarized in Table 7.2.

2.2.2.1 Tungsten Deposition

Tungsten has received increasing attention as a conductor in VLSI circuits. This is, in part, due to the success of its conventional CVD reactions (Broadbent and Ramiller, 1984; Learn and Foster, 1985). These reactions traditionally employ WF_6 as the precursor gas and H_2 as a reducing agent. Attempts to reduce the deposition temperature by PECVD techniques have not been completely successful for silicon-based ICs, because fluorine radicals generated at high concentrations during PECVD from WF_6 attack silicon and SiO_2 structures (Hess, 1984).

Table 7.2 Examples of UV laser photochemically assisted CVD processes with the laser directed parallel to the substrate, listed in order of deposited material, proceeding from left to right across the periodic table.

Deposit	Precursor	Laser	Comments
$TiSi_x$	$TiCl_4$, SiH_4	ArF*	Substrate 400–500°C[a]
Cr	$Cr(CO)_6$	Pulsed dye laser, 280–350 nm	Substrate 25°C, C, O contamination[b]
W	WF_6, H_2	ArF*	Substrate 200–500°C,[c,de]
Hg_xCd_yTe	$Hg(CH_3)_2$, $Cd(CH_3)_2$, $Te(CH_3)_2$	ArF*	Substrate 150°C, epitaxial growth on CdTe[f]
Al	$Al(CH_3)_3$	ArF*, KrF*	Substrate 200°C[g]
Al_2O_3	$Al(CH_3)_3$, N_2O	ArF*, KrF*	Substrate 25–500°C[g,h,i]
AlO_xN_y	$Al(CH_3)$, NH_3, air	ArF*	Substrate 25–500°C[j]
$Al(Ga)As_x$	$Al(CH_3)_3As(CH_3)_3$, $Ga(CH_3)_3As(CH_3)_3$	ArF*, KrF*	Substrate 25–300°C[k]
a-Si:H	Si_2H_6	ArF*	Substrate 35–400°C[m,l,n]
SiN_x	SiH_4, Si_2H_6, NH_3	ArF*	Substrate 225–625°C, hot wall reactor[o]
SiN_x	SiH_4, NH_3	ArF*	Substrate 200–600°C[p]
SiN_x	SiH_4, N_2O, N_2	ArF*	Substrate 250–400°C[q]
SiO_2	SiH_4, N_2O	ArF*	Substrate 20–600°C[r,s]
SiO_2	SiH_4, O_2	KrF*	Substrate 175–250°C[t]
SiO_2-GeO_2	SiH_4, GeH_4, N_2O	ArF*	Substrate 200–300°C[u]
Ge	GeH_4	ArF*, KrF*	Substrate 25–600°C[v]
PN_x	PH_3, NH_3	ArF*	Substrate 100–300°C[w]

[a] West et al. (1985). [b] Mayer et al. (1982). [c] Deutsch and Rathman (1984). [d] Shintani et al. (1987). [e] Adams et al. (1984). [f] Morris (1986). [g] Solanki et al. (1983). [h] Minkata and Furakawa (1986). [i] Deutsch et al. (1984). [j] Demiryont et al. (1986). [k] Zinck et al. (1987). [l] Yoshikawa and Yamaga (1984). [m] Yamada et al. (1985). [n] Zarnani et al. (1986). [o] Jasinski et al. (1987). [p] Deutsch et al. (1983). [q] Emery et al. (1984). [r] Boyer et al. (1982). [s] Tate et al. (1985). [t] Nishino et al. (1986). [u] Tate et al. (1986). [v] King et al. (1987). [w] Hirota and Makami (1985).

A reduction in deposition temperature has been achieved photochemically for tungsten CVD from WF_6 and H_2 by carrying out the process in the presence of unfocused ArF* laser radiation directed parallel to the substrate (Deutsch and Rathman, 1984; Shintani et al., 1987). In the study by Shintani and coworkers, the laser fluence was on the order of 80 mJ/cm^2. At this fluence, only single-photon dissociation of WF_6 is likely, since the absorption cross section of WF_6 at 193 nm is very low; the first absorption maximum is a weak peak at ~160 nm (Tanner and Duncan, 1951; McDiarmid, 1974). Furthermore, dissociation will not be

Table 7.3 Examples of UV laser photochemically assisted CVD processes with the laser directed onto the substrate. Listed in order of deposited material, proceeding from left to right the periodic table.

Deposit	Precursor	Laser	Comments
TiO_2	$TiCl_4$, O_2	ArF*	Substrate 25°C[a]
Cr, Mo, W	$Cr(CO)_6$, $Mo(CO)_6$, $W(CO)_6$	ArF*, KrF*, XeCl*	Substrate 25°C, C and O contamination[b]
Cr, Mo, W	$Cr(CO)_6$, $Mo(CO)_6$, $W(CO)_6$	ArF*, KrF*, XeCl* (?)	Substrate 25–40°C, inaccurate vapor pressures and cross sections[c]
Cr	$Cr(CO)_6$	KrF*	Substrate 25°C, C and O contamination[d]
Fe	$Fe(CO)_5$	ArF*, KrF*	Substrate 77°K and O contamination[e]
Fe/Ni	$Fe(C_5H_5)_2$, $Ni(C_5H_5)_2$	ArF*, N_2	Substrate 25°C, carbon contamination[f]
Pt	$Pt(PF_3)_4$	KrF*	Substrate 25°C[g]
Au	$(CH_3)_2Au(acac)$	ArF*, KrF*, XeCl*	Substrate 25°C, projection deposition, C and O contamination[h]
Au	$(CH_3)_3AuP(CH_3)_3$	KrF*	Substrate 10–25°C, projection deposition[i]
Zn	$(CH_3)_2Zn$,	ArF*, KrF*	Substrate 25–220°C[j]
ZnO_x	$(CH_3)_2Zn$, NO_2, N_2O	ArF*, KrF*	Substrate 25°C[j]
Al	$Al(CH_3)_3$	ArF*, KrF*	Substrate 25°C[k]
Al	$Al(CH_3)_3$	KrF*	Substrate 25–350°C, laser incident at 5°[l]
Al	$Al(i\text{-}C_4H_9)_3$	KrF*	Substrate 250°C, projection deposition[m]
Al_2O_3	$Al(CH_3)_3$, N_2O	ArF*, KrF*	Substrate 300–500°C[n]
GaAs	$Ga(CH_3)_3$, $As(CH_3)_3$	ArF*	Substrate 25°C[o]
InP, InO_x	$(CH_3)_3InP(CH_3)_3$, $P(CH_3)_3$, H_2	ArF*, KrF*, XeF*	Substrate 25–400°C[p,q]
a-Si:H	Si_2H_6	ArF*	Substrate 100–350°C[r]
Si	SiH_4	KrF*	Plasma-assisted, substrate 25°C[s]
Si	SiH_4	ArF*, KrF*	Substrate 25°C[t]
Ge	GeH_4	ArF*, KrF*	Substrate 25°C[t,u]

[a] Kawai et al. (1987). [b] Solanki et al. (1982). [c] Flynn et al. (1986). [d] Yokayama et al. (1985). [e] Love et al. (1984). [f] Armstrong et al. (1987). [g] Schroeder et al. (1984). [h] Baum et al. (1986). [i] Aylett (1986). [j] Solanki and Collins (1983). [k] Higashi and Rothberg (1985). [l] Motooka et al. (1986). [m] Blonder et al. (1987). [n] Deutsch et al. (1984). [o] Donnelly et al. (1987). [p] Donnelly et al. (1985). [q] Donnelly et al. (1984). [r] Yoshikawa and Yamaga (1984). [s] Gee et al. (1984). [t] Andreatta et al. (1982). [u] Osmundsen et al. (1985).

Table 7.4 Examples of CO_2 laser photochemically assisted CVD processes. Laser was operated in cw mode and directed parallel to the substrate, except where noted. Listed in order of deposited material, proceeding from left to right across the periodic table.

Deposit	Precursor	Comments
$TiSi_x$	$TiCl_4$, SiH_4	Substrate 300–500°C[a]
a-Si:H	SiH_4	Substrate 200–500°C[b,c,d]
a-Si:H, P and B doped	SiH_4, PH_3, B_2H_6	Substrate 200–500°C[e,f]
Si	SiH_4	Laser directed onto substrate, no external heating[g,h]
Si	SiH_4	Laser directed onto substrate, external heating to 550–750°C[i]
Si	SiH_4	Pulsed TEA laser directed onto substrate, no external heating[j]
Si	SiH_4	Pulsed TEA laser, parallel to and incident on substrate maintained at 25–400°C[k,l]
SiN_x	SiH_4, NH_3	[m]
Ge	R_xGeH_{4-x}	[n]

[a] West et al. (1985). [b] Bilenchi et al. (1982). [c] Meunier et al. (1983). [d] Meunier et al. (1983). [e] Bilenchi et al. (1985). [f] Branz et al. (1986). [g] Hanabasu et al. (1983). [h] Hanabasu and Kikuchi (1983). [i] Meguro et al. (1986). [j] Hanabasu et al. (1980). [k] Pauleau et al. (1984a). [l] Pauleau et al. (1984b). [m] West and Gupta (1984). [n] Stanley et al. (1986).

extensive, since the first tungsten-fluorine bond-dissociation energy is 120 kcal/mole (Dispert and Lachman, 1977), whereas the energy of one 193-nm photon is 148 kcal/mole. ArF*-laser dissociation of WF_6 reduces the activation energy for tungsten CVD from 16 kcal/mole to 9 kcal/mole, however (Shintani et al., 1987). Deposits with a resistivity of 9 $\mu\Omega$-cm were obtained at a substrate temperature of 300°C. Kinetic studies of the deposition rate versus WF_6 and H_2 partial pressure suggest that the rate-determining step in the presence of laser radiation (Shintani et al., 1987) is different from that in the purely thermal process (McConica and Krishnamani, 1986), although the nature of this step has not been identified.

2.2.1.2 SiO_2 Deposition

High-quality SiO_2 films have been deposited at temperatures below 450°C from SiH_4 and N_2O mixtures in the presence of ArF*-laser radiation directed parallel to the substrate (Boyer et al., 1982). At this wavelength, only N_2O absorbs appreciably under the process conditions (substantial excess N_2O, laser fluence ~400 mJ-cm^{-2}), giving ground electronic state

N_2 and excited $O(^1D)$ via a single-photon dissociation process (Preston and Barr, 1971); subsequent thermal reactions of $O(^1D)$ with N_2O yield N_2, O_2, and NO in their ground electronic states (Zavelovich, 1981). The oxygen-containing products of N_2O photodissociation can react with SiH_4 to drive deposition of SiO_2. The deposition rate is independent of substrate temperature over the range of 20–600°C and is as high as 300 Å-min^{-1}. At substrate temperatures ≥350°C, highly adherent, transparent SiO_2 films free of pinholes were produced via the laser-assisted process without subsequent annealing. Other properties of the films, such as hydrogen and nitrogen content, etch rate in buffered HF, and dielectric breakdown voltage, compared well with those of SiO_2 films produced by PECVD.

2.2.2 CVD Processes Assisted by Electronic Photodissociation—Perpendicular Illumination

Pulsed-laser deposition has also been performed with the laser directed onto the substrate. Several effects may be expected in this illumination geometry, in addition to the photochemical enhancement effects achieved in the case of parallel illumination. For example, laser-induced desorption (Stair and Weitz, 1987) or photoreactions of surface species may occur. Photochemical reactions occurring within the deposit may alter its composition. In addition, laser-induced surface temperature increases may cause some annealing or thermal curing of the deposit, or may alter the deposition kinetics. Pulsed-laser heating will not necessarily yield results similar to those obtained upon cw heating to the same temperature, however, because of the tremendous heating and cooling rates of solids under pulsed laser illumination. We will concentrate on the examples of Group 6 metal CVD and InP CVD, since these studies clearly illustrate the effect of laser-beam interactions with the substrate surface on the deposition process. Additional examples are summarized in Table 7.3.

2.2.2.1 Group VI Metal Deposition

The excimer laser-induced deposition from the hexacarbonyls of Cr, Mo, and W provides an interesting contrast of deposition performed under perpendicular versus parallel illumination of the substrate. Continuous, adherent films of submicrometer grain size and a relatively high metal content (50–90 atm %) were reported by two groups (Solanki et al., 1982; Flynn, 1986) for deposition onto quartz and silicon substrates under direct illumination of the substrate at rather low fluences (<50 mJ-cm^{-2}), while large, poorly adherent, dispersed grains of lower metal content ($<$50 atm %) are deposited under parallel illumination. The authors did not specifically address the causes behind this difference. There is also some disagreement between the two reports as to the level of C and O impurities in the films deposited with the laser directed onto the substrate. The results of Yokoyama et al. (1985) suggest that the metal content may actually be closer to 50 atm %.

Yokayama et al. (1985) studied KrF* laser-induced deposition from $Cr(CO)_6$ at several laser fluences under direct illumination of the quartz substrate. A dramatic change in deposit morphology, adhesion, and metal content was observed as the laser fluence increased. At high fluences ($>$40 mJ-cm^{-2}), continuous films of submicron grain size were formed. The Cr content of the films was ~50 atm %, with C and O impurities making up the remainder. At low fluences, the deposits resembled the poorly adherent, grainy deposits observed under parallel illumination. The chromium content of the deposits was insensitive to laser fluence, but the oxygen content dropped, while the carbon content increased with increasing fluence. The authors attributed this to an increasing thermal component to the deposition process. Other effects are difficult to exclude, however, particularly since the concentration and composition of the gas-phase photoproducts as well as the surface temperature vary with laser fluence. A more definitive separation of thermal and photochemical effects could be obtained by studying the wavelength dependence of the deposition kinetics and of the deposit morphology and composition.

2.2.2.2 InP Deposition

Photochemically assisted CVD of InP from $(CH_3)_3InP(CH_3)_3$ and $P(CH_3)_3$ has been carried out using an ArF* laser directed onto the substrate (Donnelly et al., 1985), as reviewed more completely in Chapter 9. Epitaxial growth of InP on single crystal InP, GaAs, and InGaAs substrates was achieved at laser fluences between 100 and 200 mJ-cm^{-2} at a substrate temperature (300–350°C), below that necessary for traditional CVD from the same precursors. Reduced substrate temperatures are advantageous in CVD of InP, since elevated temperatures lead to defects caused by phosphorus deficiency. Material deposited within the laser-illuminated area of the substrate grew epitaxially and was virtually free of carbon contamination, while material deposited outside the laser-illuminated area was amorphous and highly contaminated by carbon. The high degree of crystallinity of the irradiated films was attributed to laser annealing. These results demonstrate clearly that dramatic effects can be achieved in laser-assisted CVD where the laser is directed onto the substrate surface.

2.2.2.3 Laser Desorption of Surface Contaminants

The effect of a laser interacting with a bare substrate in the absence of photochemistry is demonstrated in an experiment on evaporation of metal films during and after irradiation of the substrate by a KrF* or XeCl* excimer laser at modest fluences ($\leq 15\,\text{mJ-cm}^{-2}$). Zinc and magnesium films evaporated onto quartz could be deposited selectively onto the area of the substrate directly illuminated by the laser (Coombe and Wodarczyk, 1980). Selectivity was maintained on a quartz substrate even when the metal evaporation source was turned on after the laser had been turned off for several minutes. This result suggests that selective deposition was caused by laser-induced desorption of surface contaminants, which increases the sticking coefficient of the metal atoms. This is consistent with the excellent adhesion of the films produced under laser illumination. It is less likely that laser radiation damages the substrate, creating nucleation sites for growth of the metal film, since selectivity was demonstrated on substrates that where highly transparent at the laser wavelength and the laser fluence was not required to be especially high to observe selective deposition.

2.2.3 CVD Processes Assisted by Electronic Photodissociation–Projection Deposition

The interconnection circuitry levels on an IC are typically formed by a multistep process involving blanket deposition of metal, definition of the circuit pattern on the metal layer by photolithography, and etching of metal from the regions uncovered in the photolithographic step. The laser-assisted deposition processes we have described thus far offer improved methods of performing the blanket deposition process, but not a reduction in the number of process steps. The entire process can be reduced to a single step, however, if patterned films can be deposited directly via photochemically assisted CVD in the perpendicular illumination geometry by imaging the laser beam onto the substrate surface with a mask and projection optics. Projection deposition has in fact been achieved with limited success. It was first demonstrated many years ago at low resolution for Hg lamp-induced deposition from $Cd(CH_3)_2$ (Jones et al., 1966). More recent experiments on laser deposition of gold and aluminum using image projection will be discussed below. Additional examples are noted in Table 7.4.

2.2.3.1 *Gold Deposition*

Gold patterns were formed by projection deposition via excimer laser-induced dissociation of dimethyl gold acetylacetonate (Baum et al., 1986). Deposition of a 0.2-μm-thick patterned film was performed on the inside of the front quartz cell window with a resolution of ~2 μm, which approached the resolution of the optics used to form the image. Little material was deposited outside of the laser-illiminated regions of the substrate. The gold deposits were relatively impure (50–70% Au).

Aylett (1986) used a modification of this one-step projection deposition process to produce patterned films of high-purity gold. In these experiments, trimethyl gold trimethyl phosphine vapor was admitted to a cell containing a substrate that was cooled below ambient temperature. Cooling of the substrate results in condensation of the precursor vapors on the substrate surface. The intent was to confine the photochemistry to dissociation of the precursor on the surface, rather than in the gas phase, thereby enhancing the resolution of the deposition process. Deposition was induced by focusing the image of a mask on the substrate, using a KrF* laser and a single-lens projection optical system. Patterned, 0.1-μm thick deposits of gold free of detectable levels of contamination (Auger) were produced. The optical system was not capable of producing images on a micrometer scale, so the ultimate resolution of this approach is unknown.

2.2.3.2 *Aluminum Deposition*

Single-step deposition of aluminum from trisobutylaluminum has been demonstrated using 193-nm light (Ehrlich and Tsao, 1985). A two-step projection deposition process has been described by Blonder et al. (1987) for producing patterned, highly conducting aluminum films. In their process, which is related to a process described by Tsao and Ehrlich (1984a), a substrate heated to 250°C in the presence of tri-isobutyl aluminum vapor is irradiated with an imaged KrF* laser beam, using a mask and projection optics. Al-film growth begins during laser irradiation and the film continues to grow by thermal CVD (see Green et al., 1985) after laser illumination ceases. The thermal growth phase is selective for the regions of the substrate that were irradiated by the laser. The result is patterned deposition of highly conducting aluminum films. Since the film can grow laterally as well as vertically during the thermal growth step, the resolution achievable by the two-step process is, in principle, more limited than that of the one-step, purely photochemical process. Thus far, however, the two-step method has achieved superior results, although the full potential of each method has not been realized.

2.2.4 CVD Processes Assisted by Vibrational Photochemistry

IR-laser-assisted CVD of both silicon and silicide films has been performed by directing the output of a cw CO_2 laser parallel to a heated substrate in a cell containing SiH_4 gas. The P(20) line (944.19 cm^{-1}) of the CO_2 laser is nearly resonant with a rotational transition of the SiH_4 ν_4 band at 944.213 cm^{-1} (Deutsch, 1979). Collisionless dissociation of SiH_4 cannot be induced even at very high laser fluences (Deutsch, 1979), because the relatively sparse rotational structure of SiH_4 and the anharmonicity of the ν_4 band do not permit absorption by SiH_4 molecules excited to $v > 0$ of this band. As a result, SiH_4 cannot be directly pumped into the vibrational quasicontinuum (see Section 2.1) by CO_2 laser irradiation, even at high laser fluences. Collisionally enhanced dissociation does occur at SiH_4 pressures above about 5 torr, however (Deutsch, 1979; Longeway and Lampe, 1981). The energy absorbed at a fixed laser fluence depends roughly on the square of the SiH_4 pressure above 5 torr, which is indicative of a collisionally enhanced absorption process (Deutsch, 1979). The primary dissociation products are SiH_2 and H_2 (Longeway and Lampe, 1981); these are also the primary products observed upon gas-phase pyrolysis of SiH_4 (Jasinski and Estes, 1985). Most importantly, gas-phase dissociation of SiH_4 to SiH_2 and H_2 is the rate-determining step of Si CVD from SiH_4 at temperatures <400°C (Scott et al., 1981), but thermal dissociation of SiH_4 at these temperatures is very slow (Jasinski and Estes, 1985). IR laser-induced dissociation of SiH_4 thus permits deposition at a substrate temperature well below that required in traditional CVD deposition of amorphous hydrogenated silicon (a-Si:H) and titanium silicide. IR-laser processes will be discussed in detail below. Additional examples of IR-laser-assisted processes are summarized in Table 7.4.

2.2.4.1 a-Si:H Deposition

IR-laser-assisted deposition has been used to great advantage in the production of a-Si:H films (Bilenchi et al., 1982; Meunier et al., 1983 Meunier et al., 1983; Bilenchi et al., 1985; Branz et al., 1986). Preparation of a-Si:H films by traditional CVD from SiH_4 at acceptable deposition rates requires temperatures >500°C. At this temperature, relatively little hydrogen is incorporated into the films; hydrogenation must be performed by subsequent exposure of the film to a hydrogen plasma at a lower temperature. Films with high levels of hydrogen incorporation can be prepared by IR-laser-assisted CVD in a single step at substrate temperatures between 300° and 400°C. SiH_4 pressures are typically in the range of 5 torr, where collisionally enhanced dissociation occurs upon CO_2-laser irradiation. An equal pressure of argon is typically added to improve the laser absorption coefficient via modest pressure broadening. Two groups (Bilenchi et al., 1985; Branz et al., 1986) have produced n-type and p-type a-Si:H by adding low levels of PH_3 and B_2H_6 to the gas mixture, although there is disagreement as to the effect of these added gases on the deposition rate. Musci and coworkers have circumvented the need for a highly uniform, large-area laser beam by rastering a small-diameter Gaussian cw CO_2 laser beam rapidly above the substrate during deposition (see Fig. 7.1). Glow discharge methods (Brodsky, 1979) and traditional CVD in a hot wall reactor (Scott et al., 1981) have also been used to prepare a-Si:H films in a single step with equal success at relatively low substrate temperatures.

2.2.4.2 Titanium Silicide Deposition

Titanium silicide films have been formed via CVD processes by mixing SiH_4 with the vapor of $TiCl_4$ (Roslera and Engle, 1984). IR-laser-assisted CVD can be used to promote this process at substrate temperatures below those accessible in the traditional CVD process. West et al. (1985) used the P(20) line of a cw CO_2 laser directed parallel to the substrate to deposit titanium silicide films from SiH_4 and $TiCl_4$ vapor. The process is analogous to the process described above for IR laser-assisted deposition of a-Si:H films from pure SiH_4, in that SiH_4 absorbs the laser radiation and dissociates by collisionally enhanced processes in the gas phase. Titanium silicide films can be produced at substrates temperatures down to 400°C, as compared to 650°C required in traditional CVD from the same precursors. The Ti/Si ratio in the films is proportional to the $TiCl_4/SiH_4$ ratio in the gas phase, although an excess of SiH_4 is required. Stoichiometric $TiSi_2$ films are produced at a $TiCl_4/SiH_4$ ratio of 0.06. The best as-deposited film resistivity is about five times that of bulk $TiSi_2$. The resistivity can be reduced to nearly the bulk value by annealing at 800°C, but at this temperature, the advantage of IR laser-assisted CVD is lost. PECVD from the same precursors yields $TiSi_2$ films with bulk resistivity after annealing to 650°C (Roslera and Engle, 1984).

In a process similar to the CO_2-laser-enhanced CVD of $TiSi_2$, West al. (1985) deposited $TiSi_2$ from the same precursors in the presence of ArF*-laser radiation directed parallel to the substrate surface. This provides an interesting comparison between laser-enhanced CVD by vibrationally induced dissociation with an IR laser and electronic photodissociation with a UV laser. In the CO_2-laser process, enhancement of the deposition rate is achieved via vibrational photodissociation of SiH_4, while in the ArF*-laser process, enhancement of the deposition rate is achieved via electronic photodissociation of $TiCl_4$ to $TiCl_3$ and Cl. The gas-phase reactions are very different in the two processes, but, interestingly, the results are remarkably similar. Each process gives reasonably high quality $TiSi_2$ at substrate temperatures between 400 and 500°C, where the as-deposited resistivity is about five times that of bulk $TiSi_2$. The main difference is that the annealing temperature necessary to achieve near-bulk resistivity is lower for films deposited by the ArF*-laser process. $TiSi_2$ films produced by this process compare favorable to films obtained via PECVD from the same precursors (Roslere and Engle, 1984).

2.3. Localized Photochemical Deposition

The first reports of localized, one-step laser deposition on a micrometer scale generated a great deal of enthusiasm in the microelectronics industry because of the number of applications where such a process can be useful. Two applications for localized laser deposition in microelectronics were mentioned in the introduction. Localized laser photochemical deposition is typically carried out by scanning a focused laser beam across a substrate in the presence of a precursor vapor, leading to deposition only in the regions of the substrate that are irradiated by the beam. Deposition often involves a combination of gas-phase and surface photochemical reactions. Localized photochemical deposition has been extensively studied recently, with much of the work aimed at understanding the high degree of localization and the relative importance of surface and gas-phase photoprocesses. Virtually all of this work has concentrated on deposition of metals. In the following sections, we will examine kinetic models for the deposition process, the experimental studies that have addressed deposition mechanisms, and the materials properties of deposits obtained by focused cw UV laser photochemical deposition.

2.3.1 Kinetic Models of Localized Deposition Processes

Focused cw laser photochemical deposition processes can be quite difficult to model, since photochemical reactions of the precursor as well as secondary reactions of the photoproducts in the gas phase and on the surface can contribute to the deposition rate. Photoproducts are created only within the laser beam, so the photoproduct concentration will be highly nonuniform. In certain limiting cases, however, focused cw laser photochemical deposition kinetics can be modeled. Many idealized cases have been treated by Chen (1987). A few models will be discussed below, including models where surface photoreactions and gas-phase photoreactions contribute to the deposition kinetics. The physical kinetics of atomic species, once formed by gas-phase or surface-phase processes, but prior to film coalescence, are equally complex. The latter can be critical in deposition localization (e.g. prenucleation) but is not treated here. Some discussion of these important processes can be found in Chapters 6 and 9.

2.3.1.1 Processes Controlled by Surface Photoreactions

A simple model has been developed to describe the deposition kinetics when the deposition rate is controlled by single-photon dissociation of a surface-adsorbed precursor. Under conditions where the deposition rate is controlled by the photon flux at the surface rather than by the flux of reacting species, the deposition flux, i.e., the number of product species deposited per unit area per unit time, can be written for a Gaussian beam as (Wood et al., 1983):

$$J(r) = N_a \sigma_a \phi \alpha \frac{2P}{\pi w^2 h\nu} \exp\left(\frac{-2r^2}{w^2}\right) \tag{7.12}$$

where N_a is the number of reactant molecules adsorbed per unit area, σ_a is the absorption cross section of the adsorbed reactant at the frequency of the incident light ν, ϕ is the quantum yield for product formation, α is the fraction of products that stick on the surface, P is the total incident power, r is the radial distance from the center of the laser beam, w is the $1/e^2$ beam diameter, and h is Planck's constant. Equation 7.12 shows that if the rate-determining step is a single-photon surface photochemical reaction, then the deposition rate should be linearly proportional to the concentration of adsorbed reactant, i.e., the deposition rate versus precursor pressure should follow the adsorption isotherm of the precursor. Isotherms for physical adsorption typically follow Type-II behavior, as depicted in Fig. 7.3a, while isotherms for chemical adsorption might follow Langmuir behavior (Adamson, 1982), as depicted in Fig. 7.3b. In each case, the concentration of adsorbed species is highly nonlinear in precursor pressure and dependent on substrate temperature. Measurements of deposition rate versus precursor pressure showing Type-II behavior have been reported for frequency-doubled argon-ion-laser-induced deposition from $TiCl_4$ (Tsao et al., 1983), as shown in Fig. 7.4. Equation 7.12 also shows that the deposition rate depends linearly on the photon flux at the surface, given by $2P/\pi w^2 h\nu$; thus, the deposition rate will increase linearly with an increase in laser power and, at constant power, will decrease quadratically with increasing laser beam radius at the surface.

A limitation of this model is that it does not account for differences in the photochemistry and in contributions to the deposition kinetics

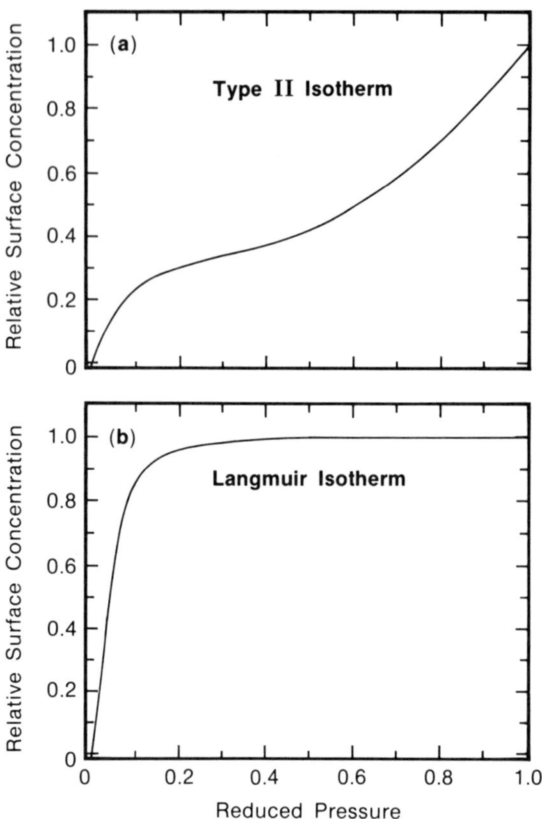

Fig. 7.3 Representative adsorption isotherms for physically adsorbed (Type II) and chemically adsorbed (Langmuir) species. The reduced pressure is normalized with respect to the vapor pressure of the species.

between photolysis of molecules in the first adsorbed layer and photolysis of molecules adsorbed in overlayers for a physically adsorbed system. Such differences could be common, since the heat of adsorption of the first monolayer is usually higher than that of the overlayers. Qualitatively, the first monolayer can be described as having a relatively strong interaction with the surface, while the overlayers more resemble a bulk liquid phase. An illustration of the difference in photochemical behavior that can be observed for molecules adsorbed in the first monolayer and in overlayers was demonstrated for photolysis of $Fe(CO)_5$ physically adsorbed on SiO_2 particles (Jackson and Trusheim, 1982). UV irradiation of

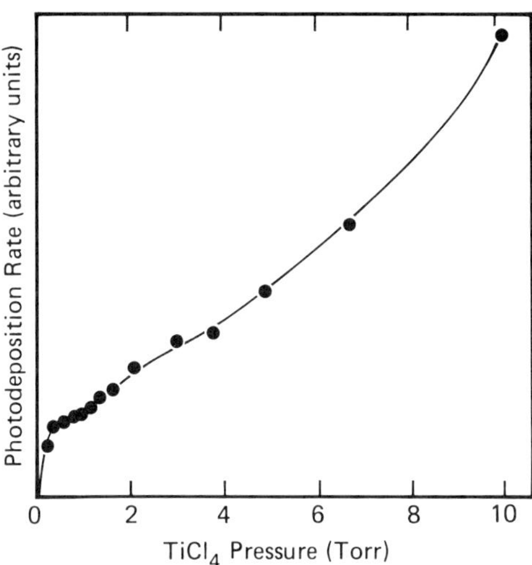

Fig. 7.4 Deposition rate versus precursor pressure for frequency-doubled argon-ion laser-induced deposition from $TiCl_4$. (Tsao et al., 1983).

adsorbed $Fe(CO)_5$ at submonolayer coverages yields $Fe_3(CO)_{12}$, while irradiation under conditions where several overlayers are present yields predominantly $Fe_2(CO)_9$. The difference was attributed to the role of surface functional groups in affecting the secondary reactions of photoproducts formed in the first monolayer and the absence of such effects in overlayers. Equation 7.12 may thus be too simple in many cases. For example, the deposition rate would not be proportional to the adsorption isotherm above monolayer coverage if the photoproducts desorbed more readily from overlayers then from the first monolayer due to a weaker interaction with the substrate. Tsao et al. (1985) indeed observed that the rate of frequency-doubled argon-ion-laser-induced deposition from $Pb(C_2H_5)_4$ was insensitive to $Pb(C_2H_5)_4$ pressure above 0.02 torr (see Fig. 7.5), even though the adsorption isotherm shows an increasing concentration of adsorbed $Pb(C_2H_5)_4$ at higher pressures (Rigby, 1969). This behavior is not anticipated by Eq. 7.12.

A unique facet of laser photochemical deposition often clearly indicates the importance of surface photodissociation. Photodeposition processes induced by (and most easily seen with polarized) laser beams frequently yield deposits that show ripples oriented perpendicular to the electric field vector of the incident beam (Osgood and Ehrlich, 1982; Brueck and Ehrlich, 1982; Wilson and Houle, 1985; Houle et al., 1986;

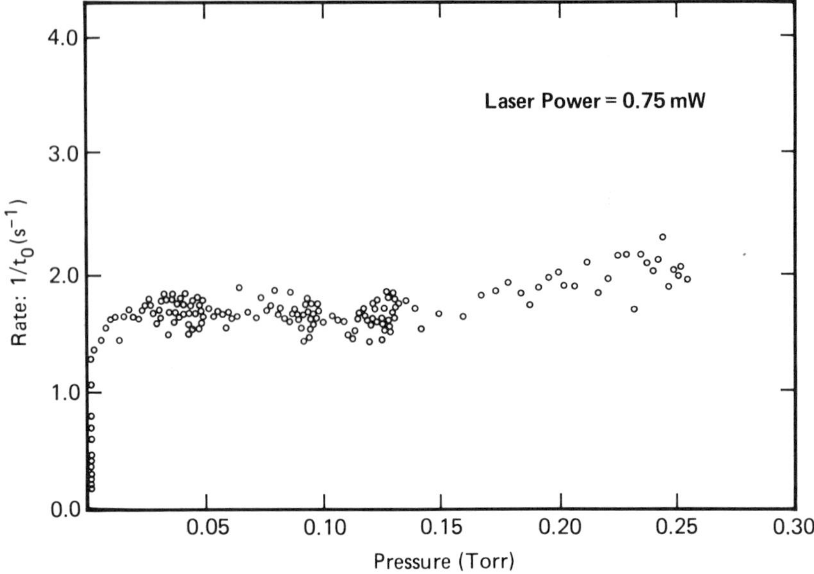

Fig. 7.5 Deposition rate versus precursor pressure for frequency-doubled argon-ion laser-induced deposition from $Pb(C_2H_5)_4$ (Tsao et al., 1985).

Yokayama et al., 1985; Jackson and Tyndall, 1988). An example of this is shown in Fig. 7.4 for a deposit prepared by photodeposition from $Cr(CO)_6$ induced by a focused, cw frequency-doubled argon ion laser. The ripple spacing is ~0.5–1.0 × the laser wavelength. In Fig. 7.6, the ripple spacing is ~160 nm. The ripples have been attributed to coherent interference between the incoming beam and surface plasma waves, the latter having the same frequency as the observed ripples (Brueck and Ehrlich, 1982; Wilson and Houle, 1985). If the deposition rate depends strongly on the light intensity at the surface, this will result in ripple formation. The appearance of ripples is thus indicative of surface photodissociation processes. The gas-phase mean free path is many times the ripple spacing in typical photochemical deposition experiments, so it is not possible for deposits formed solely as a result of a gas-phase photodissociation processes to exhibit ripples of this kind. (High laser powers can, of course, induce ripples by surface heating and annealing effects.) Ripples have been observed in deposits formed from the metal alkyls of cadmium, zinc, and aluminum (Osgood and Ehrlich, 1982), copper bis(hexafluoroacetylacetonate) (Wilson and Houle, 1986; Houle et al., 1986), and the hexacarbonyls of chromium, molybdenum, and

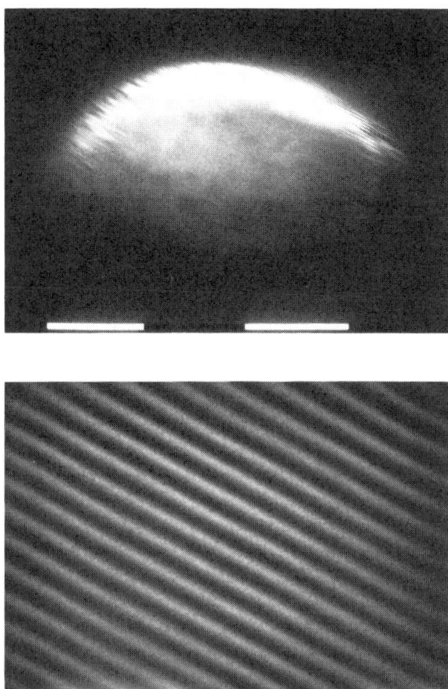

Fig. 7.6 Scanning electron micrographs at two different magnifications of a deposit formed via frequency-doubled argon-ion laser-induced deposition from $Cr(CO)_6$. The white bars at the bottom left represent 10 μm and 1 μm in the top and bottom micrographs, respectively. The ripples are oriented perpendicular to the electric field vector of the incoming polarized laser beam used to produce the deposits (Jackson and Tyndall, 1988).

tungsten (Yokayama et al., 1985; Jackson and Tyndall, 1988), indicating the importance of surface photochemical processes in each case. The importance of laser polarization is illustrated by the lack of ripples in a deposit formed via laser photochemical deposition from $Cr(CO)_6$ using a dual-plane-polarized He/Cd laser (325 nm), which has no preferential polarization direction (Jackson and Tyndall, 1988).

2.3.1.2 Processes Controlled by Gas-Phase Photoreactions

Mathematical descriptions of the deposition rate, for cases where the deposition kinetics are controlled by gas-phase photodissociation processes, can be very difficult to derive. The rate of such a process can be written exactly in some cases as a function of radial distance from the center of the laser beam and of the distance from the surface, but creation of photoproducts is only part of the deposition process. The products must migrate to the surface and stick to form the observed deposit. Photoproduct mass transport is difficult to describe accurately, except in the limits of low and high pressure. At low pressures, where the gas-phase mean free path is much longer than the radius of the deposit, a ballistic model may describe the transport of photoproducts to the substrate surface. If a photoproduct formed at any point in the cell has a randomly oriented velocity vector, the probability that the product will strike the substrate surface within a given area is equal to the solid angle subtended by that area, from the point of photoproduct creation, divided by 4π (Wood et al., 1983; Chen, 1987). Combining this probability with the equation for the rate of formation of photoproducts gives an expression for the flux of photoproducts to the deposit surface. The rate of formation of products via a one-photon dissociation process induced by a focused Gaussian light beam in an optically thin sample is described by an equation analogous to Eq. 7.12, where N_a and σ_a are replaced by the concentration of the precursor in the gas phase and the absorption cross section of the precursor in the gas phase, respectively. The beam radius w is a function of distance from the substrate z and is given by

$$w = w_0\left[1 + \left(\frac{z}{z_0}\right)^2\right]^{1/2}, \tag{7.13}$$

where w_0 is the $1/e^2$ beam diameter at the focal point and z_0 is a constant determined by the focusing optics (Demtroeder, 1982). These equations neglect contributions from light reflected or scattered from the substrate, which will lead to an underestimation of the rate if the deposit reflects or scatters a significant fraction of the incident beam. If the deposition rate is controlled by ballistic transport of products formed in a gas-phase, one-photon dissociation process, then the deposition rate will increase linearly with laser power and precursor pressure. It can also be shown that the deposition rate will decrease linearly with increasing laser-beam radius (Wood et al., 1983; Chen, 1987). The ballistic model is limited to cases where the gas-phase mean free path is large compared to the deposit radius, but in this limit, the model has some ability to predict deposit shapes (Jackson and Tyndall, 1988). At high pressures, where the mean free path in the gas phase is much smaller than the radius of the deposit, two-dimensional (r, z) diffusion equations can be used to model the gas-phase transport processes occurring during deposition. In practice, it can be difficult to model photochemical deposition processes using a diffusion equation approach, since boundary conditions that accurately describe the behavior of photoproducts at the cell walls, the substrate surface, and the deposit surface must be accurately specified and all gas-phase reactions must be explicitly taken into account. This requires a detailed knowledge of the chemistry occurring in the deposition process. The boundary conditions can dramatically effect the form of the deposition rate equation, as shown by the following two examples.

In the first example, Kodas et al. (1986) considered the problem of a cylindrical light beam of uniform intensity passing through a gas, resulting in the formation of photoproducts. The gas-phase mean free path was much shorter than the beam radius, so the steady-state concentration of photoproducts could be found by solving a one-dimensional radial diffusion equation. This problem is analogous to a photochemical deposition process induced by a loosely focused beam where the gas-phase photoproduct has a very low sticking coefficient on the substrate surface so that the surface does not disturb the photoproduct concentration in the gas phase. It was assumed that secondary gas-phase reactions of the photoproducts were negligible. The results of the model calculation show that the steady-state photoproduct concentration is directly proportional to the light-source power and the precursor pressure, but is independent of the radius of the light beam. Since the deposition rate is proportional to the flux of photoproducts at the surface, the deposition rate will show the same dependence on power and beam diameter. This situation might describe deposition from the dialkyls of Cd and Zn, since the predominant metal-containing products formed upon gas-phase UV photolysis of these dialkyls are the bare metal atoms, which have very low sticking probabilities on many substrates.

In the second example, Chen (1987), as well as Krchnavek et al. (1987), also considered the problem of a cylindrical beam passing through a gas, producing photoproducts under conditions where the gas-phase mean free path is much smaller than the beam radius, but they further assumed that the substrate was a perfect sink for the photoproducts. The substrate was assumed to be an infinite plane and the gas phase above the substrate was allowed to extend to infinity. Gas-phase secondary reactions of the photoproducts were neglected. The solution indicates that the deposition rate depends linearly on light source power and precursor pressure, but inversely on the light beam radius. The photoproduct flux was shown to be identical to that predicted by the ballastic model at all points on the substrate surface, except for a factor of two or greater *increase* in magnitude (Chen, 1987). This model might apply to photochemical deposition of a species with a very high sticking coefficient, such as chromium. The inverse dependence of the deposition rate on beam radius at constant power predicted by this model contrasts with the lack of dependence predicted by the model of Kodas et al.

Each of these models is appropriate only under idealized conditions.

For example, each model is valid only when the precursor photodissociation rate scales linearly with light-source power. At sufficiently high pressures or high light intensities, this assumption will not hold, because the photodissociation rate will be limited by the diffusion of precursor into the gas-phase volume illuminated by the light beam, rather than by the light-source power. Also each model is valid only when the sample is optically thin, which may not hold at high precursor pressures. Most importantly, each model neglects gas-phase secondary reactions of the photoproducts. It is rarely possible to neglect such reactions. In some cases, such as in photochemical deposition from the Group 6 metal hexacarbonyls, secondary reactions regenerate the starting precursor. Jackson and Tyndall (1988) showed that this can result in a decrease in the deposition rate with increasing buffer-gas pressure. In other cases, secondary reactions do not regenerate the precursor, such as in photochemical deposition from the alkyls of cadmium and zinc. Even here, however, gas-phase reactions cannot be neglected, since nucleation and growth of metal particles in the gas phase has been observed when deposition is performed at high pressure. To be general, models of photochemical deposition must account explicitly for each of these effects.

The different dependence of deposition rate on laser-beam radius has been suggested as a means for distinguishing between processes where the rate-determining step involves gas-phase photodissociation (rate depends on w^{-1}) or surface photodissociation (rate depends on w^{-2}) (Wood et al., 1983). This criterion is useful when the deposition process is adequately described by the ballistic model but may not be useful when the deposition process takes place in the diffusive transport regime. The model of Chen (1987) and Krchnavek et al. (1987) shows that the photodeposition rate also scales as w^{-1} for a process involving gas-phase photodissociation of the precursor in the diffusive-transport regime, but the model of Kodas et al. (1986) predicts that the deposition rate should be independent of beam diameter. For a deposition process proceeding by photodissociation of adsorbed species at high laser intensities, the deposition rate may be limited by the flux of precursor to the deposit surface; the deposition rate will then scale as w^{-1} in the diffusive-transport regime, as discussed in Section 3.3.2.

Other criteria may thus be more generally useful for distinguishing between deposition processes controlled by surface photochemical reactions and gas-phase transport of photoproducts. We have already pointed out that the presence of ripples in the deposit clearly indicate the dominance of surface photoreactions in deposition processes driven by a polarized laser (see Section 2.3.1.1). A criterion based on the behavior of the deposition rate versus precursor pressure is also generally useful. For a process controlled by gas-phase transport of photoproducts, the deposition rate will increase linearly with the concentration of photoproducts in the low-pressure limit, if the mean free path is much greater than the deposit radius, or in the high-pressure limit, if the concentration of photoproducts is very small compared to the concentration of other gases in the cell. The concentration of gas-phase photoproducts depends linearly on the precursor pressure if the sample is optically thin. For a deposition process controlled by surface photodissociation reactions, the deposition rate will be nonlinear in precursor pressure. The rate may scale with the adsorption isotherm of the precursor or may show more complex nonlinear behavior, as discussed in Section 2.3.1.1.

2.3.1 Experimental Studies of Deposition Mechanisms

Localized cw laser photochemical deposition has been studied to the greatest extent for two different types of precursors: the group 6 metal hexacarbonyls (Cr, Mo, and W) and the group 12 metal alkyls (Cd and Zn). Other precursors have also been used, as shown in Table 7.5, but mechanistic work for the metal carbonyls and alkyls is more complete. We will thus focus our mechanistic discussion on kinetic studies of deposition from these precursors.

Various methods have been employed to measure photochemical deposition rates, including measurements of the transmission of the laser beam through the deposit versus time as deposition proceeds (Tsao and Ehrlich, 1984b), measurements of the thickenss of deposits prepared at different irradiation times by stylus profilometry (Krchnavek, 1987), and measurements of the frequency change versus time for deposition on a quartz-crystal microbalance (QCM) where the quartz crystal is used as the substrate (Jackson and Tyndall, 1987, 1988). Each method has its advantages and limitations. The transmission method is simple and useful for weakly absorbing deposits, but is useful for strongly absorbing deposits (metallic films for example) only when the deposit is extremely thin, i.e., during the nucleation stage of growth. During nucleation, however, light scattering and interference effects as well as the film refractive index can vary strongly with time. Optical-transmission kinetic data for strongly absorbing films should thus be interpreted cautiously. Stylus profilometry has the advantage that it can give a real measure of both the deposit thickness and shape, from which one can determine the volume of material deposited per unit time as well as the deposition flux at a given point with respect to the center of the laser beam. Its disadvantage is that each deposition run represents only one point on a

Table 7.5 Examples of localized photochemical deposition processes induced by a focused UV laser (SH denotes second harmonic) listed in order of deposited material, proceeding from left to right across the periodic table. Lasers were operated in cw mode, except where noted. An asterisk notes where there is no indication of deposit purity.

Deposit	Precursor	Laser	Comments
Ti*	$TiCl_4$	Ar^+ (SH)	[a]
Ti/Al*	$TiCl_4$, $Al(CH_3)_3$	Ar^+(SH)	[b,c]
Cr	$Cr(CO)_6$	Ar^+ (SH)	Deposits contaminated by C, O[d,e,f]
Mo	$Mo(CO)_6$	Ar^+ (SH)	Deposits contaminated by C, O[d,f]
Mo	$Mo(CO)_6$	Ar^{++}, 351–364 nm	Laser thermal and photochemical[g]
W	$W(CO)_6$	Ar^+ (SH)	Deposits contaminated by C, O[d,e,f,h,i]
W	$W(CO)_6$	Ar^{++}, 351–364 nm	Laser thermal and photochemical[g]
Mn	$Mn_2(CO)_{10}$	Kr^{++}, 337–356 nm	[j]
Fe*	$Fe(CO)_5$	Ar^+ (SH)	[e]
CrO_x	CrO_2Cl_2	Ar^+, 488/514 nm	Laser thermal and photochemical[k]
Cu	$Cu(hfac)_2$	Ar^+ (SH)	Films contaminated by carbon[l,m,n]
Pt	$Pt(hfac)_2$	Ar^{++}, 351–364 nm	Laser thermal and photochemical[g,o]
Zn*	$Zn(CH_3)_2$	Ar^+ (SH)	[p,q,r,s]
Zn*	$Zn(C_2H_5)_2$	ArF*, KrF*	Focused pulsed laser[t]
Zn*	$Zn(C_2H_5)_2$	Ar^+ (SH)	[u]
Cd	$Cd(CH_3)_2$	Ar^+ (SH)	[p,q,r,s,v,w,x,y]
Al*	$Al(CH_3)_3$	Ar^+ (SH)	[v,x,y,z,aa]
Al*	$Al(i\text{-}C_4H_9)_3$	Ar^+ (SH)	[bb]
Ga*	$Ga(CH_3)_3$	Ar^+ (SH)	[z]
In	$In(CH_3)_3$	Ar^+ (SH)	[aa]
Sn	$Sn(CH_3)_4$	Ar^+ (SH)	[cc,dd]
Sn*	$Sn(C_2H_5)_4$	ArF*, KrF*	Focused pulsed laser[t]
Pb*	$Pb(C_2H_5)_4$	Ar^+ (SH)	[dd,ee]

[a] Tsao et al. (1983). [b] Ehrlich and Tsao, (1985). [c] Tsao and Ehrlich (1984b). [d] Jackson and Tyndall (1988). [e] Ehrlich et al. (1981). [f] Gluck et al. (1987). [g] Gilgen et al. (1987). [h] Jackson and Tyndall (1987). [i] Chiu et al. (1985). [j] Kitai and Wolga (1983). [k] Arnone et al. (1986). [l] Wilson and Houle (1985). [m] Houle et al. (1986). [n] Jones et al. (1985). [o] Braichotte and van den Berg (1985). [p] Osgood and Ehrlich (1982). [q] Brueck and Ehrlich (1982). [r] Ehrlich et al. (1982). [s] Ehrlich et al. (1981). [t] Aylett and Haigh (1984). [u] Krchnavek et al. (1987). [v] Deutsch et al. (1979). [w] Wood et al. (1983). [x] Ehrlich et al. (1980). [y] Ehrlich and Osgood (1981). [z] Rytz-Froidevaux et al. (1983). [aa] Osgood and Gilgen (1985). [bb] Mingxin et al. (1984). [cc] Tsao and Ehrlich (1984a). [dd] Braichotte et al. (1985). [ee] Tsao et al. (1985).

plot of deposit thickness versus irradiation time, and many points are needed to determine a single deposition rate. A QCM has the advantage that it gives a real-time measure of deposition rates and the QCM crystal frequency change can be calibrated to give an absolute determination of the mass deposited per unit time. A limitation is that the deposition flux at a specific point with respect to the center of the laser beam cannot be determined. The best general method for measuring the rates of localized photochemical deposition processes thus appears to be the QCM method combined with profilometry measurements of the deposit shape and thickness.

2.3.2.1 Localized Deposition from the Group 6 Hexacarbonyls

Photochemical deposition from Cr, Mo, and W hexacarbonyls induced by a focused, cw UV laser can be used to form localized deposits on a submicrometer scale (Jackson and Tyndall, 1988). Most of the studies performed to data have employed a frequency-doubled argon-ion laser operating at 257 nm (Ehrlich et al., 1981; Chiu et al., 1985; Jackson and Tyndall, 1987, 1988). All three carbonyls have a strong absorption band in the UV at approximately 200–220 nm, with a weaker band at slightly longer wavelengths. Although the absorption cross sections at the shorter wavelength maximum follows the trend W > Mo > Cr, the cross sections at 257 nm follow the reverse order (Jackson and Tyndall, 1988). Single-photon dissociation of $Cr(CO)_6$ in the gas phase at 248 nm yields predominantly $Cr(CO)_4$ (Tumas et al., 1982; Fletcher and Rosenfeld, 1985; Seder et al., 1986). Single-photon dissociation of $Mo(CO)_6$ and $W(CO)_6$ at ~250 nm in the gas phase has not been reported, but each has a significantly higher average metal-CO bond-dissociation energy than that for $Cr(CO)_6$ (see Table 7.1), so the degree of dissociation achieved upon 257-nm excitation of $Mo(CO)_6$ and $W(CO)_6$ will be less than or equal to that achieved upon 257-nm excitation of $Cr(CO)_6$. At least two 257-nm photons are required to form Cr atoms from $Cr(CO)_6$, while at least three photons are required to form Mo and W from their respective hexacarbonyls. It is thus extremely doubtful that metal atoms are formed in the vapor phase during photochemical deposition, since the laser intensities are usually too low ($<10\,\text{kW-cm}^{-2}$) to drive multiphoton processes.

Jackson and Tyndall (1987; 1988) studied the kinetics and mechanism of frequency-doubled argon-ion-laser induced deposition from each of the Group 6 hexacarbonyls. In their experiments, the substrate was located at the laser beam focal point and the $1/e^2$ beam radius at the surface was 15 μm. QCM measurements show that the rate of deposition from each hexacarbonyl is linear in laser power over the range of 0–12 mW, and the rate is linear in hexacarbonyl pressure up to each hexacarbonyl's ambient vapor presure. The deposition rate declines rapidly upon adding argon buffer gas to pressures beyond the point where the mean free path is smaller than the deposit radius (~2 torr), as shown in Fig. 7.7. These data indicate that the rate-determining step of the deposition process involves gas-phase transport of single-photon dissociation products to the substrate surface. Photodissociation of adsorbed hexacarbonyl species can be ruled out, since the hexacarbonyls adsorb to less than 0.1 monolayers under the conditions of the deposition experiments, even at pressures approaching the vapor pressure.

Additional evidence illustrates the importance of secondary processes, however. The deposits exhibit ripples, indicating that surface photochemical reactions make a major contribution to the deposition process. Also, the deposit composition as measured by Auger-electron spectroscopy shows significant contamination by carbon and oxygen, but much less than that expected for a deposit formed simply by condensation of gas-phase, single-photon dissociation products: one 257-nm photon does not have enough energy to remove the number of carbonyl ligands necessary to yield the levels of carbon and oxygen observed in the deposits. These observations indicate that reactions occur on the surface

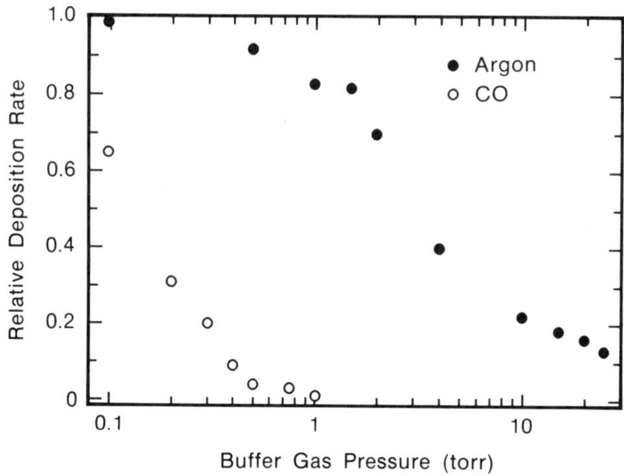

Fig. 7.7 Deposition rate versus added CO and argon pressure for frequency-doubled argon-ion-laser-induced deposition from $Cr(CO)_6$. The laser power was 10 mW, the $Cr(CO)_6$ pressure was 50 mtorr, and the $1/e^2$ beam radius was 15 μm (Jackson and Tyndall, 1988).

subsequent to the initial gas-phase, single-photon dissociation of the hexacarbonyl precursor. The presence of ripples indicates that these reactions are initiated photochemically within the focal spot of the laser beam.

To account for all of these results, Jackson and Tyndall proposed that deposition is initiated by one-photon dissociation of the hexacarbonyl to form a $M(CO)_x$ photoproduct. Some fraction of the $M(CO)_x$ species are transported to the substrate surface where they adsorb. Those $M(CO)_x$ species that adsorb on the substrate within the laser beam or that diffuse along the surface into the laser beam undergo further, photochemically-initiated dissociation to produce the final deposit. $M(CO)_x$ species that adsorb outside of the focal spot of the laser probably undergo room-temperature thermal dissociation.

2.3.2.2 Localized Deposition from the Group 12 Alkyls

Photochemical deposition from the dialkyls of Cd and Zn induced by a focused, cw frequency-doubled argon-ion laser has also been studied in detail (Deutsch et al., 1979; Ehrlich et al., 1980; 1982; Krchnavek, 1987). The dimethyl derivatives of each species absorb weakly at 257 nm, but dimethyl cadmium absorbs significantly more strongly than dimethyl zinc (Chen and Osgood, 1984). For each species, absorption at 257 nm excites the $^1A_1$ band. This state is dissociative, producing the monomethyl species and a methyl radical; each is formed in its ground electronic state, but both appear to retain a high degree of vibrational excitation (Chu et al., 1985; Young et al., 1973). The monomethyl species is not particularly stable (see Table 7.1), although it has been observed spectroscopically for both Cd and Zn following photodissociation of their respective dimethyl compounds (Young et al., 1973). Methyl zinc is somewhat more stable than methyl cadmium (see Table 7.1). In the absence of collisions, the majority of the monomethyl products will dissociate further to the ground-state metal atom and a methyl radical. At high buffer gas pressures, or even at modest pressures where the buffer gas has a large number of low-frequency vibrations, the monomethyl species can be efficiently stabilized (Young et al., 1973).

Optical transmission measurements of the deposition rate performed at very low laser intensities ($\lesssim 10$ W-cm^{-2}) indicate that the rate of deposition from dimethyl cadmium on quartz is linear in laser intensity and dimethyl cadmium pressure over the range reported (0–1.6 W-cm^{-2}, ≤ 3 torr, Ehrlich et al., 1980). The deposition rate decreases slightly upon increasing the substrate temperature over the range 25–42°C for a dimethyl cadmium pressure of 5 torr (Ehrlich and Osgood, 1981). The deposit is confined to the area of the substrate that is illuminated by the laser beam. Ripples arising from surface wave interference are observed in the deposits (Osgood and Ehrlich, 1982).

To account for these results, Ehrlich et al. (1980, 1981) proposed a mechanism, termed *pre-nucleation,* that combines surface and gas-phase photodissociation processes. They proposed that the initial phase of deposition proceeds by photodissociation of adsorbed dimethyl cadmium, which shows a Type-II isotherm on SiO_2 (see Fig. 7.8). Since cadmium atoms have a high sticking probability on cadmium metal, but a very low sticking probability on quartz, the initial deposit serves as a nucleation site for sticking of cadmium atoms formed by gas-phase photodissociation of dimethyl cadmium. This accounts for the kinetic data, since the rate-determining step, after formation of the nucleation layer, involves transport of gas-phase photoproducts to the deposit surface. It also

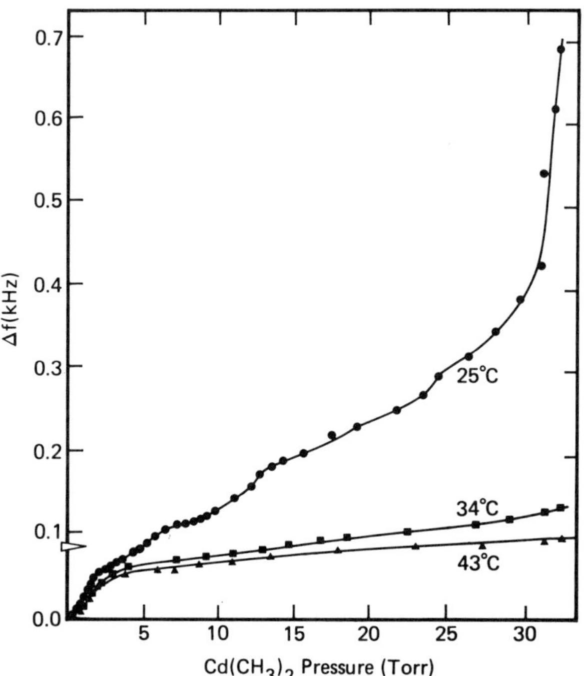

Fig. 7.8 Isotherm at three different surface temperatures for adsorption of dimethyl cadmium, measured using a quartz crystal microbalance. The frequency change Δf is proportional to the concentration of adsorbed dimethyl cadmium (Ehrlich and Osgood, 1981).

accounts for the high degree of localization of the deposits: the low sticking coefficient of cadmium atoms outside the nucleated region combined with the relatively low flux of atoms to the surface prevents formation of nucleation centers outside of the area directly illuminated by the laser beam.

More recently, Krchnavek et al. (1987) have examined the mechanism of frequency-doubled argon-ion-laser-induced deposition from diethyl zinc. This deposition process should be qualitatively similar to that obtained with dimethyl cadmium. Krchnavek et al. indicate that the deposition process is dominated by surface photodissociation processes at high laser intensities ($\gtrsim 100$ W-cm^{-2}) and by gas-phase photodissociation processes at lower intensities.

Mechanistic studies of more complex photochemical deposition processes have also been performed. For example, Tsao and Ehrlich (1984b) have discussed the mechanism of deposition induced by a frequency-doubled argon-ion laser from a mixture of $TiCl_4$ and trimethyl aluminum vapor. Deposition from each precursor alone appears to proceed largely via surface photodissociation processes. The most interesting feature of the combined system is that the deposition rate obtained at a given pressure of one of the precursors is accelerated by addition of small amounts of the other precursor. Ehrlich and Tsao interpreted their results in terms of a surface chain reaction between $TiCl_4$ and trimethyl aluminum initiated by $TiCl_3$, formed via photodissociation of adsorbed $TiCl_4$. In another study, Tsao et al. (1985) made use of the observation that frequency-doubled argon-ion laser-induced deposition from $Pb(C_2H_5)_4$ proceeds almost exclusively via surface photoreactions to measure surface diffusion rates of adsorbed $Pb(C_2H_5)_4$.

2.3.3 Material Properties of Deposits

Substrate heating by the laser or by external means is intentionally limited or absent in photochemical deposition processes induced by focused cw lasers. The lack of substrate heating is an advantage, since deposition can be performed on thermally sensitive substrates and the deposition kinetics do not depend on the substrate's optical or thermophysical properties (see Section 3.1). A major drawback to the lack of heating during deposition is that thermal processes, such as annealing and byproduct desorption, are slowed tremendously. Focused cw-laser photochemical deposition processes thus will not yield high-quality material unless potential contaminants can be removed by purely photochemical processes or by thermal processes that occur efficiently at room temperature. This generally does not occur, so it is not surprising that metals deposited by focused cw laser photochemical deposition are often highly contaminated and are poor conductors. Contaminants can be removed by subsequent thermal annealing, but then the advantage of low-temperature deposition is lost.

Metals free of contaminants can be deposited from some of the metal alkyls, however. Deposits with resistivities ~5 × that of the bulk metal have been formed via focused cw frequency-doubled argon-ion-laser-induced deposition from Cd and In alkyls (Osgood and Gilgen, 1985). One possible explanation of the high conductivity of the cadmium deposits is that 257-nm photodissociation of $Cd(CH_3)_2$ yield Cd atoms in the gas phase, so the deposits are formed by condensation of gas-phase atoms with little chance of impurity incorporation. This is an attractive explanation, since it also accounts for the low purity and conductivity of deposits formed from the metal carbonyls, which cannot photodissociate to metal atoms in the gas phase under typical focused cw-laser photochemical deposition conditions. The conducting Cd deposits showed strong surface wave ripples (Osgood and Gilgen, 1985), however, suggesting that they were formed largely by photodissociation of adsorbed species. Another possible explanation is that the conducting deposits were formed at relatively high laser intensities (1–10 kW-cm^{-2}), where laser heating could be significant, depending on the substrate. Unfortunately, the substrate was not specified in the report, so the extent of laser heating cannot be determined. Cadmium and indium are low-melting metals (m.p. = 321°C and 156°C, respectively), so laser heating to relatively low temperatures could cause significant annealing during deposition. The contribution of laser heating is suggested by the dramatic drop in resistivity with increasing laser intensity observed for indium deposits.

One can take advantage of laser heating during deposition to improve the purity and conductivity of metal deposits by carrying out the deposition process at higher laser intensities. Hybrid thermal and photochemical deposition of conducting Pt from platinum bis(hexafluoroacetylacetonate), i.e., $Pt(hfac)_2$, has been performed using the relatively high-intensity UV light available from an argon-ion laser (351–364 nm, Briachotte and van den Bergh, 1985; Gilgen et al., 1987). Although the advantage of low surface temperatures during deposition is lost via this approach, pure Pt has not been deposited from $Pt(hfac)_2$ via a totally photochemical process. Also, relatively pure Pt can be deposited by the hybrid process at temperatures lower than those required in the absence of photochemical reactions. Hybrid deposition with focused cw lasers thus shares some similarities with the photochemically assisted CVD processes discussed in Section 2.2. The light intensity and the surface temperature cannot be varied independently in the hybrid deposition process, however. Independent control of light intensity and surface temperature could be achieved by employing an external heat source, but this can result in a loss of localization, since thermal reactions of photoproducts may occur on all regions of the substrate. Better control can be achieved by using two lasers, one to induce photochemical processes and one to heat the substrate.

Limited observations have been made on properties other than composition and resistivity of deposits produced via focused cw-laser photochemical deposition. Adhesion of deposits formed from the metal alkyls, carbonyls, and β-diketonates is qualitatively good, although quantitative measurements are lacking. Detailed studies of the morphology of deposits formed from these species are also limited. Deposits formed from dimethyl cadmium and dimethyl zinc are initially fine grained, but become columnar as the thickness increases (Osgood and Ehrlich, 1982). Continuous, fine-grained deposits are obtained from $Cr(CO)_6$, while discontinuous, cracked films are formed from $Mo(CO)_6$ and $W(CO)_6$, presumably due to stress buildup as the films grow (Jackson and Tyndall, 1988; Gluck et al., 1987). The morphology of films grown from $Cu(hfac)_2$ and $Cu(hfac)_2.C_2H_5OH$ has been more extensively characterized (Wilson and Houle, 1985; Houle et al., 1986). A significant variation in morphology and composition with laser fluence was observed. This is particularly important for rippled films deposited using a polarized frequency-doubled argon-ion laser. Variations in film morphology and composition follow the light intensity variations at the surface induced by the interference of the incoming beam with surface waves, as shown by TEM and Auger-electron spectroscopic measurements. At the ripple maxima, i.e., at the points of higher light intensity, carbon contamination is lower and the films are amorphous. At the ripple minima, carbon contamination is higher, but the films are more crystalline. These results emphasize that polarization has a critical effect on the growth of films via focused cw laser photochemical deposition. Lasers with no preferential polarization direction are preferred.

3. Laser-Induced Thermal Deposition

In this section, we discuss laser thermal deposition processes, where the laser beam heats the substrate surface and is thus the energy source for the chemical reactions and physical processes that lead to deposit formation. Other terms for these processes are *laser-induced chemical vapor deposition* (LCVD), which connotes thermal processes implicitly, or *laser-induced pyrolytic deposition.* Deposition of films by laser heating of a substrate is an extension of traditional thermal CVD processes. Perhaps because of this, laser thermal deposition was explored before laser photochemical deposition. A number of fundamental new kinetic aspects arise in laser thermal deposition processes versus large-area traditional CVD processes, however, due to the enormously larger rates achievable by localized laser heating.

Since laser-induced heating is an important part of laser thermal deposition, the interaction of a laser beam with multilayered materials is discussed in the first section. The energetics of deposition reactions for both transition and main group elements that have been deposited by laser thermal deposition are then summarized. In the final part, the dynamics of laser thermal deposition are discussed, including the kinetics of nucleation and mass transport. Analytical expressions are developed for mass-transport limited rates for deposits that are either much larger or smaller than the gas mean free path. Deposition rates for deposits of arbitrary size are also described. Some consequences of diffusion-limited growth are then discussed, particularly the influence of diffusion on the deposition rates and the surface chemical kinetics taking place during laser thermal deposition.

3.1. Laser-Induced Surface Temperature Rise

The underlying driving force of laser thermal deposition is the heating of a solid substrate by absorption of light. The situation is frequently complex. Described below are expressions that have been developed for laser heating of multilayered materials for a variety of physical situations encountered in laser thermal deposition. Generally, the laser beam will have a Gaussian profile. The $1/e$ beam radius ω will typically be much smaller than the dimensions of the absorbing substrate, so the substrate may be treated as a semi-infinite solid. Heat losses above the surface as well as at the boundaries of the substrate are neglected in the expressions described below. Also, light absorption is considered to take place at the surface or within the film for a film-substrate system, and the power absorbed is given by $P(1-R)$, where P is the total laser power and R is the reflectivity of the absorbing surface. Light absorption cannot be treated as taking place entirely at the surface in many physical situations, but a small absorption coefficient in the absorbing medium complicates the analysis and therefore will not be considered here. Secondary optical interference effects will also not be considered, although they can be important for some multilayer systems. The thin-film optical interference problem has been studied both theoretically (Colinge and Van de Wiele, 1981; Calder and Sue, 1982) and experimentally (Andrew, 1986).

3.1.1 Steady-State and Time-Dependent Solutions for Multilayer Solids

The simplest multilayer case consists of a film and a substrate with similar thermophysical properties, i.e. $K_f \sim K_s = K$, where K_f and K_s are the thermal conductivities of the film and substrate, respectively. This might be used to model the surface temperature during deposition of polycrystalline silicon on a silicon substrate. The well-known expression for heating of a semi-infinite solid may be used to estimate the surface temperature rise. The steady-state solution for the temperature rise T at the center of the beam has been developed by both Sparks (1976) and Lax (1977; 1978)

$$T(r=0, z=0, t=\infty) \equiv T(0, 0, \infty) = \frac{P(1-R)}{2\sqrt{\pi}\omega K}. \tag{7.14}$$

In this equation, r is the radial coordinate, z is the depth coordinate, and t is the time from the initiation of laser heating. The time-dependent form of this solution (Ready, 1977) clearly shows that the temperature rise depends on the thermal diffusivity D of the substrate and that the temperature rises faster for smaller beam radii

$$T(0, 0, t) = \frac{P(1-R)}{\pi^{3/2}\omega K} \arctan\left[\frac{4Dt}{\omega^2}\right]^{1/2}. \tag{7.15}$$

This expression also shows that the time required to approach a steady temperature will be $t \gg \omega^2/4D$.

An entirely different situation occurs when K_f is significantly larger or smaller than K_s. For times much shorter than the thermal diffusion time $\omega^2/4D$, and when the ratio of the film depth to the laser beam radius satisfies the relation $l/\omega \gg t4D_f/\omega^2$, an analytical solution can be found (Abraham and Halley, 1987). The time-dependent temperature rise has the form

$$T(0, 0, t) = \frac{P(1-R)D_f t}{2\omega^2(K_s + K_f)}. \tag{7.16}$$

The temperature rise is a linear function of time, and the heating rate depends linearly on the power absorbed in the film, the thermal diffusivity D_f of the film, and the depth of the film. This expression shows the influence of the substrate properties on the heating rate. For example, the more insulating the substrate, the shorter the time required to reach a given temperature.

For an insulating film on a conducting substrate, an analytical expression has been developed for situations where the depth of the film is much smaller than the laser beam radius (Abraham and Halley, 1987).

An example of this case might be the heating of a surface layer of SiO_2 on silicon with a CO_2 laser. The steady-state solution for the temperature rise is

$$T(0, 0, \infty) = \frac{P(1-R)}{2\sqrt{\pi}\omega K_s}. \tag{7.17}$$

This equation shows that the temperature rise is independent of l as long as the above limits remain in effect. For a thicker layer of insulating material ($l \to \omega$), radial heat flow becomes important, and the thermophysical properties of the insulating material must be taken into account.

For radially infinite conducting films on semi-infinite insulating substrates, radial heat flow can be very important even for thin films. An example of this case would be a metal film on a glass or alumina substrate. In the limit of $K_f \gg K_s$, cooling by the substrate can be neglected as a first approximation for intermediate times, and the time-dependent temperature rise for $l/\omega \gg 1$ is given by (Lin, 1967)

$$T(0, 0, t) = \frac{P(1-R)}{8\pi K_f l} \ln\left[1 + \left(\frac{4D_f t}{\omega^2}\right)\right]. \tag{7.18}$$

This solution indicates that the surface temperature rise is dependent on the beam diameter for times when the relation $4D_f t/\omega^2 \geq 1$ is satisfied. For very long times, however, cooling by the substrate must be taken into account, and this expression is no longer valid. For conducting films where $K_f/K_s > 0.6$ and the thickness of the film is on the order of the beam size ($l/\omega \to 1$), the ratio of the radial heat transfer to the total heat transferred at $r = \omega$ is greater than 0.9 (Abraham and Halley, 1987). Because of the importance of heat transfer within the film, numerical techniques are clearly necessary to describe temperature distributions when heat cannot flow readily in all transverse directions, e.g., for conducting lines and spots on insulating surfaces.

One final situation can be described that may be applicable to deposition of metallic thin films on insulating surfaces. Heating a circular deposit of conducting material on an insulating substrate approximately models the laser-induced deposition of a metal on an insulating surface for a stationary laser beam when the deposit is cylindrical and thin. With $K_f \gg K_s$, a thin cylindrical deposit can be considered a circular surface heat source at a constant temperature. The time-dependent solution for a surface circular heat source is (Carslaw and Jaeger, 1959)

$$T(0, 0, t) = \frac{2P(1-R)(D_s t)^{1/2}}{\pi r_f^2 K_s}\left\{\frac{1}{\pi^{1/2}} - \operatorname{ierfc}\left[\frac{r_f}{2(d_s t)^{1/2}}\right]\right\}. \tag{7.19}$$

For this situation, the heating rate depends on the radius of the film r_f and the thermal diffusivity of the substrate. The temperature converges to a steady-state value

$$T(0, 0, \infty) = \frac{P(1-R)}{\pi r_f K_s.} \tag{7.20}$$

This solution is very similar to that for a Gaussian beam on a semi-infinite solid, as expected.

Numerical methods, such as finite difference or finite element techniques, have yielded some interesting results for situations that are applicable to deposition studies. Using finite element calculations, Bauerle has examined the influence of the thermal conductivity of the deposit, the geometry of the deposit, as well as the laser fluence on the surface temperature (Bauerle, 1986). These calculations show for specific situations the effects that were qualitatively described with the previous analytical expressions. The maximum surface temperature and the shape of the surface temperature profile can be a strong function of the specific geometry of the film, as well as the thermophysical and optical properties of the materials comprising the system.

3.1.2 Experimental Measurements of Surface Temperature

The comparison of experimentally measured surface temperature profiles to numerical calculations for a specific multilayer geometry is an important test of the validity of theoretical treatments of laser-induced surface temperature rises. A number of experimental techniques for measuring temperature increases induced by a highly focused laser beam have been described. These include infrared emission techniques (Salathe et al., 1981), photoluminescence from semiconductors (Salathe et al., 1981), and the use of films that undergo a phase transition as a temperature indicator (Shaapur and Allen, 1987). Thin-film thermocouples have been particularly effective in studying the influence of thin-film deposits on a laser-induced surface temperature rise (Kodas et al., 1987). Heating a micron-sized thin-film thermocouple with a focused laser beam generates a thermoelectric signal that can be calibrated to an absolute temperature scale and has a relatively fast response time ($<10\ \mu s$). One effect of applying any thin-film measuring structure is the perturbation of the underlying substrate thermal properties, particularly at fine dimensions. These perturbations, however, may be of interest in that they can apply to cases of laser heating of these geometries during LCVD of a thin film.

Temperature distributions induced by an argon-ion laser on a gold-nickel thin-film thermocouple fabricated on an SiO_2/Si substrate were measured and compared to the results of numerical finite difference calculations (Kodas et al., 1987). A comparision of the calculated to the measured profiles indicated that the measured and calculated peak temperatures with the metal thermocouple lines were much higher than the peak temperatures calculated without the presence of metal lines. This resulted because the lines were insulated from the relatively conductive silicon by the silicon dioxide layer, and could only conduct heat away along the lines. In contrast, in the case without metal lines the light was absorbed by the silicon, which was able to conduct heat away from the source in three dimensions, resulting lower surface temperatures. The temperature for the case with metal lines was also increased since the laser energy was absorbed by the gold line, which, for this experimental situation, had a lower reflectivity than the silicon at the laser wavelength used.

These experiments and calculations provide a qualitative description of the surface temperature during laser thermal deposition of metals onto substrates with transparent, insulating surface layers. When the deposit has a lower reflectivity than the substrate, the surface temperature should first increase as deposition begins and the deposited metal, insulated from the conductive substrate, absorbs the laser energy. This temperature increase can influence the deposition rate, as observed experimentally for deposition of chromium oxide films onto fused silica substrates (Arnone et al., 1986). As sufficient metal is deposited to conduct the laser energy away from the irradiated region, the temperature begins to decrease. For deposition under scanning conditions, the temperature profile becomes constant after a sufficiently long metal line is deposited. The steady-state maximum temperature after significant deposit growth can be higher or lower than the maximum temperature obtained on the bare substrate, depending on the geometry and the values of the thermal and optical properties of the deposited metal versus those of the substrate.

3.2. Thermochemistry

The chemical transformation occurring during deposition from a gaseous $ML_{(g)}$ species can be summarized by

$$ML_{n(g)} \rightarrow M_{(s)} + nL_{(g)}, \tag{7.21}$$

where $M_{(s)}$ is the solid deposit and $L_{(g)}$ is a gaseous byproduct. The

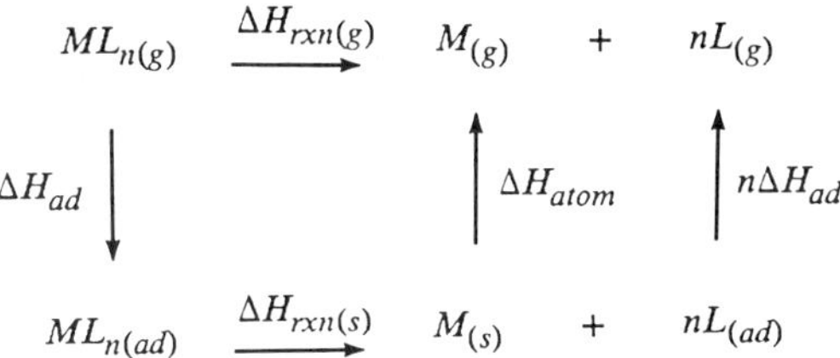

Fig. 7.9 Thermochemical enthalpy cycle for the model deposition reaction $ML_n \rightarrow M + nL$.

energetic changes can be described in more detail by the thermodynamic cycle in Fig. 7.9. This cycle shows that the enthalpy terms that make up Eq. (7.21) are $\Delta H_{\text{rxn(g)}} - \Delta H_{\text{atom}}$, i.e., the difference in the enthalpy of the gas phase reaction and the heat of atomisation of the solid deposit. Equivalently, the difference can be described by $\Delta H_{\text{rxn(s)}} - \Delta H_{\text{ad}}(\text{ML}_\text{n}) + n\,\Delta H_{\text{ad}}(\text{L})$, the heat of reaction on the surface minus the difference between the heat of adsorption of the reactant and byproducts. The free-energy change would include the entropy changes that accompany the formation of $n+1$ species from one species and the entropy of sublimation. The cycle can be used as a model for the energetic changes that must occur for this type of chemical reaction to take place. For example, Table 7.6 has been constructed for some of the typical transition metal carbonyls and metal alkyls for which the enthalpy data is known or can be easily approximated.

For the deposition reactions listed in Table 7.6, the $\Delta H_{\text{rxn(s)}}$ are all negative, indicative of the strong bonding of the metal atom in the solid, as well as the strong adsorption of the ligand to the solid. The $\Delta H_{\text{rxn(g)}}$ are generally positive, indicating the loss of the bond energy due to the formation of free atoms and ligands. The ΔH_{dep}, the enthalpy change for the deposition reaction, includes the enthalpy of desorption of the ligands from the surface and can indicate whether thermal energy must be supplied for Eq. 7.21 to take place. The values obtained for ΔH_{dep} indicate that the reaction for the metal carbonyls and acetylacetonates is endothermic, while the transformation for the metal alkyls is exothermic. This is primarily due to the large adsorption energies of the carbonyl and acetylacetonate ligands on transition metal surfaces. Typically, heats of chemisorption of ligands on transition metal surfaces are 1–6 eV (Adamson, 1982). This is particularly problematic for early transition metals. For example, elements further left and higher up in the transition metal series dissociatively adsorb CO more readily than those further to the right or bottom (Broden et al., 1976).

Table 7.6 Enthalpy Data for Metal Deposition Reactions[a].

Precursor ML_n	$\Delta H_{rxn(g)}$ [b]	$\Delta H_{rxn(s)}$ [c]	ΔH_{dep} [d]	ΔH_{atom} [e] (M)	ΔH_{vap} [f] (ML_n)	$E_{desorption}$ [g] (L from M)
$Cr(CO)_6$	153.6[h]	—	58.6	95	18.7	Dissociatively adsorbs[i]
$Mo(CO)_6$	216.8	(−234)	59.5	157.3	16.7	29.0, 54.0, 72.0[j]
$W(CO)_6$	255.4	(−288)	52.3	203.1	17.7	24.1, 53.3, 101.6[k]
$Fe(CO)_5$	140.2	(−113)	40.9	99.3	9.6	20.3, 25.1, 52.6[l]
$Ni(CO)_4$	139.3	(−77)	36.5	102.8	6.6	30.0[m]
$Sn(CH_3)_4$	36.3	(−38)	−39.9	72.2	7.9	(5)[n]
$Cd(CH_3)_2$	−19.3	(−42)	−46.0	26.72	9.1	(5)[n]
$Zn(CH_3)_2$	−3.1	(−32)	−34.3	31.25	7.1	(5)[n]
$Al(CH_3)_3$	61.9	(−9)	−16.8	78.7	15.1	(5)[n]
$Cr(acac)_3$	315	—	220	95	6.6	—
$Fe(acac)_3$	311	—	212	99.3	15.6	—
$Co(acac)_3$	283	—	181	101.6	17.9	—
$Cu(acac)_2$	170	—	88.5	81.5	15.0	—

[a] All enthalpy changes are in kcal/mol at 298°K, with the sign convention specified by Fig. 7.9. [b] Heat of reaction calculated from data in Cox and Pilcher (1970) for the metal carbonyls and alkyls and Ashcroft and Mortimer (1970) for the metal acetylacetonates. [c] Heat of reaction calculated from thermodynamic cycle in Fig. 7.9. [d] Change in enthalpy for metal deposition reaction (Eq. 7.20). [e] Heat of atomization of metals (Brewer and Rosenblatt, 1969). [f] Data for the heat of adsorption of the precursors on their respective metals are not available, therefore the value for the heat of vaporization or sublimation is used as an approximation (Cox and Pilcher, 1970). [g] The activation energy for desorption is used here as an approximation to the thermodynamic heat of adsorption. [h] Pilcher et al. (1975). [i] Shinn and Madey (1987); Baca et al. (1986); Kato et al. (1982). [j] Gillet et al. (1976). [k] King et al. (1972). [l] Benzinger and Madix (1980). [m] Nieuwenhuys (1981). [n] Ethane is considered the product ligand for the purposes of these calculations. For these metals, ethane is assumed to be physically adsorbed, and a typical value for small hydrocarbons on metals for the heat of physical adsorption is on the order of 5 kcal/mol (Madey and Yates, 1978; Steinruck et al., 1986).

If the ligand is strongly bound, i.e., chemisorbed or dissociatively adsorbed to the metal surface, then deposits with the ligand incorporated into the metal can result. A pure metal deposit from a light-induced surface reaction of a metal carbonyl, for example, would only occur in cases where energy could be supplied to remove the ligand or its constituent atoms from the deposited metal surface. Thus, even though formation of a pure deposit is favored thermodynamically ($H_{rxn(s)} \leq 0$), the reaction sequence may result in a kinetically stable deposit that is a combination of chemisorbed ligand and metal, due to large values of H_{ad}

for the ligand. For those reactions that yield pure elemental deposits, it can in some cases be important to determine what the desorbing products are, in order to better design reactive precursors that yield easily desorbed products.

Some qualitative feeling for the energetics of deposition reactions can be gained with a quantitative estimate of the relative stabilities of the various oxidation states of the elements. This is accomplished in the following section on transition elements with an oxidation-state-free energy diagram (Phillips and Williams, 1965; Mackay and Mackay, 1973). Section 3.2.1 is organized according to the types of material deposited and is thus divided in two parts with respect to the periodic table, the transition elements and the main group elements, so as to note any periodic trends in the chemistry that may be relevant to deposition reactions.

3.2.1 Transition Element Metals, Oxides, and Carbides

Laser thermal deposition of the group 3–5 transition metals has not been reported. This is likely to be due partly to the instability of these metals with respect to higher oxidation states or, from the viewpoint of the previous section, the stability of the chemisorbed ligand-metal bond (or bonds, if the ligand adsorbs dissociatively). It can be seen from Fig. 7.10 that, by using the data displayed as an indication of oxidation state stability, inferences can be drawn concerning the likelihood of metal production from higher oxidation state precursors. For example, to reduce a titanium precursor to Ti metal from any of the higher oxidation states, approximately 3 eV will need to be supplied. This indeed has been accomplished with a coreductant. The early transition metals are therefore not predisposed toward metal formation from a higher oxidation state, particularly with regard to the extreme relative stability of the II and III oxidation states. In the group IV–V elements, only TiO_2 and TiC have been reported to be deposited by laser thermal deposition (Allen, 1981; Mazumder and Allen, 1979).

The group 6 and 7 transition elements Cr and Mn also have very stable 2 and 3 oxidation states relative to the metallic state. However, for the group 6 and 7 elements, stable carbonyl complexes are available. The metal atoms in the carbonyl complexes are generally in the zero oxidation state, and in some cases thermal metal deposition, as well as oxide or carbide deposition, might be expected. As of this writing, there are no reports of the laser-induced thermal deposition of metallic Cr. Both CrO_2 and Cr_2O_3 have been reported to be deposited from

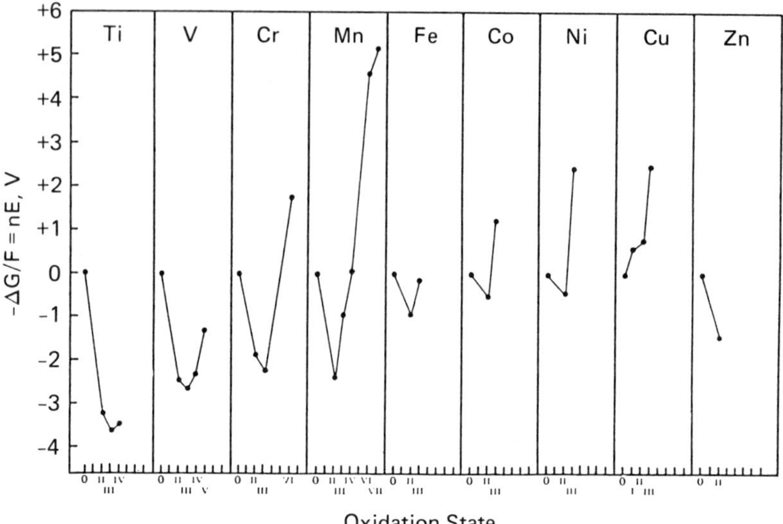

Fig. 7.10 Oxidation-state free-energy diagrams for the elements of the first transition series. The oxidation state is plotted versus the free-energy change in one electron step, with the value of the zero-valent element taken as the zero in energy (Mackay and Mackay, 1973).

CrO_2Cl_2 with an argon-ion laser (Arnone et al., 1986). The data indicate that CrO_2 is deposited by photolysis of the precursor and that Cr_2O_3 is formed by thermal elimination of oxygen. Laser thermal deposition of Mo metal has been reported from $Mo(CO)_6$ (Gilgen, et al., 1987). Tungsten deposition from $W(CO)_6$ (Gilgen et al., 1987; Allen and Trigubo, 1983) and from WF_6 with the coreactant H_2 (McWilliams et al., 1983; Mitlitsky et al., 1986; Liu et al., 1985) has been reported by a number of different laboratories.

Similar to group 6 and 7, the group 8 transition elements have stable zero oxidation state carbonyls, and again it might be expected that purely thermal dissociation may yield the pure transition metal, especially since these elements do not have extremely stable higher oxidation states, as do the earlier elements. Iron deposition from $Fe(CO)_5$ has been accomplished by several research groups. The deposits are heavily oxidized (Allen and Trigubo, 1983), or for deposition on SiO_2, are composed of both Fe and Fe_2SiO_4 (Jackman et al., 1986). The iron silicate is presumably formed from thermal reactions at the silica surface. The deposition of Co metal has not been reported. However, deposition of CoO from Co(III) tris(acetylacetonate) (Steen, 1978) has been achieved in air, which was presumably the source of oxygen in this reaction.

The deposition of metallic nickel from $Ni(CO)_4$ induced by krypton-ion (Krauter et al., 1983; Petzoldt et al., 1984), argon-ion (Herman et al., 1983), and carbon-dioxide lasers (Allen and Bass, 1979; Allen et al., 1985; Allen et al., 1986) has been studied extensively by many research groups. Deposition rate data were measured by a number of different means. In the Ni Group, the zero-valent carbonyl complexes of Pd and Pt are not thermally stable. Platinum metal, however, has been deposited from the Pt(II) bis-(hexafluoroacetylacetonate) (Gilgen et al., 1987; Braichotte and van den Bergh, 1984; Mingxin et al., 1984; Braichotte and van den Bergh, 1986). The resistivity is generally within $2\text{–}3 \times$ bulk metal, indicating that the deposits are primarily metallic.

In the Cu group, pure metals have been deposited using precursors with metal atoms in higher oxidation states. For Cu, the metallic state is stable with respect to the higher oxidation state (see Fig. 7.10). The precursor for the laser thermal deposition of metallic copper is Cu(II) bis-(hexafluoroacetylacetonate) (Houle et al., 1985; Moylan et al., 1986). For Au, a series of organometallic acetylacetonate complexes with varying numbers of fluorine atoms in the ligand has been used to deposit high-purity gold metal (Baum and Jones, 1986; Baum and Jones, 1985; Baum, 1987). These materials have been the subject of deposition rate studies (Kodas et al., 1987) discussed in the following section.

In the Zn group, the zero valent carbonyls are again not stable species. However, the metal alkyls are stable and have been used as precursors for metal deposition. Dimethyl zinc in the presence of H_2 has been used to deposit Zn metal with a Kr^+ laser (Rytz-Froidevaux et al., 1981), and dimethyl cadmium has been used to deposit Cd metal (Rytz-Froidevaux et al., 1981; Rytz-Froidevaux et al., 1982).

3.2.2 Main Group Metals

Of the main group metals, only Al and Sn have been reported to be deposited by thermal techniques. Aluminum has been deposited with a frequency-doubled argon-ion laser beam (257 nm). In this work, adsorbed layers of $HAlMe_2$ photodissociate, leaving a metal film that is heated by the laser, causing further thermal deposition of Al metal (Cacouris et al., 1987). Ultraviolet laser nucleation followed by CO_2 laser thermal deposition has been used to produce Al from tri-isobutylaluminum (Tsao and Ehrlich, 1984). Tin metal has been reported to be deposited from $Sn(Me)_4$ (Mingxin et al., 1984; Braichotte and van den Bergh, 1984; Braichotte et al., 1985).

3.2.3 Main Group Semiconductors and Insulators

The deposition of polycrystalline Si has been accomplished from $SiCl_4$ and H_2 with a CO_2 laser (Baranauskas et al., 1980), as well as from SiH_4 with a CO_2 laser (Christensen and Lakin, 1978) and an argon-ion laser (Ehrlich et al., 1981; Bauerle et al., 1982; Magnotta and Herman, 1986). The pyrolytic deposition of Si has been studied extensively by researchers in the conventional CVD field (see, for example, Breiland et al., 1986). Gallium-arsenide has been deposited from AsH_3, trimethylgallium, and hydrogen by pulsed surface heating with Nd:YAG lasers (Roth et al., 1984), and cw surface heating with argon-ion lasers (Bedair et al., 1986). InGaAs and GaAsP have been deposited on Si with argon-ion lasers (Bedair et al., 1986). Polycrystalline carbon has been deposited from C_2H_2, ethylene, and methane on tungsten rods and alumina ceramic at extremely high temperatures (Leyendecker et al., 1981; 1982; 1983).

3.3. Dynamic Processes in Laser Thermal Deposition

The deposition of thin films by a thermal laser-driven reaction can be described in an analogous manner to vacuum deposition of thin films, with the important exception that the precursor reacts on the surface, giving rise to the atoms that make up the thin films. The sequence of events that must take place involves adsorption of the precursor on the surface, followed by reaction to form adsorbed deposit atoms as well as adsorbed byproducts. The adsorbed atoms migrate to form clusters, which then grow and coalesce. Coalescence takes place to the extent that a continuous film is formed, and finally the continuous film continues to grow. These steps can be described by the following equations:
Adsorption:

$$\mathrm{ML_{n(g)}} \rightleftarrows \mathrm{ML_{n(ad)}}. \tag{7.22}$$

Reaction:

$$\mathrm{ML_{n(ad)}} \rightarrow \mathrm{M_{(ad)}} + \mathrm{nL_{(ad)}}. \tag{7.23}$$

Migration and coalescence:

$$\begin{aligned} &\mathrm{M_{(ad)}} + \mathrm{M}_{i\mathrm{(ad)}} \rightleftarrows \mathrm{M_{2(ad)}} \\ &\mathrm{M_{(ad)}} + \mathrm{M}_{i\mathrm{(ad)}} \rightleftarrows M_{i+1\mathrm{(ad)}} \equiv (i+1)\mathrm{M_{(s)}}. \end{aligned} \tag{7.24}$$

Desorption (important only for small i):

$$\mathrm{M}_{i\mathrm{(ad)}} \rightleftarrows \mathrm{M}_{i\mathrm{(g)}}. \tag{7.25}$$

In these equations, ML is the precursor to the deposit atom M and the byproduct L. The subscripts g, ad, and s indicate gas-phase species, adsorbed species, and solid-phase species, respectively, and the i index refers to a cluster of i atoms. Equation (7.22) may not necessarily be an elementary step, if the collision site of the molecule on the surface is not the adsorption site. The surface reaction Eq. 7.23 is also not necessarily an elementary step but can be a complex series of surface chemical reactions, ultimately giving rise to a metal atom.

The nucleation stage of film growth is controlled by the competition and reversibility of Eqs. 7.23 through 7.25. This stage of the deposition process is particularly complex for a laser-driven reaction. The laser-induced temperature rise, for example, can change with time because the optical and thermal properties of the substrate are being modified by the deposit, as described in Section 3.1. Secondly, although thermodynamic energy levels cannot specify exactly the rate at which an elementary process will take place, the energy terms can place limits on the rate of a particular step. For example, the adsorption energies are not constant during the nucleation stage because the nature of the surface is changing. During this early stage of deposition, the adsorption of the precursor and reaction products take place predominantly on the substrate. As clusters form, more surface area is taken up by the deposit, and adsorption will occur predominantly on the deposit. The newly deposited surface may also be catalytic towards the deposition reaction, resulting directly in a faster surface reaction rate. Because the activation energies for these processes, as well as the surface temperature, enter as exponential terms in the rate expressions, the nucleation regime is highly nonlinear in the variables that influence the time to grow deposits.

Nevertheless, there have been attempts to study heterogeneous nucleation during laser thermal deposition in order to gain some insight into film formation. For example, Herman et al. (1986) have used a Monte Carlo simulation to examine the nucleation process for Si as a function of temperature and several other parameters. This approach can be used, along with the numerous results in the literature for nucleation of various materials during CVD, to understand nucleation phenomena during laser thermal deposition (Neugebauer, 1970).

The regime where material is deposited onto an already continuous deposit is important because the bulk of the material is deposited during this time. Experimental and theoretical investigations of the growth process are therefore examined in some detail here. In contrast to the conventional CVD of thin films, the reaction zone is localized in laser-induced deposition. Therefore the conditions under which mass transport to the reaction zone is rate limiting can be substantially different for laser thermal deposition than for CVD and will require a different theoretical formulation for the mass transport kinetics. Diffusive mass transport to a surface where chemical reaction occurs can be encountered in laser thermal deposition when the gas mean free path is much smaller than the radius of the reaction zone. Of the theoretical studies of mass-transport-controlled laser thermal deposition that have been carried out (Herman et al., 1983; Ehrlich and Tsao, 1983; Kodas et al., 1987; Kodas and Comita, 1988), few have attempted to provide a general theoretical framework for the kinetics of mass transport (Kodas et al., 1987; Kodas and Comita, 1988). Because of the importance of this problem in laser thermal deposition, this work will be summarized here.

We consider two distinctly different cases for mass-transport-limited deposition. The first case is that of a single reactant transported to a surface where one or more gaseous products are formed as the result of chemical reaction. The second case is similar to the first, but a buffer gas is present in excess over the reactant and products. For both cases, two limiting situations arise depending on the total pressure in the system and the size of the reaction zone; the gas mean free path may be much smaller or much longer than the size of the reaction zone. Since the theories for each limiting case are quite different, we will discuss these cases first and then discuss the intermediate case of gas mean free path l on the order of the deposit radius r.

3.3.1 Vertical Growth Rates for $l \gg r$

The transport-limited deposition rate when the gas mean free path is much greater than the deposit radius ($l \gg r$) can be calculated using kinetic theory if the reaction probability α is sufficiently large (Kodas et al., 1987; Kodas and Comita, 1988)

$$\frac{dr}{dt} = \frac{\alpha v_1 p_1}{(2\pi m \mathrm{k} T)^{1/2}}, \tag{7.26}$$

where r is the radius of the deposit, v_1 is the volume of an atom in the deposit, p_1 is the partial pressure of the deposit precursor, m is the molecular mass of the precursor, and α is the fraction of collisions resulting in reaction. This expression differs from that used by Ehrlich and Tsao (1983) in two ways. First, a distinction is made between the buffer gas pressure and the precursor partial pressure. This is a crucial distinction for $l \ll r$ for a single-component system (see below). Secondly, this equation makes use of the mean velocity of gas molecules to calculate the molecular bombardment rate on the surface, rather than the root-mean-square velocity. For the case of $l \gg r$, the presence of the reaction zone does not disturb the velocity distribution of the incoming molecules. The deposition rate is the same in the absence or presence of buffer gas provided that the surface reaction rate does not depend on buffer-gas pressure. Equation 7.26 shows explicitly that the deposition rate is linearly dependent on p_1, but is independent of the partial pressure of the buffer gas provided that $l \gg r$.

The reaction probability, α, is a parameter that embodies the influence of the surface processes described by Eqs. 7.22 and 7.23. The reactant molecule first impinges on the deposit surface and then must stick in order to react. The fraction of molecules that stick has been called the *trapping coefficient* (Boudart and Djega-Mariadassou, 1984). The reactant molecule may then diffuse to a site where it is chemically or physically adsorbed. The adsorbed reactant may then desorb or it may react followed by desorption of the products. If the rates of the surface processes are fast relative to the rate at which molecules arrive on the surface and if desorption of reactant is negligible, the reaction probability becomes equal to the trapping probability. The trapping probability is a weak function of the surface temperature for many species and is often of order unity (Weinberg and Merrill 1971). Therefore, the deposition rate and the value of α for reactant transport-controlled deposition of a spot depend only weakly on the surface temperature. In contrast, deposition rates and values of α for surface reaction-controlled processes depend strongly on temperature, and values of α are usually much less than unity. Hence, the value of α is indicative of the rate-determining step for deposition.

3.3.2 Vertical Growth Rates for $l \ll r$

For a sufficiently high reactant pressure or with a high buffer gas pressure, the gas mean free path is much less than the deposit radius and mass transport takes place by a diffusive mechanism, if the reaction probability is sufficiently large. In the absence of buffer gas, two different cases can occur. In the first case, there are no species desorbing and the surface is an infinite sink for the reactant molecule. For single-component systems in which deposition occurs by simple condensation or by chemical reaction without product desorption, diffusive mass transport does not occur. Examples of chemically reactive systems such as this include some polymerization reactions. For these cases, mass transport to the surface for the undiluted reacting molecule occurs by molecular bombardment and the rate is described by Eq. 7.26. For condensation, the validity of Eq. 7.26 has been experimentally demonstrated (Richardson et al., 1986). In the second case for undiluted reactant gas, deposition occurs by reaction and desorbing species are produced. This is the most common situation for laser thermal deposition reactions. The ligands of the reactive precursor will in some cases form easily desorbed products, which is an important feature for the production of pure deposits, as discussed in the previous section.

The simplest example is the reaction of a precursor on the surface to form a solid species and one gaseous product

$$ML_{(g)} \rightarrow M_{(s)} + L_{(g)}. \tag{7.27}$$

For this particular case, the form of the solution to the mass transport equation is identical to that for the case with diffusion through an inert buffer gas where any number of gaseous products are produced, so long as the concentration of the products is much less than that of the buffer gas. This is because for both cases there is effectively no transport occurring due to the net molecular flux to the reaction zone. Formation of more than one species for the case without buffer gas would result in a reduction of the deposition rate.

For the simple case of diffusive mass transport of A to a hemispherical surface followed by chemical reaction to form a gaseous species B, or for the case of diffusion of A through a buffer gas to a surface where chemical reaction takes place to form any number of products, the diffusion equation in spherical coordinates can be solved (Kodas and Comita, 1988). For low surface reaction rates, the surface reaction rate constant k_1 is small and the deposition rate can be limited by the surface reaction rate. For certain boundary conditions, the rate is given by $k_1 n_A$, where n_A is the concentration of A in the gas phase at the surface. For high reaction rates the deposition rate for a hemispherical deposit is given by the familiar expression for diffusion-limited deposition (Kodas et al., 1987; Kodas and Comita, 1988)

$$\frac{dr}{dt} = \frac{D_{AB} v_1 p_1}{rkT}. \tag{7.28}$$

D_{AB} is the binary diffusion coefficient of the reactant/buffer gas system. This equation describes the deposition rate under quasi-steady-state conditions when the effects of free convection and thermal diffusion are negligible. The effect of including thermal diffusion in the theory is difficult to quantify but would probably result in a reduction of calculated deposition rates below those obtained using Eq. (7.28).

3.3.3 Vertical Growth Rates for Arbitrary Size Deposits

Only in the two limiting cases of very large and very small gas mean free path compared to the deposit radius is the theoretical description of transport simple and well understood. Equation 7.28 in particular becomes invalid as the gas mean free path increases, especially in the region of the reaction zone. In this region molecular collisions do not occur and continuum theory breaks down. The spot does not disturb the vapor concentration for $\mathrm{Kn} \gg 1$, where the Knudsen number Kn is

$$\mathrm{Kn} = \frac{3D_{\mathrm{AB}}}{r\left(\frac{8kT}{\pi m}\right)^{1/2}}, \tag{7.29}$$

the ratio of the gas mean free path, l, to the deposit radius, r.

The simplest semiempirical expression for the flux in the transition regime approaching Eqs. 7.26 and 7.28 in the limits of large and small Kn was presented by Fuchs (1959). The rate of deposit growth is given by the following expression incorporating an interpolation formula for the transition region (Kodas et al., 1987; Fuchs, 1959):

$$\frac{dr}{dt} = \frac{\alpha v_1 p_1}{(2\pi mkT)^{1/2}}\left(\frac{4\mathrm{Kn}/3\alpha}{1 + 4\mathrm{Kn}/3\alpha}\right). \tag{7.30}$$

The behavior of the rate with respect to the total pressure p_{tot} and the partial pressure of the precursor p_1 can be easily ascertained with these equations. For example, in the diffusion regime ($r \gg l$), the rate is proportional to p_1 as in the $r \ll l$ regime, but because of the dependence on D_{AB}, the rate is inversely proportional to p_{tot}.

The deposit growth equations for mass-transport-controlled laser thermal deposition are summarized in Table 7.7. The validity of these

Table 7.7 Summary of deposition rate expressions.

Mechanism	Deposition Rate (dr/dt)
Molecular bombardment ($r \ll l$)	$\frac{\alpha v_1 p_1}{(2\pi mkT)^{1/2}}$
Surface reaction[a] ($\alpha \ll 1$)	$\frac{\alpha v_1 p_1}{(2\pi mkT)^{1/2}}$
Diffusion ($r \gg l$)	$\frac{D_{\mathrm{AB}} v_1 p_1}{rkT}$
Molecular bombardment, transition ($r \sim l$), and diffusion	$\frac{\alpha v_1 p_1}{(2\pi mkT)^{1/2}}\left(\frac{4\,\mathrm{Kn}/3\alpha}{1 + 4\,\mathrm{Kn}/3\alpha}\right)$

[a] This expression is valid for all deposit sizes.

equations has been experimentally demonstrated for the deposition of Au from $Me_2Au(hfac)$ in the absence and presence of buffer gas (Kodas et al., 1987; Kodas and Comita, 1988). In this work, the rates of thermal laser deposition of gold onto alumina substrates were measured by determining the rate at which the deposit grew through the focal spot of a He/Ne laser probe beam directed parallel to the substrate. The rate of gold deposition depends linearly on the gold precursor partial pressure and, for a wide variety of conditions, does not depend on the laser power or focal spot radius. These experiments also showed that the deposition rate decreases with increasing buffer gas pressure above a critical pressure.

3.3.4 The Influence of Diffusion on Laser Thermal Deposition

In order to understand the influence of diffusion on the growth rate, the rate obtained under various conditions can be compared to the maximum possible vertical growth rate, given by Eq. 7.25. The ratio of the actual rate to the maximum possible deposition rate when $\mathrm{Kn}/\alpha \gg 1$ is given by F (Kodas et al., 1987)

$$F = \frac{dr/dt}{\dfrac{\alpha v_1 p_1}{(2\pi m \mathrm{k} T)^{1/2}}} = \frac{4\mathrm{Kn}^*/3}{1 + 4\mathrm{Kn}^*/3} \tag{7.31}$$

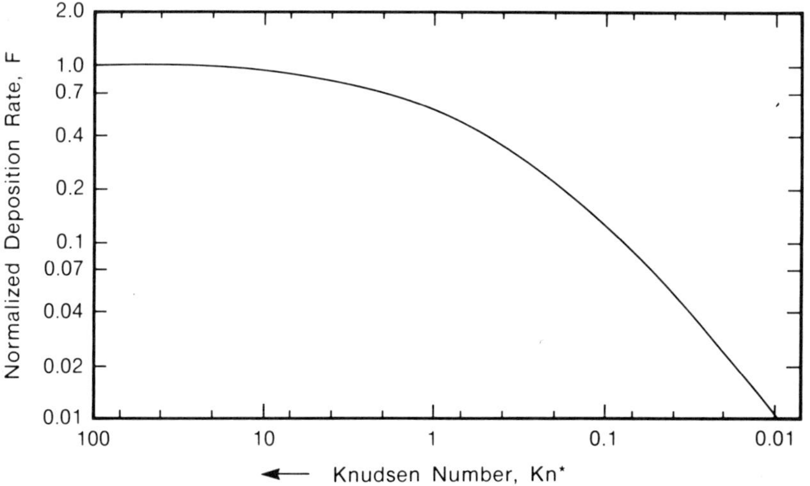

Fig. 7.11 Vertical growth rate normalized with respect to maximum possible vertical growth rate as a function of Knudsen number and reaction probability α.

where $Kn^* = Kn/\alpha$. Figure 7.11 shows how the normalized deposition rate, F, varies with Kn^*. The normalized deposition rate also depends on $D_{AB}/r\alpha$ or on $(p_{tot}r\alpha)^{-1}$ since the diffusion coefficient D_{AB} depends inversely on the total pressure. Thus, the total pressure at which diffusional effects begin to decrease the deposition rate depends on α and the size of the deposit. For systems with $\alpha \ll 1$, Kn^* is large and the deposition rate is described by Eq. 7.26.

A second consequence of diffusive growth results from the coupling of gas-phase mass transport to surface chemical kinetics. When diffusive transport occurs, the gas-phase concentration of reactant at the surface is depleted and the concentrations of products are increased, due to the surface chemical reactions. If the laser beam is modulated, the surface temperature can be modulated, thereby modulating the gas-phase reactant (and product) concentrations. These effects have been the basis for a relaxation technique for studying laser-induced surface reactions during laser-induced chemical vapor deposition (Comita and Kodas, 1987; Kodas and Comita, 1987).

References

Abraham, E., and Halley, J.M. (1987). Some calculations of temperature profiles in thin films with laser heating. *Appl. Phys.* **A42,** 279–285.

Adams, A.E., Lloyd, M.L., Morgan, S.L., and Davis, N.G. (1984). Laser photochemical deposition of metals. In *Laser Processing and Diagnostics* (Bauerle, D., ed.). *Springer Ser. Chem. Phys.* **39,** 269–273.

Adamson, A.W. (1982). *Physical Chemistry of Surfaces,* 4th ed. J. Wiley and Sons, New York.

Allen, S.D., and Bass, M. (1979). Abstract: Laser chemical vapor deposition of metals and insulators. *J. Vac. Sci. Technol.* **16,** 431.

Allen, S.D. (1981). Laser chemical vapor deposition: A technique for selective area deposition. *J. Appl. Phys.* **52,** 6501–6505.

Allen, S.D., and Tribugo, A.B. (1983). Laser chemical vapor deposition of selected area Fe and W films. *J. Appl. Phys.* **54,** 1641–1643.

Allen, S.D., Jan, R.Y., Mazuk, S.M., and Vernon, S. (1985). Real time measurement of deposition initiation and rate in laser chemical vapor deposition. *J. Appl. Phys.* **58,** 327–331.

Allen, S.D., Goldstone, J.A., Stone, J.P., and Jan, R.Y. (1986). Transient nonlinear laser heating and deposition: A comparison of theory and experiment. *J. Appl. Phys.* **59,** 1653–1657.

Andretta, R.W., Abele, C.C., Osmundsen, J.F., Eden, J.G., Lubben, D., and Greene, J.E. (1982). Low-temperature growth of polycrystalline Si and Ge films by ultraviolet laser photodissociation of silane and germane. *Appl. Phys. Lett.* **40,** 183–185.

Andrew, R. (1986). Cw Ar^+ laser-induced oxidation of Cu layers. *Appl. Phys.* **41,** 205–212.

Armstrong, J.V., Burk, A.A. Jr., Coey, J.M.D., and Moorjani, K. (1986). Wavelength control of iron/nickel composition in laser induced chemical vapor deposited films. *Appl. Phys. Lett.* **50,** 1231–1233.

Arnone, C., Rothschild, M., Black, J.G., and Ehrlich, D.J. (1986). Visible-laser photodeposition of chromium oxide films and single crystals. *Appl. Phys. Lett.* **49,** 1018–1020.

Ashcroft, S.J., and Mortimer, C.T. (1970). *Thermochemistry of Transition Metal Complexes.* Academic Press, London.

Aylett, J.R., and Haigh, J. (1984). Laser photolytic deposition on indium phosphide. In *Laser Processing and Diagnostics* (Bauerle, D. ed). *Springer Ser. Chem. Phys.* **39,** 263–268.

Aylett, M.R. (1986). Laser photolytic deposition of patterned gold and tungsten films. *Chemtronics* **1,** 146–149.

Baca, A.G., Klebanoff, L.E., Schulz, M.A., Paparazzo, E., and Shipley, D.A. (1986). Dissociative adsorption of CO and O_2 on Cr(100), Cr(110), and Cr(111) in the temperature range 300–1175°K. *Surf. Sci.* **173,** 215–233.

Baranauskas, V., Mammana, C.I.Z., Klinger, R.E., and Greene, J.E. (1980). Laser-induced chemical vapor deposition of polycrystalline Si from $SiCl_4$. *Appl. Phys. Lett.* **36,** 930–932.

Bauerle, D. (1986). *Chemical Processing with Lasers,* Vol. 1, Springer Ser. Mater. Sci. Springer-Verlag, New York.

Bauerle, D. (1982). Ar^+ laser induced chemical vapor deposition of Si from SiH_4. *Appl. Phys. Lett.* **40,** 819–821.

Baum, T.H., and Jones, C.R. (1985). Laser chemical vapor deposition of gold. *Appl. Phys. Lett.* **47,** 538–540.

Baum, T.H., and Jones, C.R. (1986). Laser chemical vapor deposition of gold: Part II. *J. Vac. Sci. Tech.* **A4,** 1187–1191.

Baum, T.H., Marinero, E.E., and Jones, C.R. (1986). Projection printing of gold micropatterns by photochemical decomposition. *Appl. Phys. Lett.* **49,** 1213–1215.

Baum, T.H. (1987). Laser chemical vapor deposition of gold: The effect of organometallic structure. *J. Electrochem. Soc.* **134,** 2616.

Bedair, S.M., Whisnant, J.K., Karam, N.H., Tischler, M.A., and Katsuyama, T. (1986). Laser selective deposition of GaAs on Si. *Appl. Phys. Lett.* **48,** 174–176.

Bedair, S.M., Whisnant, J.K., Karam, N.H., Griffis, D., El-Masry, N.A., and Stadelmaier, H. (1986). Laser selective deposition of III-V compounds on GaAs and Si substrates. *J. Cryst. Growth* **77,** 229–234.

Benziger, J., and Madix, R.J. (1980). the effects of carbon, oxygen, sulfur and potassium monolayers on CO and H_2 adsorption on Fe(100). *Surf. Sci.* **94,** 119–153.

Bilenchi, R., Gianinoni, I., and Musci, M. (1982). Hydrogenated amorphous silicon growth by CO_2 laser photodissociation of silane. *J. Appl. Phys.* **53,** 6479–6481.

Bilenchi, R., Gianinoni, I., Musci, M., Murri, R., and Tacchetti, S. (1985). CO_2 laser-assisted deposition of boron and phosphorus-doped hydrogenated amorphous silicon. *Appl. Phys. Lett.* **47,** 279–281.

Blonder, G.E., Higashi, G.S., and Fleming, C.G. (1987). Laser projection patterned aluminum metallization for integrated circuit applications. *Appl. Phys. Lett.* **50,** 766–768.

Boudart, M., and Djega-Mariadassou, G. (1984). *Kinetics of Heterogeneous Catalytic Reactions.* Princeton University Press, Princeton, New Jersey.

Boyer, P.K., Roche, G.A., Ritchie, W.H., and Collins, G.J. (1982). Laser-induced chemical vapor deposition of SiO_2. *Appl. Phys. Lett.* **40,** 716–719.

Braichotte, D., and van den Bergh, H. (1984). Structure of platinum and tin films formed by laser-induced chemical vapor deposition. In *Laser Processing and Diagnostics.* (Bauerle, D., ed.) Springer Ser. in Chem. Phys. **34,** 183–187.

Braichotte, D., Ernst, K., Monot, R., Philippoz, J.-M., Qiu, M., and van den Bergh, H.

(1985). Laser induced surface reactions in microelectronic technology. *Helv. Phys. Acta* **58,** 879–882.

Braichotte, D., and van den Bergh, H. (1985). Schottky diodes and ohmic contacts formed by thermally assisted photolytic laser chemical vapor deposition. *Proc. Int. Conf. Lasers '85,* 688–696.

Braichotte, D., and van den Bergh, H. (1986). Laser chemical vapor deposition of Pt: Conductivity measurements and schottky diodes. *Proc. Eur. Mater. Res. Soc. Mtg.,* 95–99.

Branz, H.M., Fan, S., Flint, J.H., Fiske, B.T., Adler, D., and Haggerty, J.S. (1986). Doped hydrogenated amorphous silicon films by laser-induced chemical vapor deposition. *Appl. Phys. Lett.* **48,** 171–173.

Breiland, W.G., Coltrin, M.E., and Ho, P. (1986). Comparisons between a gas-phase model of silane chemical vapor deposition and laser-diagnostic measurements. *J. Appl. Phys.* **59,** 3267–3273.

Brewer, L., and Rosenblatt, G.M. (1969). In *Advances in High Temperature Chemistry,* Vol. 2 (Eyring, L., ed.). Academic Press, New York.

Broadbent, E.K., and Ramiller, C.L. (1984). Selective low pressure chemical vapor deposition of tungsten. *J. Electrochem. Soc.* **131,** 1427–1433.

Broden, G., Rhodin, T.N., Brucker, C., Benbow, R., and Hurych, Z. (1976). *Surf. Sci.* **59,** 593–611.

Brodsky, M.H. (1979). *Amorphous Semiconductors.* Springer, Berlin.

Brueck, S.R.J., and Ehrlich, D.J. (1982). Stimulated surface-plasma-wave scattering and growth of a periodic structure in laser-photodeposited metal films. *Phys. Rev. Lett.* **48,** 1678–1681.

Cacouris, T., Scelsi, G., Beach, R., Osgood, R.M., and Krchnavek, R.R. (1987). Laser direct writing of aluminum interconnects. *Conference on Lasers and Electrooptics Extended Abstracts,* 289–291.

Calder, I.D., and Sue, R. (1982). Modeling of cw laser annealing of multilayer structures. *J. Appl. Phys.* **53,** 7545–7550.

Calloway, A.R., Galantowicz, T.A., and Fenner, W.R. (1983). Vacuum ultraviolet driven chemical vapor deposition of localized aluminum thin films. *J. Vac. Sci. Technol.* **A1,** 534–536.

Carslaw, H.S., and Jaeger, J.C. (1959). *Conduction of Heat in Solids,* 2nd ed. Clarendon Press, Oxford, 264.

Chen, C.J., and Osgood, R.M. (1984). A spectroscopic study of the excited states of dimethylzinc, dimethylcadmium, and dimethylmercury. *J. Chem. Phys.* **81,** 327–334.

Chen, C.J. (1987). Kinetic theory of laser photochemical deposition. *J. Vac. Sci. Technol.* **A5,** 3386–3398.

Chiu, M.S., Shen, K.P., and Ku, Y.K. (1985). Lead films produced by laser vapor deposition. *Appl. Phys.* **B37,** 63–65.

Chiu, M.S., Tseng, Y.G., and Ku, Y.K. (1985). Pressure-dependent rate saturation in photodeposition from $W(CO)_6$. *Opt. Lett.* **10,** 113–115.

Christensen, C.P., and Lakin, K.M. (1978). Chemical vapor deposition of silicon using a CO_2 laser. *Appl. Phys. Lett.* **32,** 254–256.

Chu, J.O., Flynn, G.W., Chen, C.J., and Osgood, R.M. Jr. (1985). Infrared emission studies of vibrational excitation in CH_2 fragments produced from ArF and KrF laser photolysis of $Cd(CH_3)_2$ and $Zn(CH_3)_2$. *Chem. Phys. Lett.* **119,** 206–212.

Colinge, J.P., and Van de Wiele, F. (1981). Laser light absorption in multilayers. *J. Appl. Phys.* **52,** 4796–4771.

Comita, P.B., and Kodas, T.T. (1987). Real time optical profilometry as a probe of rates of laser induced chemical vapor deposition. *J. Appl. Phys.* **62,** 2280–2285.

Comita, P.B., and Kodas, T.T. (1987). Modulated laser beam relaxation spectrometry of laser-induced chemical vapor deposition. *Appl. Phys. Lett.* **51,** 2059–2061.

Coombe, R.D., and Wodarczyk, R.J. (1980). UV laser-induced deposition of metal films. *Appl. Phys. Lett.* **37,** 846–848.

Cox, J.D., and Pilcher, G. (1970). *Thermochemistry of Organic and Organometallic Compounds.* Academic Press, New York.

Delahoy, A.E. (1985). High rate photochemical deposition of amorphous silicon from higher silanes. *J. Non-Cryst. Solids* **77/78,** 833–836.

Demiryont, H., Thompson, L.R., and Collins, G.J. (1986). Optical and electrical characterizations of laser-chemical-vapor-deposited aluminum oxynitride films. *J. Appl. Phys.* **59,** 3235–3240.

Demtroeder, W. (1982). *Laser Spectrocopy. Basic Concepts and Instrumentation.* Springer Ser. Chem. Phys. **5,** 270. Springer-Verlag, Berlin.

Deutsch, T.F. (1979). Infrared laser photochemistry of silane. *J. Chem. Phys.* **70,** 1187–1192.

Deutsch, T.F., Ehrlich, D.J., and Osgood, R.M. Jr. (1979). Laser photodeposition of metal films with microscopic features. *Appl. Phys. Lett.* **35,** 175–177.

Deutsch, T.F., Silversmith, D.J., and Mountain, R.W. (1983). UV laser-initiated formation of Si_3N_4. In *Laser Diagnostics and Photochemical Processing for Semiconductor Devices* (Osgood, R.M., Brueck, S.R.J., and Schlossberg, H.R., eds.). Mater. Res. Soc. Symp. Proc. **17,** 129–134.

Deutsch, T.F., Silversmith, D.J., and Mountain, R.W. (1984). UV laser-initiated deposition of Al_2O_3 Films: The effect of surface irradiation. In *Laser-Controlled Chemical Processing of Surfaces* (Johnson, A.W., Ehrlich, D.J., and Schlossberg, H.R., eds.). Mater. Res. Soc. Symp. Proc. **29,** 67–72.

Deutsch, T.F., and Rathman, D.D. (1984). Comparison of laser-initiated and thermal chemical vapor deposition of tungsten films. *Appl. Phys. Lett.* **45,** 623–625.

Dispert, H., and Lachman, K. (1977). Formation of WF_6 and its dissociative products by collisional ionization. *Chem. Phys. Lett.* **45,** 311–315.

Donnelly, V.M., Geva, M., Long, J., and Karlicek, R.F. (1984). Excimer laser induced deposition of InP and indium-oxide films. *Appl. Phys. Lett.* **44,** 951–953.

Donnelly, V.M., Brasen, D., Appelbaum, A., and Geva, M. (1985). Excimer laser-induced deposition of InP: Crystallographic and mechanistic studies. *J. Appl. Phys.* **58,** 2022–2035.

Donnelly, V.M., McCrary, V.R., Appelbaum, A., Brasen, D., and Lowe, W.P. (1987). ArF excimer-laser-stimulated growth of polycrystalline GaAs thin films. *J. Appl. Phys.* **61,** 1410–1414.

Duncan, M.A., Dietz, T.G., and Smalley, R.E. (1979). Efficient multiphoton ionization of metal carbonyls cooled in a pulsed supersonic beam. *Chem. Phys.* **44,** 415–419.

Ehrlich, D.J., Osgood, R.M. Jr., and Duetsch, T.F. (1980). Laser microphotochemistry for use in solid-state electronics. *IEEE J. Quant. Electron.* **QE-16,** 1233–1243.

Ehrlich, D.J., and Osgood, R.M. Jr. (1981). UV photolysis of van der Waals molecular films. *Chem. Phys. Lett.* **79,** 381–388.

Ehrlich, D.J., Osgood, R.M. Jr., and Deutsch, T.F. (1981). Direct writing of refractory metal thin film structures by laser photodeposition. *J. Electrochem. Soc.* **128,** 2039–2041.

Ehrlich, D.J., Osgood, R.M. Jr., and Deutsch, T.F. (1981). Spatially delineated growth of metal films via photochemical pronucleation. *Appl. Phys. Lett.* **38,** 946–948.

Ehrlich, D.J., Osgood, R.M. Jr., and Deutsch, T.F. (1981). Laser microreaction for deposition of doped silicon films. *Appl. Phys. Lett.* **39,** 957–959.

Ehrlich, D.J., Osgood, R.M. Jr., and Deutsch, T.F. (1982). Photodeposition of metal films with ultraviolet laser light. *J. Vac. Sci. Technol.* **21,** 23–32.

Ehrlich, D.J., and Tsao, J.Y. (1983). A review of laser-microchemical processing. *J. Vac. Sci. Tech.* **B1,** 969–984.

Ehrlich, D.J. and Tsao, J.Y. (1985). Laser microchemical reactions for maskless device processing. *J. Vac. Sci. Technol.* **A3,** 904–905.

Ehrlich, D.J., and Tsao, J.Y. (1985). UV laser photodeposition of patterned catalyst films from adsorbate mixtures. *Appl. Phys. Lett.* **46,** 198–200.

Emery, K.A., Thompson, L.R., Bishop, D., Zarnani, H., Boyer, P.K., Moore, C.A., Rocca, J.J., and Collins, G.J. (1984). Beam-assisted CVD of microelectronic films. In *Laser-Controlled Chemical Processing of Surfaces* (Johnson, A.W., Ehrlich, D.J., and Schlossberg, H.R., eds.). Mater. Res. Soc. Symp. Proc. **29,** 81–92.

Eres, D., Mootooka, T., Gorbatkin, S., Lubben, D., and Greene, J.E. (1987). Time-resolved spectroscopic studies of the ultraviolet-laser photolysis of Al alkyls for film growth. *J. Vac. Sci. Technol.* **B5,** 848–852.

Figueira, J.F., and Thomas, S.J. (1982). Generation of Surface Microstructure in Metals and Semiconductors by Short Pulse CO_2 Lasers. *Appl. Phys.* **B28,** 267–268.

Fisanick, G.J., Gedanken, A., Eichelberger, T.S. IV, Kuebler, N.A., and Robin, M.B. (1981). Multiphoton ionization spectroscopy of organometallics: The $Cr(CO)_6$, $Cr(CO)_3D_6H_6$, Cr $(C_6H_6)_2$ series. *J. Chem. Phys.* **75,** 5215–5225.

Fisanick, G.J., Gross, M.E., Hopkins, J.B., Fennell, M.D., Schnoes, K.J., and Katzir, A. (1985). Laser-initiated microchemistry in thin films: Development of new types of periodic structure. *J. Appl. Phys.* **57,** 1139–1142.

Fletcher, T.R., and Rosenfeld, R.M. (1985). Studies on the photochemistry of chromium hexacarbonyl in the gas phase: primary and secondary processes. *J. Am. Chem. Soc.* **107,** 2203–2212.

Flynn, D.K., Steinfeld, J.I., and Sethi, D.S. (1986). Deposition of refractory metal films by rare-gas halide laser photodissociation of metal carbonyls. *J. Appl. Phys.* **59,** 3914–3917.

Foord, J.S., and Jackman, R.B. (1984). Chemical vapour deposition on silicaon: In-situ surface studies. *Chem. Phys. Lett.* **112,** 190–194.

Fuchs, N.A. (1959). *Evaporation and Droplet Growth in Gaseous Media.* Pergamon Press, New York.

Gee, J.M., Hargis, P.J., Carr, M.J., Tallant, D.R., and Light, R.W. (1984). Plasma-initiated laser deposition of polycrystalline and monocrystalline silicon films. In *Laser-Controlled Chemical Processing of Surfaces* (Johnson, A.W., Ehrlich, D.J., and Schlossberg, H.R., eds.). Mater. Res. Soc. Symp. Proc. **29,** 15–20.

Gerrity, D.P., Rothberg, L.J., and Vaida, V. (1980). Multiphoton ionization of metal atoms produced in the photodissociation of group VI hexacarbonyls. *Chem. Phys. Lett.* **74,** 1–5.

Gerrity, D.P., Rothberg, L.J., and Vaida, V. (1983). Ultraviolet-visible multiphoton dissociation of $Cr(CO)_6$: Experimental evidence for statistical fragmentation. *J. Phys. Chem.* **87,** 2222–2225.

Gilgen, H.H., Cacouris, T., Shaw, P.S., Krchnavek, R.R., and Osgood, R.M. (1987). Direct writing of metal conductors with near-UV light. *Appl. Phys.* **B42,** 55–66.

Gillet, E., Chiarena, J.C., and Gillet, M. (1976). Etude par spectrometrie Auger et par spectrometrie de masse de la chimisorption de l'oxyde de carbone sur le molybdene polycristallin. *Surf. Sci.* **54,** 601–616.

Gluck, N.S., Wolga, G.J., Bartosch, C.E., Ho, W., and Ying, Z. (1987). Mechanisms of carbon and oxygen incorporation into thin metal films grown by laser photolysis of carbonyls. *J. Appl. Phys.* **61,** 998–1005.

Green, M.L., Levy, R.A., Nuzzo, R.G., and Coleman, E. (1984). Aluminum films prepared by metal-organic low pressure chemical vapor deposition. *Thin Solid Films* **114,** 367–377.

Gross, M.E., Fisanick, G.J., Gallagher P.K., Schnoes, K.J., and Fennell, M.D. (1985). Laser-initiated deposition reactions: Microchemistry in organogold polymer films. *Appl. Phys. Lett.* **47,** 923–925.

Gupta, A., West, G.A., and Beeson, K.W. (1985). Excimer laser-induced chemical vapor deposition of titanium silicide. *J. Appl. Phys.* **58,** 3573–3582.

Haigh, J. (1985). Mechanisms of metallo-organic vapor phase epitaxy and routes to a ultraviolet-assisted process. *J. Vac. Sci. Technol.* **B3,** 1456–1459.

Hamano, K., Numazawa, Y., and Yamazaki, K. (1984). Structural and electrical properties of photo-CVD silicon nitride film. *Jpn. J. Appl. Phys. Lett.* **23,** 1209–1215.

Hanabusa, M., Namika, A., and Yoshihara, K. (1980). Laser-induced vapor deposition of silicon. *Appl. Phys. Lett.* **35,** 626–627.

Hanabusa, M., Moriyama, S., and Kikuchi, H. (1983). Laser-induced deposition of silicon films. *Thin Solid Films* **107,** 227–234.

Hanabusa, M., and Kikuchi, H. (1983). Coherent anti-Stokes Raman spectroscopic study of CO_2 laser CVD of silane. *Jpn. J. Appl. Phys.* **22,** L712–L714.

Harriott, L.R., Cummings, K.D., Gross, M.E., and Brown, W.L. (1986). Decomposition of palladium acetate films with a microfocused ion beam. *Appl. Phys. Lett.* **49,** 1661–1662.

Herman, I.P., Hyde, R.A., McWilliams, B.M., Weisberg, A.H., and Wood, L.L. (1983). Wafer-scale laser lithography: I. Pyrolytic deposition of metal microstructures. In *Laser Diagnostics and Photochemical Processing for Semiconductor Devices* (Osgood, R.M., Brueck, S.R.J., and Schlossberg, H.R., eds.). Mater. Res. Soc. Symp. Proc. **17,** 9–18.

Herman, I.P. (1984). Laser fabrication of integrated circuits. In *Laser Processing and Diagnostics* (Bauerle, D., ed.). Springer Ser. in Chem. Phys. **39,** 396–416.

Herman, I.P., Magnotta, F., and Kotecki, D.E. (1986). Direct-laser writing of silicon microstructures: Raman microprobe diagnostics and modeling of the nucleation phase of deposition. *J. Vac. Sci. Technol.* **A4,** 659–664.

Hess, D.W., Jensen, K.F., and Anderson, T.J. (1986). Chemical vapor deposition: A chemical engineering perspective. *Rev. Chem. Eng.* **3,** 97–186.

Hess, D.W. (1984). Plasma-enhanced CVD: Oxides, nitrides, transition metals, and transition metal silicides. *J. Vac. Sci. Technol.* **A2,** 244–252.

Highashi, G.S., and Rothberg, L.J., (1985). Surface photochemical phenomena in laser chemical vapor deposition. *J. Vac. Sci. Technol.* **B3,** 1460–1463.

Hirota, Y., and Mikami, O. (1985). *Electron. Lett.* **21,** 77–78.

Hosenlopp, J.M., Samoriski, B., Rooney, D., and Chaiken, J. (1986). Intramolecular vibrational relaxation in n-alkyl benzene chromium tricarbonyls: State selective production of chromium atoms. *J. Chem. Phys.* **85,** 3331–3337.

Houle, F.A., Jones, C.R., Baum, T., Pico, C., and Kovac, C.A. (1985). Laser chemical vapor deposition of copper. *Appl. Phys. Lett.* **46,** 204–206.

Houle, F.A., Wilson, R.J., and Baum, T.H. (1986). Surface processes leading to carbon contamination of photochemically deposited copper films. *J. Vac. Sci. Technol.* **A4,** 2452–2458.

Inoue, T., Konagai, M., and Takahashi, K. (1983). Photochemical vapor deposition of undoped and n-type amorphous silicon films produced from disilane. *Appl. Phys. Lett.* **43,** 774–776.

Iskizaka, S., Simpson, J., and Williams, J.O. (1986). Laser-induced luminescence (LIL) and the photodecomposition of dimethylzinc (DMZ). *Chemtronics* **1,** 175–177.

Jackman, R.B., Foord, J.S., Adams, A.E., and Lloyd, M.L. (1986). Laser chemical vapor deposition of patterned Fe on silica glass: Observation and origins of periodic ripple structures. *J. Appl. Phys.* **59,** 2031–2034.

Jackson, R.L., and Trusheim, M.R. (1982). Surface-modified photochemistry. Preparation of silica-supported $Fe_3(CO)_{12}$ via irradiation of adsorbed $Fe(CO)_5$. *J. Am. Chem. Soc.* **104,** 6590–6596.

Jackson, R.L., and Tyndall, G.W. (1987). Quartz crystal microbalance measurements of absolute laser photodeposition rates: Application to 257-nm deposition from $W(CO)_6$. *J. Appl. Phys.* **62,** 315–317.

Jackson, R.L., and Tyndall, G.W. (1988). Kinetics and mechanism of laser-induced photochemical deposition from the Group 6 hexacarbonyls. *J. Appl. Phys.* **64,** 2092–2102.

Jasinski, J.M., and Estes, R.D. (1985). Laser powered homogeneous pyrolysis of silane. *Chem. Phys. Lett.* **117,** 495–499.

Jasinski, J.M., Meyerson, B.S., and Nguyen, T.N. (1987). Excimer laser-induced deposition of silicon nitride thin films. *J. Appl. Phys.* **61,** 431–433.

Johnson, W.E., and Schile, L.A. (1982). Photodeposition of Zn, Se, and ZnSe thin films. *Appl. Phys. Lett.* **40,** 798–801.

Jones, C.R., Houle, F.A., Kovac, C.A. and Baum, T.H. (1985). Photochemical generation and deposition of copper from a gas phase precursor. *Appl. Phys. Lett.* **46,** 97–99.

Jones, M.W., Rigby, L.J., and Ryan, D. (1966). Delineated Cd films produced by the photolysis of dimethylcadmium. *Nature* **212,** 177.

Kakolowicz, W., and Giera, E. (1983). Standard enthalpies of formation of the chelate complexes of some 3d-electron elements with pentane-2,4-dione. Metal-oxygen bond energies and ligand-field stabilization energies. *J. Chem. Thermodynamics,* **15,** 203–210.

Karny, A., Naaman, R., and Zare, R.N. (1978). Production of excited metal atoms by UV multiphoton dissociation of metal alkyl and metal carbonyl compounds. *Chem. Phys. Lett.* **59,** 33–37.

Kato, H., Sakisaka, Y., Miyano, T., Kamei, K., Nishijima, M., and M., and Onchi, M. (1982). ELS study on chemisorbed state of CO on Cr(110). *Surf. Sci.* **114,** 96–108.

Kawai, T., Choda, T., and Kawai, S. (1987). Laser-induced formation of thin TiO_2 films from $TiCl_4$ and oxygen on a silicon surface. In *Photon, Beam, and Plasma Stimulated Chemical Processes at Surfaces* (Donnelly, V.M., Herman, I.P., and Hirose, M., eds.). Mater. Res. Soc. Symp. Proc. **75,** 289–295.

King, D.A., Goymour, C.G., and Yates, J.T. (1972). Chemisorption of carbon monoxide on tungsten. *Proc. R. Soc. Lond.* **A331,** 361–376.

King, K.K., Tavitian, V., Geohegan, D.B., Cheng, E.A.P., Piette, S.A., Scheltens, F.J., and Eden, J.G. (1987). Laser photochemical vapor deposition of Ge films (300 & le. 873 K) form GeH_4: Roles of Ge_2H_6 and Ge. In *Photon, Beam, and Plasma Stimulated Chemical Processes at Surfaces* (Donnelly, V.M., Herman, I.P., and Hirose, M., eds.). Mater. Res. Soc. Symp. Proc. **75,** 189–194.

Kitai, A., and Wolga, G.J. (1983). Selective Mn doping of thin film ZnS:MN electroluminescent devices by laser photochemical vapor deposition. In *Laser Diagnostics and Photochemical Processing for Semiconductor Devices* (Osgood, R.M., Brueck, S.R.J., and Schlossberg, H.R., eds.). Mater. Res. Soc. Symp. Proc. **17,** 141–147.

Kodas, T.T., Pratsinis, S.E., and Friedlander, S.K. (1986). Aerosol formation and growth in a laminar core reactor. *J. Colloid and Interface Sic.* **111,** 12–111.

Kodas, T.T., Baum, T.H., and Comita, P.B. (1987). Surface temperature rise in multilayer solids induced by a focused laser beam. *J. Appl. Phys.* **61,** 2749–2753.

Kodas, T.T., Baum, T.H., and Comita, P.B. (1987). Kinetics of laser-induced chemical vapor deposition of gold. *J. Appl. Phys.* **62,** 281–286.

Kodas, T.T., and Comita, P.B. (1987). A diffusive transport relaxation technique for studying laser-induced chemical vapor deposition reactions of high pressure. *J. Appl. Phys.*, accepted for publication.

Kodas, T.T., and Comita, P.B. (1988). to be published.

Krauter, W., Bauerle, D., and Fimberger, F. (1983). Laser induced chemical vapor deposition of Ni by decomposition of $Ni(CO)_4$. *Appl. Phys.* **A31,** 13–18.

Krchnavek, R.R., Gilgen, H.H., Chen, J.C., Shaw, P.S., Licata, T.J., and Osgood, R.M. Jr. (1987). Photodeposition rates of metal from metal alkyls. *J. Vac. Sci. Technol.* **B5,** 20–26.

Lax, M. (1977). Temperature rise induced by a laser beam. *J. Appl. Phys.* **48,** 3919–3924.

Learn, A.J., and Foster, D.W. (1985). Resistivity, grain size, and impurity effects in chemically vapor-deposited tungsten films. *J. Appl. Phys.* **58,** 2001–2007.

Lewis, K.E., Golden, D.M., and Smith, G.P. (1984). Organometallic bond dissociation energies: Laser pyrolysis of $Fe(CO)_5$ $Cr(CO)_6$, $Mo(CO)_6$, and $W(CO)_6$. *J. Am. Chem. Soc.* **106,** 3905–3912.

Leyendecker, G., Bauerle, D., Geittner, P., and Lydtin, H. (1981). Laser induced chemical vapor deposition of carbon. *Appl. Phys. Lett.* **39,** 921–923.

Leyendecker, G., Bauerle, D., Geittner, P., and Lydtin, H. (1982). Laser induced chemical vapor decomposition of C and S. *Appl. Phys.* **B28,** 267–268.

Leyendecker, G., Noll, H., Bauerle, D., Geittner, P., and Lydtin, H. (1983). Rapid determination of apparent activation energies in chemical vapor deposition. *J. Electrochem. Soc.* **130,** 157–160.

Lin, T.P. (1967). Estimation of temperature rise in electron beam heating of thin films. *IBM J. Res. Develop.* **11,** 527–536.

Liou, H.T., Ono, Y., Engelking, H.R., and Moseley, J.T. (1986). Two-color multipohoton dissociation and ionization of ferrocene. *J. Phys. Chem.* **90,** 2888–2892.

Liu, Y.S., Yakymyshyn, C.P., Philipp, H.R., Cole, H.S., and Levinson, L.M. (1985). Laser-induced selective deposition of micron-size structures on silicon. *J. Vac. Sci. Tech.* **B3,** 1441–1444.

Longeway, P.A., and Lampe, F.W. (1981). Infrared multiphoton decomposition of monosilane. *J. Am. Chem. Soc.* **103,** 6813–6818.

Love, P.J., Loda, R.T., La Roe, P.R., Green, A.K., and Rehn, V. (1984). Laser-induced photodeposition of Fe films from iron carbonyl. In *Laser-Controlled Chemical Processing of Surfaces* (Johnson, A.W., Ehrlich, D.J., and Schlossberg, H.R., eds.). Mater. Res. Soc. Symp. Proc. **29,** 101–103.

Mackay, K., and Mackay, R. (1973). *Introduction to Modern Inorganic Chemistry.* Intertext, London.

Madey, T.E., and Yates, J.T. (1978). The adsorption of cycloparaffins on Ru(001) as studied by temperature programmed desorption and electron stimulated desorption. *Surf. Sci.* **76,** 397–414.

Magnotta, T.F., and Herman, I.P. (1986). Raman microprobe analysis during the direct laser writing of silicon microstructures. *Appl. Phys. Lett.* **48,** 195–197.

Matsui, S., and Mori, K. (1987). In situ observation on electron beam induced chemical vapor deposition by Auger electron spectroscopy. *Appl. Phy. Lett.* **51,** 646–648.

Mayer, T.M., Fisanick, G.J., and Eichelberger, T.S. IV (1982). Deposition of chromium films by multiphoton dissociation of chromium hexacarbonyl. *J. Appl. Phys.* **53,** 8462–8467.

Mazumder, J., and Allen, S.D. (1979). Laser chemical vapor deposition of titanium carbide. *SPIE Proceedings* **198,** 73–80.

McConica, C.M., and Krishnamani, K. (1986). The kinetics of LPCVD tungsten deposition in a single wager reactor. *J. Electrochem. Soc.* **133,** 2542–2548.

McDiarmid, R. (1974). Assignments in the ultraviolet spectra of MoF_8 and WF_6. *J. Chem. Phys.* **61,** 3333–3339.

McWilliams, B.M., Herman, I.P., Mitlitsky, F., Hyde, R.A., and Wood, L.L. (1983). Wafer-scale laser pantography: Fabrication of n-metal-oxide-semiconductor transistors and small-scale integrated circuits by direct-write laser-induced pyrolytic reactions. *Appl. Phys. Lett.* **43,** 946–948.

Meguro, T., Ishihara, Y., Itoh, T., and Tashiro, H. (1986). Low-temperature silicon epitaxial growth by CO_2 laser CVD using SiH_4 gas. *Jpn. J. Appl. Phys.* **25,** 524–527.

Meunier, M., Gattuso, T.R., Adler, D., and Haggerty, J.S. (1983). Hydrogenated amorphous silicon produced by laser induced chemical vapor deposition of silane. *Appl. Phys. Lett.* **43,** 273–275.

Meunier, M., Flint, J.H., Adler, D., and Haggerty, J.S. (1983). Hydrogenated amorphous silicon produced by laser induced chemical vapor deposition of silane. *J. Non-Cryst. Solids* **59/60,** 699–702.

Meyerson, B.S., and Jasinski, J.M. (1987). Silane pyrolysis rates for the modeling of chemical vapor deposition. *J. Appl. Phys.* **61,** 785–787.

Minakata, M., and Furakawa, Y. (1986). ArF excimer laser induced CVD of aluminum oxide films. *J. Electron. Mater.* **15,** 159–164.

Mingxin, Q., Monot, R., and van den Bergh, H. (1984). Structure and formation of metal film by deposition of laser chemistry. *Scientia Sinica* (*Ser. A*) **27,** 531–539.

Mishima, Y., Hirose, M., Osaka, Y., Nagamine, K., Ashida, Y., Kitagawa, N., and Isogaya, K. (1983). Silicon thin-film formation by direct photochemical decomposition of fisilane. *Jpn. J. Appl. Phys.* **22,** L46–L48.

Mishima, Y., Hirose, M., Osaka, Y., and Ashida, Y. (1984). Direct photochemical deposition of SiO_2 from the $Si_2H_6 + O_2$ system. *J. Appl. Phys.* **55,** 1234–1236.

Mitchell, S.A., Hackett, P.A., Rayner, D.M., and Humphries, M.R. (1985). Multiphoton dissociation of $Ga(CH_3)_3$. *J. Chem. Phys.* **83,** 5028–5042.

Mitlitsky, F., Whitehead, J.C., Bernhardt, A.F., and McWilliams, B.M. (1986). Fabrication of gate array interconnect structures using direct-write deposition processes. In *The Physics and Fabrication of Microstructures and Microdevices* (Kelly, M.J., and Weisbuch, C., eds.). Springer-Verlag, New York, 443–452.

Morris, B.J. (1986). Photochemical organometallic vapor phase epitaxy of mercury cadmium telluride. *Appl. Phys. Lett.* **48,** 867–869.

Motooka, T., Gorbatikin, S., Lubben, D., Eres, D., and Greene, J.E. (1986). Mechanisms of Al film growth by ultraviolet laser photolysis of trimethylaluminum. *J. Vac. Sci. Technol.* **A4,** 3146–3152.

Moylan, C.R., Baum, T.H., and Jones, C.R. (1986). LCVD of copper: Deposition rates and deposit shapes. *Appl. Phys.* **A40,** 1–5.

Mullin, J.B., and Irvine, S.J.C. (1986). Ultraviolet assisted growth of II-VI compounds. *J. Vac. Sci. Technol.* **A4,** 700–705.

Nagano, Y., Achiba, Y., and Kimura, K. (1986). Laser photoelectron spectroscopic determination of electronic states of Fe atoms produced in multiphoton dissociation of $Fe(CO)_5$ in the gas phase. *J. Chem. Phys.* **84,** 1063–1070.

Neugebauer, C.A. (1970). In *Handbook of Thin Film Technology* (Maissel, L.I., and Glang, R., eds.). McGraw-Hill, New York.

Nieuwenhuys, B.E. (1981). Correlation between work function change and degree of electron back-donation in the adsorption of carbon monoxide and nitrogen on Group VII metals. *Surf. Sci.* **105,** 505–516.

Nishino, S., Honda, H., and Matsunami, H. (1986). SiO_2 film deposition by KrF excimer laser irradiation. *Jpn. J. Appl. Phys.* **25,** L87–L89.

Numasawa, Y., Yamazaki, K., and Hamano, K. (1983). Photochemical vapor deposition of silicon nitride film by direct photolysis. *Jpn. J. Appl. Phys.* **22,** L792–L794.

Osgood, R.M. Jr., and Ehrlich, D.J. (1982). Optically induced microstructures in laser-photodeposited metal films. *Opt. Lett.* **7,** 385–387.

Osgood, R.M., and Gilgen, H.H. (1985). Laser direct writing of materials. *Ann. Rev. Mater. Sci.* **15,** 549–576.

Osmundsen, J.F., Abele, C.C., and Eden, J.G. (1985). Activation energy and spectroscopy of the growth of germanium films by ultraviolet laser-assisted chemical vapor deposition. *J. Appl. Phys.* **57,** 2291–2930.

Pauleau, Y., Stawski, R., Lami, Ph., and Auvert, G. (1984a). Chemical vapor deposition of silicon films by pulsed CO_2 laser irradiation of silane. In *Laser-Controlled Chemical Processing of Surfaces* (Johnson, A.W., Ehrlich, D.J., and Schlossberg, H.R., eds.). Mater. Res. Soc. Symp. Proc. **29,** 41–46.

Pauleau, Y., Tonneau, D., and Auvert, G. (1984b). Deposition of silicon films by photodissociation of silane under IR laser irradiation. In *Laser Processing and Diagnostics* (Bauerle, D., ed.). Springer Ser. Chem. Phys. **39,** 215–228.

Perkins, G.G.A., Austin, E.R., and Lampe, F.W. (1979). The 147-nm photolysis of monosilane. *J. Am. Chem. Soc.* **101,** 1109–1115.

Petzoldt, F., Piglmayer, K., Krauter, W., and Bauerle, D. (1984). Lateral growth rates in laser CVD of microstructures. *Appl. Phys.* **A35,** 155–159.

Phillips, S.C.G., and Williams, R.J.P. (1965). *Inorganic Chemistry.* Oxford University Press, New York.

Pilcher, G., Ware, M.J., and Pittam, D.A. (1975). The thermodynamic properties of chromium, molybdenum and tungsten hexacarbonyls in the gaseous state. *J. Less-Common Met.* **42,** 223–228.

Preston, K.F., and Barr, R.F. (1971). Primary processes in the photolysis of nitrous oxide. *J. Chem. Phys.* **54,** 3347–3348.

Price, S.J.W., (1972). The decomposition of metal alkyls, aryls, carbonyls, and nitrosyls. In *Decomposition of Inorganic and Organometallic Compounds* (Bamford, C.H., and Tipper, C.F.H., eds). Elsevier, Amsterdam, Compr. Chem. Kin., 197–257.

Ready, J.F. (1977). *Effects of High-power Laser Radiation.* Academic Press, New York.

Richardson, C.B., Lin, H.B., McGraw, R., and Tang, I.N. (1986). Growth rate measurements for single suspended droplets using the optical resonance method. *Aerosol. Sci. and Technol.* **5,** 103–112.

Rigby, L.J. (1969). Photodeposition from tetraethyl lead. *Trans. Faraday Soc.* **65,** 2421–2429.

Rosler, R.S., and Engle, G.M. (1984). Plasma-enhanced CVD of titanium silicide. *J. Vac. Sci. Technol.* **B2,** 733–737.

Roth, W., Beneking, H., Krings, A., and Krautle, H. (1984). GaAs mesa diodes made by direct-writing laser stimulated MOCVD. *Microelectronics J.* **15,** 26–29.

Rothman, L.B. (1984). Process for forming passivated metal interconnection system with a planar surface. *J. Electrochem. Soc.* **130,** 1131–1136.

Rytz-Froidevaux, Y., Salathe, R.P., and Gilgen, H.H. (1981). Laser-initiated metal deposition on GaAs substrates. *Phys. Lett.* **84A,** 216–218.

Rytz-Froidevaux, Y., Salathe, R.P., Gilgen, H.H., and Weber, H.P. (1982). Cadmium deposition on transparent substrates by laser induced dissociation of $Cd(CH_3)_2$ at visible wavelengths. *Appl. Phys.* **A27,** 133–138.

Rytz-Foidevaux, Y., Salathe, R.P., and Gilgen, H.H. (1983). Laser-initiated Ga-deposition on quartz substrates. In *Laser Diagnostics and Photochemical Processing for Semiconductor* Devices (Osgood, R.M., Brueck, S.R.J., and Schlossberg, H.R., eds.). Mater. Res. Soc. Symp. Proc. **17,** 29–34.

Saitoh, T., Muramatsu, S., Shimada, T., and Migitika, M. (1983). *Appl. Phys. Lett.* **42,** 678–680.

Salathe, R.P., Gilgen, H.H., and Rytz-Froidevaux, Y. (1985). Efficient luminescence band created in (Al, Ga)As multilayers by athermal laser processing. *IEEE J. Quant. Electronics* **17,** 1989–1995.

Shroeder, H., Kompa, K.L., Masci, D., and Gianinoni, I. (1985). Investigation of UV-laser induced metallization: Platinum from $Pt(PF_3)_4$ *Appl. Phys.* **A38,** 227–233.

Scott, B.A., Plecenik, R.M., and Simonyi, E.E. (1981). Kinetics and mechanism of amorphous hydrogenated silicon growth by homogeneous chemical vapor deposition. *Appl. Phys. Lett.* **39,** 73–75.

Seder, T.A., Church, S.P., and Weitz, E. (1981). Wavelength dependence of excimer laser photolysis of $Cr(CO)_6$ in the gas phase. A study of the infrared spectroscopy and reactions of the $Cr(CO)_x$ ($x = 5, 4, 3, 2$) fragments. *J. Am. Chem. Soc.* **108,** 4721–4728.

Shaapur, F., and Allen, S.D. (1987). Experimental determination of laser heated surface temperature distributions. *Appl. Phys. Lett.* **50,** 723–724.

Shedd, G.M., Lazec, H., Dubner, A.D., and Melngailis, J. (1986). Focused ion beam induced deposition of gold. *Appl. Phys. Lett.* **49,** 1584–1586.

Shinn, N.D., and Madey, T.E. (1987). Stimulated desorption from CO chemisorbed on Cr(110). *Surf. Sci.* **180,** 615–632.

Shintani, A., Tsuzuku, S., Nishitani, E., and Nakatani, M. (1987). Excimer laser initiated chemical vapor deposition of tungsten films on silicon dioxide. *J. Appl. Phys.* **61,** 2365–2373.

Smith, G.P., and Patrick, R. (1983). Pyrolysis studies of main group metal-alkyl bond dissociation energies: VLPP of $GeMe_4$, $SbEt_3$, $PbEt_4$, and PEt_3. *Int. J. Chem. Kinet.* **15,** 167–185.

Solanski, R., Boyer, P.K., and Collins, G.J. (1982). Low-temperature refractory metal film deposition. *Appl. Phys. Lett.* **41,** 1048–1050.

Solanski, R., and Collins, G.J. (1983). Laser induced deposition of zinc oxide. *Appl. Phys. Lett.* **42,** 662–663.

Solanski, R., Ritchie, W.H., and Collins, G.J. (1983). Photodeposition of aluminum oxide and aluminum thin films. *Appl. Phys. Lett.* **43,** 454–456.

Sparks, M. (1976). Theory of laser heating of solids: Metals *J. Appl. Phys.* **47,** 837–849.

Stair, P.C., and Weitz, E. (1987). Pulsed-laser-induced desorption from metal surfaces. *J. Opt. Soc. Am.* **B4,** 255–260.

Stanley, A.E., Johnson, R.A., Turner, J.B., and Roberts, A.H. (1986). FT-IR analysis of the photolytic laser-induced chemical vapor deposition of germanium films. *Appl. Spectrosc.* **40,** 374–378.

Steen, W.M. (1978). Surface coating using a laser. In *Int. Conf. on Advances in Surface Coating.* Steen Welding Institute, Cambridge, U.K. 175–187.

Steinfeld, J.I. (1974). *Molecules and Radiation.* Harper and Row, New York.

Steinruck, H.P., Hamza, A.V., and Madix, R.J. (1986). A molecular beam investigation on the kinetic energy dependence of the activation of ethane on the reconstructed Ir(110)-(1 × 2) surface. *Surf. Sci.* **173,** L571–L575.

Sudboe, A.S., Schultz, P.A., Shen, Y.R., and Lee, Y.T. (1986). Molecular-beam studies of laser-induced multiphoton dissociation. *Top. Curr. Phys.* **39,** 95–122.

Tanner, K.N., and Duncan, A.B.F. (1951). Raman effect and ultraviolet absorption spectra of molybdenum and tungsten hexafluorides. *J. Am. Chem. Soc.* **73,** 1164–1167.

Tarui, Y., Sorimachi, K., Fujii, K., Aota, A., and Saitoh, H. (1983). Photochemical vapor deposition of amorphous silicon through 185 nm excitation of monosilane. *J. Non-Cryst. Solids* **59/60,** 711–714.

Tate, A., Jinguji, K., Yamada, T., and Takato, N. (1985). Theoretical and experimental investigations on the deposition rate and processes of parallel incident laser-induced CVD. *Appl. Phys.* **A38,** 221–226.

Tate, A., Jinguji, K., Yamada, T., and Takato, N. (1986). Photodeposition of GeO_2-SiO_2 glass and application to planar optical waveguides. *J. Appl. Phys.* **59,** 932–936.

Tedrow, P.K., Ilderem, V., and Reif, R. (1985). Low pressure chemical vapor deposition of titanium silicide. *Appl. Phys. Lett.* **46,** 189–191.

Tison, J.K., and Cohen, M.G. (1987). Lasers in mask repair. *Solid State Technol.* **30**(2), 113–114.

Tsao, J.Y., Ehrlich, D.J., Silversmith, D.J., and Mountain, R.W. (1982). Direct-write metallization of silicon MOSFETs using laser photodeposition. *IEEE Electr. Dev. Lett.* **EDL-3,** 164–166.

Tsao, J.Y., Becker, R.A., Ehrlich, D.J., and Leonberger, F.J. (1983). Photodeposition of Ti and application to direct writing of Ti:$LINBO_3$ waveguides. *Appl. Phys. Lett.* **42,** 559–561.

Tsao, J.Y., and Ehrlich, D.J. (1984a). Patterned photonucleation of chemical vapor deposition of Al by UV-laser photodeposition. *Appl. Phys. Lett.* **45,** 617–619.

Tsao, J.Y., and Ehrlich, D.J. (1984b). UV-laser photodeposition from surface-adsorbed mixtures of trimethylaluminum and titanium tetrachloride. *J. Chem. Phys.* **81,** 4620–4625.

Tsao, J.Y., Zeiger, H.J., and Ehrlich, D.J. (1985). Measurement of surface diffusion by laser-beam-localized surface photochemistry. *Surf. Sci.* **160,** 419–442.

Tumas, W., Gitlin, B., Rosan, A.M., and Yardley, J.T. (1982). Olefin rearrangement resulting from the gas-phase KrF laser photolysis of $Cr(CO)_6$. *J. Am. Chem. Soc.* **104,** 55–59.

Tyndall, G.W., and Jackson, R.L. (1987). UV multiple-photon dissociation of $Cr(CO)_4$ to Cr and CO: Evidence for direct and sequential dissociation processes. *J. Am. Chem. Soc.* **109,** 582–583.

von Gutfeld, R.J. (1987). Laser-enhanced patterning using photothermal effects: Maskless plating and etching. *J. Opt. Soc. Am.* **B4,** 272–280.

Weinberg, W.H., and Merrill, R.P. (1971). A simple classical model for trapping in gas-surface interactions. *J. Vac. Sci. Technol.* **8,** 718–724.

West, G.A., and Gupta, A. (1984). Laser-induced chemical vapor deposition of silicon nitride films. In *Laser-Controlled Chemical Processing of Surfaces* (Johnson, A.W., Ehrlich, D.J., and Schlossberg, H.R., eds.). Mater. Res. Soc. Symp. Proc. **29,** 61–66.

West, G.A., Gupta, A., and Beeson, K.W. (1985). CO_2 laser-induced chemical vapor deposition of titanium silicide films. *Appl. Phys. Lett.* **47,** 476–478.

West, G.A., Gupta, A., and Beeson, K.W. (1985). Laser-induced chemical vapor deposition of titanium-silicide films. *J. Vac. Sci. Technol.* **A3,** 2278–2282.

Whetten, R.L., Fu, K.-J., and Grant, E.R. (1983). Photodissociation dynamics of $Fe(CO)_6$: Excited state lifetimes and energy disposal. *J. Chem. Phys.* **79,** 4899–4911.

Wilson, R.J., and Houle, F.A. (1985). Composition, structure, and electric field variations in photodeposition. *Phys. Rev. Lett.* **55,** 2184–2187.

Wood, T.H., White, J.C., and Thacker, B.A. (1983). Ultraviolet photodecomposition for metal deposition: Gas versus surface phase processes. *Appl. Phys. Lett.* **42,** 408–410.

Yamada, A., Konagai, M., and Takahashi, K. (1985). Excimer-laser chemical vapor deposition of hydrogenated amorphous silicon. *Jpn. J. Appl. Phys.* **24,** 1586–1587.

Yokayama, G., Uesugi, F., Kishida, S., and Washio, K. (1985). Photothermal effect contribution on film quality improvement in excimer-laser induced metal CVD. *Appl. Phys.* **A37,** 25–30.

Yoshikawa, A., and Yamaga, S. (1984). Growth of hydrogenated amorphous silicon films by ArF excimer laser photodissociation of disilane. *Jpn. J. Appl. Phys.* **23,** L91–L93.

Young, P.J., Gosavi, R.K., Connor, J., Strausz, O.P., and Gunning, H.E. (1973). Ultraviolet absorption spectra of $CdCH_3$, $ZnCH_3$, and $TeCH_3$. *J. Chem. Phys.* **58,** 5280–5283.

Zarnani, H., Demiryont, H., and Collins, G.J. (1986). Optical properties of UV laser photolytic deposition of hydrogenated amorphous silicon (a-Si:H). *J. Appl. Phys.* **60,** 2523–2529.

Zavelovich, J., Rothschild, M., Gornik, W., and Rhodes, C.K. (1981). VUV fluorescence following photodissociation of N_2O at 193 nm. *J. Chem. Phys.* **74,** 6787–6781.

Zinck, J.J., Brewer, P.D., Jensen, J.E., Olson, G.L., and Tutt, L.W. (1987). Excimer laser assisted deposition of GaAs, AlAs, and (Al, Ga)As from Lewis acid base adducts. In *Photon, Beam, and Plasma Stimulated Chemical Processes at Surfaces* (Donnelly, V.M., Herman, I.P., and Hirose, M., eds.). Mater. Res. Soc. Symp. Proc. **75,** 233–240.

CHAPTER 8

Laser Deposition: Experimental Approaches

J. HAIGH AND M.R. AYLETT
British Telecom Research Laboratories
Martlesham Heath
Ipswich, United Kingdom

1. Introduction

The controlled deposition of thin metal films has very great technological importance. Sputtering and evaporation techniques, which use high vacuum technology, find applications in many areas such as semiconductors and optical components. A more recent development is the application of chemical vapor deposition, CVD, using the pyrolysis (thermal decomposition) of organometallics. Laser-induced deposition using the same or related organometallics can be viewed as a branch of CVD and, when it it pyrolytic, obviously has closely similar mechanisms.

This chapter, which is based on Chapter 2 of a wider survey of photoprocesses (*Progress in Quantum Electronics,* Pergamon, Oxford, 1988) will survey the deposition of metals by laser and other photonic

Laser Microfabrication
ISBN 0-12-233430-2

techniques. The photogeneration of the volatile elements of groups V–VII is omitted since although these reactions have been much studied for their mechanistic implications, they are of lesser technological significance in our present context (except in regard to the generation of the III–V and II–VI semiconductors, reviewed separately).

The elements are grouped according to the periodic table, using Fig. 8.1 as an overall guide. Within these groups the individual precursors (italic type), usually but not exclusively organometallic, are given their own subdivision. Within each such subdivision, the arrangement is usually more-or-less historical, since this tends to reflect the development of the understanding of the processes involved.

Deposition of metals using lasers or other sources has been included in reviews of the laser chemical-processing field several times. Ehrlich, Osgood, and Deutsch (1980) were concerned with applications relevant to microelectronics. Ehrlich and Osgood (1982), in a brief review of photodeposition, concentrated on nucleation mechanisms. Ehrlich and Tsao (1983) provided a comprehensive review of laser microchemical processing, with processes classified according to the temperature rise induced by the laser, and subdivided according to the location of the phase absorbing the radiation. McGrath (1983) reviewed applications of excimer lasers in microelectronics, including their use for metal deposition. Ehrlich (1985) gave a review of metal deposition using direct patterning with excimer lasers, including both direct writing and mask projection patterning.

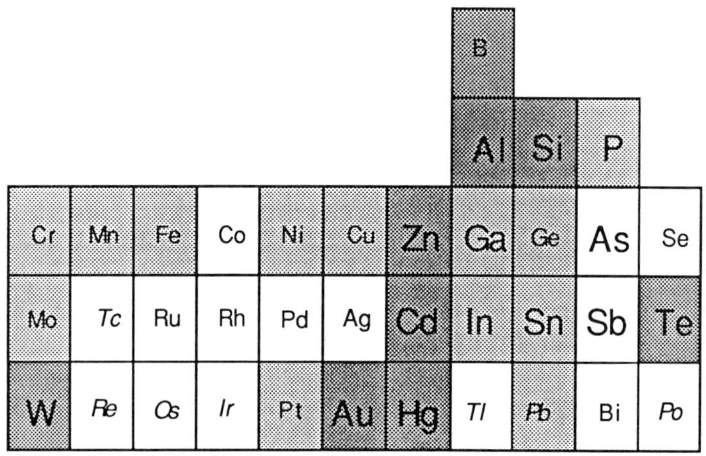

Fig. 8.1 The range of elements (shaded) that have been deposited by light-assisted processes to date. The darker shading indicates the more highly developed techniques. Bold type indicates elements of present importance in semiconductor technology.

2. Group IIB: Zinc, Cadmium, and Mercury

The role of *zinc* in semiconductor technology is as a component of the compound semiconductors zinc sulphide and selenide, and as a dopant in III–V materials. It also has a limited role as a metallization component in III–V technology, usually alloyed with gold to make low-resistance contacts to p-type material. *Zinc dimethyl* [dimethylzinc, DMZ, DMZn, $Zn(CH_3)_2$], and to a lesser extent the diethyl, are useful organometallic precursors, and the former has predominated in photodeposition work to date.

Rytz-Froidevaux, Salathe, and Gilgen (1981) used a pyrolytic laser technique (to be described in more detail below in the context of cadmium deposition) to prepare localized deposits of zinc from DMZ. Deutsch, Ehrlich, and Osgood (1979) included zinc among their early demonstrations of photolytically produced spatially localized metallizations. Deutsch, Ehrlich, Osgood, and Liau (1980) diffused Zn and Cd into InP to form ohmic contacts; the atomic species were generated photolytically at 193 nm (also at 351 nm, possibly pyrolytically), but the indiffusion seemed to require some surface melting. Johnson and Schlie (1982) photolyzed DMZ with a Hg/Xe arc lamp—largely broadband in output—to obtain a deposition rate of 0.1 nm/sec from the 0.4 W/cm^2 of available UV. They showed by using filters that the active wavelength region was that below 245 nm. This accords with the gas-phase UV spectrum. Ishizaka, Simpson, and Williams (1986) examined the laser-induced luminescence of DMZ photolyzed in the gas phase at 193 nm; the process of photolysis involves the absorption of two photons, finally generating excited Zn atoms.

Ehrlich, Osgood, and Deutsch (1981) used DMZ to demonstrate fundamental prenucleation effects and were able to deposit aluminium on surfaces prenucleated in lines with zinc, demonstrating that the effect is not highly specific chemically. They noted in a later paper (Ehrlich, Osgood, and Deutsch 1982) the high grain density of photodeposited zinc films, attributed to the high surface adsorbed-layer coverage, among other factors. Brueck and Ehrlich (1982) discussed periodic (ripple) structures obtained in the deposition of Zn, Cd, and Al. Solanki and Collins (1983) reported on zinc deposition mainly in the context of the conditions for reaction to form zinc oxide.

Krchnavek and coworkers (1987) made a thorough study of the UV spectrum and photolytic decomposition of *zinc diethyl* on quartz and fused silica, using the frequency-doubled argon-ion laser line. The UV absorption spectrum of the compound shows a peak between 200 and 300 nm, as would be expected; its variants in the gas phase and adsorbed states were examined, and the latter was further resolved into physisorbed and chemisorbed components. The physisorbed spectrum shows an enhancement at longer wavelengths. The chemisorbed spectrum—that of the first adsorbed layer, which would be expected to be most strongly perturbed by effects of strong bonding and surface hydroxyl groups—is slightly shifted to shorter wavelengths. Detailed analysis of the photolysis rates indicated that the reaction can be both surface- and gas-phase controlled, depending on temperature and on spot size (see work of Wood and coworkers on cadmium dimethyl, discussed below).

Coombe and Wodarczyk (1980) irradiated a variety of surfaces with KrF and XeCl excimer laser lines and showed that this enhanced the sticking of *zinc* and other metal *atoms* to the surface. The atoms were generated in resistively heated ovens. They surmised that the laser radiation acted by desorbing organic contaminants from the surfaces; in view of the known ability of 248-nm radiation to break organic bonds and ablate polymeric material very rapidly, this seems highly probable. *Magnesium* was also used and similar effects obtained.

Rose and coworkers (1983) examined the tungsten-halogen light-assisted deposition of zinc from *zinc sulphate* solution. This work is discussed in the section on cadmium.

The spectroscopy and photodissociation dynamics of *cadmium dimethyl* [dimethylcadmium, DMCd, $Cd(CH_3)_2$] have been extensively studied. In particular the interplay of gas-phase and surface-phase photolysis has been a continuing object of study. This must to a large extent be because the dimethyls of *cadmium* and of zinc have for some time been among the more readily available of organometallic compounds. Naturally, therefore, their photodecomposition has provided the route of choice to metallic films. Cadmium is required as one of the constituents of cadmium mercury telluride, a material useful for long-wavelength optoelectronic devices; some of the reports detailed here are of cadmium deposition as a stage in the development of this compound—see Irvine and coworkers (1984, 1985). In addition, cadmium is a valuable dopant in III–V technology, but this application has so far attracted less attention.

Anderson and Taylor (1952) made a mass spectrometric analysis of the products of the photolysis of cadmium dimethyl with Hg lamps. In the absence of hydrogen ethane C_2H_6 was the main product, with smaller amounts of methane, CH_4, and of C_2H_4, C_3H_8, and C_4H_{10}. Adding hydrogen increased the rate of production of methane considerably, but solely at the expense of the higher hydrocarbons, leaving the amount of ethane produced substantially unchanged. They concluded that the H_2 molecule is dissociated by CH_3' radicals

$$CH_3' + H_2 = CH_4 + H',$$

and the hydrogen atom then reacts with the metal alkyl

$$H' + Cd(CH_3)_2 = CH_4 + CH_3' + Cd.$$

Jones, Rigby, and Ryan (1966) photolyzed cadmium dimethyl with UV from an arc lamp, and were able to obtain patterned deposits by focusing images onto the substrate. Because the rate of deposition was linearly proportional to the partial pressure of the alkyl, the reaction was taken to be gas phase, but they suggested that the patterning was possible because of rapid deactivation of the gas-phase intermediates; only those species generated close to the surface were able to reach it and form a coherent deposit.

Jonah, Chandra, and Bersohn (1971) used a mercury arc UV source to photolyze cadmium dimethyl, and studied the angular distribution of the products as they deposited on the interior of a hemispherical reaction chamber. Using metallic film formation as means of monitoring the metal atom generation, and reaction with a molybdenum trioxide deposit to monitor the methyl radicals, they showed that when polarized UV was used the product species—both Cd atoms and methyl radicals—appeared with momentum perpendicular to the transition dipole. Assuming the transition dipole to be perpendicular to the linear molecule, this implies that the two methyls do not exit from the photolyzing molecule at equal velocities, since if they did the cadmium atom would have no net momentum. The authors concluded that excitation occurred to an antisymmetric, rather than a symmetric, stretching vibrational mode in the electronic transition. Later Pattengill (1983) used a classical trajectory method to study this asymmetry and showed that it should increase with decreasing photon energy. See also papers cited within, and Tamir, Halavee, and Levine (1974) for further theoretical analysis. See also Chu, Flynn, Chen, and Osgood (1985) for studies of vibrational excitation in the CH_3' photofragment from excimer laser excitation of $Cd(CH_3)_2$ and $Zn(CH_3)_2$.

After their early exploratory papers (Deutsch, Ehrlich, and Osgood 1979), the first technical application for semiconductors seems to have been reported by Ehrlich, Osgood, and Deutsch in 1980. $Cd(CH_3)_2$ was photolyzed with 257-nm radiation from an argon-ion laser/frequency doubler combination, while the direct 514-nm radiation was used to heat an n-type indium phosphide substrate at the point where the photolytic deposition was taking place. With a judicious combination of photolytic rate and surface heating, cadmium atoms could be produced and caused to diffuse into the substrate to form a locally doped region of dimensions of the order of 20 μm. Electrically well-behaved p-n junctions were produced. The method was contrasted with conventional alloying techniques for producing p-n junctions, in the course of which a solid or liquid cadmium film is often produced on the surface. In the photolytic technique it was shown that the specific contact resistance to the Cd-indiffused InP remained rate controlled by the UV flux (Fig. 8.2), indicating that the in-diffusion was also UV controlled and that in consequence no buildup of a condensed cadmium phase was occurring.

Rytz-Froidevaux, Salathe, and Gilgen (1981) included dimethylcadmium among the species whose thermal dissociation with a krypton laser they examined at 521–568 nm, using a spot focused to about 10 μm diameter. Crystalline deposits of the pyramidal structure expected for single-crystal cadmium were obtained.

Ehrlich, Osgood, and Deutsch (1981) used cadmium deposition from the dimethyl in their studies of photochemical prenucleation. They also showed (1982) that the ArF laser line at 193 nm dissociated the dimethyl in the gas phase to cadmium atoms that were long lived and could be collected on the reactor wall. A freshly deposited Cd film on the wall enhanced the condensation rate. Brueck and Ehrlich (1982) studied the production of periodic structures in the deposition reaction. Chen and

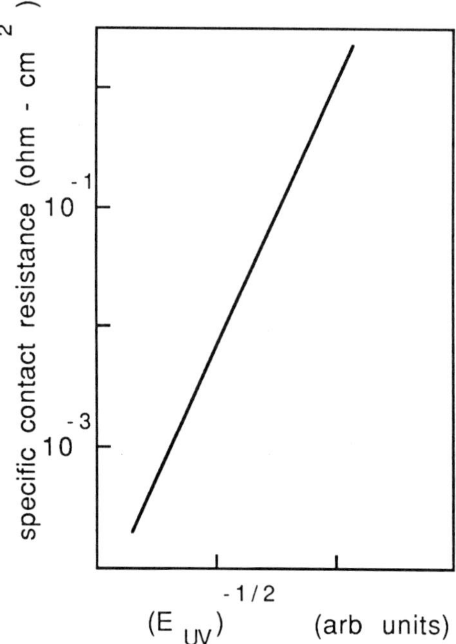

Fig. 8.2 Form of the plot of specific contact resistance versus UV exposure, for ohmic contacts made to p-InP (initially Zn-doped to 10^{18} cm^{-3}) by photolysis of cadmium dimethyl and subsequent in-diffusion of Cd. From Ehrlich, Osgood and Deutsch (*Appl. Phys. Lett.* 1980).

Osgood examined deposition of cadmium in the presence of spheres of the metal. They identified a plasma resonance intensification of the electromagnetic field close to the spheres, which enhanced the deposition.

Ehrlich and Osgood (1981) examined the initial stages of UV photolytic deposition of cadmium in the presence of a stack of silica substrates using UV spectroscopy from the point of view of comparing gas-phase and adsorbed-phase photolysis mechanisms. Although adsorption on the surface takes place strongly and generates a broadening in the UV absorption line out to wavelengths beyond 257 nm, the gas-phase photolysis at 257 nm remains dominant over the adsorbed-phase reaction because the gas-phase UV absorption remains the stronger and more effective process in collecting the radiant energy. See also Osgood (1983).

Rytz-Froidevaux, Salathe, Gilgen, and Weber (1982) examined the dissociation of dimethylcadmium adsorbed on quartz substrates, as induced by various laser wavelengths between 337 and 676 nm. By repeating the experiment of Ehrlich and Osgood described above they showed that broadening of the absorption extended further than the latter had reported—well into the visible region. They postulated that a two-photon absorption process occurred in the adsorbed layer at these longer wavelengths and that this served to nucleate a film. This film subsequently grew by a pyrolysis mechanism, the initial deposit being, of course, opaque to the radiation and so heated by it. By observing a nucleation time for the formation of the deposit at 515 nm and showing that this time was inversely proportional to the square of the laser power, they obtained evidence for this mechanism, and the observed dependence of the nucleation time on wavelength also supported this view.

Wood, White, and Thacker (1983) examined the decomposition of cadmium dimethyl at 257 nm by a spot focused from the outside of a reaction chamber onto the inner surface of a quartz window. The attenuation of the beam provided a measure of the rate of growth of the film. They derived formulae for the rate of deposition in the two cases of adsorbed-phase reaction and gas-phase reaction. The first formula assumed monolayer coverage. The two formulae lead to different dependencies of rate of film thickening on the focused spot size—an effect derived from the geometries of, respectively, the area and the volume within which photolysis is to be assumed to take place. The experimental result indicated a gas-phase reaction on this basis.

Liquid-phase processes have also been examined. Rose, Longendorfer, and Rauh (1983) used an electroless process with *cadmium sulphate* solution and tungsten-halogen lamp radiation to deposit cadmium on p-type silicon. The process, like the corresponding reaction with zinc sulphate, takes place with high quantum efficiency because of the large band bending that occurs.

For methods for the production of *mercury,* reference should be made to the work of Irvine and coworkers (1984, 1985), where the photodeposition of mercury telluride from *dimethyl-* and *diethylmercury* is extensively discussed.

3. Group IIIA: Aluminium, Gallium, and Indium

Aluminium has found favour as an interconnect metallization in silicon technology. This is because of its high electrical conductivity, good adherence, and low rectifying contact resistance, and also because it is relatively easy to deposit by vacuum evaporation. More recently, chemical vapor deposition (CVD) from organo-aluminium precursors has become prominent. Laser deposition of aluminium may be looked upon as a variant of this CVD process.

Trimethylaluminium [TMA, $Al(CH_3)_3$ or $Al_2(CH_3)_6$], like a number of other potential organometallic precursors for aluminium, is a strong electron acceptor species, and accordingly has a tendency to bond to other species that can donate electrons. [Its existence as a dimer $Al_2(CH_3)_6$ is one consequence of this.] The surfaces of semiconductors and of materials such as silica fall into this category. Thus it is not surprising that much work on the photodeposition of aluminium from TMA has involved the adsorbed species as a precursor. Silicon technology makes high demands upon the metallization, particularly as regards the production of interconnects between different levels in a device structure, where a smooth uniform transition is required across a step. Its success in producing such transitions is greatly responsible for the rise of aluminium CVD technology.

Deutsch and coworkers (1979) were perhaps the first to demonstrate the laser deposition of aluminium from trimethylaluminium, using a frequency-doubled argon-ion laser operating at 257 nm. Rytz-Froidevaux, Salathe, and Gilgen (1981) were also early in the field, using a krypton laser at 521–568 nm focused onto a GaAs substrate to bring about the decomposition of TMA. Because reaction at these wavelengths is likely to be pyrolytic, it is important to determine the temperature at which the reaction proceeds. This was done by measuring the shift in the low-energy edge of the GaAs luminescence. Experiments were carried out at temperatures of between 200 and 400°C. Twenty to ninety μm diameter "spikes" of aluminium projecting from the substrate surface and collinear with the beam were produced. These strongly resemble features described by Bauerle and coworkers and obtained during similar experiments on nickel and silicon pyrolytic deposition (qv). Analysis showed them to consist of an amorphous form of aluminium mixed with the carbide Al_4C_3. The high growth rate of the spikes (of the order of microns per second) and the relatively low temperature probably account for the amorphous nature of the deposit. Later, Rytz-Froidevaux and Salathe (1984) presented an analysis of the temperature rise at the focus of the laser, applicable to these experiments. Cross and coworkers (1986) showed how the morphology of laser-deposited aluminium on Si or Si_3N_4 was dependent on the power density and that with Si substrates at high powers reaction occurs with the surface oxide. As long as the surface oxide thickness exceeds 300 nm the Al lines remain electrically isolated from the substrate.

Andreatta and coworkers (1982) briefly reported codepositing aluminium and germanium as Al/Ge alloys from the photodissociation of TMA/GeH_4 gas mixtures, using 193-nm radiation.

Ehrlich and Osgood (1981) measured the photodissociation rate of $Al_2(CH_3)_6$ adsorbed on several parallel wafers of silica using 257-nm radiation as part of an investigation of the properties of adsorbed organometallic layers. The rate was measured by the rate of increase in attenuation of the photolyzing beam that passed through the stack of substrate wafers to a detector behind. Because of the accompanying broadening of the characteristic UV absorption to longer wavelengths, the 257-nm line is more strongly absorbed by the surface-bound TMA than by TMA in the gas phase; thus adsorbed-layer photolysis predominates. The rate showed sensitivity to temperature, reflecting the temperature dependence of the thickness of the adsorbed layer.

Tsao and Ehrlich (1984) studied the effect on the photolysis of adsorbed surface layers of trimethylaluminium of reducing the substrate temperature. It was found that the initial rate of photolysis (i.e., of the first few molecular layers) is only weakly dependent on the temperature of the substrate, while the subsequent "steady-state" deposition rate shows a strong temperature dependence. As the temperature decreases, the surface coverage increases so that the deposition rate also increases. As the temperature is lowered beyond the freezing point of the adsorbed species, however, surface mobility is reduced, and the deposition rate falls again.

The preference of TMA to photolyze in the adsorbed state at 257 nm allows it to demonstrate the stimulated-plasma-wave scattering effects described by Brueck and Ehrlich (1982). Ehrlich, Osgood, and Deutsch (1981) and Tsao and Ehrlich (1984) used the photolysis with 257-nm radiation to described "prenucleation" effects.

Calloway, Galantowicz, and Fenner (1983) showed that TMA could be photolyzed to aluminium using vacuum UV from a rare-gas lamp with microwave excitation. The wavelength range was 100–200 nm. Vacuum UV photon energies are sufficient to ionize the species by reactions such as

$$(CH_3)_3Al = (CH_3)_3Al^+ + e^- \qquad h\nu = 9.1\ \text{eV}.$$

Positive aluminium-containing ions were indeed shown to be present by an increase in the rate of deposition that was observed when a negative potential was applied to the substrate. This carries the implication that the process is gas-phase controlled.

From the technological point of view, the demonstration by Solanki and coworkers (1983) that aluminium deposited by photolysis has the same propensity to good step coverage as CVD aluminium—and other valuable properties—is of great interest. The demonstration was made using the photolysis of TMA with 248-nm excimer radiation. The beam passed parallel to the heated substrate at about 1 mm distance, and was 25 mm by 1 mm in cross section. TMA was admitted at 25% molar concentration in hydrogen, with flows of helium arranged to prevent deposition on the windows of the cell. The substrate temperature was 200°C. The electrical resistivity of the resulting film was about 9 $\mu\Omega$-cm—about three times that of bulk aluminium—and the film had excellent flatness. Values for adhesion and residual tensile stress are quoted, and conformality (smoothness of deposition over a step in the substrate) appears to be excellent. It might be expected for such a strongly electropositive and readily oxidizable metal that there would be a residual oxygen content; this was, however, stated to be less than 5%, accompanied by less than 7% of carbon. It is not entirely clear whether these figures represent analysis resolution limits or a proven level. See also Boyer and coworkers (1983).

In view of other reported work, the ability to deposit at a high rate (100 nm/min) by what appears to be exclusively a gas-phase, rather than a surface-phase, photolysis, is noteworthy. In the photodeposition of metals, generally some degree of heating is always necessary to obtain well-consolidated deposits. This must be antagonistic to the maintenance of an adsorbed layer (see Ehrlich and Osgood 1981). Thus the ability to perform the deposition from the gas phase is a significant advantage.

Tri-isobutylaluminium, [$((CH_3)_2.CH.CH_2)_3Al$, TIBA], has found favor in aluminium CVD over TMA because its pyrolysis appears to produce films with a lower carbon content. Tsao and Ehrlich examined the prenucleation of Al using TIBA followed by patterned deposition from a non-spatially selective pyrolysis. Higashi and Fleming (1986) have shown that the prenucleation of Al films may be carried out with TIBA and

248-nm pulsed excimer laser radiation, obtaining similar results, with 4-μm resolution of the patterns. Rectifying contacts were obtained to GaAs substrates. The substrate temperature was 240–250°C. They investigated a range of substrate materials for their resistance to spontaneous nucleation during the pyrolysis stage (Ehrlich and Tsao's figure of merit). In general, semiconductor surfaces showed a higher level of spontaneous nucleation when they could be shown to be free of native oxides. Film resistivities of about 50 μ-Ω-cm were obtained; these rather high values were attributed to imperfect consolidation of the film, caused by the presence of the H_2 carrier gas.

Geohegan and Eden (1984) used a laser ionization process to deposit aluminium from *aluminium tri-iodide,* AlI_3. The ions were steered by a uniform electric field on to the substrate (ion beam epitaxy). An excimer laser was used for the photoionization stage, operating at 193 or 248 nm. Film deposition rates appear to have been very low, about 42 nm per hour, due possibly to the low metal-halide vapour pressure used, but few details were given. The work is to be compared with that of Calloway and coworkers (above) on $Al(CH_3)_3$ in the vacuum UV. In both cases complete stripping of the species to Al atoms is not achieved before impingement on the substrate and so residual halogen or carbon impurities are to be expected in the film. (However, see same paper for indium below).

Gallium trimethyl [$Ga(CH_3)_3$, trimethylgallium, TMG] has been photolyzed and pyrolyzed extensively as one component of reaction mixtures used in the preparation of gallium arsenide. The only report of deposition of metallic *gallium* from this or any other species is the use of a high-pressure xenon arc lamp (producing a broad featureless emission throughout the UV) by Aylett and Haigh (1983) to photolyze *triethyl gallium,* deposition taking place onto a GaAs substrate. The gallium formed into droplets on the surface, and Auger analysis indicated that some carbon was also present.

Aylett and Haigh (1983) deposited *indium* by photolysis of *indium trimethyl* with a xenon arc lamp broad-band UV source. They also examined the photolysis of *cyclopentadienylindium*(*I*), $(C_5H_5)In$. Te deposits contained large amounts of carbon, and this was attributed to the photolytic sensitivity of the liberated cyclopentadienyl radical.

Geohegan and Eden (1984) exploited the fact that *indium monoiodide* (InI) has an ionization energy

$$InI = In^+ + I^-$$

of 6.24 eV, which corresponds to a wavelength very close to that of the ArF laser emission at 193 nm. (Compare AlI_3, where naked metal ions are not produced, above. No aluminium monoiodide exists at room temperatures.) Ion pairs are generated in a growth cell according to the equation above, by a laser beam passing transversely over the substrate. A negative voltage applied to the nickel substrate led to the formation of an indium film upon it. The collection efficiency for the ions was at its highest when the vapor pressure was such as to make the absorption length for the radiation ($1/e$) roughly equal to the length of the path through the vapor. The films were free of iodine (Auger spectral analysis). Some diffusion of indium into the nickel substrate occurred. The collection voltage used for this experiment unfortunately was not stated so that it is not possible to assess how much of this interdiffusion might have been caused by excessive ion impact energies. Deposition rates were low (42 nm/hr is quoted). Similar experiments were carried out with *thallium* and aluminium (qv).

Letokhov and coworkers (1983) used laser radiation to ionize *indium atoms* in a beam. These ions could then be focused onto a substrate to form a crystalline deposit. The method was advocated primarily as a purification procedure.

4. Group IV Silicon

4.1. Introduction

The deposition of silicon has been the most popular single area by far among those presently being reviewed, and the one where photofabrication has to compete against the best-established and most commercially important alternative. This alternative is thermal chemical vapor deposition, under which heading two reactions are commonly employed. The first is the reduction of the tetrachloride

$$SiCl_4 + 2H_2 \rightarrow Si + 4HCl.$$

This is a labile reaction carried out comparatively close to equilibrium and controllable at high temperatures, the extent of reaction as a function of temperature being governed largely by the thermodynamics. The second reaction is the pyrolysis of the tetrahydride (monosilane, SiH_4).

$$SiH_4 \rightarrow Si + 2H_2.$$

Silane is a much less thermodynamically stable species than $SiCl_4$ so that the reaction takes place some distance from equilibrium and its rate as a function of temperature is kinetically controlled, proceeding at a practical speed at considerably lower temperatures. The $SiCl_4$ reaction, because of its high temperature, can produce single-crystal epitaxial silicon, and has had a large role in the production of silicon devices. With the increasing use of implantation techniques for fabricating microcircuitry, its role, however, has somewhat diminished, and there has been a corresponding increase in emphasis on the areas of polycrystalline silicon deposition on insulators for interconnecting components in large-scale integration. At the same time "amorphous" silicon—material in which crystalline domains are vanishingly small or absent, and in which there is therefore random orientation of the bonds on an atomic scale—has grown in importance in its applications for photovoltaic devices. These applications require lower processing temperatures than does silicon epitaxy, and speed of deposition and cost economy are important for their mass-market applications. Plasma deposition, where energetic electrons assist the reaction, has proved able to give electrically satisfactory material at even lower temperatures and at a faster deposition rate.

During this evolution, the potential contribution of photostimulated deposition has become clear, especially to the Japanese research and development establishments. Much of the published work has emphasized the comparison of the properties of the material with those of that from thermal CVD and plasma CVD. Areas important for technology are stressed, such as trap density, film uniformity, and ease of chemical etching. The temperature ranges used have been comparable with those in plasma or thermal silane CVD, although the reported work suggests, on the whole, that material can be deposited photolytically at lower temperatures than is possible with the latter techniques.

4.2. Lamp-Driven Ultraviolet Photolytic Processes

The UV electronic excitation of silane as part of a technological deposition process was probably first described by van der Brekel and Severin (1972), who used it as a precursor reaction to silicon nitride. The well-established photolysis technique of mercury (Hg) sensitization formed an important part of their process. This involves the introduction of a small amount of Hg vapor into the reactant gas. The excited species Hg ($^3P_1$), often simply written Hg*, is then generated by irradiating with UV light from a low-pressure mercury lamp at either of its two wavelengths, 185 or 254 nm. These excited Hg atoms are generated in good yield and, because of the spectral resonance, can often pass on their energy to the precursor molecule in radiationless processes, so stimulating chemical reactions.

Soon workers were showing that silane may be photolyzed directly to a form of silicon by the Hg sensitization process. Niki and Mains (1964) found "polymeric silicon hydride" on the walls of their reactor when studying the kinetics of the process. Saito and coworkers (1983a, 1983b) used 254-nm Hg lamp radiation and, with the substrate at 200–300°C, obtained partially crystalline silicon films that were similar to the hydrogenated amorphous silicon generated by plasma or thermal CVD techniques. The same group (Tarui and coworkers 1983) reported a similar process using 185-nm radiation. Meanwhile Mishima, Ashida, Hirose, and coworkers (1983a, 1983b) reported the Hg lamp photolysis of disilane, Si_2H_6, and showed that this could be achieved without Hg sensitization. Using a 1% mixture of Si_2H_6 in helium and a UV flux of 0.08 W/cm^2 at 254 nm, they reported a deposition rate of 1.5 nm/min, with the resulting film containing 6–9% of residual hydrogen as measured by infrared spectroscopic determination of Si-H bonds. The deposition rate was independent of temperature, as would be expected for a UV-activated process, except in the presence of BH_3 or PH_3, which were introduced as sources of B and P, respectively, for p- and n-type doping (see below).

In 1984 Aota, Tarui, and Saito examined in more detail the effects of the two Hg lamp wavelengths, 185 and 254 nm, using Hg sensitization, on the photolysis of monosilane. The 185-nm wavelength proved to be much more efficient, and this was linked with the higher SiH_4 absorption cross section at this wavelength. In determining the optimal partial pressure of silane for the process, they found that there was a pressure at which the deposition rate saturated. This could be attributed to the effect of increasing transmission loss at the higher pressures.

Tsuda, Oikawa, and Nagayama (1985) compared the mechanisms of photo-CVD and plasma-CVD, in both cases with monosilane as the precursor. They suggested that the plasma process generates the lowest triplet state of monosilane, which then loses H to give 'SiH_3. This cannot happen with the 257-nm photoprocess because the photon energy is less than the bond energy and so the mechanisms of the two processes must be basically different (see below). See also Nitta (1985).

One problem with using 185-nm radiation is its high attenuation by even the best commercial grade of UV-transparent silica. To circumvent this problem Aota, Tarui, and Saito (1984) produced an Hg-line discharge directly in the reactor to stimulate 185-nm photolysis of SiH_4, with some success. Kamisako, Aota, and Tarui (1984) further examined analytically the spatial distribution of the deposition in this novel reactor.

Meanwhile lamp-driven photolysis of disilane continued to attract attention, despite the technical difficulties of using the unstable precursor. Inoue, Konagai, and Takahashi (1983) obtained amorphous silicon by its Hg-sensitized decomposition at or below 300°C substrate temperature and reported in detail on the structural and optical properties. These were close to those of films from the glow-discharge method. Ashida et al. (1984) and Mishima et al. (1983) compared the Hg lamp photo-CVD process with thermal CVD. The latter is a gas-phase-controlled process with an activation energy that above 430°C fits the reaction.

$$Si_2H_6 \rightarrow SiH_4 + {}''SiH_2.$$

The hydrogen content of the deposit falls as temperature increases. The rate of the photoprocess is independent of temperature, as would be expected. Addition of dopants produces significant changes, however. With diborane present, an activation energy comparable to that of the thermal process is observed. (See Mishima, Ashida, and Hirose, 1983).

Kenne et al. (1985) and Tanaka et al. (1984) have successfully made solar cells consisting of a p-i-n silicon carbide/silicon structure on glass with the precursors disilane, methylsilane, and acetylene. Hg sensitization was used with the 254-nm radiation but was found not to be necessary with 185 nm.

Recently Nishida, Shiimoto, and coworkers (1986) have reported the growth of silicon films at 200°C from mixtures of SiH_6 and SiH_2F_2, using 185 nm. The activation energy was 0.18 eV. The films showed some evidence of a degree of epitaxial orientation. Equally unexpectedly, Mutsukura and Machi (1986) showed by transmission electron microscopy that some "amorphous" silicon films—including one photochemically deposited at only 150°C—contained single crystal grains.

4.3. *Laser-Driven Processes*

Both Bauerle and coworkers (1982) and Ehrlich, Osgood, and Deutsch (1981) described the deposition of silicon from silane using visible or near-UV light from a powerful focused argon-ion laser. For the first time highly spatially delineated deposition of silicon became possible, with deposits consisting of lines down to a few microns in width. These lines could be doped p-type with boron from the decomposition of BCl_3. It was clear that, unlike the lamp-driven UV processes described above, this was essentially a pyrolytic process. Bauerle et al. (1982) carried out a thorough investigation of the mechanistics of their reaction, together with those of the deposition of the related element carbon, which could be produced from hydrocarbon precursors. They identified separate temperature regimes in which kinetics and diffusion, respectively, were in control. See also Bauerle et al. (1983) and Ishizu and coworkers (1984). Du and coworkers (1986) demonstrated periodic structures in the deposits Magnotta and Herman (1986) and Herman, Magnotta, and Kotecki (1986) have recently used Raman spectroscopy to examine the phase state of the deposit while it was forming. McWilliams et al. (1983) reported the full fabrication of MOS transistors and logic gates at micron sizes using the argon-ion laser, with phosphorus n-type doping. See also Ishizu et al. (1984). Parker, Liu, and Zhang (1984) used a frequency-doubled Nd:YAG laser at 532 nm to deposit polysilicon from an unstated precursor for microcircuit link-making and measured the electrical resistance of the tracks as a function of the laser pulse power. Nishida, Konagai, and coworkers (1986) compared photo- with plasma-deposited microcrystalline silicon as regards its piezoresistive properties. The comparison favored the former.

Deposition using infrared lasers has also been an active field. At an early date Christensen and Lakin (1978) and Hanabusa, Namiki, and Yoshihara (1979) had reported pyrolytic polycrystalline silicon deposition from silane using a CO_2 laser at a 10.6-μm wavelength, the former with a degree of focusing of the beam and some limited spatial delineation of deposition. Baranauskas and coworkers (1980) studied the CO_2-laser-induced reduction of $SiCl_4$ in H_2, the laser being used to provide a heated spot of defined size on a quartz substrate. Because of the labile nature of this process it has a well-defined cutoff temperature governed by thermodynamics, and as a result very sharply defined deposition features were obtained within the heated spot. Hanabusa, Namiki, and Yoshihara (1979); Hanabusa and Kikuchi (1983, 1984); and Hanabusa, Moriyama, and Kikuchi (1983a, 1983b) have used a CO_2 laser at 10.6 μm with SiH_4 as the precursor. Deutsch (1979) investigated the mechanism of this process. One motive for the use of this laser source is that it may be tuned to a line in the SiH_4 vibrational spectrum. When this is done CARS shows that energy is transferred directly to the gas-phase molecule, and an enhancement of the gas-phase temperature is observed. The deposition rate is also found to be increased. When the laser line is detuned from the absorption, deposition occurs by substrate heating only. However, in the former case it is not yet clear whether an infrared photolytic mechanism is important or whether the energy is thermalized before decomposition occurs. In this context see also Tonneau and coworkers (1986). Iwanaga and Hanabusa (1984) have examined the 10.6-μm decomposition of disilane. Meunier et al. (1983), Meunier et al. (1984), and Branz et al. (1986) have carried out similar work on the silane decomposition to generate both p- and n-type material and have reported on the electrical conductivity, hydrogen content, and mechanical stress of the resulting films for their solar cell applications. The dopants PH_3 and B_2H_6 both influence the growth kinetics (compare above, Mishima and coworkers, 1983) but in different ways. The hydrogen content is 4% when the deposition temperature is 400°C. See also Bilenchi, Musci, and Murri (1984), who also reported on p and n doping.

Meguro and coworkers (1986) have reported epitaxial growth of silicon from SiH_4 at 650°C using the CO_2 laser. The deposition rate depends on the polarization and angle of incidence of the beam, which suggests a surface activational process (compare the work of Frieser, below).

Turning to processes that may be presumed to be photolytic; in 1982 Andreatta et al. used the far UV wavelengths of the excimer laser (193, 248, and 308 nm) to deposit silicon and germanium from their monohydrides. Far ultraviolet lasers may expel energetic ions from most solids by multiphoton interactions, and, in a unique development, Lubben et al. (1984, 1985), exploited this possibility by using the 248-nm line of the excimer laser system at powers above 10^7 W/cm^2 to generate Ge and Si ions at 40 eV from wafer targets. These ions were then either allowed to fall directly on a silicon or gallium arsenide substrate or to generate secondary Ge or Si ions. Films produced by this method were epitaxial when the substrate temperature was above 300°C, as demonstrated by electron channeling methods. Some difficulties were caused by the presence of atomic clusters in the ion beam, but these could be shuttered out. Virupaksha Reddy (1986) has given a variant of the process involving a cooled silicon substrate onto which may be condensed a layer of argon or other inert gas. This serves to reduce reflection and increase the energy absorption. The method would appear to offer possibilities for the etching of silicon and other materials in situations where damage due to the energetic nature of the beam does not matter.

Both these latter developments can be said to be photolytic—the latter in the sense that the UV triggers a high-energy, nonthermalized, emission process in the target surface. That UV can influence deposition via surface excitation may have been demonstrated by the work of Frieser (1968), who reported that deposition of crystalline silicon from Si_2Cl_6/H_2 gas mixtures could be enhanced by the UV content of a mercury arc-lamp emission. He deduced that the UV photon energy was lower than the Si-Si bond-breaking activation energy, indicating that direct photofragmentation of the Si_2Cl_6 was not occurring; by elimination, therefore, the UV action must be on the Si substrate.

4.4. *Mechanisms of the Photolytic Production of Amorphous Silicon from the Hydrides*

The generation of films with semiconducting and photosensitive properties by the action of heat or of plasmas on silane and disilane has attained great technological importance over the last 10 years, and techniques such as coherent antistokes Raman spectroscopy (CARS), which enable the species present in the gas phase to be identified have been used in analytical work in this direction (Tanaka, Hata, and Matsuda 1984). The even more recent development of UV-photogenerated films, and their high electronic quality, means that there will undoubtedly soon be much work in the literature on the mechanism of their production. However, a substantial body of work on the homogeneous aspects of the mechanistics of UV photolysis of SiH_4 and Si_2H_6 already exists and will be reviewed.

The pyrolysis of SiH_4 is found to yield Si_2H_6 (Purnell and Walsh, 1966). The first point of interest seems to be the nature of the primary precursor. The *a priori* candidates are SiH_3' and SiH_2''. Walsh (1988) considers the reaction via SiH_3' to be unlikely on the basis of estimated activation energies. Thus the homogeneous thermal process is likely to involve SiH_2''. (Wall-initiated side reactions may interfere.) A mechanism propounded by Erwin, Ring, and O'Neal (1985) for the reaction involves the "extrusion" of H_2 to leave SiH_2''. This then reacts further by insertion:

$$SiH_2'' + SiH_4 \rightarrow Si_2H_6.$$

Direct photolysis with 147-nm UV also gives these primary products (Perkins, Austin, and Lampe, 1979), and Si_2H_6 and Si_3H_8 are generated in addition to the deposited film. Taguchi and coworkers (1986) report that the formation of the deposit from SiH_4 using 193 nm is enhanced by the presence of some Si_2H_6, in a manner that suggests the latter is an intermediate in the process rather than a byproduct. Recently Oikawa and coworkers (1986) appear to have shown by *ab-initio* MO configurational-interactional calculations that SiH''' may be formed in a single step from SiH_4 and so under some circumstances may be a precursor.

Hg-sensitized decomposition of silane is stated to go via SiH_3' exclusively (Perkins, Austin, and Lampe, 1979 and references therein). Pollock and coworkers (1973) examined Hg-photosensitized decomposition of disilane. They concluded that the primary step is

$$Hg^* + Si_2H_6 \rightarrow Si_2H_5' + H' + Hg.$$

This process has high quantum efficiency. The final products they were able to observe in addition to the solid polymer were H_2, SiH_4, Si_3H_8, and Si_4H_{10}. The last two may plausibly be derived by addition or abstraction reactions involving Si_2H_5'. They went on to investigate the route by which SiH_4 was produced, using an isotopic substitution method, and concluded that a displacement reaction

$$Si_2H_6 + H' \rightarrow SiH_4 + SiH_3'$$

was occurring. The SiH_3' radical then disproportionated, generating the silylene diradical

$$2SiH_3' \rightarrow SiH_2'' + SiH_4.$$

As in the thermal reactions, this diradical can then insert itself in other species and is presumed to play a role in the buildup of the molecular weight of the solid polymer. They concluded, however, that it was not the only species active in this respect and that Si_2H_5' was also involved in the molecular weight buildup by a direct chain displacement reaction

$$Si_2H_5' + Si_2H_6 \rightarrow Si_3H_8 + {}'SiH_3$$

$${}'SiH_3 + Si_2H_6 \rightarrow SiH_4 + Si_2H_5'.$$

Baggott and coworkers (1986) studied the 193-nm laser photolysis of phenylsilane and concluded that a three-center elimination of H_2 occurred, generating a silylene diradical,

$$PhSiH_3 \rightarrow PhSiH'' + H_2 \qquad \text{or}$$

$$PhSiH_3 \rightarrow {}''SiH_2 + PhH,$$

and that these silylene species seem to have solid polymer producing capabilities.

Another important factor is the establishment of the final bound hydrogen content of the film. If the solid is formed initially by a mechanism of continuing the substitutional reactions demonstrated by Pollock, it should be rich in Si-H bonds, with a stoichiometry corresponding to $(SiH_2)_n$. A mechanism must therefore exist to reduce this to the much lower Si-H bond content typically found in the films. Mishima and coworkers (1983) postulated such a mechanism based on the hydrogen atoms generated in the Hg-sensitized photochemical route propounded by Pollock. They proposed that these can abstract hydrogen atoms from the $-SiH_2-$ groups in the developing polymer chain, thereby promoting crosslinking and establishing the lower hydrogen content. This step appears to be unique to the photoprocess, where indeed very low levels of hydrogen can result.

Motooka and Greene (1986) have modeled the UV-induced deposition of Si and Ge from the monohydride. The reaction is assumed to go via SiH_2'' and GeH_2'' (photogenerated). The calculations were based on structures and energies of complexes such as Ge_6H_{16} postulated to exist—as models—on the surface of the Ge crystal; MO calculations then determined the energy of H_2 elimination from this complex. The route does not involve any photon participation in the elimination.

5. Group IV: Germanium, Tin, and Lead

Some of the published work on *germanium* deposition has already been touched upon in the section on silicon; *germanium hydride* (monogermane or germane, GeH_4) is the common starting compound. Rousseau and Mains (1966) studied the 254-nm (Hg lamp) photolysis of GeH_4 in terms of product yields. They did not report directly the absorption intensity at this wavelength. Andreatta and coworkers (1982) used excimer laser photoexcitation of GeH_4 at room temperature to deposit polycrystalline germanium. The dependence of the deposition rate on intensity was complex.

Later Osmundsen, Abele, and Eden (1985) made a thorough study of the photochemistry of this process at excimer wavelengths. The UV absorption of GeH_4 extends up in wavelength to 230–240 nm, but the cross section is low, and about 300 times lower at 248 than at 193 nm. These workers concluded that the primary reaction is multiphoton, producing the germylene, GeH_2'', diradical. They studied the fluorescence emissions attributable to atomic Ge and to GeH''' in the gaseous photolysis product from 248-nm irradiation. The intensities of bands

$$A_2 \rightarrow X_2\pi \quad \text{for GeH}''' \quad \text{and}$$

$$5s^1P_1 \rightarrow 4p^2\,{}^1D_2 \quad \text{and}$$

$$5s^3P_1 \rightarrow 4p^2\,{}^3P_0 \quad \text{for Ge}$$

were proportional to (laser intensity).^{2} This indicated that both Ge and GeH''' originated from a precursor that was formed from GeH_4 by a two-photon absorption process. A continuum emission that could be attributed to GeH_2'' was observed, supporting the conclusion that this was the initial photoproduct. The activation energy for film growth was measured as about 2 kcal/mol, much lower than for the pyrolysis of GeH_4 and consistent with the inference that the primary gas-phase photoprocess was rate limiting. It was left undecided as to which of the species studied were important intermediates in the growth of the film. However, its weak UV absorption means that GeH_4 is a somewhat unsatisfactory precursor.

Alkylgermanes as alternatives to GeH_4 have been explored by Stanley and coworkers (1986). $C_2H_5GeH_3$ and $(C_2H_5)_2GeH_2$ were the most successful. They used Fourier transform infrared spectroscopy to monitor the gas-phase reaction products. The solid germanium produced was polycrystalline, and could be doped with Cd and Al by addition of dimethylcadmium or trimethylaluminium to the vapor.

A certain amount of work has been carried out on the deposition of *tin* from the simple alkyls *tetramethyltin* and *tetraethyltin,* as much, probably, because of the relative innocuousness of these reagents (toxic but of low air sensitivity), and their convenient vapor pressures at room temperature, than from an intrinsic interest in the reactions involved or in the technical value of tin deposition. However, tin does have a useful role in semiconductor structures, being an n-type dopant in several III–V compounds and a constituent of alloys used for their metallization.

Mingxin, Monot, and van den Bergh (1984) reported both pyrolytic and photolytic deposition of tin on quartz from the vapor of the tetramethyl, using the argon-ion laser, respectively without and with frequency doubling, and a mercury lamp. The relatively low melting point of tin rendered the deposit from the pyrolysis liquid and made it difficult to obtain a uniform thin film. Much more uniform deposits were obtained by photolysis, as long as the power levels were kept low. The deposition rate under these conditions was linear both with the vapor pressure and with the radiation intensity, the latter indicating that the decomposition is a single-photon process. Striations in the deposit were reported that the authors believed corresponded to interference effects in the irradiation pattern, presumably situated near the substrate. The spacing of these striations appears to indicate a spatial resolution for the deposition of 0.2 μm, which seems to imply a surface reaction, or perhaps surface activation of absorption as invoked by Rigby in the case of rather similar experiments with lead deposition (qv).

The widespread use of tin metal as a dopant in molecular beam epitaxial growth of gallium arsenide led Kowalczyk and Miller (1985) to a novel investigation of its production in the MBE apparatus by photolysis from a range of species using xenon arc and low-pressure Hg lamps. *Tetramethyl-* and *tetrabutyltin* (no information as to which isomer), *dibutyltin dibromide* $(C_4H_9)_2SnBr_2$, and *stannic chloride* $SnCl_4$ were all used. The authors were looking for a method of incorporating tin in selected areas of the grown epitaxial layer. Tin metal incorporation in MBE is known to take place by a surface accumulation mechanism. A condensed film of tin forms from the impinging beam, and atoms then enter the growing crystal from this film. Thus if this film could be formed in a spatially delineated manner, selective area incorporation would be possible. These authors therefore wished to be able to absorb and desorb the species without pyrolytic breakdown so that unphotolyzed material could be removed from the surface by evaporative heating before subsequent crystal growth. The growth itself, of course, requires the substrate to be heated, in this case to about 580°K. The tetramethyl was found to pyrolyze too readily and to incorporate carbon in the layer (a problem also encountered with other methyl species in III–V crystal growth) and so was unsuitable. Tetrabutyltin and dibutyltin dibromide also pyrolyzed before desorption, albeit more cleanly than the tetramethyl, but also ruling themselves out. Stannic chloride also pyrolyzed and the resulting chlorine attacked the crystal surface. [See also the later paper (1986)]. Thus no suitable precursor for the selective-area incorporation was found, although much useful information was furnished on the possibility of organotin compounds for the developing technique of metallo-organic molecular beam epitaxy (MOMBE).

The work of Green and Kuku (1983) on the photolysis of stannous iodide (SnI_2) in the solid phase is not strictly within the present brief but is interesting. They used filtered quartz-halogen radiation in the range 430–550 nm. The stannous iodide was evaporated onto a quartz substrate, which formed part of a quartz crystal microbalance, so that the extent of reaction could be monitored by the weight loss. Three temperature regimes were found, and the results were similar to those found and reported in a preceding work on plumbous iodide, PbI_2. Free carriers are created in the microcrystals that constitute the film, and these migrate to the surface of the film where they generate tin and iodine atoms. The latter may be lost by evaporation, leaving tin. The formation of the opaque tin attenuates or scatters the radiation and accounts for some of the variability with temperature of the kinetics; the authors go on to suggest that there are critical densities for the formation of the tin atom clusters and that this also has an effect.

At an early date Leighton and Mortenson (1936) had measured the products of the photolyses of several tetraalkyls of *lead* and observed free methyl radicals. In general the workers examining the photolysis of organolead species have had their main interest in mechanistic studies rather than directly in the development of technology; relevant to this is the fact that organolead species share with organotin species a relatively innocuous nature (they are admittedly highly toxic, but not so air sensitive as many of their analogs with other elements). They have no applications of which the present authors are aware in semiconductor technology.

There has been much emphasis on surface diffusion effects with lead species. Possibly the lack of electron-accepting ability in lead (IV) [and tin (IV))] organometallics leads to a more loosely bound state of the molecule on the surface and enhances its migration rate, making such effects more obvious. The work of Rigby (1969) and that of Perry and Roberts (1972) gave rise to some of the first reports of surface-enhanced processes in organometallic photolysis. In 1969 Rigby investigated the deposition of lead from tetraethyl lead, $Pb(C_2H_5)_4$, using radiation from a short-arc lamp. From the consideration of the rate of formation of products, he deduced that the deposition was controlled by a gas-phase reaction. However, when collimated radiation was used a delineated deposit was obtained, the degree of delineation being better than would have been allowed if the gas-phase process, followed by diffusion to the surface, was fully controlling—the intermediate species would have diffused out of the beam. This implied that a rapid surface photolytic reaction was occurring, which in effect increased the sticking probability of the intermediate species on that part of the surface that was illuminated.

Later Perry and Roberts (1972) used both broad-band and monochromatic Hg lamps to photolyze $Pb(C_2H_5)_4$ vapor, in the presence of a glass substrate at 0°C. With the 254-nm line they were able to photolyze the $Pb(C_2H_5)_4$ when this was adsorbed on the glass, but as a lead film developed the 254-nm radiation became less effective than the slightly longer wavelengths from the broad-band source. The inference was that the lead tetraethyl adsorbed on the developing lead film was "activated" in the sense of being decomposed at longer wavelengths (lower energies) than when adsorbed on glass.

Zeiger (1985) and Tsao (1985), with their coworkers, used tetraethyllead on sapphire in their investigations on surface diffusion of photolyzable species. Similar work at 257-nm irradiation wavelength was reported by Chiu and coworkers (1985). They found, however, that the presence of helium slowed the deposition, and they attributed this to its effect on the rate-determining gas-phase diffusion of the photolyzed species to the surface. They reported some initial effects in the film formation that were attributable to surface photolysis, but unfortunately do not give the data on irradiation intensity that would allow results to be compared with those of Zeiger and of Tsao.

6. Early Transition Metals: Titanium, Manganese, Chromium, Molybdenum, and Tungsten

Despite the fairly extensive organometallic chemistry of *titanium,* the only practical precursor for its photon-assisted deposition is the tetrachloride $TiCl_4$ Organo-titanium compounds containing carbon—metal bonds are in general difficult to handle and their gas-phase photolytic properties have not been examined. Complexes such as dicyclopentadienyltitanium dichloride exist, but experience with a related compound of manganese (see below) and elsewhere suggests that carbon incorporation would be a problem. Stable compounds with Ti-N bonds exist, such as $Ti(N(CH_3)_2)_4$, but these have not afforded high-quality metal deposits because of the strong tendency of metallic titanium to occlude nitrogen, oxygen, and carbon. The disadvantage of using the tetrachloride, on the other hand, would appear to be the corrosive nature of the chlorine byproduct. This reduces its applicability for deposition in III–V technology, where sputtered and evaporated titanium films are used as contacts, but other applications remain (see below).

Titanium, and films containing titanium and chlorine, have been deposited from the tetrachloride $TiCl_4$ both by infrared (Allen, 1979) and by 257-nm UV photolysis. Tsao et al. (1983) deposited Ti onto lithium niobate, using a writing technique with a 4-μm diameter spot at 257 nm, and then heated the substrate to in-diffuse the titanium and thereby produce channel waveguide structures. No measurements of the refractive index profile of these waveguides were reported, but the mode confinement was comparable with that of a conventionally fabricated $Ti:LiNbO_3$ structure. The deposition showed a nonlinear dependence on $TiCl_4$ pressure of the type indicating photolysis of adsorbed species. Ehrlich and Tsao (1984) deposited Ti-Cl films as part of an investigation of nucleation effects in surface-adsorbed photolytic deposition; in conjunction with adsorbed alkyluminium species these films acted as catalysts analogous to those of Ziegler and Natta for olefin polymerization.

George and Beauchamp (1980) used *methylcyclopentadienyl manganese tricarbonyl* to deposit *manganese* onto a soda lime glass substrate. The light source used was a 2.5-kW Hg-Xe arc lamp passed through a monochromator. The film was nonconducting, presumably because of residual organic incorporation from the strongly bound cyclopentadienyl groups. From analogous experiments with iron deposition from $Fe(CO)_5$ the decomposition mechanism appeared to involve the reaction of low-energy electrons photoejected from the substrate surface.

Another potential precursor is *decacarbonyl dimanganese,* $Mn_2(CO)_{10}$. Karny and coworkers (1978) studied its photodissociation, and that of $Mn(CO)_5Br$, using the UV output from an excimer laser operating at 193 or 248 nm. They discussed the multiphoton mechanism by which excited manganese atoms were produced.

The group VI metals *chromium, molybdenum,* and *tungsten* have wide applications in several technologies, and so their deposition has been fairly extensively examined. Tungsten, in particular, has been used for electrical interconnections between devices, as a low-resistance gate metallization on silicon wafers, and as a contact diffusion barrier on both silicon and III–V materials. Pauleau (1984) has reviewed the chemical vapor deposition of tungsten films for IC metallization, including laser deposition. Many workers have studied the hexacarbonyls of the three elements as a group, since they possess closely related properties. *Molybdenum and tungsten hexacarbonyls,* $Mo(CO)_6$ and $W(CO)_6$, have been used for some time as precursors for conventional (thermal) CVD of Mo and W films [see Kaplan and d'Heurle (1970), for example, who report 400–500°C for the reduction of $W(CO)_6$ in H_2]. The choice of the hexacarbonyls for laser CVD experiments followed naturally from this. The work falls into three types based on the mechanisms proposed: dielectric breakdown, pyrolytic ("photothermal"), and photolytic.

Langsam and Ronn (1981) demonstrated the possibility of producing chromium atoms from $Cr(CO)_6$ using a focused CO_2 laser to provide intense infrared radiation at a point within the $Cr(CO)_6$ vapor, from the point of view of spectroscopic studies of excited atoms and the mechanisms of their production. Draper (1980) reported the laser-induced dielectric breakdown of the vapor-phase chromium and molybdenum hexacarbonyls to produce ultrafine particles of the metals. This was achieved by focusing a pulsed CO_2 laser into a gas cell containing the vapor to a local field intensity of $6 \cdot 10^{10}\,W/cm^2$. This electromagnetic field generates a localized plasma that disrupts the metal-carbonyl bonds. Condensation of the metal from the gas phase leads, as might be expected from their very low volatility, to ultrafine particles with dimensions in the sub-10-nm range. Draper (1980) then attempted to generate fine molybdenum particles by this method to act as a dispersed, high-surface-area catalyst for alkene hydrogenation. The laser decomposition was carried out in the hydrogenation vessel. However, in the presence of a straight-chained hydrocarbon it was found that rather than the fine particles being produced, molybdenum "cobwebs" resulted. These webs formed much more slowly than the particles, over time scales of tens of seconds. They were, unfortunately, not catalytically active.

Jervis and Newkirk (1986) reported the laser-induced dielectric breakdown deposition of molybdenum from $Mo(CO)_6$ in a manner similar to that of Draper, except that a substrate was placed in the vicinity of the focus of the CO_2 laser so that deposition of molybdenum films resulted rather than the production of powders. Auger analysis gave the film composition as being 40% Mo, 30% C, and 30% O. Shin and coworkers (1985) used the laser-induced dielectric breakdown method of Draper to produce chromium-molybdenum-tungsten alloy powders from mixtures of the metal carbonyls. Two types of particles were formed by the homogeneous nucleation of species generated in the CO_2 laser-induced plasma; one type, spherical with diameters of <10 nm, was apparently amorphous, with a Cr : Mo : W ratio similar to that of the precursor vapor mixture; the other type were larger and ellipsoidal, with the major axis $\sim 0.5\,\mu m$—these were found to be polycrystalline and rich in Mo but depleted in Cr. This was seen as a demonstration of a potentially powerful technique for generating otherwise difficult-to-make amorphous materials and alloys.

Allen and Trigubo (1983) used a CO_2 laser at $10.6\,\mu m$ to deposit tungsten pyrolytically onto quartz substrates. The film thickness was found to be self limiting due to the large increase in reflectivity as the metal was deposited. Problems were reported with the low-room-temperature vapor pressure of $W(CO)_6$; if the cell was heated to increase the vapor pressure, the high gas-phase spectral absorption of the precursor proved troublesome.

Solanki and coworkers (1981) reported the photolytic deposition of Cr, Mo, and W films from the vapors of the hexacarbonyls using a pulsed copper hollow cathode laser operating in the UV at 260 nm. Decomposition occurred via multiphoton absorption processes. The laser could be focused onto the quartz or silicon substrate to give localized deposition; this was done for Cr and Mo.

Ehrlich and coworkers (1981) reported the use of pulsed excimer laser radiation at 193 and 248 nm as well as the cw 257-nm output from a frequency-doubled argon-ion laser for deposition of refractory metals from a number of metal carbonyl compounds, including chromium and tungsten from the hexacarbonyls, on silicon, silica, and Pyrex glass substrates. Chromium was deposited in spots and lines using the focused 257-nm beam at its room-temperature vapor pressure. Micron-scale resolution was achieved; the width for the tungsten features, for example, was 2.5 microns. While deposition of iron from $Fe(CO)_5$ gave particulate deposits over the entire substrate surface as well as at the illuminated spot, with chromium deposition from $Cr(CO)_6$ this was not the case, presumably due to the lower vapor pressure of the chromium precursor at the temperature of the experiment.

Mayer and coworkers (1982) used a pulsed Nd:YAG-pumped dye laser, frequency-doubled to give UV radiation in the range 280–350 nm. Chromium deposition from the hexacarbonyl was monitored at a number of wavelengths in this range by the rate of mass gain on a quartz crystal microbalance near to the focus of the beam, as well as by monitoring the attenuation of the beam by the deposits on the cell window. These measurements indicated that essentially all the $Cr(CO)_6$ molecules in the laser beam path were dissociated. The decomposition was determined to be homogeneous rather than heterogeneous, since no $Cr(CO)_6$ adsorption on the quartz crystal microbalance could be observed, and there as no indication of an induction period for deposition nor any dependence of deposition rate on substrate temperature. Deposition of chromium on silicon and quartz substrates was achieved with the beam normal to, and focused on, the surface. The high-energy laser pulses used melted the silicon surface and the deposited chromium. Auger electron spectroscopic analysis of the deposited chromium revealed the presence of carbon, at up to half the abundance of the chromium, as well as oxygen. This contamination was attributed to incomplete decarbonylation of the $Cr(CO)_6$.

By contrast, Creighton (1986) studied the photodecomposition of $Mo(CO)_6$ adsorbed on silicon at low temperatures, using temperature-programmed desorption and Auger spectroscopy. $Mo(CO)_6$ was shown to be only physisorbed onto silicon and was removed at 210–230°K. Photolysis experiments with the adsorbed $Mo(CO)_6$ at 150°K using the 248-nm radiation from an excimer laser suggested that one CO is removed from each $Mo(CO)_6$, leaving a product on the surface that contains substantial levels of carbon and oxygen.

Solanki and coworkers (1982) and Boyer and coworkers (1983a, 1983b) deposited chromium over a large area of the substrate using the beam from an excimer laser at various different wavelengths in the UV; the KrF operating wavelength of 248 nm was found to give the fastest deposition in all cases (around 0.2 μm/min). The films contained some oxygen ($<7\%$), but very little carbon (below 1%). The ArF line, in the case of tungsten, gave the best adherence.

Yokoyama and coworkers (1985) studied the deposition of chromium from $Cr(CO)_6$ vapor in an argon stream using an excimer laser operating at 248 nm, monitoring the development of the film by the intensity of the transmitted UV. They concluded that for high laser intensities there was a significant contribution to the deposition process by photothermal effects, as well as photolytic decomposition. The chromium content of these deposited films was only ~5%, however. Patterned deposition was obtained by imaging a patterned mask onto the quartz substrate; resolution was ~10 μm.

Flynn and coworkers (1986) also reported deposition of Cr, Mo, and W from the hexacarbonyls, using an excimer laser operating at 193, 248, or 308 nm. In the case of Cr the findings of Boyer and coworkers were confirmed; the fastest deposition rate was found at 248 nm. Mo was deposited using 193 and 248 nm, both wavelengths apparently giving the same deposition rate of 14 pm/pulse. W was deposited faster at 193 nm (3 pm/pulse) than at 248 nm (1 pm/pulse). In all cases oxygen and carbon incorporation into the films was ascribed to the gas-phase or surface-adsorbed-phase photolysis of CO, or to the incomplete decarbonylation of the precursor. The good adherence of the films was due to the generation by the incident UV radiation of strong binding sites for metal adsorption.

Chiu and coworkers (1985) used a frequency-doubled argon-ion laser to deposit tungsten from $W(CO)_6$. The laser beam was focused to a 5-μm diameter spot. Increasing the laser intensity led to a saturation in the deposition rate, which was claimed to be due to the latter becoming limited by the rate of diffusion of precursor to the reaction zone at the substrate surface.

Aylett (1987) described patterned photolytic deposition of tungsten using the 248-nm output from an excimer laser. Patterning was achieved by imaging a mask onto the InP substrate surface. The tungsten patterns deposited were used as antidiffusion barriers for patterned gold films photolytically deposited in the same experiment. Auger analysis of the gold-on-tungsten structure showed the tungsten layer to contain ~50% tungsten and ~50% oxygen.

The general drift of the work reported above suggests that a combination of UV and heating is necessary to achieve something approaching pure metal deposits. Aylett and Haigh (1985) reported deposition of tungsten using $W(CO)_6$ as the precursor by photolysis using a high-pressure xenon arc lamp as the UV source and a controlled degree of heating. Films were deposited at 320°C on n- and p-InP, n-GaAs, and on epitaxial layers (lattice-matched to InP) of n- and p-(Ga, In)As. Ohmic contacts were obtained for tungsten films deposited onto these materials. Haigh (1986) later gave further details of these tungsten films. Auger analysis showed the films to contain some carbon, but generally very little oxygen. Contact resistance measurements were carried out that gave an upper limit of 10^{-2} Ω-cm^2. Experiments performed with the UV radiation focused above the substrate gave deposition downstream from the illuminated region, indicating that the UV photolysis takes place in the gas phase, followed by diffusion of species to the substrate surface.

Dibenzenechromium and *dibenzenemolybdenum,* $(C_6H_6)_2Cr$ and $(C_6H_6)_2Mo$, are potentially suitable for laser pyrolysis because of their low decomposition temperatures, around 300 and 100°C, respectively. However, they have vapor pressures too low for practical gas-phase deposition. Yokoyama and coworkers (1984) decomposed their solutions in benzene by laser pyrolysis. A cw argon-ion laser operating at 488 nm was focused onto the inside of a glass plate forming one wall of the solution cell. Localized deposits were formed, where the laser beam heated first the solution and then, as deposition ensued, the metal deposit itself. As might be expected from results with other species containing conjugated aromatic ligands, Auger analysis of the deposited spots showed carbon as the dominant contaminant, although generally at levels below 10%. Lines could also be formed by scanning the laser spot across the plane of the substrate. In the case of Cr a 20-μm wide and ~1-μm thick line was drawn at a lateral rate of 50 μm/s.

Jan and Allen (1984) reported the deposition of Mo from a *screen ink* coated onto a glass or silica substrate. An argon-ion laser or a CO_2 laser was used to pyrolyze this precursor film. The quality of the resulting deposit was not specified.

As stated above, tungsten has been widely used for electrical interconnections on silicon, and in this connection the reactions involving Si, hydrogen, and *tungsten hexafluoride,* WF_6, have attracted great attention. The hexafluoride is volatile and relatively easy to handle; its lack of oxygen or carbon content has been another attraction. Both thermal and photolytic enhancements by laser have been discussed.

Berg and Mattox (1973), in a very early report of laser CVD, described the use of a CO_2 laser to provide localized heating to reduce tungsten hexafluoride with H_2:

$$WF_6 + 3H_2 \rightarrow W(s) + 6HF.$$

Fused silica substrates were used, as these were identified as having several desirable characteristics for laser CVD:

1. They absorb the laser wavelength.
2. They have a high thermal shock resistance.
3. They have a high thermal stability.
4. They have a low thermal conductivity, allowing localization of the heated area.

Lines were deposited by rotating the substrate with the beam focused off-axis.

Laser DVC of tungsten films from WF_6 and H_2 was reported by Allen and Trigubo (1983); Allen, Trigubo, and Liu (1984); and Allen et al. (1984), using a focused CO_2 laser to heat the absorbing substrate. Optical self-limiting of the films was observed, with the deposition rate falling off as the reflective metal film was deposited. Resistivities were 3–10 times that for bulk metal, considered good in view of the nonoptimized nature of the process.

McWilliams and coworkers (1983) reported localized deposition of tungsten films from WF_6 and H_2 using an argon-ion laser operating on the 514.5-nm line focused to a small spot on the substrate surface. Herman and coworkers (1984) reported the deposition of tungsten on silicon substrates by the reduction of WF_6 with the silicon substrate surface itself (rather than using H_2 to reduce the WF_6). A focused argon-ion laser operating at 514.5 nm was used to locally heat the silicon substrate surface. The reaction is

$$WF_6 + 3/2\,Si(s) \rightarrow W(s) + 3/2\,SiF_4.$$

Auger analysis indicated that no fluorine was incorporated into the tungsten films. No carbon or oxygen were seen in the bulk of the deposited metal, although some surface contamination was seen, ascribed to handling of the samples. There was no evidence for any formation of WSi_2.

Liu and coworkers (1985) used a focused argon-ion laser to reduce WF_6 at a silicon substrate surface. Although the focused spot was ~20 μm in diameter, the nonlinearity of temperature behavior expected for an activated process allowed linewidths of only a few microns to be produced. Raman spectra of deposited material showed the deposit to be tungsten disilicide (WSi_2), and Rutherford backscattering and Auger spectroscopy indicated a composition of $WSi_{2.3}$.

Jervis and Newkirk (1986) reported deposition of tungsten films by hydrogen reduction of WF_6 using a focused CO_2 laser to induce dielectric breakdown in the gas mixture. Deposited tungsten films were shown by Auger analysis to contain 55% tungsten and 45% oxygen. The oxygen was thought to be incorporated into the film during deposition from contamination in the source gases.

The resistivities reported by Kaplan and d'Heurle (1970) in the reductive CVD of tungsten hexacarbonyl were about twice the value of the bulk metal, with the resistivity increasing with increasing amounts of carbon incorporated in the deposited tungsten films. In photothermal work, Deutsch and Rathman (1984) compared conventional CVD tungsten films with films deposited from WF_6 and H_2 using an excimer laser operating at 193 nm. Gas-phase photolysis only was used, the laser beam running parallel to the surface and being formed into a sheet maintained at a height of ~1 mm above the substrate surface. At deposition temperatures of 440°C, resistivities in the laser-deposited films were as low as twice the resistivity of pure bulk metal; adhesion of the films to the substrate was good only at 440°C or above. Auger analysis of these films showed no fluorine, carbon, or oxygen incorporation. Films deposited at temperatures as low as 240°C showed good conformal coverage of steps in the substrate, but did not survive a self-adhesive-tape pull test. The dependence of film properties on deposition temperature was ascribed to production at lower temperatures of the alternative, beta phase of tungsten, while at higher temperatures the higher conductivity alpha-W was formed. This was confirmed by x-ray diffraction measurements of laser-deposited films. The reaction mechanism for tungsten film deposition is changed by the addition of the 193-nm radiation, since photolysis of WF_6 releases F atoms that can then react with H_2 to give H′ atoms; these then reduce the WF_6 and WF_{6-n} to tungsten.

Foord and Jackman (1986) studied tungsten deposition at low temperatures onto silicon from WF_6 using thermal-desorption spectrometry (TDS), combined with Auger and LEED. No adsorption of WF_6 on the substrate was seen in the range 150–300°K, but at 77°K there was efficient physisorption. The adsorbed WF_6 was desorbed when the temperature was increased to 120°K. Mercury-arc-lamp photolysis of the physisorbed WF_6 at 77°K followed by TDS revealed that the species present after the photolysis desorbed at higher temperatures than the WF_6 had before photolysis. SiF_4, F, "SiF_2, and unreacted WF_6 were desorbed from the substrate after photolysis of WF_6. Auger analysis of the deposited tungsten films as they were heated showed a falling W signal as the temperature increased, presumably due to transport of the tungsten into the bulk of the silicon substrate. Because of formation of WSi_2, the surface concentration did not, however, decay away completely.

Toyama, Itoh, and Morya (1986a, 1986b) developed a novel model to account for their observation that the photolytic deposition rate of tungsten on silicon from WF_6 and H_2 is proportional to the deposition time for times up to 100 minutes. For this deposition a xenon arc lamp provided the UV illumination; deposition took place at ~300°C. The model eventually developed postulated the buildup of decomposition intermediates on the deposited tungsten film surface and includes effects caused by the collision of excited precursor molecules from the gas phase with these intermediates on the surface, leading to further decomposition.

7. Late Transition Metals and Group IB

The chemistry of the photodeposition of the elements of groups VIII, IX, X, and IB demonstrates a gradual transition from the use of the carbonyl complexes so valuable with the elements of group VI to the direction of the alkyl compounds that dominate the work with group IIB. Simple volatile molecular alkyls of the type MR_n are not known for copper and silver, however, and are very unstable in the case of gold; there is only one report of such a compound being used. Most of the gas-phase work that has been reported has involved volatile oxygenated chelate complexes. There has accordingly been a heavy emphasis on developing laser-assisted variants of standard solution deposition processes. However, the gas-phase area may well become the growth area because of a growing recognition of the stability of a small but useful number of volatile gold, and possibly silver, species that are partially alkylated and partially complexed.

Iron pentacarbonyl, ($Fe(CO)_5$, pentacarbonyliron) has been the most widely used percursor for *iron.* George and Beauchamp (1980) generated deposits from $Fe(CO)_5$ on soda-lime and silver substrate surfaces in their study of photoemission effects on deposition. The film was stated to be electrically conducting, but no other characterizing data were presented. Ehrlich and coworkers (1981) briefly touched upon photolytic deposition from $Fe(CO)_5$ in the course of a report on spatially delineated metal deposition with the frequency-doubled argon-ion laser at 257 nm.

Love and coworkers (1984a, 1984b) used the unfocused beam from an excimer laser operating at 193 or 248 nm to photolyze iron pentacarbonyl, both at room temperature and at 77°K, in an ultrahigh vacuum system. With the Al_2O_3 (sapphire) substrate cooled, the precursor had to be photolyzed as fast as it condensed onto the surface to avoid reaction intermediates being frozen into the deposit. Best results with a cooled substrate generated films containing 7% C and 4% O and were obtained with the 193-nm laser line. Problems with deposition on windows limited the film thickness obtainable to about 10 nm by progressively blocking the laser illumination. These films contained 13% C and 8% O. See also Lloyd and Ibbs (1983) for excimer laser work.

Bottka, Walsh, and Dalbey (1983) photolyzed the pentacarbonyl with a Hg lamp in the adsorbed state on a GaAs surface. Adsorption was enhanced by cooling, and the photolyzed film then contained adsorbed carbon monoxide, which desorbed on warming to room temperature.

Foord and Jackman (1986) examined the pyrolytic and UV photolytic decomposition of $Fe(CO)_5$ on the [100] face of silicon in some detail. No adsorption of the pentacarbonyl took place above 135°K. At this temperature the desorption energy was around 33 kJ/mol, a figure identical, within experimental error, with the heat of vaporization. Above 300°K some iron was formed on the surface, and this contained a few atom percent of oxygen and carbon contaminants. At 77°K a multilayer physisorption took place. In a second experiment this adsorbate at 77°K was photolyzed with UV from a low-pressure Hg lamp. Auger analysis showed that as long as photolysis was carried to completion at this temperature, all carbon monoxide was eventually removed. The thermal desorption behavior was monitored using a mass spectrometer and following the $Fe(CO)^+$ signal. As the 77°K photolysis proceeded, the peak became weaker, and shifted to progressively higher temperatures, indicating a gradual increase in the binding energy.

Moving to slightly longer wavelengths, Jackman and coworkers (1986) used a scanned focused beam of the multiline emission of an argon-ion laser to generate a deposit on silica. Higher laser powers were needed in the nucleation stage than for subsequent deposition. A ripple structure of 50-μm period was observed. The period was independent of the scan speed and of the laser power, and was attributed to the self-propagation of the decomposition reaction, which, from the wavelengths used, was likely to be thermal in nature. Iron silicates were formed by chemical reaction at the substrate-layer interface.

Allen and Trigubo (1983) carried out a pyrolytic decomposition with the CO_2 laser using a relatively large (600 μm) spot size and scanning this across the quartz substrate to produce the localized heating effect. $Fe(CO)_5$ has no spectral absorption at this wavelength. When a deposit of about 50 nm had built up, deposition effectively stopped. This effect was attributed to the change in reflectivity from the low value characteristic of quartz to the much higher value of the iron surface; this greatly reduced the power absorbed and the temperature rise. The electrical resistance was high because of oxygen (oxide) incorporation.

Deposition of *nickel* from *nickel tetracarbonyl* $Ni(CO)_4$ was carried out by Krauter, Bauerle, and Fimberger (1983) by pyrolysis with the focused krypton laser beam at 531 nm and other wavelengths. Using a spot of micron dimensions, stripe structures (on amorphous silicon/glass substrates) and rods could be produced. Rods, projecting normal to the substrate, developed from the original circular spots when the position of the spot was held steady. There was an initial nucleation period, which varied in duration with the nature of the surface, followed by steady growth of the feature. By careful estimates of the laser power density and of the temperature distribution in the irradiated region, it was deduced that the rates were thermally controlled and independent of the laser wavelength. In a later paper from this group (Petzoldt and coworkers 1984), growth rates were measured over a wide range of temperature and an activation energy of 22 ± 3 kcal/mole was estimated, in agreement with published values for the pyrolysis of $Ni(CO)_4$.

Allen and Trigubo (1983) and Jervis (1985) used CO_2 laser radiation to pyrolyze the tetracarbonyl in the gas phase. Allen (1979, 1981) related the shapes of spots deposited from $Ni(CO)_4$ to the laser power, the time of exposure, and the temperature distribution. The effects of reflection in inhibiting deposition were also covered (Allen and Bass, 1979, Allen et al. 1986). Jervis' results were discussed in terms of a plasma, induced by the high laser power, but the actual power levels were not stated, unfortunately. Chemical analysis of the film showed the presence of about 10% carbon, and X-ray diffraction linewidths indicated that there were diffracting domains of dimension about 2.5 nm.

Among the group IB elements *gold* serves as a paradigm of the trends and is the most studied element. It has been used for electrical contacts to all types of electronic and optoelectronic devices, because of its inertness, high electrical and thermal conductivity, and the ease which wires can be attached to thin gold layers by compression bonding to form the leads for the device or circuit. Conventionally it has been deposited by vacuum evaporation, screen printed as inks containing gold compounds (which then need to be fired to produce the metal film), or electroplated from gold electrolyte solutions. The latter two methods have been adapted into laser deposition processes. The relative dearth of stable organogold compounds has delayed the introduction of light-initiated processes based on CVD; these have only lately begun to be reported.

Among the first CVD workers were Baum and Jones (1985, 1986), who reported the use of *dimethylgold (III) acetylacetonate* (dimethyl gold 2,4-pentanedionate $Me_2Au(acac)$) for deposition of lines of gold by laser pyrolysis using an argon-ion laser as the localized heat source. The metal had a resistance of from 4 to 22 times that of bulk gold, for scan rates of between 5 and 40 $\mu m\, s^{-1}$. Linewidths were on the order of 30 μm. The gold deposition was compared with that of copper from copper hexafluoroacetylacetonate using the same laser pyrolysis apparatus. The copper gave better resistivities (relative to the bulk metals); this was attributed to its deposition rate being slower, giving fewer crystalline defects or occluded organics in the deposit. The resistivity of deposited gold lines was decreased to two to three times the resistivity of bulk gold by annealing at 300–350°C for 30 minutes. This decrease in resistivity was attributed to an increase in grain size of the gold films (observed in TEM of laser-deposited gold on carbon) rather than a purification of the metal. SAES and XPS showed the gold to be 96–97% pure, with about 1% oxygen and 2% carbon incorporated. Measurements of the deposition rates showed that when argon gas was added to the organogold vapor, the reaction changed from a surface-limited kinetic regime to a diffusion-limited one. This was in agreement with model calculations (Ehrlich and Tsao, 1983; Moylan, Baum, and Jones, 1986) for laser pyrolytic deposition using gas-phase precursors.

Dimethyl gold III acetylacetonate was again used, in photolytic deposition, by Baum and coworkers (1986). Here excimer lasers at 193, 248, and 308 nm were used to project the image of a mask onto the substrate surface with a 4 × projection lens. The beam was taken through the transparent quartz substrate so that as deposition occurred, the UV available for photolysis was attenuated. This self-limiting effect restricted the maximum thickness of the films obtained to 0.3 μm. XPS analysis of the films showed that they contained differing carbon and oxygen levels depending on the wavelength used for photolysis. The highest gold content was found using 248-nm irradiation, at 76%; 193-nm and 308-nm irradiation gave much more lower gold contents at only 23–28%. Up to 73% of carbon and 13% oxygen were measured (in different samples). A low-pressure mercury arc lamp was also used to deposit films with the same precursor. XPS analysis showed these to be made up of 78% gold and 22% carbon. TEM examination revealed similarities to evaporated gold films; this was taken to indicate that the deposition process was similar to that for evaporated films, namely, a nucleation of the surface followed by condensation of gold atoms from the gas phase to give a crystalline deposit.

The high contents of carbon and oxygen in these films present an obvious problem. Aylett and Haigh (1985) reported the deposition of gold by 248-nm excimer laser photolysis of the volatile organometallic compound *trimethylgold(III)-trimethylphosphine*, $(CH_3)_3Au\text{-}P(CH_3)_3$. Auger analysis of a deposited gold film indicates a clean decomposition to give a pure gold deposit. The gold was deposited either as a broad area film or as a patterned film by imaging a mask onto the substrate surface. Photolysis of precursor species adsorbed on the substrate surface rather than in the gas phase (because of the low vapor pressure) means that this process has good potential for high-resolution patterning. A micrograph of a patterned gold deposit in this way is shown in Fig. 8.3. Bunkin and coworkers (1986) reported pyrolytic deposition of micro-sized gold features from solid films of a *triphenylphosphine-gold (I) complex*

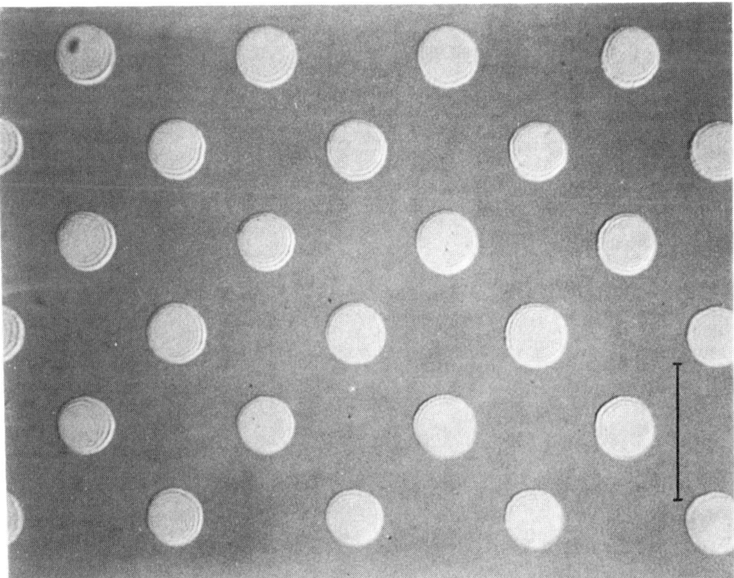

Fig. 8.3 Pattern of 100 μm diameter spots obtained by laser projection deposition of gold. Reproduced from Aylett, M.R.; *Chemtronics* (1987) **1** 146–9, by permission of the publishers, Butterworth and Co. (Publishers) Ltd ©.

$P(C_6H_5)_3AuR$ using a copper vapor laser. They did not state the chemical composition of the group R.

Turning to solution deposition, Fisanick and coworkers (1984) used an argon-ion laser to locally heat a thin film of the proprietary gold screen printing ink *Engelhard Bright Gold NW,* which had been spread, by spinning, on a silica substrate. Metallic regions were thereby produced in the ink film. Scanning the laser spot across the film produced a track of metal with a periodic structure that varied in periodicity with varying scan speed. This structure was attributed to a process analogous to that seen with "explosive recrystallization" of amorphous films. The laser heats a localized area and initiates an exothermic reaction, which then is self-sustaining until the reaction front has expanded to such an extent that heat losses due to conduction are sufficient to lower the temperature below that required for the reaction to take place. The moving laser beam then initiates another reaction front when it reaches more unreacted material. At low scan speeds and low laser powers, the periodic structure is not seen because the higher relative rate of heat dissipation means that the reaction is not fast enough to allow the formation of a self-sustaining reaction front.

Gross and coworkers (1985) provided a comparison beween this screen ink and two other blended materials, *NH_4AuCl_4-nitrocellulose* and *NH_4AuCl_4-poly(vinyl alcohol)*. The thermal decomposition of *gold p-tert-butyl phenyl mercaptide* was also investigated, as this was believed to provide a good model species for the gold component in the screen ink. Differential scanning calorimetry and thermogravimetric analysis showed the gold mercaptide decomposition to be a two-state endothermic process, while poly(vinyl alcohol) decomposed slowly exothermically above the decomposition temperature range of the gold mercaptide, and nitrocellulose decomposed strongly exothermally at a lower temperature. As expected from this, NH_4AuCl_4-nitrocellulose gave rise to "explosive" periodic features under most conditions of scan speed and laser intensity, while NH_4AuCl_4-poly(vinyl alcohol) did not.

Gross, Appelbaum, and Schnoes (1986) provided more details of the composition of lines written by laser pyrolysis of *Engelhard Bright Gold NW* screen ink and developed in dichloromethane. Scanning Auger electron spectroscopy showed that the lines were bounded by regions with high carbon and sulphur content. Further development in chromic/sulphuric acid was found to remove these edges without adversely affecting the gold in the center of the deposit. 257-nm laser irradiation also resulted in photothermal decomposition of the ink, but the resulting deposits were not adherent. This was attributed to the increased optical density of the ink at this UV wavelength, which attenuates the laser beam sufficiently to prevent reaction of the lower part of the film of ink.

Jan and Allen (1984) also reported pyrolysis of another commercial gold screen printing ink, *Engelhard Bright Gold 7621,* using either a CO_2 laser or an argon-ion laser. Silica and soda-lime glass substrates were used. Irradiation of single sites on the screen ink films led to a decrease in the film thickness, greater where the irradiation was more intense. Unirradiated areas of the ink could be removed with a solvent wash, leaving only the irradiated regions, and conventional heat treatment of the samples then resulted in localized metal features. It was suggested that the screen ink pyrolysis proceeds in two stages: an initial consolidation, followed by conversion to metallic gold at higher induced temperatures or longer irradiation times.

Karlicek and coworkers (1983) used *chlorauric acid* ($HAuCl_4$) in methanol or aqueous solution to deposit gold onto InP substrates. An excimer laser-pumped dye laser was used in the wavelength range 580–720 nm, where the solutions do not absorb. Aqueous solutions gave gold deposits with a rough, porous-looking morphology, as did methanolic solutions at higher laser power densities. At lower laser power densities—less than 0.2 J/cm^2—smooth films were obtained. The deposition mechanism was proposed to be pyrolytic or photothermal, that is, the precursor in solution was decomposed thermally by the temperature rise induced in the substrate by the incident laser beam. A chemical interaction with the surface was found. Examination of the films in transmission electron microscopy by Brasen and coworkers (1983) revealed that when thin they consisted of a mixture of mainly Au and some Au_4In. No Au-P species were observed; this is consistent with the fact that the reaction that would be required to form species such as Au_2P_3 from Au + InP is highly endothermic. Gold deposition appears to proceed both by direct decomposition of the gold salt,

$$2HAuCl_4 \rightarrow 2Au(s) + 2HCl + 3Cl_2,$$

and also by reduction of gold ions, with the corresponding oxidation of phosphorus from the InP. The gold deposits were polycrystalline, with a uniform distribution of crystallite sizes between ~5 and ~50 nm.

Kuiken and coworkers (1983) used an argon-ion laser to deposit gold from an aqueous solution of *potassium aurocyanide* $KAu(CN)_2$. Deposition onto nickel-iron substrates gave well-localized spots, but deposition onto highly thermally conductive copper-zinc ("Tombak') substrates gave poor localization of the deposit, due to spreading of the heated zone. This was overcome by adding a thin film of NiP (which has a low thermal conductivity) to the surface of the substrate, giving a well-localized heated zone at the surface, despite the underlying high conductivity material. Localization of the heated zone at the surface then resulted in localization of the deposited gold.

Commercially available gold plating solutions were used by von Gutfeld, Tynan, and coworkers (1979) for laser-enhanced localized deposition of gold films. The argon-ion laser enhanced the rate by heating the substrate, which consisted of a thin film of metal on glass. Micheels and coworkers (1981) used a standard gold/cyanide plating solution to deposit gold photoelectrochemically onto p-Si and both n- and p-GaAs. A tungsten-halogen lamp was used for illumination, patterned deposits being produced by imaging an illuminated mask onto the substrate. Positive deposition patterning (deposit only where the substrate was illuminated) was observed for deposition onto p-type material, whereas negative deposition patterning (deposit only where the substrate was not illuminated) was found to occur for the n-type substrate. The resolution of the process was limited to about 10 μm by nucleation effects, causing deposition to occur outside the illuminated area.

von Gutfeld, Acosta, and Romankiw (1982) developed a theoretical model of this laser-enhanced plating that suggested that the temperature rise induced at the substrate caused a "thermobattery" effect. This was demonstrated experimentally by exchange plating gold onto nickel, copper, and tungten films, the deposition being localized to the laser-irradiated area, while in the surrounding region the substrate metal film was etched. von Gutfeld, Gelchinski, and Romankiw (1983) used various different commercially available plating solutions to deposit gold using a focused argon-ion laser beam to locally heat the Be-Cu substrate. It was found that successful deposition could be achieved using gold-cyanide solutions if no cobalt was added, but addition of cobalt (to give hard gold) led to optical absorption in the plating solution, and the resultant heating of the solution meant that thermal blooming took place, with consequent defocusing of the beam. Good gold deposits were obtained both with low beam powers of 10^4 W/cm^2 or less, and by scanning the beam across the substrate using an oscillator mirror. Both of these factors serve to prevent the region where deposition is taking place from overheating. It was found that with high-density gold solutions deposition rates of up to 1 μm/s could be obtained.

Rose and coworkers (1983) used a gold cyanide solution to photoelectrochemically deposit patterned gold films. The light source was a tungsten-halogen lamp, giving 50 mW/cm^2 illumination at the p-Si substrate surface. Patterning was achieved by imaging a mask onto the surface. The resolution of the process was only moderate, due to inhomogeneous nucleation of the gold. Optimum resolution found for deposition onto p-GaAs was ~10-μm linewidth.

A new technique emerged in 1983 from the work of von Gutfeld, Gelchinskii, and coworkers called *laser-jet plating*. Here the laser beam is directed onto the substrate along a 0.5-mm wide jet of plating solution. The jet serves as a light guide to localize the illumination onto one region of the substrate and also provides a continuously replenished supply of fresh solution to the plating region. Nickel-plated beryllium-copper was used as the substrate material, and plating rates of up to 10 μm/s were reported, using up to 25 W of cw laser power focused into the jet. The deposits were adherent and appeared to consist of pure gold with no pores or inclusions. Repeating the laser-jet deposition experiments without the laser irradiation resulted in poorly adherent deposits that contained voids. Laser-jet plating was used by von Gutfeld (1986) to obtain very high gold deposition rates of up to 20 μm/s, still with good adherence of the deposit to the substrate.

Zahavi, Tamir, and Halliwell (1986) used a pulsed Nd:YAG laser for localized deposition of gold from either an acid plating solution or a cyanide solution. Substrates used were n- and p-type silicon and undoped gallium arsenide wafers. The deposits were analyzed by scanning and transmission electron microscopies. Auger electron spectroscopy, ESCA, Rutherford backscattering, and x-ray diffraction. The results indicated that the deposition was initially photoelectrochemical, followed by diffusion of metal atoms into the surface, probably associated with local melting of the surface.

A similar situation exists with *copper* in regard to solution-phase versus gas-phase deposition as exists with gold. Jones and coworkers in 1985 reported the use of *copper bis*(*hexafluoroacetylaceonate*) (copper bis-1,1,1,5,5,5-hexafluoropentanedionate,

$$Cu(CF_3.CO.CH.CO.CF_3)_2),$$

in the vapor and in solution in ethanol for photolytic deposition of films and localized spots of copper. A low-pressure mercury arc lamp gave no deposit with the vapor, but when ethanol vapor was added, a copper mirror was produced on the quartz window of the cell. X-ray photoelectron spectroscopy and Auger analysis of this film showed copper, carbon, oxygen, and fluorine. Accordingly, a frequency-doubled argon-ion laser operating at 257 nm was used to give a higher UV power density at the substrate. In this case, deposition of copper was observed both with and without ethanol. Spots were deposited on silicon substrates, and analysis showed that these copper films also contained carbon at between

10 and 90%. Traces of fiuorine were also observed, but no oxygen. The films with less carbon contamination were those where ethanol had been present in the photolysis cell. The deposition rate was in the range 0.1–20 nm/min. An excimer laser (operating at 193 or 248 nm) was also used for photolysis of this precursor, again with and without ethanol; films in this case were limited to thicknesses of ~50 nm. Analysis of these films also revealed the presence of copper, with carbon, oxygen, and fluorine contamination. The contamination of all these copper films with carbon, and to a lesser extent oxygen and fluorine, was attributed to the photolysis of the chelating ligand. Different contamination levels with and without ethanol may have been due to different deposition mechanisms, single-photon decomposition leading to less contamination than multiphoton decomposition.

Houle and coworkers in 1985 reported on the pyrolysis of the same complex in the vapor phase, using an argon-ion laser to locally heat the silicon substrates used. Spots and lines of copper were generated, at a volume deposition rate of up to 270 $\mu m^3/s$. X-ray photoelectron spectroscopy and Auger analysis of these showed that, apart from surface contamination with fluorine and carbon, the deposits consisted of copper and oxygen only. The oxygen level was low (much below 10%) and consistent with what might have been caused by oxidation during deposition of the originally pure copper resulting from the rough vacuum used in the deposition cell. Four-point probe resistivities were measured to be as low as twice that of pure bulk copper, and adhesion of the deposits to the substrate was good. An analysis of the kinetic factors influencing deposition rate and deposit morphology for this photothermal copper deposition was given by Moylan and coworkers (1986). This indicated that deposition was taking place in a surface-kinetics-limited regime rather than being limited by diffusion of the precursor in the gas phase.

Braichotte and van den Bergh (1985) reported using an argon-ion laser operating at 514 nm to pyrolytically deposit lines of copper onto glass substrates from the bis(hexafluoroacetylacetonate) complex. The vapor-phase absorption spectrum was measured, and this showed that the absorption of the precursor at this wavelength is low. The deposition therefore was caused by laser heating of the substrate. The deposited copper had electrical resistivities of 20 times the bulk value and was deposited at speeds of up to 500 μm/s. Photolysis of the precursor was attempted using the frequency-doubled 257-nm beam from the argon-ion laser; this gave impure deposits with high bulk resistances.

von Gutfeld and coworkers (1979) reported the use of an argon-ion laser to enhance the plating rate of copper from *cupric sulphate* plating solutions. Micheels and coworkers (1981) used 1:4 diluted *cupric pyrophosphate* and also a solution in acetonitrile obtained by the electrolytic reduction of *cupric perchlorate,* $Cu(ClO_4)_2$ for photoelectrochemical deposition of copper. A tungsten-halogen lamp was used as the light source, and patterning of the deposits formed on the substrate was achieved by forming an image of an illuminated mask on the substrate. The substrates used were p-type Si and GaAs, and also n-type GaAs. On p-Si the resolution of the deposited features was determined to be about 10 μm. The photocurrent efficiency was found to be 100% (within experimental error). Photodeposition of copper onto n-GaAs gave negative images, with deposition only taking place where the substrate was not illuminated. This was attributed to a positive photovoltage at the semiconductor/electrolyte interface, due to the applied bias voltage being insufficient to reverse the band bending.

Rose and coworkers (1983) compared electroless photochemical deposition of copper from an electroplating solution onto p-silicon with zinc and cadmium deposition. Because of the band bending induced at the silicon-electrolyte interface, Zn and Cd deposition was efficient; because the $Cu^{+2/0}$ potential is near the flat-band potential of p-Si, however, little band bending is induced in this case so that electroless photoelectrochemical deposition is not observed. (A He-Ne laser was used for Zn and Cd deposition).

von Gutfeld and Vigliotti (1985) applied the laser-jet plating technique previously demonstrated for gold deposition to the deposition of copper. The laser-jet technique uses a jet of plating solution directed at the substrate, along which the laser beam is focused. The jet serves to act as a waveguide to illuminate only a small spot on the substrate, as well as continually replenishing the supply of plating solution at the point of deposition. For gold deposition, laser illumination enhanced both the deposition rate and the morphology of the deposits; for copper deposition, this was not found to be the case, since the deposition rate remained constant, regardless or whether or not the laser was turned on. The morphology of the copper deposits was significantly enhanced, however, with severe cracks and voids present without the laser but absent with it. The resistivity of laser-jet plated lines approached the resistivity of bulk pure copper. Jan and Allen (1984) used films of a commercially available copper screen ink for deposition by laser pyrolysis of copper. The films were deposited on silica and soda-lime glass substrates, the localized laser heating being provided by either an argon-ion laser or a CO_2 laser. No details were given of the morphology or electrical properties of the copper produced.

Metev (1983) suggested that *cupric formate,* $(HCOO)_2Cu$, as a solid thin film on a substrate surface could be decomposed thermally to give metallic patterns using masked laser-beam irradiation. No experimental results for this deposition method were presented, however. Montgomery and Mantei (1986) reported deposition of copper films by UV laser irradiation of *cupric chloride,* $CuCl_2$, dissolved in acetonitrile. A He-Cd laser was used, operating at 325 nm. The properties of the deposited copper were not reported.

The noble metals *platinum* and *palladium* are growing in importance for semiconductor metallization because of their low propensity to diffuse into the semiconductor. Volatile compounds suitable for CVD exist but have been little explored as yet; work on laser deposition, as with copper and gold, has concentrated on solution techniques. Micheels and coworkers (1981) deposited palladium onto p-Si photoelectrochemically from a standard palladium/cyanide plating solution. A 100-W tungsten-halogen lamp was used to illuminate a mask, which was then projected onto the substrate, immersed in the plating solution. Montgomery and Mantei (1986) deposited palladium from a solution consisting of acetonitrile, methanol, and $Pd_2(NCMe)_6(PF_3)_2$, upon irradiation with a He-Cd laser operating at 325 nm. The mechanism of the reaction leading to the deposition was not determined; it was suggested that the methanol acts as a sacrificial reducing agent for photolytically produced palladium radicals. Details of the film properties were not disclosed. Zahavi and coworkers (1986) used a pulsed Nd:YAG laser for localized deposition of palladium-nickel on to n-type silicon substrates from a commercial palladium-nickel plating solution. Analysis of the deposits showed that palladium and nickel silicides were formed at the surface. The morphology of the surface indicated that local melting had occurred due to the intense laser irradiation.

Karlicek and coworkers (1982) used *chloroplatinic acid* H_2PtCl_6 in 0.1 M aqueous solution for deposition of platinum onto n-type, p-type, and nominally undoped InP. The beam from an excimer laser-pumped dye laser was directed onto the surface of the substrate. Wavelengths used were in the range 580–720 nm, where the solution of H_2PtCl_6 does not absorb. Deposition only occurred in the illuminated region of the substrate. The morphology of the platinum deposits at various stages in their formation was presented, showing island growth initially, these islands growing together to form a continuous film. Since the deposition took place on both n- and p-type material, as well as undoped, the deposition mechanism was determined to be photothermal rather than photoelectrochemical. Brasen and coworkers (1983) examined such platinum deposits by TEM and found that $Pt(P, In)_2$ was formed. No elemental Pt was observed; this was interpreted to indicate that the Pt deposition took place via oxidation of In rather than direct decomposition of the precursor species.

Jan and Allen (1984) reported the use of a commercially available platinum screen ink for laser pyrolytic deposition onto silica or soda-lime glass using an argon-ion or CO_2 laser. The quality of the deposits was not reported.

Mingxin and coworkers (1984) used an argon-ion laser for both pyrolytic LCVD and photolytic deposition of platinum from *platinum bis(hexafluoroacetylacetonate)*, $Pt(CH(CO.CF_3)_2)_2$, onto quartz substrates. Pyrolysis was achieved by using the 514-nm beam, while photolysis followed when this was frequency-doubled to 257 nm. Calibration of the Pt film absorption of the incident 514-nm light allowed these workers to calculate the vapor pressure of the Pt precursor by performing a series of deposition experiments with the cell containing the precursor source maintained at different temperatures. Braichotte and van den Bergh (1986) reported on the electrical properties of Pt/n-GaAs junctions prepared in this way, and in 1987 on the electrical conductivity of 10- to 20-μm wide stripes of Pt on resistive substrates.

Montgomery and Mantei (1986) deposited *silver* from various solutions by a novel process whereby silver ions in solution were reduced by UV laser-generated metal carbonyl radicals. A He-Cd laser operating at 325 nm was used to photolyze the metal–metal bonded carbonyl compound. Silver films were obtained from *silver trifluoromethanesulphonate* (AgOTf, $CF_3SO_3^- Ag^+$) and decacarbonyl dimanganese, $Mn_2(CO)_{10}$, in toluene solution. The mechanism of the process is that the UV radiation breaks the Mn-Mn bond

$$Mn_2(CO)_{10} + h\nu \rightarrow 2Mn(CO)_5'.$$

These radicals can then reduce the silver ions in the AgOTf solution to neutral silver atoms, which then plate out onto the substrate as a metallic film. In this case, the substrate was the window of the deposition cell, made of either quartz or glass. This arrangement meant that the thickness was self-limiting at ~20 nm. SIMS analysis of a deposited silver film showed only silver present. The resistivity of the silver was high, at ~100 times that of bulk pure silver, but the authors attributed this to the extreme thinness of the film.

Acknowledgements

We would like to thank several of our colleagues at British Telecom Research Laboratories for helpful comments. We also thank Ms. Fay

Owston for valuable help with literature searches. Acknowledgement is made to the Director of Research, British Telecom, for facilities granted during the preparation of this review.

References

Allen, S.D. (1979). *Proc. SPIE, Int. Soc. Opt. Eng.* **196,** 49–56.

Allen, S.D., and Bass, M. (1979). *J. Vac. Sci. Technol.* **16,** 431.

Allen, S.D., and Tringubo, A.B. (spelt elsewhere Trigubo). (1983). *J. Appl. Phys.* **54,** 1641–3.

Allen, S.D., Goldstone, J.A., Stone, J.P., and Jan, R.Y. (1986). *J. Appl. Phys.* **59,** 1653–7.

Allen, S.D., Jan, R.Y., Mazuk, S.M., Shin, K.J., and Vernon, S.D. (1984). *Laser-Controlled Chemical Processing of Surfaces,* Mat. Res. Soc. Symp. Proc. **29,** 1–8.

Allen, S.D., Trigubo, A.B., and Liu, Y.-C. (1984). *Opt. Eng.* **23,** 470–4.

Anderson, R.D., and Taylor, H.A. (1952). *J. Phys. Chem.* **56,** 498–502.

Andreatta, R.W., Lubben, D., Eden, J.G., and Greene, J.E. (1982). *J. Vac. Sci. Technol.* **20,** 740–1.

Andreatta, R.W., Abele, C.C., Osmundsen, J.F., Eden, J.G., and Greene, J.E. (1982). *Appl. Phys. Lett.* **40,** 183–5.

Aota, K., Tarui, Y., and Saito, T. (1984). *Amorphous Semiconductor Technologies and Devices* (Hamakawa, Y. ed.), Ohmsha Ltd, Tokyo and North Holland, Amsterdam, 98–107.

Ashida, Y., Kitagawa, N., Isogaya, K., Mishima, Y., Hirose, M., and Osaka, Y. (1984). *Conf. Record on the 17th IEEE Photovoltaic Specialist Conf.*, 1398–9.

Aylett, M.R., and Haigh, J. (1985). *Extended Abstracts of Materials Research Soc. Symposium, Boston, MA.*

Aylett, M.R., and Haigh, J. (1983). *Laser Diagnostics and Photochemical Processing for Semiconductor Devices,* Mat. Res. Soc. Symp. Proc. **17,** 177–82.

Aylett, M.R. (1986). *Chemtronics* **1,** 146–9.

Baggott, J.E., Frey, H.M., Lightfoot, P.D., and Walsh, R. (1986). *Chem. Phys. Lett.* **125,** 22–6.

Baranauskas, V., Mammana, C.I.Z., Klinger, R.E., and Greene, J.E. (1980). *Appl. Phys. Lett.* **36,** 930–2.

Bauerle, D., Irsigler, P., Leyendecker, G., Noll, H., and Wagner, D. (1982). *Appl. Phys. Lett.* **40,** 819–21.

Bauerle, D., Leyendecker, G., Geittner, P., and Lydtin, H. (1982). *Appl. Phys.* **B28,** 267–8.

Bauerle, D., Leyendecker, G., Wagner, D., Bauser, E., and Lu, Y.C. (1983). *Appl. Phys.* **A30,** 147–9.

Baum, T.H., and Jones, C.R. (1985). *Appl. Phys. Lett.* **47,** 538–40.

Baum, T.H., and Jones, C.R. (1986). *J. Vac. Sci. Technol.* **B4**(5), 1187–91.

Baum, T.H., Marinero, E.E., and Jones, C.R. (1986). *Appl. Phys. Lett.* **49**(18), 1213–1215.

Berg, R.S., and Mattox, D.M., *Procs of the 4th Int Conf on CVD, Sandia Labs NM USA,* (Wakefield, G.F., ed.), 196–204.

Bilenchi, R., Musci, M., and Murri, R. (1984). *Proc. SPIE, Int. Soc. Opt. Eng.*, **459,** 61–5.

Bjorkholm, J.E., and Ballman, A.A. (1983). *Appl. Phys. Lett.* **43**(6), 574–6.

Bottka, N., Walsh, P.J., and Dalbey, R.Z. (1983). *J. Appl. Phys.* **54,** 1104–9.

Boyd, I.W., Wilson, J.I.B., and West, J.L. (1981). *Thin Solid Films* **83,** L173–6.

Boyer, P.K., Moore, C.A., Solanki, R., Ritchie, W.K., Roche, G.A., and Collins, G.J. (1983). *Laser Diagnostics and Photochemical Processing for Semiconductor Devices,* Mat. Res. Soc. Symp. Proc. **17,** 119–127.

Boyer, P.K., Moore, C.A., Solanki, R., Ritchie, W.K., Roche, G.A., and Collins, G.J. (1983). *Proc. SPIE, Int. Soc. Opt. Eng.* **385,** 120–6.

Braichotte, D., and van den Bergh, H. (1986). *Proceedings of the European Mat. Res. Soc. Meeting, Strasbourg.*

Braichotte, D., and van den Bergh, H. (1985). In *Integrated Optics,* Springer Series in Optical Sciences Vol. 48, (Nolting, H.-P. and Ulrich, R., eds.). Springer Verlag, Berlin, 38–42.

Braichotte, D., and van den Bergh, H. (1987). *Appl. Phys.* **A44,** 252–9.

Branz, H.M., Fan, S., Flint, J.H., Fiske, B.T., Adler, D., and Haggerty, J.S. (1986). *Appl. Phys. Lett.* **48,** 171–3.

Brasen, D., Karlicek, R.F., and Donnelly, V.M. (1983). *J. Electrochem. Soc.* **130,** 1473–5.

Brueck, S.R.J. and Ehrlich, D.J. (1982). *Phys. Rev. Lett.,* **48,** 1678–81.

Bunkin, F.V., Grandberg, K.I., Luk'yanchuk, B.S., Perevalova, E.G., and Shafeev, G.A. (1986). *Sov. J. Quantum Electronics* **16,** 868–9.

Calloway, A.R., Galantowicz, T.A., and Fenner, W.R. (1983). *J. Vac. Sci. Technol.* **A1,** 534–6.

Chen, C.J., Gilgen, H.H., and Osgood, R.M. (1983). *Inst. Phys. Conf. Ser.* **67,** 149–55.

Chen, C.J., and Osgood, R.M. (1983). *Phys. Rev. Lett.* **50,** 1705–8.

Chen, C.J., and Osgood, R.M. (1984). *J. Chem. Phys.* **81,** 327–34.

Chiu, M.S., Shen, K.P., and Ku, Y.K. (1985). *Appl. Phys.* **B37,** 63–5.

Chiu, M.S., Tseng, Y.G., and Ku, Y.K. (1985). *Opt. Lett.* **10**(3), 113–5.

Christensen, C.P., and Lakin, K.M. (1978). *Appl. Phys. Lett.* **32,** 254–6.

Chu, J.O., Flynn, G.W., Chen, C.J., and Osgood, R.M. (1985). *Chem. Phys. Lett.* **119,** 206–12.

Chuang, T.J. (1985). *J. Vac. Sci. Technol.* **B3**(5), 1408–20.

Chuang, T.J., and Sesselman, W. (1985). *J. Vac. Sci. Technol.* **3,** 1507–12.

Chuang, T.J., and Sesselman, W. (1985). *Surface Science* **162,** 1007–10.

Coombe, R.D., and Wodarczyk, F.J. (1980). *Appl. Phys. Lett.* **37,** 846–8.

Creighton, J.R. (1986). *J. Appl. Phys.* **59**(2), 410–414.

Cross, D., John, P., Milne, D., and Wilson, J.I.B. (1986). *IEE Colloquium on Laser Processing of Materials.* IEE Digest No 1986/129 (London).

Deutsch, T.F. (1979). *J. Chem. Phys.* **70,** 1187–92.

Deutsch, T.F., Ehrlich, D.J., and Osgood, R.M. (1979). *Appl. Phys. Lett.* **35,** 175.

Deutsch, T.F., Ehrlich, D.J., Osgood, R.M., and Liau, Z.L. (1980). *Appl. Phys. Lett.* **36,** 847–9.

Deutsch, T.F., and Rathman, D.D. (1984). *Appl. Phys. Lett.* **45**(6), 623–5.

Draper, C.W. (1980). *Met. Trans.* **11A,** 349.

Draper, C.W. (1980). *J. Phys. Chem.* **84,** 2089–90.

Du, Y.C., Kempfer, U., Piglmayer, K., and Bauerle, D. (1986). *Appl. Phys.* **A39,** 167–71.

Ehrlich, D.J. (1985). *Solid State Technol.* **81-5**.

Ehrlich, D.J., and Osgood, R.M. (1981). *Chem. Phys. Lett.* **79,** 381–8.

Ehrlich, D.J., and Osgood, R.M. (1982). *Thin Solid Films* **90,** 287–94.

Ehrlich, D.J., and Tsao, J.Y. (1983). *J. Vac. Sci. Technol.* **B1**(4), 969–84.

Ehrlich, D.J., Osgood, R.M., and Deutsch, T.F. (1980). *Appl. Phys. Lett.* **36**(8), 698–700.

Ehrlich, D.J., Osgood, R.M., and Deutsch, T.F. (1980). *Appl. Phys. Lett.* **36,** 916–8.

Ehrlich, D.J., Osgood, R.M., and Deutsch, T.F. (1980). *Proc. IEEE* **QE-16,** 1233–43.

Ehrlich, D.J., Osgood, R.M., and Deutsch, T.F. (1981). *Appl. Phys. Lett.* **36,** 399–401.
Ehrlich, D.J., Osgood, R.M., and Deutsch, T.F. (1981). *Appl. Phys. Lett.* **37,** 946–8.
Ehrlich, D.J., Osgood, R.M., and Deutsch, T.F. (1981). *Appl. Phys. Lett.* **38,** 1018–20.
Ehrlich, D.J., Osgood, R.M., and Deutsch, T.F. (1981). *Appl. Phys. Lett.* **39,** 957–9.
Ehrlich, D.J., Osgood, R.M., and Deutsch, T.F. (1981). *J. Electrochem. Soc.* **128**(9), 2039–41.
Ehrlich, D.J., Osgood, R.M., and Deutsch, T.F. (1982). *J. Vac. Sci. Technol.,* **21,** 23–32.
Erwin, J.W., Ring, M.A., and O'Neal, H.E. (1985). *Int. J. Chem. Kinetics* **17,** 1067–83.
Fisanick, G.J., Gross, M.E., Hopkins, J.B., Fennell, M.D., Schnoes, K.J., and Katzir, A. (1984). *J. Appl. Phys.* **57,** 1139–42.
Flynn, D.K., Steinfeld, J.I., and Sethl, D.S. (1986). *J. Appl. Phys.* **59**(11), 3914–7.
Foord, J.S., and Jackman, R.B. (1986). *J. Opt. Soc. Amer.* **B3**(5), 806–810.
Frieser, R.G. (1968). *J. Electrochem. Soc.* **115,** 401–5.
Geohegan, D.B., and Eden, J.G. (1984). *Appl. Phys. Lett.* **45,** 1146–8.
George, P.M., and Beauchamp, J.L. (1980). *Thin Solid Films* **67,** L25–8.
Green, M., and Kuku, T.A. (1983). *J. Phys. Chem. Solids* **44,** 990–1008.
Gross, M.E., Applebaum, A., and Schnoes, K.J. (1986). *J. Appl. Phys.* **60**(2), 529–33.
Gross, M.E., Fisanick, G.J., Gallagher, P.K., Schnoes, K.J., and Fennell, M.D. (1985). *Appl. Phys. Lett.* **47,** 923–5.
Haigh, J. (1986). *Chemtronics* **1,** 134–6.
Hanabusa, M., and Kikuchi, H. (1983). *J. Appl. Phys.* **22,** L712-4.
Hanabusa, M., and Kikuchi, H. (1984). *Laser-Controlled Chemical Processing of Surfaces,* Mat. Res. Soc. Symp. Proc. **29,** 21-7.
Hanabusa, M., Moriyama, S., and Kikuchi, H. (1983). *J. Non-crystalline Solids* **59,** 703–6.
Hanabusa, M., Moriyama, S., and Kikuchi, H. (1983). *Thin Solid Films* **107,** 227–34.
Hanabusa, M., Namiki, A., and Yoshihar, K. (1979). *Appl. Phys. Lett.* **35,** 626–7.
Herman, I.P., McWilliams, B.M., Mitlitsky, F., Chin, H.W., Hyde, R.A., and Wood, L.L. (1984). *Laser-Controlled Chemical Processing of Surfaces,* Mat. Res. Soc. Symp. Proc. **29,** 29–34.
Herman, I.P., Magnotta, F., and Kotecki, D.E. (1986). *J. Vac. Sci. Technol.* **A4,** 659–64.
Higashi, G.S., and Fleming, C.G. (1986). *Appl. Phys. Lett.* **68,** 1051–3.
Higashi, G.S., Rothberg, L.J., and Fleming, C.G. (1985). *Chem. Phys. Lett.* **115,** 167–72; and *Laser Chemical Processing of Semiconductor Devices,* (Houle, F.A., Deutsch, T.F., and Osgood, R.M., ed.).
Houle, F.A., Jones, T., Baum, T., Pico, C., and Kovac, C.A. (1985). *Appl. Phys. Lett.* **46,** 204–6.
Inoue, T., Konagai, M., and Takahashi, K. (1983). *Appl. Phys. Lett.* **43,** 774–6.
Irvine, S.J.C., Giess, J., Mullin, J.B., Blackmore, G.W., and Dosser, O.D. (1985). *J. Vac. Sci. Technol.* **B3,** 1450–5.
Irvine, S.J.C., Giess, J., Mullin, J.B., Blackmore, G.W., and Dosser, O.D. (1985). *Materials Lett.* **3,** 290–3.
Irvine, S.J.C., Mullin, J.B., Robbins, D.J., and Glasper, J.L. (1985). *J. Electrochem. Soc.* **132,** 968–72.
Irvine, S.J.C., Mullin, J.B., and Tunnicliffe, J. (1984). *J. Crystal Growth* **68,** 188–93.
Ishizaka, S., Simpson, J., and Williams, J.O. (1986). *Chemtronics* **1,** 175–7.
Ishizu, A., Miki, H., Onishi, Y., Nishimura, T., and Akasaka, Y. (1984). *Extended Abstracts of the 16th Int. Conf. on Solid State Devices and Materials,* Suppl. **52-3**.
Iwanaga, T., and Hanabusa, M. (1984). *Jpn. J. Appl. Phys.* **23,** 473–5.
Jackman, R.B., Foord, J.S., Adams, A.E., and Lloyd, M.L. (1986). *J. Appl. Phys.* **59,** 2031–4.

Jackson, R.L. (1984). *Proc. SPIE, Int. Soc. Opt. Eng.* **459,** 33–9.

Jan, R.Y., and Allen, S.D. (1984). *Proc. SPIE, Int. Soc. Opt. Eng.* **459,** 71–6.

Jervis, T.R. (1985). *J. Appl. Phys.* **58,** 1400–1.

Jervis, T.R., and Newkirk, L.R. (1986). *J. Mater. Res.* **1**(3), 420–424.

Johnson, W.E., and Schlie, L.A. (1982). *Appl. Phys. Lett.* **40,** 798–801.

Jonah, C., Chandra, P., and Bersohn, R. (1971). *J. Chem. Phys.* **55,** 1903–7.

Jones, C.R., Houle, F.A., Kovac, C.A., and Baum, T.H. (1985). *Appl. Phys. Lett.* **46,** 97–99.

Jones, M.W., Rigby, L.J., and Ryan, D. (1986). *Nature* **5058,** 177.

Kamisako, K., Aota, K., and Tanui, Y. (1984). *Jpn. J. Appl. Phys.* **23,** L776–8.

Kaplan, L.H., and d'Heurle, F.M. (1970). *J. Electrochem. Soc.* **117**(5), 693–700.

Karlicek, R.F., Donnelly, V.M., and Collins, G.J. (1982). *J. Appl. Phys.* **53,** 1084–90.

Kamy, Z., Naaman, R., and Zare, R.N. (1978). *Chem. Phys. Lett.* **59**(1), 33–7.

Kenne, J., Yamada, A., Konagai, M., and Takahashi, K. (1985). *Jpn. J. Appl. Phys.* **24,** 997–1002.

Kowalczyk, S.P., and Miller, D.L. (1985). *J. Vac. Sci. Technol.* **B3,** 1534-8.

Kowalczyk, S.P., and Miller, D.L. (1986). *J. Appl. Physics* **59**(1) 287–9.

Krauter, W., Bauerle, D., and Flmberger, F. (1983). *Appl. Phys.* **A31,** 13–18.

Krchnavek, R.R., Gilgen, H.H., Chen, J.C., Shaw, P.S., Licata, T.J., and Osgood, R.M. (1987). *J. Vac. Sci. Technol.* **B5**(1), 20–26.

Kuiken, H.K., Mikkers, F.E.P., and Wierenga, P.E. (1983). *J. Electrochem. Soc.* **130,** 554–8.

Langsam, Y., and Ronn, A.M. (1981). *Chem. Phys.* **54**(2), 277–290.

Larson, C.A., and Stringfellow, G.B. (1986). *J. Crystal Growth* **75,** 247–54.

Leighton, P.A., and Mortensen, R.A. (1936). *J. Amer. Chem. Soc.* **58,** 448–54.

Letokhov, V.S., Mishin, V.I., Muchnik, M.L., Orlov, Yu.V., and Chernyuk, E.V.A. (1983). *Sov. J. Quantum Electronics* **13,** 1308–9.

Lin, J., Murphy, W.C., and George, T.F. (1984). *Ind. Eng. Prod. Res. Dev.* **23,** 334–344.

Liu, Y.S., Yakymyshyn, C.P., Philipp, H.R., Cole, H.S., and Levinson, L.M. (1985). *J. Vac. Sci. Technol.* **B3**(5), 1441–4.

Lloyd, M.L., and Ibbs, K. (1984). *Laser-Controlled Chemical Processing of Surfaces,* Mat. Res. Soc. Symp. Proc. **29,** 35.

Love, P.J., Loda, R.T., La Roe, P.R., Green, A.K., and Rehn, V. (1984). *Laser-Controlled Chemical Processing of Surfaces,* Mat. Res. Soc. Symp. Proc. **29,** 101–108.

Love, P.J., Loda, R.T., Rosenberg, R.A., Green, A.K., and Rehn, V. (1984). *Proc. SPIE,* Int. Soc. Opt. Eng. **459,** 25–32.

Lubben, D., Barnett, S.A., Suzuki, K., and Greene, J.E. (1984). *Laser-Controlled Chemical Processing of Surfaces,* Mat. Res. Soc. Symp. Proc. **29,** 359–65.

Lubben, D., Barnett, S.A., Suzuki, K., Gorbatkin, S., and Greene, J.E. (1985). *J. Vac. Sci. Technol.* **B3,** 968–74.

Magnotta, F., and Herman, I.P. (1986). *Appl. Phys. Lett.* **48,** 195–7.

Mayer, T.M., Fisanick, G.J., and Eichelberger, T.S. (1982). *J. Appl. Phys.* **53**(12), 8462–9.

McGrath, T. (1983). *Solid State Technology* **26,** 165–9.

McWilliams, B.M., Herman, I.P., Mitlitsky, F., Hyde, R.A., and Wood, L.L. (1983). *Appl. Phys. Lett.* **43,** 946–8.

Meguro, T., Ishihara, Y., Itoh, T., and Tashiro, H. (1986). *Jpn. J. Appl. Phys.* **25,** 524–7.

Menzel, D. (1982). *J. Vac. Sci. Technol.* **20,** 538–43.

Metev, S.M. (1983). *Procs. NATO Adv. Study Inst.,* Pianore Italy, 503–518.

Meunier, M., Flint, J.H., Adler, D., Haggerty, J.S., Atkins, L.V., Petti, C.J., and Wong,

F. (1984). *Laser Chemical Processing of Semiconductor Devices,* (Houle, F.A., Deutsch, T.F. and Osgood, R.M. ed.). Proc. Symp. Boston.

Meunier, M., Gattuso, T.R., Adler, D., and Haggerty, J.S. (1983). *Appl. Phys. Lett.* **43,** 273–5.

Mingxin, Q., Monot, R., and van den Bergh, H. (1984). *Scientia Sinica* **A27,** 531–9.

Michaels, R.H., Darrow, A.D., and Rauth, R.D. (1981). *Appl. Phys. Lett.* **39,** 418–20.

Mishima, Y., Ashida, Y., and Hirose, M. (1983). *J. Non-crystalline Solids* **59,** 707–10.

Mishima, Y., Ashida, Y., Hirose, M., and Osaka, Y. (1983). *Ext. Abs. 15th Conf. on Solid State Devices and Materials, Jpn. Soc. Appl. Phys.* 121–4.

Mishima, Y., Hirose, M., Osaka, Y., and Ashida, Y. (1984). *J. Appl. Phys.* **55,** 1234–6.

Mishima, Y., Hirose, M., Osaka, Y., Nagamine, K., Ashida, Y., Kitagawa, N., and Isogaya, K. (1983). *Jpn. J. Appl. Phys.* **22,** L46–8.

Montgomery, R.K., and Mantei, T.D. (1986). *Appl. Phys. Lett.* **48**(7), 493–5.

Motooka, T., and Greene, J.E. (1986). *J. Appl. Phys.* **59,** 2015–8.

Moylan, C.R., Baum, T.H., and Jones, C.R. (1986). *Appl. Phys.* **A40,** 1–5.

Mutsukura, N., and Machi, Y. (1986). *Appl. Phys. Lett.* **48,** 544–5.

Niki, H., and Mains, G.J. (1964). *J. Phys. Chem.* **68,** 304–9.

Nishida, S., Konagai, M., and Takahashi, K. (1986). *Jpn. J. Appl. Phys.* **25,** 17–21.

Nishida, S., Shiimoto, T., Yamada, A., Karasawa, S., Konagai, M., and Takahashi, K. (1986). *Appl. Phys. Lett.* **49,** 79–81.

Nitta, S. (1985). *Solid State Phys.* **20,** 564–8.

Oikawa, S., Tsuda, M., Yoshida, J., and Tisai, Y. (1986). *J. Chem. Phys.* **85,** 2808.

Okano, H., Horlike, Y., and Sekine, M. (1984). *Oyo Buturi* **53,** 979–84.

Osgood, R.M. (1983). *AIP Top Conf Proc.* (USA) No. 100. Topical Meeting on Excimer Lasers, 256–65.

Osmondsen, J.F., Abete, C.C., and Eden, J.G. (1985). *J. Appl. Phys.* **57,** 2921–30.

Parker, D.L., Lin, F.-Y., and Zhang, D.-K. (1984). *IEEE Trans.* **CHMT-7,** 438–42.

Pattengill, M.D. (1983). *Chem. Phys.* **75,** 59–66.

Pauteau, Y. (1984). *Thin Solid Films* **122,** 243–258.

Perkins, G.G.A., Austin, E.R., and Lampe, F.W. (1979). *J. Amer. Chem. Soc.* **101,** 1109–1115.

Perry, D.L., and Roberts, M.W. (1972). *J.C.S. Chem. Commun.* **147**.

Pollock, T.L., Sandhu, H.S., Jodhau, A., and Strausz, O.P. (1973). *J. Amer. Chem. Soc.* **95,** 1017–24.

Purnell, J.H., and Walsh, R. (1966). *Proc. Roy. Soc.* **A293,** 543–561.

Rigby, L.J. (1969). *Trans. Faraday Soc.* **65,** 2421–9.

Rose, T.L., Longendorfer, D.H., and Rauh, R.D. (1983). *Appl. Phys. Lett.* **42,** 193–5.

Rose, T.L., Micheels, R.H., Longendorfer, D.H., and Rauh, R.D. (1983). *Laser Diagnostics and Photochemical Processing for Semiconductor Devices,* Mat. Res. Soc. Symp. Proc. **17,** 265–272.

Rousseau, Y., and Mains, G.J. (1966). *J. Phys. Chem.* **70,** 3158–62.

Rytz-Froidevaux, Y., and Salathe, R.P. (1984). *Proc. SPIE, Int. Soc. Opt. Eng.* **459,** 55–60.

Rytz-Froidevaux, Y., Salathe, R.P., and Gilgen, H.H. (1985). *Appl. Phys.* **A37,** 121–38.

Rytz-Froidevaux, Y., Salathe, R.P., and Gilgen, H.H. (1981). *Phys. Lett.* **84A,** 216–8.

Rytz-Froidevaux, Y., Salathe, R.P., Gilgen, H.H., and Weber, H.P. (1982). *Appl. Phys.* **A27,** 133–8.

Saito, T.S., Muramatsu, S., Shimada, T., and Migitaka, M. (1983). *Appl. Phys. Lett.* **42,** 678–9.

Saito, T., Shimada, T., Migitaka, M., and Tarui, Y. (1983). *J. Non-crystalline Solids* **59,** 715–8.

Shaapur, F., and Allen, S.D. (1986). *J. Appl. Phys.* **60,** 470–2.
Shin, S.M., Draper, C.W., Mochel, M.E., and Rigsbee, J.M. (1985). *Materials Lett.* **3**(7, 8), 265–9.
Sipe, J.E., Young, J.F., Preston, J.S., and van Driel, H.M. (1983). *Phys. Rev.* **B27,** 1141–54.
Solanki, R., and Collins, G.J. (1983). *Appl. Phys. Lett.* **42,** 662–3.
Solanki, R., Boyer, P.K., and Collins, G.J. (1982). *Appl. Phys. Lett.* **41**(11), 1048–50.
Solanki, R., Boyer, P.K., Mahan, J.E., and Collins, G.J. (1981). *Appl. Phys. Lett.* **38**(7), 572–4.
Solanki, R., Ritchie, W.H., and Collins, G.J. (1983). *Appl. Phys. Lett.* **43,** 454–6.
Stanley, A.E., Johnson, R.A., Turner, J.B., and Roberts, A.H. (1986). *Appl. Spectroscopy* **40,** 374–8.
Taguchi, T., Morikawa, M., Hiratsuka, Y., and Toyoda, K. (1986). *Appl. Phys. Lett.* **49,** 971–3.
Tamir, M., Halavee, U., and Levine, R.D. (1974). *Chem. Phys. Lett.* **25,** 38–42.
Tanaka, K., Hata, N., and Matsuda, A. (1985). In *Amorphous Semiconductor Technologies and Devices,* (Hamakawa, Y. ed.). Ohmsha Ltd and North Holland, Tokyo.
Tanaka, T., Woo, Y.K., Konagai, M., and Takahashi, K. (1984). *Appl. Phys. Lett.* **45,** 865–7.
Tarui, Y., Sorimachi, K., Fujii, K., Aota, K., and Saito, T. (1983). *J. Non-crystalline Solids* **59,** 711–4.
Tonneau, D., Auvert, G., and Pauleau, Y. (1986). *J. Vac. Sci. Technol.* **A4,** 670–2.
Toyama, M., Itoh, H., and Moriya, T. (1986). *Jpn. J. Appl. Phys.* **25**(5), 679–683.
Toyama, M., Itoh, H., and Moriya, T. (1986). *Jpn. J. Appl. Phys.* **25**(5), 688–694.
Tsao, J.Y., and Ehrlich, D.J. (1984). *Appl. Phys. Lett.* **45,** 617–9.
Tsao, J.Y., and Ehrlich, D.J. (1984). *J. Chem. Phys.* **61,** 4620–6.
Tsao, J.Y., and Ehrlich, D.J. (1984). *J. Crystal Growth* **68,** 176–87.
Tsao, J.Y., and Ehrlich, D.J. (1984). *Proc. SPIE,* Int. Soc. Opt. Eng. **459,** 2–8.
Tsao, J.Y., Becker, R.A., Ehrlich, D.J., and Leonberger, F.J. (1983). *Appl. Phys. Lett.* **42,** 559–61.
Tsao, J.Y., Zeiger, H.T., and Ehrlich, D.J. (1985). *Surface Science* **160,** 419–42.
Tsuda, M., Oikawa, S., and Nagayama, K. (1985). *Chem. Phys. Lett.* **118,** 498–502.
van der Brekel, C.H., and Severin, P.J. (1972). *J. Electrochem. Soc.* **119,** 372–6.
Virupaksha Reddy, K. (1986). *J. Opt. Soc. Amer.* **B3,** 801–5.
von Gutfeld, R.J. (1986). *IEEE Circuits and Devices Magazine* **2,** 57–60.
von Gutfeld, R.J., and Vigliotti, D.R. (1985). *Appl. Phys. Lett.* **46,** 1003–5.
von Gutfeld, R.J., Acosta, R.E., and Romankiw, L.T. (1982). *IBM J. Res. Dev.* **26,** 136–44.
von Gutfeld, R.J., Gelchinski, M.H., and Romankiw, L.T. (1983). *Proc. SPIE Int. Soc. Opt. Eng.* **385,** 118–119.
von Gutfeld, R.J., Gelchinski, M.H., Romankiw, L.T., and Vigliottl, D.R. (1983). *Appl. Phys. Lett.* **43**(9), 876–878; see also (1984). *Laser-Controlled Chemical Processing of Surfaces,* Mat. Res. Soc. Symp. Proc. **29,** 325–332.
von Gutfeld, R.J., Tynan, E.E., Melcher, R.L., and Blum, S.E. (1979). *Appl. Phys. Lett.* **35,** 651.
Walsh, R. (1988). In *Proceedings of NATO Advanced Research Workshop on the Design, Activation and Transformation of Organometallics* (R.M. Laine, ed.), Martinus Nijhoff.
Wood, T.H., White, J.C., and Thacker, B.A. (1983). *Appl. Phys. Lett.* **42,** 408–10.
Yokoyama, H., Kishida, S., and Washio, K. (1984). *Appl. Phys. Lett.* **44**(8), 755–7.
Yokoyama, S., Yamakage, Y., and Hirose, M. (1985). *Appl. Phys. Lett.* **47,** 389–91.

CHAPTER 9

Epitaxy

S.J.C. IRVINE
Royal Signals and Radar Establishment
Great Malvern, Worcester,
United Kingdom

1. Introduction

Epitaxial growth can be considered as a special case of thin-film deposition whereby a single crystal layer is deposited onto a single crystal substrate. Thermally activated epitaxial processes are more common and include hydride chemical vapor deposition for Si or the use of metal organics for III–V and II–VI semiconductors in metal organic vapor phase epitaxy (MOVPE), also variously referred to as OMVPE, MOCVD (metal organic chemical vapor deposition), or OMCVD. These have been reviewed elsewhere (Ludowise, 1985). All these vapor-phase epitaxial (VPE) techniques use a hydride or metal organic precursor that contains a constituent atom of the semiconductor to be deposited. These precursors have to be sufficiently volatile, and stable at room temperature, to enable rapid transport in a stream of a suitable carrier gas such as H_2, He, or N_2. The walls of the reactor cell are normally cold, and the substrate is heated to a temperature at which these precursors will pyrolyze, yielding the constituent atoms, e.g., Si, Ga, As, etc., together with volatile reaction products H_2, CH_4, C_2H_6, etc. This completes the transport part of the process, but for epitaxial growth the atoms need to diffuse across the surface to find suitable sites for nucleation of the crystal lattice such as at step edges or kinks. In this way compound semiconductors such as GaAs can be grown epitaxially from trimethyl gallium (Me_3Ga) and arsine (AsH_3) by the generalized reaction

$$(CH_3)_3Ga + AsH_3 \xrightarrow{700°C} GaAs + 3CH_4. \tag{9.1}$$

This type of generalized reaction gives an oversimplistic view of the process and does not consider whether the pyrolsis reactions occur in the vapor or on the surface or whether the cleaving of radicals is by a stepwise process. A surface model for pyrolysis has been proposed by Schleyer and Ring (1976) and a vapor decomposition model by Leys (1981).

The Schleyer and Ring model proposes that the pyrolysis of $(CH_3)_3Ga$ is catalytic on the product surface to yield Ga and CH_4. This occurs by reaction with AsH_3 on the surface via the following reaction steps:

$$(CH_3)_3Ga + AsH_3 \rightarrow (CH_3)_2GaAsH_2 + CH_4 \tag{9.2}$$

$$(CH_3)_2GaAsH_2 \rightarrow H_3CGaAsH + CH_4 \tag{9.3}$$

$$H_3CGaAsH \rightarrow GaAs + CH_4. \tag{9.4}$$

The Leys vapor reaction model is based on observations in a GaAs reaction cell using infrared spectroscopy. A sample tube was inserted into a horizontal reactor above the substrate, and some of the gas flow was then bled off into an infrared absorption cell. Within the boundary layer, all the $(CH_3)_3Ga$ and 60% of the AsH_3 are decomposed. From these results it was concluded that Ga atoms or clusters were formed in the vapor, and As_4 species formed at the substrate. Ga atoms or clusters then diffused through the boundary layer to react with As_4 on the substrate. Photons can be used to interact with the vapor or the surface to aid epitaxy under conditions where otherwise no growth or only polycrystalline growth may occur.

Photoepitaxy encompasses all photolytic and photothermal processes that can be used to modify the above VPE reactions, either by means of decomposition of the precursors or in interaction with the surface to provide sufficient atom mobility necessary for epitaxial growth. This, of course, covers a wide range of processes using either photon sources in the infrared, visible, or UV, with the alternatives of incoherent lamps or lasers. This chapter will explore the range of both photoepitaxial processes and the range of semiconductor materials successfully prepared by photoepitaxy.

1.1. Reasons for Photoepitaxy

Photoinduced reactions can be considered as modifications to VPE processes and are therefore added complications to already complex chemical and kinetic processes. The need for added complexity is reflected in limitations in thermal VPE processes.

The possible benefits of photoepitaxy are as follows:

1. Selective area deposition or modulation of growth
2. Low-temperature epitaxy
3. Compatible with other laser deposition/etching processes for in-situ processing
4. Independent control of substrate temperature and precursor decomposition processes

All these benefits can extend the capability of existing epitaxial processes to include features such as spatial modulation of growth using photons. This can be in the form of selective area deposition or doping. Low-temperature epitaxy may be important in reducing diffusion rates, for example, in II–VI semiconductors a high metal atom interdiffusion rate can occur on the group II sublattice. In-situ processing relates to (1) and (2) in the sense that nonphotoresist processing will reduce the number of steps and lower processing temperatures will enable a wider range of materials to be used for device fabrication. The final point (4) emphasizes that the use of photoinduced reactions will enable greater flexibility in epitaxial growth, which is an important requirement in the growth of highly complex structures such as III–V optoelectronic devices, III–V/Si hybrids, and in multilevel silicon processing.

1.2. Classifications for Photoepitaxial Growth

The general description of how photochemistry can expand the capabilities of VPE, given in the previous section, opens up a large number of possible ways and degrees to which the materials scientist may modify a process with the introduction of photons. For convenience the broad topic will be divided into three main categories, which represent varying degrees to which the reactions are dependent on a photon beam. These are shown diagrammatically in Fig. 9.1.

1. *Photomodified epitaxy,* which can include laser-enhanced doping or laser-induced change in alloy composition. This was proposed by Kukimoto et al. (1986) and Ban et al. (1986a, 1986b) as a way of achieving selective area control for semiconductor device processing.

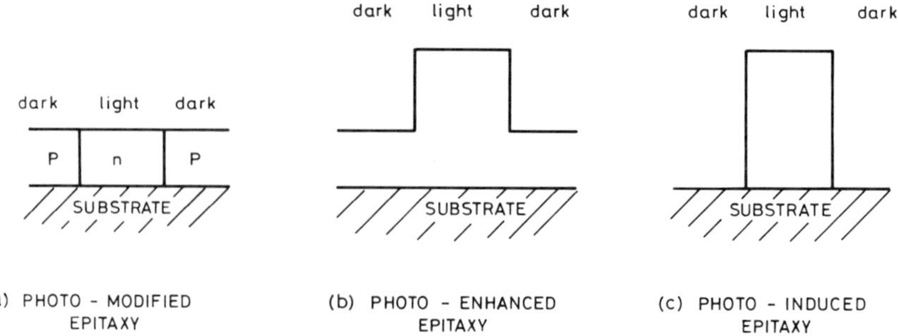

Fig. 9.1 The three main categories of photoepitaxy, identified by the effect of illumination of CVD growth rate and material properties such as conductivity or alloy composition.

2. *Photoenhanced epitaxy.* In photomodified epitaxy, no change in growth rate of the epitaxial material will occur in the illuminated region, whereas photoenhanced epitaxy brings about an enhancement of the growth rate that would have been attributed to purely thermal processes arising from the steady-state substrate temperature. Examples are the enhanced epitaxial growth of CdTe by Kisker et al. (1986) using a low-pressure Hg arc lamp and enhanced low-temperature growth of GaAs by Roth et al. (1983) using a Nd:YAG laser. The former is an example of a photolytic process, and the latter is a photothermal process. Both also appear to induce an improvement in crystalline quality.
3. *Photoinduced epitaxy.* This can be simply defined as epitaxial growth only in the illuminated region with no deposition outside this region. As with photoenhanced epitaxy, this can be as a result of photolytic or photothermal reactions. Examples are low-temperature epitaxial growth of CdTe $\leqslant 250°C$, Kisker et al. (1986), or low-temperature epitaxial growth of HgTe, Irvine et al. (1984a). Low temperatures are essential in this category, as the substrate temperature must be below that at which pyrolysis of the precursor will occur.

A list of reported photoepitaxial growth is shown in Table 9.1, indicating the semiconductor material, radiation used, and classification. The last two columns deal with whether the reactions are photothermal or photolytic; in some cases both play a role. This list includes a report of laser-modified MBE (molecular beam epitaxy) for the photoactivated doping of CdTe (Bicknell et al., 1986). No metal-organic or hydride precursors are used, but the argon-ion laser interaction with the surface activates the dopant, so it is an example of photomodified epitaxy.

Table 9.1 Semiconductor Materials Grown by Photoepitaxy.

Reference	Semi-conductor	Radiation	Photo-modified	Photo-enhanced	Photo-induced	Photo-thermal	Photo-lytic
Eden (1983)	Ge	KrF			√		√
Nishida (1986)	Si	Lp Hg arc			√		√
Hargis (1984)	Si	ArF			√	√	
Ishitani (1987)	Si	ArF, KrF, Hg-Xe		√		√	√
Karam (1986)	Ga(As, P)	Ar^+			√	√	
Balk (1987)	GaAs	Lp Hg arc, ArF, KrF			√		√
Doi (1986)	GaAs	Ar^+		√			√
Kukimoto (1986)	GaAs	ArF	√				√
Kisker (1985)	CdTe	Lp Hg arc		√			√
Donnelly (1985)	GaAs, InP	ArF			√	√	√
Irvine (1984)	HgTe, CMT	Hp Hg arc			√		√
Morris (1986)	CMT	ArF			√		√
Ando (1985)	ZnSe	LP Hg arc			√		√
Ahlgren (1987)	CdTe HgTe	HP Hg arc			√		√
Bicknell (1986)	CdTe	Ar^+	√				√
Roth (1983)	GaAs	Nd-YAG		√		√	

A large number of the epitaxial processes shown in Table 9.1 involve both photothermal and photolytic processes. Laser radiation can interact with a more conventional epitaxial process at some, or all, of these stages. Laser-surface interactions have been described in Chapter 2 and other laser-vapor and laser-solid interactions were described in Chapter 6. The relevance of these processes to epitaxy can be seen in the schematic representation of epitaxial growth shown in Fig. 9.2. Laser interaction can therefore be used to modify any part of this chain of processes by either enhancing the substrate or vapor temperature or by inducing photolytic reactions at any one or more of these stages.

The photoinduced or photoenhanced/photolytic reactions are normally intended to reduce epitaxial growth tempratures, usually by means of precursor decomposition. Examples in Table 9.1 are Ge epitaxy at 150°C

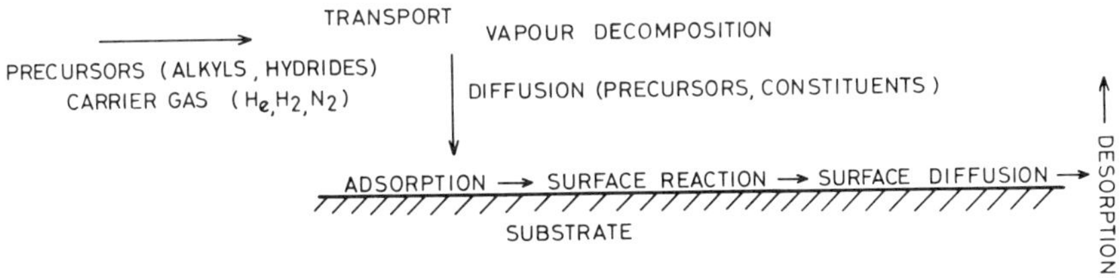

Fig. 9.2 Schematic of process steps for epitaxial growth from vapor precursors.

(Eden et al., 1983), Si epitaxy at 200°C (Nishida et al., 1986), GaAs epitaxy at 400°C (Balk et al., 1987), and HgTe epitaxy at 250°C (Irvine et al., 1984), etc. Alternatively, the laser radiation can be used as a local area or transient heating source where the baseline substrate temperature may be as low as, for example, GaAs epitaxy reported by Karam et al. (1986), where the base-line temperature was 350°C and the transient laser-induced temperature was ~600°C. More recent work by these authors has also indicated significant growth-rate enhancements for high scan rates giving estimated temperature increases of ~50°C.

Hence, in this example decomposition of the precursors at the surface is controlled entirely by the incident laser radiation, and epitaxial deposition will therefore only occur within these regions. In all these examples the deposition conditions must be suitable for epitaxial growth, which must include a suitable clean, undamaged substrate surface, sufficient surface temperature for a) surface mobility of atoms to locate on the correct sites and b) desorption of organic reaction products in order to avoid site blocking or carbon incorporation. The latter is particularly important in GaAs photoepitaxy, where high carbon levels have been detected for layers grown at ~400°C (Beneking, 1985). (Normal MOCVD growth temperatures are 700°C.) Photoassisted desorption of reaction products could be an important key to low-temperature photoinduced epitaxy, but little is known as present about these processes.

1.3. Reaction Cells for Photoepitaxy

Photoepitaxy has occurred largely as a development of more conventional CVD processes, and this is reflected in the types of reaction cells used. An MOCVD reaction cell can be mounted horizontally or vertically, and either configuration can be used for photoepitaxy with some modification. The main modification is to allow laser or lamp radiation to pass through the wall of the reactor cell and impinge on the surface, or run parallel to the surface if purely vapor excitation is required. The two basic designs of reactor cell with their modified forms are shown in Fig. 9.3. Substrate heating is normally by means of RF inductively coupled heating of a graphite susceptor, but radiant heating of the susceptor can also be used and is shown in Fig. 9.3c.

The choice of window materials depend on the wavelength of the radiation. Silica is suitable for visible wavelengths and in the UV down to the 184.9-nm mercury line if Suprasil grade is used. For the near infrared, sapphire can be used up to 6 μm; for longer wavelengths, materials such as germanium, zinc sulphide, and cadmium telluride are suitable. If the radiation is to be focused for laser writing or a mask pattern is to be imaged on the surface, then the window has to be polished flat to an optical quality of finish.

Because photoinduced reactions are likely to occur on the reactor wall or in the vapor, some form of protection of the optical window is required to avoid reaction products settling on the surface. This protection normally takes the form of an additional high-purity purge gas directed over the window region. Purges for vertical and horizontal configurations are shown in Figs 9.3c and 9.3d. Variations on these alternatives have been used, but these two examples demonstrate the principles involved. For example, in Fig. 9.3c, the purge gas enters parallel to the window, causing a downward flow towards the reactant entrance port. The ideal flow conditions at the top of the reactor cell is laminar, as this represents the minimum back diffusion rate for the reactant gases. High flows may induce vortex flow, which could induce the flow of reactants to the window. On the other hand, if the purge flow is too low, then back diffusion of the reactants will increase. Lower down

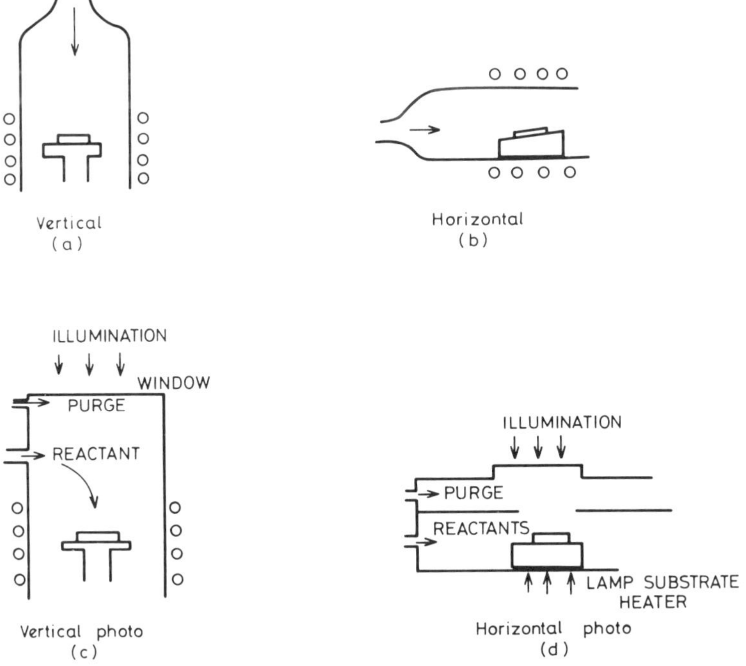

Fig. 9.3 Four different types of reactor cell configurations commonly used. (a) and (b) are used for pyrolytic growth; (c) and (d) are used for photolytic growth.

the reactor cell, the purge and reactant gases must mix homogeneously in order to avoid nonuniform deposition rates.

Ando et al. (1985) used a variation on this design with the reactants diethyl zinc and dimethyl selenium introduced via injection tubes. The high velocity of reactant gas at the injection nozzles would have aided mixing with the purge gas.

The horizontal cell purge shown in Fig. 9.3(c) relies on separating the purge and reactant flows, with the only contact occurring in the illumination zone. As in the case of the vertical reactor, vortex flow must be avoided and flow sheer between purge and reactants minimized. This reactor design was used by Kukimoto et al. (1986) for photomodified growth of GaAs and AlGaAs. The radiation source was a 193-nm ArF excimer laser, and substrate heating was by means of an RF coil. In this example the reactor was operated at reduced pressure (50–200 torr). Radiant substrate heating with a horizontal reactor was used by Irvine et al. (1985) for low-temperature epitaxial growth of HgTe (200–300°C). The work of Kukimoto required higher substrate temperatures (550–800°C), which is usually achieved by RF heating.

More complex reactor designs can be anticipated for parallel and vertical substrate illuminations or in vacuum conditions, where it is combined with in-situ diagnostics such as mass spectrometry. Laser excitation can also be combined with plasma-assisted CVD, such as in the work of Hargis et al. (1984), which requires a multi-chamber cell (see Fig. 9.5).

1.4. Vapor Phase Versus Surface Reactions

The reaction kinetics of vapor and surface photochemical reactions have been described in Chapter 6 and are applicable to all deposition processes, including epitaxy. However, in epitaxial growth these kinetic factors are not only important for determining growth rate, but also in determining the quality of epitaxial growth. For vapor decomposition of precursors, the diffusion of constituent atoms to the surface becomes analogous to vapor transport based on sublimation of source materials. The kinetics of these growth processes for compound semiconductors are well understood and have been described in detail elsewhere (see Faktor et al., 1986).

For high-temperature growth, where the reevaporation flux from the layer is significant compared with the incident flux, these processes would be expected to result in epitaxial growth on a suitable substrate. Difficulties can arise, however, when vapor diffusion of the constituents is used at low temperatures, where most of the photoepitaxial processes occur. Either very low growth rates are observed, for example, in the case of CdTe (Irvine et al., 1987), or compound formation will occur in the vapor as a result of high supersaturation. The latter may lead to polycrystalline growth as a result of secondary nucleation on vapor-nucleated "dust" or due to overgrowth on a cluster of atoms that have had insufficient time to reorder on the correct lattice sites.

In most cases, selective photoexcitation of precursors on the surface is not practical, due to the broad UV absorption bands of metal organics and the relatively small shift in wavelength attributed to adsorption on a substrate (see Chapter 2).

However, atomic UV absorption can be very narrow and a small shift of ~1 nm due to adsorption can be very much greater than the absorption half width. This has been exploited in the case of HgTe photoepitaxy. Hg was introduced as an atomic vapor, from a liquid Hg source, and the precursor was Et_2Te. A relatively high Hg partial pressure of $\sim 5 \times 10^{-2}$ atm was used at a substrate temperature of 250°C to a) maintain equilibrium partial pressure of Hg over the epitaxial layer and b) achieve a high surface coverage of Hg, shown schematically in Fig. 9.4. The

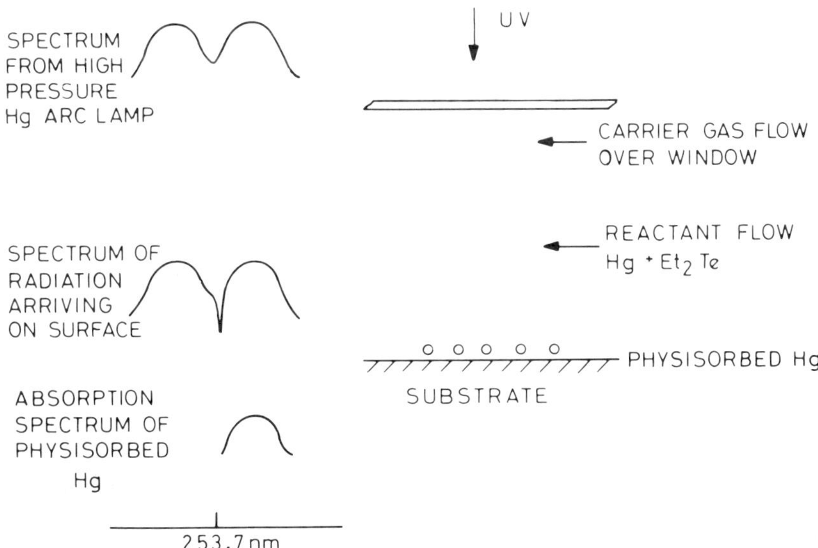

Fig. 9.4 Schematic representation of the photosensitization process for HgTe epitaxy, showing selective UV absorption in an adsorbed layer of mercury. The actual shift in the absorption spectrum of physisorbed mercury is shown to longer wavelengths, but shorter wavelength shifts are also possible.

radiation source was a high-pressure Hg arc lamp that had a broad UV emission spectrum down to 240 nm. Due to the resonant $6S_0 \rightarrow 6^3P_1$ transition at 253.7 nm, the high pressure of Hg will act as a narrow-band filter. However, a wavelength shift in this transition for adsorbed Hg atoms will absorb radiation transmitted through the vapor. The energy of the excited Hg atoms is 4.87 eV, which is greater than the energy required to cleave both $C_2H_5^{\circ}$ radicals in Et_2Te. By this photosensitization mechanism, Et_2Te will only decompose on the surface, avoiding the vapor-phase reaction of Hg and Te. This is an example of photo-stimulated epitaxial growth, as no thermal decomposition of Et_2Te will occur at substrate temperatures <400°C. Typical vapor sublimation temperatures are considerably higher, around 500–600°C. An interesting feature of the HgTe compound is that surface mobility of Te and Hg atoms is high and therefore at the relatively low temperatures of 250–300°C, good crystal growth can be achieved.

An alternative approach to stimulating epitaxial growth has been used by Donnelly et al. (1985, 1986) for epitaxial growth of InP where vapor decomposition is induced by a laser. An excimer laser (193 nm), illuminating the substrate, decomposed all the precursor molecules in the volume of the beam for the duration of the laser pulse. The precursors used were the adduct $(CH_3)_3InP(CH_3)_3$ with additional $P(CH_3)_3$, diluted with H_2. At low laser fluences (<0.03 J/cm^2) the deposited layer was not epitaxial and the estimated substrate temperature was below 300°C. With increased laser fluence, the absorbed photon energy increased the transient substrate temperature, with corresponding improvements in crystalline quality. Good-quality epitaxy was observed for fluences around 0.1 J/cm^2, where the transient substrate temperature was around the melting point for InP of 1070°C. The high transient temperature reordered atoms on the surface and overcame poor nucleation that would otherwise have arisen from vapor-nucleated InP. Excimer-laser-induced epitaxy can be viewed as a two-stage process where the first pulse decomposes the precursors and the second pulse heats the substrate and provides the energy for surface diffusion of In and P atoms that had impinged on the surface. Further discussion of this technique is given in the section on III–V photoepitaxy.

Two other examples of vapor and surface laser excitation will be mentioned here to show the scope for laser processes. The first example is illumination of the vapor parallel to the substrate so that no UV falls on the substrate—all the laser interaction is in the vapor. This approach was used by Morris (1986), who grew epitaxial $Cd_xHg_{1-x}Te$ at 150°C using Me_2Cd, Me_2Hg, and Me_2Te illuminated with a 193-nm excimer laser. Within the pulse duration, complete decomposition of the alkyls occurred to yield Cd, Hg, and Te atoms that diffused to the surface. The absence of laser irradiation of the surface meant that the substrate temperature of 150°C has to be sufficient to induce surface diffusion. As the surface is not illuminated, this technique would not be suitable for surface patterning of an epitaxial layer.

An example where there was no vapor phase reaction induced by the laser beam, but a purely surface interaction, was the laser-induced doping of CdTe with In using MBE (Bicknell et al., 1985). Cd, Te_2, and In are transported as molecular beams and do not interact in the vapor due to long mean free path lengths. A 514-nm argon-ion laser was used to illuminate the surface and induce a change in the incorporation of In dopant, making it electrically active. Without illumination, no change in the conductivity of CdTe was observed, but photoluminescence showed that In was present in the film, occupying a distorted lattice site. The mechanism for this surface change was not clear, but the laser intensity of $\sim$100 mW/cm^2 was insufficient to cause any significant substrate heating. This is an example of laser-modified growth where the growth rate was not increased by laser illumination, but the incorporation of a dopant was modified, thus making the CdTe conducting.

2. Photoepitaxy of Ge and Si

The first reported photoepitaxial growth was that of Eden et al. (1983), who used an excimer laser (KrF) to decompose GeH_4 onto sodium chloride or quartz substrates. The substrate temperature was maintained below the pyrolytic threshold of 280°C, and emission spectroscopy indicated that the decomposition process was photolytic in the gas phase. Ge and GeH species were detected, but film growth at temperatures as low as 150°C was attributed to the diffusion of liberated Ge atoms to the substrate. This was supported by the very low activation of 2 kcal/mole. The films on quartz substrates were polycrystalline, but those deposited onto (100) NaCl substrates were epitaxial with only the 400 reflections seen in x-ray diffraction.

Silicon CVD is normally carried out at substrate temperatures in excess of 1000°C using silane (SiH_4) or $SiCl_4$ in hydrogen. However, there are penalties in using such high processing temperatures where dopant diffusion may significantly degrade a junction. Temperatures below 800°C are required for abrupt junction devices, but for three-dimensional structures, temperatures below 450°C may be needed, (see Srinivasan and Meyerson, 1987). For normal CVD conditions, reduction in growth temperature from 1150°C to 1000°C causes a dramatic reduction in MOS device yields, with a corresponding increase in anodic defects. However, this deterioration in epitaxial quality is not a fundamental feature of Si epitaxy, such as a limitation to surface diffusion of Si atoms, but has been attributed to the quality of the Si substrate surface. Defects caused by poor nucleation onto a substrate can propagate through the epitaxial layer. Defects will nucleate on surface contamination, such as oxygen or carbon, which are difficult to remove thermally below 1000°C. For example, above 1000°C the very stable surface oxide SiO_2 will react with subsurface Si to form the suboxide SiO. This suboxide is volatile at temperatures above 800°C and will leave a clean Si surface provided the oxygen pressure is below 10^{-5} torr. In most CVD systems, the background oxygen and water pressure is sufficient to prevent oxide removal below 1000°C.

Robbins and Young (1987) have shown that a sufficiently clean CVD system can be realized for good epitaxy at 800°C if the reactor is baked under UHV conditions before growth. The limitation on epitaxy at these temperatures is not one of surface diffusion or SiH_4 pyrolysis but substrate surface quality.

Further reductions in epitaxial growth temperature, below 800°C, requires an additional energy source such as an RF plasma or laser. As with the higher temperature growth, the substrate precleaning phase is more critical. Donahue and Reif (1986), extended Si epitaxy to ~650°C using plasma-assisted CVD but required a minimum bias of 300 V on the substrate during the plasma precleaning.

A combined plasma/laser-enhanced CVD system has been used by Hargis and Gee (1984) (see also Gee and Hargis 1984; Anderson and Hargis, 1986), for Si deposition. The growth reactor chamber is shown schematically in Fig. 9.5. A DC discharge was sustained between the anode and cathode in 30 mtorr of SiH_4. Side arms near the cathode contained a) a window to admit laser radiation and b) mounting for the substrate, which was kept at room temperature. The substrate was therefore kept well away from the plasma discharge, and the processes of plasma and laser excitation were well separated. The plasma discharge decomposed SiH_4 and a KrF laser stimulated the surface for either deposition or etching, depending on laser power. At low laser fluences (<0.45 J/cm^2/pulse) the film on a (100) substrate was polycrystalline. A clue to the role of the surface irradiation can be found in the change in crystallinity with temperature. At 0.15 J/cm^2, cross-sectional TEM shows a 250-Å thick amorphous layer between the substrate and polycrystalline overgrowth. At 0.25 J/cm^2 the amorphous layer is not present and the deposited film consists of columnar grains of Si. Hargis and Gee proposed that initial deposition from the glow discharge produced an amorphous Si:H film that would be highly absorbing in the UV. The

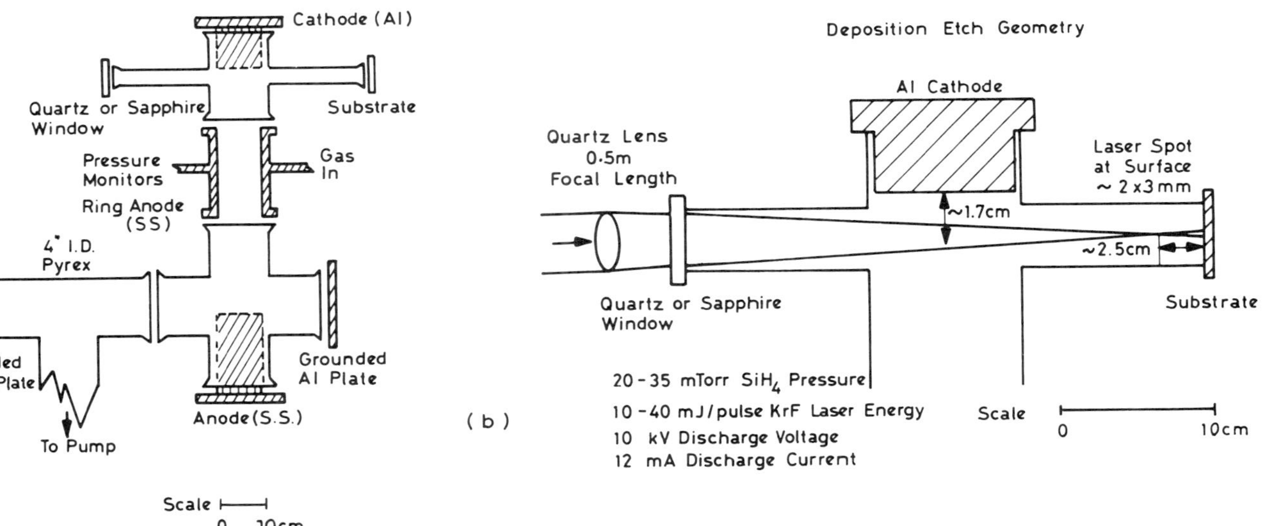

Fig. 9.5 Schematic diagram of apparatus for plasma/laser-assisted CVD of Si. (a) shows the complete apparatus with method for generating the glow discharge, and (b) shows the deposition zone and optical arrangement (Hargis et al., 1984).

resultant temperature rise could bring about recrystallization of the Si and evolution of hydrogen. The thermal effect is consistent with a critical fluence for epitaxial growth. All the pulse energies above 0.13 J/cm^2 were sufficient to take the Si:H layer to its melting point; the higher fluences would allow more time for recrystallization of the Si film. Further evidence for thermal processes can be found in the differences in etching effects on different substrates. For fluences above 0.2 J/cm^2 on quartz substrates, a net etching effect was observed. On silicon substrates a decrease in deposition rate was observed, but no etching; and on aluminum substrates, a constant growth rate above 0.2 J/cm^2 was observed. The greatest temperature rise for a given fluence would be expected for quartz, which has a lower thermal conductivity, and, conversely, the lowest temperature rise expected would be for aluminum, having a higher thermal conductivity. This is consistent with the above etching, deposition rates.

Increasing laser influence above 0.2 J/cm^2 had a dramatic effect on the morphology of the deposited film. A columnar structure at lower laser fluences coalesced at higher fluences to give a smooth film for fluences above 0.40 J/cm^2. From measurements of Raman linewidths it was concluded that films deposited at these high fluences onto (100) silicon substrates were epitaxial. The transition from columnar growth to single crystal growth was attributed to the surface-cleaning effects of high transient temperatures. This process was essentially photothermal where the absorption of the laser energy in the amorphous Si:H layer provided sufficient thermal energy to

1. Ablate the oxide
2. Desorb hydrogen from the surface
3. Enable sufficient surface atom mobility for single crystal growth

Without laser irradiation of the substrate no crystalline silicon deposition will take place so, as shown in Table 9.1, this is an example of photoinduced epitaxy.

By contrast, Shaji Nishida et al. (1986) used a photolytic process to bring about low-temperature Si epitaxy at 200°C, a temperature considerably lower than thermal or photothermal CVD. A mixture of Si_2H_6, Si_2F_2, and H_2 was used in a low-pressure (~2 torr) CVD growth reactor.

Dissociation of the disilane and fluorosilane was by means of a UV photosensitization process in which a low pressure of Hg vapor is used to absorb radiation from a low-pressure Hg arc lamp (30 mW/cm^2). The Hg atoms are excited to the $6^3\,P_1$ state, which can decay back to the ground state by transfer of energy to Si_2H_6 or SiH_2F_2 via collisions, resulting in the rupture of Si-H and Si-F bonds. The low power density from the lamp means that no significant thermal enhancement would be expected. For a substrate temperature of 200°C, the deposited silicon film was epitaxial when SiH_2F_2 flows exceeded a critical value (20 times the Si_2H_6 flow). The explanation given by Nishida was that the etching effect of the enhanced fluorine radical concentration would have been sufficient to remove the native oxide from the silicon substrate. Again, an essential part of the process is the exposure of a clean silicon surface before epitaxial growth can proceed. It is remarkable that with no additional thermal energy, silicon epitaxial growth could occur at such a low temperature compared with normal CVD temperatures.

A detailed study of photoenhanced silicon epitaxy, using UV radiation, was made by Ishitani et al. (1987). Photothermal enhancement was achieved using ArF or KrF lasers and photolytic enhancement was achieved using a Hg-Xe arc lamp. The effects of UV radiation during the crucial prebake period, as well as during growth, were investigated. The prebaking conditions were in an atmosphere of hydrogen that would normally require temperatures at or in excess of 1000°C in order to reduce the SiO_2. Using a 1-minute exposure of the surface to an ArF excimer laser at 950°C was sufficient to leave a clean surface. Similar results were observed using a Hg-Xe arc lamp, but the longer wavelength KrF was less successful. This was considered to be evidence for a photoreduction of the SiO_2 to SiO via photogenerated holes using the shorter wavelengths of ArF and Hg-Xe.

Similar results were obtained with UV photoenhanced growth of Si using SiH_2Cl_2 as the precursor. Improvements in epitaxial quality were observed for comparable temperatures between 800 and 900°C for ArF or Hg-Xe radiation. With the pulse lengths of 10 nsec and repetition rates ~100 Hz for the KrF and ArF lasers, it was estimated that the observed enhancement in growth rate could not, based on transport arguments, possibly be due to transient thermal effects enhancing growth rate. However, the time-averaged temperature increase was considered to be significant and would enhance growth rates.

The observed growth rate enhancement from 1.8 μm/min to 2.8 μm/min at 900°C with Hg-Xe radiation was attributed to a photonic dissociation of $SiCl_2$. This was the dominant species in the reactor cell arising from the thermal decomposition of SiH_2Cl_2.

3. Photoepitaxy of III–V Compounds

III–V compounds, such as GaAs and InP, are technologically important for a range of optoelectronic and microwave devices. Lasers and detectors for 0.9-μm and 1.3-μm wavelengths are used for optic fibre communications. The shorter wavelength devices are made from heterostructures of GaAs and GaAlAs. The longer wavelength devices are composed of the quaternary alloy GaInAsP. Substrates are either GaAs or InP single crystal wafers, and normally the combination of substrate and alloy composition is chosen in order to achieve lattice matching between the substrate and epitiaxial layer. By using quaternary compositions, both band gap and lattice parameter can be tailored. Multilayer structures can be grown by liquid-phase epitaxy (LPE), molecular-beam epitaxy (MBE), or MOVPE. Photoepitaxy of III–V semiconductors is mainly as a modification to MOVPE in order to achieve either lower growth temperatures (normal pyrolytic range 600–700°C) or for selective area epitaxy. At present discrete solid-state lasers comprised of double heterostructures or quantum wells are encapsulated in an optoelectronic device. Future requirements are for integrated optoelectronic devices, which may include buried laser structures and wave guides all on one chip. These structures will require selective area etching and deposition, which could be realized with a laser-based technology with either writing or projection imaging.

One such technique that could fulfill these requirements is the excimer-laser-induced deposition of InP reported by Donnelly et al. (1985). The process was briefly described in Section 1.4. An excimer laser illuminates the substrate, and, by maintaining the substrate temperature below the pyrolysis temperature ~400°C, an epitaxial film can be grown in only the illuminated regions. The process does, however, involve a thermal step of raising the substrate temperature with the laser pulse to allow recrystallization. MOCVD of III–V compounds normally requires temperatures that are considerably above pyrolysis thresholds in order to allow sufficient surface mobility for single-crystal epitaxial growth rather than polycrystalline growth. As will be seen in the next section, this is not the case with II–VI epitaxy, where higher surface diffusion rates will permit lower temperatures for single-crystal growth.

The thermal effects induced by pulsed-laser heating were well illustrated in Donnelly's experiments, the results of which are reproduced in Fig. 9.6. The quality of single crystal growth was determined by the Rutherford backscattering (RBS) parameter θ_{ch}/θ_{ran} which is the ratio of the backscattered counts in a channel direction compared with a random crystalline direction. The temperature of the illuminated surface increases with laser fluence, and the calculated curve is shown in Fig. 9.6. Over the range of laser fluences shown, complete decomposition of the vapor precursors Me_3In and Me_3P would be expected, and this represents the photochemical part of the process.

In the absence of photochemical dissociation of the precursors, pulsed

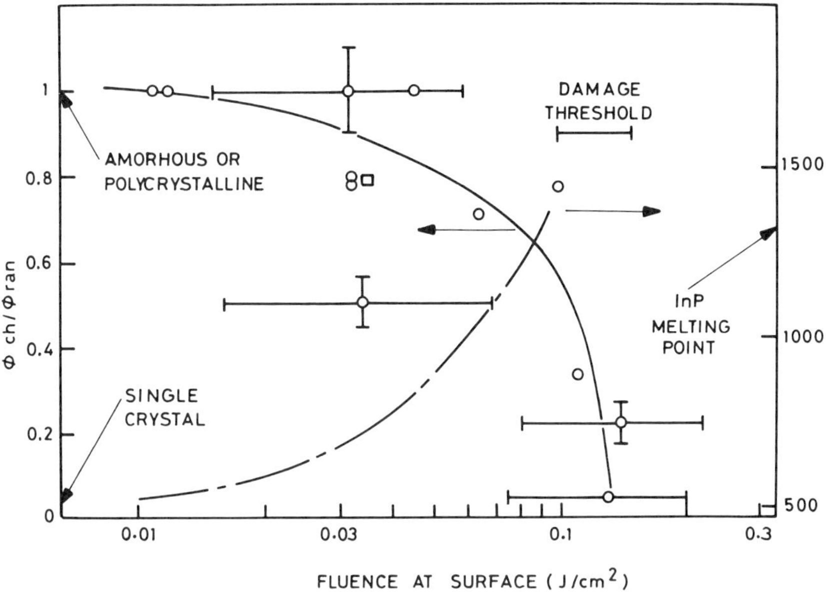

Fig. 9.6 θ_{ch}/θ_{ran} versus laser fluence incident on the substrate for InP excimer laser deposition. The calculated peak substrate temperature is also shown. After Donnelly et al., 1986.

laser irradiation of the surface at low baseline temperatures will enhance both the decomposition rate and surface mobility. Roth et al. (1983) used a Nd:YAG laser (530 nm) to irradiate the surface of GaAs for low-temperature MOCVD using Me_3Ga and AsH_3 precursors. The thermal stability of AsH_3 is such that without laser irradiation the growth rate decreases with temperatures below 520°C, with no measurable growth at 450°C. With 200-mJ/pulse and 10-Hz repetition rate, the growth rate was enhanced for baseline temperatures below 500°C, as shown in Fig. 9.7.

The growth rate enhancement in this example is photothermal, in which the pyrolysis efficiency of Me_3Ga, and in particular AsH_3, is increased as a result of absorption of laser radiation by the surface. However, there is an underlying difficulty in modeling this process that is similar to the difficulty encountered by Ishitani et al. (1987) for pulsed laser-enhanced growth of Si. The heat pulses are of sufficiently short duration that enough precursor molecules could not arrive at the substrate while it is illuminated to cause a significant enhancement of growth rate. In the example of Roth et al. (1983), the calculated heat

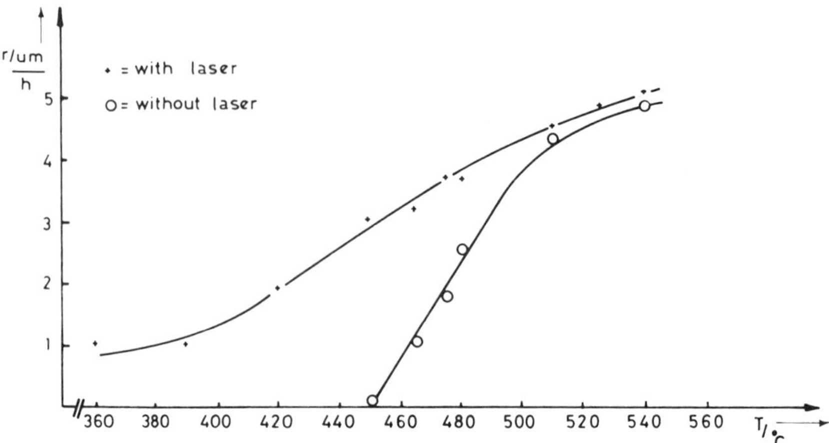

Fig. 9.7 GaAs growth rates versus temperature with and without laser irradiation. After Roth et al., 1983.

cycle would be 1 μs and the total time for enhanced thermal growth effects would be only 10 μs/s or 1 in 10^5. The explanation given for enhancement lies in the adsorption of precursor molecules during the "off" periods, which are then pyrolyzed in the "on" periods. The implication of this argument is that the adsorption is much higher than would be expected for a physisorbed Langmuir film of the precursors at these temperatures. The mechanism does provide a controlled surface reaction process, which is an important element in good photoepitaxial growth.

Surface morphology and crystalline quality are also strongly influenced by laser irradiation. At 450°C without illumination, the morphology was very poor, with whisker growth indicative of arsenic depletion. Increasing the laser pulse energy density to 120 mJ/cm^2 resulted in a smooth, single crystal layer. The band of energies suitable for epitaxial growth was found to be fairly small, as 70 mJ/cm^2 resulted in polycrystalline growth and 150 mJ/cm^2 damaged the wafer.

Another approach to transient heating of the substrate that could be used to pattern the surface with an epitaxial structure is the technique of scanning the surface with the 514-nm line from an argon-ion laser. This technique was reported by Karam et al. (1986) and Bedair et al. (1986a, 1986b). The reactor cell is shown schematically in Fig. 9.8 with a vertical configuration but with laser radiation entering via the quartz window in the side. Translation of the beam across the substrate was achieved by

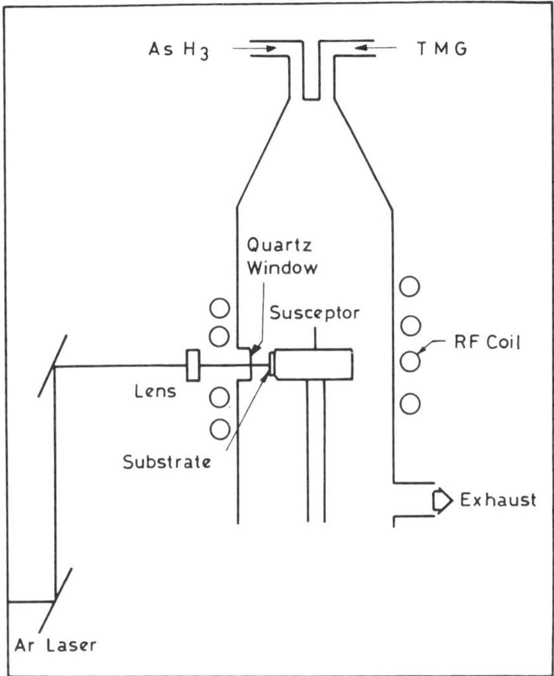

Fig. 9.8 Schematic diagram of laser CVD deposition cell for GaAs and GaAsP epitaxy. After Karam et al., 1986.

scanning the mirrors. Optimum scan rates were fairly high, in the range 100–200 μm/s. Power densities were in the range $1.5 \times 10^3 - 3.3 \times 10^4$ W/cm^2, and bias temperatures 375–500°C. At higher power densities rough surfaces were observed, although the films were thicker, of the order of several microns. The most satisfactory approach for thick films was found to be a multipass at the more intermediate power densities. The disadvantage here is in the time taken for processing. Although the above scan rates are high, processing times would be multiplied by the number of passes. An advantage of this approach is the much higher transport rates to a small deposition region, arising from three-dimensional transport, compared with the slower one-dimensional transport in more conventional systems. This was confirmed by the growth rates, which were found to be kinetically limited. Good crystalline quality of the epitaxial films, grown under optimum conditions, was confirmed by cross-sectional TEM.

The alloy GaAsP was grown by the laser scanning technique to look at effects of local changes in surface temperature. The composition of this alloy is particularly sensitive to temperature in conventional MOCVD. The phosphorus content changed with position across a photodeposited GaAsP line, with high phosphorus content in the center and low phosphorus content at the edges. This was interpreted in terms of the higher thermal stability of PH_3 compared with AsH_3. Therefore, the phosphorus content will follow the temperature profile, with more phosphorus in the hotter central region. The laser scan process is essentially one of raising the substrate temperature from a baseline temperature (below pyrolysis threshold) to a temperature at which efficient pyrolysis will occur and is classified in Table 9.1 as a photothermal, photoinduced process.

A GaAsP-GaAs superlattice structure has been grown by laser scanning, as reported by Karam et al. (1987). This approach uses laser modulation to bring about an alloy composition change, rather than a change in gas composition. It takes advantage of the composition dependence of GaAsP on substrate temperature discussed above. The baseline temperature was set at 500°C so that only GaAs would grow over the whole substrate. The laser was then scanned to locally grow the GaAsP alloy by locally enhancing the substrate temperature up to a maximum of 800°C. A sequence of scans produced a sequence of GaAs/GaAsP epitaxial layers, with a period of ~400 Å, as shown schematically in Fig. 9.9. This type of structure should have a very abrupt interfaces and may be suitable for buried structures in integrated electro-optic applications.

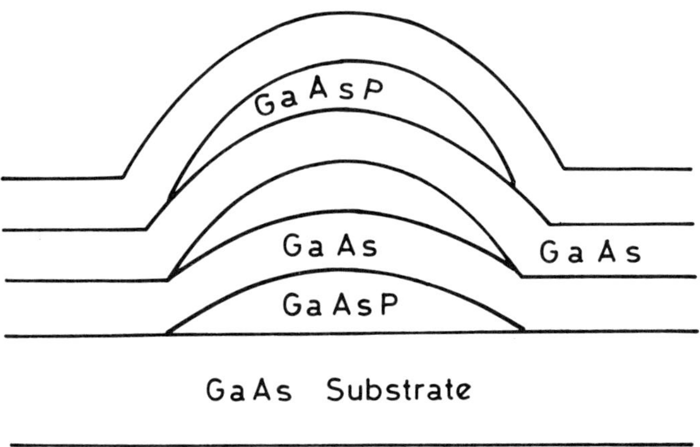

Fig. 9.9 Cross-sectional representation through a laser-stimulated GaAsP-GaAs superlattice. After Karam et al., 1986.

Laser-enhanced MOVPE can be taken a step further to chop a laser beam so that a surface-absorbed layer can be exposed to radiation resulting in decomposition. Doi et al. (1986a, 1986b, 1987) used this modulation technique to induce atomic layer epitaxy (ALE) of GaAs. The concept of ALE is to grow alternate layers of Ga and As resulting in a very-high-quality crystal lattice. In the laser-ALE technique the precursors of Me_3Ga and AsH_3 were introduced alternatively into the growth chamber. Significant growth enhancement from 0.33 μm to 1.07 μm at 500°C was observed when the laser was on during the Me_3Ga flow cycle. The alternative schemes are shown in Fig. 9.10 for illumination during Me_3Ga or AsH_3 cycles, but no enhancement was seen for the latter. Karam et al. (1988) have also shown similar laser-assisted ALE of GaAs but with a lower bias temperature of 400°C.

Doi observed that the growth enhancements were greater when n-type GaAs substrates were used than for p-type or semi-insulating. The mechanism for enhancement was described as a photoassisted catalytic effect, where Me_3Ga pyrolysis was enhanced on an arsine surface under 514-nm argon-ion laser radiation. A photoinduced ALE growth was observed around 400°C where no growth was observed without laser

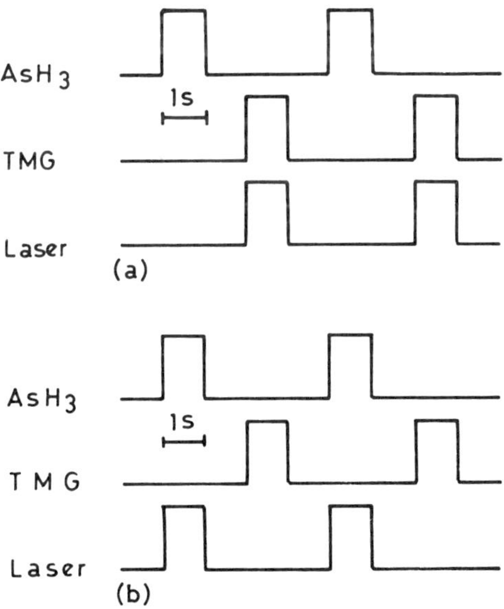

Fig. 9.10 Schematic growth procedure for switched laser MOVPE. The flows of Me_3Ga and AsH_3 are introduced alternatively into the growth chamber. Argon-ion laser irradiation can occur concurrently with the introduction of either (a) Me_3Ga or (b) AsH_3.

radiation, and the growth rate of 0.3 nm/cycle corresponded with the correct rate anticipated for a true ALE process (i.e., atomic layer growth). For pyrolytic ALE, calculated growth rates indicated that temperatures around 500°C would be needed for one monolayer/cycle growth rates.

Photolytic, photoinduced epitaxy of GaAs has been demonstrated by Balk et al. (1987). A reduced-pressure (0.1 bar) horizontal reactor cell was used for low-temperature UV-stimulated deposition. The alternative radiation sources used were a low-pressure Hg arc lamp, and excimer laser wavelengths of 193 nm or 351 nm. The low-temperature kinetically limited growth regimes for Me_3Ga, Et_3Ga, and iBu_3Ga precursors were investigated for the influence of UV stimulation. The kinetically limited growth regimes were different for the three alternative precursors, corresponding with temperatures below 580°C for Me_3Ga, below 430°C for Et_3Ga, and below 380°C for iBu_3Ga. These variations reflect different thermal stabilities for these precursors and indicate that the reductions in growth rates at lower temperatures are due to a slowing of the thermal decomposition rate on the substrate.

Small enhancements in growth rates were observed with the UV lamp on. It was noted that the UV illumination improved surface morphology, with a reduction of growth hillocks and whiskers. This would be consistent with improved surface kinetics arising from UV illumination.

Using the 193-nm excimer radiation (173 mJ/cm^2), enhancements of a factor of two were observed when using Me_3Ga, but no enhancement for 351 nm was observed. The UV absorption is at its maximum at 193 nm in Me_3Ga but is very weak at 351 nm. Therefore, the main mechanism for enhanced growth was UV photodissociation and not thermal effects.

For the laser-induced epitaxial layers, the enhanced deposition was confined to the area of illumination and would therefore be suitable for patterning. Reactions close to or on the substrate surface were considered important for epitaxial growth and in order to avoid deposition in the unilluminated regions. To avoid thermal deposition, the growth temperature would have to be at or below 400°C for Me_3Ga and even lower for the less stable precursors.

Laser modulation of AlGaAs alloys used the same experimental setup but with best results at a lower substrate temperature of 630°C. An example of the modification to alloy composition used a vapor composition ratio $[Me_3Al]/([Me_3Al] + [Me_3Ga])$ of 0.7. In the laser irradiation region the composition in the alloy ($Al_xGa_{1-x}As$) was $x = 0.60$, and in the unirradiated regions it was $x = 0.56$. The aluminium content increased at higher temperatures, and the composition change on irradiation was reduced. The increased Al content was explained by laser enhancement of Me_3Al decomposition at the lower substrate temperatures and thermal enhancement at the higher temperatures.

These laser-modified epitaxial processes could be used for the fabrication of buried optical waveguide structures and possibly buried laser structures.

A different approach to selective area control of III–V semiconductor layers is photomodified epitaxy, which will bring about a change in the doping concentration or alloy composition by illumination. The principle of this approach, proposed by Kukimoto et al. (1986), is that the substrate is held at a temperature where efficient pyrolysis of the precursors will occur and the growth is vapor-transport limited.

Under these conditions no overall change in growth rate is expected, but some of the difficulties encountered in going to lower temperatures—such as poor surface morphologies and carbon incorporation—are avoided.

Kukimoto et al. (1986) used a horizontal reactor cell with a suprasil window and hydrogen window flush. The radiation source was a 193-nm excimer laser with 50-mJ/cm^2 pulse energy at a 50-Hz repetition rate. Laser-enhanced doping of GaAs, grown at 700°C used the dopant precursor tetramethylsilane, which has a strong absorption band around 200 nm. In the irradiated areas, the carrier concentration increased from 10^{16} to 10^{17} cm^{-3}. The background carrier concentration in the unirradiated, undoped GaAs was 2×10^{15} cm^{-3}.

4. Photoepitaxy of II–VI Compounds

4.1. General Considerations

The application of photoepitaxy to the II–VI's has gained as much attention as the more widely used semiconductor materials described in Sections 2 and 3. The reasons for this interest lie in the difficulties in preparing epitaxial layers of these materials by more conventional processes. The incentives for low-temperature growth are very strong where, for example, the interdiffusion in the infrared detector material $Cd_xHg_{1-x}Te$ can be in excess of 100 μm at 600°C for a normal growth period and would have to be grown at 400°C or below for <1 μm interdiffusion. Other effects of high diffusion rates in II–VI compounds are a) dopant or impurity diffusion, b) complex formation with associated deep levels, c) dislocation loops, d) precipitates, and e) subgrain boundaries.

The wide band-gap II–VI compounds such as ZnSe are important for electroluminescent displays and, if the material quality could be adequately controlled, would be efficient light-emitting diodes (LED). Difficulties encountered with doping and with deep-light-emitting centres can both be related to growth temperature. The objectives for photoepitaxy of these materials is to grow high-quality epitaxial layers at the lowest possible temperatures. Other advantages of photoepitaxy, such as selective area deposition and integration with other photoprocessing technologies, are as attractive to II–VI optoelectronic devices as for processing of silicon or III–V devices. A detailed review of II–VI photoepitaxy has been given by Irvine (1988).

A limitation to reducing growth temperatures using the MOCVD technique has been the stability of the alkyl precursors. Table 9.2 lists some of the group II and group VI precursors with the II–VI compound pyrolytic growth temperatures. However, the minimum growth temperatures for the compounds can be lower than the separate pyrolysis temperatures, due to vapor phase reactions. For example, ZnSe can form in the vapor at room temperature from a mixture of Me_2Zn and H_2Se. Both these precursors are stable before mixing and would require temperatures in excess of 300°C for efficient pyrolysis. Another example is the growth of CdTe from the alkyls Me_2Cd and Et_2Te, which can proceed efficiently at 350°C, whereas temperatures in excess of 400°C are required for the efficient pyrolysis of just Et_2Te (see Mullin et al., 1981).

Table 9.2 Examples of Growth Temperatures for MOVPE-Grown II–VI Compounds.

Compound	Precursors	Substrate
ZnO	Et_2Zn-O_2	325–400°C
ZnO	Me_2Zn-C_4H_8O	350–450°C
ZnS	Me_2Zn-H_2S	350–750°C
ZnSe	Me_2Zn-H_2Se	250–350°C
CdS	Me_2Cd-Me_2S	400–500°C
CdSe	Me_2Cd-Me_2Se	375°C
CdTe	Me_2Cd-Et_2Te	350–500°C
CdTe	Me_2Cd-Me_2Te_2	250–500°C
CdTe	Me_2Cd-$(t\mathrm{Bu})_2Te$	220–260°C
HgTe	Hg(v)-Et_2Te	410°C
HgTe	Hg(v)-$(\mathrm{iPr})_2Te$	350–410°C
HgTe	Hg(v)-$(t\mathrm{Bu})_2Te$	230–330°C
HgTe	Me_2Hg-Et_2Te	220–270°C

Vapor-phase reactions can bring about their own difficulties in handling the vapor mixing and in reactor design. For ZnSe epitaxy this led to the use of an injection tube arrangement (see Stutius, 1982) to avoid ZnSe dust forming in the entrance zone. The objective with pyrolytic, as with photolytic, epitaxy is to bring about controlled reactions on the substrate. With photoepitaxy, lower temperatures can be achieved without going to less stable precursors that create their own handling problems, but thermally stable precursors can be decomposed in the desired reaction zone.

Ando et al. (1985) used the alkyl precursors Et_2Zn and Me_2Se in order to avoid the vapor phase reaction. Pyrolytic growth would only occur for substrate temperatures of 400°C or above, which confirms the absence of unwanted gas-phase reactions and is consistent with the thermal stability of these alkyls. Photodissociation was achieved with a low-pressure Hg arc lamp with emission at 253.7 nm and 184.9 nm. A vertical reactor cell was used, operated at low pressures (~1 torr). A total UV intensity of 40 mW/cm^2 was sufficient to enhance growth rates at 400°C by a factor of two to 0.2 $\mu m/h$ and to induce epitaxy of ZnSe at temperatures down to 200°C (200°C below pyrolysis temperatures).

The substrate material used for these experiments was (100) GaAs. The lattice parameters of GaAs and ZnSe are similar, 5.6532 and 5.6676 Å, respectively, which is suitable for high-quality epitaxy. However, Ga is a donor dopant in ZnSe, and therefore the growth temperature must be sufficiently low to avoid Ga diffusion into the ZnSe layer.

A more widely studied II–VI compound for photoepitaxy is CdTe. The interest in this material is in its application for infrared detectors. Although CdTe has a moderately large band gap of 1.6 eV, when alloyed with the semimetal HgTe, a range of band gaps down to zero can be achieved by choice of alloy composition x in $Cd_xHg_{1-x}Te$. The lattice parameter of CdTe is 6.4829 Å, which closely matches the HgTe lattice parameter of 6.4605 Å. CdTe can therefore be used as a substrate material over the whole compositional range, being chemically compatible and satisfying the condition of a small lattice mismatch.

However, non-lattice-matched substrates have been sought as alternatives in order to circumvent the poor crystalline quality of currently available CdTe substrates. Alternative substrates such as sapphire and GaAs have been successfully used where an epitaxial layer of CdTe is grown as a "buffer" between the non-lattice-matched substrate and the "active" $Cd_xHg_{1-x}Te$ epitaxial layer. This type of multilayer epitaxial structure is shown schematically in Fig. 9.11. The object of the buffer layer is to confine the misfit dislocation structure and potential dopant

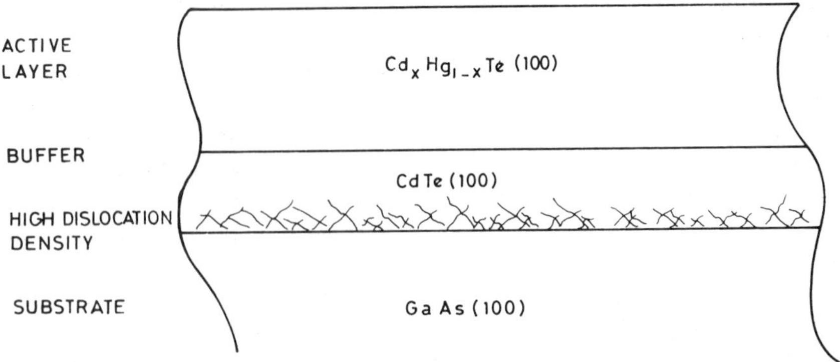

Fig. 9.11 Cross-sectional representation of an ideal buffer structure for the growth of $Cd_xHg_{1-x}Te$ onto GaAs substrates. The misfit dislocation structure and residual strain should be confined to the region of the buffer/substrate interface.

impurities such as Ga, As, and Al to the region of the buffer/substrate interface.

CdTe has been shown to fulfill these requirements, provided that the buffer layer is grown at sufficiently low temperatures (see, for example, Giess et al., 1987). A further advantage of this approach is that the substrate, e.g., GaAs could be an active part of the device with, for example, readout circuitry to address a two-dimensional array of infrared detectors. These highly complex applications would require some selective area epitaxy to make optimum use of the device.

Examples of photoepitaxial CdTe are those reported by Kisker et al. (1985) and Ahlgren et al. (1986). In both examples Hg arc lamps were used as the UV sources, but with the difference of the former using a low-pressure lamp. The low-pressure lamp differed in having a lower UV intensity ($\sim 10\,mW/cm^2$), but also a narrower distribution of intensity. Kisker et al. (1985) estimated from CdTe growth rates that photoenhancement was greater than expected from purely photodissociation, giving an apparent photochemical quantum efficiency of 3. This was interpreted by Mullin et al. (1981) as a radical-induced chain reaction initiated by photodissociation and is shown in the following set of reaction steps:

$$h\nu + (CH_3)_2Cd \rightarrow CH_3^\circ + CH_3Cd^\circ \rightarrow 2CH_3^\circ + Cd \tag{9.5}$$

$$CH_3^\circ + H_2 \rightarrow CH_4 + H^\circ. \tag{9.6}$$

Due to the reactive nature of monatomic hydrogen radicals, one possible quenching reaction is

$$(CH_3)_2Cd + H^\circ \rightarrow Cd + CH_4 + CH_3^\circ. \tag{9.7}$$

Hence one photon can release more than one Cd atom and similarly for Te with reactions between the Te precursor and hydrogen radicals.

In the case of a high-pressure Hg arc lamp, more power is available and at shorter wavelengths than the 253.7-nm resonant line, where the alkyl precursors absorb more strongly. The increased primary photodissociation in the case of Irvine et al.'s experiments, using a high-pressure lamp, showed the necessity for using a helium carrier gas instead of hydrogen in order to reduce hydrogen quenching reactions for the alkyl radicals and thus permit longer free-radical lifetimes. Either approach has been used to successfully grow epitaxial layers onto CdTe and GaAs substrates. The major constraint in the choice of growth conditions is in avoiding nucleation of CdTe in the vapor that would fall to the surface as a CdTe dust. An example of the polycrystalline structures that can nucleate on dust particles is shown in Fig. 9.12. The parameters of precursor concentration, UV intensity, and substrate temperature can be chosen to avoid dust nucleation and to allow CdTe nucleation on the substrate at a rate determined by the surface kinetic processes.

The conditions for which homogeneous vapor nucleation or heterogeneous surface nucleation would occur were calculated by Irvine et al. (1988), based on a model for critical supersaturation conditions. The

Fig. 9.12 SEM micrograph of a cross section through a CdTe polycrystalline layer. The crystallites, randomly oriented with respect to the substrate, had nucleated onto CdTe "dust" particles. After Irvine et al., 1987.

conditions of vapor concentration and UV absorption from a broad-band source can be complex and difficult to determine, but each of these processes can be related to vapor-transport-limited growth rate. The latter can readily be measured, and for each temperature a maximum heterogeneous growth rate could be related to different combinations of vapor concentrations, UV intensity, etc. Assuming that the vapor concentration close to the surface is very small compared with the free stream concentrations, the transport-limited growth rate is as follows.

$$R = \frac{DC\eta I_0(\lambda)(1 - \exp(-\sigma(\lambda)p/RT))\,d\lambda}{\delta} \tag{9.8}$$

where D is the diffusion coefficient of the atomic species (Cd or Te), C is the inlet alkyl concentration, $\sigma(\lambda)$ is the alkyl cross section at wavelength λ, p is the alkyl partial pressure, η is the photochemical quantum efficiency, $I_0(\lambda)$ is the UV intensity at λ, T is the vapor temperature, and δ the boundary-layer thickness. The maximum growth rate for heterogeneous reaction can be related to a critical free energy for homogeneous nucleation $\Delta\mu_{crit}$. If this free energy is exceeded then homogeneous nucleation will occur, but lower free energies will permit heterogeneous nucleation. The critical growth rates versus vapor temperature for two values of $\Delta\mu_{crit}$ are shown in Fig. 9.13. The simple transport-limited growth rates, assuming no homogeneous reaction for a range of conditions, are also plotted. Experimental results identified as being single crystal or polycrystalline are shown, and a general agreement with the model can be seen.

To confirm the crystalline quality of CdTe photoepitaxy grown under heterogeneous nucleation conditions, double-crystal x-ray diffraction rocking curve widths were measures for the 004 reflection. A layer grown at 250°C had a rocking curve width of 42 arc secs on a CdTe substrate of 25 arc secs. The layer rocking curve width compares well with other epitaxial growth processes for CdTe at comparable temperatures.

Epitaxial growth onto GaAs substrates does not occur as a natural consequence of growth onto CdTe but often requires a more specific range of growth parameters. Kisker et al. (1985) demonstrated epitaxial growth of CdTe onto (100) GaAs down to 250°C, provided that the ratio of $Me_2Cd:Et_2Te$ precursor concentrations was Cd rich in a ratio of 1.5:1. A 1:1 ratio at 250°C resulted in polycrystalline growth.

A Cd-rich stoichiometry ratio has been confirmed by Haq et al. (1988) for epitaxial growth below 350°C. These authors were able to demonstrate that in order for the growth to be epitaxial it was necessary to attenuate UV power to the point where dust nucleation was avoided and

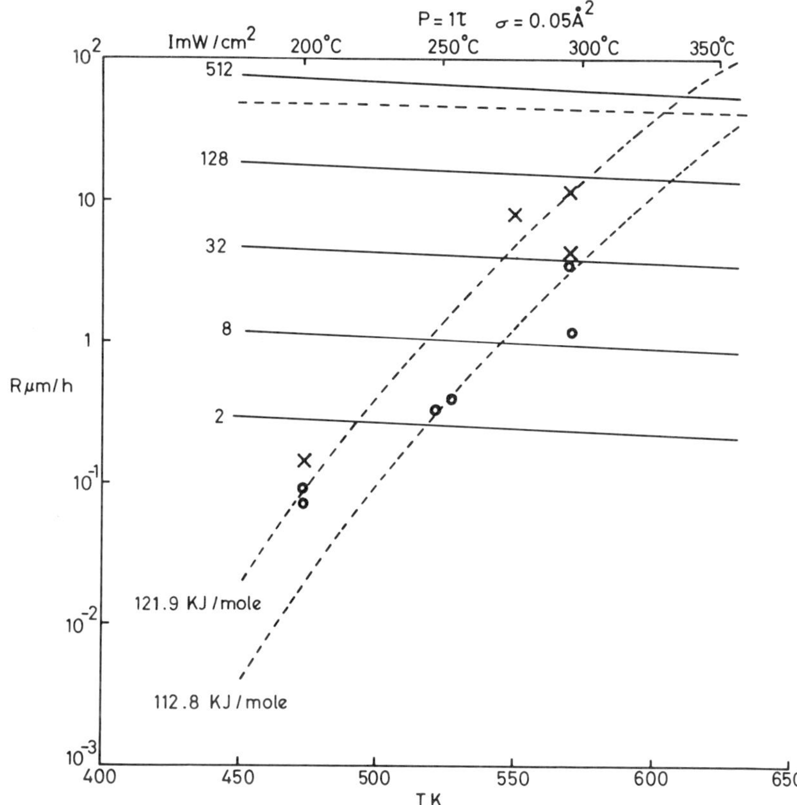

Fig. 9.13 Predicted critical growth rates for critical excess chemical potentials of 121.9 and 112.8 kJ/mole. Experimental results are shown for polycrystalline (x) and epitaxial (o) CdTe layers. The horizontal lines are the predicted growth rates for a transport model for conditions of 1-cm thick boundary layer, 1 torr partial pressures of Me_2Cd and Et_2Te, absorption cross section of 0.05 $Å^2$, and for UV intensities between 2 and 512 mW/cm^2. The horizontal dashed line shows the maximum growth rate under beam saturation conditions. After Irvine et al., 1988.

to maintain a Cd-rich vapor concentration. Cross-sectional TEM of these films grown at 300°C showed that the misfit dislocation structure was confined to a very narrow band at the interface, with no eividence for dislocations threading through to the bulk of the layer.

This would suggest that UV-enhanced epitaxial growth of CdTe actually improves epitaxial quality compared with pyrolytic growth. Kisker et al. investigated this by measuring photoluminescence spectra from photoenhanced CdTe and pyrolytically grown CdTe. The ratio in intensity of the sharp near-band-edge luminescence, associated with bound excitons, to the broad deeper luminescence, associated with structural defects, has been used as a figure of merit for the crystalline quality of CdTe. Comparing photoenhanced and pyrolytically grown CdTe at 350°C, the ratios were 20 and 0.5, respectively, indicating the superior quality of the photoenhanced material.

An example of photomodified CdTe is the photodoping of molecular beam epitaxy (MBE) CdTe reported by Bicknell et al. (1986). Undoped CdTe films, grown within the temperature range of 250–275°C, are electrically insulating. This is not unusual for undoped CdTe, due to deep traps, and for lightly donor-doped CdTe because of vacancy-donor complexes. The addition of an indium source in the MBE growth chamber can lead to incorporation of indium in the film, but with poor activation. Illumination of the substrate with an argon-ion laser, with an intensity of 150 mW/cm^2, during growth yielded conducting CdTe films with good In activation. These intensities would not be sufficient to raise the substrate temperature; the process is probably photolytic, but the exact mechanism is not clear. Indium-doped CdTe has been grown at 230°C with carrier concentrations as high as $1 \times 10^{17}\ cm^{-3}$. Without illumination, indium degrades the low-temperature photoluminescence spectra with strong broad-band luminescence at 1.44 eV, associated with defects, and little or no near-band-edge liminescence. However, the low-temperature photoluminescence for the photoassisted doped layers was dominated by near-band-edge luminescence with little or no luminescence at 1.44 eV. This photomodified growth of CdTe could be suitable for patterning with islands or conducting tracks of activated indium. As the photochemical reaction appears to be on the surface, the resolution of these features would be determined by the beam resolution.

HgTe photoepitaxy contrasts with CdTe in that the heterogeneous surface reaction is strong due to the surface decomposition of Et_2Te, while the vapor reaction is weak. This has enabled epitaxial growth over a wide temperature range from 180°C to 300°C with no evidence for dust nuclei settling on the substrate. The mechanism of surface photosensitization was described more fully in Section 1.4; in this section the quality of these epitaxial layers will now be considered.

Single crystal x-ray diffraction was used by Irvine et al. (1984b) to assess HgTe layers on InSb substrates. InSb has a similar lattice parameter to HgTe, 6.4798 Å compared with 6.4605 Å. InSb (001) substrates were etched in 25:4:1, lactic acid:nitric acid:HF prior to loading into the reactor cell. The substrates were heated to the growth temperature without further surface treatment. Subsequent x-ray examination of the films (~1 μm thick) using 400 reflections showed them to be epitaxial over the growth temperature range of 200–310°C. The diffractometer resolution was 3.6 minutes of arc, and all the layers gave peak widths close to this. However, a systematic trend in $K_{\alpha 1}$ and $K_{\alpha 2}$ peak separations was observed, indicating poorer films at higher temperatures. This would not be expected on the basis of surface atom mobility, which would reduce crystalline defects at higher temperatures, essentially because the Hg and Te atoms would be more likely to find the correct sites. Examination of these layers in cross section using transmission electron microscopy shows that the layer/substrate interface is considerably disrupted. This has been attributed to the preferential loss of antimony from the InSb substrate prior to growth. The excess indium is then free to react with tellurium to form indium tellurides as an interfacial layer. Therefore, the epitaxial quality of photoepitaxial HgTe in this case was limited by the substrate/layer interface rather than fundamental kinetic problems during growth.

The epitaxial quality of HgTe has been further investigated using double-crystal x-ray diffraction, which has peak width resolution of ~10 arc seconds. The interface problem was avoided by using $Cd_{0.96}Zn_{0.04}Te$ substrates and selected for their relatively narrow rocking curves (see Irvine et al., 1988). The rocking curve of x-ray intensity versus Bragg angle θ from one such layer, grown at 250°C, is shown in Fig. 9.14. A 1-μm thick HgTe layer had a rocking curve width of 46.6 arc seconds, and the underlying substrate rocking curve width was 23.2 arc seconds. This result is comparable, with best results reported for HgTe grown by molecular beam epitaxy (MBE) or by thermal MOVPE.

The electrical properties of HgTe epitaxial layers have been assessed by low-temperature Hall measurements (see Irvine et al., 1987; Ahlgren et al., 1987). For 1-μm thick epitaxial layers, the 77 K Hall mobilities were in the range 3.2×10^4 to $5.5 \times 10^4\ cm^2\ V^{-1}\ s^{-1}$. Similar Hall mobilities for photolytic HgTe have been reported by Ahlgren et al. (1986). These authors also used a high-pressure arc lamp with Et_2Te and Hg as the reactants. A comparison of 77 K Hall mobilities between photolytic and pyrolytic HgTe is shown in Table 9.3. The photolytic layers compare well with pyrolytic layers and are mostly grown at lower temperatures. This further demonstrates that a reduction in epitaxial growth temperature by using photoepitaxy does not necessarily lead to poorer crystal quality but can give a lower equilibrium defect concentration. However, the highest 77 K Hall mobility of $1.1 \times 10^5\ cm^2\ V^{-1}\ s^{-1}$ reported by Bhat and Ghandhi (1984) is twice that of the best photoepitaxial result and shows the scope for further improvements.

Epitaxial growth of the alloy $Cd_xHg_{1-x}Te$ has also been demonstrated

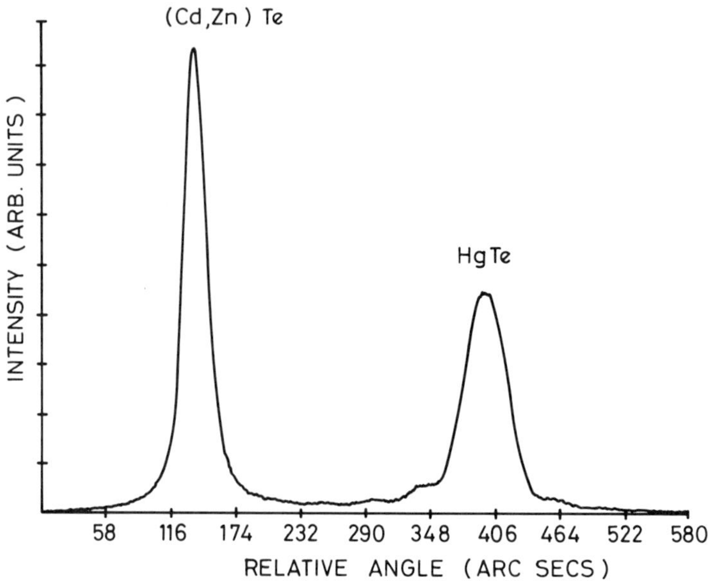

Fig. 9.14 Double-crystal x-ray diffractometer trace for 004 reflection from a 1-μm HgTe layer deposited by photoepitaxy at 250°C onto a $Cd_{0.96}Zn_{0.04}Te$ substrate.

using Et_2Te and Me_2Cd precursors in a vapor of mercury in helium carrier gas by both Irvine et al. (1985) and Ahlgren et al. (1986). The method is an extension of the binary growth of CdTe and HgTe, and control of alloy composition was by means of adjusting Me_2Cd and Et_2Te partial pressures. An alternative photoinduced technique has been used by Morris (1986). An excimer laser (193 nm) was used to photodecom-

Table 9.3 A Comparison of Hall Mobilities and Carrier Concentrations for Photolytic and Pyrolytic Epitaxy of HgTe.

Method	Substrate Temperature °C	77 K Mobility $cm^2 V^{-1} s^{-1}$	Carrier Concentration cm^{-3} 77 K	4.2 K
Pyro (Irvine et al., 1982)	410	50,000		2×10^{16}
Pyro (Bhat & Ghandhi, 1984)	440	110,000	3×10^{16}	
Pyro (Hoke et al., 1986)	270	63,900	7×10^{16}	
Photo (Irvine et al., 1986)	240	55,000	9×10^{16}	3×10^{16}
Photo (Ahlgren et al., 1987)	280	35,000→50,000	8×10^{16}	7×10^{16}

pose a mixture of Me_2Cd, Me_2Te, and Me_2Hg in the vapor. The beam was directed parallel to the substrate and the pulse energy of 384 mJ/pulse was sufficient to decompose all the alkyls. An epitaxial film was grown at 150°C onto a CdTe substrate; the layer composition was 80% CdTe and 20% HgTe. This technique relies on vapor-phase decomposition, and, although suitable for reducing growth temperature, it would not be suitable for patterning applications. However, by using a parallel illumination geometry, laser heating of the substrate was avoided.

4.2. Photoepitaxy of Heterostructures

The growth of heterostructures in the CdTe-HgTe system is particularly difficult because of interdiffusion between cadmium and mercury, which will blur the interface. The graded interface width for structures grown at 410°C is typically 0.3 μm, reducing to 0.1 μm for layers grown at 350°C (see Irvine et al., 1986). An advantage of reducing growth temperatures is the reduction of the inderdiffusion region at heterostructure interfaces. MBE growth between 180°C and 200°C has been successful in reducing this region to ~10 Å and is therefore suitable for the growth of superlattices of HgTe and CdTe.

Photoepitaxy has the flexibility of growing CdTe/HgTe heterostructures over a wider range of temperatures than MBE because of greater flexibility in Hg flux pressure. A CdTe/HgTe structure was grown by photo-MOVPE at 230°C onto a (100)CdTe substrate by Irvine et al. (1986) to investigate interdiffusion at this temperature. The SIMS sputter

Fig. 9.15 A transmission electron microscope (TEM) micrograph of 70 Å/30 Å, HgTe/CdTe superlattices grown at 182°C (after Ahlgren et al., 1987).

depth profile for mercury gave an interface width of 105 Å. Extrapolation to lower temperatures predicted that growth temperatures of 150°C or lower would be required for superlattice growth. However, Ahlgren et al. (1986) have successfully demonstrated superlattice growth at 180°C using photoepitaxy. Composition modulation was achieved by switching the Me_2Cd flow on and off; the Hg-Xe arc lamp was run continuously. Thirty-five periods were grown with 70 Å of HgTe and 30 Å of CdTe in each period. A micrograph of the angle-lapped structure is shown in Fig. 9.15.

Photoepitaxy can offer flexibility in the growth of heterostructures by controlling the growth with the UV source rather than by switching gas flows. For example, the UV could be switched off during the critical phase of switching gas flows, introducing a growth pause when otherwise pressure transients would disturb the growth. Complex three-dimensional structures could be fabricated by lateral patterning at the same time as growing multilevel heterostructures with appropriate acceptor and donor doping.

References

Ahlgren, W.I., James, J.B., Ruth, R.P., Patten, E.A., and Staudenmann, J.-L. (1987). HgTe-CdTe superlattices grown by photo-MOCVD. *Mat. Res. Soc. Symp.* **90,** 405–415.

Anderson, H.M., and Hargis Jr, P.J. (1986). A model for silicon dendrite growth during laser/plasma deposition from a silane discharge. *Mat. Res. Soc. Symp. Proc,* **75**.

Ando, H., Inuzuka, H., Konagai, M., and Takahashi, K. (1995). Photoenhanced metalorganic chemical vapour deposition of ZnSe films using diethylzinc and dimethylselenide. *J. Appl. Phys.* **58,** 802–805.

Balk, P., Fischer, M., Grundmann, D., Lückerath, R., Lüth, H., and Richter, W. (1987). Ultraviolet-assisted growth of GaAs. *J. Vac. Sci. Technol.* **5,** 1453–1459.

Ban, Y., Takechi, M., Ishizaka, M., Murata, H., and Kukimoto, H. (1986a). Laser-irradiation effects on the incorporation of impurities in GaAs during MOVPE growth. *Jpn. J. Appl. Phys.* **25,** L967–L969.

Ban, Y., Takechi, M., Ishizaki, M., Murata, H., and Kukimoto, H. (1986b). Selective area formation of semi-insulating GaAs by laser-assisted metalorganic vapour phase epitaxy. In *Semi-Insulating III–V Materials,* Ohmsha Ltd, 521–524.

Bedair, S.M., Whisnant, J.K., Karam, N.H., Griffis, D., El-Masry, N.A., and Stadelmaier, H.H., (1986a). Laser selective deposition of III–V compounds on GaAs and Si substrates. *J. Crystal Growth* **77,** 229–234.

Bedair, S.M., Whisnant, J.K., Karam, N.H., Tischler, M.A., and Katsuyama, T. (1986b). Laser selective deposition of GaAs on Si. *Appl. Phys. Lett.* **48,** 174–176.

Beneking, H. (1985). Laser deposition of single crystalline GaAs and stimulated sheet doping. *Springer Ser. Chem. Phys.* **39,** 188–196.

Bhat, I., and Ghandhi, S.K. (1984). The growth and characterisation of HgTe epitaxial layers made by organometallic epitaxy. *J. Electrochem. Soc. Sol. Stat. Sci. Technol.* **131,** 1923–1926.

Bicknell, R.N., Giles, N.C., and Schetzina, J.G. (1986). Growth of high mobility n-type CdTe by photoassisted molecular beam epitaxy. *Appl. Phys. Lett.* **49,** 1095–1097.

Doi, A., Aoyagi, Y., and Namba, S. (1986a). Growth of GaAs by switched laser metalorganic vapour phase epitaxy. *Appl. Phys. Lett.* **48,** 1787–1789.

Doi, A., Aoyagi, Y., and Namba, S. (1986b). Stepwise monolayer growth of GaAs by switched laser metalorganic vapour phase epitaxy. *Appl. Phys. Lett.* **49,** 785–787.

Doi, A., Aoyagi, Y., and Namba, S. (1987). Private communication.

Donahue, T.J., and Reif, R. (1986). Low temperature silicon epitaxy deposited by very low pressure chemical vapour deposition. *J. Electrochem. Soc. Sol. Stat. Sci. Technol.* **133,** 1691–1697.

Donnelly, V.M., Brasen, D., Appelbaum, A., and Geva, M. (1985). Excimer laser-induced deposition of InP; Crystallographic and mechanistic studies. *J. Appl. Phys.* **58,** 2022–2035.

Donnelly, V.M., Brasen, D., Appelbaum, A., and Geva, M. (1986). Excimer laser induced deposition of InP. *J. Vac. Sci. Technol.* **A4,** 716–721.

Donnelly, V.M., McCrary, V.R., and Brasen, D. (1987). Production of electronically excited P_2 and In from ArF excimer laser irradiation of InP. *Mat. Res. Soc. Symp. Proc.,* **75**.

Eden, J.G., Greene, J.E., Osmundsen, J.G., Lubben, D., Abele, C.C., Gorbatkin, S., and Desai, H.D. (1983). Semiconductor thin films grown by laser photolysis. *Mat. Res. Soc. Symp. Proc.* **17,** 185–192.

Faktor, M.M., Garrett, I., and Heckingbottom, R. (1971). Diffusional limitations in gas phase growth of crystals. *J. Crystal Growth* **9,** 3–11.

Gee, J.M., Hargis Jr, P.J., Carr, M.J., Tallant, D.R., and Light, R.W. (1984). Plasma-initiated laser deposition of polycrystalline and monocrystalline silicon films. *Mat. Res. Soc. Symp. Proc.* **29,** 15–20.

Gee, J.M., and Hargis Jr, P.J. (1984). Laser-induced etching of insulators using a DC glow discharge in silane. *Proc. SPIE* **459,** 132–137.

Giess, J., Gough, J.S., Irvine, S.J.C., Mullin, J.B., and Blackmore, G.W. (1987). Orientation effects on the heteroepitaxial growth of $Cd_xHg_{1-x}Te$ onto CdTe and GaAs. *Mat. Res. Soc. Symp. Proc.* **90,** 389–395.

Haq, S., Dobson, P.S., and Irvine, S.J.C. (1987). Growth of CdTe on GaAs by photo-MOVPE. *Les Editions de Physique, E-MRS* **15,** 199–204.

Hargis Jr, P.J., and Gee, J.M. (1984). Laser-plasma interactions for the deposition and etching of thin-film materials. *Sol. Stat. Technol.* 127–133.

Hoke, W.E., and Lemonias, P.J. (1985). Metalorganic growth of CdTe and HgCdTe epitaxial films at a reduced substrate temperature using diisopropyltelluride. *Appl. Phys. Lett.* **46,** 398–400.

Irvine, S.J.C., Mullin, J.B., and Royle, A. (1982). Epitaxial growth of HgTe by a MOVPE process. *J. Crystal Growth* **57,** 15–20.

Irvine, S.J.C., Mullin, J.B., and Tunnicliffe, J. (1984a). Photosensitisation: A stimulant for the low temperature growth of epitaxial HgTe. *J. Crystal Growth* **68,** 188–193.

Irvine, S.J.C., Mullin, J.B., and Tunnicliffe, J. (1984b). Low temperature growth of HgTe by a UV photosensitisation method. *Springer Ser. Chem. Phys.* **39,** 234–238.

Irvine, S.J.C., Mullin, J.B., Blackmore, G.W., and Dosser, O.D. (1985). Photo-metal organic vapour phase epitaxy: A low temperature method for the growth of $Cd_xHg_{1-x}Te$. *J. Vac. Sci. Technol.* **B3,** 1450–1455.

Irvine, S.J.C., Giess, J., Gough, J.S., Blackmore, G.W., Royle, A., Mullin, J.B., Chew, N.G., and Cullis, A.G. (1986). The potential for abrupt interfaces in $Cd_xHg_{1-x}Te$ using thermal and photo-MOVPE. *J. Crystal Growth* **77,** 437–451.

Irvine, S.J.C., Mullin, J.B., Blackmore, G.W., Dosser, O.D., and Hill, H. (1987). Photochemical processes in photo-assisted epitaxy of $Cd_xHg_{1-x}Te$. *Mat. Res. Soc. Symp. Proc.* **90,** 153–161.

Irvine, S.J.C. (1988). UV photo-assisted crystal growth of II–VI compounds. *CRC Critical Reviews in Sol. Stat. and Mat. Sci* **13,** 279–309.

Ishitani, A., Oshita, Y., Tanigaki, K., and Takada, K. (1987). Prebaking and silicon epitaxial growth enhanced by UV radiation. *J. Appl. Phys.* **61,** 2224–2229.

Karam, N.H., El-Masry, N.A., and Bedair, S.M. (1986). Laser direct writing of single crystal III–V compounds on GaAs. *Appl. Phys. Lett.* **49,** 880–882.

Karam, N.H., Bedair, S.M., El-Masry, N.A., and Griffis, D. (1987). Laser stimulated deposition of GaAs, GaAsP and GaAsP-GaAs superlattices. *Mat. Res. Soc. Symp. Proc.* **75,** 241–248.

Karam, N.H., Liu, H., Yoshida, I., and Bedair, S.M. (1988). Direct writing of GaAs monolayers by laser-assisted atomic layer epitaxy. *Appl. Phys. Lett.* **52,** 1144–1146.

Kisker, D.W., and Feldman, R.D. (1985). Photolysis-assisted OMVPE growth of CdTe. *Materials Letters* **3,** 485–488.

Kisker, D.W., and Feldman, R.D. (1985). Photon-assisted OMVPE growth of CdTe. *J. Crystal Growth* **72,** 102–107.

Kukimoto, J., Ban, Y., Komatsu, H., Takechi, M., and Ishizaki, M. (1986). Selective area control of material properties in laser-assisted MOVPE of GaAs and AlGaAs. *J. Crystal Growth* **77,** 223–228.

Leys, M.R., and Veenvliet, H. (1981). A study of the growth mechanism of epitaxial GaAs as grown by the technique of metal organic vapour phase epitaxy. *J. Crystal Growth* **55,** 145–153.

Ludowise, M.J. (1985). Metalorganic chemical vapour deposition of III–V semiconductors. *J. Appl. Phys.* **58,** R31–R55.

Morris, B.J. (1986). Photochemical organometallic vapour phase epitaxy of mercury cadmium telluride. *Appl. Phys. Lett.* **48,** 867–869.

Mullin, J.B., Irvine, S.J.C., and Ashen, D.J. (1981). Organometallic growth of II–VI compounds. *J. Crystal, Growth* **55,** 92–106.

Nishida, S., Shiimoto, T., Yamada, A., Karasawa, S., Konagai, M., and Takahashi, K. (1986). Epitaxial growth of silicon by photochemical vapour deposition at a very low temperature of 200°C. *Appl. Phys. Lett.* **49,** 79–81.

Robbins, D.J., and Young, I.M. (1987). A new approach to the kinetics of silicon vapour phase epitaxy at reduced temperature. *Appl. Phys. Lett.* **50,** 1575–1577.

Roth, W., Kräutle, H., Krings, A., and Beneking, H. (1983). Laser stimulated growth of epitaxial GaAs. *Mat. Res. Soc. Symp. Proc.* **17,** 193–198.

Schlyer, D.J., and Ring, M.A. (1976). An examination of the product-catalyzed reaction of trimethylgallium with arsine. *J. Organometallic Chem.* **114,** 9–19.

Srinivasan, G.R., and Meyerson, B.S. (1987). Current status of reduced temperature silicon epitaxy by chemical vapour deposition. *J. Electrochem. Soc.: Sol. Stat. Sci. and Technol.* **134,** 1518–1523.

Stutius, W. (1982). Growth and doping of ZnSe and ZnS_xSe_{1-x} by organometallic chemical vapour deposition. *J. Crystal Growth* **59,** 1–9.

CHAPTER 10

Doping and Oxidation

IAN W. BOYD
Electronic and Electrical Engineering
University College London
London, United Kingdom

This chapter will deal with the use of photon beams to locally modify the structural and electronic properties of various materials by the controlled introduction of a range of atomic species. As with conventional processing techniques for doping and oxidation, these laser-based methods are capable of selectively altering the electronic conductivity of semiconductor surfaces by up to 20 orders of magnitude. Since the reaction mechanisms inherent to these processes are quite different they will be treated separately here.

1. Laser Doping

1.1. Introduction

For the production of microelectronic devices, the controlled incorporation of dopant atoms into semiconductors is extremely important.

0-12-233430-2

Indeed, the ability to tailor the semiconducting properties of silicon by doping is a major reason for its successes in large-scale device integration. Two major approaches are traditionally employed to this end, namely, diffusion, and ion implantation.

Impurities used for doping can be obtained from vapor-phase chemical compounds at high temperature, from solid sources placed on or adjacent to the wafer surface, or from highly energetic beams of ions directed at the material. Key goals in these technologies are precise control of the final dopant profile and concentration in the semiconductor, and the ability to fabricate large areas reproducibly and cost effectively. In order to optimize these doping methods, several processing steps must be performed. Firstly, the impurity species must be created and then introduced into the material. Subsequent steps must ensure that the required distribution and concentration has been obtained in the solid. Finally, the atoms must be incorporated and electrically activated within the lattice, to contribute to the conduction of charge within the layer. Diffusion techniques were an early development in integrated microelectronic technology, and, historically, the dopant for these methods has been provided in a number of ways. Solid-phase sources such as vacuum-evaporated or paint- and spin-on coatings can be applied directly to the sample. Alternatively, in a heated environment, solid sources placed near to the substrate can liberate evaporant that settles onto the sample surface.

The use of gas-phase compounds (or precursors) for directly doping the substrate, has been very popular in recent years. In this method, the gaseous molecules pyrolytically decompose upon hitting a sufficiently hot sample surface. The dopant atoms are henceforth released on to the substrate, whilst the reaction by-products are swept away in the exhaust flow. At the high temperatures required to crack these molecular species and dissolve the dopant into the near-surface region, limited diffusion into the bulk invariably occurs. For specific applications, the dopant layer can even be co-deposited with the host material onto a variety of substrates.

In ion implantation, ions are created from a gaseous source and are filtered through a resolving slit, accelerated to energies between 3 and 500 keV, and then scanned across the sample to achieve a spatially uniform dose. With this technique, the depth of penetration of the atoms as well as the dose implanted can be precisely controlled. The energy scattered by the decelerating ions, however, leads to crystal damage by introducing point defects and the formation of amorphous material.

An important and usually unavoidable feature of these doping procedures is the requirement of an additional high-temperature processing step. In ion implantation, a high-temperature annealing cycle is often essential to return the sample surface to its original high degree of crystalline perfection; it also assists in the placement of the dopant atoms onto electrically active interstitial sites. With the other systems, the dopant is usually not initially located at the required depth within the material, being usually deposited at the surface, and again a separate higher temperature diffusion cycle is often necessary. In this drive-in process, no additional dopant enters the substrate, and the high temperatures enable the impurities already present to diffuse to the required depth and concentration. Mathematical models of isothermal diffusion have been developed over the years to predict the dopant profiles induced under a variety of processing conditions.

It is not only the inconvenience of the additional high-temperature processing step that is a disadvantage in these doping operations, it is the fact that the complete wafer, and not just the area to be doped, is isothermally heated to the predetermined annealing or drive-in temperature, inducing undesirable dopant diffusion in any previously prepared underlayers. Since the trends of VLSI are to produce devices with submicron features, it is important to minimize the total amount of high-temperature processing on the production line. A further disadvantage to current doping procedures is that photolithographic patterning is necessary to define the areas across the sample to be selectively doped. This usually requires up to 15 individual processing steps. Techniques allowing direct doping without the requirement of prepatterning could therefore become economically desirable.

Laser processing offers advantages in these areas, as has been pointed out many times already in this book. Firstly, it can locally confine chemical and physical reactions, in some cases to areas as small as $10^{-13}\,m^2$ (Ehrlich and Tsao, 1982). Recently the general concept of in-situ processing has emerged from several major research laboratories. Such a technology may involve an interlocking multi-chambered arrangement and include directed ion, electron, and laser beam modification of materials as well as large-area film fabrication, using rapid thermal or plasma-related techniques. A wide range of microchemical reactions for pattern definition would be required, including not only doping and oxidation, but also etching and deposition. Throughout this book it is clear that progress has already been made in these directions.

By providing unprecedented heating and cooling rates, laser-induced heating can also significantly reduce the heating up and cooling down times associated with traditional furnace processing (Boyd, 1983a). Often too, significant reductions in processing temperature are available by inducing photolytic or photochemical reactions on or near the sample surface, rarely above temperatures of 300–400°C (Micheli and Boyd, 1987a). For the fabrication of ultrashallow n^+-p and p^+-n junctions, therefore, which severely test ion-implantation capabilities by requiring 10–20 keV implants without significant planar or axial ionic channeling, in-situ laser doping could prove advantageous. The precise depth of diffusion, as well as the shape of the dopant profile, can be accurately controlled by adjusting the amount of laser radiation entering the sample. The various laser-based approaches to doping will be reviewed in the following sections.

1.2. Mechanisms

Lasers offer a variety of mechanisms by which solids can be doped, as shown in Table 10.1. Whilst the introduction of dopant into the material can be achieved in a number of ways, the drive-in step invariably requires thermal diffusion. In some instances, however, the deposition of the

Table 10.1 Summary of Laser-Induced Doping Methods.

Laser	Precursor/Dopant	Substrate/Film	Reference
Nd:YAG, CO_2	Surface Film of B, P	Si	Affolter et al. (1978)
Ruby	Evaporated B	Si	Narayan et al. (1978)
Ruby	Implanted B, As	Si	Young et al. (1980)
Ruby	Implanted B, Evaporated Sb	Si	Muller et al. (1980)
Ar (257, 514 nm)	DMCd	InP	Ehrlich et al. (1980)
XeCl, ArF	DMCd, DMZn	InP	Deutsch et al. (1980)
Alexandrite	PH_3	Si	Turner et al. (1981)
XeCl/ArF	BCl_3, PCl_3	Si	Deutsch et al. (1981)
Ruby	Evaporated Sb, Ga, Bi, In	Si	Fogarassy et al. (1981)
Ar (257, 514 nm)	DMZn	Al	Ehrlich et al. (1980)
Nd:YAG (532 nm)	Spin-on As	Si	Kiang et al. (1982)
XeCl/ArF	H_2S	GaAs	Deutsch et al. (1982)
Ar (257, 514 nm)	BCl_3, PCl_3	Si	Ehrlich et al. (1982)
ArF/XeF	TEB	Si	Ibbs & Lloyd (1983)
Ar	$SiH_4 + PH_3$	Si	McWilliams et al. (1983)
N_2-dye	P-doped glass	Poly-Si	Miyauchi et al. (1983)
ArF	$PH_3 + Si_2H_6$	glass	Yoshikawa et al. (1984)
XeCl	B_2H_6	Si	Carey et al. (1985)
CO_2	B_2H_6, $PH_3 + SiH_4$	Glass	Bilenchi et al. (1985)
Nd:YAG (532 nm)	H_2Se, DEZn	GaAs	Kräutle et al. (1985)
CO_2	Evaporated B	Glass	Branz et al. (1986)
ArF	TMSi	GaAs	Ban et al. (1986)
XeCl	AsH_3	Si	Carey et al. (1986)
XeCl	PH_3	Si	Sigmon et al. (1987)
XeCl	RF-deposited B	Si	Sameshima et al. (1987)

complete layer, doped appropriately, on top of the substrate can eliminate the requirement for the drive-in procedure. Here, several of these techniques will be described.

1.2.1 Creation of Dopant Species

In Chapter 8 of this book, a large range of laser-induced deposition reactions has been reviewed. The laser radiation can liberate the necessary atomic species for doping by absorption in, and dissociation of, gas-phase precursors close to the required substrate. The donor-bearing molecules can also adsorb onto the sample surface and there dissociate either by direct photon absorption or by vibrational energy transfer. Alternatively, the precursors can decompose, or crack, upon random collision with the laser-excited surface. However, the earliest work in this field made use of the more traditional modes of dopant supply. Laser annealing, for example, involved heating the substrate sufficiently to redistribute and activate ion-implanted species, while simultaneously recrystallizing the damaged semiconductor surface (Boyd, 1983; Cullis, 1985). Similar work on laser doping involved precoating the substrate with commercially available "paint-on" (Affolter et al., 1978), evaporated layers (Narayan et al., 1978) containing the required dopant species and then subjecting it to laser-induced heat treatment.

A glance at Table 10.1 will indicate that the vast majority of precursors used for laser doping have been specifically designed for more established thin-film technologies such as vapor- and liquid-phase epitaxy (VPE and LPE) or chemical vapor deposition (CVD). It is largely coincidental that these are also useful for laser chemical processing and somewhat surprising that there is a lack of new precursors designed specifically for this technology.

For boron (and phosphorous) doping, the use of BCl_3 (PCl_3) and B_2H_6 (PH_3) has been prevalent. The metal organics, dimethyl cadmium [DMCd or $Cd(CH_3)_2$], trimethylaluminum [TMA or $Al_2(CH_3]$, trimethylboron [TMB or $B(CH_3)_3$], dimethyl zinc [DMZn or $Zn(CH_3)_2$], and trimethyl silicon [TMSi or $Si(CH_3)_3$] have also attracted some attention. Production of the required dopant species can occur via one or more of the mechanisms outlined above. Traditionally, it has been difficult to isolate and identify the kinetics controlling the dopant production. Usually, however, one can reasonably determine the dominant reaction mechanism by performing a range of wavelength-dependent experiments, by investigating the effect of plane or parallel incidence of the laser beam, and by studying the pressure dependence of the induced reactions.

1.2.2 Gas-Phase Doping

UV radiation can be used to photodissociate many precursor compounds (Ehrlich and Tsao, 1983; Solanki et al., 1985; Rytz-Froidevaux et al., 1985; Micheli and Boyd, 1987a). Some of the first work in this area was reported by the MIT group in the late 1970s (Deutsch et al., 1979) and led to the proposed application towards localized doping when combined with a drive-in process (Deutsch et al., 1980; Ehrlich et al., 1980). Using DMCd and the 257-nm radiation from a frequency-doubled argon-ion laser, a strong linear relationship was found between Cd incorporation in InP and the DMCd pressure, as well as the laser flux, consistent with linear photodeposition rates obtained on an unheated quartz substrate (Deutsch et al., 1980; Ehrlich et al., 1980). Evidence of photochemical effects was also found in laser deposition of Zn from DMZn (Deutsch et al., 1980; Ehrlich et al., 1981). While almost no deposition was observed when using up to 10^6 W/cm^2 of 514-nm radiation, an appreciable deposit was obtained when less than 10 W/cm^2 of 257-nm light was used.

In contrast to these results, irradiation by 193-nm radiation from an ArF laser did not alter the overall carrier concentration obtained during MOVPE growth of GaAs at 700°C in arsine (AsH_3) and TMGa and DMZn (Ban et al., 1986). For the flow rates and gas concentrations used, it appears that thermal decomposition was sufficiently strong at 700°C to swamp any photochemical effects introduced by the UV photons. On the other hand, when TMSi replaced DMZn during growth at 700°C a 10-fold increase in the carrier concentration was found, indicating that although the precursor was thermally stable, it readily photodecomposed with 193-nm radiation. No change in the overall carrier concentrartion was found between 600 and 800°C when silane (SiH_4) replaced TMSi during irradiation and growth. Although it is the more traditional source of Si dopant, it does not absorb efficiently at 193 nm.

Films of p-type hydrogenated amorphous silicon carbide (a-SiC:H) have also been prepared by direct photodeposition from disilane (Si_2H_6), ethylene (C_2H_2), and diborane (B_2H_6) mixtures using Hg lamps (Yamada et al., 1985; Takei et al., 1985). Interestingly, it was found that the optical band gap of the films was narrower than in layers of identical conductivity prepared using photosensitizing reactions (Yamada et al., 1985). Such reactions will be discussed in the following section.

1.2.3 Photosensitized Dopant Creation

Where gas-phase absorption is not sufficient to dissociate the parent species and liberate the dopant atoms, indirect methods of excitation may be used, principally, through photosensitization of an intermediary species. With the aid of the well-known mercury photosensitized reaction, incorporation of dopant has been achieved during the growth of hydrogenated amorphous silicon (a-Si:H) film by using a low-pressure mercury lamp to decompose silane/diborane mixtures (Inoue et al., 1983, 1984). By this method, Hg atoms were introduced into the reaction chamber and photoexcited by the UV radiation; the diborane molecules are then dissociated by collisional energy transfer from the Hg atoms. The photoinduced reaction increased the overall deposition rate by two orders of magnitude, and the photostimulated reaction became twofold slower when the 0.2–0.4% diborane was not present.

Boron-doped p-type a-SiC:H films have also been prepared via photosensitized deposition from acetylene, disilane, and diborane mixtures, using a low-pressure Hg lamp (Inoue et al., 1983; Takei et al., 1985). In this case the reaction rate was not enhanced by the addition of diborane (Inoue et al., 1983). Phosphine (PH_3), rather than diborane, was chosen by Yoshikawa and Yamaga (1984) to laser photodeposit doped a-Si:H films because of its higher optical absorption cross section at 193 nm. Although their electrical conductivity had increased by more than four orders of magnitude, the overall doping efficiency was less than that reported for the plasma CVD process.

CO_2 lasers have been employed for gas-phase doping, but, unlike UV photons, IR quanta are individually not sufficiently energetic to rupture chemical bonds. However, in the case of SiH_4, the P(20) emission of the laser is resonantly absorbed by a fundamental vibrational mode of the molecule. The excitation of such a mode in a single molecule will not usually lead to dissociation, and it is only if many photons are absorbed—either by the same molecule or by the others in the cell—that enough energy can be accumulated to break up the molecule. Collisional energy transfer amongst the molecules can be important in this energy collection mechanism.

When phosphine or diborane—both of which are transparent to CO_2 laser radiation—are added to silane, they too can be dissociated via collisional energy transfer from the absorbing SiH_4 molecules (Bilenchi et al., 1985; Branz et al., 1986). However, once created, the B and P species can have differing influences on the growth of the film (Ashida et al., 1984; Bilenchi, 1985). While adsorbed B atoms or radicals encourage further growth activity because of their catalytic behavior, P species tend to poison the growing Si surface, thereby restricting further reactivity. Adsorption effects are further discussed in the next section.

1.2.4 Surface Adsorption Effects

The role of adsorption during laser deposition has often been examined. For example, in studies of p and n doping of Si from BCl_3 and PCl_3 using

193-nm and 351-nm laser radiation, Deutsch et al. (1981) found very little wavelength dependence to the dopant incorporation. Since the optical properties of BCl_3 are markedly different at these two wavelengths, strong photolytic mechanisms were ruled out. Furthermore, the dependence of sheet resistance of the films on gas pressure was much weaker than expected for pyrolysis. By filling and evacuating the sample cell prior to irradiation, dopant incorporation was still achieved, however, and the presence of an adsorbed layer, confirmed independently by isotherm measurements, was proposed. It was therefore suggested that both adsorption and gas-phase pyrolytic processes played an important role in this supply of dopant into the material. In contrast to this observation, no adsorbed layer was detected when H_2S was used to dope GaAs (Deutsch et al., 1982). A similar two-wavelength study showed that again the shorter wavelength radiation gave rise to slightly, though not vastly, superior electrical properties. In this case, the additional photochemical effects induced by the ArF laser over the pyrolytic mechanisms induced by both was thought to be important. For doping of GaAs with Se or Zn, using a frequency-doubled Nd:YAG laser, Kräutle et al. (1985) found that the sticking coefficient of H_2Se and DEZn must approach unity.

Sigmon (1987) has studied the role of adsorption during XeCl laser (308 nm) doping of silicon with diborane, arsine, or phosphine. Two models were examined. The first considered the amount of dopant impinging upon the sample surface only during the molten phase induced by the laser beam, while the second assumed that the source of impurity was an adsorbed layer that accumulated on the surface over a longer period of time. Once the laser heated the sample to a sufficiently high temperature, thermal diffusion into the bulk ensued. The latter model most readily fitted the experimentally obtained dopant profiles, supporting monolayer adsorption of the hydrides. The next section will deal with the thermal diffusion step following dopant incorporation into the material.

1.3. Drive-In and Diffusion

In addition to providing dopant, laser processing has been more commonly used to diffuse impurities. By controlling the beam intensity and the duration of exposure, the irradiated material can be precisely heated in a predetermined fashion (Cullis, 1985). In many of the cases outlined above, and in those summarized below, such a drive-in or diffusion step was provided by the laser beam creating the dopant. Ehrlich and Tsao (1982), for example, used 514-nm radiation both to pyrolize BCl_3 and simultaneously diffuse B atoms into Si. There have also been occasions when a different laser wavelength was used for each process. Ehrlich et al. (1981) and Ehrlich and Tsao (1982) have employed a dual-beam coaxial system to separately initiate photodeposition and diffusion of Zn and Cd.

The confinement of laser light can be readily achieved through a range of lensing systems. Present optical technology enables beams of radiation as small as their own wavelength to be produced, and, thus, spot sizes less than 0.5 μm can be projected onto sample surfaces. Also, within optically dense media laser radiation of the appropriate wavelength can be absorbed well within 1 μm. Therefore, extremely large amounts of energy can be delivered to usefully miniature volumes of material for confined three-dimensional material processing. For practical purposes, however, particularly when the sample must be enclosed in a cell for gas immersion processing, it is not always possible to achieve optimum high resolution, and spot sizes several times the diffraction limit may only be achievable.

Since high temperatures are necessary for diffusion, strong optical absorption is required in the substrate. In the case of semiconductor processing, laser wavelengths in the visible or UV are generally used to achieve this. Where pulsed lasers are applied, surface melting promotes rapid liquid-phase indiffusion of the dopant (Deutsch et al., 1980, 1981, 1982; Turner et al., 1981). For beam irradiation, solid-state diffusion, characterized by a diffusion coefficient seven orders of magnitude smaller than in the liquid phase, can more slowly redistribute the atoms to the required depth (Ehrlich et al., 1980, 1981).

Optical energy is absorbed in semiconductors by the electronic system and, on a timescale of the electron-phonon scattering events ($\sim 10^{-13}$ s), starts to be redistributed into vibrational modes of the atomic system. Therefore, heating rates approaching 10^{15} K/s are achievable with femtosecond and picosecond pulses. After fsec or psec irradiation, the thermal energy in the heated zone (α^{-1}, the absorption depth) diffuses into the bulk according to $L = (Dt)^{1/2}$, where D is the characteristic diffusivity, t is time, and L the diffusion length. For nanosecond pulses, one can consider that heating initially occurs within α^{-1}. In this regime, if $\alpha^{-1} < L$, energy delivered during the initial portion of the pulse can diffuse a distance L during the remainder of the pulse. Thus, the temperature profile induced is influenced by heat diffusion during the nsec pulse. If $\alpha^{-1} > L$, diffusion does not influence the temperature induced. For even longer pulse exposure, or cw irradiation, energy is usually delivered at such a slow rate that thermal diffusion determines the temperature profile. However, if sufficiently high energies are used at a particular pulse-width, thermal diffusion effects can be minimized to some extent. Nevertheless, it is generally well known that laser-induced heating leads to a strongly peaked temperature profile, both laterally, of the order of the size of the incident laser spot, and into the sample, to some depth between the absorption depth and the thermal diffusion length.

A large number of models have been developed to estimate the laser-induced temperature rise in semiconductors, and the subsequent behavior of the thermally excited system (see, for example, Baeri et al., 1979; Meyer et al., 1980; Nissim et al., 1980; Wood, 1982; Thompson et al., 1984; and Schvan and Thomas, 1985). Consequently, and with a prolific amount of experimental detail also published, the temperature profile within many laser-irradiated materials, especially semiconductors, can be readily estimated to a high degree of accuracy.

Simple models of laser-induced diffusion have also been developed over the years by Kiang et al. (1982), Ehrlich and Tsao (1982), Fogarassy et al. (1983), and Kräutle et al. (1985). With a knowledge of the temperature profile induced by the laser beam, one can calculate the spatial variation of the diffusion coefficient for a known dopant, D_d. In the solid phase, D_d is typically described by

$$D_d \alpha \exp\left(\frac{-E_a}{kT}\right), \tag{10.1}$$

where k is Boltzman's constant, t is temperature (in K), and E_a is the usual isothermal activation energy for diffusion; e.g., $E_a = 3.69$ eV for B in Si.

Since the temperature profile can be steep, the strong temperature dependence of D_d will lead to an even steeper diffusivity profile within the heated zone. For example, a temperature change of 100°C around 1000°C produces a 10-fold difference in D_d for the common dopant impurities in silicon. Figure 10.1 shows a typical calculated laser-induced temperature profile, together with the corresponding spatial variation of D_d, and the diffusion length for boron. Thus, it is readily obvious that submicron resolution should be achievable laterally with this technique. Using this property, doped linewidths as narrow as 300 nm have indeed been achieved (Ehrlich and Tsao, 1982).

In short-pulse (nsec) laser heating of semiconductors, it is generally quite straightforward to controllably melt very shallow depths into the substrate. This property of laser melting presents a possibility for producing extremely shallow doped layers. Figure 10.2, for example, shows spreading resistance measurements and Rutherford backscattering spectra (RBS) of laser-doped Si, where junction depths less than 0.1 μm

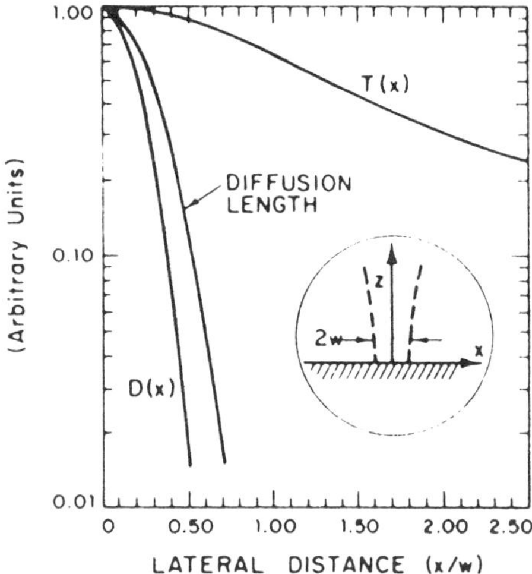

Fig. 10.1 Normalized lateral distribution of surface temperature $T(x)$ and the corresponding spatial variation of boron diffusion coefficient $D(x)$ and diffusion length for typical experimental conditions. With spotsizes between 1.5 and 2 μm, submicron diffused linewidths are predicted. From Ehrlich and Tsao, 1982.

have been produced (Sigmon, 1987). The technique, known as *gas immersion laser doping,* GILD, used a XeCl laser to simultaneously dissociate adsorbed species and drive them into the sample—in the case of As^+ doping, to a depth of 600 Å (Carey et al., 1986). Junction depths as shallow as 500 Å have been reported for ArF laser doping of Si with triethylboron [$B(C_2H_5)_3$ or TEB] by Ibbs and Lloyd (1983).

Comprehensive modeling of laser doping by liquid-phase diffusion has been performed by Fogarassy et al. (1983) for 20-nsec pulsed ruby-laser irradiation of Si. Figure 10.3 compares the modeling results with RBS obtained after a thin evaporated layer of dopant was irradiated by various beam fluences. The agreement for these relatively low-impurity doses is clearly excellent and shows the gradual increase in doped depth with the higher laser fluences, as expected by the increased melting depth. The formation of impurity-rich cellular structures was observed when larger dopant doses were used. Solid solubility limitations were thought to initiate this solid-liquid interfacial instability; although the effect of small variations of dopant concentration on the solidification temperature may also have been important. Therefore, although the impurities can be redistributed by laser diffusion, a fraction of them may not become

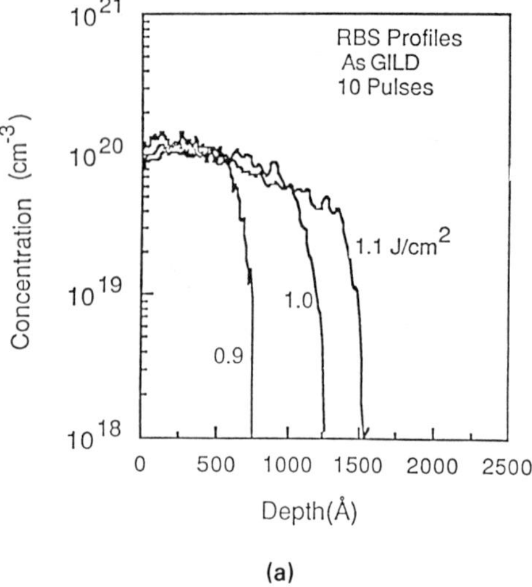

(a)

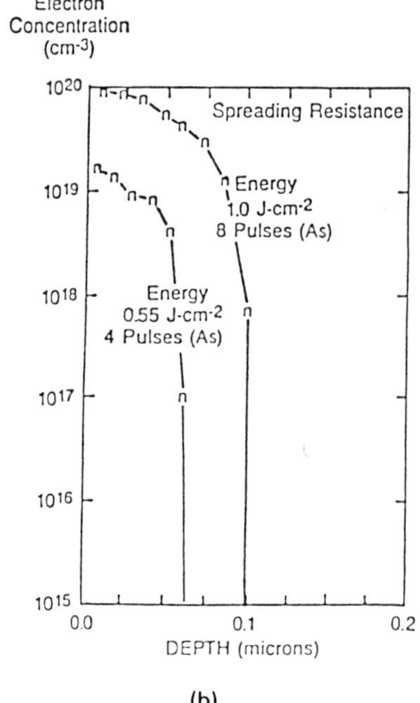

(b)

Fig. 10.2 (a) RBS profiles of As in laser-doped Si for 10 pulses and three different energies; (b) Spreading resistance profiles for Si excimer laser doped with As after 8 pulses at 1 J/cm^2. From Sigmon, 1987.

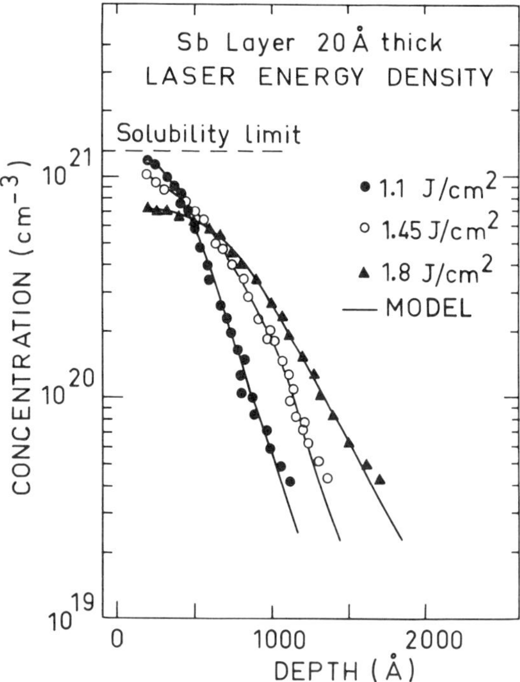

Fig. 10.3 Comparison of RBS spectra and modeling results ($k = 0.8$) of ruby-laser-diffused Sb films on Si. From Fogarassy et al., 1983.

incorporated into the usual lattice structure. Figure 10.4 shows the relative ratio of atoms substitutionally and nonsubstitutionally bonded in Si after laser doping from a 60-Å thick surface layer. Clearly, nearest the surface a sizeable proportion of the dopant has not been incorporated chemically into the lattice and therefore does not contribute to efficient charge conduction. More recently, Sameshima and Usui (1987) found excellent agreement between diffusion calculations and SIMS measurements after excimer laser doping of Si from a predeposited boron film. For laser pulse energies above 0.6 J/cm^2, melt front velocities faster than 8 m/s were calculated, and these were found to be so fast that the dopant could not diffuse away from the surface before resolidification occurred.

1.4. Device Fabrication

Several groups have applied selective laser doping to device manufacturing. Affolter et al. (1978), for example, first produced ohmic contacts by

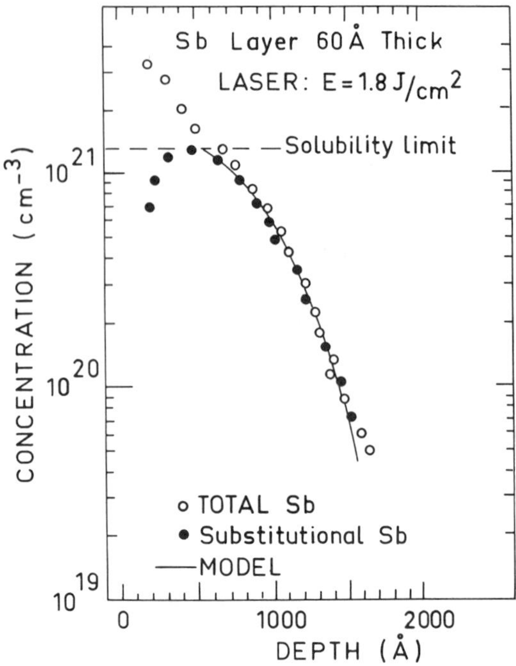

Fig. 10.4 Plot of the total Sb concentration and the substitutionally bonded Sb profile after ruby-laser irradiation of a 60-Å thick Sb film on Si with 1.8 J/cm^2 pulses, along with the atomic distribution predicted by modeling. From Fogarassy et al., 1983.

locally melting-in doped layers coated onto Si surfaces. Using an aqueous suspension of red phosphorous (for n-type doping) and diluted solutions of boric acid (for p-type doping), the Si was "painted" and dried before irradiation with pulses from either a CO_2 or Nd:YAG laser. However, relatively poor contact resistances were obtained, of the order of 0.1–1 Ω-cm^2 compared with 10^{-6}–10^{-3} Ω-cm^2 manufactured conventionally, and particular growth phenomena were suggested to be responsible. Narayan et al. (1978) similarly used a ruby laser to diffuse a pre-evaporated boron layer 0.3 μm into n-type Si to produce a p-n junction mesa diode. The reverse current of about 10^{-7} A/cm^2, with a forward current-voltage ideality factor of 1.39, was said to be of a superior diode and comparable with those obtained by traditional furnace processing. Kräutle and Wachenschwanz (1985) have shown that specific contact resistances of $3–5 \times 10^{-5}$ Ω-cm^2 for nonalloyed ohmic n- and p-type contacts on GaAs can also be obtained by laser doping in H_2Se and

DEZn, respectively. After alloying, up to one order of magnitude was further obtained in these values.

Carey et al. (1985, 1986) used diborane and arsine sources to provide B and As atoms during XeCl irradiation of n- and p- type Si. Junction depths around 0.1 μm with peak B concentrations of $2 \times 10^{20}\,\text{cm}^{-3}$ were obtained. The diodes manufactured showed regions of ideal-type behavior limited by series resistance at high forward bias and excess generation current at low forward bias, as shown in Fig. 10.5. A subsequent annealing step reduced both of these features and provided an overall ideality factor close to 1. Leakage currents as low as $47\,\text{nA/cm}^{-2}$ were obtained on $7 \times 10^{14}\,\text{cm}^{-3}$ doped substrates at a bias voltage of -10 V. More recently, even shallower p^+-n junctions (70 nm) have been produced by excimer-laser-induced melting and diffusion of RF-deposited boron into Si, producing reverse bias currents around $50\,\text{nA/cm}^2$ (Sameshima et al., 1987).

With the As doping, submicron gate length MOSFETs with laser-doped drain and source regions have also been produced (Carey et al.,

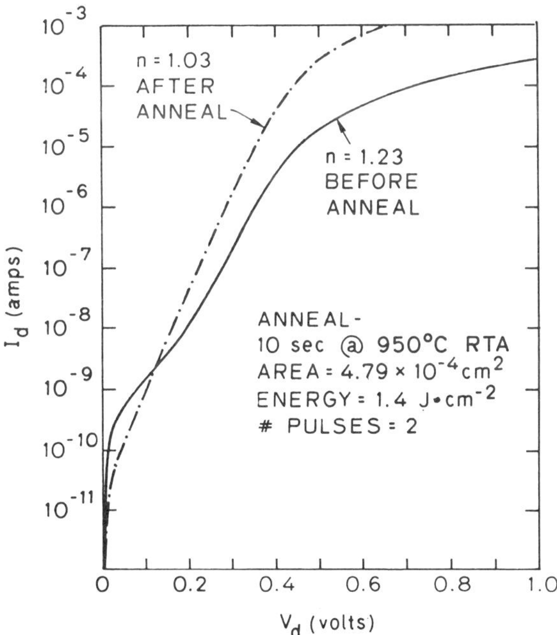

Fig. 10.5 Current-voltage characteristics of a diode fabricated by laser doping, before and after annealing. From Carey et al., 1985. Copyright © 1985 IEEE.

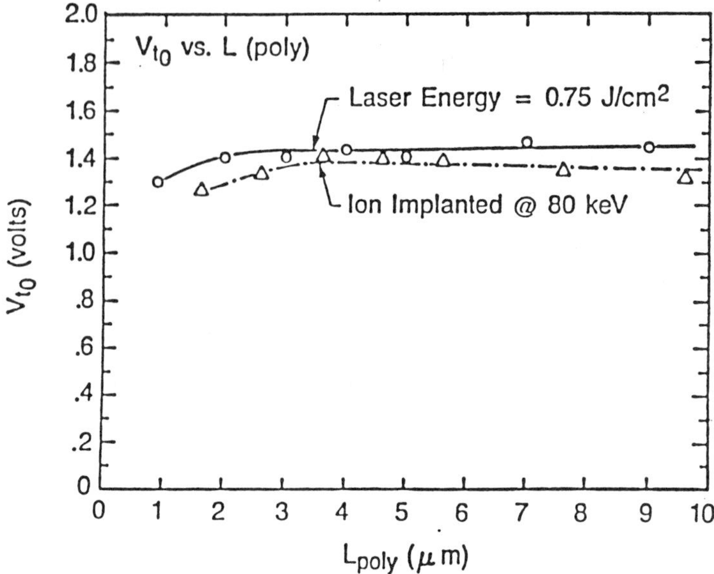

Fig. 10.6 MOSFET threshold variation versus poly gate length for laser-doped and ion-implanted devices. From Carey et al., 1986. Copyright © 1986 IEEE.

1985, 1986). Again, annealing produced slightly superior results, with leakage currents around 44 nA/cm^2 being achieved. Figure 10.6 compares the threshold voltage of the devices as a function of gate length (L_{poly}) with standard ion-implanted devices. Laser-doped MOSFETs gave better short-channel behavior than the best ion-implanted devices, and they also generally exhibited more uniform short-channel behavior because of the restricted diffusion of dopant under the gate region. McWilliams et al. (1983) have already shown that laser doping can be readily included within a completely laser-based technology for *ab initio* fabrication of n-MOS transistors and small area ICs.

The fabrication of p-n junctions in GaAs structures using laser doping from the gas phase has also led to successful solar cell manufacturing. Deutsch et al. (1982), using H_2S an the source for S dopant, found the ArF laser produced better devices than the XeCl laser. The best reported air mass 1 efficiencies were 10.8%, which reportedly compared favorably with the smaller area 12% obtained for ion-implanted and cw-laser-annealed devices.

As Table 10.1 shows, the doping process has not been extensively applied to the III–V materials. This is principally because of the requirement for high temperatures for diffusion, and such materials are notoriously difficult to handle under such conditions. However, in contrast to junction formation on InP, there was found to be less limitation on processing parameters for the fabrication of ohmic contacts (Deutsch et al., 1980). By depositing Cd or Zn from their parent organometallic compounds using an ArF laser, resistance-area products as low as $2 \times 10^{-4}\ \Omega/\mathrm{cm}^2$ have been obtained. These values are as good as the furnace-manufactured contact characteristics, and the $0.5–2 \times 10^{-4}\ \Omega/\mathrm{cm}^2$ values achieved by laser annealing of ion-implanted material (Deutsch, 1984). Figure 10.7 shows the progressive decrease in contact resistance of a photodeposited Cd contact as a function of increasing number of laser shots.

Many types of Si solar cells have been made using laser doping techniques. In fact, laser annealing of ion-implanted layers has produced efficiencies around 14–15% (with anti-reflection, AR, coatings), while laser diffusion of deposited layers has produced similar results (Young et al., 1980; Fogarassy et al., 1981; Muller et al., 1980). Early devices laser doped by gas-phase pyrolysis and diffusion compared favorably. Using PH_3 and a Q-switched alexandrite laser, efficiencies of 8.6% were

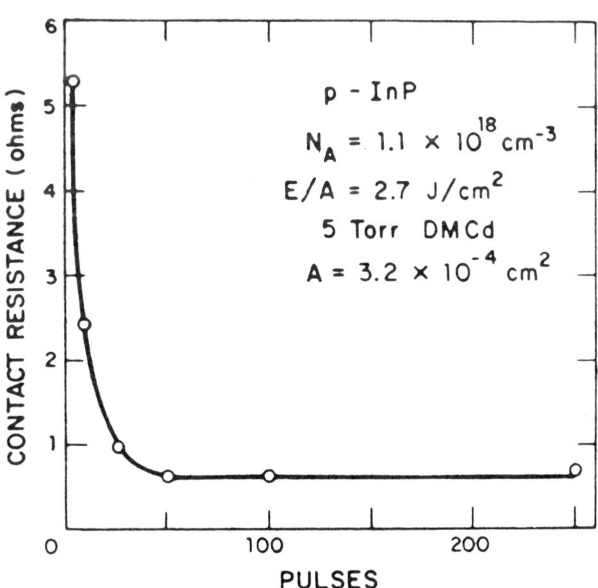

Fig. 10.7 Resistance of a photodeposited ohmic contact on p-type InP as a function of the number of ArF laser pulses at 2.7 $\mathrm{J/cm^2}$, using 5 torr of DMCd. From Deutsch et al., 1980.

obtained by Turner et al. (1981), while ArF laser photolysis of TMA or TMB produced device efficiencies up to 9.6% (Deutsch et al., 1981), without AR coatings, but equivalent to AR efficiencies up to 14%.

Most laser doping applications to solar cells has involved a-Si:H and a a-SiC:H technologies. Some of the first devices used photosensitized deposition reactions initiated with the Hg lamp, as discussed above. There were p-i-n cells, where the i and n layers were photodeposited on glow discharge (GD) layers, with efficiencies around 4.4% (Inoue et al., 1983). This is compared with typical solar cell efficiencies around 11% obtained using GD or plasma CVD prepared layers. More recent work involved completely photodeposited p-i-n devices in a single reaction chamber, with efficiencies of 8.29% (Tanaka et al., 1984), and by performing the three deposition tasks in separate reaction chambers efficiencies were further improved to 9.64% (Takei et al., 1985).

Direct photodeposition of p-type a-SiC:H layers allowed Yamada et al. (1985) to fabricate cells of 9.46% efficiency. By using a "super chamber," with 10^{-9} torr background pressure, low gas leakage, and impurity concentrations, Nakano et al. (1986) have incorporated p-SiC layers produced by photo-CVD into a-Si solar cells, thus obtaining operating efficiencies of 11.7%. The first photo-deposition of new amorphous superlattice-structure films by the same group has also led to a p-layer cells with efficiencies of 10.5%. These developments indicate the promise of LCVD of doped films for low-cost solar energy cells.

Probably the most original application of laser doping thus far is towards VLSI programming. Miyauchi et al. (1983) have shown, by employing a nitrogen-pumped dye laser, that intrinsic polysilicon can be selectively doped from an overlying passivation layer containing phosphorous. By irradiating a region between two memory cells, a conducting path is created by reducing the intrinsic resistivity by seven orders of magnitude to 1000 Ω. This technique has been applied to 2-μm high-density CMOS memory devices successfully without any sign of damage.

2. Laser Oxidation

2.1. Introduction

As can be seen in this volume, a wide variety of reactions can be induced by laser radiation. This section will review one particular class of reaction that has received considerable attention, namely oxidation. Because of its unquestionable technological importance, silicon oxidation has been extensively studied and therefore will form the bulk of this section, although oxidation of other semiconductors and metals will also be discussed. This review will focus on the kinetics involved in the reactions, parcicularly photonic-induced mechanisms. A more complete discussion of photoformation of dielectrics and their properties is reported elsewhere (Boyd, 1987).

2.2. *Silicon Oxidation*

Silicon dioxide (SiO_2) has contributed significantly to the development of the planar process and MOS technology as well as the high level of performance of many other advanced microelectronic devices. The natural oxide of silicon is an excellent insulator, can withstand very large electrical fields, and is chemically very stable. In fact, the band-gap electronic states normally present at a clean surface of crystalline silicon (c-Si) as a result of free radicals, or dangling bonds, can be reduced by more than five orders of matnitude under ideal conditions by the growth of a thin oxide layer on the surface.

Traditionally oxidation has been performed in RF-heated quartz-tube furnaces at temperatures between 700°C and about 1200°C. In recent years, intense and directed optical radiation has been investigated as a means of heating Si to the desired temperatures (Boyd, 1986). Heating rates can be as fast as 10^{15} K/s using picosecond laser radiation or as modest as 500 K/s using incoherent light sources. Beam technology can also give rise to localized and confined processing, presenting the opportunity of fine-scale pattern production, either by direct writing or by projection methods. Optical technology also presents new processing conditions under which to study the oxidation reaction. Thus, for the first time, we can study Si oxidation in extremely short timescales, or initiate or alter specific reaction pathways via photon-induced excitation.

2.3. *Thermal Oxidation of Silicon*

The mechanism by which c-Si oxidizes has for several decades been the subject of intensive investigation. Yet, remarkably, the simple thermal reaction of solid Si and dry oxygen gas, $Si + O_2 \rightarrow SiO_2$, is not yet fully understood. Whilst many models can predict reaction rates under certain conditions, none can explain the observed behavior under a wider set of experimental environments, as discussed in detail elsewhere (Boyd, 1986, 1987).

Radioactive tracer experiments have confirmed that for the growth of oxides thicker than a few hundred angstroms, the oxidizing species diffuses through any pre-existing film to the Si/SiO_2 interface, where it reacts with the available Si atoms. As oxidation proceeds, Si is consumed below the oxide as the Si/SiO_2 interface retreats further into the bulk.

The nature and charge state of the oxidizing species has been the source of much controversy over the years. However, it presently appears from the work of Modlin and Tiller (1985) that neutral species are responsible for oxidation when the oxide layer already formed is thicker than 300 Å. The role of charged species, however, such as O_2^- or O_2^{2-}, may yet be important during the reaction, when the intervening oxide layer is thinner. It is known that oxides of 200 Å and less will be routinely required to enable the drive towards ever-diminishing geometries to continue, and it is becoming more crucial to understand the optimum criteria for producing high-quality oxide layers at these thicknesses and less.

2.4. Optically Induced Oxidation of Silicon

Intense optical beams can provide unique environments for the growth of thin oxide layers. These include gas-phase deposition (photochemical and pyrolytic) (Boyer et al., 1982; Szikora et al., 1984), photooxidation of SiO in O_2 (Blum et al., 1983; Fogarassy et al., 1987), O-implantation and annealing (Chiang et al., 1981), or solid- or liquid-phase annealing of amorphous samples in O_2-rich environments (Garulli et al., 1980; Boyd, 1983b). These particular methods are reviewed elsewhere (Boyd, 1984). Here we will discuss the application of pulsed and cw optical sources to induce or enhance thermal oxidation on c-Si. Several regimes will be examined, namely, rapid thermal oxidation in Section 2.4.1, photonically enhanced oxidation in Section 2.4.2, and liquid phase oxidation in Section 2.4.3.

2.4.1 Rapid Thermal Oxidation

Thermal oxidation of c-Si in an RF-heated furnace usually requires around 10 minutes heating up-time and 10–25 minutes of ramping-down time on either side of the actual reaction period. Oxide growth normally occurs at a rate of 0.3–3 Å/s in production, depending upon temperature, oxygen purity, substrate doping and orientation, etc. However, intense 10-μm radiation from CO_2 lasers can induce similar reaction rates with heating and cooling times between 10^{-3} and 10 s (Boyd et al., 1981), introducing the regime of rapid thermal processing (RTP).

Although 10-μm radiation is not strongly absorbed by c-Si, preheating to 400°C increases its absorption by several orders of magnitude (see Fig. 10.8). The Si can then be readily heated by the laser such that it glows red around 700°C, and progressively orange and yellow at higher temperatures, and then brilliant yellow when melting is achieved. Since the reaction mechanism in this case is principally thermal, no modifica-

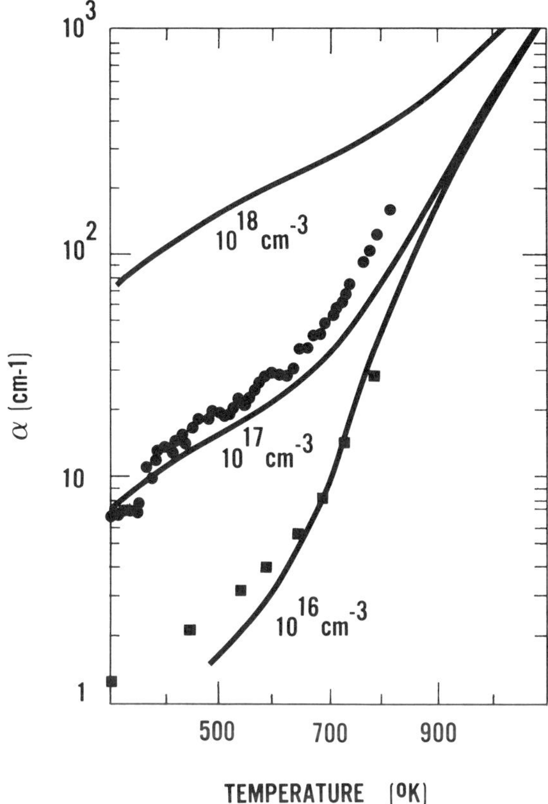

Fig. 10.8 Temperature dependence of the 10.6-μm absorption coefficient of c-Si for 10^{15} cm^{-3} doped samples (■, ●) compared with theory (solid lines). From Boyd et al., 1984).

tions to the oxide growth process should be expected. Because of the Gaussian beam shape applied, the temperature rise across the surface is radially variant, so this technique lends itself to the study of the physical and chemical properties of oxides grown at many different temperatures simultaneously but under otherwise precisely identical conditions. A recent study (Boyd and Wilson, 1987) has taken advantage of this technique to study the IR absorption properties of ultrathin oxide layers grown on c-Si.

RTO by visible radiation is also possible, and indeed ruby and argon lasers, and intense incoherent lamp sources, have recently been used to this effect. One of the earliest studies in the field used 30-nsec pulses up to 1.3 J/cm^2 to heat in short periods of time, and apparently not melt, c-Si in well-controlled environments. Thus, gradual oxide coverage on freshly cleaned c-Si surfaces could be monitored (Cros et al., 1982). It was found that after approximately 27 pulses, a saturated thickness of about 6.5 Å (i.e., two monolayers) developed beyond which oxidation barely progressed. It is remarkable that growth rates in this regime are around 6 mm/s. The number of atoms striking the surface during irradiation cannot account for the total oxide growth, and since any contribution from interstitial oxygen in the bulk Si was effectively ruled out, adsorption prior to irradiation was suggested to be the controlling mechanism. However, in high vacuum, laser radiation can also desorb impurities from the c-Si surface. Since there are no comparable rates for thermal oxidation in the initial few monolayers, it is not clear whether the radiation itself introduces additional nonthermal mechanisms that may enhance the reaction in some way. The emergence of processing equipment incorporating intense lamp sources in the past few years, primarily for work in the annealing and alloying areas, has opened up new possibilities for rapid thermal processing in general. This is a simple extension of the earlier work using focused laser beams to achieve rapid heating, although for large area processing it yields clear advantages (Nulman et al., 1985). Very few groups have studied the growth rates of the layers in detail, but Ponpon et al. (1986) have found a possible enhancement in their measured oxidation rate, while Chan Tung et al. (1986) reported a rapid initial oxidation rate of up to 11 Å/s during the first 5 seconds of the reaction and just over 3 Å/s during the next 5–60 seconds at 1250°C. Activation energies of 0.9 eV (Chan Tung et al., 1986) and 1.2 eV (Nulman et al., 1985) have been extracted from the growth data for the rapid oxidation regime, while 1.4 eV was calculated for the slower post-5-second oxidation (Chan Tung et al., 1986). These observations do not agree with extrapolations of the data for conventional oxidation. For example, Massoud et al. (1985) find a 3.2 eV activation energy for conventional thermal oxidation at temperatures above 1000°C. Clearly, further work is needed to establish whether this RTP procedure is purely thermal in nature. Possible mechanisms that could induce nonthermal or "photonic" effects are discussed in the following section. Whether or not these enhancements are operative, it is clear that this "optical furnace" will become an important piece of silicon processing apparatus in the near future.

2.4.2 Photonically Enhanced Oxidation

More than 20 publications have shown that optical radiation can not only induce oxidation of various materials, but that it can also *enhance* the reaction somewhat (Boyd, 1987). This has been observed not only for Si, but also for GaAs (Petro et al., 1982; Bermudez, 1983; Siejka et al., 1985; Bertrand, 1985) and other III–V binary, ternary, and quaternary compound semiconductors (Fukuda and Takahei, 1985), InP (Fathipour et al., 1985), and SiO (Blum et al., 1983), as well as metals (Ursu et al., 1984a,b) and the not-unrelated reactions of laser enhanced nitridation (Sugii et al., 1984) and etching (Chuang, 1984), and the reaction of hydrogen with graphite (Ashby, 1983). Whilst some of the photonic mechanisms may apply to one or more of the reactions, we shall concentrate in this section only on the enhanced oxidation of c-Si.

The term *photonic enhancement* (PE) used here embraces all reported effects in a light-induced reaction that cannot be attributed to simple thermal mechanisms. Figure 10.9 summarizes most of the proposed PE mechanisms in the literature in terms of spectral response. An attempt has been made to estimate the relative contributions of the various effects for a constant incident beam power, but the reader is warned that not only is the relative importance of each open to debate, but in several cases, so are some of the enhancement effects, as will be discussed below. It is also important to recognize that the photon flux directed at the c-Si surface during the enhancement will inevitably also induce some thermal effects, since none of the reactions reported to date are 100% quantum effcient. Therefore, in almost every experiment, there will be a background oxidation enhancement due to an increase in sample temperature.

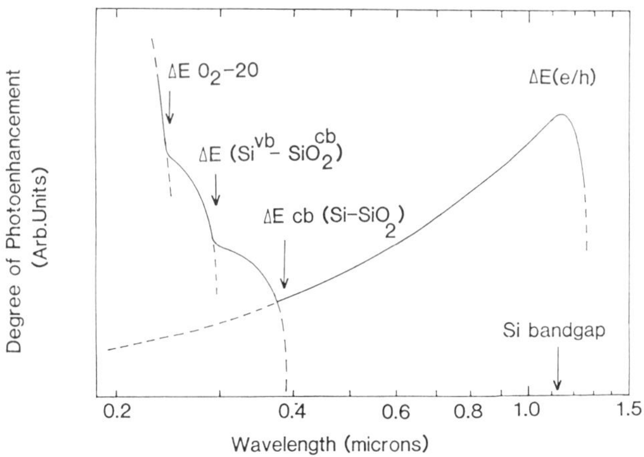

Fig. 10.9 Spectral summary of the various enhancement mechanisms proposed for photonically induced silicon oxidation involving energy differences between several electronic transitions in Si and SiO_2.

The first report of light-enhanced Si oxidation was by Oren and Ghandi (1971), who used UV radiation during a conventional thermal oxidation reaction. They attributed the observed increase in reaction rate to the photogeneration of electrons, which are then liberated into the oxide leaving the c-Si in a highly charged state. The energy threshold for this was suggested to be that required by electrons to make a transition from the valence band of the Si into the conduction band of the oxide (4.25 eV). This corresponds to a wavelength of 292 nm and a prominent line of their mercury arc lamp near 254 nm. A similar proposition was made by Schafer and Lyon (1982), who applied low-power optical radiation to a Si sample that was already undergoing thermal oxidation in a furnace at elevated temperatures. Owing to the fluence levels used, however, allowances had to be made for the increase in temperature, and only small PE rates were extracted, typically of the order of 20–50%. A major problem with this and related techniques is the satisfactory isolation of thermal and non-thermal effects. As a consequence of their study, it was noted that there was a threshold to the enhancement around 3–3.5 eV, which was attributed to the photoexcitation of conduction-band electrons from the Si into the conduction band of the oxide. Since this does not require the Si band-gap energy (1.1 eV), it closely corresponds to the threshold proposed earlier.

The majority of the present PE models are based principally on the indirect laser photogeneration of conduction electrons in the Si from the valence band, a process that occurs at photon energies near to and above 1.1 eV (Gibbons, 1981; Boyd et al., 1981; Boyd, 1983; Schafer and Lyon, 1981). In chemical terms, this may be likened to antibonding. Gibbons (1981) also proposed photoionization. In these models, the enhancement is related to the new equilibrium concentration of photogenerated carriers above that established by the usual intrinsic thermal generation, which may affect the reaction-limited process.

Increased laser powers enable the furnace thermal contribution to be minimized and the incident photon flux to be maximized, and thus oxidation can be seen to proceed by factors of 2–5 faster than the furnace-induced rates (Gibbons, 1981; Boyd et al., 1981; Boyd, 1983). However, isolation and analysis of the photonic contributions to the reaction are difficult. Because direct methods of temperature monitoring cannot easily be used here, heating effects caused by the relaxation of the photoexcited carriers have to be estimated using laser heating models. In these, a small uncertainty in the beam energy or spot size can translate into a larger error in estimated growth rate. For example, Fig. 10.10 shows how an uncertainty in the measurement of the laser beam parameters of $\pm 10\%$ leads to an error in estimated temperature of 50 K

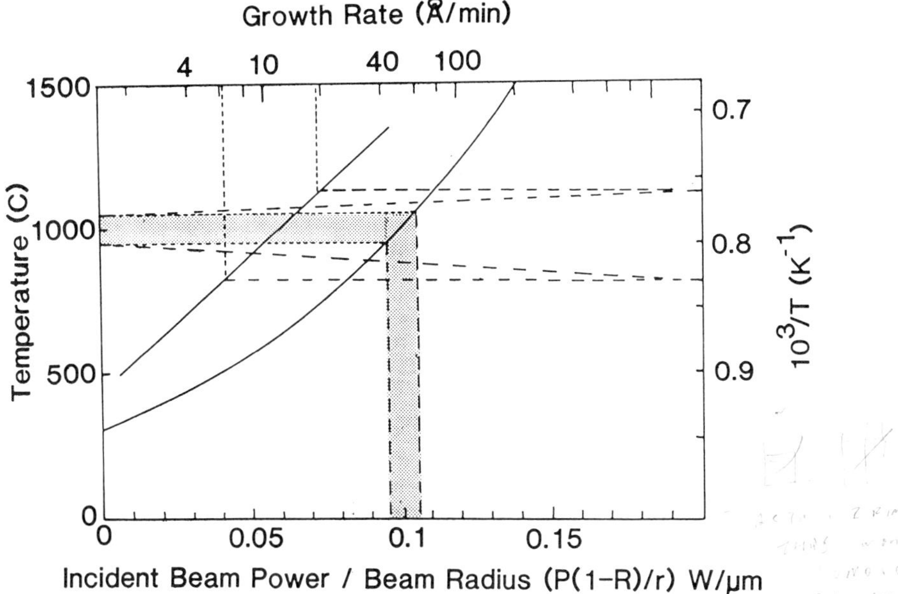

Fig. 10.10 Example of how an uncertainty of ±5% in the laser beam parameters can lead to an "error" in predicted temperature, which when transformed gives a factor of three spread in estimated oxidation rate. From Boyd, 1986.

using, for illustrative purposes, a perfect theoretical heating model. Such a spread in temperature represents, by transforming to reciprocal units, an uncertainty of a factor of three in the thermal growth rate. This also indicates that any successful temperature measurement must be capable of accuracies significantly better than ±25 K. Nevertheless, from the calculations presented (Boyd, 1983; Schafer and Lyon, 1982), it is clear that the PE effects in the visible are secondary to the thermal contributions induced by electron-phonon scattering.

Several qualitative experiments have recently been performed to highlight PE in the visible. Young and Tiller (1983) found that a larger enhancement occurred for (100)Si than with (111)Si. Because of the different density of atoms in each crystal plane, this was contrary to expectations for a purely thermal process. The enhancement was also found to be directly proportional to the number of photons incident during the reaction, rather than the power density of the Ar laser beam. Thus, by reducing the power of the 514 nm beam to less than that of the 488-nm light, in order to equate the flux at each wavelength, the same rate of growth could be obtained. Similarly, Boyd and Micheli (1987)

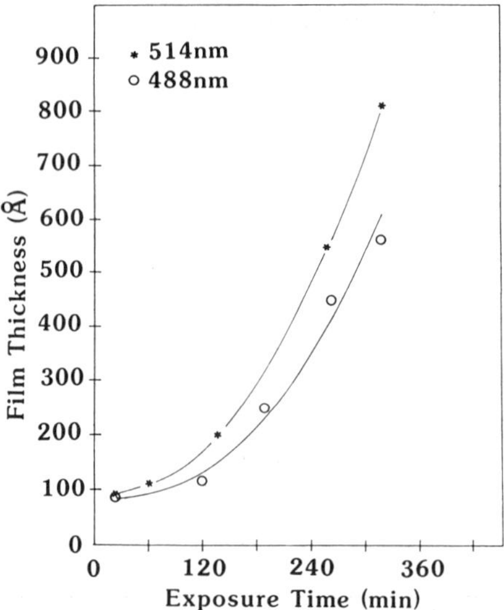

Fig. 10.11 Oxide thickness grown by equal (initial) absorbed power levels of 488- and 514-nm radiation from a cw Ar laser. From Boyd and Micheli, 1987.

used the multiple interference characteristics of the thin oxide to amplify the small differences the growth rate induced by identical absorbed powers of 488 and 514 nm laser radiation. Figure 10.11 shows the oxide thickness as a function of irradiation time, indicating that the *longer* wavelength light induces faster film growth. Consideration of the slight differences in absorption and thin film reflectivity properties between the two wavelengths would tend to favor increased thermal contributions from the 488 nm radiation. Clearly, the extra photon flux presented by the 514 nm radiation more than compensates for this. Simple estimates reveal that the difference in photonic contributions between the 488 and 514 nm light corresponds to an increase of about 20% in the thermal reaction rate (Micheli and Boyd, 1987b).

The question remains as to what effect photoexcited carriers have on the overall reaction. Since the early stages of Si oxidation are presently not well understood, it is difficult to define a precise photonic mechanism. An initial attempt at this (Boyd, 1983) was based upon the model proposed by Blanc (1978). Since this model is essentially a special case of the more general model of Ghez and van der Meulen (1972), the latter remains a possible starting point for analysis. Basically, the proposed mechanism involves O_2 diffusion to the Si/SiO_2 interface, where an equilibration reaction produces some O atoms from the O_2, both of which react with the silicon. Similarly, the more recent model of Hu (1983), which also encompasses the Blanc theory, may also be valid. Hu assumes the intermediate step in the reaction is the chemisorption of O_2 molecules so that the pressure dependence of the linear rate constant takes the form of the chemisorption isotherm, taken to be the Freundlich isotherm. For ultrathin films, Hu (1983) has suggested that the transported oxidant may no longer be molecular oxygen. Nevertheless, it would not be be difficult to associate PE with photostimulated charged species effects. In fact most of the models requiring the presence of electrons, or holes, or partially ionized silicon could predict PE effects in the visible.

The final category of enhancement modes depicted in Fig. 10.9 is the photo-dissociation of molecular oxygen close to the Si substrate. The bond energy of O_2 is close to 5.1 eV, corresponding to a wavelength of 240 nm. Photons with energy greater than this will crack the molecule into two extremely reactive atoms. If they do not react immediately with the Si or the oxide film, they may alternatively react with any remaining O_2 to form ozone (O_3), a much more powerful oxidant than its allotropic cousin. Further reactions may also be induced between ozone and the UV radiation. Consequently, oxidation will proceed extremely quickly. Shorter UV radiation can also interact with the O_2, or directly with substrate, but this regime has not yet been explored.

Finally, we discuss the phenomenon of UV (5 eV)-induced bond reorientation within an adsorbed layer of oxygen on a Si surface (Fiori, 1984). It has been found that after exposure to 248 nm radiation an energetically unstable oxide is formed on the Si. After annealing at 949°K, the oxide relaxes into a more conventional amorphous SiO_2 structure. No oxide could be formed when 308 nm radiation was used. Although these effects were induced with an absorbed power density of 0.5 W/cm^2, films as thick as 200 Å were grown when 1.5 W/cm^2 was applied for 300 minutes to c-Si in an oxygen pressure of 1 kg/cm^2.

2.4.3 Liquid-Phase Oxidation

Silicon oxidizes when gaseous oxygen dissolves into the thin pregrown oxide layer on the Si and diffuses to the oxide-Si interface where it reacts with the available Si atoms. As the temperature is increased, so does the solubility and the diffusivity of the oxygen, as well as the rate of production of thermally excited Si. During the growth of thin oxides, the latter process is usually considered to most important. However, Si can be readily melted by lasers, thus introducing completely novel oxidation regimes (Garulli et al., 1980; Hoh et al., 1980; Cohen et al., 1986).

The kinetics of the process are extremely complicated in this rapid transient situation and by no means yet fully understood. During irradiation, the Si heats up in the solid phase until 1685 K, and once the latent heat has been absorbed, it melts, transforming both the optical and thermal properties. Strong temperature gradients will induce thermal convection within the liquid, and at sufficiently high beam intensities the liquid will enter a further latent heating regime and evaporate, creating significant losses of material. Eventually it would be reasonable to assume that the evaporated material would shield the Si from the tail end of the pulse. The amount of O_2 incorporated would thus be limited at some saturation fluence, as has been experimentally observed (Cohen et al., 1986). During this time, species from the surrounding ambient will impinge upon, and dissolve and react with, the surface. Products such as SiO may leave the surface, and any nearby gas-phase reactions between thermally or photoexcited species and the ambient could also be important. Upon resolidification, the quantity of atoms dissolved, together with their precise location and the solubility and segregation characteristics of the species, will determine the amount and quality of oxide formed. Thus the precise temperature reached and the nature and pressure of the ambient, as well as the presence of additonal impurities, will affect the overall reaction.

Another important criterion is the presence of any surface oxide on the Si before irradiation. SiO_2 melts at 1980 K, whereas c-Si melts at 1685 K. If the laser beam induces temperatures between these points the Si will melt, and one would expect increased oxygen dissolution and reactivity. Estimates have shown that only a small energy increase is required to achieve this for an insulating film on top of a very efficient bulk metallic conductor (Cohen, et al., 1986). Extrapolating the data of Hirata and Hoshikawa (1980) reveals that just below the melting point of SiO_2, the dissolution rate into Si would be around 500 Å/s, and a native oxide layer of 20 Å would take 40 msec to dissolve, which is several orders of magnitude longer than the known time duration of the molten phase induced by nanosecond pulses. Any carbon in the melt (unavoidably present at levels up to 5×10^{17} atoms/cm^3 in Czochralski Si) will accelerate the dissolution of SiO_2 only slightly by encouraging the formation of SiO and CO (Bathy et al., 1980). Other impurities could also have similar effects on the reaction.

Neglecting these influences for the moment, it is instructive to estimate the appropriate rate constants for the reaction at a temperature just below the oxide melting point. The impingement rate of gas-phase molecules onto a surface can be estimated, assuming the ideal gas law, by

$$M_i = \frac{3.513 \times 10^{22} P}{\sqrt{MT}}, \tag{10.2}$$

where P is the equilibrium gas pressure (in torr) at temperature T and M is the molecular weight of the gas. For pure oxygen at room temperature and one atmosphere this is $2.7 \times 10^{23}\ \mathrm{s^{-1}\ cm^{-2}}$. It can be calculated that around 2.2×10^{14} oxygen atoms are present per Å in SiO_2. Therefore by assuming a sticking coefficient of unity, sufficient oxygen could be incorporated to grow one angstrom of oxide every nanosecond. This, of course, assumes no oxygen evaporation from the melt and a rapid diffusivity of the oxygen in the liquid silicon. We estimate the oxygen diffusivity in molten Si to be approximately $10^{-4}\ \mathrm{cm^2/s}$, while for diffusion in solid SiO_2 at 1975 K it is calculated using (Tsai, 1983)

$$D = 2.7 \times 10^{-4} \exp\left(\frac{-1.16}{kT}\right) \tag{10.3}$$

and found to be $3 \times 10^{-7}\ \mathrm{cm^2/s}$, some 100 times faster than at 900°C. Thus, oxygen can diffuse through 100 Å of liquid silicon in only 5 nsec, or through 1000 Å in 500 nsec. If a 20 Å film of native oxide is present, the oxygen will take nearly 70 nsec to travel across into the liquid. Therefore in this regime the maximum oxidation rate could be limited by solid-phase transport.

Resolidification processes will also affect the mechanism of O incorporation. At low levels of oxygen incorporation, the Si may solidify either in crystal form or in an amorphous state, depending upon the cooling rate. The oxygen can be incorporated in many ways, e.g., interstitially, as precipitates, or in polymerized complexes, sometimes involving additional impurity atoms. If the concentration of oxygen is sufficiently greater than the solubility level at a particular temperature, then the supersaturated condition is reached and precipitation will occur. The preferred location for the nucleation of these oxides or conglomerates is not yet known. If a thin film is formed with one pulse, this could introduce a diffusion barrier to the oxidant if a second pulse attempts to repeat the process, highlighting important considerations between single-shot and multishot liquid-phase oxidation.

Cohen et al. (1986) have argued that as much as 80% of the O_2 impinging upon the molten Si is not incorporated, introducing questions regarding the balance between adsorption and loss of oxygen under these transient heating conditions. Westendorp et al. (1982) found no evidence for oxygen diffusion into Si (with a native oxide layer) exposed to 20-nsec ruby-laser pulses at fluences ~1.5 J/cm^2. These observations can be explained by the diffusion arguments outlined above, providing the native oxide does not melt. Hoh et al. (1980) have similarly found that a 905 Å oxide layer inhibited further oxidation that could otherwise be induced by 150-nsec pulses at 0.53 or 1.06, μm sufficient to melt but not vaporize the c-Si. However, Garulli et al. (1980) found, in contradiction to Westendorp et al., that for identical irradiation conditions, an oxygen supersaturation could be obtained midway in depth through the laser-melted region. A possible explanation is that Garulli et al. underestimated the laser fluence that caused oxide melting, atomic evaporation, convective flow, or boiling. This is supported in part by the more recent data of Bentini et al. (1984) showing that oxygen is only mildly incorporated for fluences just above the melting threshold, while a massive uptake occurs for fluences around three times the melting threshold. Thicknesses up to 500 Å have been grown this way. Cohen et al. (1986) have considered many possible incorporation and loss schemes for the reaction, as well as convection, under these conditions.

UV laser radiation has also been used to induce liquid-phase oxidation of Si. Liu et al. (1981) used 15-nsec pulses at 266 nm to induce a crystal-amorphous phase transition in c-Si by rapid quenching, and in an air or O_2 ambient, oxide layers up to 300 Å were produced. Fogarassy et al. (1986) have also shown that oxygen incorporation by multishot melting in the UV is very much enhanced by the presence of specific impurities during the melting. In fact, by contrast to As- and Sb-doped Si, both In and Bi enabled appreciably higher rates of incorporation, and this was partly related to surface segregation effects. Richter et al. (1984) have used multiple shots from an XeCl laser at 308 nm to grow oxides as thick as 3000 Å at rates of more than 100 Å/s. The rapid reaction rate was achieved with fluences up to 0.85 J/cm^2, just above the single-shot melting threshold (Unamuno, 1984). It is not yet understood why these experimental conditions are conducive to such efficient oxide growth.

2.5. Oxidation of Compound Semiconductors

Early experiments involving pulsed-laser processing of compound semiconductors induced melting with segregation of the constituent atoms. However, when GaAs was irradiated in air or O_2, both hexagonal Ga_2O_3 and cubic As_2O_3 were formed, although the top 500 Å was found to be Ga rich (Matsuura et al., 1981). As with Si, experiments showed that before any oxygen became incorporated, the incident beam intensity had to be nearly three times that required to melt the GaAs, E_{th} (Bentini et al., 1982). Indeed, between E_{th} and $3E_{th}$ no oxides were formed. It has since been shown that the native oxide must be evaporated at this higher intensity before significant new oxide formation can occur by oxygen penetration through the melted region (Cohen et al., 1986).

The formation of oxides on GaAs without surface melting is clearly a much more attractive proposition. Petro et al. (1982) have shown that low-intensity argon laser irradiation ($<3\,W/cm^2$) of cleaved GaAs substrates increases the sticking coefficient of O_2 by a factor of 1000. Heating effects and oxygen excitation were ruled out as possible causes of the enhancement in favor of electronic excitations of the GaAs surface. It was suggested that oxidation proceeded via absorption by a loosely bound oxygen molecule, followed by breakup of the molecule into reactive species (Bertness et al., 1984). Evidence for this model indicated that in the dark the uptake of oxygen by the GaAs was strongly temperature dependent, while under laser illumination the chemisorption was totally independent of thermal activation. Coverage of up to two monolayers was found, with no evidence of saturation.

Bermudez et al. (1983) have shown that irradiation with similar intensity incoherent radiation also produced photoenhanced solid-state oxidation of GaAs in the presence of O_2, but not in CO and H_2O atmospheres. Since the latter molecules do not chemisorb, this suggests that the intramolecular bonding in the oxidizing species must be weakened for the reaction to occur. The bonding in the adsorbate is weakened by charge transfer from the bulk, and oxidation proceeds with the photoenhanced breaking of Ga-As bonds. Enhancement factors up to 20 have been observed in the times required to form the first monolayer of oxide, while the rate of photoenhancement does not strongly depend upon the type of excess carriers present in the material. Whilst experiments with NO indicated a much smaller sticking coefficient and subsequent oxide growth than with O_2, Bertrand (1985) found that oxidation of GaAs irradiated by 254 and 185 Å lamp wavelengths occurred in the presence of N_2O. In this case the As-rich surfaces oxidized much more slowly, and saturation occurred once all the adsorbed oxygen became incorporated to form the oxide.

Siejka et al. (1985) have similarly observed preferential oxidation of Ga, in addition to some As loss during 193-nm laser irradiation in an ^{16}O atmosphere of GaAs ^{18}O enriched native oxides. Additionally, very efficient exchange of oxygen was observed after 1000 laser pulses, with the total oxygen content of the films being conserved. Since thermal heating was considered negligible, the authors appear to have found a new mechanism of laser-assisted formation of highly mobile oxygen species in GaAs.

An independent study using cw argon laser radiation also found optically enhanced oxidation rates for GaAs and for other III–V and ternary compounds (Fukuda and Takahei, 1985). It was found that after initial rapid oxidation, film growth proceeded in proportion to the logarithm of illumination time when intensities of about 1 kW/cm^2 were used. Semiconductors containing Ga oxidized easily during the initial stage of irradiation, while those containing As oxidized more readily than the others only after a certain amount of film growth or at temperatures near 300°C. While calculations showed that the laser increased the temperature by only around 20°C, the growth rate was enhanced as if the temperature had increased by 50°C. Since the logarithmic oxidation behavior did not change under laser illumination, it was suggested that the enhancement was due to accelerated transport of the elements necessary for oxidation.

The oxidation of InP has been achieved using an ArF laser in an atmosphere of N_2O (Fathipour et al., 1985). The process is most likely initiated by the photodissociation of the N_2O molecules, since it was found that for a given constant laser power the oxide thickness increased with increasing N_2O pressure. When the laser beam impinged directly on the sample surface during the film growth, the oxidation rate was enhanced by approximately 20%. In agreement with the photon-enhanced Si oxidation models, and some of the related ideas connected with enhanced reaction rates also discussed throughout this chapter, it was suggested that the increased concentration of electron-hole pairs could assist in the breakage of the lattice bonding, thereby increasing the availability of atoms for the reaction. Films as thick as 160 Å have been grown using this method, and these appear to be amongst the thickest laser-grown oxides on compound semiconductors reported so far.

2.6. Oxidation of Metals

The CO_2 laser-induced thermochemical reaction of oxygen with the surface atoms of metal mirrors has for many years been a severe problem in IR laser studies when using high-intensity CO_2 lasers. In this section we will discuss the use of lasers to deliberately grow metallic oxides using a wide variety of laser-initiated reactions (see Table 10.2).

Controlled heating of the substrate within an O_2-rich environment offers unique opportunities to study not only the early stages of the oxidation reaction, but also the structural evolution of the oxide layers themselves. Such has been the drive behind the investigation of cw CO_2 laser oxidation of vanadium (Ursu et al., 1984a) and copper (Wautelet and Baufay, 1983, Ursu et al., 1984b). For example, it has been shown for the case of copper (Ursu et al., 1984c) that there is generally good

Table 10.2 Summary of Methods for Laser Oxidation of Metals.

Substrate	Oxide Species	Laser	Reference
Cu	Cu_2O	Ar	Baufay et al. (1987)
Cu	Cu Oxide	CO_2	Arzuov et al. (1979)
Cu	Cu_2O, CuO	Ar	Andrew (1986)
Cu, Cd	Cu_2O, CdO	Kr	Wautelet & Baufay (1983)
Zr	ZrO_2	CO_2	Ursu et al. (1986b)
V	V_2O_5, VO_2	CO_2	Ursu et al. (1984a)
Cu	Cu_2O, CuO	CO_2	Ursu et al. (1984b)
Ti	TiO_2	Nd:YAG	Thuillard & von Allmen (1985)
Ti	TiO, Ti_2O_3, TiO_2	Nd:glass	Akomov et al. (1982)
$Ti_{50}Zr_{10}Be_{40}$	Ti_2O_3	Kr	Merlin & Perry (1984)
Cr	Cr_2O_3	Nd:glass	Metev et al. (1981)
Cr	Cr_2O_3	Nd:glass	Vieko et al. (1973)
Te	TeO_2	Kr	Wautelet & Andrew (1987)
W	WO_3	CO_2	Arzuov et al. (1976)
Al, Cu, Au	various	CO_2	Arzuov et al. (1976)
Cr	Cr_2O_3	CO_2	Birjega et al. (1985)

agreement between experimentally inferred values of oxide layers thickness grown in the early stages and the theoretical predictions of the Cabrera–Mott cubic growth rate law in the 100–250°C range

$$\frac{dx}{dt} = \left(\frac{d_1}{x^2}\right) \exp\left(\frac{-T_1}{T}\right) \tag{10.4}$$

and of the Wagner relationship for layers in the 500 to 1000-Å range

$$\frac{dx}{dt} = d_0 \exp\left(\frac{-T_0}{T}\right), \tag{10.5}$$

where dx/dt is the rate of change of oxide thickness x with time t; d_0, T_0, d_1, and T_1 are thermodiffusive constants; and T is temperature. The laser-grown Cu oxides adhered poorly to the metal and often cracked into small plates when scratched. Only Cu_2O was observed for the thinner films, while small quantities of CuO were present in thicker layers. Cu_2O was similarly formed by cw krypton laser radiation (Wautelet and Baufay, 1983). However, more recently Wautelet (1983) reported that as well as the usual red- and green-colored zones associated with Cu_2O, a black external halo is present. Since this appears *before* Cu_2O, CuO may indeed be formed prior to Cu_2O and is most certainly a necessary precursor to the rapid formation of Cu_2O.

Over the years, there have been several observations during laser oxidation of metals, especially Cu, that could not be adequately explained by purely thermal models (Ursu et al., 1986; Andrew et al., 1985). Principally because a variety of wavelengths—from 488 nm to 10.6 μm—were used in these experiments, an equally wide range of "photonic" mechanisms were proposed to explain these early results. Ursu et al. (1986b), for example, suggested that the 10-μm radiation from the cw CO_2 laser altered the effective diffusion of charged particles during the reaction and successfully modeled these effects. Andrew et al. (1985) similarly proposed that visible radiation may modify the space-charge part of diffusion. Birjega et al. (1986) have also perceived a laser enhancement to cw CO_2 laser oxidation of Cr films. The heating conditions, however, are extremely complicated, and are presently much better understood. Much more sophisticated recent studies (Andrew, 1986; Baufay et al., 1987) have taken into account many of the time-dependent physical properties of the films grown, including multiple interference effects, and have now concluded that for Ar-laser oxidation of metals the "oxidation, which proceeds at rates up to μm/s, is due to only heating effects of the laser beam" (Andrew et al., 1986). It is not yet clear, however, why the laser-induced diffusion mechanism proposed for the ions during the laser oxidation is not required in the models pertaining to visible oxidation of Cu.

Continuous-wave Kr-laser irradiation of other metals in air or O_2 has led to the controlled formation of SnO, ZnO, CdO, and TeO_2 (see Table 10.2). By taking into account the thickness dependence of the reflectivity and the reaction's heat of formation, it has been concluded that the rapid oxidation rates ($\sim 10^5$ Å/s for CdO) would only be attainable if the metal actually melted. Similar irradiation (Merlin and Perry, 1984) of $Ti_{50}Zr_{10}Be_{40}$ produced amorphous films of Ti_2O_3 more than 1000 Å thick with no evidence of any crystalline phases, such as TiO_2, which are often dominant in conventional thermally oxidized Ti and Ti alloys. Titanium oxidation using pulsed lasers was first reported by Akimov et al. (1980), who formed TiO_2 on the surface and various lower oxides deeper into the Ti when irradiating Ti sheet with millisecond pulses from a Nd:glass laser. Continuous lasers have also been used by Akimov et al. (1982) to oxidize Ti to form TiO_2. Novel oxide structures have been produced in thin-film Ti by Thuillard and von Allmen (1985) using a Nd:YAG laser. It was found that the oxygen content and structure of the insulators produced was strongly dependent upon the incident fluence. Additionally, the electrical resistance of the films could be reversibly and continously varied by up to 10 orders of magnitude.

The oxidation of chromium to Cr_2O_3 in air using Nd:glass laser pulses

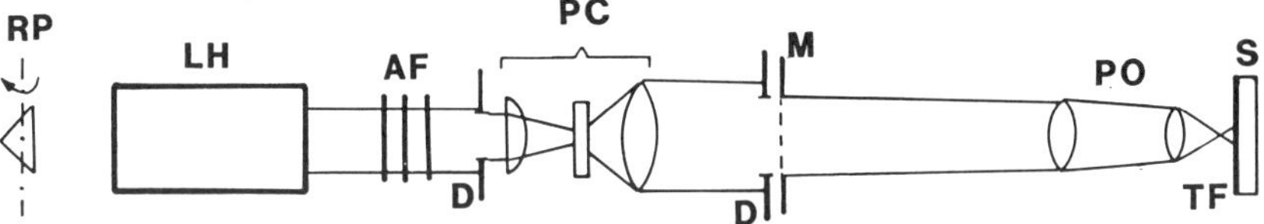

Fig. 10.12 Optical arrangement used to locally oxidize Cr in the fabrication of masks (Metev, 1980). RP: rotating prism, LH: laser head, AF: absorbing filters, D: diaphragm, PC: phase corrector, M: mask, PO: projection objective, S: sample, TF: thin film.

has potential applications in mask making (Veiko et al., 1973). It has been shown (Metev et al., 1980) that such laser-induced reactions permit direct generation of patterns without any intermediate photolithographic processing. Basically, an optical image of a mask is formed on the chromium surface by projection using 50-nsec duration pulses to selectively oxidize the material (see Fig. 10.12). Reductions of between 3x and 10x were used, and since the beam intensity on the actual mask itself was only 100 Wcm^{-2}, emulsion masks could be used. Studies showed that one pulse of 50-nsec duration produced a 5 to 7 Å thick oxide layer on the 2000 Å thick Cr film. Although this layer was thick enough to permit selective etching of the unirradiated (i.e., masked) zones, up to five successive laser pulses were used to make slightly thicker oxide films. Figure 10.13 shows a mask obtained by this photo-oxidation technique and etching in solutions of $K_3Fe(CN)_6$, NaOH, and H_2O. The total dimensions of the projected pattern were 9×7 mm, while the smallest

Fig. 10.13 Patterned oxide obtained by selective laser oxidation. From Metev, 1980.

feature resolvable was 3 μm, closely corresponding to the diffraction limit of the wavelength used. Clearly, visible or UV radiation should reduce the minimum feature size further.

2.7. Summary

This chapter has reviewed the current status of laser applications in the complementary fields of doping and oxidation. The various mechanisms by which the appropriate dopant or dielectric forming reactants are created and introduced into the appropriate have been reviewed, as have several recent device applications. Although the processing conditions available with these photon-assisted methods bring about a totally new and exciting approach towards in-situ device fabrication, much work remains to be done to fully develop the potential capabilities inherent with this manufacturing philosophy.

References

Affolter, K., Lüthy, W., and von Allmen, M. (1978). Properties of laser-assisted doping in silicon. *Appl. Phys. Lett.* **33,** 185–187.

Akimov, A.G., Gagarin, A.P., Dugarov, V.G., Makin, V.S., and Pudkov, S.D. (1980). Composition of an oxide film formed after pulsed heating of a metal. *Sov. Phys. Tech. Phys.* **25,** 1439–1441.

Akimov, A.G., Bonch-Bruevich, A.M., Gargarin, A.P., Dorofeev, V.G., Zimin, P.A., Ivanova, I.N., Libenson, M.N., Makin, V.S., and Pudkov, S.D. (1982). The relationship of oxide film thickness and composition of titanium to the absorption coefficient on laser irradiation in an oxidizing atmosphere. *Bull. Acad. USSR Phys. Ser.* **46,** 145–151.

Andrew, R., Baufay, L., Wautelet, M., and Pigeolet, A. (1985). Non-thermal effects in laser assisted oxidation of thin metal processing. (Nguyen, V.T., and Cullis, A.G., eds.). Les Editions de Physique, Les Ulis, 163–168.

Andrew, R. (1986). CW Ar^+ laser induced oxidation of Cu layers. *Appl. Phys.* **B41,** 205–212.

Ashby, C.I.H. (1983). Photoenhancement of the reaction of graphite with atomic hydrogen by UV radiation. *J. Vac. Sci. Technol.* **1,** 935–936.

Ashida, Y., Mishima, Y., Hirose, M., and Osaka, Y. (1984). Impurity doping in chemically vapour-deposited amorphous hydrogenated silicon from disilane. *J. Appl. Phys.* **55,** 1425–1428.

Baeri, P., Campisano, S.U., Foti, G., and Rimini, E. (1979). A melting model for pulsing-laser annealing of implanted semiconductors. *J. Appl. Phys.* **50,** 788–797.

Ban, Y., Takechi, M., Ishizaki, M., Murata, H., and Kukimoto, H. (1986). Laser-irradiation effects on the incorporation of impurities in GaAs during MOVPE growth. *Jpn. J. Appl. Phys.* **25,** L967–L969.

Bathy, B., Bates, H.E., and Cretella, M. (1980). Effect of carbon on the dissolution of fused silica in liquid Si. *J. Electrochem. Soc.* **128,** 771.

Baufay, L., Houle, F.A., and Wilson, R.J. (1987). Optical self regulation during laser-induced oxidation of copper. *J. Appl. Phys.* **61,** 4640–4651.

Bentini, G.G., Berti, M., Cohen, C., Drigo, A.V., Iannitti, E., Pribat, D., and Siejka, J. (1982). Effect of oxygen pressure on oxygen incorporation in Si and GaAs during Q-switched laser irradiation. *J. de Phys C1* **43,** 229–234.

Bentini, G.G., Berti, M., Drigo, A.V., Iannitti, E., Cohen, C., and Siejka, J. (1984). Silicon-oxygen reaction induced by laser pulse irradiation in oxidizing atmosphere. In *Laser Chemical Processing of Semiconductor Devices,* Extended Abstracts of MRS Symp. B (Houle, F.A., Deutsch, T.F., and Osgood, R.M., eds.). MRS, Pittsburgh, 126–128.

Bermudez, V.M. (1983). Photoenhanced oxidation of gallium arsenide. *J. Appl. Phys.* **54,** 6795–6798.

Bertness, K.A., Petro, W.G., Silberman, J.A., Friedman, D.J., and Spicer, W.E. (1984). Optically enhanced low temperature oxygen chemisorption on GaAs(110). *J. Vac. Sci. Technol.* **A3,** 1464–1467.

Bertrand, P.A. (1985). The photochemical oxidation of GaAs. *J. Vac. Sci. Tehcnol.* **132,** 973–976.

Bilenchi, R., Gianinoni, I., and Musci, M. (1985). CO_2 laser-assisted deposition of boron and phosphorous-doped hydrogenated amorphous silicon. *Appl. Phys. Lett.* **47,** 279–281.

Birjega, M.I., Nanu, L., Mihailescu, I.N., Dinescu, M., Popescu-Pogrion, N., and Sarbu, C. (1986). Oxidation of thin polycrystalline sputtered Cr films by CW CO_2 laser radiation. *Optica Acta* **33,** 1073–1082.

Blanc, J. (1978). A revised model for the oxidation of Si with oxygen. *Appl. Phys. Lett.* **33,** 424–426.

Blum, S.E., Brown, K.H., and Srinivasan, R. (1983). Photo-oxidation of silicon monoxide to silicon dioxide with pulsed far-ultraviolet (193 nm) laser radiation. *Appl. Phys. Lett.* **43,** 1026–1027.

Boyd, I.W., Wilson, J.I.B., and West, J.L. (1981). Laser induced oxidation of silicon. *Thin Solid Films* **83,** L173–L176.

Boyd, I.W. (1983a). A review of laser beam applications for processing silicon. *Contemporary Physics* **24,** 461–490, and references therein.

Boyd, I.W. (1983b). Incorporation of oxygen atoms into As^+ implanted silicon during cw CO_2 laser annealing in O_2. *Appl. Phys. Lett.* **A31,** 71–74.

Boyd, I.W., (1984). Laser assisted pyrolytic growth and photochemical deposition of thin oxide films. In *Laser Processing and Diagnostics* (Bäuerle, D., ed.). Springer-Verlag, Heidelberg, 288–299.

Boyd, I.W., Binnie, T.D., Wilson, J.I.B., and Colles, M.J. (1984). Absorption of infrared radiation in silicon. *J. Appl. Phys.* **55,** 3061–3063.

Boyd, I.W. (1986). Photoformation of silicon dielectrics. In *Dielectric Layers on Semiconductors* (Bentini, G.G., Fogarassy, E., and Golanski, A., eds.). Les Editions de Physique, Les Ulis, 177–190.

Boyd, I.W., and Wilson, J.I.B. (1987). Structure of ultrathin silicon dioxide films. *Appl. Phys. Lett.* **50,** 320–322.

Boyd, I.W. (1987). *Laser Processing of Thin Films and Microstructures.* Springer-Verlag, Berlin, 1987.

Boyd, I.W., and Micheli, F. (1987). Wavelength dependence of optical-oxidation of silicon. *Electronics Letts.* **23,** 298–300.

Boyer, P.K., Roche, G.A., Ritchie, W.H., and Collins, G.J. (1982). Laser induced chemical vapour deposition of SiO_2. *Appl. Phys. Lett.* **40,** 716–718.

Branz, H.M., Fan, S., Flint, J.H., Fiske, B.T., Adler, D., and Haggerty, J.S. (1986). Doped hydrogenated amorphous silicon films by laser-induced chemical vapour deposition. *Appl. Phys. Lett.* **48,** 171–173.

Carey, P.G., Sigmon, T.W., Press, R.L., and Rahlen, T.S. (1985). Ultra-shallow high-concentration boron profiles for CMOS processing. *IEEE Electron Dev. Lett.* **EDL-6,** 291–293.

Carey, P.G., Bezjian, K., Sigmon, T.W., Gildea, P., and Magee, T.J. (1986). Fabrication of submicrometer MOSFETs using gas immersion laser doping (GILD). *IEEE Electron Dev. Lett.* **EDL-7,** 440–442.

Chan Tung, N., Caratini, Y., and Liauzu, L. (1986). Oxidation kinetics and electrical characteristics of thin gate dielectrics obtained by rapid thermal processes. In *Dielectric Layers in Semiconductors* (Bentini, G.G., Fogarassy, E., and Golanski, A., eds.). Les Editions de Physique, Les Ulis, 247–253.

Chiang, S.W., Liu, Y.S., and Reihl, R.F. (1981). Oxidation of silicon by ion implantation and laser irradiation. *Appl. Phys. Lett.* **39,** 752–754.

Chuang, T.J., Hussla, I., and Sesselman, W. (1984). Laser assisted chemical etching of inorganic materials: Mechanistic studies. In *Laser Processing and Diagnostics* (Bäuerle, D., ed.). Springer-Verlag, Heidelberg, 300–314, and references therein.

Cohen, C., Siejka, J., Bentini, G.G., Berti, M., Dona Dalle Rose, L.F., and Drigo, A.V. (1986). Loss and incorporation mechanisms during pulsed laser irradiation of silicon in gaseous atmosphere. In *Dielectric Layers in Semiconductors* (Bentini, G.G., Fogarassy, E., and Golanski, A., eds.). Les editions de Physique, Les Ulis, 197–207.

Cros, A., Salvan, F., and Derrien, J. (1982). Laser induced oxidation of the Si(111) surface. *Appl. Phys.* **A28,** 241–245.

Cullis, A.G. (1985). Transient annealing of semiconductors by laser, electron beam and radiant heating techniques. *Rep. Prog. Phys.* **48,** 1155–1233.

Deal, B.E., and Grove, A.S. (1965). General relationship for the thermal oxidation of silicon. *J. Appl. Phys.* **36,** 3770–3776.

Deutsch, T.F., Ehrlich, D.J., and Osgood, R.M. (1979). Laser photodeposition of metal films with microscopic features. *Appl. Phys. Lett.* **35,** 175–177.

Deutsch, T.F., Ehrlich, D.J., Osgood, R.M., and Liau, Z.L. (1980). Ohmic contact formation on InP by pulsed laser photochemical doping. *Appl. Phys. Lett.* **36,** 847–849.

Deutsch, T.F., Ehrlich, D.J., Rathman, D.D., Silversmith, D.J., and Osgood, R.M. (1981). Electrical properties of laser chemically doped silicon. *Appl. Phys. Lett.* **39,** 825–827.

Deutsch, T.F., Fan, J.C.C., Ehrlich, D.J., Turner, G.W., Chapman, R.L., and Gale, P. (1982). Efficient GaAs solar cells formed by UV laser chemical doping. *Appl. Phys. Lett.* **40,** 722–724.

Deutsch, T.F. (1984). Applications of excimer lasers to semiconductor processing. In *Laser Processing and Diagnostics* (Bäuerle, D., ed.). Springer-Verlag, Berlin, 239–251.

Ehrlich, D.J., Osgood, R.M., and Deutsch, T.F. (1980). Direct writing of regions of high doping on semiconductors by UV laser photodeposition. *Appl. Phys. Lett.* **36,** 916–918.

Ehrlich, D.J., Osgood, R.M., and Duetsch, T.F. (1981). Laser photochemical microalloying for etching of aluminium thin films. *Appl. Phys. Lett.* **38,** 399–401.

Ehrlich, D.J., and Tsao, J.Y. (1982). Submicrometer-linewidth doping and relief definition in silicon by laser controlled diffusion. *Appl. Phys. Lett.* **41,** 297–299.

Ehrlich, D.J., and Tsao, J.Y. (1983). A review of laser-microchemical processing. *J. Vac. Sci. Technol.* **B1,** 969–978, and references therein.

Fathipour, M., Boyer, P.K., Collins, G.J., and Wilmsen, C.W. (1985). Photoenhanced thermal oxidation of InP. *J. Appl. Phys.* **57,** 637–642.

Fiori, C. (1984). Far ultraviolet laser induced oxidation at the Si(111) surface by bond rearrangement. *Phys. Rev. Lett.* **52,** 2077–2080.

Fogarassy, E., Stuck, R., Grob, J.J., and Siffert, P. (1981). Silicon solar cells realized by laser induced diffusion of vacuum deposited dopants. *J. Appl. Phys.* **52,** 1076–1082.

Fogarassy, E., Stuck, R., Toulemonde, M., Salles, D., and Siffert, P. (1983). A model for laser induced diffusion. *J. Appl. Phys.* **54,** 5059–5063.

Fogarassy, E., White, C.W., Lowndes, D.W., and Narayan, J. (1986). Laser induced oxidation of heavily doped silicon. In *Beam-Solid Interactions and Phase Transformations* (Kurz, H., Olson, G.L., and Poate, J.M., eds.). MRS, Pittsburgh, 1986, 173–178.

Fogarassy, E., Unamuno, S., Regolini, J.L., and Fuchs, C. (1987). Analysis of the thermal contribution to UV laser induced oxidation of silicon and silicon dioxide. *Phil. Mag.* **B55,** 253–260.

Fukuda, M., and Takahei, K. (1985). Optically enhanced oxidation of III–V compound semiconductors. *J. Appl. Phys.* **57,** 129–134.

Garulli, A., Servidori, M., and Vecchi, I. (1980). In-depth oxygen contamination produced in silicon by pulsed laser radiation. *J. Phys.* **D13,** L199–202.

Ghez, R., and Van Der Meulen, Y.J. (1972). Kinetics of thermal growth of ultrathin layers of silicon dioxide on Si: II Theory. *J. Electrochem. Soc.* **119,** 1100–1107.

Gibbons, J.F. (1981). CW laser processing of silicon. *Jpn. J. Appl. Phys. Suppl.* **19,** 121–127.

Hirata, H., and Hoshikawa, K. (1980). The dissolution rate of silica in molten silicon. *Jpn. J. appl. Phys.* **19,** 1573.

Ho, K., Koyama, H., Uda, K., and Miura, Y. (1980). Incorporation of oxygen into silicon during pulsed laser radiation. *Jpn. J. Appl. Phys.* **19,** L375–L378.

Hu, S.M. (1983). New oxide growth law and the thermal oxidation of silicon. *Appl. Phys. Lett.* **42,** 872–874.

Ibbs, K.G., and Lloyd., M.L. (1983). Ultraviolet laser doping of silicon. *Optics and Laser Technology* **11,** 35–39.

Inoue, T., Konagai, M., Takahashi, K., and Migitaka, M. (1983). Photochemical vapour deposition of undoped and n-type amorphous silicon films produced from disilane. *Appl. Phys. Lett.* **43,** 774–776.

Inoue, T., Tanaka, T., Konagai, M., and Takahashi, K. (1984). Electronic and optical properties of p-type amorphous silicon and wide bandgap amorphous silicon carbide films prepared by photochemical vapour deposition. *Appl. Phys. Lett.* **44,** 871–873.

Kiang, Y.C., Moulic, J.R., Chu, W.-K., and Yen, A.C. (1982). Modification of semiconductor device characteristics by lasers. *IBM J. Res. Develop.* **26,** 171–176.

Kräutle, H., Roentgen, P., Maier, M., and Beneking, H. (1985). Laser-induced doping of GaAs. *Appl. Phys.* **A38,** 49–56.

Kräutle, H., and Wachenschwanz, D. (1985). Ohmic contacts on n- and p-layers of GaAs using laser-induced diffusion. *Solid State Electron.* **28,** 601–603.

Liu, Y.S., Chiang, S.W., and Bacon, F. (1981). Rapid oxidation via adsorption of oxygen in laser-induced amorphous Si. *Appl. Phys. Lett.* **38,** 1005–1007.

Matsuura, M., Ishida, M., Suzuki, A., and Hara, K. (1981). Laser oxidation of GaAs. *Jpn. J. Appl. Phys.* **20,** L726–L728.

Massoud, H.Z., Plummer, J.D., and Irene, E.A. (1985). Thermal oxidation of silicon in dry oxygen. *J. Electrochem. Soc.* **132,** 1745–1753.

McWilliams, B.M., Herman, I., Mitlitsky, F., Hyde, R.A., and Wood, L.L. (1983). Wafer-scale laser pantography: Fabrication of n-metal-oxide-semiconductor transistors and small-scale integrated circuits by direct-write laser-induced pyrolytic reactions. *Appl. Phys. Lett.* **43,** 946–948.

Merlin, R., and Perry, T.A., (1984). Growth of amorphous Ti_2O_3 layers by laser-induced oxidation. *Appl. Phys. Lett.* **45,** 852–853.

Metev, S.M., Savtchenko, S.K., and Stamenov, K. (1980). Pattern generation by laser induced oxidation of thin metal films. *J. Phys.* **D13,** L75–L76.

Meyer, J.R., Kruer, M.R., and Bartoli, F.J. (1980). Optical heating in semiconductors: Laser damage in Ge, Si, InSb, and GaAs. *J. Appl. Phys.* **51,** 5513–5522.

Micheli, F., and Boyd, I.W. (1987a). Laser microfabrication of thin films: Part III. *Optics and Laser Technology* **19,** 75–82, and references therein.

Micheli, F., and Boyd, I.W. (1987b). Photon controlled oxidation of silicon. *Appl. Phys. Lett.* **51,** 1149–1151.

Miyauchi, T., Hongo, M., Masuhara, T., Minato, O., Kawanabe, T., and Nagasawa, K. (1983). Laser diffusion connection technology for VLSI programming. *Annals CIRP* **32,** 141–144.

Modlin, D.N., and Tiller, W.A. (1985). Effects of corona discharge-induced oxygen ion beams and electric fields on silicon oxidation kinetics. *J. Electrochem. Soc.* **132,** 1659–1663.

Muller, J.C., Fogarassy, E., Salles, D., Stuck, R., and Siffert, P.M. (1980). Laser processing in the preparation of silicon solar cells. *IEEE Trans. Electron. Dev.* **ED-27,** 815–821.

Nakano, S., Kuwano, Y., and Ohnishi, M. (1986). High performance a-Si solar cells and new fabrication methods for a-Si solar cells. *Appl. Phys.* **A41,** 267–274.

Narayan, J., Young, R.T., Wood, R.F., and Christie, W.H. (1978). P-N junction formation in boron deposited silicon by laser induced diffusion. *App. Phys. Lett.* **33,** 338–340.

Nissim, Y., Lietoila, A., Gold, R.B., and Gibbons, J.F. (1980). Temperature distributions produced in semiconductors by a scanning elliptical or circular CW laser beam. *J. Appl. Phys.* **51,** 274–276.

Nulman, J., Krusius, J.P., and Gat, A. (1985). Rapid thermal processing of thin gate dielectrics: Oxidation of silicon. *IEEE Electron Dev. Lett.* **EDL-6,** 205–207.

Oren, R., and Ghandi, S.K. (1971). Ultraviolet enhanced oxidation of silicon. *J. Appl. Phys.* **42,** 752–756.

Petro, W.G., Hino, I., Eglash, S., Lindau, I., Su, C.Y., and Spicer, W.E. (1982). Effect of low intensity laser radiation during oxidation of the GaAs(110) surface. *J. Vac. Sci. Technol.* **21,** 405–408.

Ponpon, J.P., Grob, J.J., Grob, A., and Stuck, R. (1986). Formation of thin silicon oxide films by rapid thermal heating. *J. Appl. Phys.* **59,** 3912–3923.

Richter, H., Orlowski, T.E., Kelly, M., and Margaritondo, G. (1984). Ultrafast UV laser induced oxidation of silicon: Control and characterisation of the Si-SiO_2 interface. *J. Appl. Phys.* **56,** 2351–2355.

Rytz-Froidevaux, Y., Salathé, R.P., and Gilgen, H.H. (1985). Laser generated microstructures. *Appl. Phys.* **A37,** 121–138.

Sameshima, T., Usui, S., and Sekiya, M. (1987). Laser induced melting of predeposited impurity doping technique used to fabricate shallow junctions. *J. Appl. Phys.* **62,** 711–713.

Sameshima, T., and Usui, S. (1987). Analysis of dopant diffusion in molten silicon induced by a pulsed excimer laser. *Jpn. J. Appl. Phys.* **26,** L1208–L1210.

Schvan, P., and Thomas, R.E. (1985). Time dependent heat flow calculation of CW laser-induced melting of silicon. *J. Appl. Phys.* **57,** 4378–4741.

Schafer, S.A., and Lyon, S.A. (1981). Optically enhanced oxidation of semiconductors. *J. Vac. Sci. Technol.* **19,** 494–497.

Schafer, S.A., and Lyon, S.A. (1982). Wavelength dependence of laser enhanced oxidation of silicon. *J. Vac. Sci. Technol.* **21,** 422–425.

Siejka, J., Perriere, J. and Srinivasan, R. (1985). Ultraviolet (193 nm) laser-induced oxygen exchange and bulk diffusion in gallium arsenide native oxides: ^{18}O experiments. *Appl. Phys. Lett.* **48,** 773–775.

Sigmon, T. (1987). UV laser processing of semiconductor devices. In *Photon, Beam and Plasma Stumulated Chemical Processes at Surfaces*. (Donnelly, V.M., Herman, I.P., and Hirose, M., eds.). Vol. 75, MRS Proceedings, Pittsburgh.

Solanki, R., Moore, C.A., and Collins, G.J. (1985). Laser-induced chemical vapour deposition. *Solid State Technology* **29,** 220–227.

Sugii, T., Ito, T., and Ishikawa, H. (1984). Low temperature nitridation of silicon by excimer laser irradiation. *Extended abstracts of the 16th International Conference on Solid-State Devices and Materials,* Kobe, 433.

Szikora, S., Kräuter, W., and Bäuerle, D. (1984). Laser induced deposition of SiO_2. *Mater. Lett.* **2,** 263–264.

Takai, H., Tanaka, T., Kim, W.Y., Konagai, M., and Takahashi, K. (1985). High-energy conversion efficiency amorphous silicon solar cells by photochemical vapour deposition. *J. Appl. Phys.* **58,** 3664–3666.

Tanaka, T., Kim, W.Y., Konagai, M., and Takahashi, K. (1984). Amorphous silicon solar cells fabricated by photochemical vapour deposition. *Appl. Phys. Lett.* **45,** 865–867.

Thompson, M.O., Galvin, G.J., Mayer, J.M., Peercy, P.S., Poate, J.M., Jacobson, D.C., Cullis, A.G., and Chew, N.G. (1984). Melting temperature and explosive crystallisation of amorphous silicon during pulsed laser irradiation. *Phys. Rev. Lett.* **52,** 2360–2363, and references therein.

Thuillard, M., and Von Allmen, M. (1985). Laser produced Ti/Ti-oxide thin film structures. *Appl. Phys. Lett.* **47,** 936.

Tsai, J.C.C. (1983). Diffusion. In *VLSI Technology*. (Sze, S.M., ed.). McGraw-Hill, London, 169–217.

Turner, G.B., Tarrant, D., and Pollock, G. (1981). Solar cells made by laser-induced diffusion directly from phosphine gas. *Appl. Phys. Lett.* **39,** 967–969.

Unamuno, S., Toulemonde, M., and Siffert, P. (1984). Melting model for UV lasers. In *Laser Processing and Diagnostics* (Bäuerle, D., ed.). Springer-Verlag, Berlin, 35–39.

Ursu, I., Nanu, L., Dinescu, M., Hening, Al., Mihailescu, I.N., Nistor, L.C., Teodorescu, V.S., Szil., E., Hevesi, I., Kovacs, J., and Nanai, L. (1984a). Vanadium oxidation as a result of CW CO_2 laser irradiation in atmospheric air. *Appl. Phys.* **A35,** 103–108.

Ursu, I., Nistor, L.C., Teodorescu, V.S., and Nanu, L. (1984b). Early oxidation of copper during CW CO_2 laser irradiation. *Appl. Phys. Lett.* **44,** 188–189.

Ursu, I., Nanu, L., Mihailescu, I.N., Nistor, L.C., Teodorescu, V.S., Prokhorov, A.M., Konov, V.I., and Chaplieu, N.I. (1984c). On the theoretical description of the early oxidation of copper by CW CO_2 laser irradiation. *J. Physique. Lett.* **45,** L737–L740.

Ursu, I., Mihailescu, I.N., Nistor, L.C., Popescu, M., Teodorescu, V.S., Prokhorov, A.M., Ralchenko, V.G., and Konov, V.I. (1986a). On the mechanism of surface compounds formation by CO_2 laser irradiation of zirconium samples in air. *J. Appl. Phys.* **59,** 668–672.

Ursu, I., Nanu, L., and Mihailescu, I.N., (1986b). Diffusion model for the laser oxidation of metallic samples in air. *Appl. Phys. Lett.* **49,** 109–111.

Veiko, V.P., Kotov, G.A., Libenson, M.N., and Nikitin, M.N., (1973). Thermochemical action of laser radiation. *Sov. Phys. Doklady.* **18,** 83–85.

Wautelet, M. (1983). Oxidation of thin Cd, Cu, and Te films under continuous laser irradiation. *Mater. Lett.* **2,** 20–22.

Wautelet, M., and Baufay, L. (1983). Laser induced oxidation of thin cadmium and copper films. *Thin Solid Films* **100,** L9–L11.

Wautelet, M., and Andrew, A. (1987). Laser induced oxidation of metals. *Phil. Mag.* **B55,** 223–227.

Westendorp, J.F.M., Wang, Z.L., and Saris, F.W. (1982). An ion scattering study of indiffusion during pulsed laser annealing/cleaning of silicon. In *Laser and Electron Beam Interactions with Solids* (Appleton, B.R., and Celler, G.K., eds.). North-Holland, Amsterdam, 255–260.

Wood, R.F. (1982). Macroscopic theory of pulsed laser annealing III. Nonequilibrium segregation effects. *Phys. Rev.* **B25,** 2786–2811, and references therein.

Yamada, A., Kenne, J., Konagi, M., and Takahashi, K. (1985). Wide band-gap, fairly conductive p-type hydrogenated amorphous silicon carbide films prepared by direct photolysis: Solar cell application. *Appl. Phys. Lett.* **46,** 272–274.

Yoshikawa, A., and Yamaga, S. (1984). Growth of hydrogenaged amorphous silicon films by ArF excimer laser photo-dissociation of disilane. *Jpn. J. Appl. Phys.* **23,** L91–L93.

Young, R.T., Wood, R.F., Narayan, Y., White, C.W., and Christie, W.H. (1980). Pulsed laser techniques for solar cell processing. *IEEE Trans. Electron. Dev.* **ED-27,** 807–815 (1980), and references therein.

Young, E.M., and Tiller, W.A. (1983). Photon enhanced oxidation of silicon. *Appl. Phys. Lett.* **42,** 63–65.

Index

G

H

I

K

U

V

W

X

Z